Biochemistry

Reactions in biochemistry can be explained by an understanding of fundamental organic chemistry principles and reactions. This paradigm is extended to biochemical principles and to a myriad of biomolecules. The text provides a framework for understanding various topics in biochemistry, including the chemical behavior of biomolecules, enzyme activity, and more. It goes beyond mere memorization. Using several techniques to develop a relational understanding, including homework, this text helps students fully grasp and better correlate essential organic chemistry concepts with those at the root of biochemistry. The goal is to achieve a better understanding of the fundamental principles of biochemistry.

Professor Michael B. Smith was born in Detroit, Michigan, in 1946 and moved to Madison Heights, Virginia, in 1957, where he attended Amherst County High School. He received an A.A. from Ferrum College in 1967 and a B.S. in chemistry from Virginia Polytechnic Institute in 1969. After working for three years at Newport News Shipbuilding and Dry Dock Co. in Newport News, VA, as an analytical chemist, he entered graduate school at Purdue University. He received a Ph.D. in Organic Chemistry in 1977, under the auspices of Professor Joe Wolinsky. He spent one year as a faculty research associate at Arizona State University with Professor G. Robert Pettit, working on the isolation of cytotoxic principles from plants and sponges. He spent a second year of postdoctoral work with Professor Sidney M. Hecht at the Massachusetts Institute of Technology, working on the synthesis of bleomycin A2. He began his academic career at the University of Connecticut in 1979, where he achieved the rank of professor of chemistry. In 1986, he spent a sabbatical leave in the laboratories of Professor Leon Ghosez at the Université Catholique de Louvain in Louvain-la-Neuve, Belgium, as a visiting professor. He retired as a full professor from the University of Connecticut on January 1, 2017, was granted the rank of Professor Emeritus, and is currently pursuing his interests as an author.

Biochemistry
An Organic Chemistry Approach

Second Edition

Michael B. Smith

CRC Press
Taylor & Francis Group
Boca Raton London New York

CRC Press is an imprint of the
Taylor & Francis Group, an **informa** business

First edition published 2026
by CRC Press
2385 NW Executive Center Drive, Suite 320, Boca Raton FL 33431

and by CRC Press
4 Park Square, Milton Park, Abingdon, Oxon, OX14 4RN

CRC Press is an imprint of Taylor & Francis Group, LLC

ISBN: 9781032996202 (hbk)
ISBN: 9781032996196 (pbk)
ISBN: 9781003605157 (ebk)

DOI: 10.1201/9781003605157

Typeset in Times
by codeMantra

Contents

Preface

The focus of the first edition of this book was to use the principles and reactions of organic chemistry to explain the fundamentals of biochemistry. The inspiration for this book came from teaching the second semester of organic chemistry to many students who were taking introductory biochemistry at the same time. Many students expressed that the topics discussed in organic chemistry helped explain and better understand the concepts they were discussing or had discussed in biochemistry. On many occasions, topics were discussed in organic chemistry that had been covered previously in the biochemistry course. An understanding of the organic chemistry reactions and mechanisms helped the students better understand many of those biochemical principles. It is hoped that understanding the organic chemistry foundations of biochemistry will provide similar assistance to students today. Where feasible, organic chemical reactions, such as those found in a typical undergraduate organic chemistry course, are included in this book to bring their biological chemistry counterparts into perspective. The intent has not changed, nor has the putative focus of this book. However, this second edition has made biochemistry the focus rather than organic chemistry.

The first chapter discusses the importance of water in chemistry and acid-base chemistry in organic and biochemical processes. Next, amino acids are introduced and then their use in the formation of peptides and proteins, as well as the importance of these biologically important compounds. The next chapter introduces enzymes, enzyme kinetics, and classes of enzymes. Pertinent organic chemical reactions are included for each enzyme class for direct comparison. After amino acids, proteins, and enzymes are introduced, there is a review of the fundamentals of an undergraduate organic chemistry course. For those who have not had a full organic chemistry course, this chapter will not suffice. This chapter is intended only as a supplement for an organic chemistry course and to function as a review for the biochemical principles to follow. Nucleophilic substitution and elimination reactions are introduced here, followed by the chemistry of radicals and radical reactions. Dienes and conjugated systems are discussed along with sigmatropic rearrangements. Enols and enolate reactions are discussed, including enolate condensation reactions. Lipid chemistry is discussed along with the chemistry of carboxylic acids and acid derivatives. Aromatic chemistry and heterocyclic aromatic compounds are also discussed. Chapter 10 introduces organometallic compounds, including biologically relevant metals and chelating reagents. Carbohydrates are introduced, and the chemistry of carbohydrate derivatives is elaborated, followed by a discussion of the importance of glycosides. This book concludes with a discussion of nucleosides, nucleotides, DNA, and RNA, followed by the answers to Homework.

I thank all my former students who inspired this book with the sincere hope that this approach will help those students interested in biochemistry. I also thank my editor, Hilary Lafoe, and my publisher, Dr. Fiona McDonald, at Taylor & Francis, for their support. This book would not have been possible without them.

All structures and reactions were drawn using ChemDraw Professional 18.0.0.231. I thank PerkinElmer Informatics, Inc. for a gift of this software. All 3-D drawings and molecular models were prepared using Spartan'18 software, version 1.2.0 (181121). I thank Warren Hehre and Sean Ohlinger of Wavefunction, Inc. for a gift of this software. I thank Ms. Christine Elder (https://christineelder.com), graphics design artist, for her graphic arts expertise to render the drawings in Figures 1.4, 1.9, 1.11, 1.15, 1.16, 1.20, 1.22, 1.29, 1.32, 1.37, 1.44, 1.45, 2.7, and 9.1.

Every effort has been made to keep this manuscript free of errors. Where there are errors, however, I take full responsibility and encourage readers to contact me at the email provided with questions, comments, and corrections. Thank you.

Michael B. Smith
Professor Emeritus
University of Connecticut
michael.smith@uconn.edu
June 04, 2025

Shutterstock Acknowledgments

Figure 1.6 General Phospholipid Structure of Membrane Cells. Stock Vector ID: 2144223119.
Figure 3.1 Protein Structure Primary Secondary Tertiary Quaternary Amino Acid Residues Helix Polypeptide Chain Assembled Subunits. Reprinted from Shutterstock vector ID: 1474657079.
Figure 3.9 Quaternary Structure of Rat AQUAPORIN-4, 3D. Shutterstock Stock Illustration ID: 1881163822.
Figure 3.11 The Standard Genetic Code. Shutterstock. Image ID 2052705743.
Figure 10.4 Structure of Phospholipid Bilayers of Cell Membrane or Cytoplasmic Membrane. Stock Vector ID: 1163460010.
Figure 12.10 Magnetic Resonance Image (MRI) of a Human Brain: Comparison MRI of the Brain Axial and Coronal. Reprinted from Shutterstock Asset id: 1390007888.
Figure 14.3 (a) The Structure Of Plant Cell. Stock Vector ID: 2465117249. (b) Diagram of Animal Cell Anatomy Illustration. Stock Vector ID: 1117954796.

Common Abbreviations

Other, less common abbreviations are given in the text when the term is used.

Ac	Acetyl	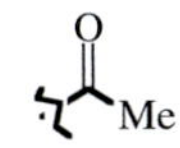
AIBN	Azobisisobutyronitrile	
aq	Aqueous	
ACP	Acyl carrier protein	
ADP	Adenosine 5′-diphosphate	
AMP	Adenosine monophosphate	
ATP	Adenosine triphosphate	
Ax	Axial	
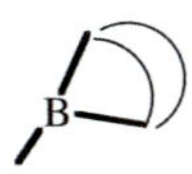	9-Borabicyclo[3.3.1]nonylboryl	
9-BBN	9-Borabicyclo[3.3.1]nonane	
Bn	Benzyl	-CH_2Ph
Boc	*tert*-Butoxycarbonyl	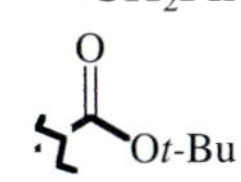
Bu	*n*-Butyl	–$CH_2CH_2CH_2CH_3$
BUN	Blood urea nitrogen	
Bz	Benzoyl	
°C	Temperature in Degrees Celsius	
13C NMR	Carbon Nuclear Magnetic Resonance	
cat	Catalytic	
Cbz	Carbobenzyloxy	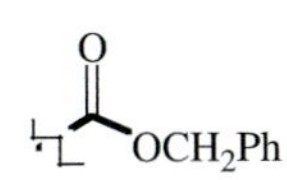
CIP	Cahn–Ingold-Prelog	
CoA	Coenzyme A	
mCPBA	3-Chloroperoxybenzoic acid	
DCC	1,3-Dicyclohexylcarbodiimide	*c*-C_6H_{11}-N=C=N-*c*-C_6H_{11}
DDT	Dichlorodiphenyltrichloroethane	
DEA	Diethylamine	$HN(CH_2CH_3)_2$
DMAP	4-Dimethylaminopyridine	
DME	Dimethoxyethane	$MeOCH_2CH_2OMe$
DMF	*N,N′*-Dimethylformamide	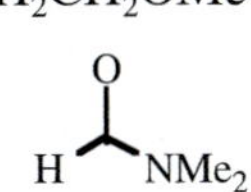
DMSA	Dimercaptosuccinic acid	
DMSO	Dimethyl sulfoxide	
DNA	Deoxyribonucleic acid	
DNP	2,4-Dinitrophenol	
EA	Electron affinity	
EDTA	Ethylenediaminetetraacetic acid	
ee or % ee	% Enantiomeric excess	
Equiv	Equivalent(s)	
Er	Enantiomeric ratio	
Et	Ethyl	–CH_2CH_3
Ether	Diethyl ether	$CH_3CH_2OCH_2C_3$
Eq	Equatorial	

FC	Formal charge	
FDNB	Sanger's reagent, 1-fluoro-2,4-dinitrobenzene	
FMO	Frontier molecular orbitals	
FVP	Flash vacuum pyrolysis	
GC	Gas chromatography	
GMP	Guanosine 5′-monophosphate	
h	Hour (hours)	
1H NMR	Proton Nuclear Magnetic Resonance	
HDL	High-density lipoprotein	
HDPE	High-density poly(ethylene)	
HIV	Human immunodeficiency virus	
HMPA	Hexamethylphosphoramide	
HOMO	Highest occupied molecular orbital	
HPLC	High-performance liquid chromatography	
hν	Irradiation with light	
IP	Ionization potential	
iPr	Isopropyl	$–CH(Me)_2$
IR	Infrared	
IUPAC	International Union of Pure and Applied Chemistry	
K	Temperature in kelvin	
LCAO	Linear combination of atomic orbitals	
LDA	Lithium diisopropylamide	LiN(*i*-Pr)$_2$
LDL	Low-density lipoprotein	
LTA	Lead tetraacetate	
LUMO	Lowest unoccupied molecular orbital	
MAL	beta-Methylaspartate ammonia-lyase	
mCPBA	*meta*-Chloroperoxybenzoic acid	
MDPE	Medium-density poly(ethylene)	
Me	Methyl	$–CH_3$ or Me
min	Minutes	
MIT	Monoiodotyrosine	
MO	Molecular orbital	
MRI	Magnetic resonance imaging	
mRNA	Messenger ribonucleic acid	
MS	Mass spectrometry	
NMR	Nuclear magnetic resonance	
N.R.	No reaction	
NAD+	Nicotinamide adenine dinucleotide	
NADH	Reduced nicotinamide adenine dinucleotide	
NADP+	Nicotinamide adenine dinucleotide phosphate	
NADPH	Reduced nicotinamide adenine dinucleotide phosphate	
NBS	*N*-Bromosuccinimide	
NCS	*N*-Chlorosuccinimide	
Ni(R)	Raney nickel	
NMO	*N*-Methylmorpholine *N*-oxide	
Nu (Nuc)	Nucleophile	
OLDA	*N*-Oleayldopamine	
Oxone®	$2\ KHSO_5KHSO_4{\bullet}K_2SO_4$	
PCB	Polychlorobiphenyl	
PCC	Pyridinium chlorochromate	
PDC	Pyridinium dichromate	

PEG	Poly(ethylene glycol)	
PES	Photoelectron spectroscopy	
Ph	Phenyl	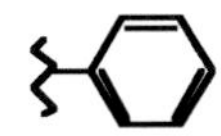
PhMe	Toluene	
PPA	Polyphosphoric acid	
Ppm	Parts per million	
Pr	*n*-Propyl	$-CH_2CH_2CH_3$
PTFE	Poly(tetrafluoroethylene)	
PVC	Poly(vinyl chloride)	
Py	Pyridine	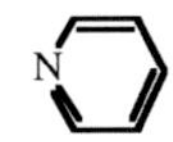
rf	Radio frequency	
RNA	Ribonucleic acid	
ROS	Reactive oxygen species	
rt	Room temperature	
s	Seconds	
SCF	Self-consistent field	
$(Sia)_2BH$	Disiamylborane (Siamyl is *sec*-Isoamyl)	
sBuLi	*sec*-Butyllithium	$CH_3CH_2CH(Li)CH_3$
SEAr	Electrophilic aromatic substitution	
SET	Single electron transfer	
SNAr	Nucleophilic aromatic substitution	
SOMO	Singly occupied molecular orbital	
T	Temperature	
t-Bu	*tert*-Butyl	$-CMe_3$
TBHP (t-BuOOH)	*t*-Butylhydroperoxide	Me_3COOH
TFA	Trifluoroacetic acid	CF_3COOH
$ThexBH_2$	Thexylborane (*tert*-hexylborane)	
THF	Tetrahydrofuran	
THP	Tetrahydropyran	
TMEDA	Tetramethylethylenediamine	$Me_2NCH_2CH_2NMe_2$
Tol	Tolyl	$4\text{-}(Me)C_6H_4$
TS	Transition state	
Ts(Tos)	Tosyl = *p*-Toluenesulfonyl	$4\text{-}(Me)C_6H_4SO_2$
TTP	Thiamine pyrophosphate	
UHMWPE	Ultrahigh molecular weight poly(ethylene)	
UTP	Uridine 5′-triphosphate	
UV	Ultraviolet spectroscopy	
VIS	Visible	
VDW	van der Waals	

1 Water, Acids, and Bases

Without water, life as we know it would not exist. Indeed, life likely arose in an aqueous medium and the adult human body has a water content of up to 60%.[1] The brain and heart are about 73% water, the lungs about 83% water, the skin contains 64% water, and the muscles and kidneys are 79% water. Blood is a body fluid in the circulatory system of humans that delivers necessary substances to the cells and transports metabolic waste products away from those same cells. Blood is composed of blood cells suspended in blood plasma, which is about 55% of blood fluid, but 92% water by volume. Typically, less than 10% of blood plasma consists of dissolved substances such as proteins, electrolytes, glucose, mineral ions, vitamins, amino acids, and hormones.[2] Water is involved in many essential functions, and is important for myriad biomolecules, both in terms of structure and function.

1.1 SOLUBILITY

As just mentioned, an important function of blood is to transport dissolved substances around the body. The extent to which compounds are dissolved in blood is a function of the solubility of those moieties in the blood, which is mostly water. It is therefore important to understand solubility in aqueous media. The term "solubility" refers to one molecule *dissolving* in another, which means that the molecules mix such that one phase (one layer) is formed. If two things are not mutually soluble (like oil and water), two or more phases are formed. Solubility in water, for example, is a measure of the amount of a chemical substance that can dissolve in water at a specific temperature. The term "dissolve" is formally defined as "to cause to pass into solution" or "to break up." If there are two compounds and at least one of them is a liquid, one dissolves in the second compound to form a solution. In chemistry, a unit of solubility is measured in mg/L (milligrams per liter) or ppm (parts per million).

The hydrophilic and hydrophobic nature of different compounds or molecules is important. Indeed, "like-dissolves-like" is an old axiom for solubility that is pertinent to chemistry. This axiom means that polar compounds are likely to dissolve in other polar compounds, but not very well in nonpolar compounds. Similarly, a nonpolar compound will dissolve in a nonpolar liquid, but it is usually not very soluble in a polar liquid. For example, water is polar and an alkane is nonpolar, so they are not expected to be mutually soluble. One alkane should dissolve in another alkane or in most hydrocarbons, but not in the polar molecule water. In biochemistry, highly lipophilic molecules should have poor solubility or be sparingly soluble in aqueous media. Blood is largely water, and most lipophilic molecules are sparingly soluble.

When dealing with typical organic molecules, one functional group in a molecule of less than five carbons renders that compound polar. A molecule with one functional group in a molecule with more than eight carbons is usually considered to be nonpolar. An alcohol with only two carbons is sufficiently polar so that alcohol will likely dissolve in water. Compounds of 5–7 carbon atoms may not fit into one category, which means that it is difficult to predict their solubility in water.

Water tends to separate and solvate ions, so Na^+, K^+ and NH_4^+ salts are usually soluble, as are ionic compounds with chloride, bromide, and iodide anions. In biological systems, water-soluble compounds often have counterions such as acetate ($CH_3CO_2^-$) and sulfate salts (SO_4^{2-}). Conversely, many carbonate salts (CO_3^{2-}) and phosphate salts (PO_4^{3-}) are insoluble in neutral aqueous media.

[1] The Franklin Institute Inc. "Blood-The Human Heart". Archived from the original on 5 March **2009**.

[2] (a) The Water in You: Water and the Human Body By Water Science School, May 22, **2019**. U.S. Geological Survey's (USGS) Water Science School. https://www.usgs.gov/special-topics/water-science-school/science/water-you-water-and-human-body; (b) Mitchell, H.H., Hamilton, T.S.; Steggerda, F.R.; Bean, H.W. *J. Biol. Chem.* **1945**, *158*, 625.

DOI: 10.1201/9781003605157-1

*Biocompatible mate*rials are those that function with a specific application within a living organism. Most biocompatible polar materials are soluble in blood, such as wettable biomaterial surfaces and protein-coated substrates. The products of digestion (e.g., glucose, amino acids, mineral salts, and vitamins) are sufficiently soluble to be carried from the final section of the small intestine (the ileum) to other organs. Vitamins and essential minerals are also absorbed into the blood plasma in the ileum and carried around the body.

Proteins tend to have polar, hydrophilic amino acid residues (see Section 2.2) on their surface that interact with water and increase solubility. Typical hydrophilic amino acids include arginine, lysine, aspartic acid, and glutamic acid. These amino acids tend to be on the exterior surfaces of proteins. Amino acids with nonpolar, hydrophobic side chains, such as alanine, valine, leucine, or isoleucine, are usually located in the interior of the protein. In the tertiary structures of proteins (see Section 3.1), amino acid residues inside a protein are kept away from water, and internal interactions tend to keep them aggregated together. The solubility of proteins in blood requires a pH in the range of 7.35–7.45. Note that this essential pH of blood is maintained by the bicarbonate-carbonic acid buffer system, where the bicarbonate concentration is greater than the carbonic acid, as discussed in Section 1.5.

1.2 HYDROGEN BONDING

When a hydrogen atom is connected by a covalent bond to a more electronegative atom with a lone electron pair, such as N, O, or F, there is an electrostatic force of attraction between that hydrogen atom and a polarized heteroatom. This interaction is called a *hydrogen bond. Hydrogen bonding* is important for both the solubility of materials and the structural integrity of biomolecules. Hydrogen bonds are important in biological systems such as DNA and proteins, dyes, and pigments, as well as in organic semiconductors.[3] In a hydrogen bond, there is a sharing between a "H-bond donor" and a "H-bond acceptor." A H-bond donor character increases with acidity, while a H-bond acceptor character increases with basicity. Hydrogen bonds are stronger than a van der Waals interaction but weaker than covalent or ionic bonds. Hydrophilic (water-loving) materials are generally polar, capable of hydrogen bonding, and water-soluble.

Two properties of water make it an excellent solvent for polar molecules: cohesiveness due to its H-bonding ability and polarity. The cohesiveness and polarity of water lead to weaker intramolecular forces between polar molecules as the water molecules "pull" on other molecules. Cohesion is essentially the fact that water molecules attract other water molecules by hydrogen bonding. Water is known to form hydrogen bonds with other water molecules, as well as with other molecules that contain heteroatoms. The δ^- oxygen of water can form dipole-dipole interactions with heteroatom-containing molecules. Likewise, the δ^+polarized hydrogen atom of a water molecule can also form dipole-dipole interactions with heteroatoms. Hydrophobic (water-hating) materials are generally non-polar, incapable of hydrogen bonding, poorly soluble or insoluble in water, and they tend to aggregate or cluster together. In other words, hydrophobic materials are repelled by water.

Since water (H–O–H) is such a common solvent and is frequently used in reactions with organic molecules, it is important to understand that water can form strong hydrogen bonds to itself, as well as to molecules that contain a polarized X–H bond (X is usually O, S, N, etc.). An example is a solution of methanol in water. Methanol has a polarized covalent O–H bond, where the hydrogen is $\delta^+$and the oxygen is δ^-. Methanol is an *amphiphilic* (amphipathic) molecule, which is a molecule having both hydrophobic (nonpolar) and hydrophilic (polar) regions. When the δ^-oxygen atom of a methanol molecule is near the $\delta^+$of a hydrogen atom of a water molecule, this proximity establishes an intramolecular *dipole-dipole interaction*, a *hydrogen bond*. This interaction is illustrated for methanol and water in Figure 1.1. Since an O–H bond is polarized and a hydrogen atom is rather small, such an O------H interaction is much stronger than a common dipole-dipole interaction. A hydrogen bond is relatively strong and requires a relatively large amount of energy to disrupt.

[3] Glowacki, E.D.; Irimia-Vladu, M.; Bauer, S.; Sariciftci, N.S. *J. Mater. Chem. B* **2013**, *1,* 3742.

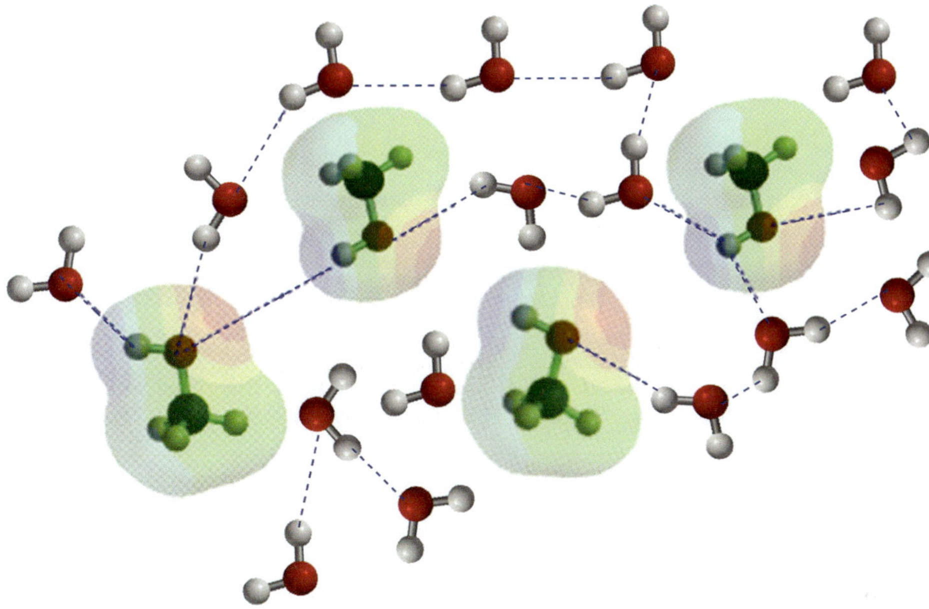

FIGURE 1.1 Hydrogen bonding of methanol molecules in water.

Hydrogen bonding is important for many of the non-biological physical and chemical properties of organic compounds that contain N, O, and F. Intermolecular hydrogen bonding is responsible for the high boiling point of water, for example.

In most discussions of the role of water in biochemistry, there are two traditional classes of water molecules in the hydration shell. One class is the "bound" waters that are hydrogen-bonded to a protein at specific locations and typically remain in position even in an anhydrous environment. The second class is "free" water, which is mobile even in the immediate hydration layer of the protein. Mobile or flexible regions of the protein have fast-moving waters, while stable or less mobile regions generally are surrounded by a slower hydration layer of water molecules. For a given protein, the dynamics of the protein and the solvation shell can be localized and somewhat different for secondary structures such as α-helix, β-sheet, and loop or connector region.

The hydrophilic and hydrophobic interactions and the properties of water lead to a hydration shell around many biomolecules that will influence both the structure and the function of that molecule. The hydration layer is composed of water molecules that cover a biomolecule. The properties of water molecules in the hydration layer differ from those of free water molecules that are proximate to the biomolecule. The spatial arrangement of water molecules in the first hydration layer is different from that in bulk water. The electronic forces and hydrogen bonding of the polar and/or charged groups at the surface contribute to the structural heterogeneity of the hydration layers and to hydration shell dynamics in the conformational dynamics of biomolecules.

The hydration shell model is oversimplified. Indeed, there is a continuum between "bound" molecules and "free" molecules. The hydration shell may be dominated by just a small number of hydration sites in deep surface pockets and clefts, which may mediate the folding of proteins. Water molecules may be trapped in the folded structures of a protein as an element of the secondary structure that is not easily expelled. Water molecules can bridge two proteins and carry important information content that determines a specific interaction. Such water-mediated interactions may be most important when they involve two oppositely charged groups, such as an acid-base pair. Note that many proteins contain hydrophobic cavities, but they do not have to be water-repelling. For proteins, the amino acid sequence is the primary structure of the protein, and the secondary structure focuses on the local structural conformation, which is dependent on the hydrogen bonding available

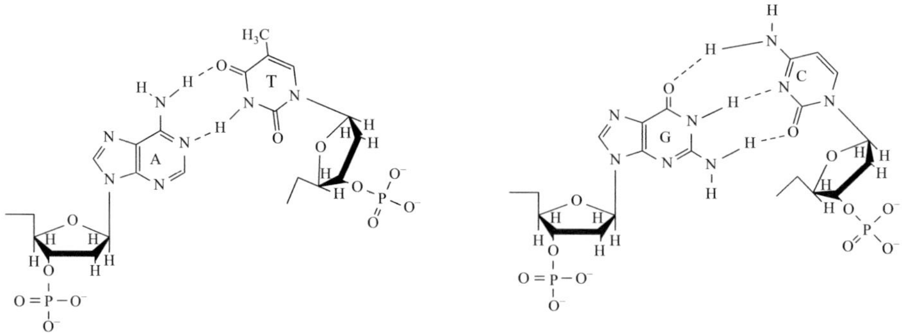

FIGURE 1.2 Watson-Crick base pairs. The C–G and T–A interactions. (Reprinted with permission and adapted from Figure 2 in Rajski, S.R.; Williams, R.M. *Chem Rev.* **1998**, *98,* 2723–2796. Copyright 1998 American Chemical Society.)

to the various amino acid residues of the protein (Section 3.1). Amino acids with hydrophobic side chains are usually found in the interior of the protein, and hydrophilic amino acid residues are found on the surface, where they interact with the aqueous environment.

The previous discussion makes it clear that hydrogen bonds are important for the structural maintenance of biomolecules. Deoxyribonucleic acid (DNA) contains many polarized functional groups, and both intramolecular and intermolecular hydrogen bonding are important. Intermolecular hydrogen bonding between base pairs is responsible for the double-stranded helix of DNA (Sections 15.1 and 15.2). An adenosine (**A**) nucleotide (called a *base* in DNA and RNA) can hydrogen bond with a thymine nucleotide (**T**) of the other strand (an **A–T** base pair), and a guanine nucleotide (**G**) base in one strand can hydrogen bond with a cytosine nucleotide (**C**) base in the other strand (a **G–C** base pair). The strands must be antiparallel (one strand extends 3′→5′ while the other strand aligns 5′→3′) to maximize hydrogen bonding. These hydrogen-bonding base pairs are called *Watson-Crick base pairs*. An adenine moiety (**A**, a 2′deoxyribose bearing the heterocyclic base adenine) can hydrogen bond with a thymine moiety (**T**) of the other strand (an A–T base pair), as shown in Figure 1.2.[4] Similarly, a guanine moiety (**G**) in one strand can hydrogen bond with a cytosine moiety (**C**) in the other strand (a G–C base pair). The stereogenic nature of the ribofuranose and deoxyribofuranose units leads to a "twist" in the polynucleotide backbone.[5]

As mentioned, hydrogen bonding leads to a twist, which is one of the unique structures of a peptide, the so-called α-helix structure, which is another *secondary structure* of a peptide (Section 3.1). In a long-chain peptide composed of L-amino acid residues, intramolecular hydrogen bonding leads to a right-handed helix (called an *α-helix*), where the hydrogen atom on the amide nitrogen is hydrogen-bonded to the oxygen of the carbonyl on the fourth amino acid residue. The dodecapeptide, dodeca(L-alanine), shown in Figure 1.3, can assume a simple α-helix structure, where the intramolecular hydrogen-bonding interactions are indicated by the red "dashed" lines. These hydrogen bonds stabilize the α-helix structure. However, some amino acid residues destabilize an α-helix, including glutamate, aspartate, lysine, arginine, glycine, serine, isoleucine, and threonine. One amino acid residue (proline) creates a bend in the α-helix (Figure 1.4).

Apart from an α-helix, there are two important secondary structures observed with peptides (Section 3.1). One is called a β-pleated sheet, and the other is called a random coil. Poly(aspartic acid) forms a random coil structure, which *does not assume a regular structure* (e.g., the α-helix) because hydrogen bonds are not easily formed. *The β-pleated sheet, on the other hand, involves intermolecular hydrogen bonds between two different peptide chains* rather than intramolecular hydrogen bonds within a single peptide chain. Beta sheets are connected by at least two or three

---

[4] Rajski, S.R.; Williams, R.M. *Chem. Rev.* **1998**, *98,* 2723.

[5] Hecht, S.M. (Ed.), *Bioorganic Chemistry: Nucleic Acids*, Oxford University Press, New York, **1996**, p. 6.

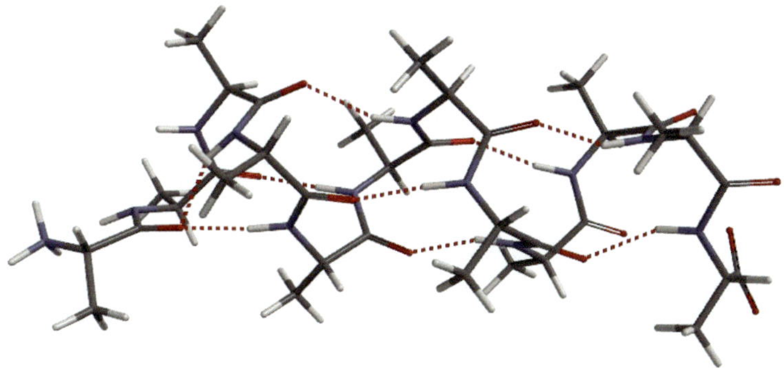

FIGURE 1.3 α-Helical peptide structure for dodeca(alanine) with internal hydrogen bonds marked.

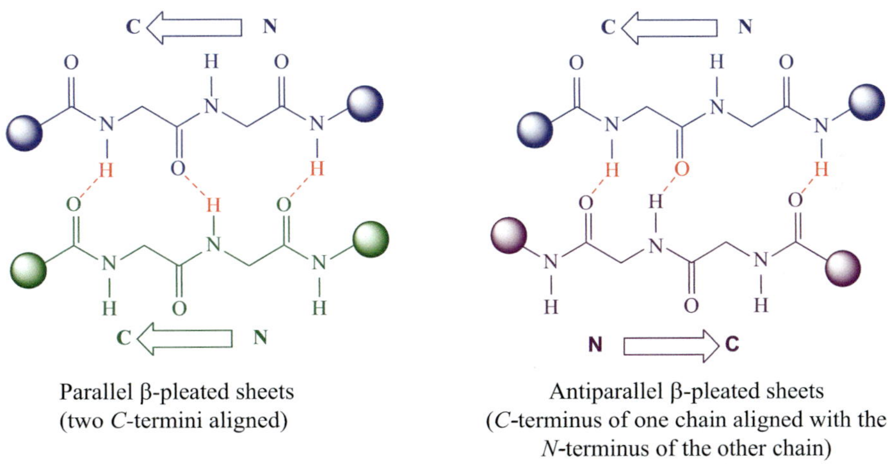

FIGURE 1.4 Parallel and antiparallel β-pleated sheets.

backbone hydrogen bonds, which leads to a generally twisted, pleated sheet. A β-strand is a stretch of polypeptide chain, usually 3–10 amino acids long, with a backbone in an extended conformation. There are parallel[6] or antiparallel β-pleated sheets, and both are shown in Figure 1.4. These variations are defined by having the two *C*-termini aligned or the *C*-terminus of one chain aligned with the *N*-terminus of the other chain.[6] The oxygen of a carbonyl in one peptide chain forms a hydrogen bond to the proton on the amide nitrogen of a second tetrapeptide unit.

Intermolecular hydrogen bonding is important for the biological activity of many enzymatic processes. In the enzyme *haloalkane dehalogenase* from *Xanthobacter autotrophicus* GJ10 (DhlA), a chlorine moiety is first converted to an ester, and hydrolysis liberates 2-chloroethanol. In this transformation, the chloride ion is bound in the active site. In a two-step reaction process, Asp-124, as the initial nucleophile, generates an ester intermediate. As shown in Figure 1.5, there is extensive intermolecular hydrogen bonding in the active site. Once 1,2-dichloroethane is bound to the active site, the carboxylate group displaces a chloride moiety to generate the intermediate ester.[7] In a second step, the His and the carboxylic acid (Asp/Glu) activate a water molecule for hydrolysis of the ester intermediate. The hydrolysis step releases 2-chloroethanol.

---

[6] Smith, C.K.; Regan, L. *Acc. Chem. Res.* **1997**, *30*, 153.

[7] O'Hagan, D.; Schmidberger, J.W. *Nat. Prod. Rep.* **2010**, *27*, 900. Also see Chovancova, E.; Kosinski, J.; Bujnicki, J.M.; Damborsky, J. *Proteins: Struct. Funct. Bioinf.* **2007**, *67*, 305; Lightstone, F.C.; Zheng, Y.-J.; Maulitz, A.H.; Bruice, T.C. *Proc. Natl. Acad. Sci. USA* **1997**, *94*, 8417.

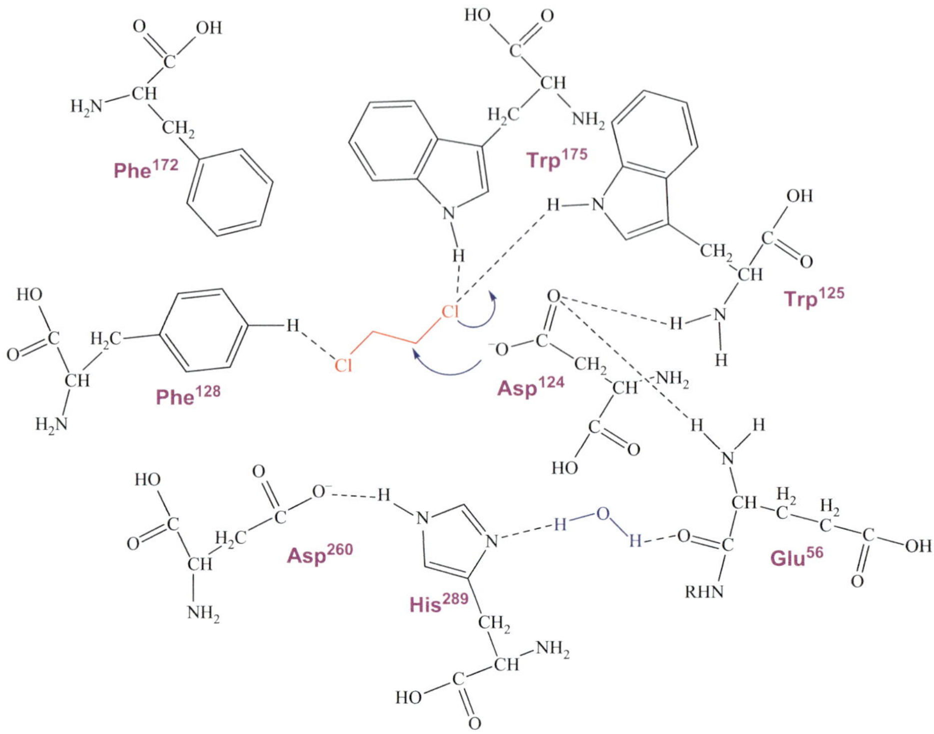

FIGURE 1.5 Partial reaction mechanism of the haloalkane dehalogenase from *Xanthobacter autotrophicus* GJ10 (DhlA) showing the reaction progression, as determined under various crystallization conditions. (Reproduced from O'Hagan, D.; Schmidberger, J.W. *Nat. Prod. Rep.* **2010**, *27*, 900, Fig. 15, p. 909, with permission from the Royal Society of Chemistry.)

1.3 CELLULAR SYSTEMS

Solubility in biological systems is often complicated by the presence of cellular structures. Cells have a phospholipidic membrane, illustrated in Figure 1.6, that tends to be insoluble in aqueous media, and fats are generally insoluble. In both cases, lipids are part of the structure, and the long-chain hydrophobic part of the lipid is responsible for the insolubility of these moieties. When fatty acids are in an aqueous medium, they tend to form micelles with the nonpolar portion of the fatty acid on the interior, exhibiting hydrophobic interactions that lead to aggregation, and the polar carbonyl groups on the surface. Biomolecules such as phospholipids and glycolipids contain both strongly nonpolar and polar groups and are *amphiphilic substances* because they exhibit both hydrophilic and hydrophobic properties. An *organelle* is a specialized subunit within a cell with a specific function, and lipid bilayers typically separate different organelles. Another component of a cell membrane is a liposome. A *liposome* is a spherical vesicle composed of phospholipids and compatible with a lipid bilayer structure. An example is shown in Figure 1.6. In an aqueous medium, lipids tend to form a film at least two molecules thick, where each molecule is arranged with its hydrophobic end directed inward toward the opposite side of the film and its hydrophilic end directed outward toward the aqueous medium. This "film" is called a *bilayer* (see Figure 1.6). Bilayers are formed by the aggregation of amphiphilic lipids. In living cells, a bilayer is a flexible two-layer membrane composed of phospholipids that surrounds the cell. In a bilayer, the long hydrophobic lipid chains aggregate away from water, and the ionic heads are the hydrophilic moieties at the surface to enhance solubility.

Oleic acid, (9Z)-octadecenoic acid, is a typical fatty acid lipid with a 17-carbon hydrocarbon chain that is hydrophobic, but it has a hydrophilic –COOH head. The carboxylic acid head is

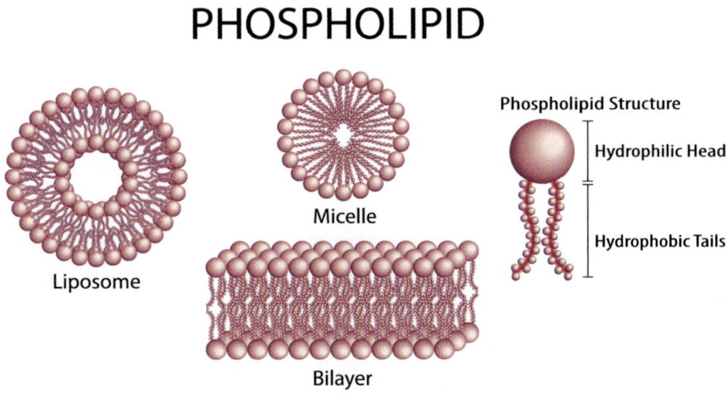

FIGURE 1.6 General phospholipid structure of membrane cells. Shutterstock Vector ID: 2144223119.

expected to have some water solubility, whereas the hydrocarbon tail is expected to be insoluble in aqueous media. A *micelle*, illustrated in Figure 1.6, is an aggregate of surfactant lipid molecules dispersed in a liquid, forming a colloidal suspension. A typical micelle in water forms an aggregate with the hydrophilic "head" regions in contact with the surrounding aqueous medium, accompanied by the sequestering of the hydrophobic "tail" regions in the micelle center. In the case of a fatty acid like oleic acid, the polar –COOH group can hydrogen bond with water, enhancing the water solubility of the micelle.

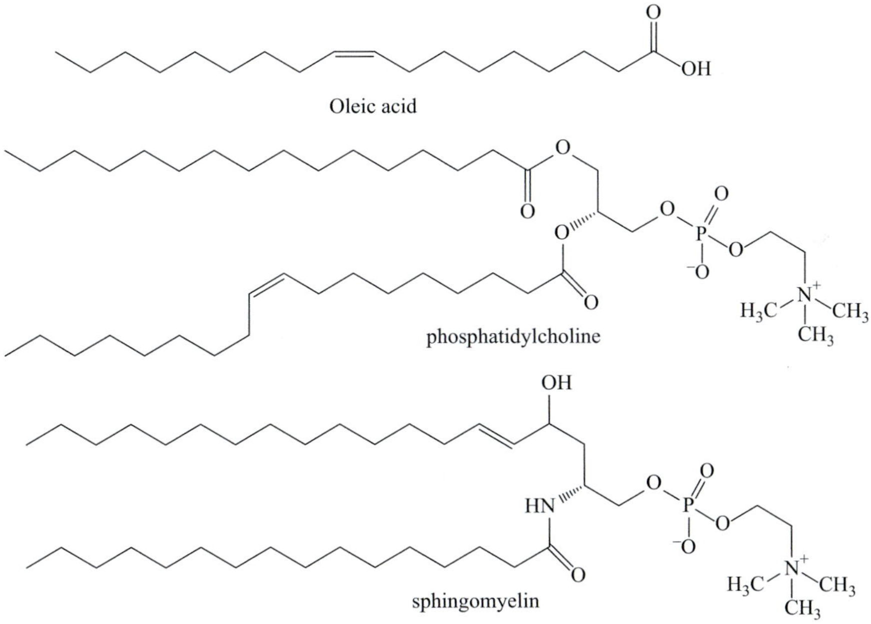

Phospholipid molecules are a group of polar lipids that consist of two fatty acids, a glycerol unit, and a phosphate group, which is esterified with an organic molecule. A phospholipid bilayer structure is part of a cell membrane, as shown in Figure 1.6. In water, the phosphate end tends to be on the surface, with the lipophilic fatty ends pointing toward each other in the middle. Two examples of phospholipids are phosphatidylcholine and sphingomyelin. Phosphatidylcholine is a major part of lecithin, which is essential to cells in the human body. Acetylcholine, which is important for memory, is made from phosphatidylcholine. Phosphatidylcholine is also found in soy, sunflower, mustard, and eggs. Sphingomyelin is a dominant sphingolipid in membranes of mammalian cells. Sphingomyelins are a lipid class found in the plasma membrane, the endocytic recycling compartment, and the trans-Golgi network. These and other lipids are discussed in Section 10.2.

Hydrophobic and hydrophilic interactions are also important for relatively small molecules, including pharmaceutical compounds. When pharmaceutical compounds are ingested, their distribution in the human body is dependent on polarity and hydrogen bonding. The penicillins are important for the treatment of bacterial infections like strep throat, ear infections, and urinary tract infections. They work by attaching to and damaging the cell walls of bacteria. The optimization of dosing is of critical importance to ensure therapeutic success and minimize harmful side effects, including toxicity and drug-resistant infections. Optimal antimicrobial therapy requires an understanding of pharmacokinetics and pharmacodynamics.[8]

Pharmacokinetics is the science that concerns itself with what the body does to a drug. The *distribution coefficient* is an important factor since a drug must first pass through lipid bilayers in the intestinal epithelium in order to be absorbed after oral ingestion. Polarity and solubility are clearly important issues in this process. *Pharmacodynamics* is the science of what a drug does to the body. It is known that hydrophobic drugs tend to be more toxic because they tend to be retained longer, have a wider distribution within the body, are somewhat less selective in their binding to proteins, and are often extensively metabolized. Penicillins tend to have low protein binding in plasma, but the bioavailability depends on the type of penicillin. Penicillin G has low bioavailability, below 30%, whereas penicillin V has higher bioavailability, between 60% and 70%. Penicillins have a short half-life and are excreted via the kidneys.[9] The potassium salt of penicillin V has a distinct advantage over penicillin G in that it is resistant to inactivation by gastric acid. Average blood levels are two to five times higher than the levels following the same dose of oral penicillin G and show much less individual variation.[9] Once absorbed, penicillin V is about 80% bound to serum protein. Tissue levels are highest in the kidneys, with lesser amounts in the liver, skin, and intestines.[9] The drug is excreted as rapidly as it is absorbed in individuals with normal kidney function. However, recovery of the drug from the urine indicates that only about 25% of the dose given is absorbed.[9]

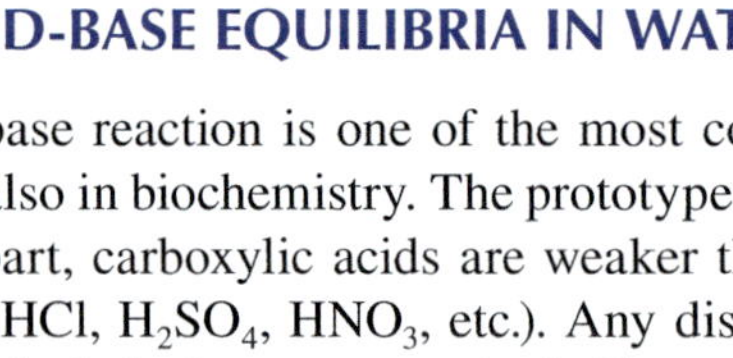

Penicillin G

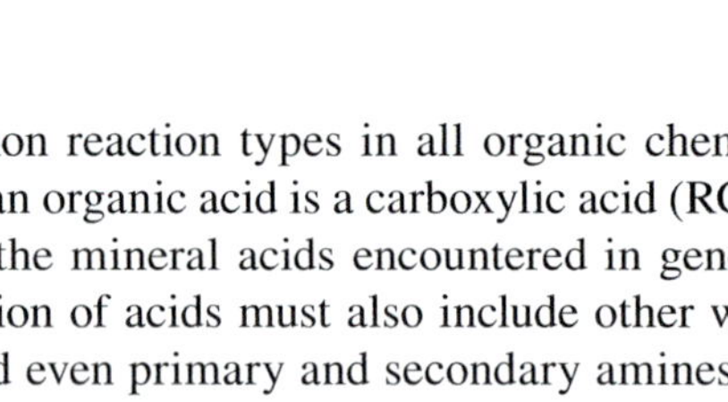

Penicillin V

1.4 ACID-BASE EQUILIBRIA IN WATER

The acid-base reaction is one of the most common reaction types in all organic chemistry and, arguably, also in biochemistry. The prototype for an organic acid is a carboxylic acid (RCO_2H). For the most part, carboxylic acids are weaker than the mineral acids encountered in general chemistry (e.g., HCl, H_2SO_4, HNO_3, etc.). Any discussion of acids must also include other weak acids, including alcohols, ketones, terminal alkynes, and even primary and secondary amines. All these molecules react as acids in the presence of a suitable base to give the corresponding conjugate base. Inorganic bases include NaOH, KOH, $NaNH_2$, $CaCO_2$, $CsCO_3$, etc. Weaker bases include NaOAc, NaOR, amines (R_2NH), and phosphines (R_2PH). The oxygen of alcohols (ROH) and the sulfur of thiols (RSH) and disulfides (RSR) react as the electron donors, and so are bases. A weak base may react with a very strong acid, generating the corresponding conjugate acid; in these cases,

[8] Levison, M.E.; Levison, J.H. *Infect. Dis. Clin. North America* **2009**, *23*, 791.

[9] Wilson, R.C.; Riezk, A.; Arkell, P.; Ming, D.; Armiger, R.; Latham, V.; Gilchrist, M.J.; Märtson, A.-G.; Hope, W.W.; Holmes, A.H.; Rawson, T.M. *Sci. Rep.* **2024**, *14*, 16762.

carboxylic acids, alcohols, ammonium salts, and phosphonium salts. In biochemical systems, a carboxylic acid unit (–COOH) often functions as an acid, and a carboxylate anion ($–COO^-$) often functions as a base.

When discussing a Brønsted-Lowry acid, what is the "acid?" Normally, a generic acid is written as H^+. This representation is simply a proton, of course, but a bottle of "proton" is not possible. An acid is a real molecule where a hydrogen atom is attached to a heteroatom, taking the form H–X. In HX, the hydrogen atom is polarized δ^+. The X group in H–X is usually a heteroatom, particularly O or S, or a halogen. Carboxylic acids contain an O–H unit, and they are well-known "organic acids." Alcohols (ROH) and thiols (RSH) are weak acids. An amine such as R_2NH is a very weak acid, so it requires a strong base for an acid-base reaction. Amines (RNR_2) are well-known organic bases, but their conjugate acids, ammonium salts $RNHR_2^+$ are weak organic acids. The K_a for the acid-base reaction of an amine to form an ammonium salt is shown, along with the K_a formula for this reaction.

$$\text{A—H} \quad + \quad NR_3 \quad \underset{}{\overset{K_a}{\rightleftharpoons}} \quad \text{H—}\overset{+}{N}R_3 \quad + \quad A^-$$

$$K_a = \frac{[R_3NH]\,[A^-]}{[HA]\,[NR_3]}$$

A classical acid-base reaction reacts an acid (A:H) with a base (B:) to yield a conjugate acid (B:H) and a conjugate base (A:) as the products (Figure 1.7). This process is an equilibrium reaction, and the equilibrium constant, K, is given the special designator K_a. If K_a is used for this reaction, then $pK_a = -\log K_a$. Since pK_a and K_a are inversely proportional, $K_a = 10^{-pK_a}$.

The Brønsted–Lowry definition states that an acid (H–X) is a proton donor, but the proton does not "fly off" an acid. A base is defined as a proton acceptor, but how does the base accept a proton? The base forms a covalent bond to the hydrogen atom, and it donates two electrons to form that bond. In other words, a Brønsted-Lowry base donates two electrons to a proton to form a new covalent bond, breaking the X–H bond at the same time. Rather than the acidic proton being "donated," the acidic proton in A:H is "pulled off" by the base. This reaction leads to cleavage of the covalent bond between A and H with transfer of those two electrons from the A–H bond to A, forming A^- (the conjugate base). Therefore, the acid-base reaction shown is simply a chemical reaction in which the electron-rich base donates two electrons to the electron-poor proton, forming a new covalent bond and breaking the covalent bond between A–H with transfer of those two electrons to A.

An acid A–H is usually a molecule with an O–H, S–H, or N–H unit, where each bond is polarized with a δ^+ hydrogen atom. Polarization of an X–H bond generally makes it easier to transfer electrons to the more electronegative atom when the bond is broken. The proton in A–H is polarized δ^+, and the electron-rich base usually has a formal charge of –1 or has a dipole δ^-. The curved, double-headed reaction arrow in Figure 1.7 shows electron donation *from* the electron-rich base *to* the electron-deficient hydrogen atom. *The arrow also indicates the formation of a new covalent bond between B and H in the conjugate acid.*

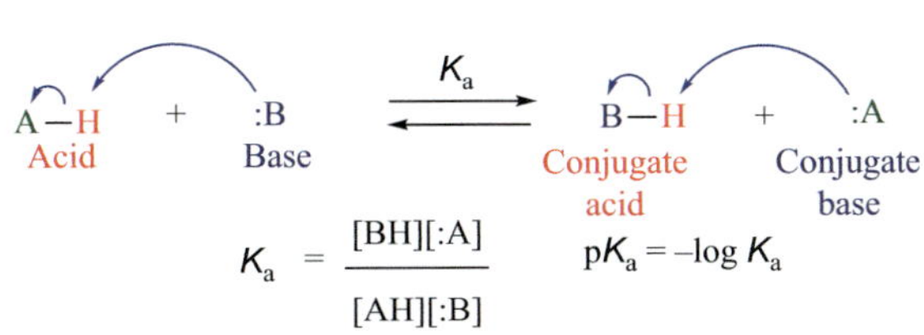

FIGURE 1.7 The generalized acid–base reaction.

By definition, an acid-base reaction has the acid–base pair (the reactants) on the *left* and the conjugate acid-conjugate base pair (the products) on the *right*. In the formula $K_a = \left[\frac{\text{products}}{\text{reactants}}\right]$, the products are therefore the conjugate acid and base, and the reactants are the initial acid and base. If H–A has a weak bond, it is easier for the base to react with the proton, and it is therefore more reactive. If A–H reacts with the base to a greater extent, the equilibrium lies to the right, there is more product, and K_a is larger; HA is more acidic. A larger K_a (smaller pK_a) correlates with A:H as a stronger acid.

If the H–A bond is rather strong, it is more difficult to break that bond, which correlates with a weaker acid. In an acid-base reaction, a stronger H–A bond means a lower concentration of the products, so K_a will be smaller. If K_a is small (large pK_a), the equilibrium lies to the left, which essentially means that A:H does not react with the base. A smaller K_a (a larger pK_a) indicates a weaker acid.

When weak acids dissociate in solution, the *Henderson-Hasselbalch equation* can be used to measure the acidity. In this context, the equation is used to calculate the pH of solutions created by mixing known amounts of acids and conjugate bases. In other words, the *Henderson-Hasselbalch* equation is used to estimate the pH of buffer solutions (Section 1.5) and to calculate the isoelectric point of proteins (Section 1.7). This equation can also be used to estimate the amount of bicarbonate in blood, as it influences pH. The equation is: $pH = pK_a + \log_{10}([A^-]/[HA])$, where p$K_a$ is the negative log of the dissociation constant, [HA] is the molar concentration of the weak acid that is not dissociated, and ([A^-] is the molar concentration of the conjugate base of that acid.

In any acid-base reaction, examination of the stability of the conjugate bases derived from HA is essential, as well as the relative bond strength in the acid. If the conjugate base (A^-) is more stable, it is less reactive so A^- does not react with the conjugate acid HB to yield AH+B^-. *If A^- is less reactive*, there will be a higher concentration of A^- and HB (products) and a lower concentration of AH and B^- (reactants), so K_a is larger, consistent with HA as a stronger acid. This observation effectively explains why a strong acid yields a weak conjugate base.

If the conjugate base (A^-) is less stable, it reacts faster with the conjugate acid HB. This statement means that the reaction of the products is more favorable than the reaction of HA with B in Figure 1.7. If the concentration of products is lower and the concentration of reactants is greater in Figure 1.7, K_a is smaller, and HA is categorized as a weaker acid. This observation correlates with the statement that a weaker acid yields a stronger conjugate base.

If the pK_a of A:H is 2.5 and the pK_a of B:H is 12.1, A:H is the stronger acid, and the equilibrium constant is relatively large for the "forward" reaction, so the reaction proceeds to the right. If the pK_a of A:H is 4.6 and the pK_a of B:H is 4.7, then K_a is close to unity and there will be a mixture of both reactants and both products. In the first case, A:H is categorized as a relatively strong acid, but in the second case, A:H is a relatively weak acid or an acid of moderate strength. Several factors influence the relative reactivity of AH vs. BH.

A convenient measure of an acid-base reaction in water is pH, which is a measure of the acidity or basicity of an aqueous solution. The magnitude of pH depends on the concentration of hydrogen ions or hydroxide ions in the solution. The pH scale is logarithmic and typically ranges from pH 0 (strongly acidic) to pH 7 (neutral) to pH 14 (strongly basic). Acidic solutions have a higher concentration of hydrogen ions, whereas basic solutions have a higher concentration of hydroxide ions. Since the pH scale is logarithmic, a pH of 2 has ten times the concentration of hydrogen ions than a pH 1, and so on.

Actually, pH is the logarithm of the reciprocal of the hydrogen ion activity, a_{H+}, or

$$pH = -\log_{10}\left(a_{H^+}\right) = \log_{10}\left[1/a_{H^+}\right]$$

However, the most common form of pH in this equation is the negative logarithm of the hydrogen ion concentration, measured in moles/liter.

$$pH = -\log\left[H^+\right]$$

The pH of aqueous solutions is important for the concentration of nutrients such as nitrogen compounds and phosphorus compounds in biological systems. It is useful to be able to correlate pH with familiar items or processes. For example, battery acid typically has a pH of 0, and stomach acid typically has a pH of around 1.[10] Lemon juice or vinegar has a pH around 2, and commercially available colas have a pH of around 2–3.[10] Beer has a pH of about 4 and acid rain typically has a pH around 5, whereas black coffee has a pH between 5 and 6 with milk having a pH of around 6.[10] The pH of blood ranges from 7.36 to 7.44, eggs have a pH around 8 and toothpaste has a pH around 9.[10] Aqueous detergent has a pH around 10, household ammonia around 11, and oven cleaner around 13.1. Finally, liquid drain cleaners have a pH of around 14.[10]

Identification of the pH of common items allows the pH values of different body fluids to be placed in their proper perspective. The pH of gastric juice is 1.0–2.0, saliva is 6.7–7.4, bile is 7.4–8.5, urine is 5.0–7.5, cerebrospinal fluid is 7.35–7.45, tears are 7.4, and stool is 7.0–7.5.[11] Changes in pH are noteworthy in human digestion. The pH in the human digestive tract begins with saliva, which is usually between 6.5 and 7.5.[12] Upon swallowing, food travels to the upper portion of the stomach (pH between 4.0 and 6.5). As the stomach secretes HCl and pepsin, the pH changes to 1.5–4.0. From the stomach, food enters the duodenum, where the pH is 7.0–8.5 and absorption of nutrients occurs.

In general, decreasing the pH increases the solubility of sparingly soluble bases and basic salts, whereas increasing the pH has the opposite effect. The solubility of the acid and/or base will affect their reactivity and hence the equilibrium constant. Likewise, the solubility of the conjugate acid and/or conjugate base will influence the equilibrium. The solubility of a substance is affected not only by temperature but also by the presence of other solutes. Therefore, the solubility of a conjugate base may be influenced and/or changed by the concentration of the initial acid or by the conjugate acid as it is formed. Changes in solubility of a drug candidate, for example, can be correlated with the pH, and solubility can be predicted by the pH_p equation. The pH_p is the pH below which an acid, or above which a base, will begin to precipitate. For a weak acid, $pH_p = pK_a + \log (S - S_o)/(S_o)$. For a weak base, $pH_p = pK_w - pK_b + \log (S_o)/(S - S_o)$. In these equations, S_o is the molar solubility of the undissociated acid or base, and S is the molar concentration of the salt form of the initially added drug. The term pK_w (w = water) is usually defined as $-\log [H^+][OH^-]$, or 14.00 at 25°C. While this definition is true in pure water, the ionization of pure water is rarely important compared with the ionization of solutions containing dissolved solutes. Solubility in blood is an example. "Blood has presumably the same ionic strength as normal saline (9 g/L NaCl) or $\mu = 0.15$, but it is unlikely that the pK_w is the same as in normal saline because blood serum also contains 8% of organic matter. The presence of this organic matter will change the structure, the concentration, and therefore the activity of the water in the serum."[13] This calculation determines the percentage of a reactant that will precipitate at or below a certain pH.

Acid-base chemistry is important for many common processes in biological systems. One useful example of acid-base equilibrium is metabolic acidosis and metabolic alkalosis. The normal range for blood pH is 7.36–7.44. If pH levels drop below 6.9, it can lead to coma. The body can generate acidic compounds that change this pH, making blood or tissue more acidic because of the natural breakdown of fats. This statement is particularly true when fats are used for energy rather than carbohydrates (see Sections 10.1 and 10.2), although other processes can also generate acids. Bicarbonate is generated in biological systems and reacts as a base. If there is insufficient bicarbonate to compensate for the extra acid, acidosis can occur. Formally, acidosis is defined as a significant decrease in the pH of extracellular fluid.[14] This condition can result from both respiratory and metabolic abnormalities. Respiratory acidosis occurs when breathing abnormalities cause CO_2 retention and an elevation in P_{CO2} in the alveoli and arterial blood (known as hypercapnia).[14] The term P_{CO2} refers to the partial pressure of CO_2 in the pulmonary alveoli during respiration.[14] Retention of CO_2

---

[10] https://www.epa.gov/sites/production/files/2015-10/documents/1622624.pdf

[11] Gupta, P.P., *Textbook of Biochemistry with Significance*, 2nd ed., CBS Publishers, New Delhi.

[12] https://www.news-medical.net/health/pH-in-the-Human-Body.aspx

[13] Hawkes, S.J. *J. Chem. Ed.* **1995**, *72*, 799.

[14] Meyers, F.H.; Jawetz, E.; Goldfien, A. *Review of Medical Pharmacology*, Lange Medical Pub., Los Altos, CA, **1972**, p. 431.

can result from inadequate ventilation during anesthesia, certain central nervous system (CNS) diseases, or drug use, and it is also observed in emphysema.[14] Metabolic acidosis occurs with starvation, uncontrolled diabetes mellitus with ketosis, and electrolyte and water loss due to diarrhea.[14]

Formally, alkalosis is a significant increase in the pH of extracellular fluid.[14] Both respiratory and metabolic alkalosis are recognized. Respiratory alkalosis results from hyperventilation, which produces a lowered P_{CO2} and a higher pH in extracellular fluids.[14] Anxiety is the usual cause of hyperventilation. Metabolic alkalosis can result from the loss of gastric juices that are rich in HCl, from excessive sodium bicarbonate ingestion, and it is also associated with potassium ion deficiency.[14]

An important and fast-acting influence on the regulation of pH is the respiratory rate, through which we excrete more carbon dioxide. Carbon dioxide is in equilibrium with carbonic acid, and it is excreted with increased respiration. Breathing more deeply and quickly can move the blood pH toward the alkaline side. The kidneys can regulate the reabsorption of carbonic acid in the tubule, increasing or reducing acid secretion. So, urine that is more acidic than normal may mean the body is ridding itself of excess dietary acid and thus making blood pH more alkaline. Ammonia produced by the kidneys is another way to regulate pH balance.

Local anesthetics are commonly used to block pain in medical procedures (e.g., filling a tooth by a dentist). Many local anesthetics block nerve conduction by reducing the membrane permeability to sodium ions.[15] A common local anesthetic is lignocaine (also called lidocaine). This compound is clearly an amine, and the nitrogen atom marked in blue is a weak base.[15] The other nitrogen atom (marked in blue) is part of an amide functional group but is not basic due to the proximal carbonyl unit and delocalization of the nitrogen lone pair. Lidocaine is usually distributed as an aqueous solution of the HCl salt (lidocaine HCl), which is the conjugate acid formed by the reaction of lidocaine with HCl,[15] as shown in the generic reaction given. At physiological pH, which is usually between 7.36 and 7.44, most of the drug will exist in its ionized form (lidocaine •HCl), which is believed to combine with the excitable membrane to inhibit sodium permeability.[15] Some of the drug will exist in the unionized form since, in the acid–base equilibrium both the base (lidocaine) and its conjugate acid (lidocaine ·HCl) will be present. It is believed that lidocaine more easily penetrates the lipid barrier around and within the nerve tissues.[15]

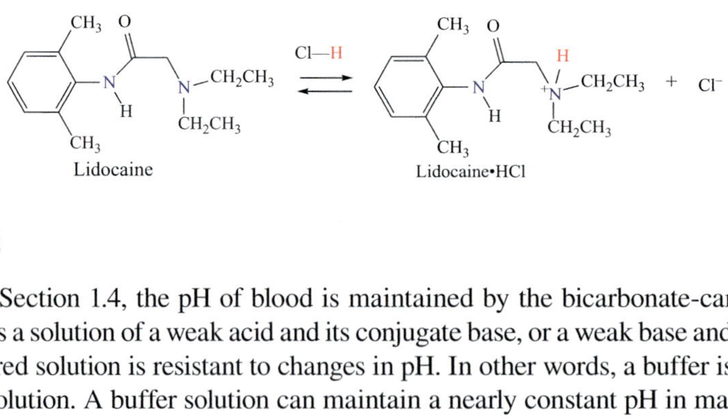

1.5 BUFFERS

As mentioned in Section 1.4, the pH of blood is maintained by the bicarbonate-carbonic acid buffer system. A *buffer* is a solution of a weak acid and its conjugate base, or a weak base and its conjugate acid salt. Such a buffered solution is resistant to changes in pH. In other words, a buffer is used to maintain a stable pH in a solution. A buffer solution can maintain a nearly constant pH in many chemical reactions, for example, by reacting with small amounts of an acid as it is produced in a chemical reaction. The Henderson-Hasselbalch equation (Section 1.4) is commonly used to approximate the pH of a buffer solution. More importantly, this equation can be used to choose a buffer in the desired pH range.

In organic chemistry, buffers are typically used when a planned reaction is known to generate an acidic or basic medium that would be deleterious to the outcome of the reaction. Similarly, the purpose of a buffer in a biological system is to maintain intracellular and extracellular pH within a very narrow range and to resist changes in pH in the presence of internal and external influences.

[15] Grahame-Smith, D.G.; Aronson, J.K. *The Oxford Textbook of Clinical Pharmacology and Drug Therapy*, Oxford University Press, Oxford, UK, **1984**, pp. 551–552.

In vitro enzymatic activity is highly dependent on the reaction medium, and buffers are an important parameter to keep the pH stable. Preventing a severe pH change prevents possible denaturation of the enzyme, for example. However, buffer agents may have negative effects on enzymatic activity, such as competitive substrate inhibition.[16] Other problems include (i) the buffering capacity of the buffer agent may be unsuitable; (ii) the buffering agent may have poor solubility; (iii) the buffer may have high ionic strength; (iv) there may be destabilizing interactions between the buffer agent and the enzyme; and (v) there may be competition between the buffering agent and the reaction substrate for the active site.[17]

A method has been introduced to assess enzymatic activity in different buffer systems prior to experimental buffer screenings.[17] Finding the best compromise of the reaction optima for each enzyme includes finding the best buffer compound. A preassessment of probable appropriate or improper buffers uses a proposed method that computationally predicts possible enzymatic activity reduction due to competitive binding of the buffering agent to the active site. Well-established computational modeling techniques have been adapted to this biotechnological problem. A docking and scoring approach is used, similar to the methods used to find protein inhibitors in drug design. The strength of binding of an enzyme substrate is only weak during the enzymatic conversion step compared to the tight binding of protein inhibitors used as drugs. If a buffering compound can be docked in silico with a good interaction pattern into the active site, a reduction of enzymatic activity in this buffer medium is very likely. This is reasonable because the buffer competes with the substrate for binding to the active site. Since the buffer concentration in the reaction medium mostly exceeds the substrate concentration, this effect should not be neglected. With this computational approach, certain buffering agents can be ruled out from a list of possible buffers without testing them experimentally. The method for computational prediction first prepares the three-dimensional enzyme structure, and the active site is determined. The compound docking library is then compiled. For each compound, up to 100 docking poses are generated. These poses are scored with the HYDE[18] scoring function. The best-scored pose of each compound is evaluated and compared to the scores of the other compounds. Finally, the scores are analyzed in comparison to the score of the substrate to assess the inhibition potential of the buffer compounds.[18]

It is possible to choose buffers that maintain the pH in a relatively narrow range at the pH of choice, and this statement applies to traditional chemical reactions as well as protein reactions.[19] The bicarbonate buffering system, for example, helps maintain the pH of blood. Proteins also form part of the buffer system that regulates pH levels, acting as both H^+ acceptors and donors since they contain both basic and acidic groups. Sodium acetate is commonly used as a buffer, and the equilibrium involves the acid-base reaction of acetate (ethanoate anion) with H^+, which establishes an equilibrium with the conjugate acid, acetic acid (ethanoic acid). As more acid is added, it reacts with the basic acetate, but the equilibrium is maintained, keeping the pH relatively constant. In such a situation, the addition of acetate to an acid-forming reaction maintains the pH at a relatively constant level by neutralizing small amounts of excess H^+ (acid) as it is produced during a chemical reaction. Phosphate buffers are widely used to moderate pH levels.

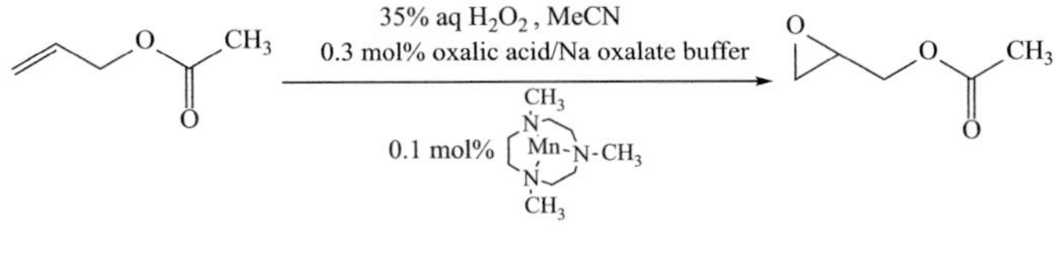

[16] Schomburg, K.T.; Ardao, I.; Götz, K.; Rieckenberg, F.; Liese, A.; Zeng, A.-P.; Rarey, M. *J. Biotechnol.* **2012**, *161*, 391.
[17] (a) Desmarais, W.T.; Bienvenue, D.L.; Byzmek, K.P.; Holz, R.C.; Petsko, G.A.; Ringe, D. *Structure* **2002**, *10*, 1063; (b) Scorpio, R. *Fundamentals of Acids, Bases, Buffers & Their Application to Biochemical Systems*, Kendall/Hunt Pub. Co., Dubuque, **2000**; (c) Zaks, A.; Klibanov, A.M. *Proc. Natl. Acad. Sci. U.S.A.* **1985**, *82*, 3192.
[18] Reulecke, I.; Lange, G.; Albrecht, J.; Klein, R.; Rarey, M. *ChemMedChem* **2008**, *3*, 885.
[19] https://www.applichem.com/fileadmin/Broschueren/BioBuffer.pdf

Apart from the use of buffers in enzyme-catalyzed reactions, there are many examples of organic reactions that are modulated by the addition of a buffer. One example is the peroxide-induced epoxidation of terminal alkenes.[20] Such an epoxidation reaction is often slow, and reversibility can be a problem. In the peroxide oxidation, an acid is produced as a byproduct, which can react with the epoxide product with deleterious consequences, especially since terminal alkenes react rather slowly. Therefore, a buffer is often required. In the example shown, allyl acetate reacts with 35% aqueous hydrogen peroxide in acetonitrile, with 0.1% of a Mn-trimethyltriazacyclonane complex and 0.3% of an oxalic acid/sodium oxalate buffer. Under these conditions, oxirane-2-ylmethyl acetate was formed in >99%.[20] The 1:1 oxalic acid-oxalate buffer gave the optimum balance of rate and efficiency.

1.6 STRUCTURAL FEATURES THAT INFLUENCE ACID STRENGTH

A carboxylic acid has a COOH unit that constitutes the functional group, with a polarized carbonyl group (C=O) attached to an O–H unit. The electron-withdrawing effect of the carbonyl oxygen on the acidic hydrogen of the OH unit is called an *inductive effect.* This inductive effect makes the attached hydrogen more positive, which means that the molecule is more acidic. However, bond polarization of the O–H is not enough to explain the large difference in acidity of carboxylic acids compared to other O–H functional groups with an acidic proton, such as alcohol ROH. The complete acid-base reaction must be considered, which means that the structure and stability of the conjugate acid, and conjugate base pair must be considered as well.

Methanoic acid (formic acid, HCOOH) is a specific example of a carboxylic acid, with an experimentally measured pK_a of 3.75. The relative acidity of a carboxylic acid can be examined in the acid-base reaction of formic acid with sodium amide to yield sodium formate and ammonia, as shown in Figure 1.8. The reaction of methanol (CH_3OH; pK_a of 15.2) with sodium amide to form the conjugate base (methoxide ion) and the conjugate acid (ammonia) is also shown in Figure 1.8. When compared to the O–H unit of an alcohol, the O–H unit in formic acid is more polarized, has a larger δ^+ dipole, and the proton is more acidic. Experimental data show that formic acid is much more acidic than methanol, but bond polarization of the O–H unit, while an important parameter, does not fully explain the difference in acidity. Turning attention to the conjugate base, the formate anion is resonance-stabilized, whereas the methoxide ion is not. In other words, the charge in methoxide resides entirely on oxygen, whereas the charge on the formate anion is dispersed over three atoms (see Figure 1.8). If the charge on a conjugate base is dispersed over a greater area, *it is more stable*

$$HCOOH + NaNH_2 \longrightarrow HCOO^- + NH_3$$
$$CH_3OH + NaNH_3 \longrightarrow CH_3O^- + NH_3$$

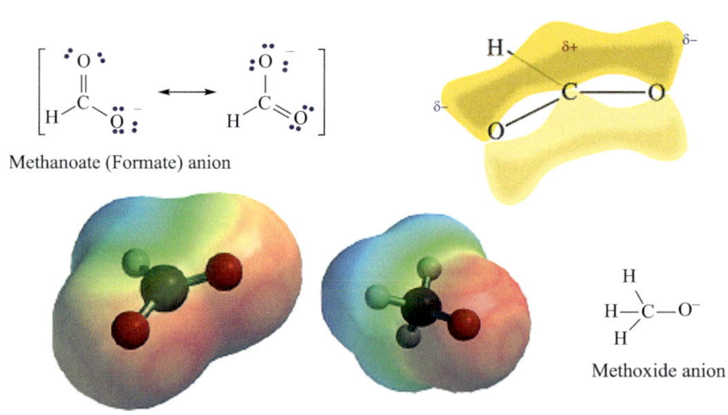

FIGURE 1.8 Structures of the formate anion and the methoxide anion.

[20] De Vos, D.E.; Sels, B.F.; Reynaers, M.; Subba Rao, Y.V.; Jacobs, P.A. *Tetrahedron Lett.* **1998**, *39*, 3221.

and therefore less reactive with the conjugate acid. The greater stability of the formate anion means that ionization of formic acid is greater, so the concentration of the conjugate acid/conjugate base is greater. In other words, the equilibrium of the reaction will be shifted toward the products; i.e., formic acid is more acidic. The methoxide product is not resonance-stabilized, so it is less stable and more reactive. The acid-base equilibrium is not shifted as far to the right, so methanol is not as acidic.

Phosphoric acid derivatives, especially phosphates ($ROPO_3^{2-}$) are important in biological systems. It is known that phosphoric acid (pK_a, 2.16) has three acidic protons, and it is more acidic than formic acid (pK_a, 3.75). Several phosphorous-based acids are important, each with multiple pK_a values. Phosphonic acid is $HP(=O)(OH)_2$, and the first pK_a is 2.48 and the second pK_a is 7.29.[21] Methyl phosphonic acid ($CH_3PO_3H_2$, pK_a, 2.48 and 7.29) is more acidic than ethanoic acid (pK_a, 4.76).[21] Deprotonation of these two acids generates the corresponding conjugate base, the first conjugate base, methyl phosphonate $CH_3PO_3^-$ and then the second conjugate base, dianion $CH_3PO_3^{2-}$.

The dianion of methylphosphonic acid (the second conjugate base) is resonance-stabilized to a greater extent than the mono-anion (the first conjugate base), which is also resonance-stabilized, as is the acetate anion. Note that phosphorus is larger than carbon, which contributes to greater dispersal of charge in the methylphosphonate monoanion. Greater charge dispersion leads to increased stability and increased acidity. Lastly, the O–P bond is more polarized than the O–C bond, and the increased size of phosphorus relative to carbon suggests that the O–H bond is more reactive in methylphosphonic acid. In other words, it is more acidic. Conversely, the alkylphosphonate dianion is very stable and has diminished reactivity due to resonance and the larger size of the phosphorus atom.

1.7 ACID-BASE EQUILIBRIA OF AMINO ACIDS

As discussed above, an amine is a base, and a carboxylic acid is an acid. A basic amine reacts with the acidic proton of acetic acid in an intermolecular reaction to form a conjugate base and a conjugate acid. If methanamine (methylamine) and acetic acid are mixed, for example, the product is methylammonium acetate and the acetate anion in a normal Brønsted-Lowry acid-base reaction. Amino acids have an amine moiety and an acid moiety in the same molecule. In other words, an amino acid contains both a base and an acid. If 2-aminopropanoic acid (alanine) is examined, the molecule does not exist as a neutral acid and a neutral amine, but rather the basic amine unit reacts with the carboxylic acid unit in an *internal acid-base reaction that yields 2-ammoniopropanoate*, as shown in Figure 1.9. This salt has a positively charged ammonium unit ($-NH_3^+$) and a negatively charged carboxylate unit ($-CO_2^-$) in the same molecule, so the molecule is electrically neutral. Such molecules are known as dipolar ionic molecules or *zwitterions*. The zwitterion form is used for all amino acids discussed in this book, unless otherwise noted.

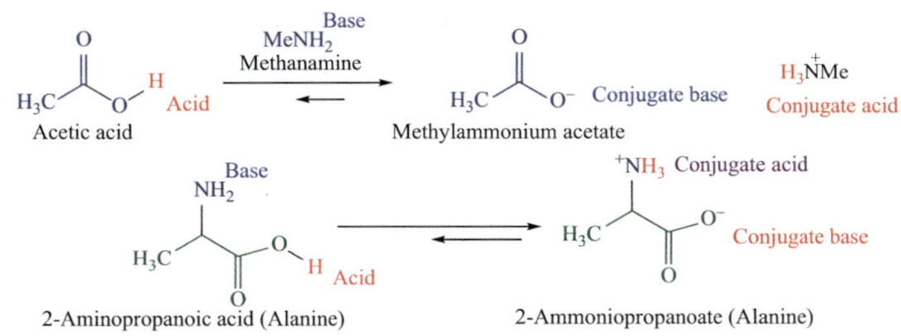

FIGURE 1.9 Internal acid–base reaction of an amino acid.

[21] Zhang, S.; Baker, J.; Pulay, P. *J. Phys. Chem. A* **2010**, *114*, 432.

Examination of the zwitterion form of alanine, ammoniopropanoate, shows that it has an ammonium salt moiety, which is a weak acid. In the acid-base equilibrium shown, both the COOH unit and the ammonium salt moiety are available, so there are *two acidic species.* The salient point of this observation is that an amino acid will have two K_a values and, therefore, two pK_a values.

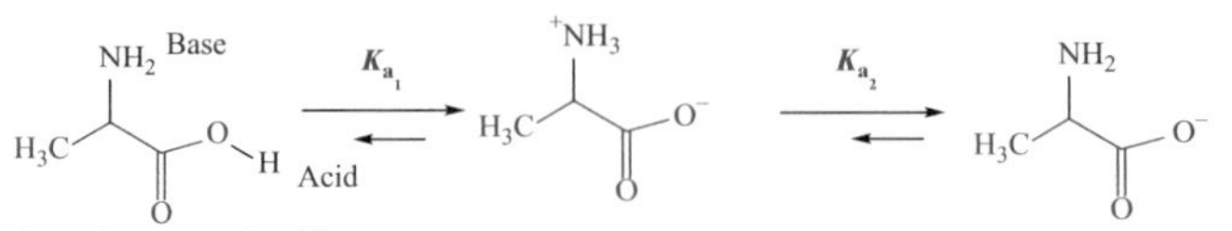

2-Aminopropanoic acid (Alanine) 2-Ammoniopropanoate (Alanine) 2-Aminopropanoate (Alanine)

Using the amino acid alanine as an example, an internal acid-base reaction of the amine unit and the carboxylic acid unit will lead to the ammonium carboxylate in an equilibrium reaction labeled K_{a1}. A second acid-base reaction is possible that converts the ammonium carboxylate to the amine carboxylate (2-ammoniopropanoate) in the equilibrium reaction labeled K_{a2}. Therefore, there are two acid-base reactions andtwo acidity constants, K_{a1} and K_{a2}. The value of K_{a1} is the equilibrium constant for the interconversion of 2-aminopropanoic acid and 2-ammoniopropanoate. The value of K_{a2} is the equilibrium constant for the interconversion of 2-aminopropanoate and 2-ammoniopropanoate. The values of pK_{a1} and K_{a2} will change somewhat as the nature of the substituents attached to the α-carbon of the amino acid is varied. Formally, the values of K_{a1} and K_{a2} are defined as follows, consistent with the standard definition of K_a for each acid-base reaction.

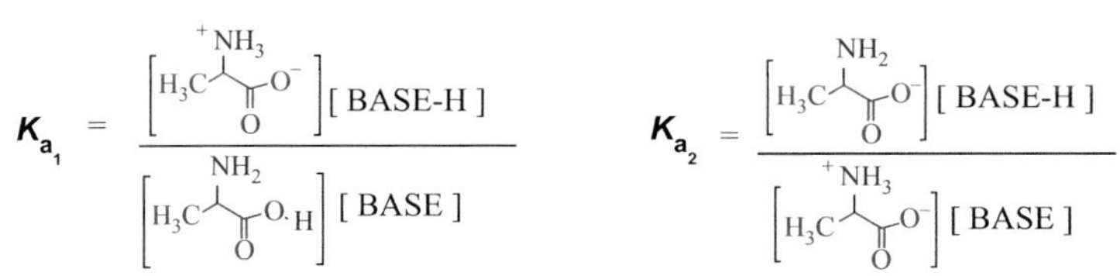

All reactions are equilibrium reactions, as indicated, and one of the species in that equilibrium will be a neutral species. The point in the equilibrium when this neutral species is formed is the *isoelectric point, pI*, which is defined as the pH at which the structure carries no net electrical charge. If p*I* is used to represent the isoelectric point, then p*I* is defined by the following equation:

$$pI = \frac{pK_{a_1} + pK_{a_2}}{2}$$

A pH curve versus the number of equivalents of base for a generic amino acid shows the isoelectric point and both K_{a1} and K_{a2} (Figure 1.10). The values of K_{a1} and K_{a2} vary with the substituents attached to the amino acid. The point of this figure is to show that K_{a1} and K_{a2} can be experimentally determined for any amino acid and, therefore, the isoelectric point can be determined.

As will be discussed in Chapters 2 and 3, the side chains attached to the amino acid residues in a protein may be positive, negative, neutral, nonpolar, or polar in nature, which will determine the overall charge of a protein. Proteins will have “free” carboxyl units and “free” amino units, depending on the nature of the side chains of the amino acid residues.

It is important to point out that each of the individual amino acids has pK_a values and an isoelectric point. Table 1.1 shows the name of the amino acid as well as the three-letter code and the one-letter code.[22] The isoelectric point of each amino acid is also shown in Table 1.1.

[22] *CRC Handbook of Chemistry and Physics*, 94th ed., CRC Press, Boca Raton, FL, **2014**, pp. 7-1–7-2.

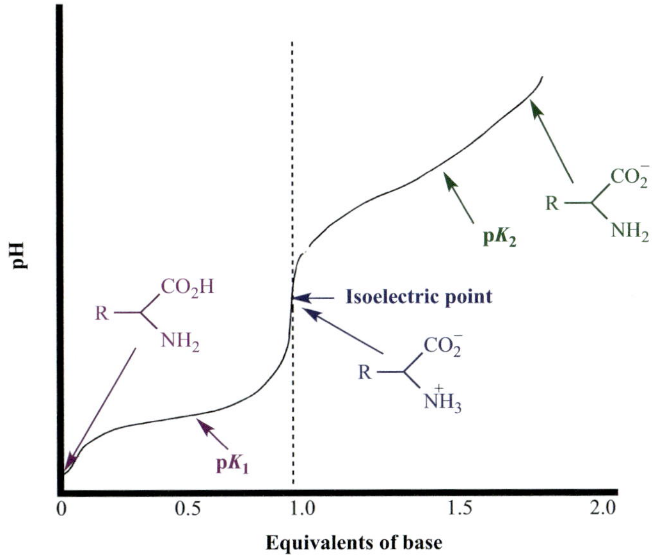

FIGURE 1.10 The position of p*K* and the isoelectric points for a generic amino acid.

TABLE 1.1
The p*K* Values of Amino Acid at The Isoelectric Point in Water At 25°C

R in $^+NH_3CHRCO_2^-$	Name	Isoelectric Point	Three-Letter Code	One-Letter Code
H	Glycine	5.97	gly	G
Me	Alanine	6.00	ala	A
$CHMe_2$	Valine	5.96	val	V
$CHMe_2$	Leucine	5.98	leu	L
CH(Me)Et	Isoleucine	6.02	ile	I
CH_2Ph	Phenylalanine	5.48	phe	F
CH_2OH	Serine	5.68	ser	S
CH(OH)Me	Threonine	5.60	thr	T
CH_2(4-hydroxy-C_6H_4)	Tyrosine	5.66	tyr	Y
CH_2SH	Cysteine	5.07	cys	C
CH_2CH_2SMe	Methionine	5.74	met	M
CH_2CONH_2	Asparagine	5.41	asn	N
$CH_2CH_2CONH_2$	Glutamine	5.65	gln	Q
CH_2COOH	Aspartic acid	2.77	asp	D
CH_2CH_2COOH	Glutamic acid	3.22	glu	E
$CH_2CH_2CH_2CH_2NH_2$	Lysine	9.74	lys	K
CH_2(2-indolyl)	Tryptophan	5.89	trp	W
CH_2(4-imidazolyl)	Histidine	7.59	his	H
$CH_2NHC(=NH)NH_2$	Arginine	10.76	arg	R
2-Pyrrolidinyl	Proline	6.30	pro	P

The neutral amino acid glycine (2-ammonioacetate) is shown in the zwitterionic form in Figure 1.11, and it is in equilibrium with carboxylmethanaminium and 2-aminoacetate. Note that a zwitterion is simply a term for a dipolar ion. The equilibrium for an amino acid with an acidic side chain is more complex because of the extra acidic group. When the side chain has an acidic proton that is less acidic than the ammonium proton or the carboxyl proton (e.g., the phenolic proton in tyrosine), the

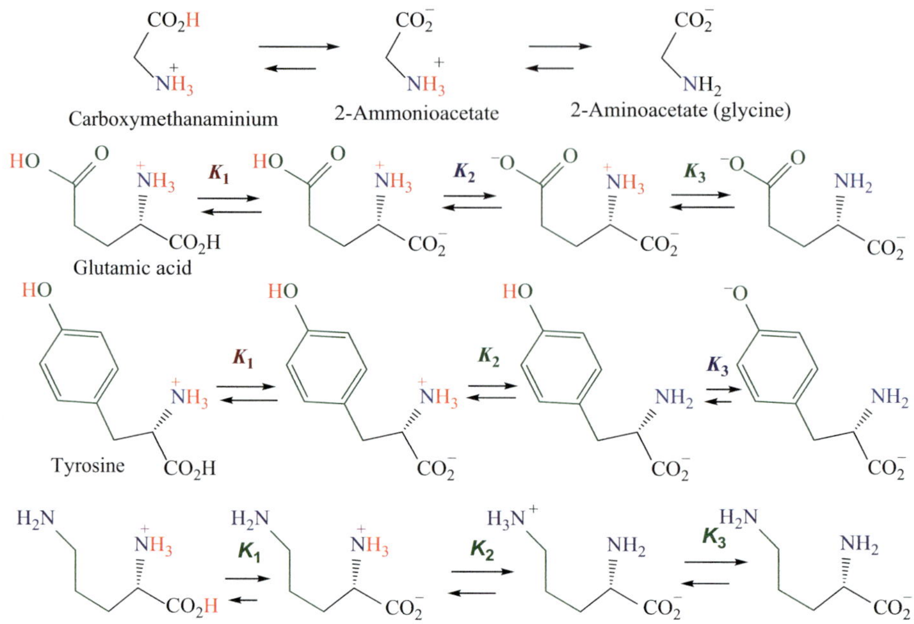

FIGURE 1.11 The K_a equilibria for glutamic acid and tyrosine.

equilibrium is slightly different. As shown in Figure 1.11, equilibration of the COOH side chain of glutamic acid and of the phenolic OH in tyrosine is represented by pK_{a3}. Similarly, in Figure 1.11, lysine has a third pK value for the ammonium salt of the amine unit on the side chain, as shown.

A few amino acids have attached functional groups that are either acidic or basic. Some amino acids have hydrophobic side chains, and several amino acids have polar side chains that are neither strongly hydrophilic nor hydrophobic. In an aqueous environment, bending and twisting about the bonds of the peptide lead to conformations (Section 5.12) with many of the hydrophilic amino acid residues on or near the surface, allowing maximum interaction of those groups with water. Likewise, the hydrophobic amino acid residues tend to aggregate away from contact with water at the surface, and they tend to be in the interior of the protein.

Protein folding is largely determined by the interactions between water and amino acids near the protein surface. The rearrangement of water molecules around unfolded protein chains releases energy that drives the folding process. If hydrophobic amino acid residues were on the surface of a protein, the large surface area with water would be reduced. Therefore, hydrophobic amino acid residues tend to aggregate in the interior of the protein, away from water, as noted. Accessible surface areas of a protein are determined by the polar, nonpolar, charged, positively charged, and negatively charged moieties of the amino acid residues. Therefore, the interaction of charged amino acids has a great influence on the final tertiary structure of the protein, and the negative surface charge had the strongest correlation with increased protein solubility.[23] Indeed, the amino acid on the protein surface largely determines the solubility of the protein.[24]

The free energy change on folding or unfolding is due to the combined effects of both protein folding/unfolding and hydration changes. Folding of a protein is further stabilized by the formation of disulfide linkages when folding brings two cysteine residues into proximity. When the carbonyl of one amino acid is brought into proximity to the NH unit of another amino acid, hydrogen bonding is possible to stabilize the tertiary structure of a protein. The carboxylic acid side chain moiety

---

[23] Cheng, B.; Cui, S.-X. *Chinese J. Polymer Sci.* **2018**, *36,* 379.

[24] Kramer, R.M.; Shende, V.R.; Motl, N.; Pace, C.N.; Scholtz, J.M. *Biophys. J.* **2012**, *102*, 1907.

can react intramolecularly with the amino unit side chain moiety of another amino acid to form ammonium carboxylate moieties. All these factors influence the tertiary structure of a protein (see Section 3.1), the solubility of the protein, and its interaction with other biomolecules.

1.8 ACID-BASE REACTIONS IN BIOLOGICAL SYSTEMS

The two acid-base reactions that occur with amino acids such as alanine were described in Section 1.7. The reaction of alanine with sodium amide is shown, and the conjugate base is 2-ammoniopropanoate with an ammonium unit, which is a weak acid. The ammonium salt reacts as an acid to give 2-aminopropanoate. Therefore, there are two equilibrium constants, K_{a1} and K_{a2}, as discussed previously for alanine. For alanine, the pK_{a1} is about 2.3 and the pK_{a2} is about 9.7. To generate a reasonable equilibrium concentration of the pK_{a1} conjugate base, 2-ammoniopropanoate, the base chosen to react with the amino acid should be a stronger base than the ammonium carboxylate product, and the conjugate acid should be a weaker acid than the neutral amino acid. To generate the amino carboxylate, a weaker base can be used.

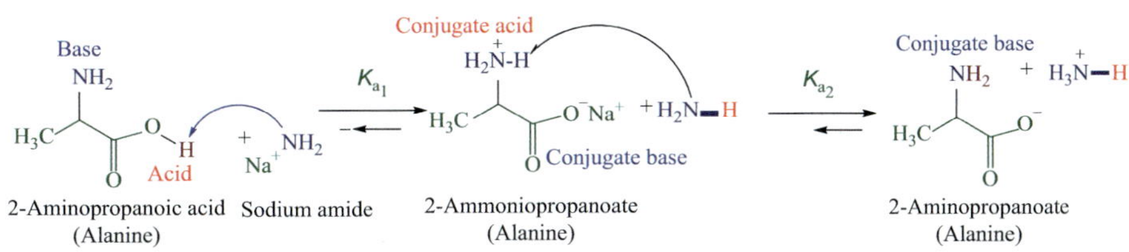

In biochemical systems, typical bases are carboxylate ($-CO_2^-$), carbonate (CO_3^{2-}), and bicarbonate (HCO_3^-). But the amine units in amino acid residues are also important in biochemical reactions. The primary amine unit on the side chain of the amino acids lysine and tryptophan, or the secondary amine unit of the imidazole side chain of histidine, for example, reacts as a base in biological acid-base reactions. A lysine residue in a peptide may be proximal to an aspartic acid residue. The acid-base reaction will generate an ammonium carboxylate as shown. The pK_a of an ammonium salt derived from an amine is typically ~ 10–11. As noted in Section 1.7, the ammonium salt generated from an amino acid has a pK_a of about 9.

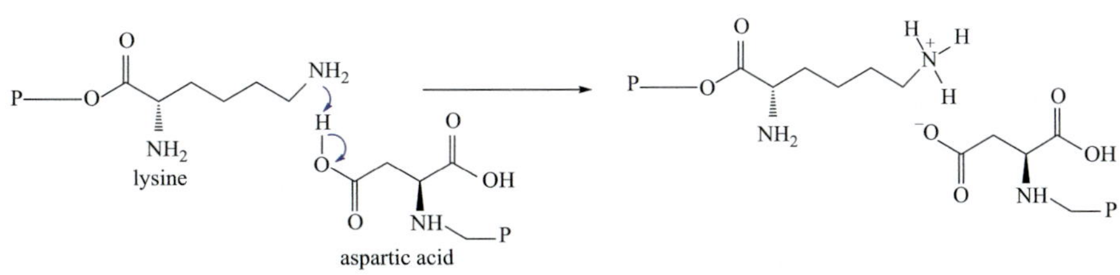

Proton transfer is one of the most common reactions of enzymes. Acid-base reactions may stabilize developing charges in the transition state of reactions that involve nucleophilic and electrophilic groups. Leaving groups can also be stabilized. In enzyme reactions, only weak acids and bases are available to hydrogen bond to reactants or to transfer protons in response to developing charges. Most enzymes show maximum activity at a neutral pH value. An example is shown in Figure 1.12.[25] An initial acid-base interaction of the carboxyl group of aspartic acid[102] in an enzyme with the basic nitrogen atom of the amine in lysine[201] in **1** to give **2** occurs via acyl addition of the nitrogen of lysine[167] to the carbonyl of the aldehyde. Note that a water molecule is involved in this process.

[25] (a) Weiss, H.M. *J. Chem. Ed.* **2007**, *84*, 440; (b) Heine, A.; DeSantis, G.; Luz, J.G.; Mitchell, M.; Wong, C.-H.; Wilson, I.A. *Science* **2001**, *294*, 369.

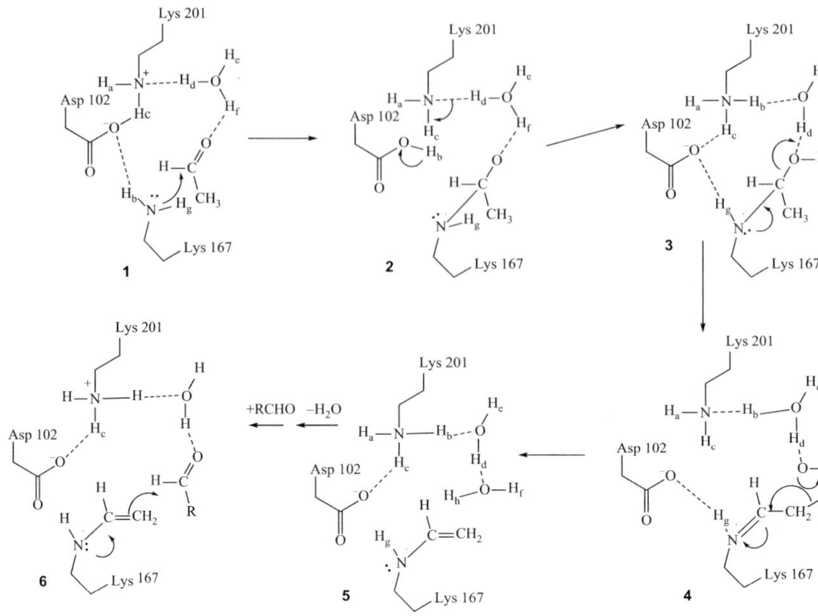

FIGURE 1.12 Mechanism of D-2-deoxyribose-5-phosphate aldolase. (Scheme II, p 442 in Weiss, H.M. *J. Chem. Educ.* **2007**, *84*, 440. Copyright 2007 American Chemical Society.)

A second acid-base reaction of aspartic acid[102] with lysine[167] leads to **3** via transfer of a proton to water. Transfer of a proton to a second water molecule leads to **4** with the formation of an enamine. Internal proton transfer of the aspartic acid[102] residue leads to **5**, and loss of water and reaction with additional aldehyde leads to **6**. This example illustrates the importance of water in the biopathway as well as the importance of acid-base reactions.

1.9 ACID AND BASE CHARACTER OF ALCOHOLS, THIOLS, AMINES, AND CARBONYLS

1.9.1 Acids

Molecules with an OH, SH, or NH unit are putative Bronsted-Lowry acids. In other words, in addition to the carboxylic acids discussed in Section 1.6, alcohols, thiols, or primary and secondary amines are acids. Alcohols are weak acids with pK_a values of 16–18. Methanol, however, has a pK_a of 15.2, whereas water has a pK_a of 14.0 (an older pK_a is reported as 15.7).[26] This fact appears to be the result of greater stabilization of the methoxide ion in water relative to the hydroxide ion. Ethanol has a pK_a of 15.9, but longer chain alcohols such as pentan-1-ol (pK_a, 16.84) are less acidic. Thiols are more acidic than alcohols.[27] Methanethiol (CH_3SH) has a pK_a of about 10.4. Ethanethiol has a pK_a of about 10.6, and pentanethiol has a pK_a of about 10.5.

The acid-base reaction of alcohols with a suitable base is a typical equilibrium reaction. As an example, sodium amide is a strong base that readily reacts with methanol. In this reaction, sodium methoxide is the conjugate base and ammonia is the conjugate acid. In the reverse reaction, sodium methoxide reacts with ammonia to regenerate methanol and sodium amide. Sodium amide is a stronger base than methoxide and methanol (pK_a 15.2) is a stronger acid than ammonia (pK_a 41), so the K_a is larger and the equilibrium lies to the right, as shown by the equilibrium arrows. In other

[26] Silverstein, T.P.; Heller, S.T. *J. Chem. Ed.* **2017**, *94*, 690.

[27] Ugur, I.; Marion, A.; Parant, S.; Jensen, J.H.; Monard, G. *J. Chem. Inf. Model.* **2014**, *54*, 2200; Yu, H.-Z.; Yang, Y.-M.; Zhang, L.; Dang, Z.-M.; Hu, G.-H. *J. Phys. Chem. A* **2014**, *118*, 606.

words, there is a higher equilibrium concentration of the conjugate base, methoxide, and the conjugate acid, ammonia. In biological systems, the base is often an amino unit from another amino acid, such as that in lysine, histidine, arginine, or asparagine.

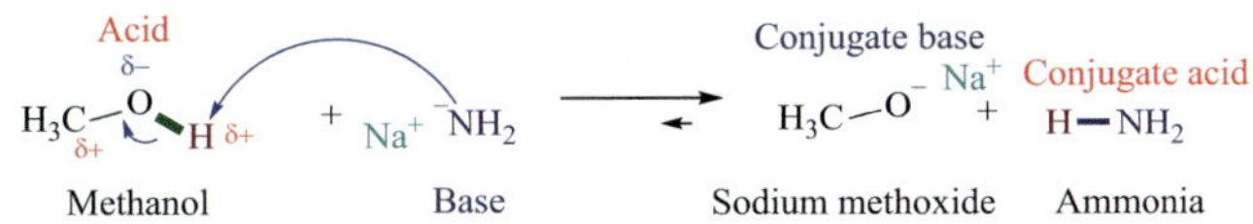

In common acid-base reactions, a weaker acid yields the stronger base. If the stronger acid and the stronger base in the equilibrium are on the right side of the equation, the equilibrium is forced back to the left, so K_a is much smaller. In water, the acid-base reaction of methanol is expected to give methoxide as the conjugate base and the hydronium ion as the conjugate acid. In such an equilibrium, the hydronium ion (pK_a, −1.7) is a stronger acid than methanol (pK_a 15.2), and methoxide is a stronger base than water. The equilibrium therefore lies far to the left (a greater concentration of methanol and water at equilibrium), so methanol is a very weak acid in water. Amino acids typically do not react with water for the same reason.

Some molecules are classified as both an acid and a base, which means that they react as an acid in the presence of a suitable base or as a base in the presence of a suitable acid. The property of a compound to react as either an acid or a base is called *amphoterism*. Certain structural features must be present, such as an atom that has an excess of electrons (oxygen) and the presence of a polarized bond to a proton (e.g., O–H). Alcohols fit these criteria perfectly. Alcohols, water, and other compounds that function in a similar manner are referred to as *amphoteric compounds*. It is possible to generalize and say that an amphoteric compound (e.g., ethanol) will function as a base in the presence of an acid with a pK_a significantly lower than its own. In the presence of a base that is much stronger than the conjugate base of ethanol, the alcohol reacts as an acid. If an alcohol reacts with a base that is much weaker than an alkoxide, however, the value of K_a may be quite small or approach 1 (a 50:50 mixture).

The N–H bond of an amine is polarized because nitrogen is more electronegative than carbon, so primary and secondary amines are acids. However, amines are weak acids, requiring very strong bases for the acid-base reaction. Strong bases such as butyllithium (Section 12.3) are required. An example is the reaction of butyllithium with diethylamine. The conjugate acid is the alkane, butane, and the conjugate base is lithium diethylamide. The pK_a of ammonia is reported to be 36, and *N*-ethylethan-1-amine is >36. In general, aliphatic amines have pK_a values between 36 and 42.

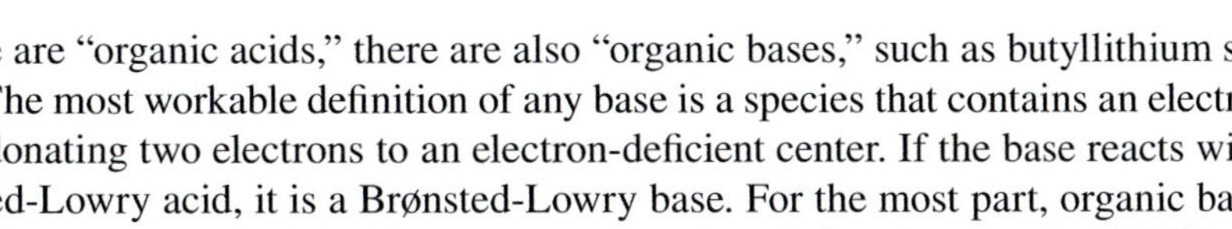

1.9.2 Bases

Just as there are "organic acids," there are also "organic bases," such as butyllithium shown in section 1.7.A. The most workable definition of any base is a species that contains an electron-rich atom capable of donating two electrons to an electron-deficient center. If the base reacts with the proton of a Brønsted-Lowry acid, it is a Brønsted-Lowry base. For the most part, organic bases contain a heteroatom such as N, O, P, or S. The most common organic base is an amine, RNH_2 or R_2NH, but phosphines (R_3P), thiols (RSH), or sulfides (RSR) are also bases. An anion, which has a negative charge localized on an atom, is a base. Therefore, amide anions (R_2N^-), phosphine anions (R_2P^-), and sulfide anions (RS^-) are stronger bases.

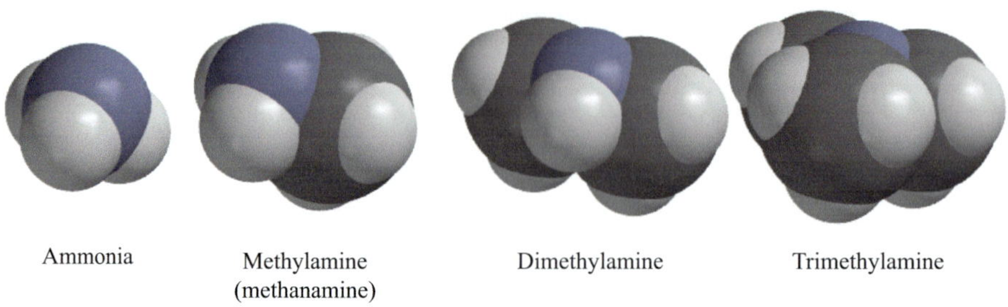

FIGURE 1.13 Steric effects on nitrogen in amines.

Amines are common bases in organic chemical reactions. The lone electron pair on the nitrogen atom of an amine is easily donated to a Bronsted-Lowry acid to generate an ammonium ion as the conjugate acid. Common amine bases are diethylamine, pyrrolidine, and pyridine. The reaction of propanoic acid and diethylamine to give the propanoate salt and diethylammonium is shown. Differences in basicity between primary, secondary, and tertiary amines can be attributed to inductive effects. When attached to nitrogen, alkyl groups are classified as electron-releasing groups. This effect increases the electron density on nitrogen, thereby increasing the basicity of the amine. Ammonia has no substituents to give an inductive effect. The methyl group in methanamine leads to more electron density on nitrogen, so methanamine is a stronger base. A secondary amine has two carbon substituents and is more basic than a primary amine, but the presence of three alkyl groups in a tertiary amine does *not* enhance basicity. Trimethylamine is a *weaker* base than dimethylamine or methylamine. The space-filling molecular models in Figure 1.13 show that less of the nitrogen atom is exposed in trimethylamine due to the "steric effect" of the three methyl groups. The steric effect here means that the three methyl groups on nitrogen block access to the nitrogen as the amine approaches another atom. For an acid-base reaction, anything that makes it more difficult for nitrogen to approach another atom diminishes the base strength of the amine based on accessibility to the nitrogen. Therefore, the tertiary amine is less basic. Based on inductive and substitution effects at nitrogen in amines, it is possible to categorize the base strength of amines, and the trend in solution is $NH_3 < R_3N < RNH_2 < R_2NH$, where secondary amines are the strongest and ammonia is the weakest base.

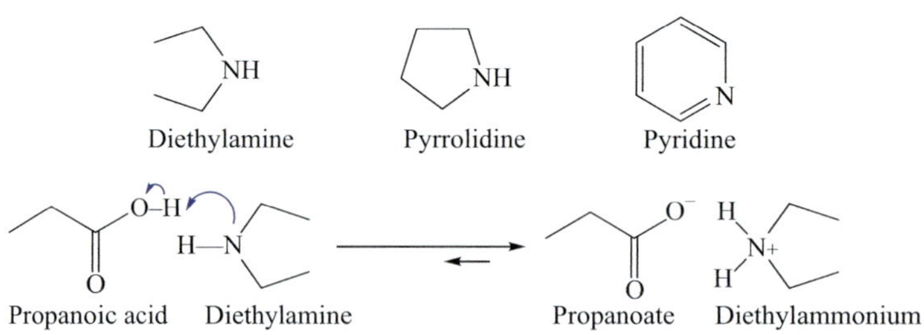

If an electron-withdrawing group is attached to nitrogen, the electron-withdrawing inductive effects make the molecule a weaker base because electron density is removed from the nitrogen. The most common electron-withdrawing group attached to nitrogen is probably a carbonyl group, which leads to a functional group called an amide. An example is ethanamide, shown in Figure 1.14, which has the common name acetamide. The δ^+ carbonyl carbon leads to an inductive effect that pulls electron density away from nitrogen, as shown, so the nitrogen cannot donate electrons to an acid as effectively as in an amine. In other words, a primary amide (NH_2 is attached to the carbonyl) is a weaker base than a primary amine because of the electron-withdrawing carbonyl group. Examination of the electron-potential map for acetamide in Figure 1.14 shows that the higher concentration of electron density (red) is on the oxygen rather than on the nitrogen. This model

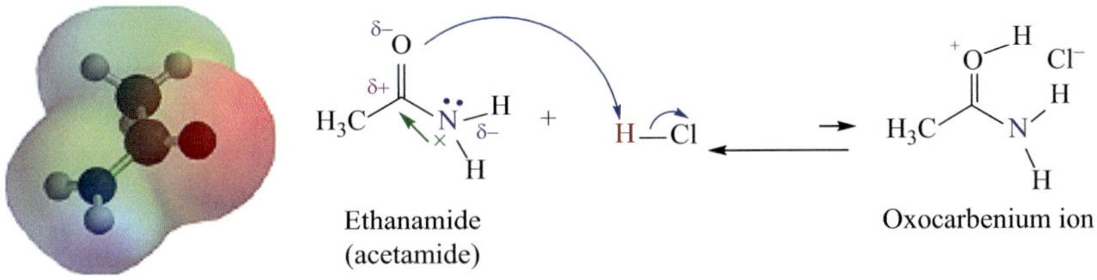

FIGURE 1.14 Reaction of acetamide with HCl.

illustrates the withdrawal of electron density from nitrogen toward the oxygen by the carbonyl, making the nitrogen less basic.

Because of the inductive effect and the delocalization of the electron pair, facilitated by the carbonyl oxygen, an amide is a much weaker base than an amine. If the nitrogen atom of an amide reacts as a base, a stronger acid is required to protonate the amide than is required to protonate an amine. If acetamide did react with HCl, the product (the conjugate acid) would be an acyl ammonium salt, ethanamide ammonium chloride. Compare this reaction with the reaction of methylamine and HCl. The diminished basicity of the amide nitrogen is quite important for the chemistry of peptides and proteins. In other words, the amide units of a peptide are relatively stable, and most of the chemical reactions occur at the more reactive side-chain amine units (see Section 12.3).

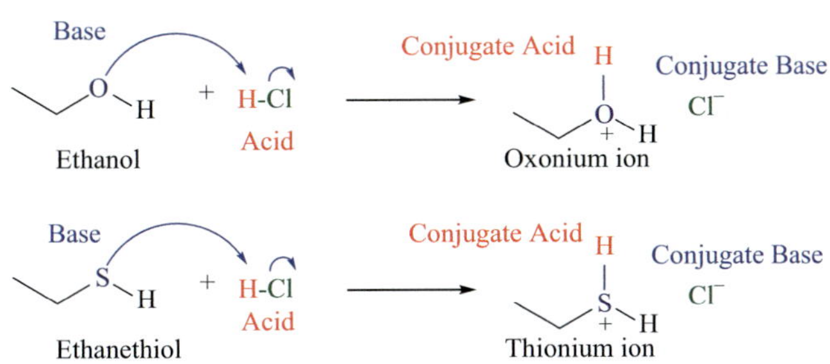

The amphoteric nature of alcohols was discussed in Section 1.9.A. In the presence of strong mineral acids like HCl, alcohols such as ethanol react as bases to form an *oxonium ion*, a transient and highly reactive *intermediate*. Oxonium ions are rather strong acids, and K_a in this reaction generally lies to the left with acids other than mineral acids because alcohols are relatively weak acids. Indeed, in an aqueous solvent, alcohols are very weak bases. Thiols react as bases like alcohols, as shown by the reaction of ethanethiol.

The oxygen of carbonyl compounds (ketones, aldehydes, carboxylic acid derivatives) has unshared electrons and reacts as a Brønsted-Lowry base or a Lewis base. When an aldehyde, ketone, or carboxylic acid derivative, such as an ester, reacts with a Brønsted-Lowry acid (H^+), the conjugate acid that is formed is a "protonated carbonyl," an *oxocarbenium ion* intermediate, as shown in Figure 1.15. As expected, oxocarbenium ions serve as highly reactive and are intermediates that undergo further reactions. The carbonyl oxygen is a relatively weak base in these acid-base reactions, requiring a rather strong acid before reaction occurs.

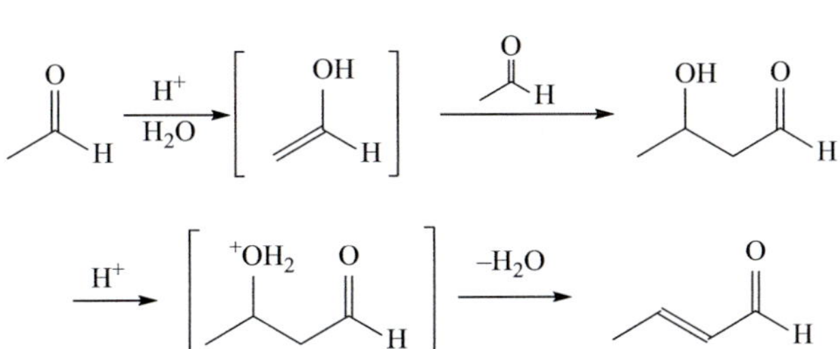

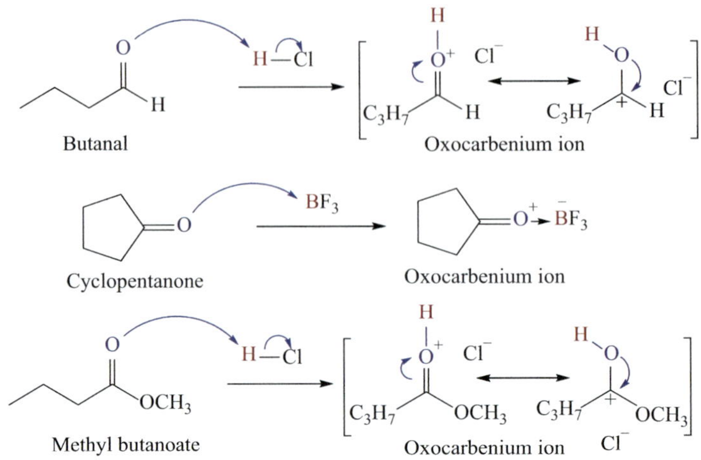

FIGURE 1.15 Acid–base reactions of the carbonyl group of aldehydes, ketones,0, and esters.

The oxygen atom in butanal reacts with HCl, for example, to form the conjugate acid, which is an *oxocarbenium ion* that is resonance-stabilized, as shown in Figure 1.15. Ketones react similarly. The formal name of this oxocarbenium ion is 1-hydroxybutan-1-ylium chloride. Oxocarbenium ions can also be generated in biochemical reactions. The carbonyl unit of an ester reacts to form an oxocarbenium ion with a pendant OR group (*O*-methyl for methyl butanoate), so the intermediate has a somewhat different structure and reactivity. These reactions are discussed in Sections 6.5 and 6.6.

1.10 DIRECTIONALITY

Acid-base reactions are equilibrium reactions, so they are reversible. Since many, if not most, biological reactions are reversible, it is appropriate to discuss the factors that shift the direction of a reaction one way or the other. An equilibrium is effectively controlled by application of *Le Chatelier's principle,* which states that *changes in concentration, temperature, volume, or partial pressure in a chemical system at equilibrium will shift the equilibrium to counteract that change.* While changes in temperature and pressure in biological systems are minimal, changes in concentration are important. In addition, initially formed products may react further, thus shifting an equilibrium reaction, and changes in the binding of starting materials and/or products can influence the direction of the reaction. Perhaps unique to enzyme catalyzed reactions, changes in the conformation of the enzyme and/or the substrate/product can have a significant influence on the directionality of the reaction.

The groups on side chains of amino acid residues in a protein or an enzyme behave differently at different pH values. The COOH unit of aspartic acid or glutamic acid, for example, may exist as COOH or COO^-, depending on the pH and the proximity of other functional groups nearby. Amino acid residues with less acidic OH or SH groups, such as serine, threonine, tyrosine, or cysteine, may exist as OH or O^-, SH or S^-· depending on the pH, or the OH units may be converted to oxonium species ($-OH_2^+$). Amino acids that have amine units on the side chains, such as lysine, may exist as the free amine unit ($-NH_2$) or as the ammonium salt ($-NH_3^+$). These changes may affect the solubility of a protein, the conformation of a protein, and reactivity at the active site. In large part, the acidic and basic groups on the side chains of the amino acid residues act as weak acids and bases, with K_a values that determine the extent of dissociation of the group depending on the pH of the system. For *enzymes*, which are proteins, catalytic activity occurs only within a certain pH range. Metabolism occurs only at a pH near neutrality, and organisms must maintain a specific and constant pH for optimum activity of enzymes. For this reason, buffering is important, and phosphate buffer systems as well as carbonate or bicarbonate buffer systems are important. The pH of blood is controlled by the bicarbonate buffer system, for example.

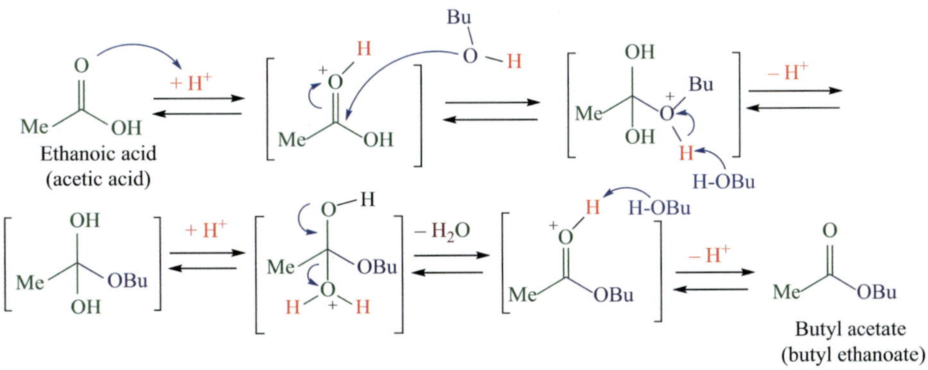

FIGURE 1.16 Acid-catalyzed esterification of acetic acid.

The importance of pH, acid catalysis, and equilibrium in chemical reactions is perhaps best illustrated by reactions that have a central place in organic chemistry. Heating a carboxylic acid (acetic acid) and an alcohol (butan-1-ol) in the presence of an acid catalyst gives the corresponding ester as well as water, as shown in Figure 1.16. Every step of the reaction sequence shown in Figure 1.16 is reversible, and using butanol as the solvent (a large molar excess of the alcohol) helps shift the equilibrium toward the ester per Le Chatelier's principle. *If one product is removed from this reversible process, such as water, the equilibrium is also shifted toward the ester product.* By adding a chemical reagent that reacts with water, such as $CaCl_2$, or one that binds water, such as molecular sieve 4 Å, water is removed from the equilibrium, shifting the equilibrium to the right, giving the product. Esters are hydrolyzed with aqueous acid under acidic conditions, and the mechanism of ester hydrolysis is the exact reverse of the esterification reaction, except that an excess of water is provided to shift the equilibrium toward the acid and away from the ester.

The acid-catalyzed aldol condensation is an example of a reaction product that undergoes further reaction that affects the directionality of the reaction. Treatment of acetaldehyde with acid gives the enol (see Section 9.2), which reacts with a second molecule of acetaldehyde to give the aldol (a β-hydroxy aldehyde) as the initial product.[28] Under the acidic conditions, however, the hydroxyl group is labile and quickly reacts with the acid catalyst to form the oxonium ion, and loss of water generates the conjugated aldehyde that is the isolated product. The instability of the initially formed product in the reaction medium helps to shift the direction of the reaction toward the conjugated product.

Chemical reactions are driven by changes in energy. Heating imparts energy to the molecules in a reaction. Molecules absorb energy from their environment by collision with the sides of the reaction vessel (a flask) or with another atom or molecule, and this energy is utilized in bond making and breaking. Therefore, monitoring changes in energy is important for following the progress of a chemical reaction. The energy of the entire system ($\Delta G°$, called the free energy) is the important parameter, and it is a function of enthalpy as well as entropy and temperature. Whether or not a chemical reaction will proceed spontaneously is determined by the change in standard free energy ($\Delta G°$). The standard free energy is calculated from the change in enthalpy ($H°$) and the change in entropy ($S°$) by the *Gibbs Free Energy equation*: $\Delta G° = \Delta H° - T\Delta S°$. This calculation *assumes that the reaction is done under standard conditions*: in a solution with a concentration of 1 M and at 298.15 K. The entropy term ($\Delta S°$) measures the "disorder" of a given system. In practical terms, if the number of particles for a reaction remains the same or decreases, the magnitude of the entropy term is usually quite small. If the number of particles greatly increases during a reaction, then entropy usually increases. In the reaction that converts chloromethane to iodomethane, replacing

[28] See Baigrie, L.M.; Cox, R.A.; Slebocka-Tilk, H.; Tencer, M.; Tidwell, T.T. *J. Amer. Chem. Soc.* **1985**, *107*, 3640.

chlorine with iodine does not change the overall number of particles, and the size and shape are not greatly changed. The change in entropy is therefore expected to be small.

For most reactions, relatively small structural changes in the molecule lead to small changes in entropy, so the assumption that the change in entropy is small for most reactions in organic chemistry is usually correct. A negative sign for $\Delta G°$ means that energy is *released* during the reaction, and this is an *exothermic* (exergonic) *process*. Such a reaction is said to be spontaneous because it produces enough energy during the reaction to be self-sustaining, once it has started. If $\Delta G°$ has a positive value, the process is *endothermic* (endergonic), and energy must be put into the system for it to continue. An endothermic process is usually not spontaneous because less energy is produced during the reaction than is required to keep it going. A negative value of $\Delta G°$ is associated with a spontaneous or exothermic reaction (heat is generated during the reaction by the bond-making and bond-breaking processes).

In general, enzymes work by lowering the activation energy of a reaction, making the reaction faster and easier. Biological systems rarely have changes in temperature or pressure, so the direction of a chemical reaction is regulated by changes in the potential energy (free energy, G). In general, the minimization of free energy determines the direction of a biochemical reaction. In other words, reactions tend to proceed in the direction for which ΔG is negative. Many chemical reactions in cells have a positive $\Delta G°$ *and so* are energetically unfavorable (non-spontaneous). The synthesis of proteins from amino acids is possible if cells couple it to a reaction that has a negative ΔG so that the sum of the two reactions has a negative ΔG. An important reaction is the hydrolysis of one or both phosphoanhydride bonds in adenosine triphosphate (ATP). Adenosine triphosphate has two high-energy phosphoanhydride bonds that link the three phosphate groups. Hydrolysis of a phosphoanhydride bond has a negative $\Delta G°'$. The removal of a phosphate or a pyrophosphate group from ATP leaves adenosine diphosphate (ADP) or adenosine monophosphate (AMP), respectively. Cells can use the free energy released to carry out reactions that would otherwise be energetically unfavorable. Indeed, the hydrolysis of ATP is important for protein formation from amino acids.

The role of water and hydrophobic and hydrophilic interactions in protein folding was noted previously. The tertiary structure of a protein, due to folding, is a conformational issue for the macromolecule. It is therefore appropriate to comment on how conformation may influence biochemical reactions. Indeed, conformational changes of molecules (Section 5.12) are important in many biochemical processes and in molecular recognition.

Conformational changes generally have a positive contribution to the free energy, which disfavors the association process. Conformational changes to the free energy thought to be represented by the conformational entropy of the involved biomolecules.[29] The relationship between ligand binding and protein conformational change is critical for biological interactions. The active site of an enzyme, and its spatial conformational change, takes place. In other words, the binding of a ligand molecule to a protein is often accompanied by conformational changes in the protein to a specific conformation. Different substrates induce different degrees of conformational change.[30]

HOMEWORK

1.1. Briefly explain how hydrogen-bonding influences the structure of a polypeptide. Also see Section 3.1.
1.2. Briefly explain how hydrogen-bonding differs between an α-helix structure of a polypeptide and β-pleated sheet structures of a polypeptide.
1.3. Draw the structure of four different amino acids that have hydrophobic side chains.
1.4. Draw a generic structure for a phospholipid and for a glycolipid. See Chapter 10 if there are questions about the structures.

---

[29] Ahmad, M.; Helms, V.; Kalinina, O.V.; Lengauer, T. *J. Phys. Chem. B* **2016**, *120*, 2138.
[30] Pan, R.; Zhang, X.-J.; Zhang, Z.-J.; Zhou, Y.; Tian, W.-X.; He, R.-Q. *J. Biol. Chem.* **2010**, *285*, 22950.

1.5. Give a reasonable definition of the "hydration layer" that is associated with a biomolecule.
1.6. Briefly explain how an increase in entropy in water influences the aggregation state of non-polar molecules inside a protein.
1.7. Draw the acid-base equilibrium for leucine and sodium acetate in water. Briefly explain how the addition of an excess of sodium acetate helps to keep the reaction from being too acidic.
1.8. Briefly explain why phenylalanine has a smaller pK_{a1} when compared to alanine.
1.9. Draw the oxocarbenium ion formed by the reaction of a catalytic amount of sulfuric acid with (a) butan-2-one (b) pentanal (c) methyl butanoate
1.10. The pK_a1 and pK_a2 for glycine, valine, and cysteine as given. Calculate the pI for each amino acid: glycine: 2.34, 9.60; valine: 2.32,9.62; cysteine: 1.96, 10.28.
1.11. Give the complete mechanism for the acid-catalyzed esterification of glycine and propan-2-ol.
1.12. What is the structure of choline, and what are the important biological properties?
1.13. What are the structures of penicillin F, penicillin X, and penicillin N?

2 Amino Acids

Amino acids are important difunctional molecules critical to an understanding of biology processes. Amino acids comprise the backbone of peptides, and thereby of enzymes. This chapter will discuss the structure, nomenclature, and characteristics of amino acids.

2.1 CHARACTERISTICS OF AMINO ACIDS

An amino acid, as the name implies, has one amine unit ($-NR_2$) and one carboxylic acid unit (a carboxyl group, COOH). The nomenclature for a generic amino acid is determined by the carboxyl group, so the parent name is "acid," and the NR_2 unit is treated as a substituent. When an amine unit is a substituent, the name "amino" is used, so these compounds are amino carboxylic acids, or simply amino acids. Amino acids are easily named using IUPAC nomenclature, and the carboxylic acid is the parent for each compound. Two examples are 2-aminopropanoic acid (known as alanine) and 5-amino-3,5-dimethylheptanoic acid.

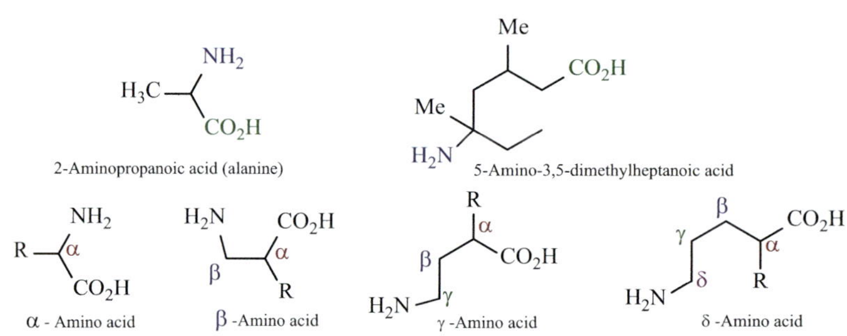

There are a variety of structural variations for amino acids. If the amine unit is attached to C2, the α-carbon of the carboxylic acid chain, the compound is an α-amino acid. If the amine group is on C3, the β-carbon, it is a β-amino acid. Similarly, there are γ-amino acids, δ-amino acids, and so on. Due to their biological importance, α-amino acids will be the focus of this discussion. The common names of α-amino acids are presented in Table 1.1 in Section 1.7. To distinguish α-amino acids from other amino acids, the term non-α-amino acids is used. 5-Amino-3,5-dimethylheptanoic acid is a non-α-amino acid, for example.

Perhaps the most important chemical feature of an amino acid is the presence of an amine, which is a base, *and* a carboxyl group (an acid) in the same molecule. The basic amine reacts intramolecularly with the acidic proton of acetic acid to form a negatively charged conjugate base and a positively charged conjugate acid in the same molecule. An intermolecular example of this reaction occurs when methanamine (methylamine) and acetic acid are mixed, and the product is methylammonium acetate. 2-Aminopropanoic acid (alanine) yields 2-ammoniopropanoate, with a positively charged ammonium unit ($-NH_3^+$) and a negatively charged carboxylate unit ($-CO_2^-$). Such molecules are known as dipolar ionic molecules or *zwitterions*. As seen in Figure 2.1, both the neutral molecule and the zwitterion are alanine.

The ammonium salt in 2-ammoniopropanoate is a weak acid, so there are two acidic species in the equilibrium reaction shown for alanine: the carboxylic acid unit and the ammonium salt unit. The internal acid-base reaction of the amine unit and the carboxylic acid unit leads to the ammonium carboxylate in an equilibrium reaction labeled K_{a1} (see Figure 2.2). A second acid-base reaction is possible, which converts the ammonium carboxylate to the amine carboxylate

DOI: 10.1201/9781003605157-2

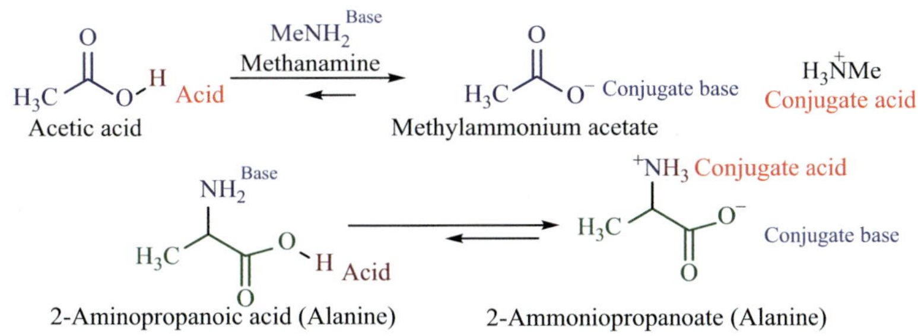

FIGURE 2.1 Intermolecular and intramolecular acid–base reactions.

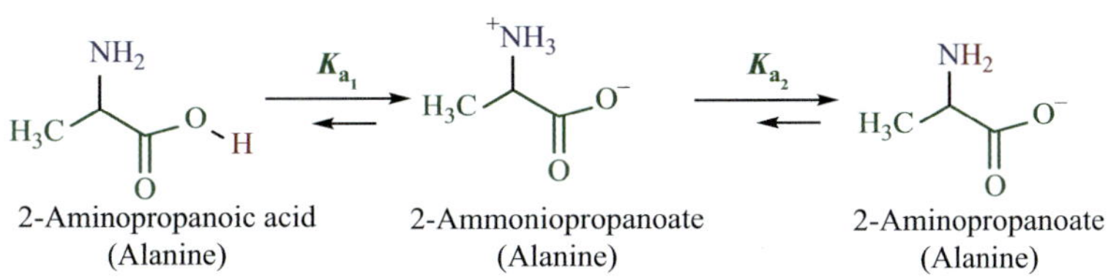

FIGURE 2.2 Internal acid–base reactions of alanine.

(2-ammoniopropanoate) in the equilibrium reaction labeled K_{a2}. Therefore, the two acid-base reactions and two acidity constants are K_{a1} and K_{a2}.

Formally, the values of K_{a1} and K_{a2} follow the standard definition of K_a for each acid-base reaction. As with other acid-base reactions, pK values are used to discuss acidity in amino acids rather than the values of K_{a1} or K_{a2}. As a reminder, $pK_a = -\log K_a$, and $K_a = 10^{-pKa}$. The values of K_{a1} and K_{a2} for alanine are derived from the equilibrium equations shown. The values for different amino acids vary with the nature of the substituents attached to the α-carbon of the amino acid. In other words, different amino acids will have different values of K_{a1} and K_{a2} and therefore different values of pK_{a1} and pK_{a2}.

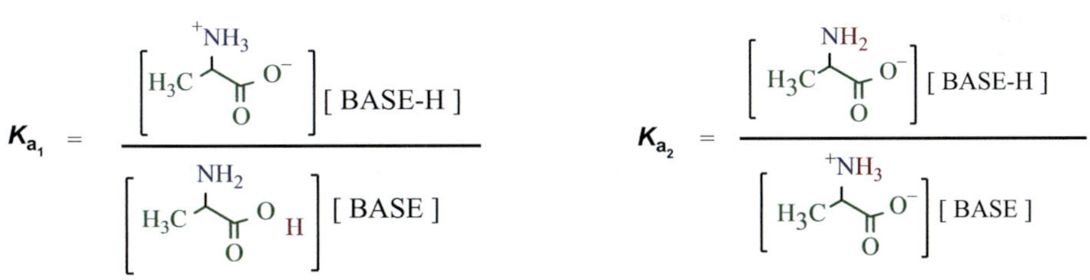

One of the species in the equilibrium will be a neutral species. The point in the equilibrium at which this neutral species is formed is the *isoelectric point, pI*, which is defined as *the pH at which the material carries no net electrical charg*e (see Section 1.7). If pI is used to represent the isoelectric point, then pI is defined by the following equation:

$$pI = \frac{pk_{a_1} + pK_{a_2}}{2}$$

Figure 2.3 shows a pH curve for a generic amino acid in which the isoelectric point and both K_{a1} and K_{a2} are marked. The values of K_{a1} and K_{a2} vary with the substituents attached to the amino acid. The purpose of this figure is to show that K_{a1} and K_{a2} can be experimentally determined for any amino acid, and the isoelectric point can then be calculated.

As will be seen in Section 3.3, amino acid units constitute the backbone of proteins. Amino acids that make up proteins may be positive, negative, neutral, or polar in nature, depending on the side

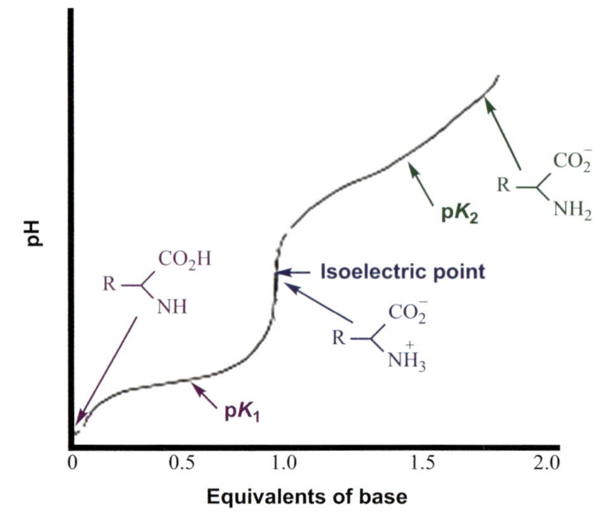

FIGURE 2.3 The position of p*K* and the isoelectric points for a generic amino acid.

chain, and the nature of these side chains determines the overall charge of a protein. At a pH below their p*I*, proteins carry a net positive charge; above their *pI*, they carry a net negative charge. As the groups on the α-carbon of an amino acid change, each new amino acid residue may have a *different* pK_{a1} value and, therefore, a different isoelectric point. As a practical matter, this *means the isoelectric point of each different amino acid is a function of its structure.*

2.2 STRUCTURE OF α-AMINO ACIDS

Proteins are made up of amino acids linked together in a way that forms a polymeric structure, as discussed in Section 3.4. The genetic code that provides a set of instructions for the biosynthesis of proteins will also be discussed in Section 3.4. The genetic code utilizes primarily ~ 20 α-amino acids, so this introduction to amino acids is limited to those compounds. Examination of all but one of the 20 α-amino acids shows that the α-carbon is a stereogenic center, so those α-substituted amino acids are chiral molecules. These chiral amino acids are known to have the (*S)* absolute configuration at the α-carbon. Note that the so-called nonessential amino acids are synthesized by mammals, while the essential amino acids must be obtained from dietary sources.

All 20 amino acids commonly found in proteins, except glycine (H_2NCH_2COOH), contain a hydrogen atom and one alkyl group at the α-carbon rather than two alkyl groups. There is a compound called glyceraldehyde [$HOCH_2CH(OH)CHO$], drawn using the usual line notation in Figure 2.4 and in Fischer projection (Section 5.15).

Glyceraldehyde has one stereogenic center, so there are two enantiomers: one with a (+) specific rotation [(*R*)-2,3-dihydroxypropanal] and one with a (–) specific rotation [(*S*)-2,3-dihydroxypropanal]. Before the R/S system (Section 5.16) was developed, the concept of the Fischer projection was based on the structure of (+)-glyceraldehyde, with the CHO unit on "top" and the OH unit on the "right." The Fischer projection shown for glyceraldehyde is drawn so that the hydroxyl group and the hydrogen atom are projected in front of the page (on the horizontal line), with the CHO group and the CH_2OH group projected behind the page (on the vertical line). Note that the four groups appear to be attached to a "cross" (see Figure 2.4). The distinguishing OH unit is on the right (dexter in *Latin*), and this stereoisomer was labeled as the D-configuration. D-(+)-glyceraldehyde is now identified as (2*R*),3-dihydroxypropanal. The other enantiomer is (2*S*),3-dihydroxypropanal, drawn with the OH unit on the left (lever in *Latin*), so this is the L-configuration. Therefore, L-(–)-glyceraldehyde is (2*S*),3-dihydroxypropanal. Using Fischer projections, (D) was assigned to the (+) enantiomer, which

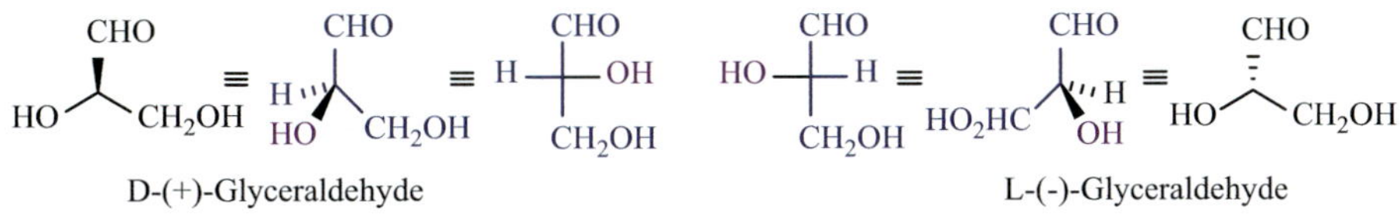

FIGURE 2.4 D- and L-Glyceraldehydes.

happens to be dextrorotatory, but this was an arbitrary choice (a guess that is properly called an assumption). As will be seen in Section 5.17, (+) and (–) refer to specific rotation, which is a physical property, whereas *D*/*L* are names.

There are two enantiomers for each amino acid, based on the zwitterionic forms of a generic amino acid with one alkyl group R and one hydrogen atom. Stereoisomers are drawn in line notation and in Fischer projection in Figure 2.5. In the Fischer projections, the COOH unit is on the "top," just as the CHO unit is on top in glyceraldehyde. The distinguishing feature is the NH_3unit, so if the amino group is on the "right", it is a D-amino acid, whereas an L-amino acid has the NH_3unit on the left. In the line drawings, the (*R*) configuration has the amino group projected forward, and the (*S*) configuration has the amino group projected to the rear. For the most part, the amino acids will be drawn using line notation.

The structures of the 20 amino acids most often used in protein biosynthesis can be shown by changing the R group in Figure 2.5, and all are L-amino acids. Imagine three categories of substituent "R": R is neutral, R has an acidic group as part of its structure, or R contains a basic group as part of its structure. The subcategories of α-amino acids are therefore neutral, acidic, or basic. Table 2.1 (see Table 1.1) shows the structures of the 20 protein amino acids, the "R" group, the name of the amino acid, and *the three-letter and one-letter codes used to abbreviate each compound.*

The first class of amino acids includes those where R is a simple alkyl fragment, which is neutral and nonpolar. This classification includes *glycine* (R=H), *alanine* (R=Me), *valine* (R=isopropyl; $CHMe_2$), *leucine* (R=isobutyl), *isoleucine* (R=*sec*-butyl), and *phenylalanine* (R=benzyl; CH_2Ph). A second class of amino acids has a substituent that is considered neutral but has a polarized group. These include *serine* (R=CH_2OH), *threonine* (R=CH[OH]Me), and *tyrosine* (R=CH_2-[4-OH)phenyl), which contain hydroxyl substituents. *Cysteine* (R=CH_2SH) has a thiol unit, and *methionine* (R=CH_2CH_2SMe) has a thioether unit. Note that the presence of the SH in cysteine leads to an (*R*)-configuration, although it remains an L-amino acid. Also note that the alcohol unit, the thiol unit in cysteine, and the phenolic hydroxyl in tyrosine are weak acids. Two amino acids have an amide unit as part of the side chain, but they are neutral amino acids: *asparagine* (R=CH_2CONH_2) and *glutamine* (R =2).

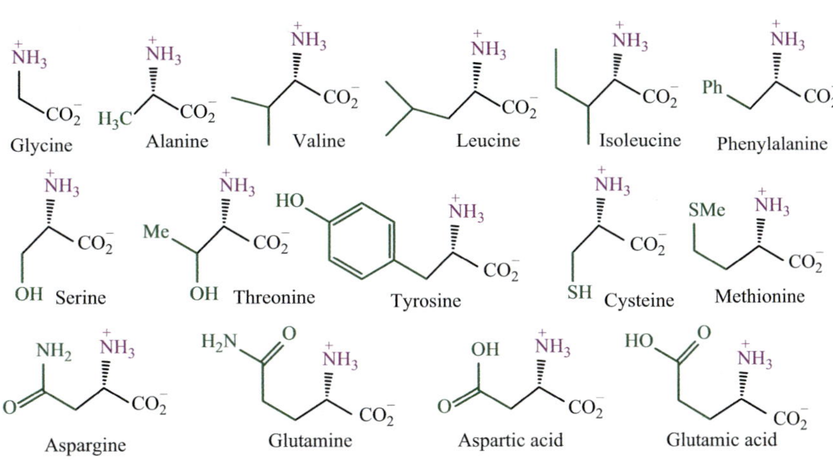

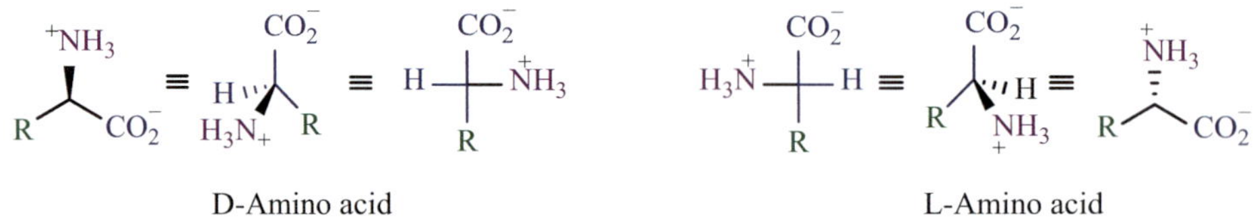

FIGURE 2.5 Fischer projections for enantiomers of a generic amino acid.

TABLE 2.1
Structures, Names, Three-Letter Code, and One-Letter Code of the 20 Essential Amino Acids, Based on the Structure in Figure 2.5

R	Name	Three-Letter Code	One-Letter Code
H	Glycine	gly	G
Me	Alanine	ala	A
$CHMe_2$	Valine	val	V
$CHMe_2$	Leucine	leu	L
CH(Me)Et	Isoleucine	ile	I
CH_2Ph	Phenylalanine	phe	F
CH_2OH	Serine	ser	S
CH(OH)Me	Threonine	thr	T
CH_2(4-hydroxy-C_6H_4)	Tyrosine	tyr	Y
CH_2SH	Cysteine	cys	C
CH_2CH_2SMe	Methionine	met	M
CH_2CONH_2	Asparagine	asn	N
$CH_2CH_2CONH_2$	Glutamine	gln	Q
CH_2COOH	Aspartic acid	asp	D
CH_2CH_2COOH	Glutamic acid	glu	E
$CH_2CH_2CH_2CH_2NH_2$	Lysine	lys	K
CH_2(2-indolyl)	Tryptophan	trp	W
CH_2(4-imidazolyl)	Histidine	his	H
$CH_2NHC(=NH)NH_2$	Arginine	arg	R
2-Pyrrolidinyl	Proline	pro	P

There are acidic amino acids that have a carboxyl group (COOH) as part of the side chain: *aspartic acid* (R=CH_2COOH) and *glutamic acid* (R=CH_2CH_2COOH). Note that asparagine is the amide derivative of aspartic acid, and glutamine is the amide derivative of glutamic acid. There are also amino acids with an amine as part of the R group, and such groups are basic. *Lysine* [R=$(CH_2)_4NH_2$] has a primary amine, *tryptophan* (R=CH_2-indole) has an indole unit, and *histidine* (R=CH_2-imidazole) has an imidazole unit. *Arginine* [R=$(CH_2)NHC(=NH)NH_2$] has a guanidine unit. *Guanidine* itself is essentially the imine derivative of *urea*, and the NH_2unit is basic. The remaining amino acid is proline, which is unique for several reasons. First, it contains a pyrrolidine ring that effectively connects the amine unit to the α-carbon. Second, proline is the only α-amino acid that has a secondary amine unit.

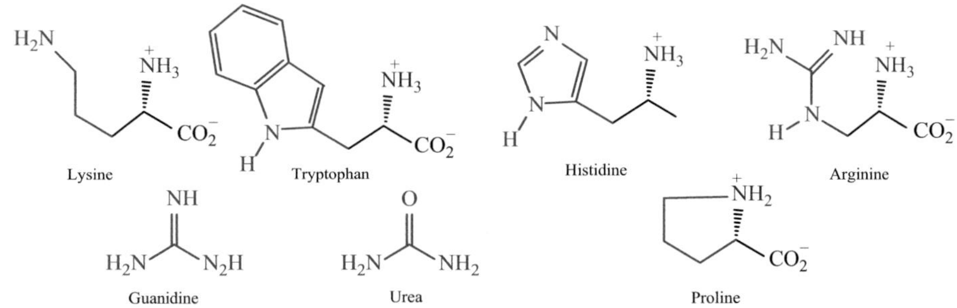

Each of the amino acids just discussed has pK_a values and an isoelectric point. Table 2.2 shows the name of the amino acid, pK_{a1}, pK_{a2}, the isoelectric point, and pK_{a3}.[1]

The neutral amino acid glycine (2-ammonioacetate) is shown in the zwitterionic form in Figure 2.6, in equilibrium with carboxylmethanaminium and 2-aminoacetate. When the amino acid has an acidic side chain, the equilibrium is more complex because of the third acid group. When the side chain has an acidic proton that is less acidic than the ammonium proton or the carboxyl proton (e.g., the phenolic proton in tyrosine), the equilibrium is slightly different. As shown in Figure 2.6, the equilibrium for the COOH side chain of glutamic acid is also represented by pK_{a3}, and the equilibrium for the phenolic OH in tyrosine is represented by pK_{a3}. These values are shown in Table 2.2. Similarly, lysine has a third pK value for the ammonium salt of the amine unit on the side chain.

TABLE 2.2
The p*K* Values of Amino Acid at The Isoelectric Point in Water At 25°C

Amino Acid	pK_{a1}	pK_{a2}	Isoelectric Point	pK_{a3}
Glycine	2.34	9.60	5.97	
Alanine	2.34	9.69	6.00	
Valine	2.32	9.62	5.96	
Leucine	2.36	9.60	5.98	
Isoleucine	2.36	9.60	6.02	
Phenylalanine	1.83	9.13	5.48	
Serine	2.21	9.15	5.68	
Threonine	2.09	9.10	5.60	
Tyrosine	2.20	9.11	5.66	10.07
Cysteine	1.96	10.28	5.07	8.18
Methionine	2.28	9.21	5.74	
Asparagine	2.02	8.80	5.41	
Glutamine	2.17	9.13	5.65	
Aspartic acid	1.88	9.60	2.77	3.65
Glutamic acid	2.19	9.67	3.22	4.25
Lysine	2.18	8.95	9.74	10.53[a]
Tryptophan	2.83	9.39	5.89	
Histidine	1.82	9.17	7.59	6.00[a]
Arginine	2.17	9.04	10.76	12.48[a]
Proline	1.99	10.60	6.30	

[a] *This pKa_3 value is for the ammonium salt of the amine side chain.*

[1] *CRC Handbook of Chemistry and Physics*, 94th ed., CRC Press, Boca Raton, FL, **2014**, pp. 7-1–7-2.

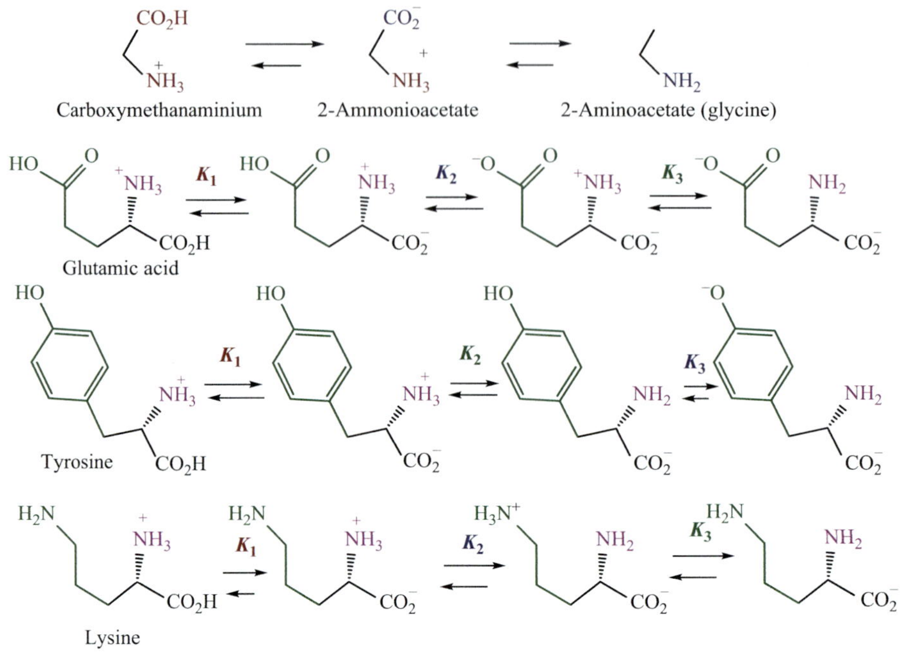

FIGURE 2.6 The K_a equilibria for glutamic acid, tyrosine, and lysine.

2.3 BIOSYNTHESIS OF AMINO ACIDS

Some microorganisms can reduce nitrogen gas to two molecules of ammonia in a process called *nitrogen fixation*.[2] The carbon skeletons of amino acids are biosynthetically available via the glycolytic pathway, the pentose phosphate pathway, or the citric acid cycle. Human beings cannot synthesize eight of the 20 amino acids most often used. These amino acids must be supplied in the diet. Amino acids that are required in the diet because they are not produced in the body are called *essential amino acids*.[3]

There are examples of the biosynthesis of myriad amino acids, but a detailed discussion is far beyond the scope of this book. Amino acid synthesis is discussed online.[4] The biopathways that generate amino acids and many of their bioreactions are discussed in an article published by the *Medical Biochemistry* page.[5] "All tissues have some capability for synthesis of non-essential amino acids via amino acid remodeling, conversion of non-amino acid carbon skeletons into amino acids and other derivatives that contain nitrogen. Nitrogen metabolism occurs in the liver, where the potentially toxic nitrogen of amino acids is eliminated via transamination, deamination, and urea formation. The carbon skeletons are generally conserved as carbohydrate via gluconeogenesis or as fatty acids via fatty acid synthesis pathways."[5]

"Amino acids fall into three categories: glucogenic, ketogenic, or glucogenic and ketogenic. Glucogenic amino acids give rise to a net production of pyruvate or TCA (tricarboxylic acid) cyclic intermediates, such as 2-oxoglutarate (α-ketoglutarate) or oxaloacetate, precursors to glucose via gluconeogenesis. All amino acids except lysine and leucine are at least partly glucogenic. Lysine and leucine are the only amino acids that are solely ketogenic and give rise only to acetyl-CoA or acetoacetyl-CoA, neither of which can bring about net glucose production."[5] Note that glucogenic

[2] Bernhard, A. *Nat. Educ. Knowl.* **2010**, *3*, 25; Vitousek, P.M.; Cassman, K.; Cleveland, C.; Crews, T.; Field, C.B.; Grimm, N.; Howarth, R.W.; Marino, R.; Martinelli, L.; Rastetter, E.B.; Sprent, J.O. *Biogeochemistry* **2002**, *57–58*, 1.

[3] See https://www.healthline.com/nutrition/essential-amino-acids#benefits-of-supplementing

[4] https://en.wikipedia.org/wiki/Amino_acid_synthesis

[5] See https://themedicalbiochemistrypage.org/amino-acid-biosynthesis-and-catabolism/

amino acids can be converted into glucose through gluconeogenesis, whereas ketogenic amino acids can be converted into ketone bodies

"A few of amino acids, isoleucine, phenylalanine, threonine, tryptophan, and tyrosine, give rise to both glucose and fatty acid precursors and are characterized as being glucogenic and ketogenic. Note that during times of starvation the reduced carbon skeleton is used for energy production, with the result that it is oxidized to CO_2 and H_2O."[5] The evolution of amino acids has also been discussed.[6]

Most amino acids are synthesized from α-keto acids or by transamination from another amino acid, such as glutamate. The conversion to other amino acids is catalyzed by an *aminotransferase* (EC 2.6; see Section 4.2). In the enzyme-catalyzed reaction, an α-keto acid is converted to α-ketoglutarate. Activated tetrahydrofolate is a coenzyme that is important in the biosynthesis.

An example of the biosynthesis of an amino acid is shown in Figure 2.7[7] for the biosynthesis of L-serine from 3-phosphoglycerate. In the first step, 3-phosphoglycerate is converted to 3-phosphohydroxypyruvate, facilitated by *3-phosphoglycerate dehydrogenase* (EC 1.1.1.95). The next step led to 3-phosphoserine, mediated by 3-phosphoserine aminotransferase (EC 2.6.1.52). In the final step, L-serine is produced by the enzyme 3-phosphoserine phosphatase (EC 3.1.3.3).

As will be introduced in Chapter 4, *transaminases*, or *aminotransferases*, are *transferase* enzymes that catalyze a transamination reaction between an amino acid and an α-keto acid. The transaminase *aspartate aminotransferase* (EC 2.6.1.1) will be described in Section 4.4.A for the catalysis of the reversible transfer of the (*R*)-amino group between aspartate and 2-oxoglutarate.[8] A biosynthesis of L-lysine from aspartate is shown in Figure 2.8,[9] which includes the *2L,6L-diaminopimelate aminotransferase* (EC 2.6.1.83)-catalyzed reaction that converts 2,3,4,5-tetrahydropicolinate to (2*S*),(6*S*)-diaminopimelate. The first reaction is catalyzed by *aspartate kinase* (EC 2.7.2.4) to form 4-phosphoaspartate, followed by the NADPH-dependent *aspartate semialdehyde dehydrogenase* (EC 1.2.1.11) reaction to give aspartate 4-semialdehyde. In the next step, aspartate 4-semialdehyde reacts with pyruvate, mediated by *dihydrodipicolinate synth*ase (EC 4.2.1.52), to give 2,3-dihydropicolinate. Reduction to 2,3,4,5-tetrahydrodipicolinate by *dihydrodipicolinate reductase* (EC 1.3.1.26) allows the transamination reaction catalyzed by *LL-diaminopimelate aminotransferase* (EC 2.6.1.83) with glutamate as the amino group donor. Epimerization, catalyzed by *diaminopimelate epimerase* (EC 5.1.1.7), gives the direct precursor of lysine, *meso*-2,6-diaminopimelate, and reaction with *diaminopimelate decarboxylase* (EC 4.1.1.20) produces L-lysine.

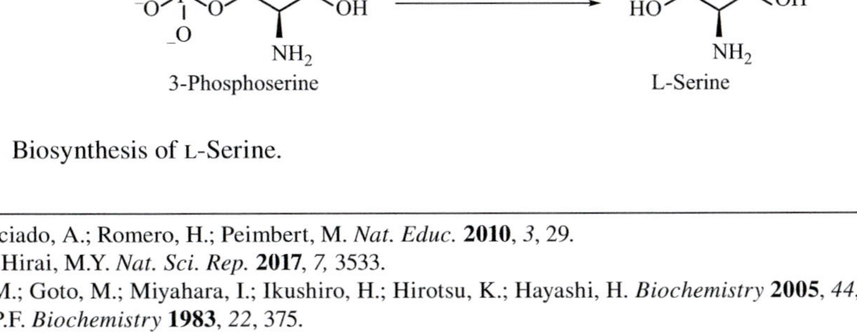

FIGURE 2.7 Biosynthesis of L-Serine.

[6] Gutiérrez-Preciado, A.; Romero, H.; Peimbert, M. *Nat. Educ.* **2010**, *3*, 29.

[7] Okamura, E.; Hirai, M.Y. *Nat. Sci. Rep.* **2017**, *7*, 3533.

[8] (a) Islam, M.M.; Goto, M.; Miyahara, I.; Ikushiro, H.; Hirotsu, K.; Hayashi, H. *Biochemistry* **2005**, *44*, 8218; (b) Kiick, D.M.; Cook, P.F. *Biochemistry* **1983**, *22*, 375.

[9] Bromke, M.A. *Metabolites* **2013**, *3*, 294.

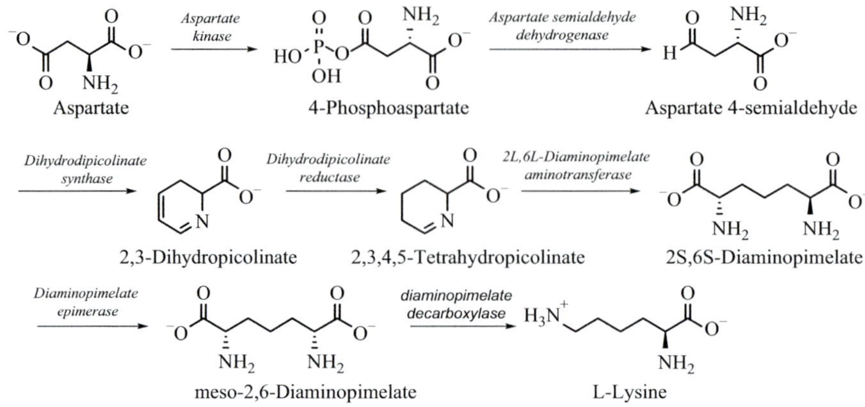

FIGURE 2.8 Biosynthesis of L-lysine. (© 2013 by the authors; licensee MDPI, Basel, Switzerland. This article is an open-access article distributed under the terms and conditions of the Creative Commons Attribution license (http://creativecommons.org/licenses/by/3.0/).)

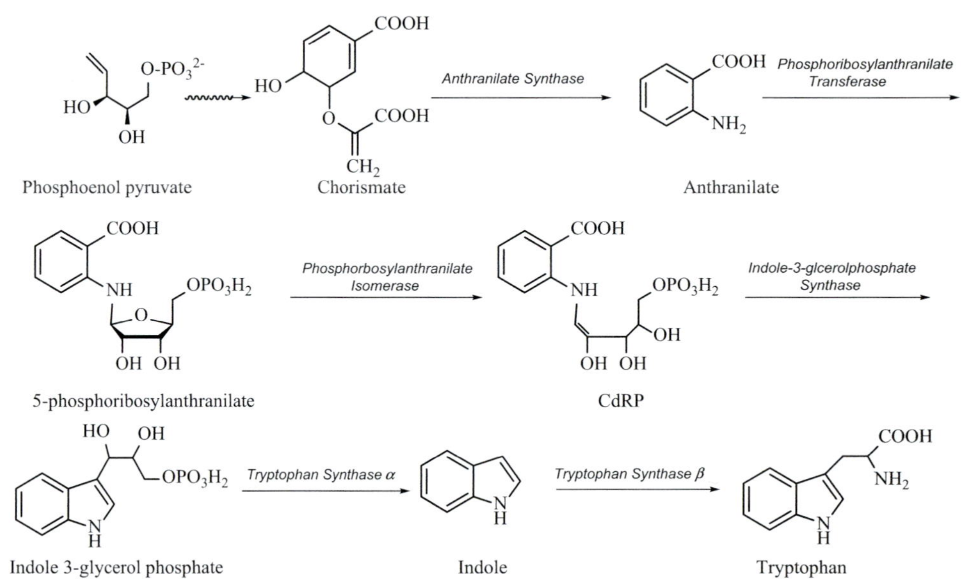

FIGURE 2.9 Biosynthesis of tryptophan. (From Mannino, G.; Pernici, C.; Serio, G.; Gentile, C.; Cinzia M. Bertea,. M. *Int. J. Mol. Sci.* **2021**, *22*, 9996. https:// doi.org/10.3390/ijms22189996. Figure 3, p. 10000. MDPI: © 2021 by the authors. Licensee MDPI, Basel, Switzerland. This article is an open-access article distributed under the terms and conditions of the Creative Commons Attribution (CC BY) license (https://creativecommons.org/licenses/by/4.0/).)

Several synthase enzymes, along with a transferase and an isomerase, are prominent in the biosynthetic pathway for tryptophan.[10] The sequence begins with the conversion of phosphoenol pyruvate to chorismate via several steps, which is then converted to anthranilate, mediated by *anthranilate synthase* (EC 4.1.3.27). Anthranilate is converted to 5-phosphoribosylanthranilate by *5-phosphoribosylanthranilate transferase* (EC 2.4.2.18), which catalyzes the transfer of a phosphoribosyl moiety from phosphoribosylpyrophosphate to anthranilate (Figure 2.9). The ribose ring is opened by *phosphosylanthranilate isomerase* (EC 5.3.1.24) to give CdRP [(carboxyphenylamino)-1-deoxyribulose

[10] Mannino, G.; Pernici, C.; Serio, G.; Gentile, C.; Cinzia M.; Bertea, M. *Int. J. Mol. Sci.* **2021**, *22,* 9996. Also see Mannino, G.; Gentile, C.; Ertani, A.; Serio, G.; Bertea, C.M. *Agriculture* **2021**, *11*, 212.

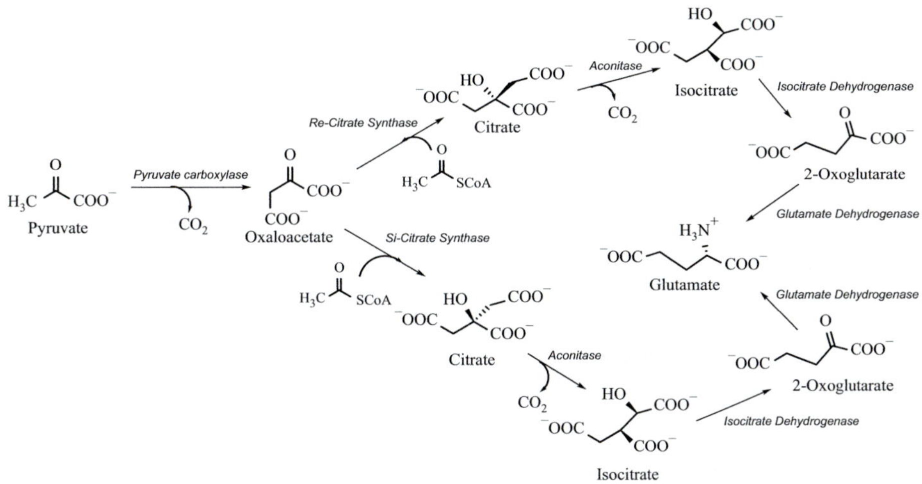

FIGURE 2.10 Two pathways for glutamate biosynthesis via *Si-citrate synthase* and *Re-citrate synthase*. (From Kim, M.; Le, H.M.; Xie, X.; Feng, X.; Tang, Y.J.; Mouttaki, H.; McInerney, M.J.; Buckel, W.; *Appl. Environ. Microbiol.* **2015**, *81*, 8434. Figure 1, p. 8435. Open Access by https://creativecommons.org/licenses/by/4.0/.)

5-phosphate]. Conversion to the indole derivative, indole 3-glycerol phosphate, by *indole-3-glycerol phosphate synthase* (EC 4.1.1.48) is followed by conversion to indole with *tryptophan synthase α* (EC 4.2.1.20). The final conversion to tryptophan is catalyzed by *tryptophan synthase β* (EC 4.2.1.20). Note that α and β refer to the subunits of *tryptophan synthase*.[11]

A study elucidated the pathway for glutamate biosynthesis in *S. aciditrophicus*, which uses a *Re-citrate synthase* and the oxidative branch of the Krebs cycle (see Section 14.3) to make glutamate.[12] It was suggested that acetyl-CoA and bicarbonate (carbon dioxide) are major precursors for the biosynthesis of cellular components. Acetyl-CoA is first converted to pyruvate.[13] Pyruvate is decarboxylated with the loss of carbon dioxide to give oxaloacetate via *pyruvate carboxylase* (EC 6.4.1.1) (Figure 2.10). The next step is the coupling with acetyl-CoA to give citrate, showing the transformation for both *Re-* and *Se-citrate synthase* (E.C. 2.3.3.1, previously known as 4.1.3.7). The enzyme *aconitase* (*aconitate hydratase*, EC 4.2.1.3) leads to isocitrate with the loss of carbon dioxide. Isocitrate is converted to 2-oxoglutarate with *isocitrate dehydrogenase* (EC 1.1.1.42), and *glutamate dehydrogenase* (EC 1.4.1.2) gives glutamate. Glutamate is the anion of glutamic acid, a potent neurotransmitter.

2.4 REACTIONS AND SYNTHESIS OF A-AMINO ACIDS

Amino acids undergo several important reactions that are dominated by the amine group and/or the carboxyl group. The fact that the amine and the carboxyl can react with each other complicates things, because one functional group can influence the reactivity of the other. The conversion of a simple carboxylic acid to an ester is quite easy, either by reaction of the carboxylic acid with an alcohol under acidic conditions or by first converting the acid to an acid chloride followed by reaction with an alcohol. However, amines also react with acid chlorides or with esters to yield the corresponding amide. In other words, trying to form an ester in the presence of an amine may be

[11] Raboni, S.; Bettati, S.; Mozzarelli, A. *Cell Mol. Life Sci.* **2009**, *66*, 2391.

[12] Kim, M.; Le, H.M.; Xie, X.; Feng, X.; Tang, Y.J.; Mouttaki, H.; McInerney, M.J.; Buckel, W. *Appl. Environ. Microbiol.* **2015**, *81*, 8434.

[13] McInerney, M.J.; Rohlin, L.; Mouttaki, H.; Kim, U.; Krupp, R.S.; Rios-Hernandez, L.; Sieber, J.; Struchtemeyer, C.G.; Bhattacharyya, A.; Campbell, J.W.; Gunsalus, R.P. *Proc. Natl. Acad. Sci. U.S.A.* **2007**, *104*, 7600.

problematic. To target one functional group in an amino acid, the other must be considered and usually blocked (*protected*) with a functional group that can be removed after the planned reaction.

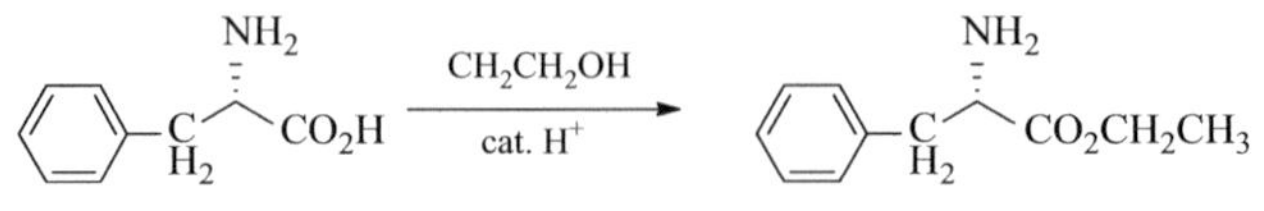

Phenylalanine — Phenylalanine ethyl ester

The carboxylic acid group of an amino acid can be converted to an ester by reaction with an alcohol in the presence of aqueous acid. Using the zwitterionic form of phenylalanine as an example, treatment with acid generates the protonated form, (1*S*)-carboxyethanaminium. Subsequent reaction with aqueous ethanol and an acid catalyst converts the COOH unit to the corresponding ester, phenylalanine ethyl ester. The amine unit exists as its ammonium salt in the aqueous acid medium.

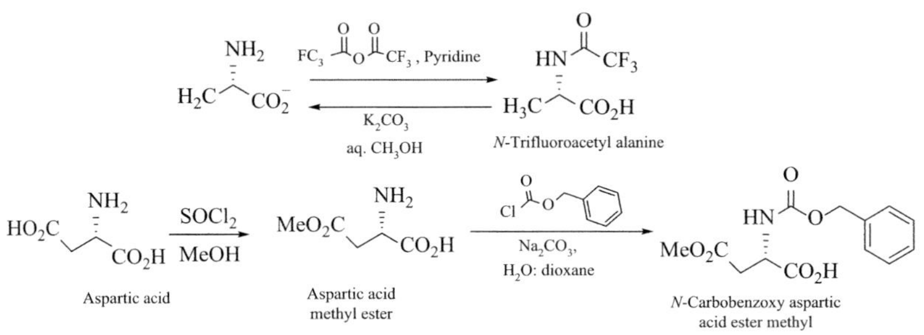

The amine unit of an amino acid can be functionalized. The amine group of alanine can be treated with trifluoroacetic anhydride and pyridine to give the *N*-trifluoroacetamide derivative.[14] If desired, the COOH unit can also be functionalized, and the $COCF_3$ unit can be easily converted back to the amine by reaction with K_2CO_3 in aqueous methanol. As shown for aspartic acid, the distal carboxyl group can be converted to the methyl ester by reaction with thionyl chloride followed by methanol. Subsequent reaction with benzyloxy chloride (Cbz-Cl) gives the corresponding *N*-carbobenzoxy derivative.[15] This is a common protecting group for amino acid manipulation and is called the N-Cbz derivative. This protecting group can be removed by catalytic hydrogenation.

Once the amine unit is blocked (protected) as an amide, other reactions of the carboxyl unit are possible. For example, the reaction of N-benzyl alanine with ethanol and dicyclohexylcarbodiimide (DCC) forms the ethyl ester, N-benzylalanine ethyl ester. However, reaction with ethanamine and DCC yields the amide–amide, N-benzyl-N-ethylalanine amide. These examples are far from exhaustive, but they illustrate the ability to functionalize the amine unit and the carboxyl unit. Both options are necessary for the functionalization and elaboration of existing amino acids, are also important for the synthesis of amino acids from non-amino acid substrates.

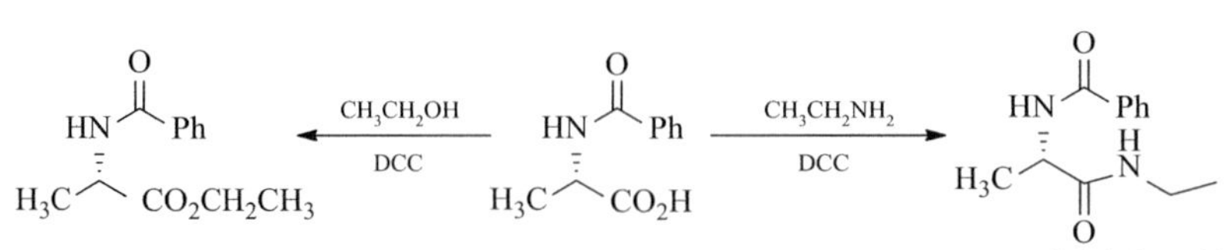

N-Benzyl alanine ethyl ester — *N*-Benzyl alanine — *N*-Benzy-*N*-ethyl alanine amide

The N-alkylation of amino acids is a crucial reaction, as many of its derivatives have applications in medicinal chemistry and drug development. This reaction is illustrated in Figure 2.11

[14] Fones, W.S.; Lee, M. *J. Biol. Chem.* **1954**, *210*, 227.
[15] Nirmalahrudaya, Ch. *J. Chem. Pharm. Sci.* **2014**, *5*, 34.

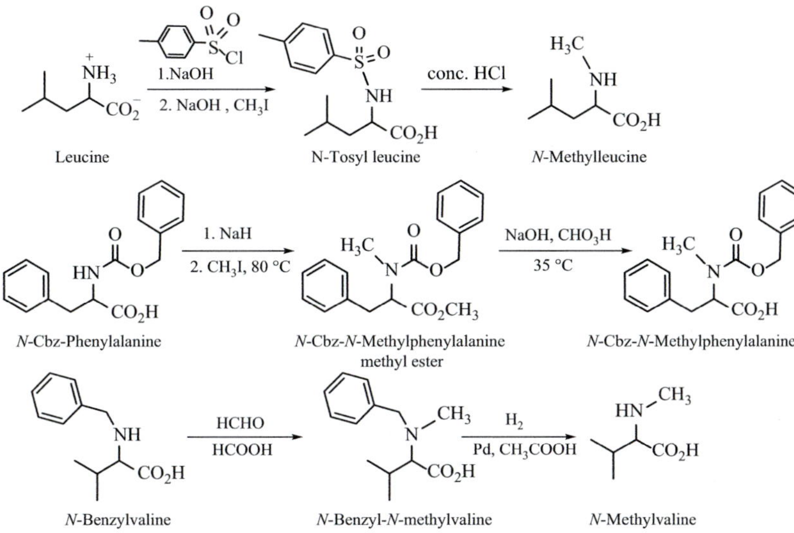

FIGURE 2.11 Methylation of leucine and synthesis of N-methylvaline from phenylalanine.

by the N-methylation of amino acids.[16] In the first example, the nitrogen of leucine is protected by conversion to the *N*-tosyl (4-toluenesulfonyl) derivative, and then deprotonation of the nitrogen proton allows the reaction with iodomethane to give the N-methyl derivative, as shown in Figure 2.11.[17,18] The N-tosyl group is recovered by reaction with concentrated HCl. In the second example, N-Cbz phenylalanine is converted to the methyl ester by reaction with sodium hydride (NaH) followed by reaction with iodomethane, which also generates the N-methyl moiety, as shown in Figure 2.11. Saponification of the ester to the amino acid uses NaOH in methanol. In the third example, *N*-benzylvaline is converted to the N-methyl derivative by reaction with formaldehyde in formic acid.[19] The benzyl group is removed by catalytic hydrogenation with a palladium catalyst. Apart from simple N-alkylation reactions, a variety of groups can be attached to nitrogen.

Conversion to an ester is arguably the most common method for protecting the COOH unit of an amino acid. Methyl or ethyl esters are prepared most of the time, but benzyl, *p*-nitrobenzyl, and *tert*-butyl esters are also used. The *p*-nitrobenzyl ester is a common protecting group, and an example is the reaction of the silver salt of valine with 4-bromomethyl nitrobenzene to give 4-nitrophenyl valinate.

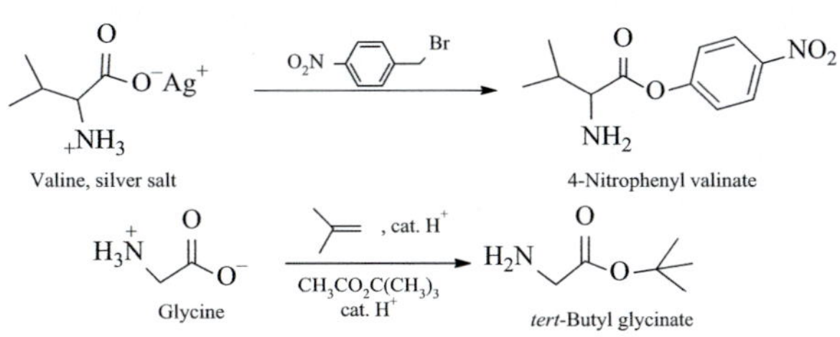

A *tert*-butyl ester is another common protecting group, usually prepared by reacting an amino acid such as glycine with 2-methylprop-2-ene (isobutylene) in the presence of an acid catalyst. The

[16] Aurelio, L.; Brownlee, R.T.C.; Hughes, A.B. *Chem. Rev.* **2004**, *104*, 5823.

[17] (a) Hlaváček, J.; Poduška, K.; **Šorm**, F.; Sláma, K. *Coll. Czech. Chem. Commun.* **1976**, *41*, 2079; (b) Hlaváček, J.; Frič, I.; Buděšínský, M.; Bláha, K. *Coll. Czech. Chem. Commun.* **1988**, *53*, 2473.

[18] (a) Benoiton, N.L.; Kuroda, K.; Cheung, S.T.; Chen, F.M. F. *Can. J. Biochem.* **1979**, *57*, 776; (b) McDermott, J.R.; Benoiton, N.L. *Can. J. Chem.* **1973**, *51*, 2555; 2562.

[19] (a) Ebata, M.; Takahashi, Y.; Otsuka, H. *Bull. Chem. Soc. Jpn.* **1966**, *39*, 2535; (b) Brockmann, H.; Lackner, H. *Chem. Ber.* **1967**, *100*, 353.

alkene reacts with the acid catalyst to give a carbocation, which traps the carboxylate oxygen to form the ester, *tert*-butyl glycinate. An alternative method reacts the acid unit of an amino acid (glycine) with *tert*-butyl acetate in the presence of an acid catalyst. Transfer of the ester unit via a transesterification reaction (see Section 10.6) gives *tert*-butyl acetate.

Many groups are available for the protection of the nitrogen unit of an amino acid. Sulfonyl chlorides are commonly used, as shown for *N*-tosylleucine. Another *N*-protecting group is a tertiary halide called triphenylmethyl chloride (Ph_3CCl; trityl chloride), and when it reacts with valine in the presence of diethylamine, the product is *N*-trityl valine. Apart from the N-Cbz group introduced in previous examples, another important *carbamate* group is the *tert*-butoxycarbonyl group (Boc or *t*-Boc; *tert*-butyl carbamate). *tert*-Butoxycarbonyl chloride is too unstable to be isolated, so either *tert*-butoxycarbonyl azide or di-*tert*-butyl anhydride (Boc anhydride) is used. The reaction of isoleucine with *tert*-butoxycarbonyl azide gives the *N*-Boc derivative, *N*-Boc-isoleucine.

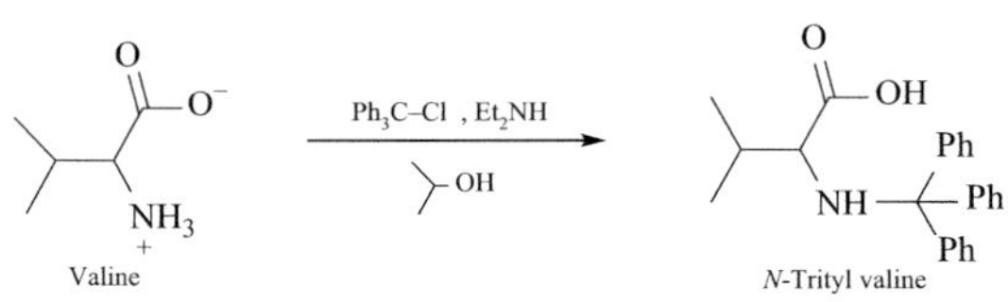

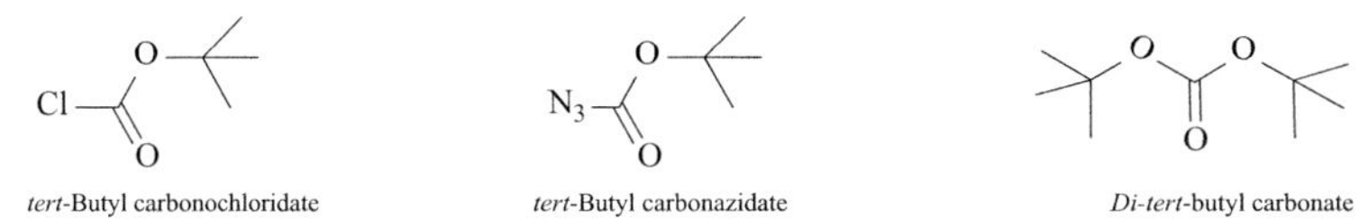

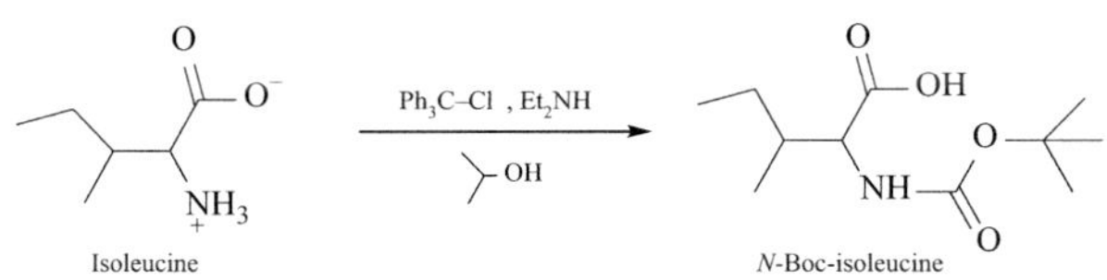

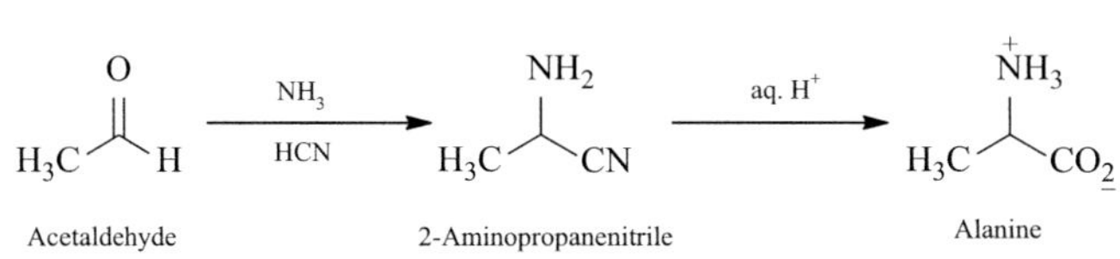

These reactions of amino acids are relatively normal with respect to the amine and the carboxyl unit, considering the presence of both functional groups in the same molecule. Several methods are used for the synthesis of amino acids. The syntheses shown here give *racemic* amino acids. An early preparation of an amino acid, the *Strecker synthesis*, reacts an aldehyde with ammonia in the presence of HCN to form an amino nitrile. Acid hydrolysis of the nitrile unit gives an amino acid. In a specific example, heating acetaldehyde with ammonia and HCN gives 2-aminopropanenitrile, and subsequent acid hydrolysis leads to alanine. *The alanine produced by this method is racemic, as is any other amino acid made this way.* It is a general procedure, however, and many different amino acids can be prepared. The enantiopure amino acid is obtained by resolution.

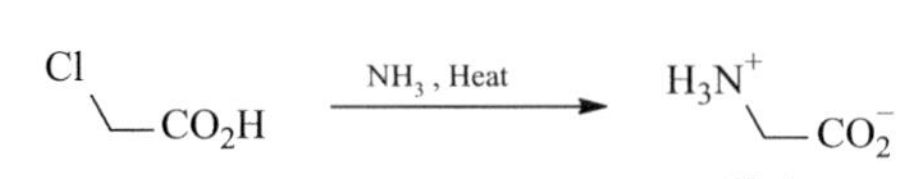

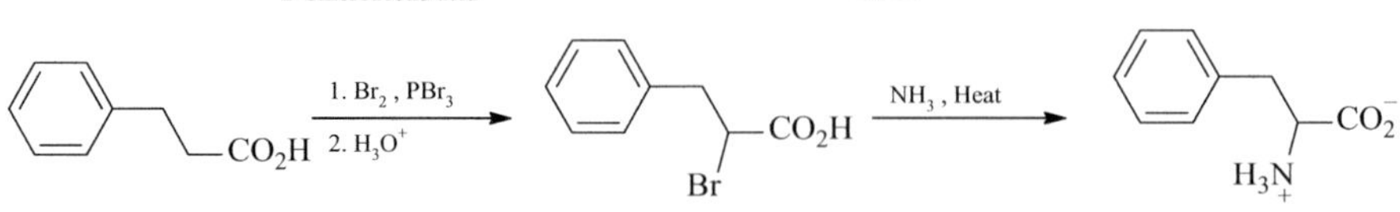

As early as 1858, glycine was prepared by heating chloroacetic acid with ammonia. The preparation of an α-bromo acid by the so-called *Hell-Volhard-Zelinskii reaction*[20] involves the reaction of a carboxylic acid such as 3-phenylpropanoic acid with bromine and phosphorus tribromide to give the α-bromo acid bromide. A mild aqueous workup gives the α-bromo derivative, 2-bromo-3-phenylpropanoic acid. Subsequent heating with ammonia and neutralization gives the amino acid phenylalanine. Many amino acids can be prepared by this method, but as noted, they are racemic.

Ammonia is not always the best reagent to generate amines from halides, and amine surrogates such as azides or phthalimide usually give better yields in S_N2-type reactions (see Chapter 6). However, their use requires an extra step to *liberate* the amine unit. Esters of malonic acid are very reactive at C3, and if diethyl malonate is treated with bromine/phosphorus tribromide, the 3-bromo derivative is readily formed. Reaction of this α-bromo ester with ammonia should lead to an amino acid precursor, but ammonia also reacts with esters to form amides. Reaction of a base such as potassium hydride (KH) with the amine surrogate phthalimide gives potassium phthalimide, which reacts with bromomalonate to give **1**, as shown in Figure 2.12. If **1** is treated with sodium metal (or NaH, LDA, etc.), the enolate anion is formed, and reaction with benzyl bromide gives **2**. Subsequent heating with aqueous NaOH and then aqueous HCl leads to deprotection and decarboxylation, leading to phthalic acid and phenylalanine.

Amino acids that have an aromatic group as part of the side chain are prepared from aromatic aldehydes by the reaction of glycine with acetic anhydride to give *N*-acetylglycine. Subsequent reaction with potassium thiocyanate (KS–C≡N) gives acyl thiourea **3** after hydrolysis, as shown in Figure 2.13. Cyclization gives thiazolone **4**, which rearranges under the reaction conditions to give a thiohydantoin, a 2-thioxo-4-imidazolidinone (3-acetyl-2-thiohydantoin, **5**). Treatment with 4-hydroxybenzaldehyde and pyridine leads to an enolate condensation reaction to give a new thiohydantoin, **6**. Subsequent heating with thiourea at 100°C for several hours leads to the loss of S=C=NH and the amino acid tyrosine (Figure 2.13).

The so-called *azlactone synthesis* is another route to amino acids. Remember from above that an azlactone is an oxazolone. Glycine is first converted to its *N*-benzoyl derivative (known as hippuric acid) by reaction with benzoyl chloride, and subsequent treatment with acetic anhydride gives the azlactone **7** (hippuric acid azlactone), as shown in Figure 2.14. An enolate anion condensation reaction (see Section 6.2) with 2-methylpropanal in the presence of pyridine gives azlactone **8**. Catalytic hydrogenation of the alkene unit and acid hydrolysis lead to the amino acid leucine.

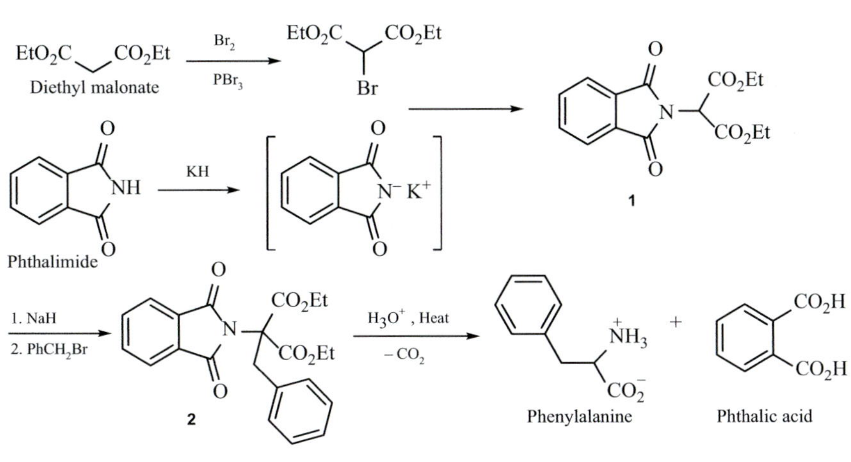

FIGURE 2.12 Synthesis of phenylalanine from phthalimide and diethyl malonate.

[20] Harwood, H.J. *Chem. Rev.* **1962**, *62*, 99; see pp. 102–103.

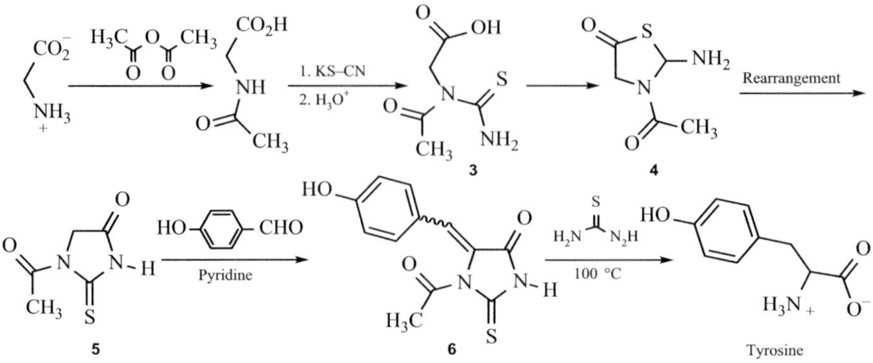

FIGURE 2.13 Synthesis of tyrosine from glycine.

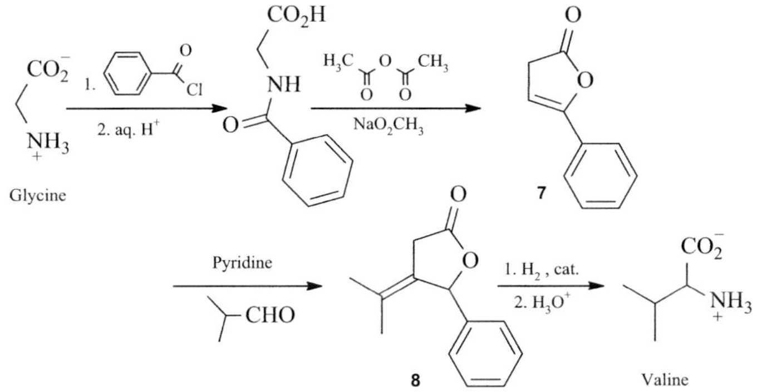

FIGURE 2.14 Synthesis of valine from glycine.

As pointed out above, all the amino acids prepared in this section are *racemic*. However, there are methods available today that generate amino acids with good enantioselectivity. The examples shown illustrate the challenges an organic chemist faces when preparing amino acids, as well as some of the chemical reactions that are possible.

To obtain an enantiopure amino acid requires separation of the enantiomers via resolution. The physical properties of enantiomers are identical except for specific rotation (Section 5.17). Since separation methods rely on differences in physical properties, this is a problem that can be overcome if the racemic amino acid mixture reacts with a reagent that has a stereogenic center. The resulting product will be a mixture of diastereomers, which have different physical properties and may be separated. A second chemical reaction is required to remove the first reagent and "release" the enantiopure amino acid. This process is called *optical resolution*. Chemicals obtained from nature as a single enantiomer are most often used as the reactive agent, but they must have a functional group that is able to react with one of the functional groups in a racemic molecule. Most of these chiral compounds have an amine, a carboxylic acid, or an alcohol unit that can interact with or react with the racemic mixture. Such compounds are called chiral *resolving agents*. Three common resolving agents that may be used for optical resolution are shown. The first two reagents are naturally occurring alkaloids brucine and strychnine. The third is an optically pure amine, (*R*)-phenylethylamine. If a racemic amino acid reacts with an enantiopure resolving agent, the salt of each enantiomer is formed. These diastereomeric salts are separated, and each is isolated and purified. Subsequent treatment with dilute base regenerates the enantiopure (*R*)- and (*S*)-amino acids.

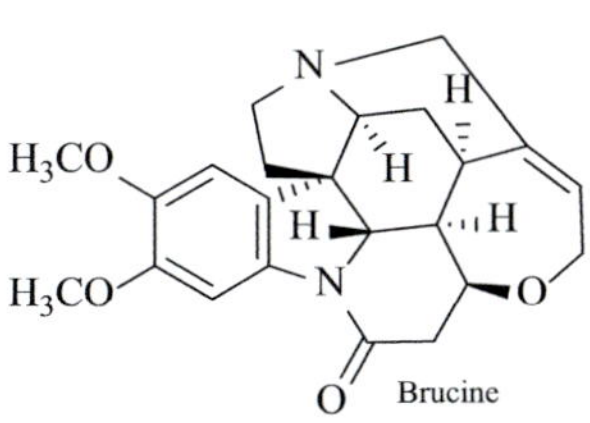

Brucine

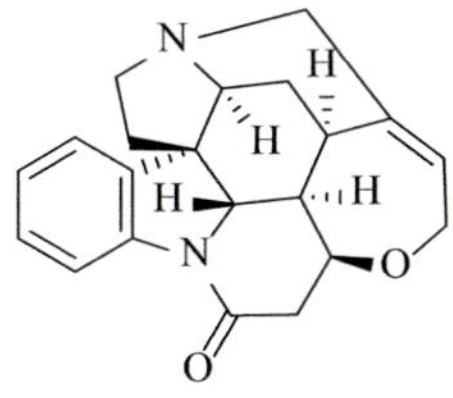

Strychnine

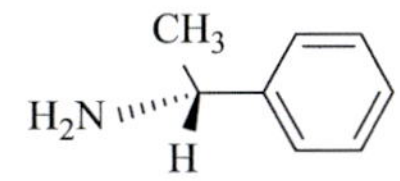

(*R*)-Phenylethylamine

HOMEWORK

11.1. Draw the structures of (a) 5-amino-2-ethylhexanoic acid, (b) 3-amino-3-phenylpentanoic acid, (c) *N*-methyl 4-aminobutanoic acid, (d) 3-amino-4,4-diethylheptanoic acid, and (e) 2-amino-4-(2-methylethyl)pentanoic acid.

11.2. Draw leucine in the zwitterionic form.

11.3. Calculate the isoelectric point for tyrosine: pK_{a1} 2.20 and pK_{a2} 9.11; Calculate the isoelectric point for asparagine: pK_{a1}2.20 and pK_{a2} 8.80.

11.4. Draw the structure of the following in their zwitterionic form: L, F, D, and W.

11.5. The enzymes *3-phosphoglycerate dehydrogenase* and 3-phosphoserine aminotransferase (in Figure 2.7) were shown to be important for the biosynthesis of serine. Give the chemical transformation that each enzyme mediates.

11.6. In the first edition of this work, Figure 11.8 stated that epimerization was catalyzed by *diaminopimelate epimerase* to give meso-2,6-diaminopimelate, and a subsequent reaction with *diaminopimelate epimerase* gave L-lysine. This is incorrect. What is the error?

11.7. In Figure 2.9, anthranilate is converted to 5-phosphoribosylanthranilate by *5-phosphoribosylanthranilate transferase*. What type of reaction is this?

11.8. What is the product of the reaction of *N*-benzylphenylalanine with formaldehyde in formic acid?

11.9. Give the major product of each of the following.

(a)

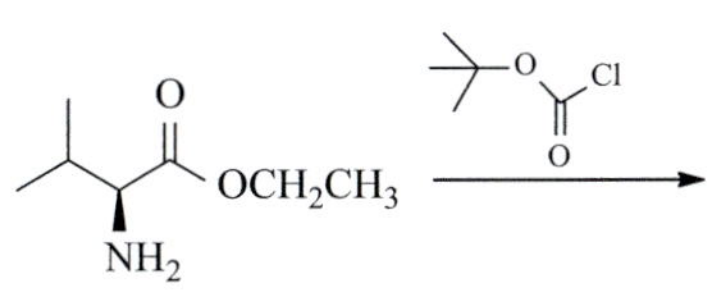

(b)

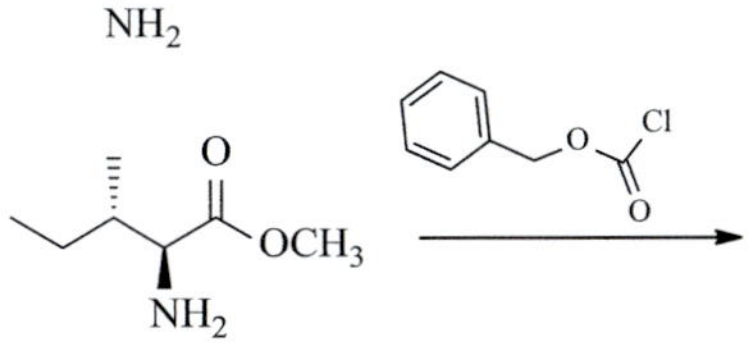

3 Peptides and Proteins

Polymeric chains of amino acids are peptides, and proteins are poly(peptides). Proteins are essential to living systems, and the critically important proteins called enzymes catalyze processes that are essential to life. This chapter will discuss peptides, their characteristics, and how they are formed. In addition, the structure and function of proteins will be discussed.

3.1 PEPTIDES ARE POLY(AMIDES) OF AMINO ACID RESIDUES

Poly(amides) derived from amino acids are known as *peptides*, and the amide bonds within the peptide are called *peptide bonds*. The amino acid components of a peptide are known as *amino acid residues*. Serine-tryptophan is composed of two amino acid residues and is a dipeptide. A peptide with three amino acid residues, such as leucine-phenylalanine-arginine, is a tripeptide, and a peptide with 15 Amino acid residues are called a pentadecapeptide. A decapeptide has ten amino acid residues. Large peptides (hundreds or thousands of amino acid residues) use primarily the 20 α-amino acids identified in Chapter 2 to prepare proteins. Proteins control many life processes. There are four major types of proteins: globular proteins (spheroproteins), fibrous proteins, disordered proteins, and membrane proteins. *Globular proteins* are water soluble and named for their spherical shape, and an example is ovalbumin, the protein in egg whites. The polypeptide chains in *fibrous proteins* are organized approximately in parallel with long fibers or large sheets and are resistant to solubilization in water. A *disordered protein* lacks an ordered three-dimensional structure. *Membrane proteins*, divided into peripheral and integral proteins, are associated with or attached to the membrane of a cell or an organelle inside the cell.

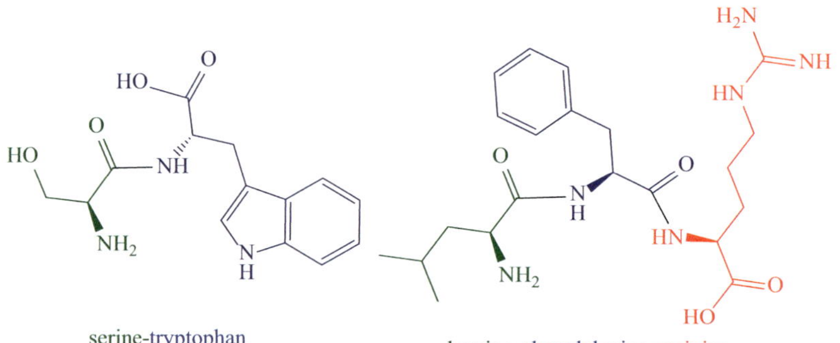

Four structural parts have been identified for proteins: primary, secondary, tertiary, and quaternary, as shown in Figure 3.1. The *primary structure* is the connectivity of the amino acids used to make the protein. The *secondary structure* focuses on the local structural conformation, which is dependent on the hydrogen bonding available to the various amino acid residues of the protein. The α-helix and the ß-sheet constitute the two main types of secondary structure. The *tertiary structure* is the overall three-dimensional shape of the protein, which is essentially the lowest energy state that gives the protein the maximum stability. The tertiary structure is due to the stabilizing bonding interactions of the amino acid side chains. The *quaternary structure* refers to how these protein subunits interact with each other and arrange themselves to form a larger aggregate protein complex. A protein subunit is a single protein molecule that assembles (or "co-assembles") with other protein molecules to form a protein complex. The quaternary structure is how protein subunits interact with each other and arrange themselves to form a larger aggregate protein complex. Different subunits are composed of more than one poly(peptide) chain.

DOI: 10.1201/9781003605157-3

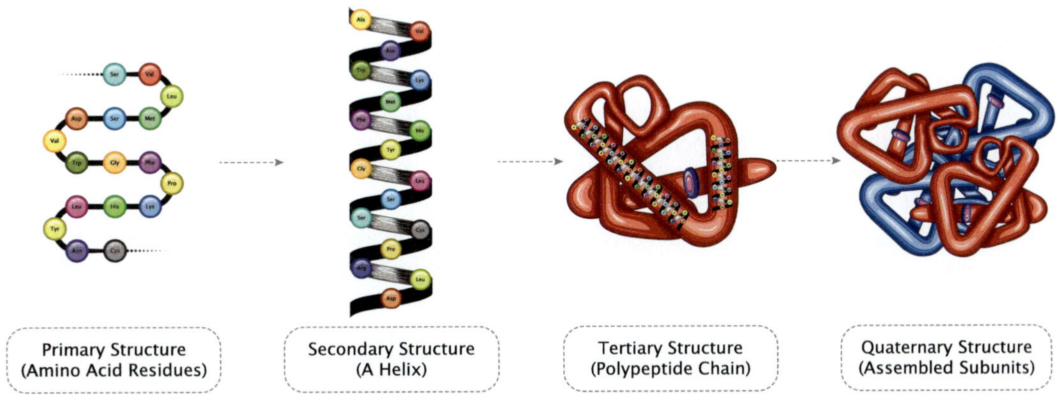

FIGURE 3.1 Protein structure, primary, secondary, tertiary, and quaternary amino acid residues helix polypeptide chain assembled sub units. (Reprinted from Shutterstock vector ID: 1474657079.)

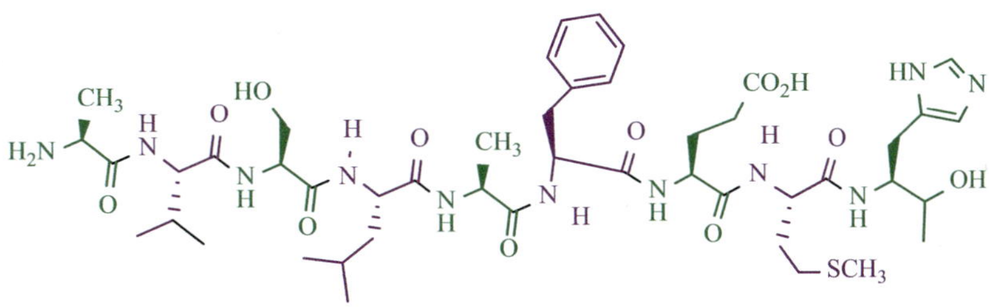

Alanine–valine–serine–leucine–alanine–phenylalanine–glutamic acid–methionine–histidine

If the subunits are the same, the subunit is called a homodimer, whereas if the subunits are different, it is called a heterodimer. Many proteins are made up of multiple polypeptide chains, often referred to as *protein subunits*. These subunits may be the same (as in a homodimer) or different (as in a heterodimer).

An example of a peptide is the nonapeptide (nine amino acid residues) composed (reading from left to right) of alanine–valine–serine–leucine–alanine–phenylalanine–glutamic acid–methionine–histidine, using only (*S*)-amino acids. This name is rather long, and to facilitate communicating the structure of peptides, a shorthand notation assigns a three-letter code or a one-letter code to each amino acid residue, as introduced in Table 2.1 (Section 2.2). These codes take the first three letters of the name of the amino acid in most cases, but isoleucine is abbreviated Ile, glutamine is abbreviated Gln, and tryptophan is abbreviated trp. A one-letter code is used for particularly large peptides, and those codes are also listed in Table 2.1. Using the three-letter codes, this nonapeptide shown is ala-val-ser-leu-ala-phe-glu-met-his, and it is A-V-S-L-A-F-E-M-H using the one-letter codes.

Closer examination of ala-val-ser-leu-ala-phe-glu-met-his shows that there is one carboxyl group on one end (*the carboxyl terminus or C-terminus*) of the peptide and an amino group at the other end (*the amino terminus or N-terminus*). In the case of ala-val-ser-leu-ala-phe-glu-met-his, an alanine residue occupies the *N*-terminus and a histidine residue occupies the *C*-terminus. *By convention, the N-terminus is always drawn on the left, and the C-terminus is drawn on the right.* This convention is important because the name ala-val-ser-leu-ala-phe-glu-met-his indicates that alanine is the *N*-terminus and histidine is the *C*-terminus. The order in which the amino acids are connected is called the *primary structure of a peptide.*

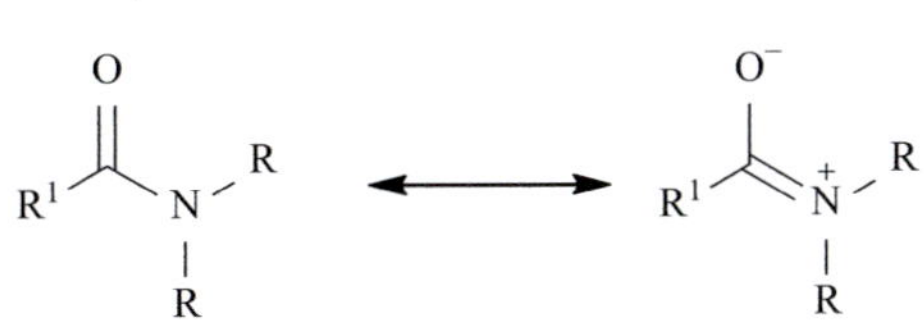

An amide (peptide) bond connects two amino acid residues. The amide unit is quite interesting in that it is essentially planar. The O=C–N bond of an amide has two resonance structures, one with the carbonyl unit and the other with an alkoxy–iminium salt unit, so the C–N unit has a "partial double-bond character," which is normal for the C–N unit in simple amides. As a practical matter, rotation around the C–N bond depends on several factors, and the significant amount of sp^2 hybridization leads to a geometry for the amide unit that is rather planar. Figure 3.2[1] shows rotation about the C–N bond (between the carbonyl carbon and the nitrogen). As mentioned, the amide unit is essentially planar, and Figure 3.2[1] suggests that, due to rotation about the carbonyl carbon–N bond, the groups attached to the carbonyl and the nitrogen may have different stereochemical relationships, and that relationship will vary with the nature of the R groups. There is hindered rotation about the C–N bond due to the resonance forms shown, and it is known that the groups on carbon and nitrogen can be trans to each other. The trans relationship is lower in energy than the arrangement of groups in conformations resulting from the other rotational angles.

If all amide bonds in a peptide are taken to have a planar geometry, the attached groups in a peptide will have a cis or trans relationship. Therefore, the nature of the substituent at the α-carbon has an important influence on the conformation of the entire molecule. A tripeptide example has three amino acids ala-val-ser. If the C–N unit of the amide bond is planar, rotation occurs only around the C–N bond and the C–C bond. Each $C_{carbonyl}$–N–C_α unit defines a plane, and because of the restricted rotation a rotational angle is defined for the atoms connected to the green and to the blue bonds in Figure 3.3. The rotational angle for the C^α–C=O bond is labeled by ψ, which is the angle defined by

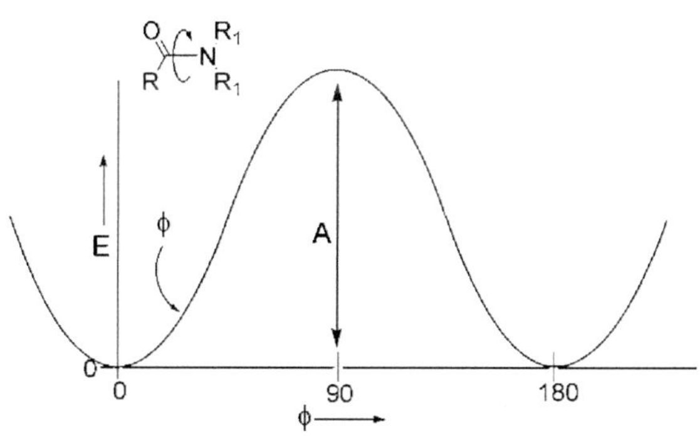

FIGURE 3.2 Approximation of the potential energy surface for amide bond rotation as a function of the twist angle φ. (Reprinted with permission from Pluth, N.D.; Bergman, R.G.; Raymond, K.N. *J. Org. Chem.* **2008**, *73*, 7132. Copyright 2008 American Chemical Society.)

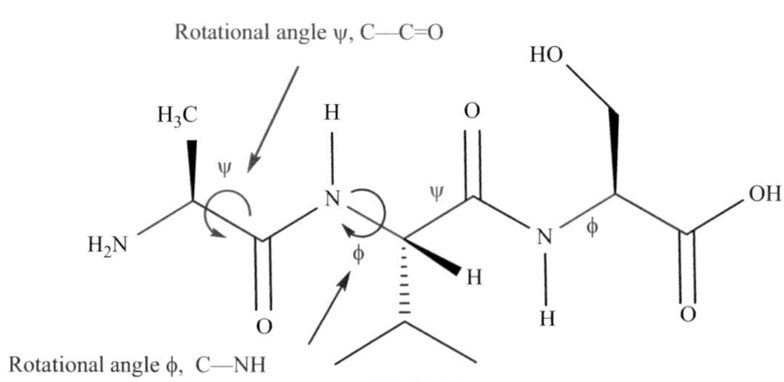

FIGURE 3.3 Rotational angles associated with the tripeptide ala-val-ser.

[1] See Figure 6 in Pluth, N.D.; Bergman, R.G.; Raymond, K.N. *J. Org. Chem.* **2008**, *73*, 7132.

rotation about that bond, where C$^\alpha$ is the carbon of the amino acid that bears the substituent (methyl, isopropyl, hydroxymethyl, etc.).

The angle ϕ is defined as the angle for rotation about the C$^\alpha$–N bond. The alkyl groups in each amino acid residue have a great influence on the magnitude of angles ψ and ϕ, and these angles of rotation define the conformation for that portion of the peptide. In Figure 3.4, the amide unit of one amino acid residue is anti to the amide unit of the adjacent amino acid residue. The carbonyl of the valine residue in this tripeptide is anti to the carbonyl of the valine residue, which is anti to the carbonyl of the serine residue. A peptide chain usually assumes this alternating, or anti, pattern.

If two dipeptide units are brought into proximity, as shown in Figure 3.4, the oxygen of one carbonyl can form a hydrogen bond to the proton on the amide nitrogen of the second dipeptide. For this to occur, however, one dipeptide must have an anti-orientation. This type of hydrogen bonding occurs in long-chain peptides, and when combined with the planar nature of the amide units, the anti-orientation of adjacent amide carbonyls, and the magnitude of the angles ψ and ϕ leads to a rather unique structure for the peptide.

Peptides can form an α-helical structure, as shown for a small portion of the dodecapeptide, dodeca(alanine), in Figure 3.5. The α-helix structure was proposed by Linus Pauling and Robert Brainard Corey in 1951. The hydrogen atoms have been omitted from the structure in Figure 3.5 so the helical structure is easier to see. In a long-chain peptide composed of L-amino acid residues, a right-handed helix (called an α-helix) is formed, where the hydrogen atom on the amide nitrogen is hydrogen-bonded to the oxygen of the carbonyl on the fourth amino acid residue. The hydrogen bonds stabilize the α-helix structure, which is an example of the *secondary structure* of a peptide.

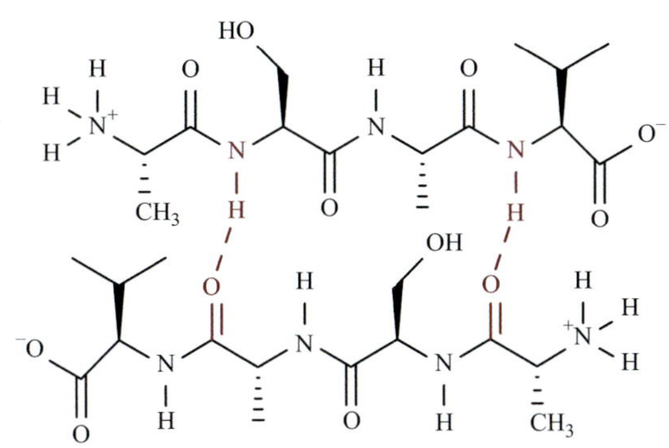

FIGURE 3.4 Intermolecular hydrogen bonding between two peptide chains.

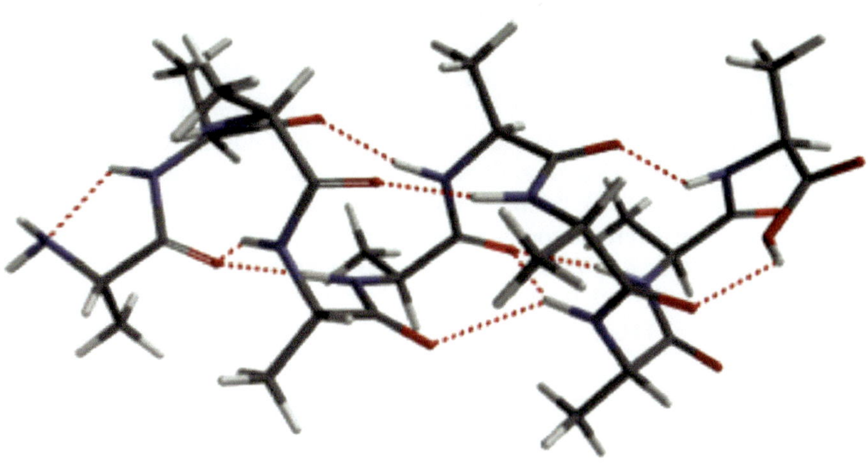

FIGURE 3.5 α-Helical peptide structure for dodeca(alanine) with the hydrogen bonding shown.

Formally, the secondary structure defines the amount of structural regularity in a peptide that results from hydrogen bonding between the peptide bonds.

Analysis of the α-helix reveals some interesting data. First, the hydrogen bonds occur between the carbonyl of one amino acid residue and an amide proton, which is usually on the fourth amino acid residue. In formal terms, *one turn of the helix represents 3.6 amino acid residues, and there are 13 atoms involved in the turn, starting with the carbonyl oxygen and ending with the amide proton.* The number of turns in the α-helix of the peptide depends on the number of amino acid residues in that peptide. The pitch of the α-helix is defined as the "rise" of the amino acid residue in the helix for each turn. Since there are 3.6 residues per turn and the rise per amino acid is measured to be ~ 1.5 Å (150 pm), the pitch of each turn is ~ 5.4 Å (540 pm). The presence of the side chains on the amino acid residues may introduce steric hindrance, which will influence the pitch of each turn.

Some peptides have a high percentage of their secondary structure as an α-helix, whereas others have a low percentage. This percentage is largely determined by the nature of the side-chain groups of the amino acid residues. Some amino acid residues destabilize the β-helix, including glutamate, aspartate, lysine, arginine, glycine, serine, isoleucine, and threonine. One amino acid residue (proline) creates a bend in the β-helix. A peptide composed entirely of leucine, poly(leucine), for example, will exist almost entirely in an α-helix secondary structure, whereas a peptide composed only of aspartic acid residues, poly(aspartic acid), will form another secondary structure. Note that a left-handed helix can be formed when D-amino acids are used to make a peptide.

As noted above, amino acids with significant branching at the beta carbon may interfere with alpha-helix formation. There are at least two other secondary structures observed with peptides. One is called a β-pleated sheet, and the other is called a random coil. Poly(aspartic acid), mentioned above, forms a random coil structure. A *random coil,* as its name implies, does not assume a regular structure (e.g., the α-helix) because hydrogen bonds are not easily formed. Rotation about the ψ and ϕ angles (see Figure 3.3) leads to a random orientation of the various amino acid residues. The β-pleated sheet, on the other hand, does involve intermolecular hydrogen bonding. In other words, there are hydrogen bonds between two different peptide chains rather than within a single peptide chain. An α-helix and a β-pleated sheet are contrasted by saying that the α-helix is held together by *intramolecular* hydrogen bonds, whereas the β-pleated sheet is held together by *intermolecular* hydrogen bonds. There are parallel[2] or antiparallel β-pleated sheets, and both are shown in Figure 3.6. These variations are defined by having the two *C*-termini aligned or the *C*-terminus of one chain aligned with the *N*-terminus of the other chain.[2]

The anti-parallel β-pleated sheet occurs when the *N*-terminal of one peptide chain is aligned with the *C*-terminal of an adjacent chain. The parallel β-pleated sheet occurs when neighboring polypeptide chains run in the same direction, meaning the *N*- and *C*-terminals of the peptide chains align, but the hydrogen bonds may be distorted. Due to this distortion of hydrogen bonds, parallel beta sheets are not as stable as anti-parallel beta sheets.

The overall structure of a long polypeptide is determined by its amino acid sequence (primary structure) and whether it forms an α-helix or a β-pleated sheet (secondary structure). Another feature of peptide structure must be considered. The peptide chain folds and coils into a very complex structure known as its *tertiary structure*, as shown in Figure 3.7[3] for the enzyme ribonuclease A in the so-called ribbon diagram or "ribbon drawing."[4] A ribbon diagram is a shorthand method for drawing the tertiary structure of complex peptide structures, in which β-strands are shown as "thick" arrows, an α-helix as a spiral ribbon, and non-repetitive structures as ropes. These shorthand structures are also called *Richardson diagrams*.[4a,b] As mentioned, ribonuclease A is shown in Figure 3.7, along with the helical ribbon and "tube" used to represent a helix. The "wide arrow"

---

[2] Smith, C.K.; Regan, L. *Acc. Chem. Res.* **1997**, *30*, 153.
[3] Richardson, J.S. *Methods Enzym. Macromol. Crystal. Part B* **1985**, *115*, 359. Figure 13 on p. 374.
[4] (a) Richardson, J.S. *Adv. Protein Chem.* **1981**, *34*, 167; (b) Richardson, J.S. *Methods Enzym. Macromol. Crystal. Part B* **1985**, *115*, 359–380; (c) Richardson, J.S. *Nature Struct. Mol. Biol.* **2000**, *7*, 624.

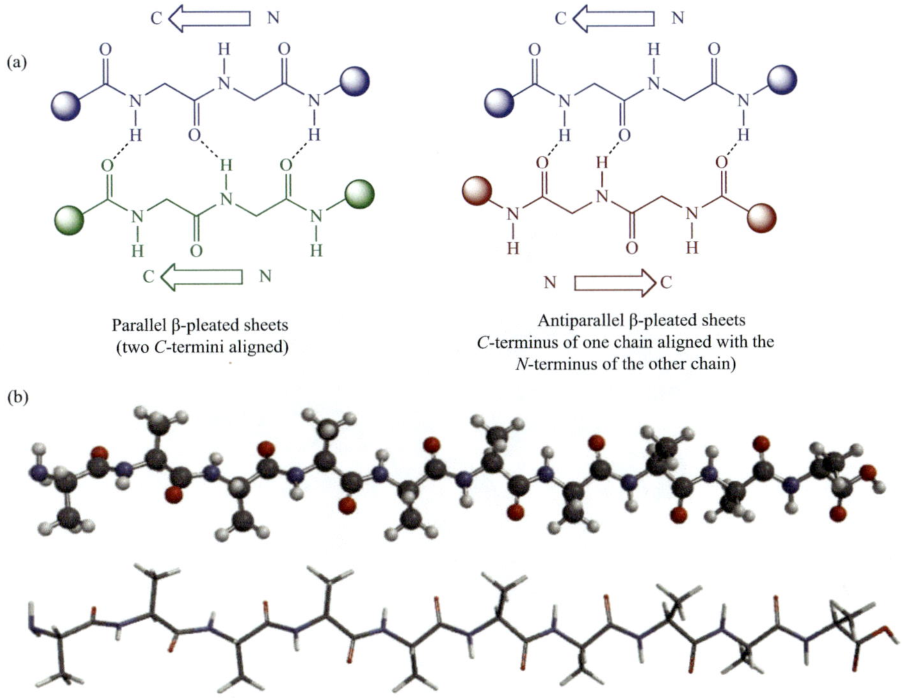

FIGURE 3.6 (a) Parallel and antiparallel β-pleated sheets. (b) β-Pleated sheet of deca(alanine).

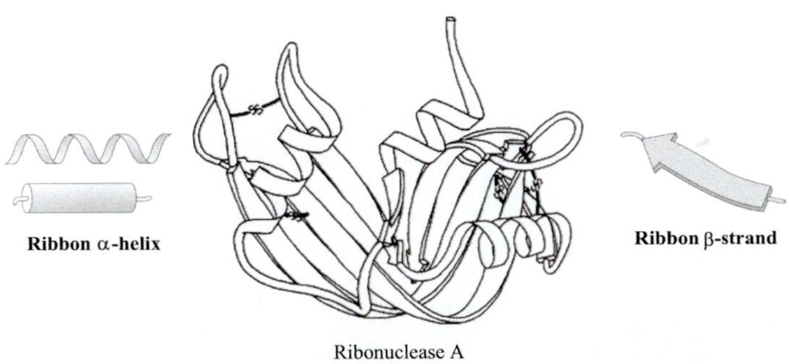

FIGURE 3.7 Richardson drawings. Ribonuclease A, illustrating a variety of junctions between loops and helices and between loops and β-strand arrows. (Reprinted from Richardson, J.S. *Methods Enzymol. Macromol. Crystallogr. B* **1985**, *115,* 359–380, Methods in Enzymology. Figure 13 on p. 374, with permission from Elsevier.)

used to represent a β-strand in a ribbon structure is also shown. An example is the ribbon structure shown in Figure 3.8[5] for YdjH, a sugar kinase in *Escherichia coli* K12.

The *quaternary structure* of a polypeptide or protein refers to two or more peptide units (called subunits) that combine to form a protein structure. An example of a quaternary structure is rat AQUAPORIN-4, shown in Figure 3.9. Note that a protein *domain* is a conserved part of a given protein sequence that can function independently of the rest of the protein chain. Domains are typically compact and globular, with a roughly spherical structure consisting of multiple motifs. The interior of a globular protein is composed primarily of hydrophobic amino acids, whereas the exterior is composed mainly of hydrophobic amino acids.

[5] Huddleston, J.P.; Raushel, F.M. *Biochemistry* **2019**, *58*, 3354.

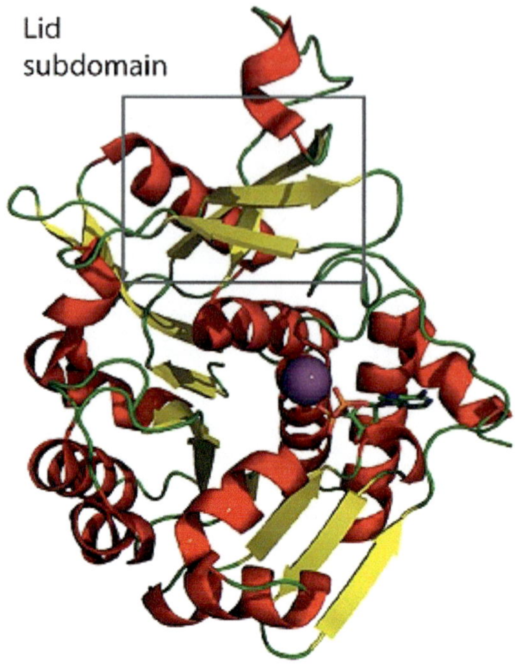

FIGURE 3.8 Ribbon diagram of YdjH in which the α-helices are colored red, β-strands are colored yellow, and the sodium is shown as a purple sphere. (Reprinted with permission from Figure 6 in Huddleston, J.P.; Raushel, F.M. *Biochemistry* **2019**, *58*, 3354–3364. Copyright 2019 American Chemical Society.)

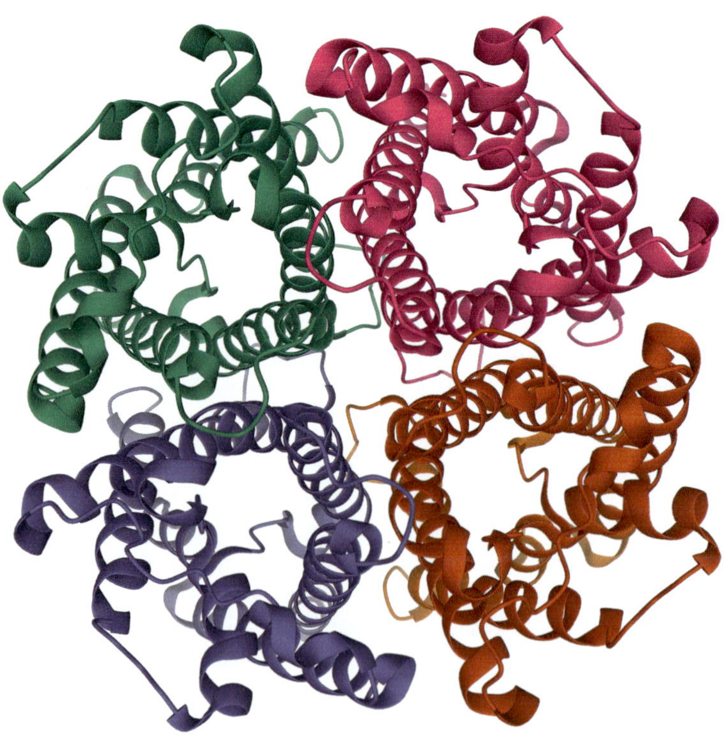

FIGURE 3.9 Quaternary structure of rat AQUAPORIN-4, 3D. Shutterstock Stock Illustration ID: 1881163822.

3.2 PEPTIDE BIOSYNTHESIS

The process of synthesizing a protein from messenger RNA (mRNA; Section 15.2) template is known as *translation*. Ribonucleic acids (RNA) are molecules that are important for peptide biosynthesis. Messenger RNA is a single-stranded nucleic acid with unique secondary structures. It carries the genetic code from deoxyribonucleic acid (DNA; Sections 15.2 and 15.5–15.7) to the ribosome, and this code is read via forming hydrogen bonds to the three base pairs of transfer RNA (tRNA; also see Section 15.2), which is called the *triplet anticodon*. A strand of mRNA may be categorized as having groups of three base pairs, where each group is the genetic code for a specific amino acid. These three base pairs (called *codons*) will match a complementary set of three base pairs on a tRNA (an anticodon, see Figure 3.10).[6] To ensure this genetic code is easily "read," mRNA is generally less structured than tRNA or rRNA (ribosomal RNA; Section 15.2).

The anticodon is at the bottom loop of the tRNA molecules shown in Figure 3.10, and *the codon and anticodon are complementary. A G–C–C codon on mRNA will match a C–G–G anticodon on tRNA*. This finding is important because the C–G, G–C, and G–C base pairs are capable of hydrogen bonding (see Figure 15.2 in Section 15.2). This hydrogen bonding allows the tRNA to attach itself to the mRNA at the proper place. The amino acid carried by the tRNA is then released to a growing peptide chain being synthesized on the ribosome. This process is illustrated in Figure 3.10, where the tRNA anticodons are shown. This constitutes the mechanism by which an organism synthesizes proteins from amino acids.

Each amino acid has a tRNA with a unique anticodon, and that amino acid will attach only to that tRNA. The ribosome binds a strand of mRNA to build a specific protein, which has a specific sequence of amino acids. The genetic code on the mRNA (the set of codons) matches the amino acid sequence of the protein. Therefore, the mRNA binds tRNA molecules in a particular sequence,

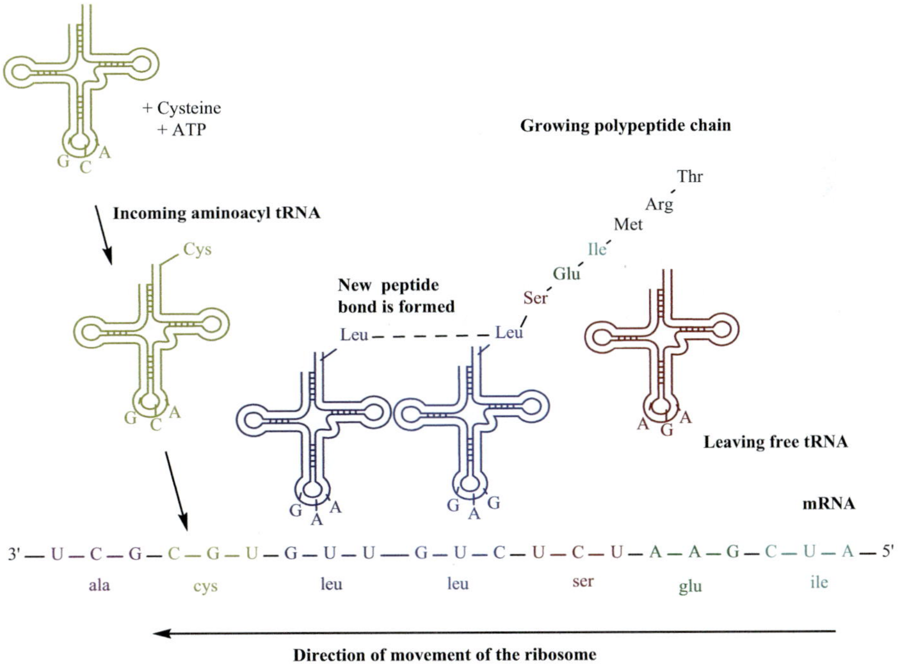

FIGURE 3.10 Protein synthesis. (Adapted from Figure 1 and reprinted with permission from Lipman, F. *Accounts Chem. Res.* **1973**, *6*, 361. Copyright 1973 American Chemical Society.)

[6] Lipman, F. *Acc. Chem. Res.* **1973**, *6*, 361.

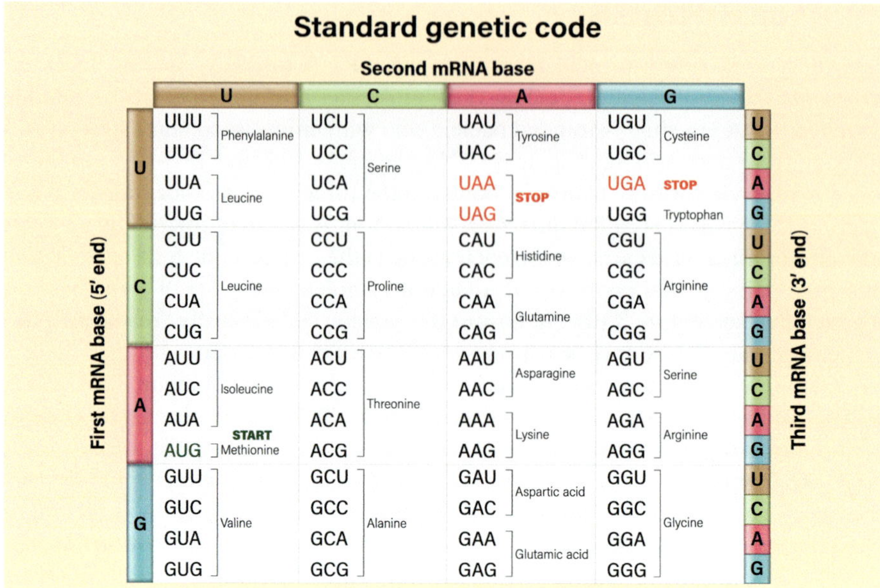

FIGURE 3.11 The Standard genetic code. Shutterstock. Image ID 2052705743.

allowing each amino acid to be released to the growing protein sequence associated with the ribosome. The genetic code for the amino acids used to make proteins is shown in Figure 3.11.[7]

Many natural peptides are produced by bacteria via nonribosomal peptide synthesis. Nonribosomal peptide synthesis means that the peptide is not produced by the transfer RNA–messenger RNA (tRNA–mRNA) mechanism just described. *Bacillus subtilis* produces the heptapeptide surfactin, for example, which has antibiotic and antifungal activity.[8] Each amino acid found in surfactin is directly selected for incorporation into the growing peptide chain by one of the domains of *surfactin synthetase*[9] (an example of a multienzymatic thiotemplate), shown with the pendant SH groups.

As shown in Figure 3.10,[6] substrate activation occurs after binding the amino acid, and the enzyme catalyzes the formation of an aminoacyl adenylate intermediate using Mg^{2+}-ATP and the release of a cofactor. Subsequently, the amino acid–O–AMP oxoester is converted into a thioester by a nucleophilic attack of the free thiol-bound cofactor of an adjacent PCP domain. Note that PCP is an abbreviation for *peptidyl carrier protein*. This reaction is effectively an acyl substitution. Also, note that ATP is adenosine triphosphate and AMP is adenosine monophosphate.

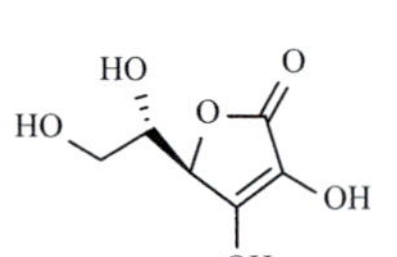

Vitamin C (ascorbic acid)

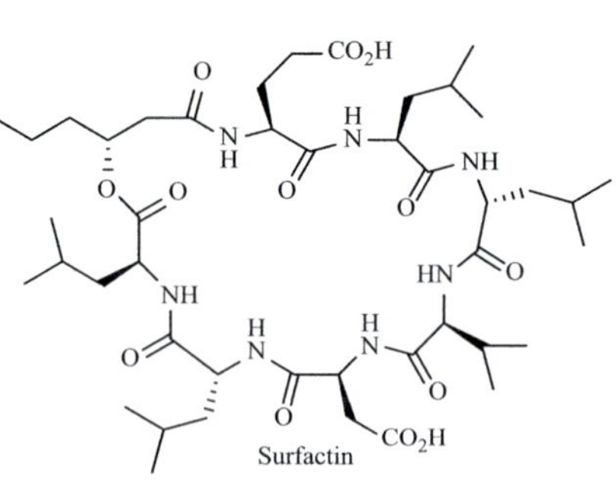

Surfactin

---

[7] Devlin, T.M. *Textbook of Biochemistry with Clinical Correlations*, 2nd Ed., Wiley, NY, **1986**, see Table 19.2, p. 738.

[8] Sieber, S.A.; Marahiel, M.A. *Chem. Rev.* **2005**, *105*, 715.

[9] See (a) Kluge, B.; Vater, J.; Salnikow, J.; Eckart, K. *FEBS Lett.* **1988**, 231, 107; (b) Shaligram, N.S.; Singhal, R.S. *Food Technol. Biotechnol.* **2010**, *48*, 119.

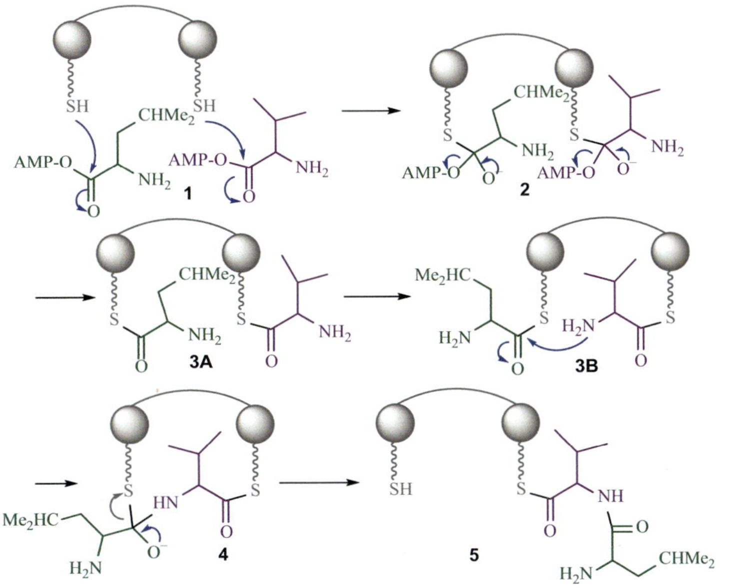

FIGURE 3.12 Biosynthesis of surfactin. (Reprinted with permission from Sieber, S.A.; Marahiel, M.A. *Chem. Rev.* **2005**, *105*, 715. Copyright 2005 American Chemical Society.)

In Figure 3.12,[8] the binding domains attach two amino acids in (**1**) by an acyl substitution reaction in which the thiol unit attacks the acyl carbon, and each coupling proceeds through the appropriate tetrahedral intermediate shown as (**2**). Loss of the O–AMP leaving group leads to (**3A**), redrawn as (**3B**) to show that the nucleophilic amine unit of one amino acid residue attacks the acyl unit of the other in an acyl substitution reaction to yield another tetrahedral intermediate (**4**). The leaving group is the thiol-bound unit, leading to the formation of the bound dipeptide in (**5**). The organism will produce the heptapeptide using similar coupling reactions and then complete the biosynthesis of surfactin. This sequence shows that organisms can produce peptides by alternative mechanisms to ribosomal peptide synthesis.

3.3 CHEMICAL SYNTHESIS OF PEPTIDES

The reaction of a carboxylic acid or an acid derivative with an amine to form an amide was discussed in Section 2.4. Since amino acids have an amine unit and an acid unit, two amino acids may be coupled together to form an amide, but there are two possible products. Alanine and phenylalanine can be coupled to form a dipeptide, for example. Since the amine unit of one amino acid may react with the carboxyl group of the other amino acid, the amine group of one amino acid and the acid group of the other must be blocked with an unspecified *protecting group* (P) to form the designated amide bond. The two possible products that can be formed by the reaction of alanine with phenylalanine are shown. Coupling one amino acid to another amino acid may generate either alanine-phenylalanine or phenylalanine-alanine. In the dipeptide alanine-phenylalanine, alanine is coupled via its amine unit to the carboxyl unit of phenylalanine. In phenylalanine-alanine, phenylalanine is coupled via its carboxyl group to the amine group of alanine. Clearly, a third, a fourth, or a larger number of new amino acids

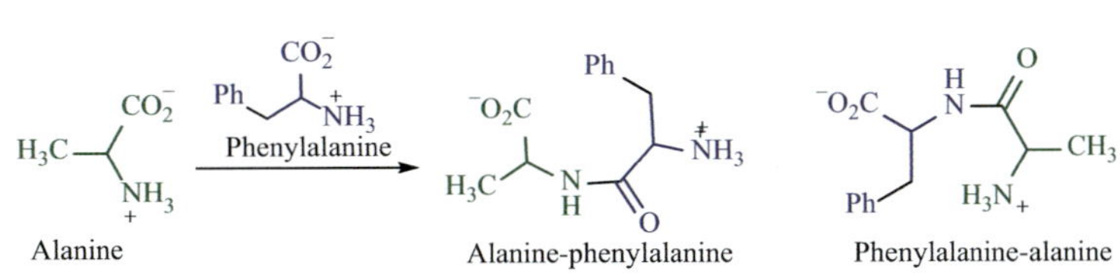

may be added to either the carboxyl end or the amino end. Dimers, trimers, tetramers, and so on, that contain several amide bonds are known as poly(amides) or just polyamides. Extension of this process will make a lengthy poly(amide), but each new reaction must be controlled to occur at the carboxyl or the amine unit. The synthesis is not trivial, however, because the amino acids must be coupled together in the correct sequence, *N*- to *C*-terminus, which is the primary structure.

P-NH-ser-CO_2H $\xrightarrow{\text{P-NH-leu-}CO_2\text{P}}$ P-NH-ser-leu-CO_2P ⟶ P-NH-ser-leu-CO_2H

$\xrightarrow{\text{P-NH-val-}CO_2\text{P}}$ P-NH-ser-leu-val-CO_2P ⟶ P-NH-ser-leu-val-CO_2H

$\xrightarrow{\text{P-NH-ala-}CO_2\text{P}}$ P-NH-ser-leu-val-ala-CO_2P ⟶ NH_2-ser-leu-val-ala-CO_2H

A tetrapeptide target such as NH_2-ser-leu-val-ala-COOH poses a more complex problem. The *N*-terminus is serine, which requires the use of an *N*-protected serine residue. The *C*-terminus is alanine, which requires the use of a *C*-protected alanine residue. To couple leucine and valine, there is a problem. To couple valine *to N*-protected serine requires a *C*-protected leucine, but after coupling, the carboxyl protecting group must be removed before the resulting peptide can react with the *C*-protected valine. To produce the tetrapeptide, further reactions require the use of an *N*-protected serine and a *C*-protected alanine. For the "middle" amino acid residues, methodology must be available to protect and deprotect either terminus of the amino acid. A discussion of peptide synthesis,0 therefore, requires a discussion of the chemical reactions and reagents used to protect the *N*- or *C*-terminus of any amino acid. Methods to couple amino acids together to form the peptide will also be discussed.

If either the nitrogen or the carboxyl unit is protected with an appropriate protecting group, there must be a reaction that will remove the protecting group (*deprotection*). The ester groups are usually removed by saponification (basic hydrolysis), but *p*-nitrobenzyl esters are removed by reaction with HBr in acetic acid, and *tert*-butyl esters are removed by treatment with trifluoroacetic acid. Removal of the nitrogen protecting groups can be more problematic. Cleavage of a trityl group requires reaction with dilute aqueous acetic acid. The Cbz group is removed by catalytic hydrogenation with a palladium catalyst, whereas the Boc group is removed by reaction with trifluoroacetic acid.

Since selective protection and deprotection of the amine and carboxyl groups of amino acids are possible, the coupling reaction of two amino acids is straightforward. The first peptide coupling reaction was reported in 1902, when an acyl azide (–CON_3) was reacted with an amino acid. A common method for making a peptide bond involves coupling an amine unit with a carboxyl unit in the presence of dicyclohexylcarbodiimide (abbreviated DCC). When *N*-Cbz-phenylalanine reacts with ethyl tryptophan in the presence of DCC and an acid catalyst, the product is the protected dipeptide. Catalytic hydrogenation and saponification yield the dipeptide phe-trp. Note that saponification is the process that cleaves esters into carboxylate salts and an alcohol using an aqueous alkali such as aqueous NaOH.

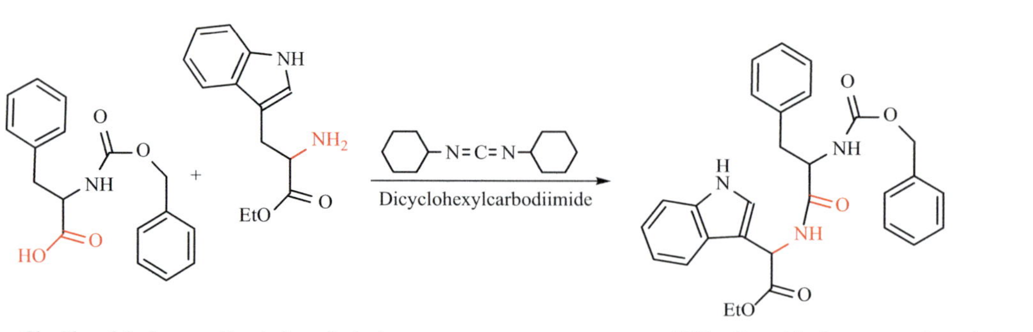

Cbz-Phenylalanine Tryptophan ethyl ester N-Cbz-Phenylalanine—tryptophan ethyl ester

An important discovery in the field of peptide synthesis was made by R. Bruce Merrifield in 1962. An amino acid was bound to a polymer bead via the *C*-terminus, and the peptide bradykinin (arg-pro-pro-gly-phe-ser-pro-phe-arg; a potent hypotensive agent) was subsequently synthesized by performing sequential chemical reactions, including the protection and deprotection steps, *on a polymer bead*. When completed, the nonapeptide was "released" from the polymer by a simple chemical reaction. This basic approach is now called the *Merrifield synthesis*, and it is one of the most important advancements in peptide synthesis.

When *N*-Boc-alanine reacts with a "Merrifield polymer," an alanine-loaded polymer is formed. The remainder of the synthesis is essentially a DCC-coupling procedure. Sequentially, the growing peptide is treated with HCl to remove the Boc group, and treatment with DCC and the next amino acid leads to the next peptide. This sequence of reactions is repeated, using other protected amino acids, until the desired peptide is prepared. At the end of the synthesis, the peptide on the polymer bead is treated with HBr and trifluoroacetic acid to release the newly synthesized peptide from the bead and isolate it. One of the more interesting features of the Merrifield synthesis is that it can be automated. Nowadays, there are peptide synthesis machines based on the Merrifield synthesis that are used to prepare peptides.

3.4 PROTEINS AND ENZYMES ARE POLY(PEPTIDES)

Proteins constitute one of the most important classes of biomolecules, which are polymeric organic molecules. Chemically, proteins are polymers of amino acid residues, linked from the amino group to the carboxyl group; large poly(peptides). Proteins are important for virtually everything connected with cell structure and cell function, and the enzymes to be described in Chapter 4 are polypeptides. Proteins differ from one another primarily in their sequence of amino acids. A linear chain of amino acid residues is called a polypeptide, and a protein contains at least one long polypeptide.

There are regulatory proteins that control the ability of other proteins to carry out their functions. Insulin regulates glucose metabolism in animals. The human insulin protein is composed of 51 amino acid residues with a mass of 5,808 Da. Note that Da is a dalton (unified atomic mass unit). The dalton is a unit of mass that is defined as 1/12 of the mass of an unbound neutral atom of carbon-12 in its nuclear and electronic ground state. The human insulin protein is a heterodimer with an A-chain and a B-chain linked together by disulfide bonds. Somatotropin, known as growth hormone, is a polypeptide chain containing about 190 amino acid residues. It is produced by the pituitary gland in mammals and has two disulfide bridges, with four alpha helices. Release of thyrotropin-stimulating hormone is stimulated by the hypothalamus. It is a tripeptide, pyroglutamyl-histidyl-proline amide. Hemoglobin is an example of a transport protein, which transports specific substances from one place to another. Specifically, hemoglobin transports oxygen from the lungs to tissues. Hemoglobin is composed of four subunits, each having one polypeptide chain and one heme group. Some nutrients are stored in special proteins called storage proteins. An example is casein, the most abundant protein found in milk. Casein is a family of related phosphoproteins that are commonly found in mammalian milk. Casein contains a high number of proline amino acids, which hinder the formation of common secondary structures. There are also no disulfide bridges, so there is little tertiary structure. Ferritin stores iron in animal tissues. Ferritin is an iron storage protein formed by 24 polypeptide chains that self-assemble into a hollow, roughly spherical protein cage.

The very structure of cells and tissues is due to structural proteins (e.g., the α-keratins, collagen, elastin, and fibroin). The α-keratins make up hair, horn (e.g., rhinoceros horn), and fingernails. α-Keratin is a polypeptide that is characterized by many alanine, leucine, arginine, and cysteine residues. Fibrous keratin molecules supercoil to form a very stable, left-handed superhelix that forms filaments of multiple copies of the keratin monomer. Collagen is found in bone and tendons. In collagen, there is a peptide chain structure with three interlocking strands, a molecular weight of 300 kilodaltons, and about 3,000 amino acid residues. Fibroin is a major component of the silk used to make cocoons, and it is a major component of spider webs. Silk fibroin consists of a heavy chain (~390 kDa) and a light chain (~26 kDa) linked together via a single disulfide bond.

Some proteins are important for cell protection. The immunoglobulins (antibodies) are produced by lymphocytes and defend the body against bacteria, viruses, and so on. Thrombin and fibrinogen are important blood clotting proteins. There are glycoproteins that contain carbohydrates (see Chapter 13), lipoproteins that contain lipids (see Chapter 10), and nucleoproteins that are important for storage and transmission of genetic information. Proteins are critical to life processes in plants, insects, and animals and, as a group, may well be the most important organic chemicals known.

3.5 PEPTIDASES

Proteolytic enzymes (proteases) are available to cleave the amide bond (peptide bond) of a peptide linkage, as in an enzyme, to yield the amino acid. They often inactivate small peptides. Such enzymes are known as *peptidases*.[10] The proteases are classified as *exopeptidases* or *endopeptidases*. An *exopeptidase* is an enzyme that cleaves an amino acid from the terminal position (near the N- or C-terminal ends) of a peptide whereas an enzyme that cleaves an amide bond at an internal position (within the peptide chain rather than at the ends) is called an *endopeptidase*. *Aminopeptidases* (EC 3.4.11) can liberate single amino acids, dipeptides (*dipeptidyl peptidases*, EC 3.4.14), or tripeptides (*tripeptidyl peptidases*, EC 3.4.14) from the N-terminal end of their substrates.[10] See Section 4.2 for the EC nomenclature. Examples of *exopeptidase* enzymes include *carboxypeptidases Y, A, and B* (EC numbers 3.4.16–3.4.18); *Cathepsin C* (EC 3.4.14.1); and *aminodipeptidase* (*DAPase*; EC 3.4.11.XX). Examples of *endopeptidase* enzymes include *trypsin* (EC 3.4.14.4), *chymotrypsin* (EC 3.4.21.1), *enterokinase* (EC 3.4.21.9), *elastase* (EC 3.4.21.36, pancreatic), *subtilisin* (EC 3.4.21.62), and *thrombin* (EC 3.4.21.5).

Proteolysis is the breakdown of proteins into smaller polypeptides or amino acids. Proteolysis is important for digestive enzymes to break down proteins in food and provide amino acids for the organism. Proteolysis is also important for the regulation of some physiological and cellular processes, as well as for preventing the accumulation of unwanted or abnormal proteins in cells.

Leucine aminopeptidase (EC 3.4.11.1; isolated from hog kidney) cleaves *N*-terminal amino acids. This enzyme is specific for leucine and other *nonpolar amino acid residues*, allowing them to be isolated and identified. If proline is the *N*-terminal amino acid residue, however, this enzyme will *not* cleave it. In other words, the enzyme is used to cleave and identify most amino acid residues, but it does *not* work if a proline residue is at the *N*-terminus.

An enzymatic procedure can be used to identify the *C*-terminal amino acid residue. Four common enzymes cleave the *C*-terminal amino acid residue from a peptide. They are *carboxypeptidase A* (from bovine pancreas), *carboxypeptidase B* (from hog pancreas), *carboxypeptidase C* (from citrus leaves), and *carboxypeptidase Y* (from yeast). *Carboxypeptidase A* cleaves all *C*-terminal amino acid residues *except* proline, arginine, and lysine. *Carboxypeptidase B* cleaves *only* arginine or lysine. *Carboxypeptidase C* cleaves *all* amino acid residues from the *C*-terminus, as does *carboxypeptidase Y*. In principle, a combination of these enzymes can be used to gain information about the *C*-terminus.

The enzyme trypsin, a digestive enzyme, cleaves peptide bonds, but only when the amino acid residue has a carbonyl unit that is part of arginine or lysine. In other words, it cleaves specifically on the *C*-side of arginine or lysine of a peptide to yield fragment peptides that have an arginine or lysine at the *C*-terminus. *Chymotrypsin cleaves amino acid residues that have a carbonyl that is part of an aromatic amino acid such as phenylalanine, tyrosine, or tryptophan.* This enzyme produces fragment peptides that have one of these three amino acids at the *C*-terminus. Since chymotrypsin can cleave many other amino acids if given enough time to react, care must be exercised when using this analysis. *Staphylococcal protease* (EC 3.4.24.19) cleaves acidic amino acid residues (e.g., aspartic acid and glutamic acid), so the fragment peptides have aspartic acid or glutamic acid at the *C*-terminus.

[10] Mótyán, J.A.; Tóth, F.; Tőzsér, J. *Biomolecules* **2013**, *3*, 923.

Using hydrolysis and enzymatic cleavage, along with the end-group identification techniques described in Section 3.6, it is possible to sequence long and complex proteins. Modern technology includes automated protein sequencers to identify amino acid residues in protein sequences by tagging and removing one amino acid at a time, which is then analyzed and identified. Modern NMR techniques also provide a powerful tool for polypeptide and protein structure determination, and they are probably used more often than many of the chemical techniques described here. However, the chemical techniques remain a powerful tool in protein research and are part of automated sequencers.

3.6 PEPTIDE DEGRADATION AND END GROUP IDENTIFICATION

The process of identifying the chemical structure of a protein (or any peptide) usually begins by heating it in 6N HCl at 105°C for 24 hours. This reaction should completely hydrolyze the peptide into its constituent amino acids, which are then separated. Hydrolysis leads to the *denaturation* of proteins, which disrupts and possibly destroys the secondary and tertiary structures. In other words, denaturation unfolds the protein into a random shape. A denatured protein usually loses its biological function, and denatured proteins can form aggregates under certain conditions. Denaturation of proteins is not strong enough to break the peptide bonds, so the primary structure (sequence of amino acids) remains the same. Therefore, denaturation is the first logical step in a process that will identify the primary structure of a protein.

There are some problems with the denaturation process, depending on the reaction conditions. Tryptophan has an indole unit, which is acid sensitive, and it is partially destroyed by harsh acidic conditions. Glutamine, asparagine, glutamic acid, and aspartic acid decompose with the loss of ammonia when heated in 6N HCl. When the amino acid residue has a sterically hindered side chain, as in valine or isoleucine, hydrolysis may be incomplete, and heating for a longer period of time may be necessary. These problems mean that after the 24-hour hydrolysis, 100% of the amino acid residues in the original peptide or protein may not be available. Once all of the amino acids have been obtained, however, there are several techniques that can be used to provide structural information.

The first approach to identifying the primary structure is to identify each individual amino acid and then calculate the percentage of each residue in the protein. For example, if the protein consists of 21% alanine and 5% methionine, this knowledge is important for determining the primary structure. There are chromatography columns that will separate individual amino acids, which makes possible the identification of each constituent amino acid. As these separated amino acids are isolated, one method for their identification relies on heating them to 100°C in the presence of ninhydrin.

Ninhydrin can exist in the hydrate or tri-ketone forms as the amount of water in the medium changes. Reaction with an amino acid such as serine (Figure 3.13) initially produces an imine along with ammonia. Under the reaction conditions, decarboxylation occurs, and subsequent reaction with a second molecule of ninhydrin gives 2-(3-hydroxy-1-oxo-1*H*-inden-2-ylimino)-1*H*-indene-1,3(2*H*)-dione, which has a characteristic purple color (λ_{max}=570nm) that is easily detected by UV spectroscopy. All amino acids react with *ninhydrin except proline*. Proline reacts with ninhydrin to yield a different compound (**6**) that has a distinctive yellow color (λ_{max}=440nm).

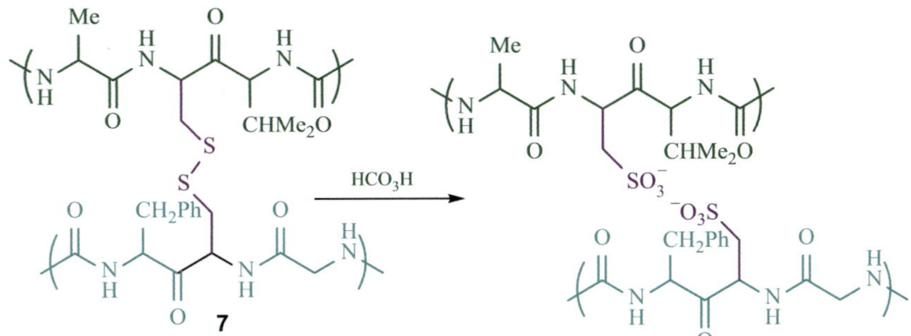

7

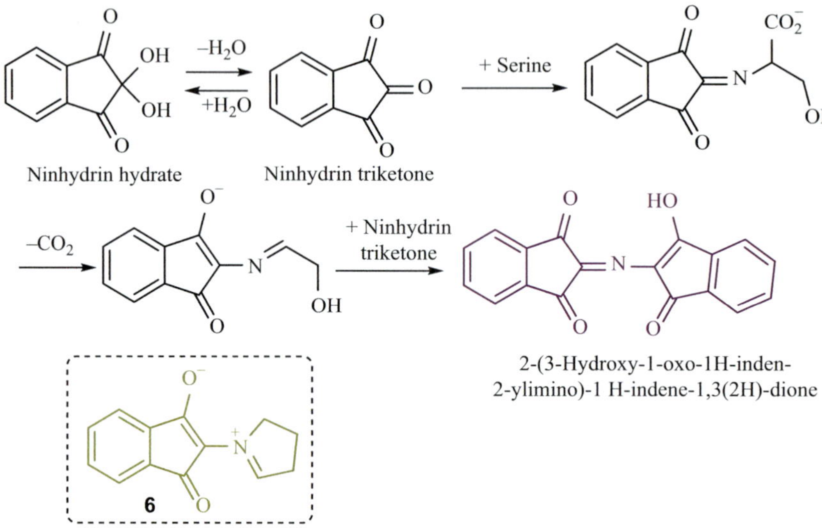

FIGURE 3.13 Identification of an amino acid with ninhydrin.

A general strategy for identifying the sequence of amino acids is to first identify the *N*- and *C*-terminal amino acid residues. The protein is then treated with specialized reagents that cleave the peptide chain at known amino acid residues, requiring a test that will identify the terminus of each fragment. When two cysteine residues are in proximity, they can exist as two different strands or in one strand that has coiled around in a manner that brings them together. In either case, they react to form a disulfide bond, as in **7**. Disulfide bonds are cleaved if the peptide is treated with peroxyformic acid, and in the case of **7**, the products are two cysteic acid residues as shown. Alternatively, **7** reacts with 2-mercaptoethanol to yield two free cysteine units as well as the disulfide, 2,2′-disulfandiyldiethanol. To prevent the cysteine residues from recombining to form a new disulfide linkage, reaction with iodoacetic acid gives **8, 9,** and the thiol fragments. Once the disulfide bridges (if any) are removed, the process of identifying the termini of the peptide can begin, which is called *end-group analysis.*

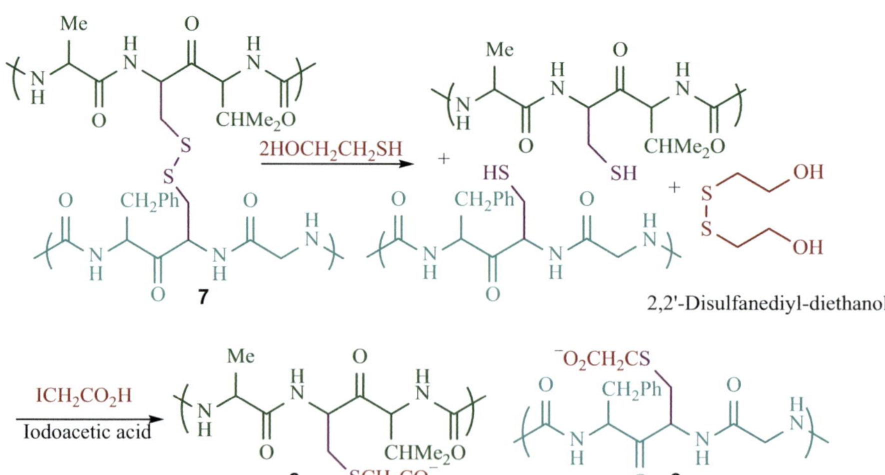

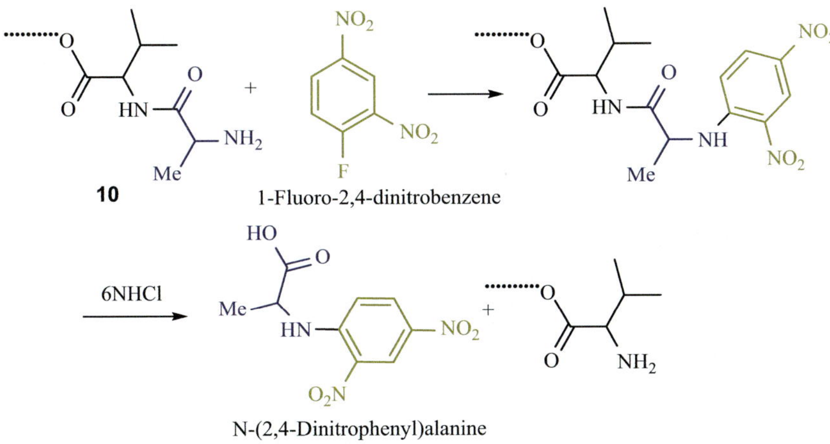

The *N*-terminal amino acid of a peptide will react with *1-fluoro-2,4-dinitrobenzene* (*FDNB*), known as *Sanger's reagent*. If there is a peptide that terminates in an alanine residue, such as (**10**), the reaction at the fluoroaryl position gives the FDNB product. Subsequent hydrolysis with 6 N HCl releases the *N*-terminal amino acid, the FDNB amino acid. The FDNB amino acid is yellow and easily identified and separated once the peptide is completely hydrolyzed. FDNB can also react with the amine unit on the side chain of lysine, the imidazole nitrogen on the side chain of histidine, and the sulfur unit on the side chain of cysteine, which can cause problems in the analysis.

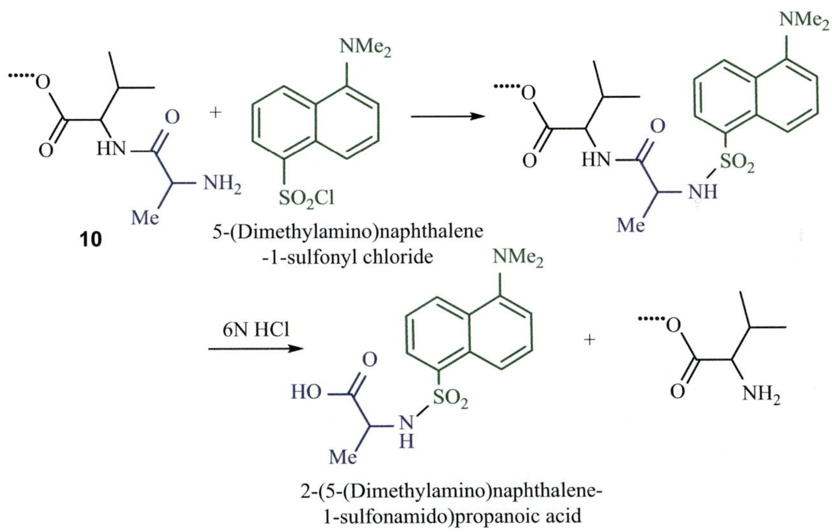

Another method for determining the identity of the *N*-terminal amino acid is to react the peptide with *dimethylaminonaphthalenesulfonyl chloride* [5-(dimethylamino)naphthalene-1-sulfonyl chloride, known as *dansyl chloride*]. When dansyl chloride reacts with a peptide such as (**10**), the amine unit reacts with the sulfonyl chloride to yield the sulfonamide. Acid hydrolysis of this *N*-dansyl peptide leads to the release of the amino acids, in this case valine and *N*-dansyl alanine, 2-(5-(dimethylamino)naphthalene-1-sulfonamido)propanoic acid. The dansyl product is easily isolated, and the presence of the dansyl group makes this sulfonamide highly *fluorescent*.

Another *N*-terminus identification technique is more versatile than Sanger's reagent or dansyl chloride in that the *N*-terminus can be identified, and the sequence of the remainder of the peptide is also possible. This reagent is *phenyl isothiocyanate*, which is known as *Edman's reagent*. This method of identification is known as *Edman degradation*.[11] The alanine residue in peptide (**10**) reacts with phenyl isothiocyanate at pH 8 to 9 to give an intermediate that cyclizes to release the phenylthiohydantoin derivative of alanine (2-methyl-4-phenyl-5-thioxopyrrolidin-3-one) and the original peptide minus the alanine residue, with a new *N*-terminus. The phenylthiohydantoin is soluble in organic solvents, easily removed from the peptide, and can be identified using various techniques. The remaining peptide has another amino unit, and it is treated with more phenyl isothiocyanate and then cleaved to examine and identify the next amino acid. Since the process begins at the *N*-terminus and progresses toward the *C*-terminus, both the identity of the amino acids and their exact sequence in the peptide can be determined. This method can be used to identify 30–60 amino acid residues in a long peptide under the right conditions.

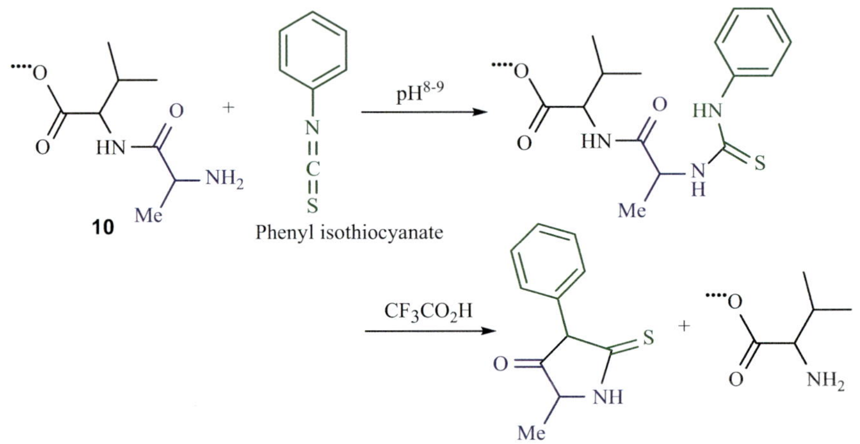

Fewer methods are available to identify the *C*-terminal amino acid residue of a peptide. When the peptide is heated to 100°C (for ~ 12 h) with hydrazine (NH_2NH_2), the amide bond of each residue is attacked and cleaved. The products are *amino hydrazides*. The *C*-terminal amino acid is *not* converted to the hydrazide because hydrazine reacts with the amide, not with the free carboxyl group. This approach can be applied to large peptides, but the peptide is destroyed, and it is difficult in some cases to separate one amino acid from many amino acid hydrazides. To overcome this problem, the mixture of hydrazides and amino acid is treated with Sanger's reagent, which converts the hydrazide compounds to a bis(dinitrophenyl) derivative. The free amino acid is also converted to its FDNB derivative, which is easily removed from the mixture, making identification of this terminal amino acid relatively easy.

An alternative method for *C*-terminus identification reduces the carboxyl group (–COOH) of a peptide to an alcohol (–CH_2OH) with lithium aluminum hydride ($LiAlH_4$). Subsequent acid hydrolysis of the peptide liberates all of the amino acids along with the one amino alcohol that was generated by reduction of the *C*-terminal amino acid residue.

HOMEWORK

3.1. What is a peptide bond?
3.2. Give a brief explanation of why a peptide bond is relatively planar.
3.3. Two amino acid residues in a peptide tend to assume a trans-relationship of the side chains. Offer a brief explanation of this observation.
3.4. What is the pitch of the α-helix in a peptide?

[11] Edman, P. *Acta Chem. Scand.* **1950**, *4*, 283.

3.5. If the statement were made that poly(isoleucine) has more difficulty forming an α-helix structure than poly(alanine), offer a brief explanation for this statement.
3.6. In a Richardson drawing of an enzyme, what do the flattened arrow structures represent?
3.7. What is the quaternary structure of a protein? What is the tertiary structure?
3.8. Using Figure 3.11, what amino acid is coded by each of the following: UUC, AUA, CAC, CCA, AGA?
3.9. What product is formed when ninhydrin reacts with tryptophan? With cysteine?
3.10. Draw the structure of the product formed when a generic peptide with an *N*-terminal serine reacts with Sanger's reagent.
3.11. Draw the structure of the product formed when a generic peptide with an *N*-terminal phenylalanine reacts with dansyl chloride.
3.12. Draw the structure of the product formed when a generic peptide with an *N*-terminal glutamic acid reacts with phenyl isothiocyanate, but before the reaction of this product with trifluoroacetic acid.

4 Enzymes

As applied to organic chemical reactions, a *catalyst* is defined as a substance added to a reaction in sub-stoichiometric amounts that increases the rate of a chemical reaction but is not consumed by the reaction and is present at the end of the reaction. Acid catalysts used in organic chemistry include sulfuric acid or *p*-toluenesulfonic acid, which are commonly added to reactions to serve as sources of H^+. An example is the acid (H^+)-catalyzed reaction of an alkene with water, as shown in Figure 4.1. An alkene such as 1-methylcyclopent-1-ene is too weak of a base to react directly with the weak acid water, so there is no reaction. However, adding a small amount of a strong mineral acid such as sulfuric acid allows a facile acid-base reaction with the alkene to generate a tertiary carbocation intermediate, which reacts quickly with water to give an oxonium ion. Loss of a proton in an acid-base reaction gives the alcohol product, 1-methylcyclopentan-1-ol, by loss of the H^+ catalyst.

In biological systems, such catalytic reactions are mediated by enzymes. This chapter describes the function and classification of enzymes in different reactions. The discussion begins with an explanation of reaction kinetics.

4.1 KINETICS

4.1.1 Kinetics in Organic Chemistry

Before discussing catalysts in biological systems, a discussion of catalysts in organic chemistry is useful. Understanding the key characteristics of chemical reactions provides insight into how reactions work, why they work, what energy characteristics drive a reaction, whether there is an intermediate, whether the reaction proceeds in one step or several steps, and what the final isolated product is. Is there more than one isolated product? If so, which is major and which is minor?

For a given chemical reaction, it is reasonable to question how fast it will proceed. In other words, how long does it take for the reaction to be completed as measured by the consumption of all the starting material? A minute? An hour? A day? A week? Apart from the obvious information about how long one will have to monitor the reaction, additional insights can be obtained from knowledge of how fast a reaction proceeds and how long it takes. In many cases, such knowledge can give insight into the mechanism of a reaction and help identify intermediates in the reaction. The number of intermediates, which defines the number of chemical steps in the transformation, is clearly important since it is part of the mechanism of a reaction. Indeed, a working knowledge of the mechanism is essential to understanding the process. In biological systems, enzymes act as catalysts, and while the rate of reaction is greatly increased, the specificity is usually quite high.

The *rate of reaction* is a parameter that measures how quickly a reaction proceeds. In other words, it indicates how long it takes to consume a starting material and convert it into a product. For a chemical reaction, the change in molar concentration of either the starting materials (reactants) or the products can be monitored. A simple reaction in which A is transformed into B ($A \rightarrow B$), with no

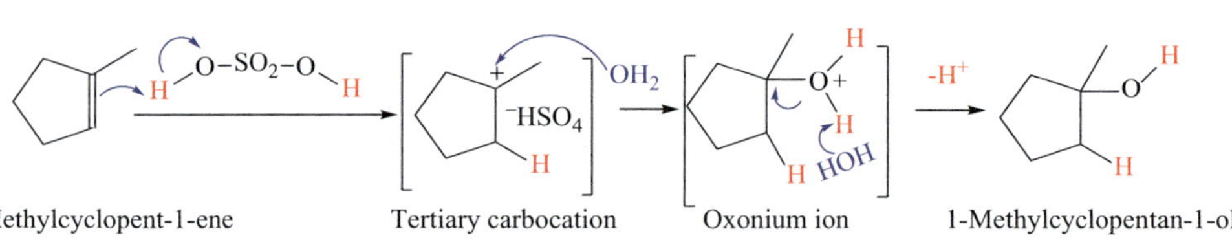

FIGURE 4.1 Acid-catalyzed hydration reaction of 2-methylbut-2-ene.

DOI: 10.1201/9781003605157-4

intermediate, will be used to illustrate the concept. The initial concentration of the starting material (how many molar equivalents) is an experimentally derived parameter, and as the starting material is consumed to form product *B*, the number of molar equivalents *decreases*. In other words, *A* is consumed to form the product *B*. Experimentally, the change in molar equivalents of *A* is measured as a function of time. In a simple case, the rate is determined by measuring the initial concentration (in mol/L) of a reactant (*A*) and then measuring the concentration of *A* at certain time intervals as it is consumed (see eq. 4.1). The rate of a reaction is a dynamic property since the concentration of *A* changes during the reaction as time progresses. In other words, the change in concentration is a function of time. Solving this problem requires a differential equation, where the rate is proportional to the concentration of *A* and to time. In eq. 4.1, $[A]_o$ is the concentration of *A* at time = 0, and $[A]_t$ is the concentration of *A* at any specified time.

(4.1)
$$Rate = -\int_{[A]_0}^{[A]_t} \frac{d[A]}{A} = k\int_0^t dt$$

When this differential equation is integrated, with $[A]$ is $[A]_o$ at $t = 0$ and it is $[A]_t$ at $t =$ "end time," the resulting expression is shown in eq. 4.2.

(4.2)
$$\ln\frac{[A]_o}{[A]_t} = k(t_{\text{end time}} - t_o) \text{ and if } t_o = 0 \text{ (as defined), then } \ln\frac{[A]_o}{[A]_t} = k\,t$$

Since concentration is proportional to time (conc $\propto t$), setting the rate of change in concentration equal to the rate of change in time requires a proportionality constant, *k* (for time changing from $t = 0$ to $t =$ some measured time). In this case, conc $= k\,t$ in eq. 4.2. *This proportionality constant (k) is defined as the rate constant. Note that the lowercase k is used to represent the rate constant, whereas a capital K is used to indicate an equilibrium constant.* The value of *k* is determined by plotting some function of *A* versus time, and a plot of $\ln\frac{[A]_o}{[A]_t}$ vs time gives a straight line, where the slope of this line gives *k*, the rate constant. Such plots are characteristic of what is called a *first-order reaction*.

Another type of reaction is illustrated by the reaction where reactant *A* must react with *B* to form a new bond and a new product, *A–B*.[1] When [*A*] is plotted against time, the curve shows that the rate of the reaction depends on the concentrations of *both A* and *B*. In other words, simply plotting the concentration of *A* vs. time or the concentration of *B* vs. time does *not* give a curve that fully describes the reaction. The required expression must include *both* concentrations, *A* and *B*, which are plotted against time to obtain the correct result for the formation of *A–B*. In other words, to calculate the rate constant, *both* reactants are necessary for the reaction to occur, and the rate expression is shown in eq. 4.3.

$$\mathrm{A} + \mathrm{B} \longrightarrow \mathrm{A{-}B}$$

(4.3)
$$-\frac{d[A]}{dt} = k[A][B] \text{ or rate} = k[A][B]$$

The differential equation for this reaction is solved to yield the complex expression shown in eq. 4.4.

(4.4)
$$\frac{1}{[A]_0 - [B]_0}\,\mathrm{In}\,\frac{[B]_0[A]_t}{[A]_o[B]_t} = kt$$

[1] For a discussion of second-order kinetics, see Daniels, F.; Alberty, R.A. *Physical Chemistry*, 3rd ed., Wiley, **1967**, p. 331.

where the $[]_0$ terms denote the initial concentrations of A or B (at time=0), and the $[\ ]_t$ terms indicate the concentrations of A and B at a specified time. To obtain a straight-line plot, time is plotted against the integrated rate expression $\ln \frac{[A][B]_0}{[A]_0[B]}$. The slope of this line gives the rate constant, which is characteristic of a *second-order reaction*.

The *half-life* of a reaction is the amount of time required for one-half of the starting material to react. This parameter is useful for determining how long it takes for a reaction to proceed to completion. Assume that it takes 100 min for 50% of the starting material to be converted into product. If the initial concentration of a reactant X is 1.0 M, then the concentration of X after 100 min will be 0.5 M. This starting material concentration remains after one half-life. After another 100 min, 50% of the remaining X will react, and the concentration of X will be 0.25 M (after the second half-life). Another 100 min is required to bring the concentration to 0.125 M, and after a total reaction time of 500 min (five half-lives), the concentration of X will be 0.0313 M. In other words, after five half-lives, 0.9787 mol of X will have reacted (97.87%). *For a reaction to be* considered *"complete," it must be allowed to proceed for more than five half-lives.* The half-life for a first-order reaction is given the symbol $t_{1/2}$, and is calculated from the simple formula: half-life (first-order) = $t_{1/2} = \frac{\ln 2}{k} = \frac{0.693}{k}$. For a second-order reaction, the half-life formula is different because the rate equation is different. The half-life for a second-order reaction is: $t_{1/2} = \frac{1}{k[A]_o}$.

4.1.2 Catalysts and Catalytic Reactions

In chemical reactions, a catalyst can be added to a reaction to either initiate the reaction or influence the rate of the reaction, and it is regenerated during the reaction. The catalyst is usually regenerated during the reaction and is present at the end of the reaction. Often, but not always, as little as < 5% and occasionally < 1% of the substrate is required. Only a small amount of the catalyst is typically used since it is not consumed by the reaction. The structure of the catalyst often changes during the reaction, although it is returned to its original form by the end of the reaction.

A catalyst usually reacts with a substrate to form a transient intermediate that facilitates the reaction of interest. Such transient intermediates rely on chemical bonds that are strong enough to form and participate in subsequent reactions, but not so strong that they cannot be broken, allowing the catalyst to return to its original state. The ability to form a transient intermediate usually changes the mechanism of the reaction relative to the uncatalyzed process. By changing the mechanism, different transition state energies are possible that are lower than the transition state energy for the uncatalyzed reaction. Such a change allows the overall reaction rate to be faster. In other words, the activation energy for the overall reaction is lowered, and the rate of reaction is thereby accelerated.

If the catalyst is soluble in the reaction medium, the medium is homogeneous, and the catalyst is called a homogeneous catalyst. Reactions that occur with a homogeneous catalyst will change the number of ligands around the catalyst, as in the transition metal salts used in homogeneous catalytic hydrogenation. Conversely, a catalyst can be insoluble in the reaction medium, in which case it is a heterogeneous catalyst. In both cases, the catalyst can be a Brønsted-Lowry acid such as sulfuric acid or an organic sulfonic acid. Any reaction with a heterogeneous catalyst occurs at the surface of the catalyst, as in the use of nickel or palladium metal in heterogeneous hydrogenation catalysis.

4.1.3 Enzyme Kinetics

Enzymes are essentially catalysts for biological systems, and they speed up biochemical reactions. Enzymes are typically proteins (Sections 3.1 and 3.4) that can exist in conformations (see Section 5.12) that allow them to form a reactive complex with a reactant, called the *substrate*, which is converted to the *product*. The site on the enzyme where substrate binding and subsequent reaction occur is called the "*active site*" or the "*binding site*." The purpose of binding into a reactive complex is to weaken key substrate bonds to facilitate the desired chemical reactions. The active site of an enzyme contains the amino acid residues (Section 2.2) that participate in making and breaking chemical bonds. Each enzyme is optimized for a particular reaction.

Once the reaction occurs, the substrate binding is sufficiently weak that the product can dissociate from the enzyme, allowing another reaction to occur. To accomplish such specificity, there is usually an equilibrium between the free substrate and the enzyme complex. In other words, complex formation between the enzyme and the substrate is reversible. However, the product is not involved in this equilibrium, and once formed, the product is "released." To be specific for a reaction or a small subset of related reactions, the active site should react with only a specific substrate or a small subset of substrates. Such specific substrate binding and equilibrium are consistent with the observation that the mechanisms for many enzymes are similar.

A reversible chemical reaction in this context is between an enzyme and a substrate (a molecule of biological importance). It has been observed that at low substrate concentrations, the reaction velocity is proportional to the substrate concentration, and the reaction is first order with respect to the substrate.[2] Increasing the concentration of the substrate causes the reaction rate to diminish, and there is a point where it is no longer proportional to the substrate concentration. At this point, the rate of the enzyme-catalyzed reaction becomes constant and independent of substrate concentration. The reaction is then zero order with respect to the substrate, and the enzyme is said to be saturated with substrate (saturation). *A zero-order reaction is independent of the concentration of the reactants,* so a higher concentration of reactants will not increase the rate of reaction.

The fundamental rate expressions and equilibrium expressions can be applied to enzyme reactions. In effect, treating an enzyme reaction as an organic chemical reaction allows for the analysis of a given transformation, leading to a better understanding of the relevant biology.

Some enzyme reactions accelerate the biological process by forming a covalent bond between the enzyme and the substrate. A substrate (e.g., **1** in Figure 4.2) reacts with an enzyme (labeled **E**), for example, that has an attached reactive species capable of donating electrons (X). This electron-donating unit reacts with the carbonyl unit to give what is known as a tetrahedral intermediate, **2** (see Section 4.5.3). Both the C–X bond and the C–Y bond are polarized, so both X and Y may be *leaving groups*, but the reaction is usually designed so that Y is the better leaving group. To be a better leaving group, the C–Y bond should preferentially break in **2** because it is more polarized, and "Y" is better able to accommodate the electrons transferred from that bond. The "Y" group is therefore expelled as the corresponding anion, Y^-, via cleavage of the C–Y bond. In this reaction, the new product is the enzyme-bound acyl derivative **3**. Once formed, the covalently bound enzyme (C–X–**E**) is available for use in various enzymatic reactions.

The catalytic activity of many enzymes is blocked by molecules known as *inhibitors*, which mimic the substrate for binding to the active site. *Competitive inhibitors* are usually structural analogs of the substrate, and they compete for the same enzymatic binding sites. The inhibitor occupies the active site by forming a complex between the enzyme and the inhibitor. Some inhibitors react preferentially rather than competitively with the binding site. The enzyme cannot react until the inhibitor dissociates from the binding site. Both competitive binding and preferential binding effectively block the substrate from binding to the active site and inhibit the activity of the enzyme. Many pharmaceutical drugs are designed to act as inhibitors. The anticancer drug methotrexate is an example of a competitive inhibitor, and it has a similar structure to the vitamin folic acid. The enzyme *dihydrofolate reductase*

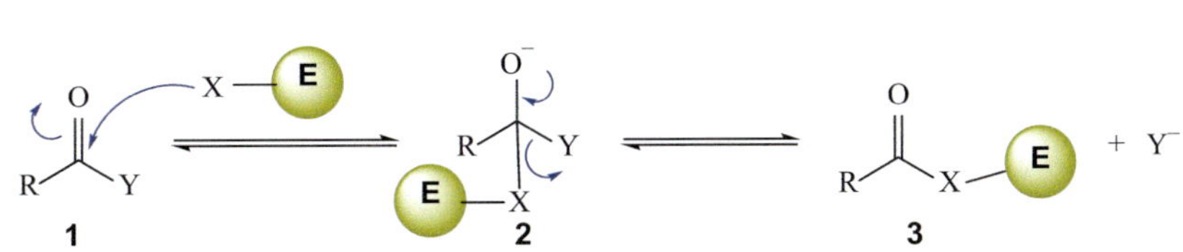

FIGURE 4.2 A reaction mediated by an enzyme.

[2] Lehninger, A.L. *Biochemistry*, Worth Publishing Inc., New York, **1970**, pp. 153–154.

(EC 1.5.1.3) cannot react, thus preventing the regeneration of dihydrofolate from tetrahydrofolate. This inhibition interferes with DNA synthesis, blocking cell division in cancer cells.[3]

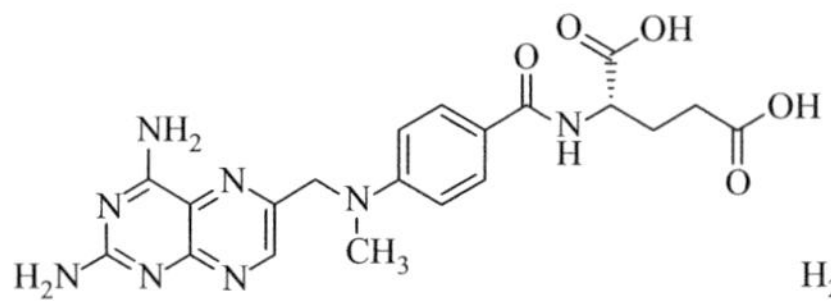

Methotrexate

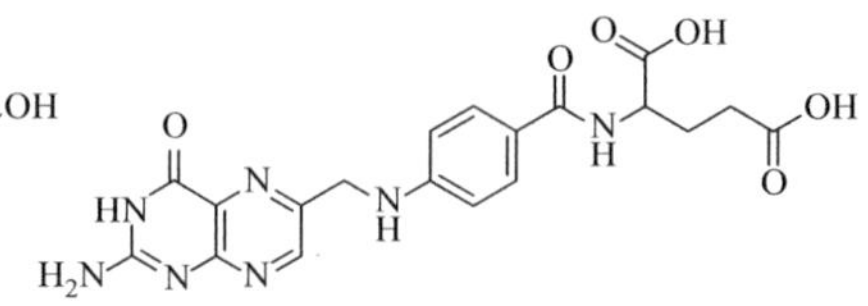

Folic acid

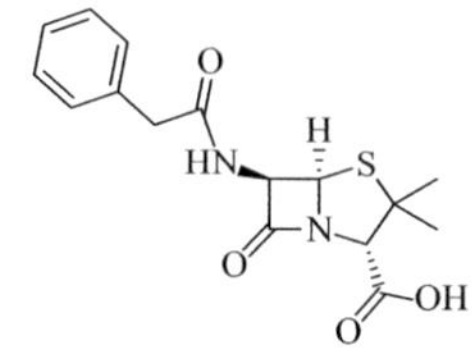

Penicillin G

The penicillins are antibacterial agents that inhibit bacterial growth by blocking cell-wall synthesis. Penicillin G (benzylpenicillin) is a competitive inhibitor of the final step in cell-wall formation, in which *glycopeptide transpeptidase* (EC 2.4.1.41) catalyzes the cross-linking reaction between peptidoglycan molecules, the main constituents of the cell wall.[4]

$$\mathrm{E} + \mathrm{S} \underset{k_2}{\overset{k_1}{\rightleftharpoons}} \mathrm{ES} \underset{k_4}{\overset{k_3}{\rightleftharpoons}} \mathrm{E} + \mathrm{P}$$

A *noncompetitive inhibitor* is a molecule that binds to the enzyme, but *away from the active site.* Typically, both the inhibitor and the substrate can bind to the enzyme. Such binding alters the shape of the enzyme so that substrate binding is less efficient and enzyme activity is reduced. An example is heavy metal poisoning, such as lead poisoning. Lead binds to enzymes such as *human porphobilinogen synthase*, causing irreversible inhibition by binding strongly to the amino acid backbone. *Porphobilinogen synthase* (PBGS; EC 4.2.1.24), also known as *5-aminolevulinic acid dehydratase* or *delta-aminolevulinic acid dehydratase*, functions in the first common step in tetrapyrrole biosynthesis (*e.g.*, heme and chlorophyll).[5]

Enzyme-catalyzed reactions are ubiquitous in biological systems. Figure 4.3 shows an energy profile of a catalyzed and an uncatalyzed reaction.[6] The catalyzed reaction appears to have more steps, but complexation with the enzyme catalyst leads to a lower activation energy for the key reaction. Indeed, catalysis by an enzyme can be correlated with the principles that govern other chemical reactions. For enzymatic reactions, the rate of a reaction is often referred to as the *velocity of the reaction*, but rate constants can be determined for each process.

There is a relationship between an enzyme-catalyzed reaction and the concentration of the substrate undergoing the reaction. Second-order and higher-order reactions are known, where the rate is proportional to the substrate. Enzyme kinetics for a single substrate with a hyperbolic rate-substrate concentration[7] are described by the *Michaelis-Menten equation*, where K_m is the *Michaelis constant*, $[S]$ is the substrate concentration, V is the rate of the reaction, and $V_{\max}$ is the reaction rate when the substrate concentration approaches infinity. Note that kinetic curves are hyperbolic; an open curve with two branches, the intersection of a plane with both halves of a double cone. Rearrangement of the *Michaelis-Menten equation* leads to *the Lineweaver-Burk equation.* The term $K_m/V_{\max}$ is obtained from the slope of a plot of $1/V$ versus $1/[S]$ in what is known as the *Lineweaver-Burk double reciprocal plot* of enzyme processes. When $V=0$, the intercept on the x-axis is $-1/K_m$. It is noted that K_m and $V_{\max}$ are characteristic of enzyme properties and the substrate system, and K measures the affinity of a substrate for that enzyme.

[3] (a) Pelley, J.W. Enzymes and Energetics, in *Elsevier's Integrated Review Biochemistry*, 2nd ed., **2012**; (b) Engelking, L.R. Enzyme Kinetics, in *Textbook of Veterinary Physiological Chemistry*, 3rd ed., **2015**.

[4] Also see Suavage, E.; Kerff, F.; Terrak, M.; Ayala, J.A.; Charlier, P. *FEMS Microbiol. Rev.* **2008,** *32*, 234.

[5] Jaffe, E.K.; Martins, J.; Li, J.; Kervinen, J.; Dunbrack Jr., R.L. *J. Biol. Chem.* **2001**, *276,* 1531.

[6] Warshel, A.; Sharma, P.K.; Kato, M.; Xiang, Y.; Liu, H.; Olsson, M.H.M. *Chem. Rev.* **2006**, *106*, 3210.

[7] Houston, J.B.; Kenworthy, K.E. *Drug Metab. Dispos.* **2000**, *28*, 246.

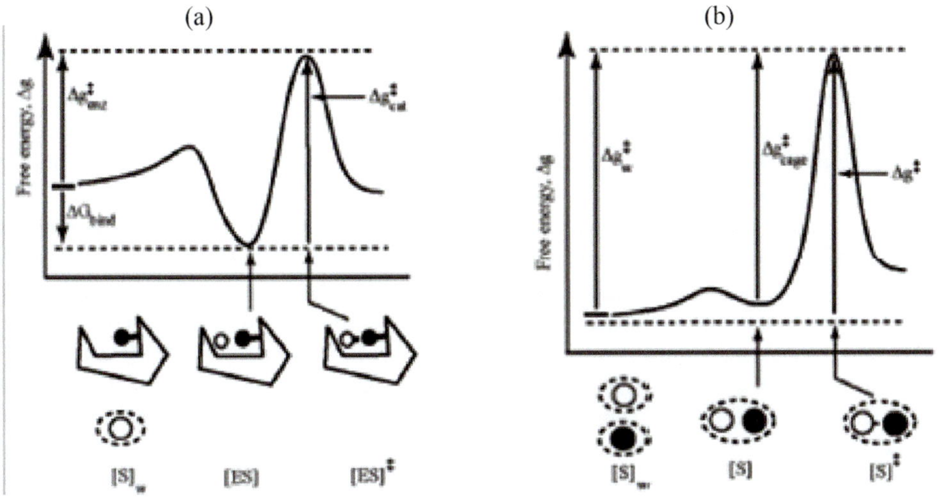

FIGURE 4.3 (a) Free energy profile for an enzymatic reaction and that for the corresponding solution reaction. The figure describes the free energies associated with k_{cat}/K_M and k_{cat}. (b) describes the energetics of a reference solution reaction that is not catalyzed. (Reprinted with permission from Warshel, A.; Sharma, P.K.; Kato, M.; Xiang, Y.; Liu, H.; Olsson, M.H.M. *Chem. Rev.* **2006**, *106*, 3210–3235. Copyright 2006 American Chemical Society.)

This effect is described by a process in which the enzyme (E) reacts with the substrate (S) to form a complex (ES), which then breaks down to regenerate the enzyme and produce the products (P). Both reactions are reversible, with the rate constants indicated as k_1_k_4. This reaction has been analyzed to give:

$$\frac{[S]([E]-[ES])}{[ES]} = \frac{k_2+k_3}{k_1} = K_M \text{ and, } \quad [ES] = \frac{[E][S]}{K_M+[S]}$$

where the second equation has a constant K_M, which replaces the rate constant term as shown, and is called the *Michaelis-Menten constant*. Further manipulation of the equation leads to the *Michaelis-Menten equation*, which defines the quantitative relationship between the enzyme reaction rate and the substrate concentration [S], if both V_{max} and K_M are known, where $V_{max}=k\,[E]$ and V_{max} is the maximum velocity for formation of the complex ES.

$$\text{Velocity} = V = \frac{V_{max}[S]}{K_m+[S]}$$

$$\textit{Michaelis} - \textit{Menton}$$

$$\frac{1}{V} = \frac{K_m}{V_{max}[S]} + \frac{1}{V_{max}}$$

$$\textit{Lineweaver} - \textit{Burk}$$

A Michaelis-Menten analysis was used in a study of the *diheme cytochrome C peroxidase* (CcP; EC 1.11.1.5; see Section 4.3 for oxidase reactions) from *Shewanella oneidensis (So)*, a cofactor in bacterial peroxidase reactions, suggesting that *So* CcP can accommodate electron donor binding in several possible orientations and that the presence of maltose-binding protein affects the availability of certain binding sites.[8] A peroxidase is a heme-containing enzyme that uses hydrogen peroxide as the electron acceptor to catalyze a number of oxidative reactions. This study is rather specific, but it illustrates how the equations discussed are used in modern-day research.

[8] Pulcu, G.S.; Frato, K.E.; Gupta, R.; Hsu, H.-R.; Levine, G.A.; Hendrich, M.P.; Elliott, S. *Biochemistry* **2012**, *51*, 974.

4.2 ENZYMES AND ENZYME CLASSES

Biochemical catalysts are known as enzymes, and they facilitate and accelerate chemical reactions in living systems. Enzymes act as catalysts for most processes in living organisms. Most enzymes are polypeptides [proteins; (poly)amino acids], as illustrated by the polypeptide structure of human putative Ga-regulatory protein acyl *thioesterase* (hAPT1), shown in Figure 4.4.[9] The ribbon structure shown is explained in Section 3.1. Note that ribozymes are RNA molecules.

The molecules that react directly with enzymes are called substrates, and the substrate is converted into a different molecule known as a product. As noted in Section 4.1, the active site of an enzyme consists of amino acid residues that form temporary bonds with the substrate and those that catalyze a reaction of the substrate. An example of an active site pocket is that of *AlbC cyclodipeptide synthase*, shown in Figure 4.5.[10] The *cyclodipeptide synthases* (CDPSs) are a family of enzymes that use aminoacyl-tRNAs (aa-tRNAs) to synthesize cyclodipeptides. This cyclization reaction is catalyzed by the *AlbC synthase* of the CDPS family. Figure 4.5[10] illustrates the interactions of a dipeptide substrate with the amino acids tyrosine, phenylalanine, histidine, glutamine, serine, and asparagine in the active site of the enzyme. The coordination of the dipeptide substrate with Glu^{182}, Tyr^{178}, and Tyr^{202} is especially important. The three-letter codes for the pertinent amino acids are defined in Section 2.2.

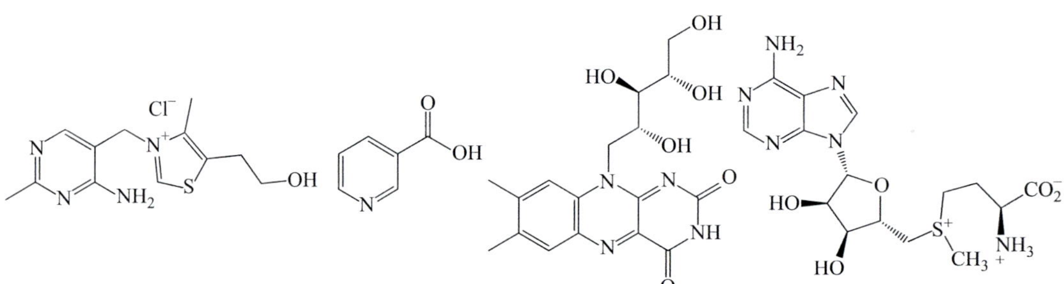

Thiamine Niacin Riboflavin S-Adenosyl thionine

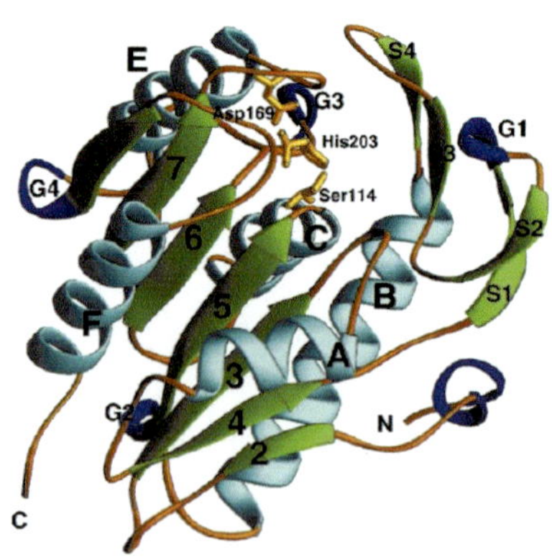

FIGURE 4.4 Structure of human acyl protein thioesterase I. (Reprinted from Devedjiev, Y.; Dauter, Z.; Kuznetsov, S.R.; Jones, T.L.Z.; Derewenda, Z.S. *Structure* **2000**, *8*, 1137–1146. Crystal Structure of the Human Acyl Protein Thioesterase I from a Single X-Ray Data Set to 1.5 Å. Copyright (2000), with permission from Elsevier.)

---

[9] Devedjiev, Y.; Dauter, Z.; Kuznetsov, S.R.; Jones, T.L.Z.; Derewenda, Z.S. *Structure* **2000**, *8*, 1137.

[10] Schmitt, E.; Bourgeois, G.; Gondry, M.; Aleksandrov, A. *Nat.: Sci. Rep.* **2018**, *8*, 7031. https://doi.org/10.1038/s41598-018-25479-5

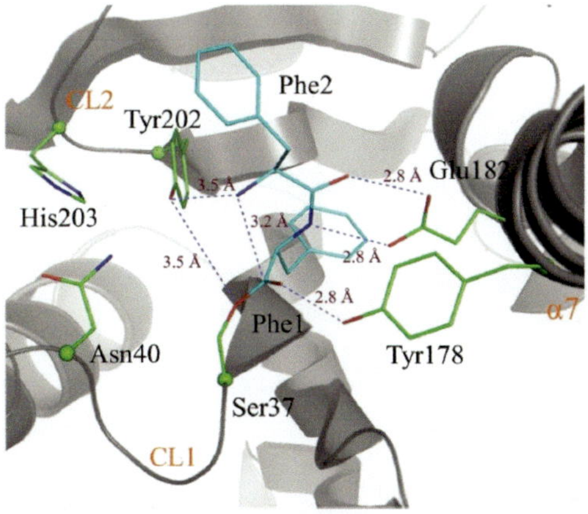

FIGURE 4.5 The active site pocket of the AlbC cyclodipeptide synthase. (Schmitt, E.; Bourgeois, G.; Gondry, M.; Aleksandrov, A. *Nat. Sci. Rep.* **2018**, *8*, 7031. https://doi.org/10.1038/s41598-018-25479-5. Open Access.)

Enzymes are generally *globular proteins*, which refers to their spherical shape, and they are often part of larger complexes. Note that the spherical shape of the protein represents the tertiary structure of that protein (Section 3.1), resulting from the aggregation of the hydrophobic portions of the amino acids toward the center of the sphere, with the hydrophilic moieties of the amino acid residues on the surface. These enzymes have some water solubility due to the polar nature of the globular protein surface.

Enzymes function as catalysts for specific transformations by lowering the activation energy of a reaction. However, enzyme catalysis can proceed by various pathways. Enzymes can stabilize the transition state of a reaction, provide an alternative pathway for the reaction, or bind to the substrate and change its ground state, as well as help transport chemical groups from one enzyme to another.

A *cofactor* is a non-protein chemical compound or metallic ion that is required for an enzyme's role as a catalyst. In effect, cofactors are "helper molecules" that assist in biochemical transformations. The cofactor may also be bound to the enzyme, in which case it is known as a *prosthetic group*. The B vitamins, such as B1 (thiamine), B2 (riboflavin), or B3 (niacin), are well-known coenzymes involved in the formation of fats, carbohydrates, or proteins. There are also non-vitamin coenzymes, such as *S*-adenosyl methionine. An organic prosthetic group can be a vitamin, a sugar, or a lipid; an example is biotin in *pyruvate carboxylase*. An example of a cofactor-independent enzyme is *orotidine 5′-monophosphate decarboxylase* (EC 4.1.1.23), which catalyzes the decarboxylation of orotidine 5′-monophosphate to uridine 5′-monophosphate during the biosynthesis of pyrimidine nucleotides.[11] The OMP (orotidine 5′-monophosphate) decarboxylase active site is shown in Figure 4.6.[11] Inorganic compounds and metal ions are often important cofactors. Examples include Zn^{+2}, Fe^{+2} or Fe^{+3}, Cu^{+} or Cu^{+2}, K^{+} or Mg^{+2} (see Section 12.2).

Ribonucleotide reductase provides deoxyribonucleotide monomers (see Section 15.1) by a radical mechanism that activates a cysteine or a glycine residue, leading to the deoxygenation of a ribonucleotide.[12] This enzyme is composed of two homodimeric proteins, subunit R1 and subunit R2. The R2 subunit has a ferric diiron cluster cofactor that stabilizes a free tyrosyl radical, which initiates cysteine activation for the enzyme to function.[13]

[11] (a) Lieberman, I.; Kornberg, A.; Simms, E.S. *J. Biol. Chem.* **1955**, *215*, 403; (b) Appleby, T.C.; Kinsland, C.; Begley, T.P.; Ealick, S.E. *Proc. Nat. Acad. Sci. U.S.A.* **2000**, *97*, 2005.

[12] (a) Follmann, H. *Chem. Soc. Rev.* **2004**, *33*, 225; (b) Stubbe, J.; Nocera, D.G.; Yee, C.S.; Chang, M.C. *Chem. Rev.* **2003**, 103, 2167.

[13] Cox, N.; Ogata, H.; Stolle, P.; Reijerse, E.; Auling, G.; Lubitz, W. *J. Amer. Chem. Soc.* **2010**, *132*, 11197.

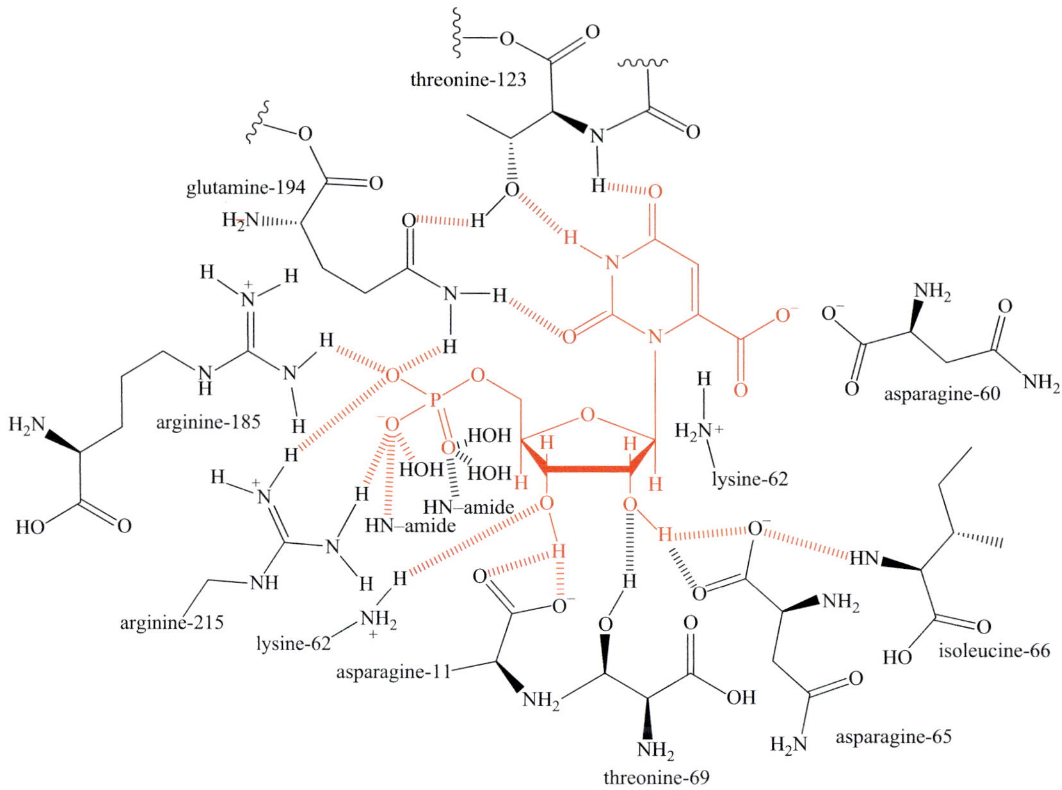

FIGURE 4.6 The orotidine 5′-monophosphate decarboxylase active site. (Appleby, T.C.; Kinsland, C.; Begley, T.P.; Ealick, S.E. *Proc. Natl. Acad. Sci. U.S.A.* **2000**, *97,* 2005, see Figure 6, p 2009. Copyright (2009) National Academy of Sciences, U.S.A.)

The flavin cofactors FMN (flavin mononucleotide) and FAD (flavin adenine dinucleotide) are examples of prosthetic groups, and they are required for the function of many proteins. These cofactors are tightly bound to proteins by cofactor-binding domains.[14] The enzyme *putrescine oxidase* (EC 1.4.3.10) from *Rhodococcus erythropolis* is an amine oxidase that contains a non-covalently bound FAD molecule.[15]

Human protein HC (also known as alpha 1-microglobulin) is a glycoprotein found in body fluids as a free monomer and as a complex with immunoglobulin A.[16] The free monomer is a polypeptide chain of 183 amino acids and carries carbohydrates and retinol.[17] The oligosaccharides of protein HC can play an active role in its physicochemical properties, turnover regulation, and interaction with macromolecules and cellular surfaces. The carbohydrates can be carried as prosthetic groups: one *O*-glycosidic linkage and two *N*-glycosidic linkages.[18]

[14] Kopacz, M.M.; Rovida, S.; van Duijn, E.; Fraaije, M.W.; Mattevi, A. *Biochemistry* **2011**, *50*, 4209.
[15] van Hellemond, E.W.; van Dijk, M.; Heuts, D.P.; Janssen, D.B., Fraaije, M.W. *Appl. Microbiol. Biotechnol.* **2008**, *78,* 455.
[16] (a) Tejler, L.; Grubb, A. *Biochem. Biophys. Acta* **1976**, *439*, 82; (b) Fernandez-Luna, J.L.; Leyva-Cobian, F.; Mendez, E. *J. Clin. Path.* **1988**, *41,* 1176; (c) Grubb, A.; Mendez, E.; Fernandez-Luna, J.L.; Lopez, C.; Mihaesco, E.; Vaerman, J.P. *J. Biol. Chem.* **1986**, *261,* 14313.
[17] Escribano, J.; Grubb, A.; Mendez, E. *Biochem. Biophys. Res. Commun.* **1988**, *155*, 1424.
[18] Escriano, J.; Lopex-Otin, C.; Hierpe, A.; Grubb, A.; Mendez, E. *FEBS Lett.* **1990**, *266*, 167.

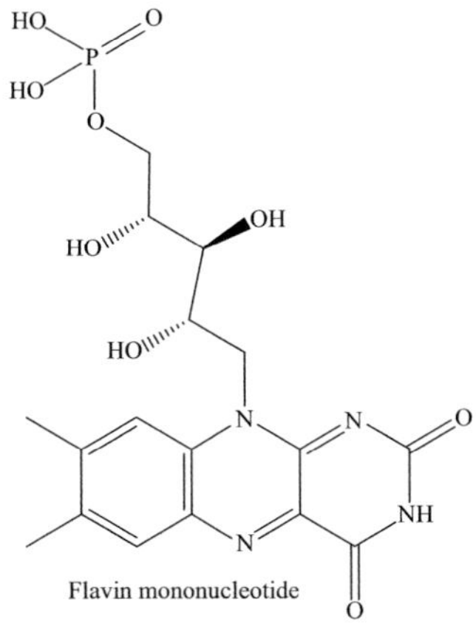

Flavin mononucleotide

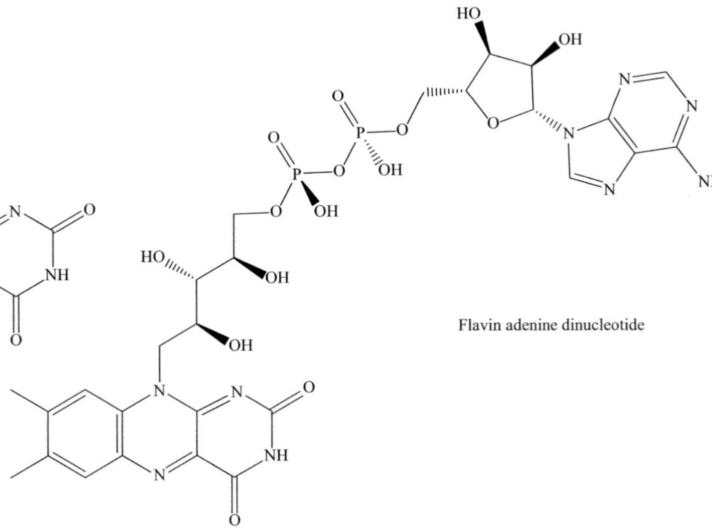

Flavin adenine dinucleotide

Some enzymes facilitate myriad chemical transformations. Two or more different enzymes that catalyze the same chemical reaction are called *isozymes*. Once the activity for a specific transformation has been confirmed, the enzyme is assigned an *EC number*: *Enzyme Commission Number*.[19] The EC number is a numerical classification scheme based on the chemical reactions that are catalyzed. *The name of an enzyme often refers to the chemical reaction it catalyzes, including oxidases, reductases, aldolases, decarboxylases, etc.* There are six 'top-level' (main) categories of enzymes, EC 1–EC 6. Oxidoreductases catalyze oxidation/reduction reactions (*EC 1*); transferases catalyze the transfer of a specific group from one molecule to another [donor to acceptor] (*EC 2*); hydrolases catalyze the hydrolysis of a chemical bond (*EC 3*); lyases cleave chemical bonds by mechanisms other than hydrolysis or oxidation (*EC 4*); isomerases catalyze the isomerization of one molecule to an isomer of that molecule (*EC 5*); and ligases catalyze the formation of a covalent bond between two molecules (*EC* 6). The actual EC number for a specific enzyme has four numbers that take the form 1.X.X.X, all beginning with the "top-level" categories 1–6. The second-level subclass and third-level sub-subclass indicate the specific bonds or functional groups involved in the reaction, and the fourth-level serial number defines the specific chemical reaction.

For the six categories of enzymes,[20] there are generalized rules for the subclasses of each main classification.[19]

- **EC 1. Oxidoreductases**. The second figure in the code number indicates the group in the hydrogen (or electron) donor that undergoes oxidation, unless it is 11, 13, 14, or 15. If the second number is 1, it denotes a –CHOH– group; 2, a –CHO or –CO–COOH group or carbon monoxide, and so on. The third number indicates the type of acceptor involved, except in subclasses EC 1.11, EC 1.13, EC 1.14, and EC 1.15. The number 1 denotes $NAD(P)^+$, 2 is a cytochrome, 3 is molecular oxygen, 4 is a disulfide, 5 is a quinone or similar compound, 6 is a nitrogenous group, 7 is an iron-sulfur protein, and 8 is a flavin. In subclasses EC 1.13 and EC 1.14, a different classification scheme is used, and sub-subclasses are numbered from 11 onwards.

[19] (a) Webb, E.C. *Enzyme Nomenclature 1992: Recommendation of the Nomenclature Committee of the International Union of Biochemistry and Molecular Biology on the Nomenclature and Classification of Enzymes*, San Diego, **1992**; (b) Also see Nath, N.; Mitchel, J.B.O. *BMC Bioinform.* **2012**, *13*, 60.

[20] Table 6.2 in Thomas, G. *Medicinal Chemistry, An Introduction*, John Wiley, Chichester, **2000**, p. 218.

- **EC 2. Transferases**. The second figure in the code number indicates the group transferred: a one-carbon group in EC 2.1, an aldehyde or ketone group in EC 2.2, an acyl group in EC 2.3, and so on. The third figure gives further information on the group transferred; *e.g.*, subclass EC 2.1 is subdivided into *methyltransferases* (EC 2.1.1), *hydroxymethyl-* and *formyltransferases* (EC 2.1.2), and so on. Note that in subclass EC 2.7, the third figure indicates the nature of the acceptor group.
- **EC 3. Hydrolases**. The second figure in the code number indicates the nature of the bond hydrolyzed. Using this strategy, EC 3.1 are the *esterases*, EC 3.2 the *glycosylases*, and so on. The third figure specifies the nature of the substrate; *e.g.*, in the esterases, there are *carboxylic ester hydrolases* (EC 3.1.1), *thiolester hydrolases* (EC 3.1.2), *phosphoric monoester hydrolases* (EC 3.1.3); in the glycosylases, the *O-glycosidases* (EC 3.2.1), *N-glycosylases* (EC 3.2.2), *etc*. The peptidyl-peptide hydrolases are an exception, and the third figure is based on the catalytic mechanism as shown by active center studies or the effect of pH.
- **EC 4. Lyases**. The second figure in the code number indicates the bond broken: EC 4.1 are carbon–carbon lyases, EC 4.2 carbon–oxygen lyases, and so on. The third figure gives further information on the group eliminated (*e.g.*, CO_2 in EC 4.1.1, H_2O in EC 4.2.1).
- **EC 5. Isomerases.** The subclasses are identified according to the type of isomerism, and the sub-subclasses according to the type of substrate.
- **EC 6. Ligases**. The second figure in the code number indicates the bond formed: EC 6.1 for C–O bonds (enzymes that acylate tRNA), EC 6.2 for C–S bonds (acyl-CoA derivatives), *etc*. Sub-subclasses are used only for the C–N ligases.

  These rules are illustrated with the *transferase* enzyme *1-glycerol-3-phosphate acyltransferase*. This enzyme has the number EC 2.3.1.51, which indicates that it is a transferase enzyme that transfers an acyl group, making it an acyltransferase specific for the transfer of unsaturated fatty acyl groups. In other words, it transfers groups other than amino-acyl groups.
- **EC7. Translocases**. A seventh category[21] comprises the *translocase* enzymes, which catalyze the movement of ions or molecules across membranes or their separation within membranes. The reactions catalyzed are designated as transfers from 'side 1' to 'side 2'. The subclasses designate the types of ion or molecule translocated. Enzymes EC 7.1 catalyze the translocation of hydrons (hydron is the general name for H^+); EC 7.2 enzymes catalyze the translocation of inorganic cations and chelates of those cations; EC 7.3 enzymes catalyze the translocation of inorganic anions; EC 7.4 enzymes catalyze the translocation of amino acids and peptides; EC 7.5 enzymes catalyze the translocation of carbohydrates and their derivatives; and EC 7.6 enzymes catalyze the translocation of other compounds. The next subclasses are: EC 7.x.1 translocations (oxidoreductase reactions); EC 7.x.2 translocations (hydrolysis of a nucleoside triphosphate); EC 7.x.3 translocations (hydrolysis of a diphosphate); and EC 7.x.4 translocations (decarboxylation reactions).

4.3 OXIDOREDUCTASES (EC 1)

As noted in the previous section, the top classes of enzymes are *oxidoreductases* (EC 1), *transferases* (EC 2), *hydrolases* (EC 3), *lyases* (EC 4), *isomerases* (EC 5), and *ligases* (EC 6). An *aconitase* is an enzyme that catalyzes the transfer of electrons from one molecule, to the reductant (the electron donor), to another, the oxidant (the electron acceptor). This group of enzymes usually utilizes NADP or NAD^+ as cofactors, although other cofactors, such as flavin adenine dinucleotide (FAD), are common. The common name is *dehydrogenase* wherever possible; alternatively, *reductase* can be used. *Oxidase* is used only in cases where O_2 is the acceptor.

[21] *Enzyme Nomenclature News*, August 2018. www.enzyme-database.org/news.php; ExplorEnz: McDonald, A.G.; Boyce, S.; Tipton, K.F. ExplorEnz: the primary source of the IUBMB enzyme list. *Nucleic Acids Res.* **2009**, *37*, D593–D597.

4.3.1 Oxidases

An *oxidase* is an enzyme that catalyzes an oxidation-reduction reaction with dioxygen (O_2) as the electron acceptor. In other words, an oxidase promotes the transfer of a hydrogen atom from a particular substrate to an oxygen molecule, forming water or hydrogen peroxide.

There are many examples. One is the enzyme *glucose oxidase*, also known as *notatin* (EC 1.1.3.4), which catalyzes the oxidation of glucose to hydrogen peroxide and D-glucono-δ-lactone. *Glucose oxidase* in *Aspergillus niger* is a β-D-glucose-specific flavoprotein *oxidase* that catalyzes the conversion to the lactone (see Figure 4.7).[22] The reaction is characterized by a concerted hydride transfer from the anomeric carbon of glucose to His^{516} and the N^5 atom of the flavin adenine dinucleotide (FAD) cofactor.[22] Both Glu^{412} and His^{559} probably act as buffers to maintain the proper acidity of the active site in order to control the reactivity, and as mentioned, His^{516} is crucial for activity.

Cholesterol oxidase is *3β-hydroxysteroid:oxygen oxidoreductase* (EC 1.1.3.6), a flavoenzyme that catalyzes the oxidation and isomerization of cholesterol to cholest-4-en-3-one. It is characterized by the flavin-binding domain and the substrate-binding domain that are fused. This enzyme is associated with the lipid bilayer, and the sterol must bind in the enzyme active site for catalysis to occur. The conversion to the corresponding 4-en-3-one product, shown in Figure 4.7, is the first and compulsory step in bacterial sterol catabolic pathways that degrade the steroid nucleus and the sterol side chain.[23] Studies with *Rhodococcus* C. *testosteroni* led to the elucidation of a biopathway for this degradation.

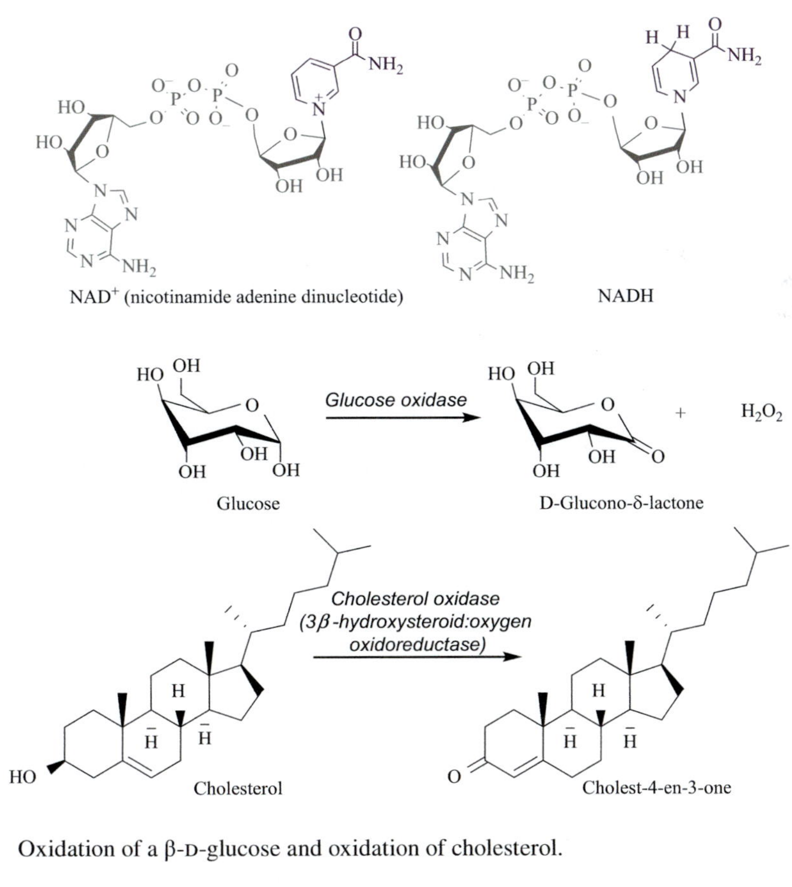

FIGURE 4.7 Oxidation of a β-D-glucose and oxidation of cholesterol.

---

[22] Petrović, D.; Frank, D.; Kamerlin, S.C.K.; Hoffmann, K.; Strodel, B. *ACS Catal.* **2017**, *7*, 6188.
[23] Kreit, J.; Sampson, N.S. *FEBS J.* **2009**, *276*, 6844.

A *dehydrogenase* is an enzyme that oxidizes a substrate by reducing an electron acceptor, usually facilitating the interconversion between alcohols and aldehydes or ketones, accompanied by the reduction of the cofactor, nicotinamide adenine dinucleotide (NAD^+ to NADH). Indeed, *alcohol dehydrogenases* (EC 1.1.1.1) are isozymes that catalyze the oxidation of primary and secondary alcohols to aldehydes or ketones, respectively, with NAD^+, and they can also catalyze the reverse reaction using NADH rather than NAD^+. In other words, NADH is the reduced form of NAD^+. Note that isozymes are enzymes that differ in amino acid sequence but catalyze the same chemical reaction. In humans, an *alcohol dehydrogenase* (EC 1.1.1.2) is found in the lining of the stomach and in the liver. The dehydrogenation reaction is an oxidation of the alcohol substrate. This enzyme catalyzes the oxidation of ethanol to acetaldehyde.[24] The conversion of ethanol to acetaldehyde with an *alcohol dehydrogenase* is shown in Figure 4.8, and further conversion to acetate uses *alcohol dehydrogenase-2*.[25] The oxidation of alcohol to acetaldehyde is reversible in vitro, but in vivo the overall reaction goes strongly toward the acetate anion due to the activity of aldehyde dehydrogenase-2, as shown in Figure 4.8. The *alcohol dehydrogenase* enzymes are responsible for most of the metabolism. The conversion of ethanol to acetaldehyde allows one to consume alcoholic beverages, but it probably exists to oxidize naturally occurring alcohols in foods or those produced by bacteria in the digestive tract. An important variant in *alcohol dehydrogenase 1B* protects people of European and Asian descent, and a different functional variant protects those of African descent.[25] A protective variant called aldehyde dehydrogenase-2 is essentially only found in Asians.[25]

Aldehyde acetals are also oxidized, as shown in Figure 4.9,[26] in which acetaldehyde hydrate is converted to acetic acid by the enzyme *horse liver alcohol dehydrogenase*. The hydrate is bound to zinc in the active site of the enzyme, and hydride transfer generates acetic acid bound to zinc. The

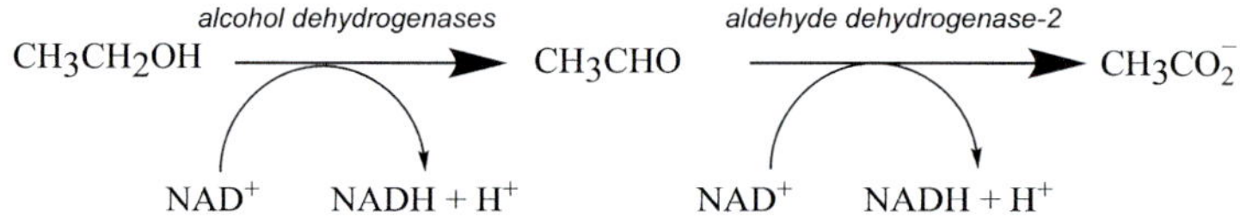

FIGURE 4.8 The oxidation of alcohol to acetaldehyde to acetate. (Edenberg, H.J.; McClintick, J.N. *Alcohol Clin. Exp. Res.* **2018**, *42*, 2281. Figure 1. Copyright © **2018**, reproduced with permission of John Wiley & Sons. Ltd.)

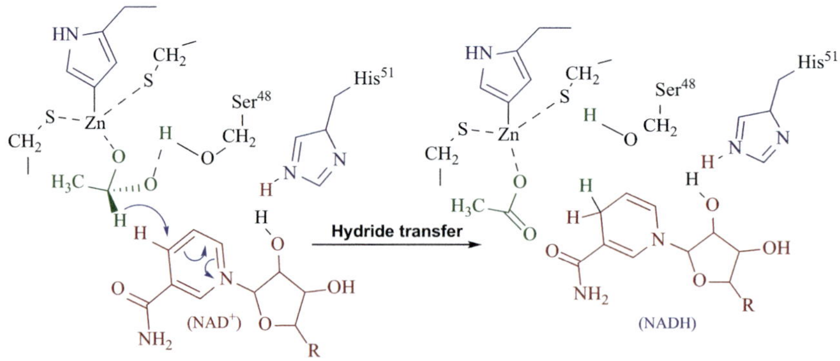

FIGURE 4.9 Mechanism of ethanol oxidation by the enzyme *liver alcohol dehydrogenase*. (Reprinted with permission from Olson, L.P.; Luo, J.; Almarssn, Ö.; Bruice, T.C. *Biochemistry* **1996**, *35*, 9782–9791. Copyright 1996 American Chemical Society.)

[24] Klund, H.; Plapp, B.V.; Samama, J.P.; Branden, C.I. *J. Biol. Chem.* **1982**, *257*, 14349.
[25] Edenberg, H.J.; McClintick, J.N. *Alcohol Clin. Exp. Res.* **2018**, *42*, 2281.
[26] Olson, L.P.; Luo, J.; Almarssn, Ö.; Bruice, T.C. *Biochemistry* **1996**, *35*, 9782.

NAD+ is also bound to the active site of the enzyme and induces a conformational change to close the active site. It is noted that the hydrate coordinates to zinc, which is stabilized by histidine and two sulfur moieties. Removal of the proton adjacent to the oxygen is accomplished by reaction with NAD+, which is reduced to NADH, and the product is the zinc-bound acetic acid, completing oxidation of the alcohol. To reiterate, the oxidation of acetaldehyde to acetic acid is accompanied by the reduction of NAD+ to NADH, as shown.

As mentioned, secondary alcohols are oxidized to ketones. In *E. coli*, glycerol enters the cell by facilitated diffusion, and enzyme-catalyzed phosphorylation in the presence of ATP (adenosine triphosphate) produces glycerol 3-phosphate.[27] This derivative is then oxidized by the membrane-associated, FAD-dependent *glycerol 3-phosphate dehydrogenase* (EC 1.1.1.8) to dihydroxyacetone 3-phosphate. Note that FAD refers to flavin adenine dinucleotide. This reaction is part of the pathway by which *E. coli* and other prokaryotes utilize glycerol as a source of carbon and energy.

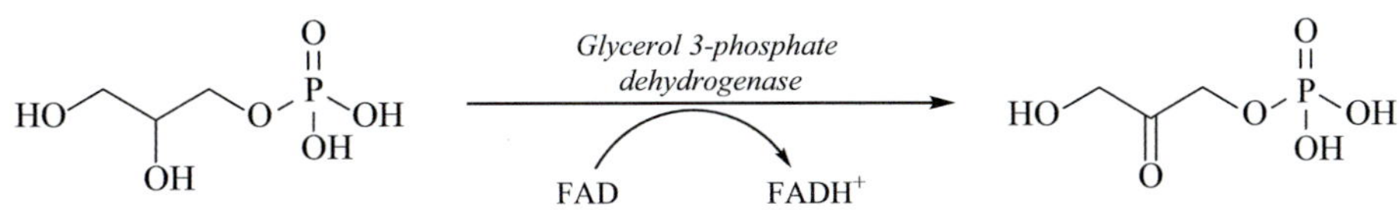

Glycerol 3-phosphate

Dihydroxyacetone 3-phosphate

4.3.2 Chemical Oxidation of Alcohols

The oxidation of alcohols in organic chemistry has traditionally relied on chromium(VI) reagents such as chromium trioxide (CrO_3) or sodium or potassium dichromate. For chromium(VI) reagents, chromic acid is prevalent in dilute aqueous media.

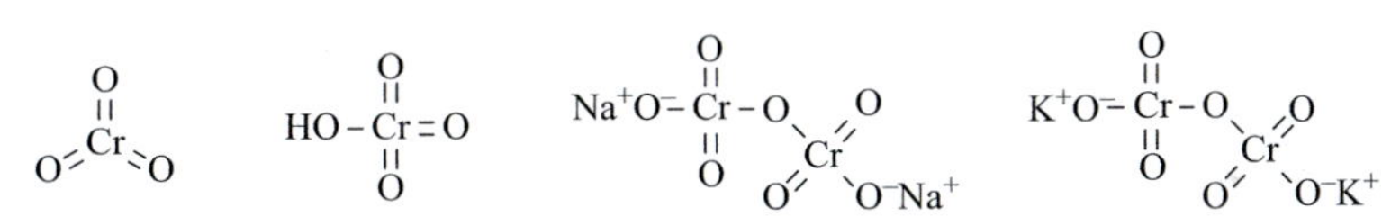

Chromium trioxide

Chromic acid

Sodium dichromate

Potassium dichromate

Most alcohols have a significant number of carbon atoms in their skeleton relative to the polar hydroxyl unit, and they are not very soluble in aqueous media. Typically, simple alcohols with fewer than five carbons are soluble in water, and with 5–8 carbon atoms they usually have partial solubility. Because of solubility issues, an organic but water-soluble or water-miscible cosolvent such as acetone is usually required. A solution of a Cr(VI) reagent in aqueous acetone (propan-2-one) in the presence of a Brønsted-Lowry acid (e.g., H_2SO_4) is called the *Jones reagent,* and the reaction of this mixture with an alcohol is called *Jones oxidation.* Cyclohexanol is oxidized to cyclohexanone, for example, by treatment with $Na_2Cr_2O_7$ and sulfuric acid in acetone.

In this example, chromium(VI) oxidation of cyclohexanol in the presence of sulfuric acid proceeds by protonation of the alcohol oxygen to give an oxonium ion. Transfer of the acidic proton of the oxonium ion to the negatively charged chromate oxygen (the base) leads to a so-called *chromate ester.* Removal of the acidic hydrogen by water is an acid-base reaction that leads to the loss of the Cr(III) leaving group. This is an elimination reaction that forms the new π-bond of a carbonyl (C=O) group in cyclohexanone. The mechanism shown in Figure 4.10 is used to explain the chromium(VI) oxidation. Note that removal of the hydrogen atom and loss of the chromium leaving group are analogous to the E2 mechanism (Section 6.7) that generates an alkene, as shown in Figure 4.10.

[27] Mallinder, P.R.; Pritchard, A.; Moir, A. *Gene* **1992**, *110*, 9.

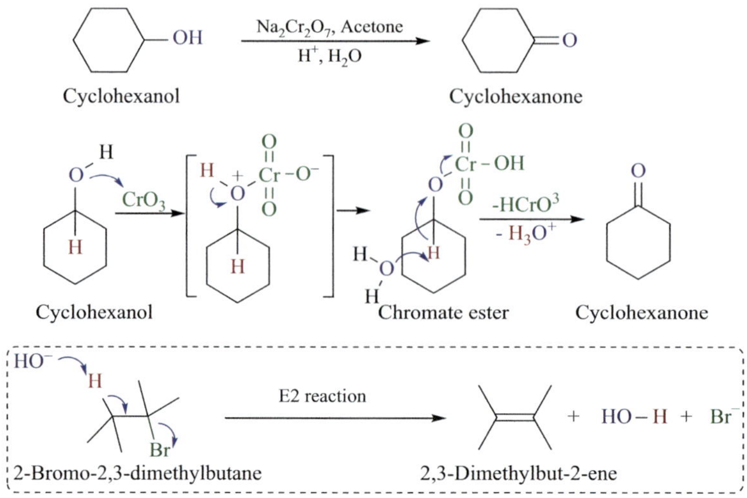

FIGURE 4.10 Mechanism of chromium trioxide oxidation.

Two modified Cr(VI) reagents that are based on the reaction of CrO_3 and pyridine are widely used for the oxidation of alcohols, especially primary and allylic alcohols. The first of these reagents is formed by the reaction of chromium trioxide with pyridine in aqueous HCl. This reaction generates a specific compound known as *pyridinium chlorochromate (PCC)*, which can be isolated and purified. In dilute aqueous acid, CrO_3 forms $HCrO_4$, which reacts with HCl to form $HCrClO_3$. Pyridine then reacts as a base with this acidic proton to form PCC. If the reaction conditions are modified to increase the amount of pyridine in the aqueous solution and the HCl is omitted, the reaction generates *pyridinium dichromate (PDC)***,** presumably by reaction of an excess of pyridine with $H_2Cr_2O_7$. In dilute solution, CrO_3 is in equilibrium with $H_2Cr_2O_7$, and pyridine reacts with both acidic hydrogen atoms to produce PDC, which can be isolated and purified.

Both PCC and PDC, in dichloromethane as a solvent, are less reactive than chromium trioxide in the Jones reagent, but they are very effective for converting primary alcohols to aldehydes and secondary alcohols to ketones, in good yield and under mild conditions. Secondary alcohols are readily converted to ketones by both reagents. An example is the reaction of 3-methylpentan-1-ol with PDC in dichloromethane, which gives 3-methylpentanal.

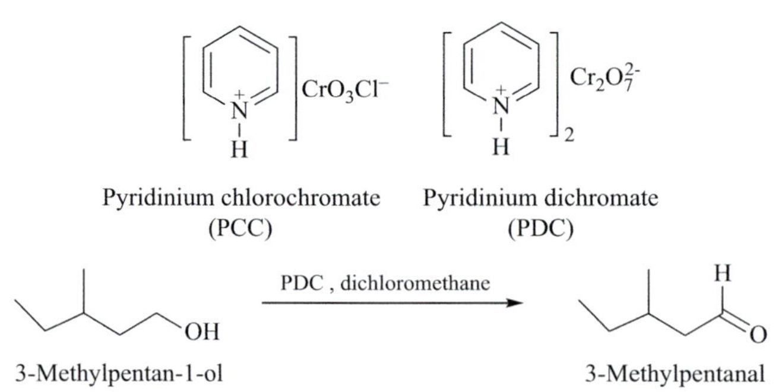

4.3.3 Reductases

A *reductase* is an enzyme that promotes the chemical reduction of various substrates, including carbinol compounds, imines, and nitro groups. Whereas *oxidase* and *dehydrogenase* enzymes catalyze an oxidation reaction in a reversible pathway, a *reductase* reduces a substrate. If a molecule can be oxidized via an oxidase, it can be reduced by a reductase.

The aldo-keto reductase family is a family of proteins that are subdivided into several categories. An example of a reductase enzyme is the NADPH-dependent *lactaldehyde reductase* (EC 1.1.1.55). It is a *dehydrogenase* enzyme that is known to be present in yeast and liver and is known to reduce L-lactaldehyde to propane-1,2-diol.[28] In this reaction, NADPH is converted to NAD^+, and this reductase is specific for L-lactaldehyde.

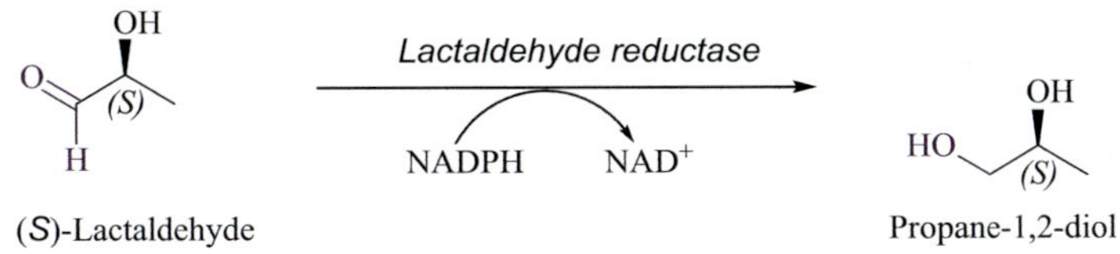

The enzymatic reduction of ester derivatives has been reported, but the reduction of the carbonyl unit of the ester gives the aldehyde, not the alcohol. The enzyme *γ-glutamyl phosphate reductase* (EC 1.2.1.41) catalyzes the second step in the biosynthesis of proline from glutamate, which is the NADP-dependent reduction of *N*-acetyl-L-glutamate 5-phosphate to *N*-acetyl-L-glutamate 5-semialdehyde and phosphate. Similarly, the enzyme *N-acetyl-γ-glutamyl-phosphate reductase* (EC 1.2.1.38; abbreviated APGR) is involved in the arginine biosynthetic pathway important for metabolism in some bacteria and plants.[29] In this example, the enzyme catalyzes the third step by reversibly converting *N*-acetyl-γ-glutamyl phosphate to *N*-acetylglutamate-γ-semialdehyde via NADPH-dependent reductive dephosphorylation (Figure 4.11).[29] Since this reaction is reversible, this enzyme is also classified as an *N-acetylglutamate-γ-semialdehyde dehydrogenase* (EC 1.2.1.88), which catalyzes the oxidative phosphorylation of the semialdehyde to produce glutamyl phosphate with $NADP^+$ as a cofactor. The crystal structure of *N*-acetyl-γ-glutamyl-phosphate reductase from *Mycobacterium tuberculosis* has been reported as a complex with NADP+.[29]

4.3.4 Chemical Reduction of Carbonyl Compounds

Typical reducing agents used in organic chemistry are discussed in this section. Lithium aluminum hydride ($LiAlH_4$) and sodium borohydride ($NaBH_4$) are versatile hydride reducing agents with wide utility in organic chemistry. Both react with ketones and aldehydes, but *$LiAlH_4$ is the more powerful reducing agent*. The reaction of 2,3,3-trimethylpentanal with $LiAlH_4$ followed by an aqueous

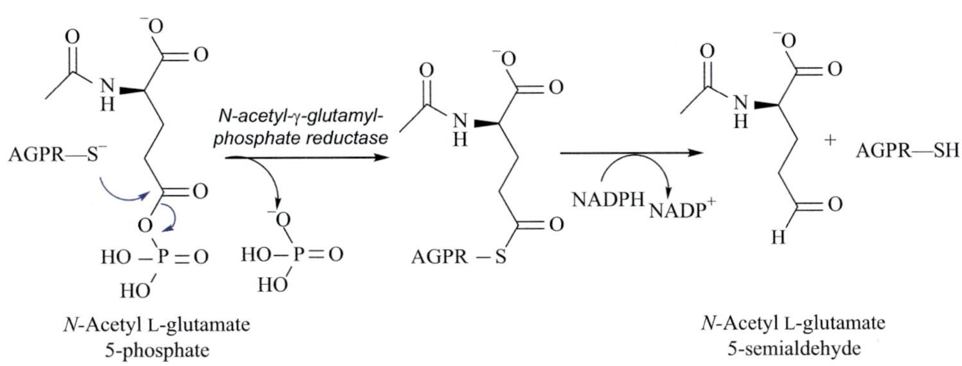

FIGURE 4.11 The reductive dephosphorylation of *N*-acetyl-γ-glutamyl-phosphate to *N*-acetylglutamate-γ-semialdehyde. (Reprinted from Cherney, L.T.; Cherney, M.M.; Garen, C.R.; Niu, C.; Moradian, F.; James, M.N.G *Journal of Molecular Biology* **2007**, *367,* 1357–1369. Copyright (2007), with permission from Elsevier.)

[28] Gupta, N.K.; Robinson, W.G. *J. Biol. Chem.* **1960**, *235*, 1609.

[29] Cherney, L.T.; Cherney, M.M.; Garen, C.R.; Niu, C.; Moradian, F.; James, M.N.G. *J. Mol. Biol.* **2007**, *367,* 1357.

workup, gives 2,3,3-trimethylpentan-1-ol, and the reaction of 4-(3,5-dimethylphenyl)butan-2-one with $NaBH_4$, followed by treatment with aqueous ammonium chloride, gives 4-(3,5-dimethylphenyl) butan-2-ol. In these examples, the first reaction gives an alkoxyaluminate or an alkoxyborate product, which requires subsequent treatment with dilute acid (H_3O^+) or aqueous ammonium chloride to yield the alcohol product.

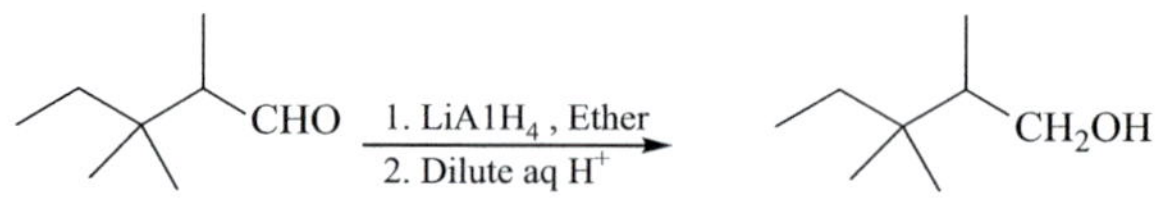

2,3,3-Trimethylpentanal 2,3,3-Trimethylpentan-1-ol

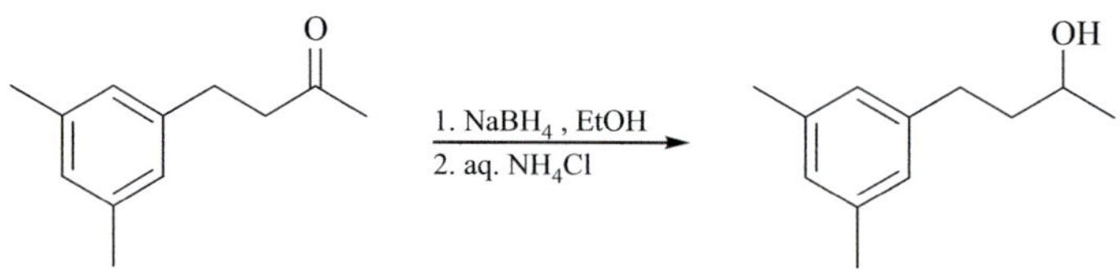

4-(3,5-Dimethylphenyl)butan-2-one 4-(3,5-Dimethylphenyl)butan-2-ol

The carbonyl moiety of carboxylic acid esters is reduced by $LiAlH_4$ to yield the corresponding alcohol, along with the alcohol derived from the OR moiety, as shown for ethyl stearate, which gives octadecan-1-ol and ethanol. Amides, such as 2-ethyl-*N*,*N*-dimethylbutanamide, are reduced with lithium aluminum hydride but do not give the alcohol as a reduction product; rather, they yield the amine 2-ethyl-*N*,*N*-dimethylbutan-1-amine. *In general, $NaBH_4$ is a much weaker reducing agent, does not reduce amides, and only reduces some esters.*

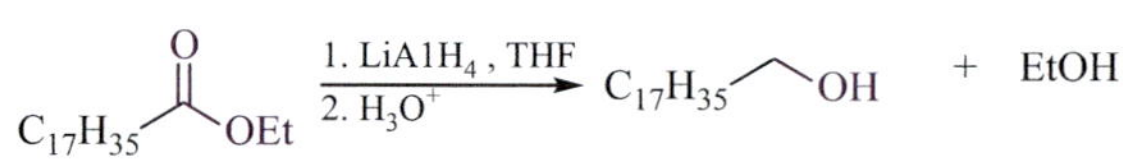

Ethyl stearate Octadecan-1-ol

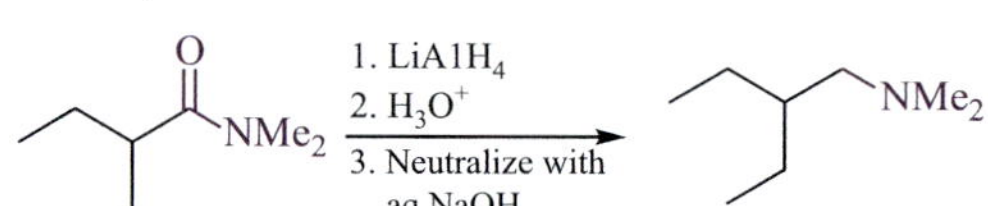

2-Ethyl-*N*,*N*-dimethylbutanamide 2-Ethyl-N,*N*-dimethylbutan-1-amine

In addition to the hydride reducing agents, the carbonyl group of an aldehyde or ketone is reduced to an alcohol by catalytic hydrogenation. As with alkenes, a transition metal catalyst is required to react first with H_2, and to coordinate the carbonyl to facilitate the transfer of hydrogen. Although palladium, nickel, ruthenium, and platinum catalyze hydrogenation of ketones and aldehydes, a platinum catalyst is usually the best choice because the yields are higher and there are fewer side reactions. Platinum oxide [PtO_2; sometimes called *Adam's catalyst*] is commonly used. The reaction of hydrogen gas with 4-phenylhexan-2-one in the presence of PtO_2 gives 4-phenylhexan-2-ol.

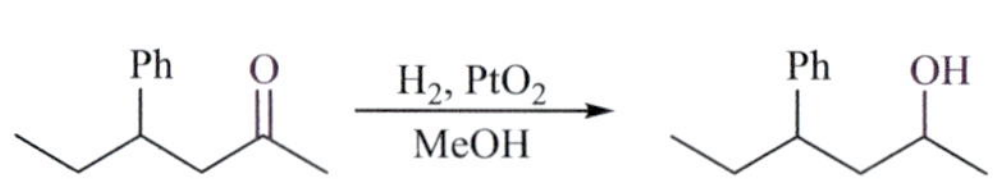

4-Phenylhexan-2-one 4-Phenylhexan-2-ol

4.4 TRANSFERASES (EC 2)

Transferases are a class of enzymes that facilitate the transfer of specific functional groups (e.g., a methyl or glycosyl group) from one molecule (called the donor) to another (called the acceptor). Their common names are usually formed according to the pattern *acceptor grouptransferase* or *donor grouptransferase*. The donor may be a cofactor (coenzyme) charged with the group to be transferred. Transaminases transfer an amino group by a well-established mechanism involving covalent substrate-coenzyme intermediates, and these enzymes are classified as a special subclass (EC 2.6.1, *transaminases*).

4.4.1 Methyl, Hydroxyl, Thiol, and Glycosyl Transferases

Several transferase enzymes transfer a methyl group to biologically important molecules. *Nicotinamide N-methyltransferase* (EC 2.1.1.7) is an enzyme first characterized in the liver that catalyzes the N-methylation of nicotinamide with the methylating agent (*S*)-adenosyl-L-methionine as the donor. This enzyme is important for the regulation of metabolic pathways, and it is expressed at high levels in several types of cancers.[30] The enzyme plays a role in nicotinamide metabolism and in the detoxification of many xenobiotics. Note that a xenobiotic is a substance, often a synthetic chemical, that is foreign to the body or to an ecological system. *N*-Methylnicotinamide is a metabolite of nicotinamide (niacin) and is found in human urine. Low levels of *N*-methylnicotinamide in the urine indicate a niacin deficiency.

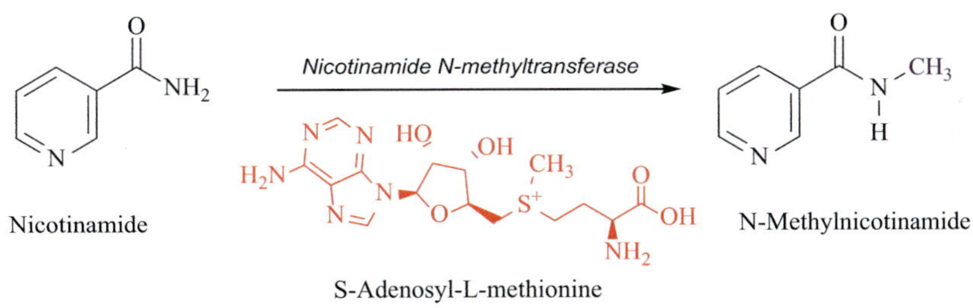

Enzymes are available for the transfer of a hydroxyl moiety from one species to another. The anaerobic bacterium *Pelobacter acidigallici* uses a *transferase* enzyme in its fermentation pathway called *pyrogallol transhydroxylase* (EC 1.97.1.2) to convert pyrogallol to phloroglucinol in the absence of O_2.[31] This enzyme is a cytoplasmic Mo-enzyme composed of a large α-subunit with 875 amino acid residues and a small β-subunit with 274 amino acid residues. The α-subunit contains a Mo ion that is coordinated to two molybdopterin guanine dinucleotide cofactors.

As shown in Figure 4.12,[31] the actual transfer of the hydroxyl group is from the co-substrate 1,2,3,5-tetrahydroxybenzene (in green) to pyrogallol (in red). This transfer does not represent a net redox reaction, but the pyrogallol is oxidized at position 5, and the co-substrate 1,2,3,5-tetrahydroxybenzene is reduced at position 2, as shown in Figure 4.12. Pyrogallol first coordinates to the histidine residue and the molybdenum(VI), and deprotonation of a phenolic moiety of pyrogallol leads to the cyclohexadienone unit via transfer of a phenolic proton to the aspartic acid residue. Loss of a proton from the pyrogallol skeleton to the coordinated dienone occurs with concomitant attack by the tetrahydroxybenzene to give the 4-aryloxy derivative. Subsequent loss of the phloroglucinol product from the tetrahydroxybenzene moiety accompanies transfer of a hydroxyl unit to the pyrogallol skeleton, and decomplexation regenerates the tetrahydroxybenzene co-substrate.

There are *transferase* enzymes that can transfer a glycosyl group from one carbohydrate unit to another. In the cytosol, maltose metabolism requires a *4-glucanotransferase* (EC 2.4.1.25), which is now referred to as disproportionating enzyme 2.[32] Note that the cytosol is the fluid portion of the cytoplasm that does not include organelles or other internal cell structures. In this pathway,

[30] Peng, Y.; Sartini, D.; Pozzi, V.; Wilk, D.; Emanuelli, M.; Yee, V.C. *Biochemistry* **2011**, *50*, 7800.
[31] Messerschmidt, A.; Niessen, H.; Abt, D.; Einsle, O.; Schink, B.; Kroneck, P.M.H. *Proc. Natl. Acad. Sci.* **2004**, *101*, 11571.
[32] Steichen, J.M.; Petty, R.V.; Sharkey, T.D. *J. Biol. Chem.* **2008**, *283*, 20797.

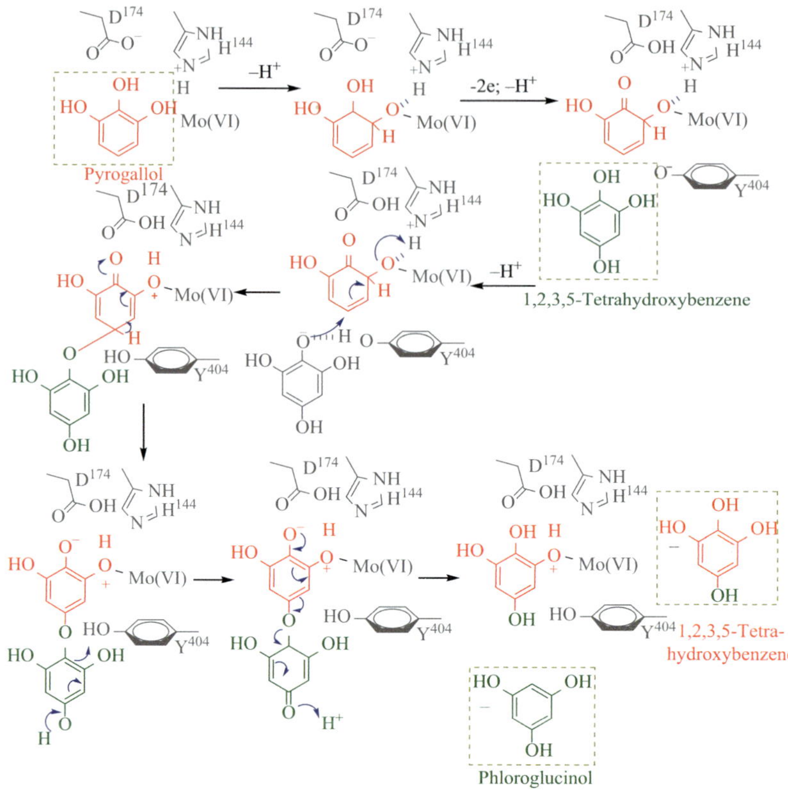

FIGURE 4.12 Biosynthetic pathway for the conversion of pyrogallol to phloroglucinol. Messerschmidt, A.; Niessen, H.; Abt, D.; Einsle, O.; Schink, B.; Kroneck, P.M.H. *Proc. Natl. Acad. Sci.* **2004**, *101,* 11571–11576. Copyright (2004) National Academy of Sciences, U.S.A.)

a maltosyl unit is transferred from one maltotriose to another maltotriose to form a dextrin long enough to be acted upon by starch *phosphorylase* or *β-amylase*. The role of 4-*glucanotransferase in vivo* is therefore presumed to be the transfer of glucosyl units during transitory starch degradation. The *4-glucanotransferase* is a modular protein that contains one domain similar to glycoside hydrolases of family 77, as well as two putative N-terminal carbohydrate-binding modules.

Transaminases, or *aminotransferases,* are transferase enzymes that catalyze a transamination reaction between an amino acid and an α-keto acid, which is important in the synthesis of amino acids that are ultimately used to synthesize proteins. In medicine, elevated transaminase levels may indicate liver damage. The transaminase *aspartate aminotransferase* (EC 2.6.1.1) is a pyridoxal 5′-phosphate-dependent enzyme that catalyzes the reversible transfer of the (*R*)-amino group between aspartate and 2-oxoglutarate.[33] The overall process requires two half-reactions, during which the (*R*)-amino group temporarily resides on pyridoxal 5′-phosphate as an imine moiety. Initially, the aldehyde group of pyridoxal 5′-phosphate is covalently linked to the ε-amino group of Lys^{258} at the active site of the enzyme, as shown in Figure 4.13.[33a] The amine unit of aspartate attacks the C=N moiety to generate the diamine intermediate, which loses the lysine unit. Loss of a proton leads to the imino unit, which reacts with water, and another loss of a proton gives the –OH moiety that facilitates the loss of the product 2-oxoglutarate and the amine on the cofactor moiety.

[33] (a) Islam, M.M.; Goto, M.; Miyahara, I.; Ikushiro, H.; Hirotsu, K.; Hayashi, H. *Biochemistry* **2005**, *44*, 8218; (b) Kiick, D.M.; Cook, P.F. *Biochemistry* **1983**, *22*, 375.

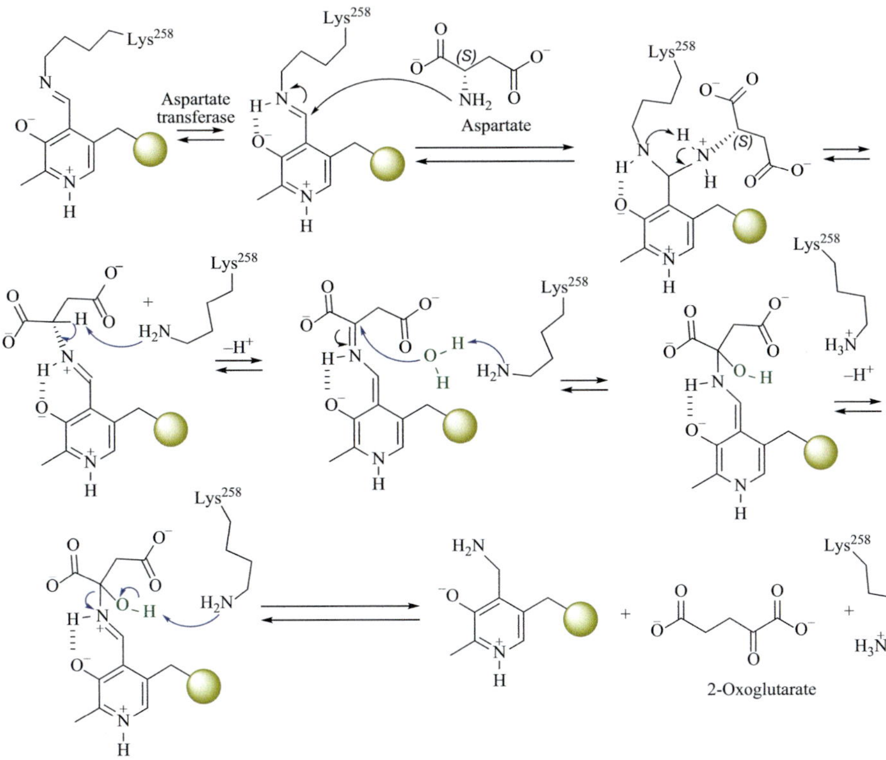

FIGURE 4.13 Reaction mechanism of *Aspartate Aminotransferase* with L-Aspartate. (Reprinted with permission from Islam, M.M.; Goto, M.; Miyahara, I.; Ikushiro, H.; Hirotsu, K.; Hayashi, H. *Biochemistry* **2005**, *44*, 8218–8229. Copyright 2005 American Chemical Society. Reprinted with permission from Kiick, D.M.; Cook, P.F. *Biochemistry* **1983**, *22*, 375–382. Copyright 1983 American Chemical Society.)

It is noted that a *kinase* is an enzyme that catalyzes the transfer of a phosphate group from a high-energy, phosphate-donating molecule such as ATP to a specified molecule. This process is known as phosphorylation, where the substrate gains a phosphate group and the high-energy ATP molecule donates a phosphate group. There are protein kinases, carbohydrate kinases, and lipid kinases. An example is the conversion of L-fucose to β-L-fucose-1-phosphate by *L-fucose kinase* (EC 2.7.1.52).[34] The conversion is mediated by the conversion of ATP to ADP.

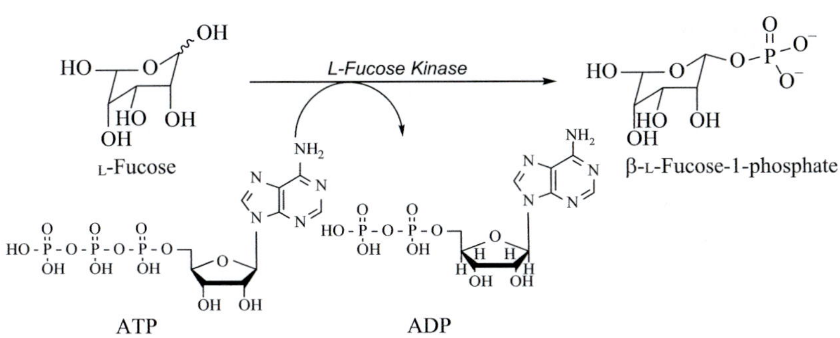

[34] Hoffmeister, D.; Yang, J.; Liu, L.; Thorson, J.S. *Proc. Natl. Acad. Sci.* **2003**, *100*, 13184; Park, S.H.; Pastuszak, I.; Drake, R.; Elbein, A.D. *J. Biol. Chem.* **1998**, *273*, 5685.

4.4.2 Chemical Reactions That Incorporate Methyl, Hydroxyl, Glycosyl or Amino Groups into New Molecules

The reaction that transfers a group from one molecule to another is essentially a substitution reaction. Incorporation of methyl groups and amines is well known, although transferring a hydroxyl group is not straightforward. It is possible to transfer a variety of organic moieties, including a glycosyl group.

A traditional method for incorporating a methyl group is to use an electrophilic methylating agent. In other words, a nucleophilic species will react with a methyl halide or a methylsulfonate ester such as iodomethane or the methyl ester of toluenesulfonic acid (methyl tosylate). Each of these electrophilic substrates is shown in a reaction with the enolate anion of acetone, which is a carbon nucleophile. In the first two cases, the methyl carbon is connected to an electron-zwithdrawing unit (halide or the sulfur of the tosylate), and the nucleophilic carbon of the enolate anion displaces the halide or $ArSO_2^-$ leaving group to give the methylation product. These reactions are S_N2 reactions, of course (see Section 6.2). If an alkoxide nucleophile is used, such as the alkoxide of pentan-3-ol, the S_N2 reaction with iodomethane gives the methyl ether in what is known as the *Williamson ether synthesis*. Amines are nucleophiles, and the reaction of *N,N*-diethylbutan-1-amine with iodomethane gives the ammonium salt S_N2 product, *N,N*-diethyl-*N*-methylbutan-1-ammonium iodide. From the standpoint of isolating the methylated amine product, this reaction is problematic in that an amine is a base that can deprotonate the acidic ammonium salt to give the new amine. This new amine may react with additional iodomethane to give a new methylated product (polymethylation). The reaction of ammonia and iodomethane can give methylation, but polymethylation is a major problem.

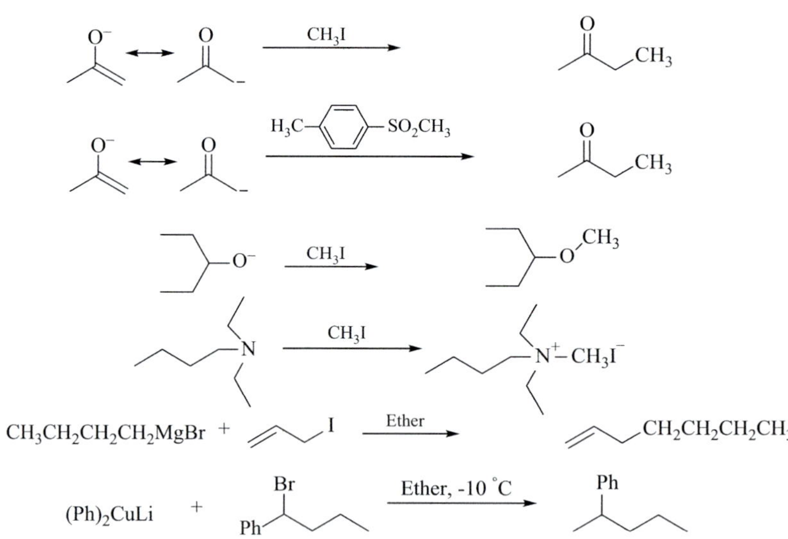

Several nucleophilic methylating agents are available with a nucleophilic carbon. Grignard reagents such as butylmagnesium bromide ($CH_3CH_2CH_2CH_2MgBr$) or organolithium reagents such as methyllithium (CH_3Li) are simple examples. These reagents will react with highly reactive alkyl halides such as allyl iodide ($CH_2{=}CHCH_2I$) or benzyl bromide ($PhCH_2Br$) to give the alkylated product. In general, the reaction of non-activated alkyl halides with these organometallic reagents is very difficult. There is a useful alternative: conversion of organolithium reagents to a corresponding organocuprate, such as lithium diphenylcuprate ($LiCuPh_2$), allows a facile reaction with alkyl halides. The reaction of butylmagnesium bromide with allyl iodide and the reaction of lithium diphenylcuprate with 2-bromopentane are shown as examples.

Incorporating hydroxyl into a molecule via an S_N2 reaction using the hydroxide ion to give an alcohol is problematic. The hydroxide ion is not a very good nucleophile with alkyl halides, in large part because it is such a good base, and elimination competes with substitution. However, it is possible to use a surrogate oxygen nucleophile such as acetate ion ($CH_3CO_2^-$). But this reaction is not very efficient because the acetate ion is not a strong nucleophile. In addition, a second reaction is required to hydrolyze the ester to give the alcohol. It is possible to convert a tertiary alkyl halide such as 1-bromo-1-methylcyclopentane to a carbocation intermediate in aqueous media, and subsequent reaction with water and loss of a proton will give 1-methylcyclopentanol via an S_N1 reaction (Section 6.4). Note that the nucleophile in this carbocation reaction is the oxygen atom of water, not hydroxide. A better way to incorporate a hydroxyl group is the reaction of an alkene such as cyclopentene with a catalytic amount of a mineral acid such as sulfuric acid in an aqueous medium, as discussed in the introduction to the chapter. The reaction of the alkene with the acid will give a carbocation, and subsequent reaction with water and loss of a proton will give the alcohol. An aromatic halide can react with sodium hydroxide under high-temperature and high-pressure conditions to give phenol, as shown, in an S_NAr reaction (nucleophilic aromatic substitution; Section 11.8).

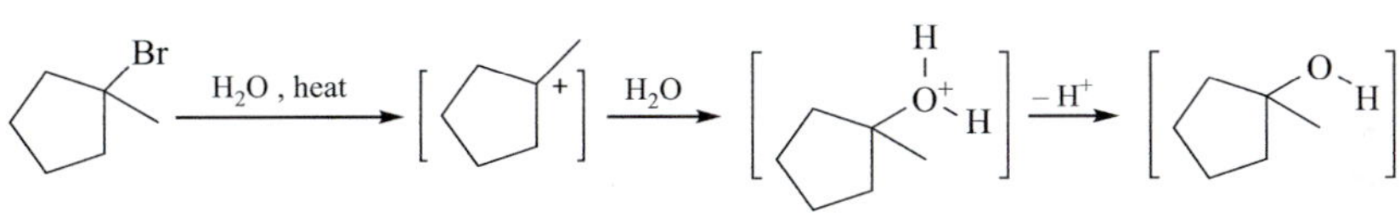

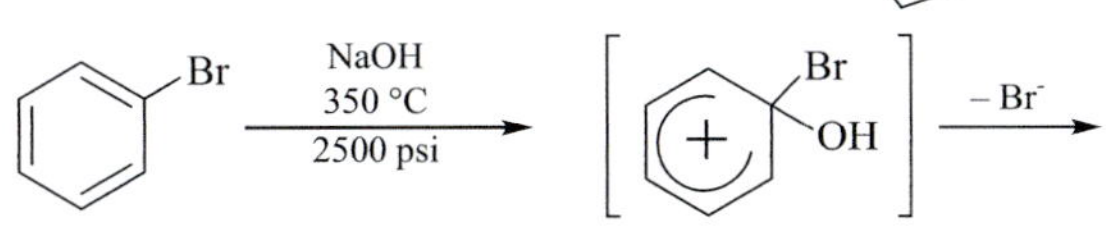

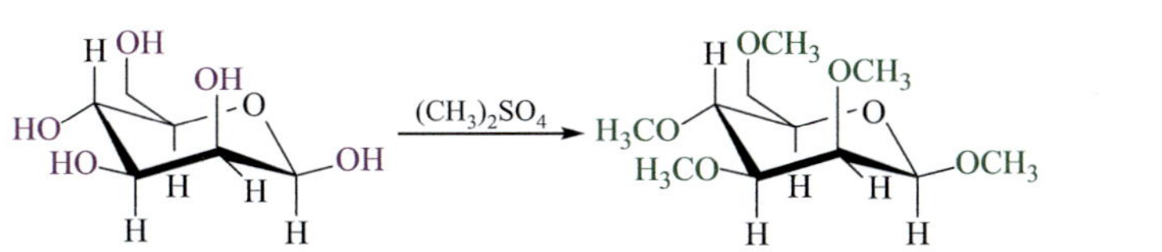

β-D-Mannopyranose | Methyl tetra-*O*-methyl-β-D-mannopyranoside

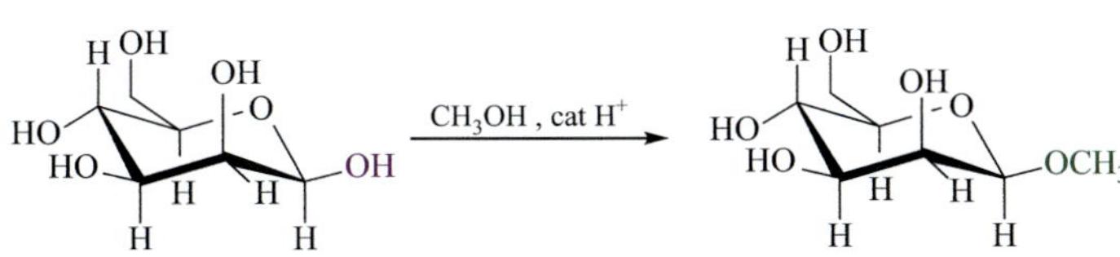

β-D-Mannopyranose | Methyl β-D-glucopyranoside

The reaction of carbohydrates with some reagents is possible, making *O*-alkylation reactions possible. It is therefore possible to transfer a glycosyl group to some alkyl groups. Two methylation reactions are shown. The reaction of a carbohydrate such as D-mannopyranose with dimethyl sulfate [$(CH_3)_2SO_4$], a common methylating reagent, gave the *O*-methyl derivative as shown. In addition, the reaction of this carbohydrate with methanol and an acid catalyst, under S_N1 conditions, likewise gave the *O*-methyl derivative.

4.5 HYDROLYASES (EC 3)

Hydrolases are hydrolytic enzymes that catalyze the cleavage of C–O, C–N, C–C, and some other bonds, including phosphoric anhydride bonds, with water. Although the systematic name always includes *hydrolase*, the common name is often the name of the substrate with the suffix *-ase*. *Hydrolases* act on ester, glycosyl, peptide, amide, or other bonds and often catalyze the hydrolytic removal of a particular group and the transfer of this group to suitable acceptor molecules. Another problem is that proteinases have "esterolytic" properties, which means that they also hydrolyze ester bonds more rapidly than natural peptide bonds. In such cases, classification may be as a peptide hydrolase based on historical priority and presumed physiological function. Several examples of hydrolase enzyme transformations will be presented.

4.5.1 Esterases

An *esterase* is a *hydrolase* enzyme that facilitates the reaction of water with esters in a hydrolytic cleavage reaction, giving an acid and an alcohol. Many different esterases differ in their substrate specificity, their protein structure, and their biological function. A specific example of an *esterase* is *tropinesterase* (EC 3.1.1.10), which is an enzyme that catalyzes the aqueous hydrolysis of atropine to give tropine and methyl tropate.

Another example is an esterase called *atropine acylhydrolase*, which allows some rabbits to hydrolyze large amounts of tropane alkaloids such as atropine or scopolamine, enabling them to consume leaves of the toxic *Atropa belladonna plant* (deadly nightshade).[35] This ability is due to the presence of the enzyme *atropine esterase* (EC 3.1.1.10) in the plasma (and other tissues), which is found only in some species of rabbits. Atropine and scopolamine are well-known muscarinic receptor antagonists. Atropine can be used to treat heart rhythm problems and stomach or bowel problems, and it is also used to dilate the pupils before an eye exam.

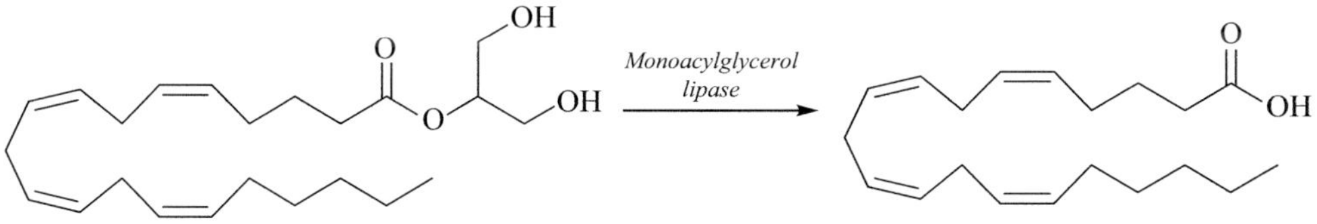

2-Arachidonoylglycerol Arachidonic acid

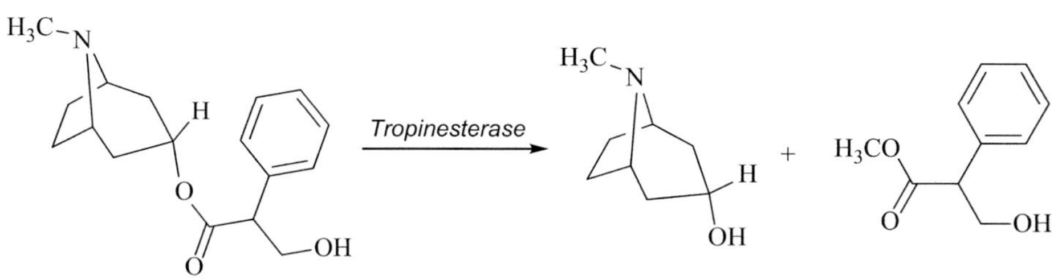

Atropine Tropine Methyl tropate

A *lipase* is a subclass of an esterase, which makes it a *hydrolase*. A lipase catalyzes the hydrolysis of fats (lipids), which is necessary for the digestion, transport, and processing of dietary lipids (e.g., triglycerides, fats, oils) in most, if not all, living organisms. In the intestine, triglycerides are split into glycerol and three fatty acids (lipolysis) with the help of *lipases*. An example of a lipase is *monoacylglycerol lipase* (or just *acylglycerol lipase*; EC 3.1.1.23), which is the primary hydrolytic enzyme for monoacylglycerols in most tissues, including the brain, liver, kidney, spleen, heart, and lung.[36] 2-Arachidonoylglycerol is an endogenous cannabinoid signaling lipid that acts as an agonist on cannabinoid receptors, and it is a substrate for this lipase. Monoacylglycerol lipase acts in the final lipolytic step in triacylglycerol breakdown.

[35] Harrison, P.K.; Tattersall, J.E.H.; Gosden, E. *Naunyn-Schmiedeberg's Archiv. Pharmacol.* **2006**, *373*, 230.
[36] Nomura, D.K.; Casida, J.E. *Chem.-Biol. Interact.* **2016**, *259*, 211.

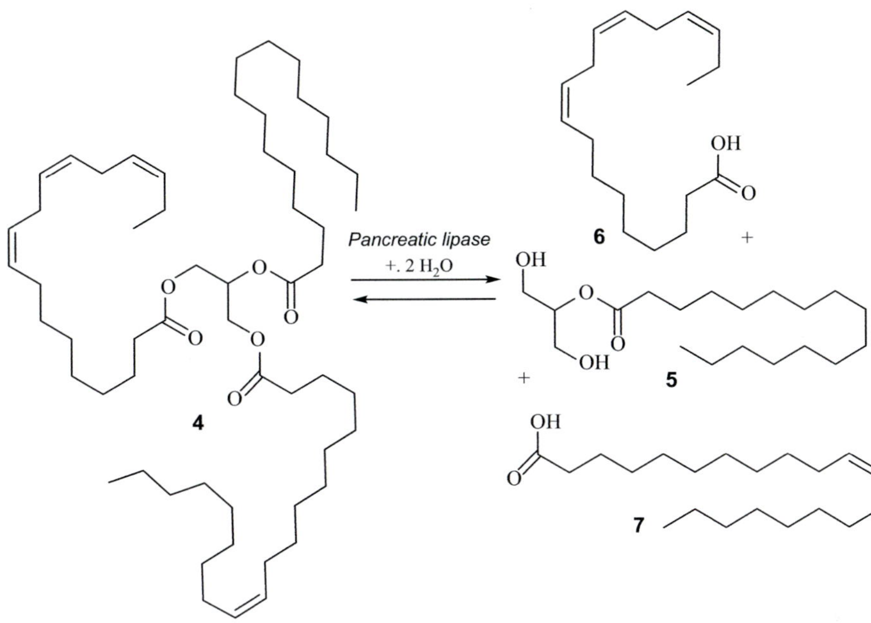

Pancreatic lipase is an enzyme secreted by the pancreas that is also known as *pancreatic triacylglycerol lipase* or *steapsin*. In the human digestive system, this enzyme hydrolyzes dietary fat molecules to convert triglyceride substrates into monoglycerides and fatty acids. An example is the hydrolysis of **4** (3-(((9Z,12Z,15Z)-octadeca-9,12,15-trienoyl)oxy)-2-(palmitoyloxy)propyl (Z)-icos-11-enoate) with *pancreatic lipase* to give monoglyceride **5** (1,3-dihydroxypropan-2-yl palmitate) plus two fatty acids, **6** ((9Z,12Z,15Z)-octadeca-9,12,15-trienoic acid) and **7** ((Z)-icos-11-enoic acid).[37]

4.5.2 Other Hydrolyases

A *phospholipase* is a hydrolase enzyme that cleaves phospholipids, including phosphoglycerol lipids, to release fatty acids and thereby remove a phosphate group from a protein. *Phospholipase A1* (EC 3.1.1.32) cleaves the C1 acyl chain. *Phospholipase A2* (EC 3.1.1.4) cleaves the C2 acyl chain; *phospholipase B* (EC 3.1.1.5) cleaves both the C1 and C2 acyl chains; *phospholipase C* (EC 3.1.4.3) cleaves before the phosphate group to yield a diacylglycerol and a phosphate-containing group; and *phospholipase D* (EC 3.1.4.4) cleaves after the phosphate to yield phosphatidic acid and an alcohol.

Phosphatidylinositol-specific *phospholipase C* hydrolyzes phosphatidylinositol 4,5-bisphosphate [PIP2; 1,2-diacyl-*sn*-glycero-3-phospho-(1-D-*myo*-inositol 4,5-bisphosphate)] via *phospholipase Cβ* (NorpA; EC 3.1.4.3) to yield two second messenger molecules: inositol trisphosphate (IP^3) and diacylglycerol.[38] This pathway is important for cellular communication. Phosphatidylinositol 4,5-bisphosphate is a minor phospholipid component of cell membranes. With light stimulation, PIP2 is hydrolyzed by *phospholipase Cβ* to yield diacylglycerol and inositol trisphosphate.[38] The two fatty acids in PIP2 are stearic acid at position 1 and arachidonic acid at position 2, as shown in Figure 4.14.[38]

Glycoside hydrolases (also called *glycosidases* or *glycosyl hydrolases*; EC 3.2.1.X) are enzymes that catalyze the hydrolysis of glycosidic bonds in complex sugars. A glycosidic bond is a covalent *bond* that joins a carbohydrate molecule to another group. *Endo-xylanases* (EC 3.2.1.8) are *glycosidases* that attack the internal β-xylosidic glycosidic linkages of the xylan backbone. Xylan, the major constituent of hemicellulose, is composed of a β-1,4-linked D-xylose backbone substituted

[37] Nuhn, P. *Naturstoffchemie*, in Hirzel, S. *Wissenschaftliche Verlagsgesellschaft*, 2nd ed., Stuttgart, 1990, pp. 308–309.
[38] Leung, H.-T. *J. Neurogenet.* **2012**, *26*, 216.

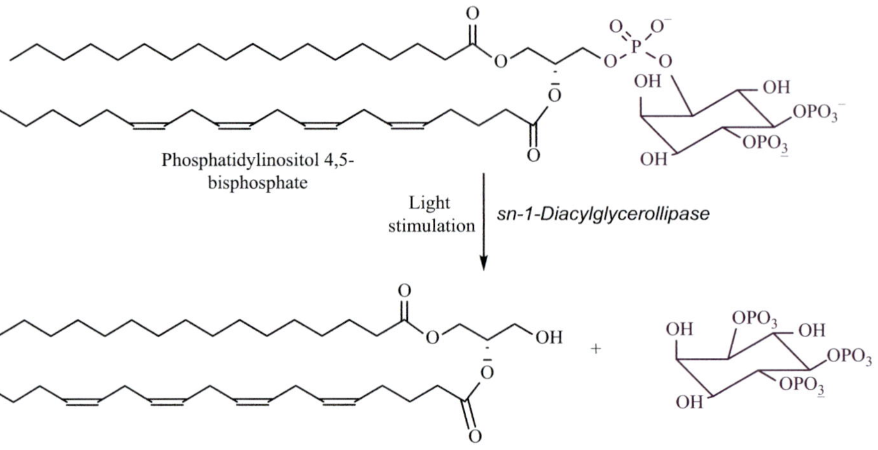

FIGURE 4.14 Cleavage of phosphatidyl linkage with a lipase.

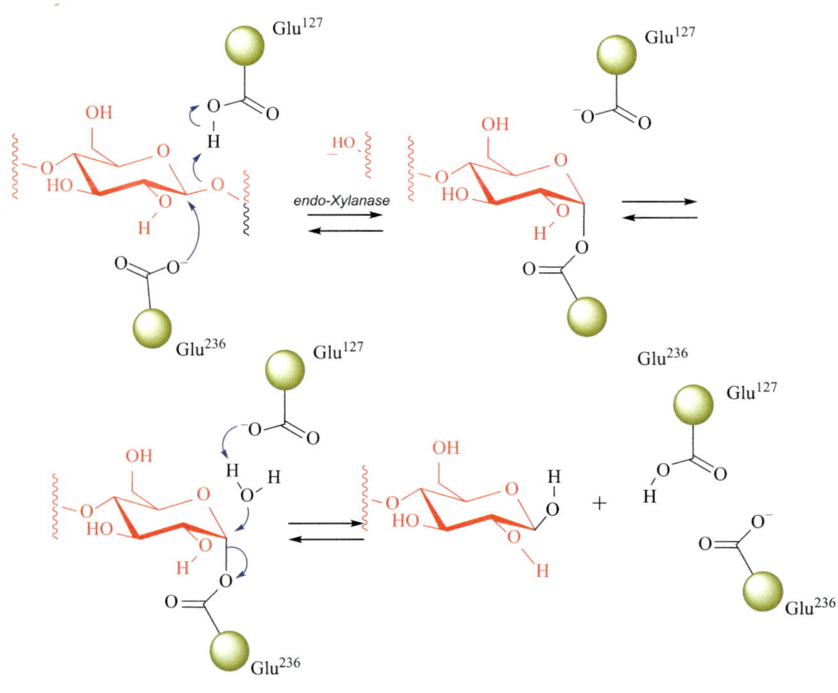

FIGURE 4.15 Double displacement reaction mechanism as applied to the *S. lividans* Xyl10A. (Reprinted from McCarter, J.D.; Withers, S.G. *Curr. Opin. Struct. Biol.* **1994**, *4,* 885–892. Mechanisms of enzymatic glycoside hydrolysis. Copyright (1994), with permission from Elsevier.)

at the 2′ and 3′ positions with L-arabinofuranose, D-glucuronic acid, and 4-*O*-methylglucuronic acid. Most *xylanases* are in families 10 and 11 of glycoside hydrolases. Xyl10A from *Streptomyces lividans* is a retaining glycoside hydrolase. An example is shown in Figure 4.15,[39] which shows the mechanism for a pyranoside hydrolysis. Initial reaction with the glutamic acid[236] residue leads to glycosidic cleavage with concomitant formation of a covalent glycosyl–enzyme acyl intermediate.

[39] McCarter, J.D.; Withers, S.G. *Curr. Opin. Struct. Biol.* **1994**, *4,* 885.

Subsequent reaction with water, assisted by glutamic acid[127], leads to displacement of the acyl group and incorporation of the hydroxyl group at the anomeric carbon.

Note that lysosomes are membrane-bound vesicles that contain digestive enzymes, such as *glycosidases*, *proteases*, and *sulfatases*. Lysosomal enzymes are synthesized in the endoplasmic reticulum, transported to the Golgi apparatus, and tagged for lysosomes by the addition of mannose-6-phosphate.

Another class of hydrolases is the *nucleosidases*, which are enzymes that promote the hydrolysis of a nucleoside. Similarly, a *phosphorylase* reversibly promotes the reaction of a nucleoside with phosphate to give a base and a phosphate of ribose or deoxyribose (a nucleoside *phosphorylase*). A purine *nucleosidase* (EC 3.2.2.1) is an enzyme that catalyzes the chemical reaction of a purine nucleoside to give D-ribose and a purine base. *A pyrimidine-5′-nucleotide nucleosidase* (EC 3.2.2.10) is an enzyme that catalyzes the hydrolysis of a pyrimidine 5′-nucleotide. Such enzymes are found on the surface membrane of several members of the protozoan family *Trypanosomatidae*, including *Leishmania* and African trypanosomes, which are pathogenic to humans.[40] These parasites are incapable of de novo purine synthesis, so nucleosidases may make purine nucleosides available to them. An example is shown in Figure 4.16 using *inosine-adenosine-guanosine-nucleoside hydrolase* to hydrolyze adenosine to D-ribofuranose and adenine.

A *protease* is a hydrolase enzyme that facilitates protein catabolism by hydrolysis of peptide bonds (proteolysis). Note that catabolism is the set of metabolic pathways that break down molecules into smaller units that can be oxidized to release energy or used in other anabolic reactions. An example is *tryptase* (EC 3.4.21.59), which is a serine proteinase found in mast cells. This enzyme is believed to be responsible for cleaving the hemagglutinin surface protein of the influenza A virus, which activates it and causes the symptoms of flu. Hemagglutinins are glycoproteins on the surface of an influenza virus. *β-Tryptase* evaluation of the substrate specificity *of β-tryptases*, by the peptide phage display technique, has revealed a strong preference for cleaving substrates with an Arg or Lys residue at the P1 position, preference for Lys/Arg in the P3 position, and some preference for Pro at the P4 position, but with little specificity at the P2 position.[41] Note that cleavage of P1-P1′ leads to incremental numbering in the N-terminal direction of the cleaved peptide bond (P2, P3, P4, etc.). Increased *β-tryptase* levels can be found in serum during extreme inflammatory conditions such as anaphylaxis.[42] Fibrinogen is an important substrate of human *tryptase*, which may display anticoagulant activity. During tissue and vascular injury, fibrinogen is converted enzymatically by thrombin to fibrin and subsequently to a fibrin-based blood clot, so this enzyme may be important for blood clotting.

In non-biochemical organic chemical reactions, both esters and amides are subject to acid-catalyzed hydrolysis and to basic hydrolysis. Esters are hydrolyzed with aqueous acid under acidic conditions, and the mechanism of ester hydrolysis, represented by ethyl hexanoate, is shown in Figure 4.17. Initial protonation of the carbonyl gives an oxocarbenium ion that reacts with water to give a tetrahedral intermediate. In the presence of an excess of water, protonation of the OEt moiety allows the loss of ethanol and the formation of hexanoic acid.

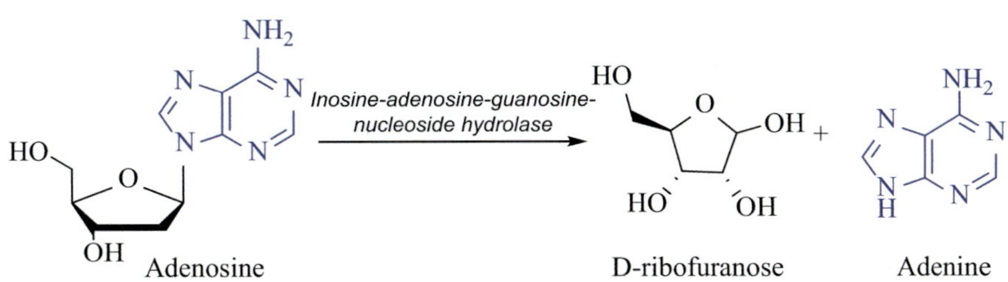

FIGURE 4.16 Reaction of a nucleoside hydrolase.

[40] Neubert, T.A.; Gottlieb, M. *J. Biol. Chem.* **1990**, *265*, 7236.
[41] Hallgren, J.; Pejler, G. *FEBS J.* **2006**, *273,* 1871.
[42] Schwartz, L.B.; Bradford, T.R.; Littman, B.H.; Wintroub, B.U. *J. Immunol.* **1985**, *135*, 2762.

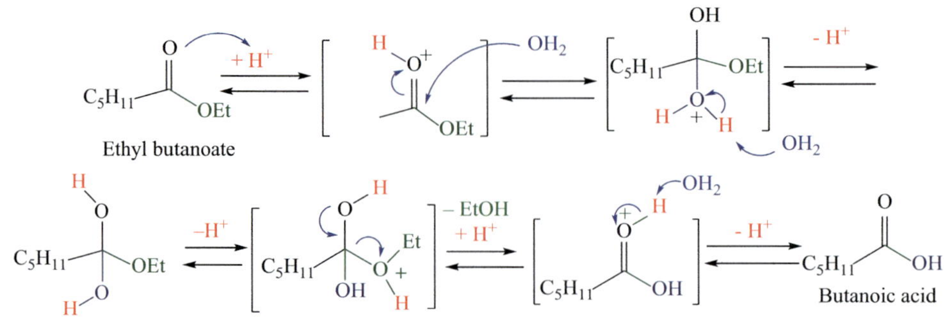

FIGURE 4.17 Acid-catalyzed hydrolysis of ethyl butanoate.

4.5.3 Chemical Hydrolysis

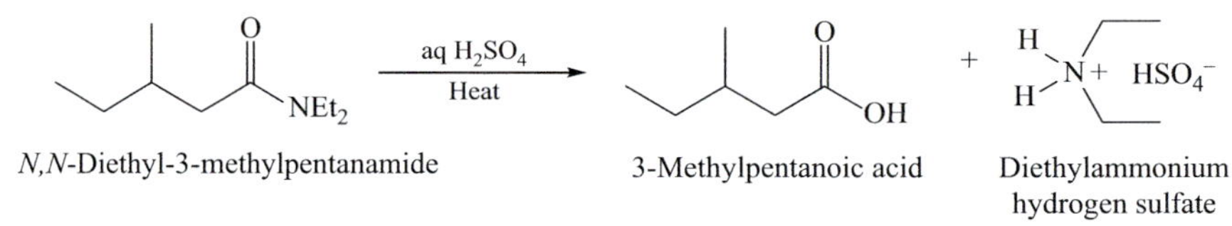

Amides are much less reactive than esters, in large part because NR_2 is a much poorer leaving group relative to OR in the tetrahedral intermediate formed via the addition of the nucleophile to the acyl carbon. However, amides do react when vigorously heated with acid to yield the parent carboxylic acid and an amine from secondary and tertiary amides, or the carboxylic acid and ammonia from a primary amide. In other words, *it is more difficult to hydrolyze an amide compared to other acid derivatives.* Note that the initially formed amine is protonated under acidic conditions to give an ammonium salt.

The process of identifying the chemical structure of a protein (or any other peptide) relies on the acid-catalyzed hydrolysis of the amide bond, in which the peptide is heated in 6 N HCl at 105°C for 24 h. This acid hydrolysis should completely hydrolyze the peptide into its constituent amino acids, which are then separated. Tryptophan has an indole unit, which is acid-sensitive, and it is partially destroyed under harsh acidic conditions. Glutamine, asparagine, glutamic acid, and aspartic acid decompose with the loss of ammonia when heated in 6 N HCl. When the amino acid residue has a sterically hindered side chain, as in valine or isoleucine, hydrolysis may be incomplete, and heating for a longer period may be necessary.

Esters are hydrolyzed under basic conditions to yield the acid, but basic hydrolysis requires two steps to convert the ester to the acid. Isopropyl acetate (1-methylethyl ethanoate), for example, was heated to reflux in aqueous NaOH to produce a tetrahedral intermediate (**8**), which loses the alkoxide to complete the acyl substitution process and give acetic acid. Under the basic conditions, acetic acid is quickly deprotonated to the acetate anion. Neutralization with aqueous H_2SO_4 in a second step gives two products: acetic acid and propan-2-ol (isopropyl alcohol), as shown in Figure 4.18.

This two-step process, shown in reaction form for isopropyl acetate to give acetic acid and propan-2-ol, is known as *saponification*. Saponification means "to make soap," and the term comes from the ancient practice of using wood ashes (rich in potassium hydroxide) to convert animal fat to soap. Animal fat, as well as vegetable oils, is usually a mixture of triglycerides, the triester derivatives of fatty acids and glycerol. Under these conditions, basic hydrolysis of all three ester units leads to the formation of glycerol and the salts of the fatty acids. The salts of these fatty acids are solids, and they are the fundamental constituents of what is known as "soap." An example is shown in Figure 4.19 for the saponification of cocoa butter (a triglyceride composed of palmitic, oleic, and stearic acid residues), which is extracted from cocoa beans. Basic hydrolysis leads to glycerol (propane-1,2,3-triol) along with the sodium salts of the three fatty acids: sodium palmitate, sodium stearate, and sodium oleate.

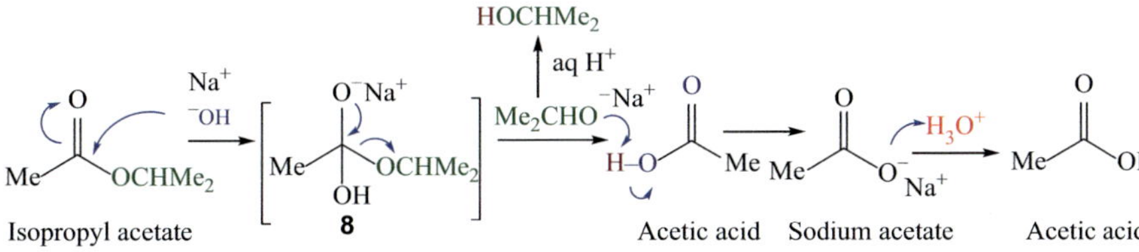

FIGURE 4.18 Base hydrolysis of an ester, isopropyl acetate.

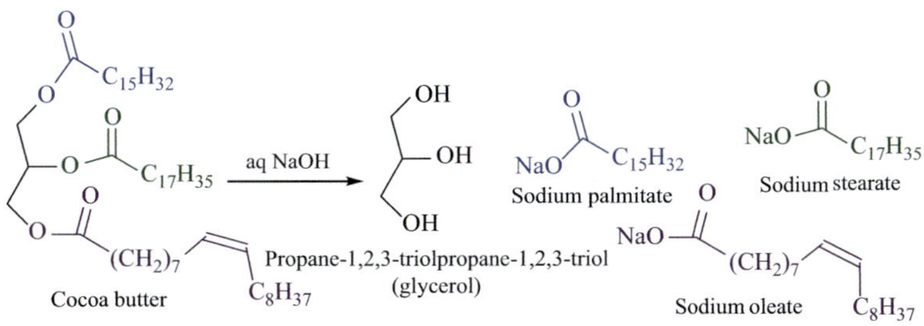

FIGURE 4.19 Saponification of cocoa butter.

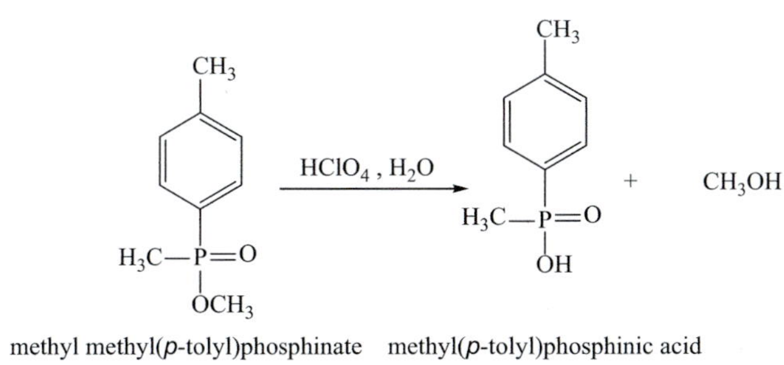

Similar acid and base conditions can be used for the hydrolysis of phosphate esters.[43] The reaction of methyl methyl(*p*-tolyl)phosphinate with aqueous perchloric acid, for example, gives methyl(*p*-tolyl)phosphinic acid and methanol.[44] The glycosidic bonds of disaccharides, trisaccharides, and polysaccharides can also be cleaved by hydrolysis facilitated by heating with dilute aqueous acid. The glycosidic bond is relatively stable to basic hydrolysis.

4.6 LYASES (EC 4)

A *lyase* is an enzyme that catalyzes the breaking of chemical bonds by means other than hydrolysis and oxidation, and the reaction is usually reversible. *Lyases* typically cleave C–C, C–O, C–N, and other bonds by an elimination reaction to give double bonds or rings. The systematic name of a lyase is formed according to the pattern *substrate group-lyase*. The *hyphen* is an important part of the name and, to avoid confusion, *should not be omitted, e.g., hydro-lyase,* not 'hydrolyase'. However, common names like *decarboxylase, aldolase, and dehydratase* (in the case of elimination of CO_2, aldehyde, or water) are often used. The enzymes are called *synthases* when the reverse reaction is more important. *Carboxylases* are better known as *decarboxylases*, and they are enzymes that cleave carbon-carbon bonds and add or remove a carboxyl group. An example is the decarboxylation of amino acids, β-keto acids, and α-keto acids.

---

[43] Kirby, A.J.; Varvoglis, A.G. *J. Amer. Chem. Soc.* **1967**, *89*, 415.
[44] Harsági, N.; Keglevich, G. *Molecules* **2021**, *26*, 2840.

4.6.1 Lyase Reactions

Pyruvate decarboxylase (EC 4.1.1.1) is a homotetrameric enzyme that catalyzes the non-oxidative decarboxylation of pyruvic acid to acetaldehyde and carbon dioxide in the cytoplasm of prokaryotes and in the cytoplasm and mitochondria of eukaryotes.[45] This enzyme is found in the yeast *Saccharomyces cerevisiae*, and it remains inactive until exposed to pyruvate. The catalytic activity of this enzyme depends on the cofactor thiamine diphosphate, which is bound mainly via a divalent metal ion (usually magnesium) to the protein moiety. It is noted that this reaction is observed in combination with the subsequent reaction of the enzyme *alcohol dehydrogenase* (EC 1.1.1.1), which reduces acetaldehyde to ethanol with the help of NADH in the metabolic pathway of alcohol fermentation.

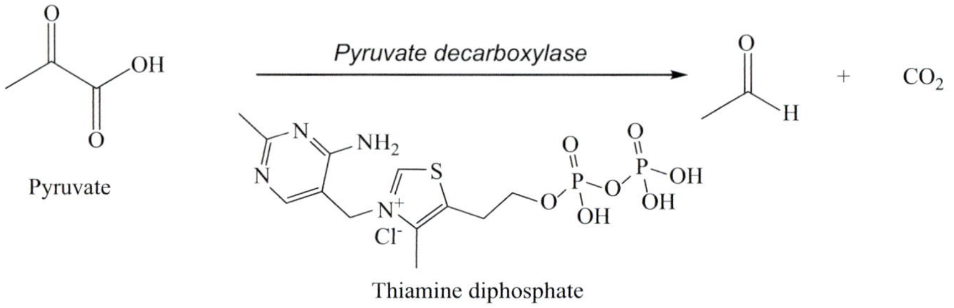

An *aldolase* is an enzyme that helps break down certain sugars by what would be a retro-aldol reaction. While an aldol condensation is the reaction of two carbonyl compounds to give a β-hydroxy carbonyl compound, a retro-aldol reaction is the reverse of the aldol reaction (Section 9.5) and cleaves a β-hydroxy carbonyl compound into two carbonyl compounds. *Aldolase* is found in high concentrations in muscle tissue. *Aldolase A*, also known as *fructose-bisphosphate aldolase* (EC 4.1.2.13), for example, is a lyase enzyme that catalyzes the reversible reaction that cleaves the aldol, fructose 1,6-bisphosphate, into the triose phosphates dihydroxyacetone phosphate and glyceraldehyde 3-phosphate (see Figure 4.20).[46]

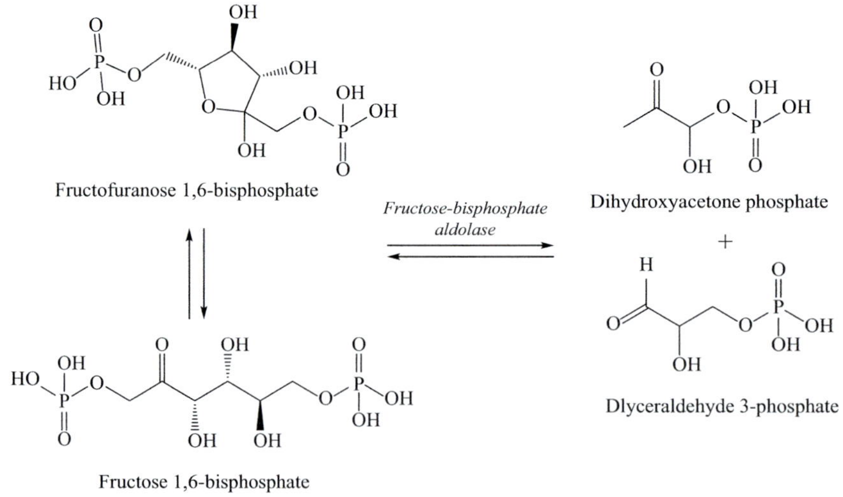

FIGURE 4.20 An aldolase reaction of fructose 1,6-biphosphate.

[45] Kutter, S.; Weiss, M.S.; Wille, G.; Golbik, R.; Spinka, M.; König, S. *J. Biol. Chem.* **2009**, *284*, 12136.
[46] Cooper, S.J.; Leonard, G.A.; McSweeney, S.M.; Thompson, A.W.; Naismith, J.H.; Qamar, S.; Plater, A.; Berry, A.; Hunter, W.N. *Structure* **1996**, *4*, 1303.

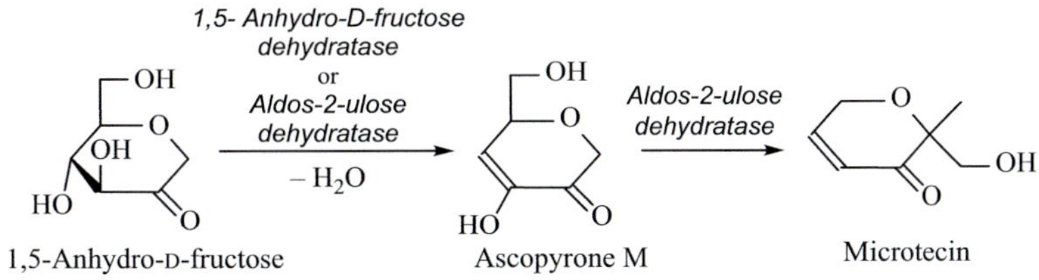

FIGURE 4.21 The proposed anhydrofructose pathway of glycogen and starch degradation.

A *hydratase* (also called a *hydro-lyase*) is any enzyme that catalyzes the addition to or removal of water from a molecule. An example is the formation of 1,5-anhydro-D-fructose, which is a metabolite of glycogen and starch after enzymatic degradation by *α-1,4-glucan lyase* (EC 4.2.2.13). The enzyme *1,5-anhydro-D-fructose dehydratase* (EC 4.2.1.111) converts 1,5-anhydro-D-fructose to ascopyrone M, which was found to be the intermediate in the formation of microthecin.[47]Microthecin (2-hydroxy-2-(hydroxymethyl)-2H-pyran-3(6H)-one) is a known fungal and red algal secondary metabolite that shows antimicrobial activity. The microthecin-forming enzyme from the fungus *Phanerochaete chrysosporium* was found to be *aldos-2-ulose dehydratase* (EC 4.2.1.110), a bi-functional enzyme that converts 1,5-anhydro-D-fructose to ascopyrone M and catalyzes the isomerization of 1,5-anhydro-D-fructose to microthecin (Figure 4.21).

The class of enzymes called *dehydratases* is *lyases* that form double and triple bonds in a substrate via the loss of water. They can be found in many places, including the mitochondria, peroxisome, and cytosol. The *dehydratase* enzyme, *3-hydroxypropionyl-CoA dehydratase* (EC 4.2.1.116) also called *3-hydroxypropionyl-CoA hydro-lyase* and catalyzes the conversion of 3-hydroxypropanoyl-CoA to acrylyl-CoA via dehydration.[48] This reaction is probably part of the 3-hydroxypropionate/4-hydroxybutyrate cycle in *M. sedula* and other members of the *Sulfolobales*.[48]

An *enolase*, a lyase enzyme also known as *phosphopyruvate hydratase* (EC 4.2.1.11), is a metalloenzyme that catalyzes the conversion of 2-phospho-D-glycerate to phosphoenolpyruvate, as shown, which is the next-to-last step of glycolysis.[49]

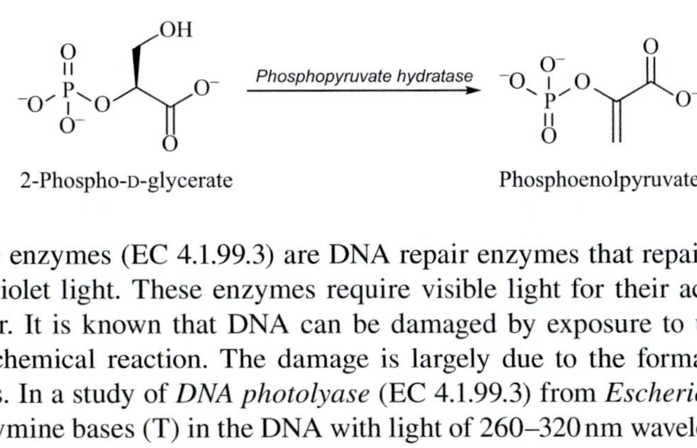

The *photolyase* enzymes (EC 4.1.99.3) are DNA repair enzymes that repair damage caused by exposure to ultraviolet light. These enzymes require visible light for their activation and for the actual DNA repair. It is known that DNA can be damaged by exposure to ultraviolet radiation, which induces a chemical reaction. The damage is largely due to the formation of cyclobutane pyrimidine dimers. In a study of *DNA photolyase* (EC 4.1.99.3) from *Escherichia coli*, irradiation of two adjacent thymine bases (T) in the DNA with light of 260–320 nm wavelengths was shown to induce a [2+2]-cycloaddition that gives the cis,syn cyclobutane thymine dimer T<>T1,[50] as shown in Figure 4.22.[51a] Such covalently linked dimers block cell replication and transcription, which

---

[47] Yu, S. *Biochim. Biophys. Acta* **2005**, *1723*, 63.
[48] Teufel, R.; Kung, J.W.; Kockelkorn, D.; Alber, B.E.; Fuchs, G. *J. Bacteriol.* **2009**, *191*, 4572.
[49] Zhang, E.; Brewer, J.M.; Minor, W.; Carreira, L.A.; Lebioda, L. *Biochemistry* **1997**, *36*, 12526.
[50] Sanders, D.B.; Wiest, O. *J. Amer. Chem. Soc.* **1999**, *121*, 5127.
[51] (a) Mitchell, D.L. *Photochem. Photobiol.* **1988**, *48*, 51. Also see (b) Brash, D.E.; Haseltine, W.A. *Nature* **1982**, *298*, 189; Taylor, J.-S.; Cohrs, M.P. *J. Amer. Chem. Soc.* **1987**, *109*, 2834.

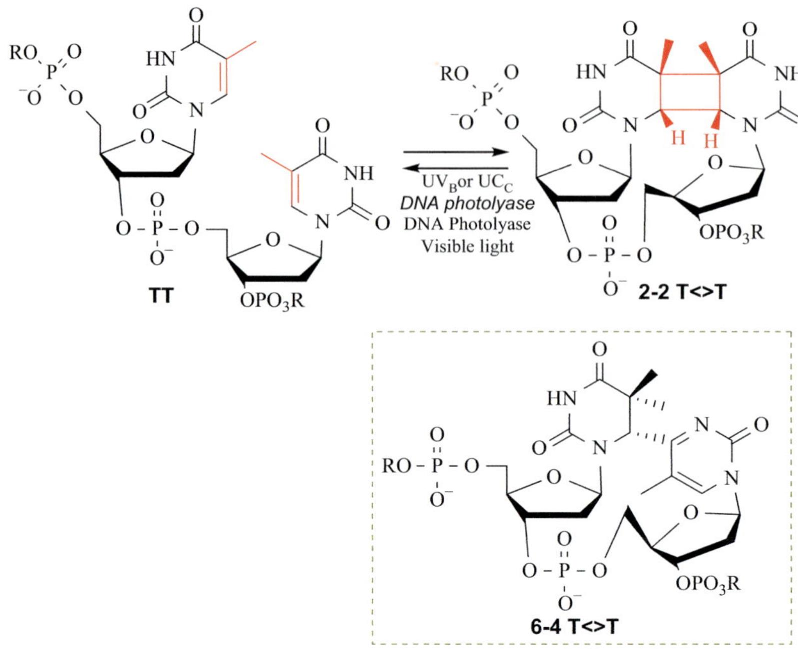

FIGURE 4.22 [2+2]-Photocycloaddition of thymine. (Reprinted with permission from Mitchell, D.L. *Photochem. Photobiol.* **1988**, *48*, 51–57. Copyright © **1988**, reproduced with permission of John Wiley & Sons. Ltd.)

leads to cell death. The presence of the T<>T2 dimer, as well as a minor photoproduct, the 6-4 photoadduct,[51a] is believed to be a major cause of basal and squamous cell skin cancers in humans.[50] The photoinduced damage of DNA can be repaired through a light-induced repair mechanism, also mediated by *DNA photolyase*. These enzymes achieve true repair through an electron-transfer-catalyzed [2+2] cycloreversion of T<>T.[52] Photolyases have been found in many organisms, but it is not clear if they occur in humans.

4.6.2 Bond Cleavage in Organic Chemistry

4.6.2.1 Decarboxylation

Carboxylic acids with a carbonyl at C3 (the carboxyl carbonyl is C1) lose carbon dioxide (CO_2) to form an enol when heated. This transformation is known as decarboxylation (loss of a carboxyl group). It is important to know that decarboxylation only occurs if a carboxylic acid has a unit that contains a π-bond attached to C3 (C=O or C=C). An example is 2-benzyl-2-methylmalonic acid, which yields 2-methyl-3-phenylpropanoic acid by the loss of CO_2 when heated, as shown in Figure 4.23. *Decarboxylation is facile for 1,3-dicarboxylic acids (1,3-propanedioic acid derivatives) and β-keto acids.*

An enol is a molecule in which an OH unit is directly attached to an sp^2 carbon of an alkene, C=C–OH. Enols are inherently unstable, and they undergo an internal acid-base reaction in which the O–H proton is transferred to the C=C unit, and the π-bond breaks with electron transfer toward oxygen to form a carbonyl. This process is called *keto-enol tautomerism* (Sections 9.2 and 9.3). Decarboxylation leads to an enol, but tautomerization of this enol gives butanoic acid. The equilibrium for this reaction lies on the side of the carboxylic acid because it is significantly more stable and because one product is a gas (CO_2), which escapes from the reaction medium. As noted, *heating 1,3-dicarboxylic acids or β-keto acids to 200°C –300°C will lead to decarboxylation.*

[52] (a) Sancar, A.; Sancar, G.B. *Ann. Rev. Biochem.* **1988**, *52*, 29. (b) Heelis, P.F.; Hartman, R.F.; Rose, S.D. *Chem. Soc. Rev.* **1995**, *24,* 289.

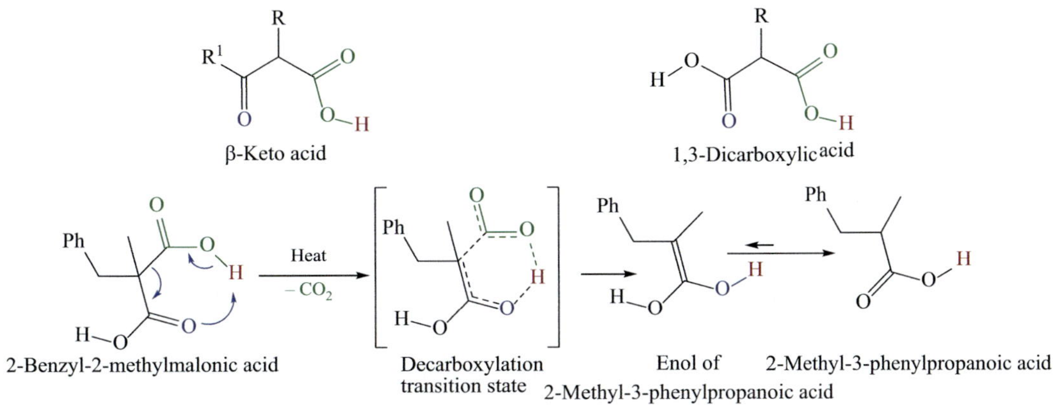

FIGURE 4.23 Decarboxylation of 2-ethyl malonic acid.

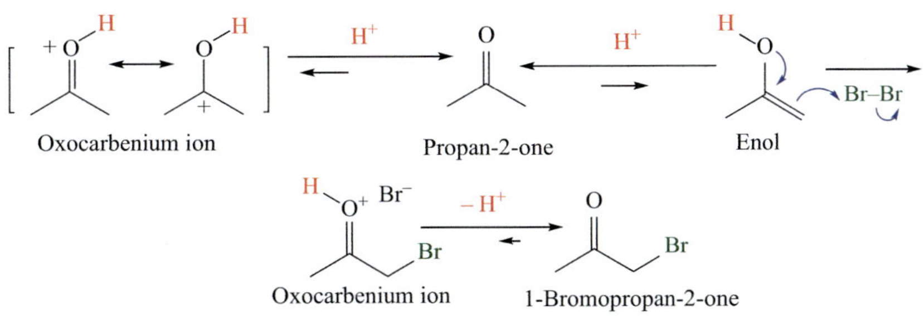

FIGURE 4.24 Keto–enol tautomerism in acetone and reaction with bromine.

4.6.2.2 Enol Formation and the Acid-Catalyzed Aldol

A ketone or an aldehyde exists with small amounts of the enol form in most cases (see Section 9.1). The carbonyl form is more stable, and for most aldehydes and ketones, only a tiny amount of enol is present, as shown by the equilibrium arrows in Figure 4.24 for acetone (propan-2-one). With an acid catalyst, there is a larger percentage of the enol, which can donate to an electrophile such as diatomic bromine, as shown, to give the protonated α-bromoacetone. Subsequent loss of a proton gives the neutral product, α-bromoacetone.

The presence of heteroatoms or electron-withdrawing functional groups at the α-carbon will increase the amount of enol at equilibrium due to stabilization of the enol form via internal hydrogen bonding. There is an increase in the enol content of 1-chloropropan-2-one and 1,1-dichloropropan-2-one, for example, relative to acetone.

If an α-carbon has more than one electron-withdrawing group, the α-proton becomes increasingly acidic as the number of attached electron-withdrawing groups increases. Pentane-2,4-dione should have an increased enol content due to the two carbonyl units, as shown. In general, dicarbonyl compounds with two carbonyl units, where one carbonyl has a 1,3-relationship with a second carbonyl ($O{=}C{-}CH_2{-}C{=}O$), will have a higher enol content. Examples include 1,3-propanedial (malonaldehyde), the diketone pentane-2,4-dione, 3-oxoesters (β-ketoesters) (e.g., methyl 3-oxobutanoate), and finally 1,3-dioic esters (malonates).

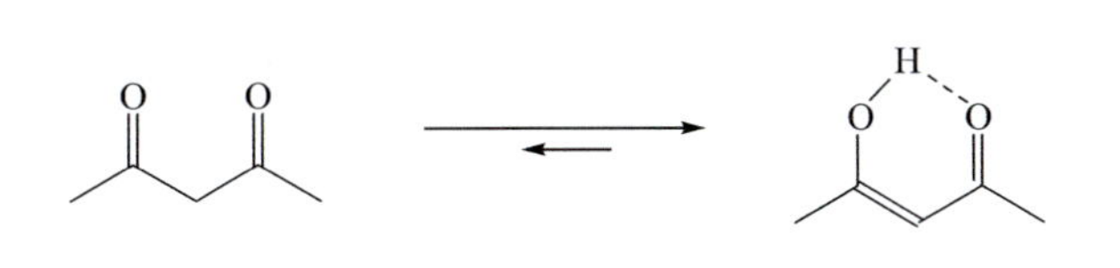

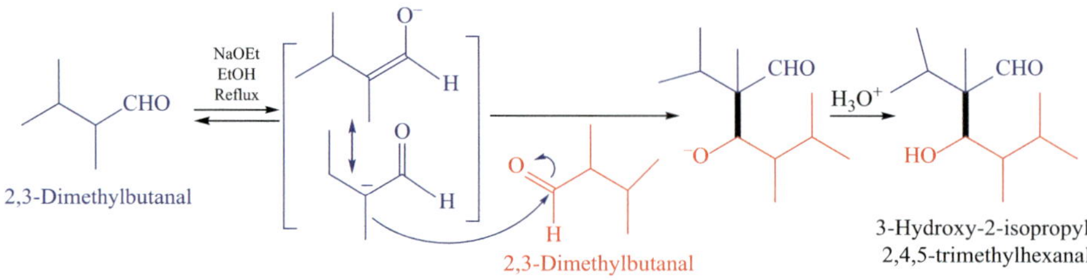

FIGURE 4.25 The aldol condensation of 2,3-dimethylbutanal.

When an aldehyde was treated with an alkoxide base in an alcohol solvent heated to reflux (heated at the boiling point of the alcohol), cooled, and then treated with dilute aqueous acid at low temperatures, *a β-hydroxy aldehyde (known generically as an aldol)* was isolated. Because of the type of product formed, this reaction has come to be called the *aldol condensation* (Section 9.5). Nowadays, the reaction is explained by deprotonation of the aldehyde (or ketone), where the α-hydrogen atom is acidic (typically, $pK_{a=}$ 19–21) to give a resonance-stabilized enolate anion. An example is the self-condensation of 2,3-dimethylbutanal by heating sodium ethoxide in ethanol to reflux, as shown in Figure 4.25. In the presence of an electrophilic species such as another molecule of the aldehyde 2,3-dimethylbutanal (or a different aldehyde or ketone that is added to the reaction), the nucleophilic enolate anion attacks the acyl carbon of the electrophilic species in an acyl addition reaction. The alkoxide product is then treated with dilute aqueous acid to give the aldol product, in this case 3-hydroxy-2-isopropyl-2,4,5-trimethylhexanal.

Aldol condensation reactions can be carried out under acidic conditions. While the base-catalyzed reaction is more common nowadays, Brønsted–Lowry acid-catalyzed condensation reactions are well known. However, the acidic conditions make the process reversible, and poor yields can result due to deleterious cationic side reactions. An example is the acid-catalyzed aldol condensation of benzophenone and benzaldehyde by heating in sulfuric acid in acetic acid.[53] The acid reacts with benzophenone to give an enol, which reacts in situ with benzaldehyde to give the aldol, 3-hydroxy-1,3-diphenylpropan-1-one, which spontaneously eliminates water in the presence of sulfuric acid to give 1,3-diphenylprop-2-en-1-one.

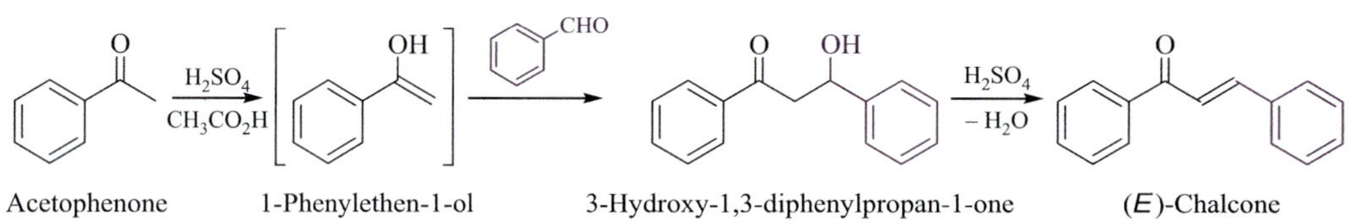

4.6.2.3 Elimination and Dehydration Reactions

Dehydration reactions are well known in organic chemistry, as shown with the acid-catalyzed dehydration reaction of 3-hydroxy-1,3-diphenylpropan-1-one to (*E*)-chalcone as discussed in the previous section. However, perhaps the best-known dehydration reaction in an undergraduate textbook occurs via a unimolecular elimination, the E1 reaction.

If 2-bromo-2-methylbutane is heated with KOH in aqueous THF, an S_N2 reaction is not possible due to the steric hindrance imposed in the transition state by the tertiary halide substrate. It is known that water facilitates ionization and solvation of the ions, thereby separating charge, so ionization will give the bromide ion and the tertiary carbocation, shown in Figure 4.26. In an aqueous solvent, formation of a stable carbocation is facile, but ionization is a slow step (rate-determining) relative to the removal of the β-hydrogen by the base. Loss of H_a by an acid–base reaction leads

[53] Noyce, D.S.; Pryor, W.A. *J. Amer. Chem. Soc.* **1955**, *77*, 1397.

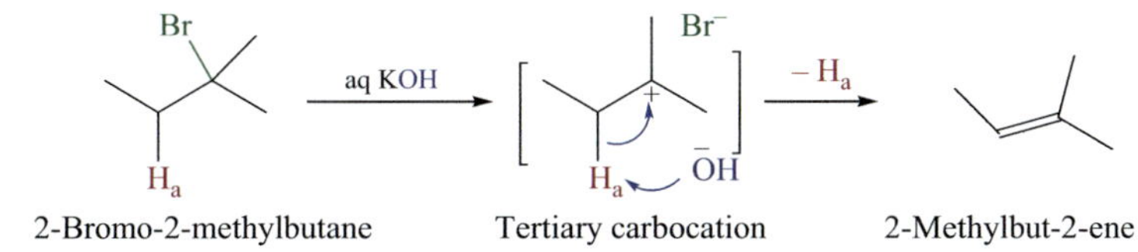

FIGURE 4.26 Ionization of an alkyl halide and an E1 reaction.

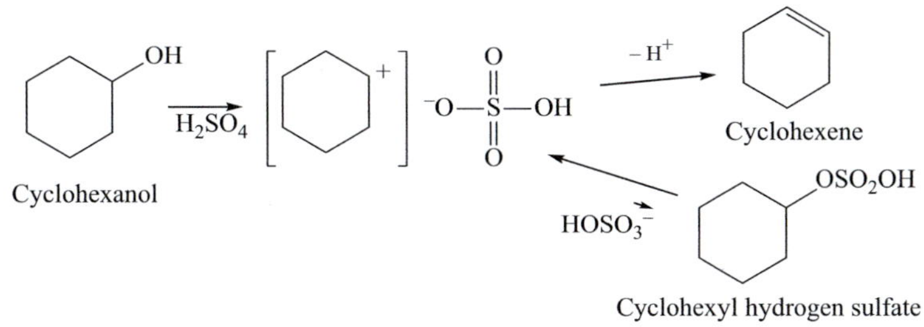

FIGURE 4.27 Dehydration of cyclohexanol.

to the formation of a conjugate acid (H–OH) and transfer of electrons from the C–H bond to the positive carbon to give an alkene, 2-methylbut-1-ene. Since the acid–base reaction that removes H_a is fast compared to ionization, *the reaction follows first-order kinetics*, and the overall reaction is a unimolecular process termed an *E1 reaction (unimolecular elimination).*

Since reaction conditions used for an E1 reaction often favor an S_N1 reaction rather than an E1 reaction, it is difficult to find a "pure" E1 reaction. In general, this statement is true, but there are exceptions when the base used in the reaction is a poor nucleophile, or if the S_N1 product is unstable and leads to a reversible reaction. If cyclohexanol is treated with concentrated sulfuric acid, as shown in Figure 4.27, the observed product is cyclohexene in a very fast reaction. The mechanism involves an acid–base reaction of the oxygen from the OH unit (the base) with sulfuric acid to form an oxonium ion, the conjugate acid, but the hydrogen sulfate counterion is a very weak nucleophile. Therefore, the substitution reaction is slow, and elimination to the alkene is favored. It is noted that the substitution product, cyclohexyl hydrogen sulfate, is very unstable in the aqueous acid medium, which also favors elimination as the major process.

4.7 ISOMERASES (EC 5)

Isomerases are a general class of enzymes that facilitate geometric or structural changes within a molecule and convert it from one isomer to another. These intramolecular rearrangements make and break bonds. Such enzymes may be called *racemases, epimerases, cis–trans isomerases, tautomerases, mutases,* or *cycloisomerases.*

4.7.1 Isomerase Reactions

An *epimerase* changes the stereochemistry of a stereogenic center in a molecule. Galactofuranose (Gal*f*) is the five-membered ring form of galactopyranose (Gal*p*), and it is found in the walls and extracellular carbohydrate sheaths of bacteria, protists, fungi, and plants. UDP-Gal*f* residues (uracildiphosophate galactofuranose) are directly produced from UDP-galactopyranose (UDP-Galp) by *UDP-galactopyranose mutase* (EC 5.4.99.9) before being incorporated into extracellular carbohydrate-containing compounds, as shown in Figure 4.28.[54] In a variety of species, ranging from

[54] Dalrymple, S.A.; Ko, J.; Sheoran, I., Kaminskyj, S.G.W.; Sanders D.A.R. *PLoS One* **2013**, *8,* e76803.

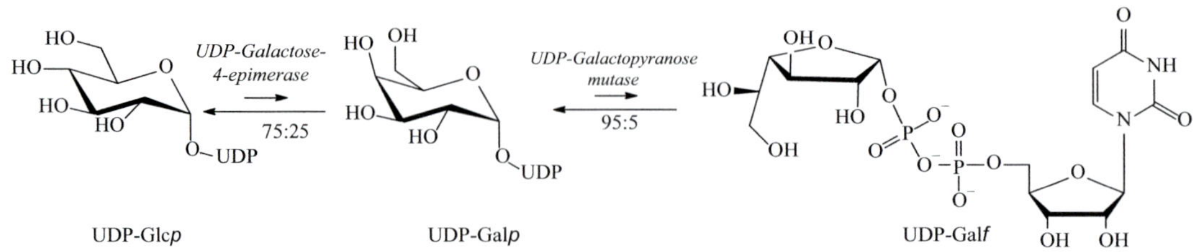

FIGURE 4.28 Proposed biosynthetic pathway of UDP-galactofuranose (UDP-Galf) in *Aspergillus nidulans*.

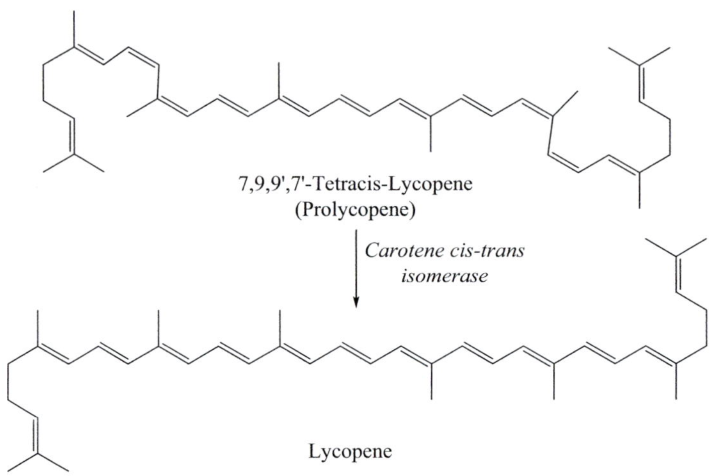

FIGURE 4.29 Isomerization of an alkene.

Escherichia coli (*E. coli*) to humans, *UDP-galactose-4-epimerase* (EC 5.1.3.2; in this study from *A. nidulans*), otherwise known as *UDP-glucose-4-epimerase*, catalyzes the interconversion of the 4′-hydroxyl configuration between UDP-glucopyranose (UDP-Glcp) and UDP-Galp. Note that both an epimerase enzyme and a mutase enzyme are used in this example. Also note the transposition of the pendant diol moiety during the transition from a six-membered ring to a five-membered ring in Figure 4.28. A *mutase* is an *isomerase* enzyme that catalyzes the shifting of a functional group from one position to another within the same molecule.

Beta-carotene isomerase (EC 5.2.1.14) is an isomerase enzyme that catalyzes the conversion of geometrical isomers in carotene and related compounds. As noted above, *isomerases* are enzymes that facilitate geometric changes, such as the isomerization of a (*Z*)-alkene to an (*E*)-alkene or an (*E*)-alkene to a (*Z*)-alkene. One such isomerase is *carotene cis-trans isomerase* (*CRTiso*; *Prolycopene isomerase*; EC 5.2.1.13), which is a constituent of the carotene desaturation pathway in cyanobacteria and in plants, and it is important in isoprenoid metabolism. Carotenoids belong to a large isoprenoid family, and some are indispensable in all photosynthetic organisms, where they play an important role in light harvesting and in the protection of cells from photo-damage caused by reactive oxygen species. In the carotenoid biosynthetic pathway, phytoene [(6*E*,10*E*,14*E*,16*Z*,18*E*,22*E*,26*E*)-2,6,10,14,19,23,27,31-octamethyldotriaconta-2,6,10,14,16,18,22,26,30-nonaene] is the first-formed polyene, but 7,9,9′,7′-tetra-cis-lycopene (commonly referred to as prolycopene) is produced later in the pathway that gives carotene (Figure 4.29).[55] The enzyme *CRTiso* isomerizes all cis double bonds in this polyene to form the all-trans-lycopene.[55] Lycopene is bright red, and it is found in tomatoes and other red fruits and vegetables such as watermelon.

[55] Yu, Q.; Ghisla, S.; Hirschberg, J.; Mann, V.; Beyer, P. *J. Biol. Chem.* **2011**, *286*, 8666.

4.7.2 Chemical Isomerization Reactions

Control of keeping isomeric integrity or changing the integrity of geometric isomers, and the control of retention or inversion of absolute stereochemistry, is important in organic chemistry. Epimerization refers to formally changing the stereochemistry of a stereogenic center from (*R*)- to (*S*)- or from (*S*)- to (*R*)-, and such a reaction is possible. More often, one plans a reaction to control the process that generates the stereogenic center to give the (*R*)-stereoisomer or the (*S*)-stereoisomer, as desired.

One way to change the stereochemistry of a stereogenic center is to take advantage of the fact that one stereoisomer may be higher in energy than the other. One type of stereocenter that can be corrected is a CHX moiety, where X is an electron-withdrawing group, and the H can be removed by treatment with base to form a planar enolate anion. In Olivo and coworkers' synthesis of callipeltoside-A,[56] the bicyclic lactone **9** was treated with a base, and a relatively planar enolate anion was generated. Subsequent protonation of this enolate anion gave a new lactone **10** in which the stereochemistry of the 3-methyl group was inverted. This inversion of configuration can be explained by examining the planar enolate, where the hydrogen can be delivered from the top face or from the bottom face. Delivery of the proton from the bottom face of the flattened enolate anion leads to the original lactone, in which there will be greater *A*-strain (Section 5.14) than if the proton is delivered from the top face, so the product of this reaction is the lower-energy product **10** shown in Figure 4.30.

Another example in Figure 4.30 is taken from Hortmann and coworkers' synthesis of occidentalol[57] and involves the α-proton of an ester group in **11**, which, when treated with a base, forms a planar enolate anion. When this enolate anion is protonated, the proton can be delivered from the top face or from the bottom face. If the proton is delivered from the bottom face of the planar enolate anion, the less stable ester is formed, in which there will be greater *A*-strain (see Section 5.14). However, if the proton is delivered from the top face, the lower-energy ester **12** is formed.

Sigmatropic rearrangements are important reactions that involve the movement of a group from one place in the molecule to another. Sigmatropic rearrangements are defined as reactions in which a σ-bond—meaning an atom or a substituent—*moves across a conjugated system to a new site.*[58] A [1,3]-sigmatropic shift involves the migration of a single atom across three carbon atoms. A 1,3-hydrogen shift, for example, involves the migration of a hydrogen atom from C3 to C1 to yield a new propene. A [1,5]-shift involves the migration of a single atom across a five-atom fragment, illustrated by the migration of a hydrogen atom from C5 in penta-1,3-diene to C1. A [3,3]-sigmatropic shift involves the migration of two fragments, each three atoms in

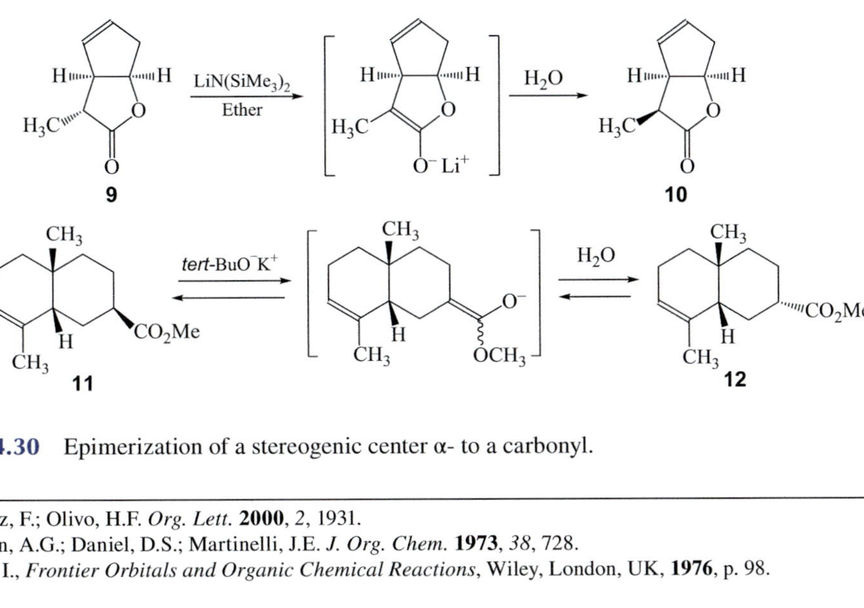

FIGURE 4.30 Epimerization of a stereogenic center α- to a carbonyl.

---

[56] Velázquez, F.; Olivo, H.F. *Org. Lett.* **2000**, *2*, 1931.
[57] Hortmann, A.G.; Daniel, D.S.; Martinelli, J.E. *J. Org. Chem.* **1973**, *38*, 728.
[58] Fleming, I., *Frontier Orbitals and Organic Chemical Reactions*, Wiley, London, UK, **1976**, p. 98.

length. Another example is the *Claisen rearrangement*,[59] illustrated by the rearrangement of allyl vinyl ether [3-(prop-1-en-2-yloxy)but-1-ene] upon heating, which gives an alkene-ketone such as (*E*)-hept-5-en-2-one.

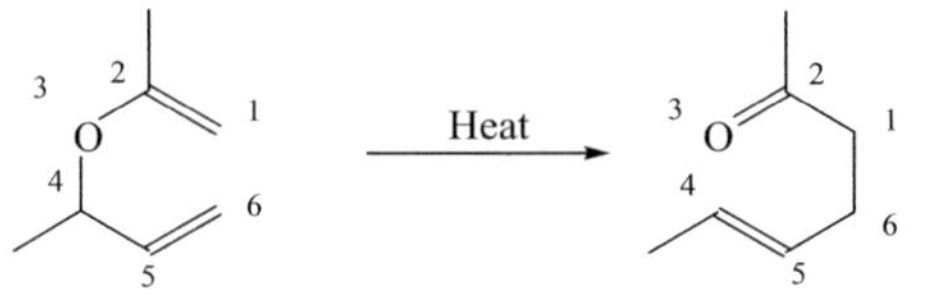

3-(Prop-1-en-2-yloxy)but-1-ene (*E*)-Hept-5-en-2-one

The example of a [1,5]-sigmatropic rearrangement with substituted cyclopentadienes occurs readily at room temperature[60] and generates three different isomers. These rearrangements can be described as a series of [1,5]-sigmatropic hydrogen shifts that lead to different isomers.

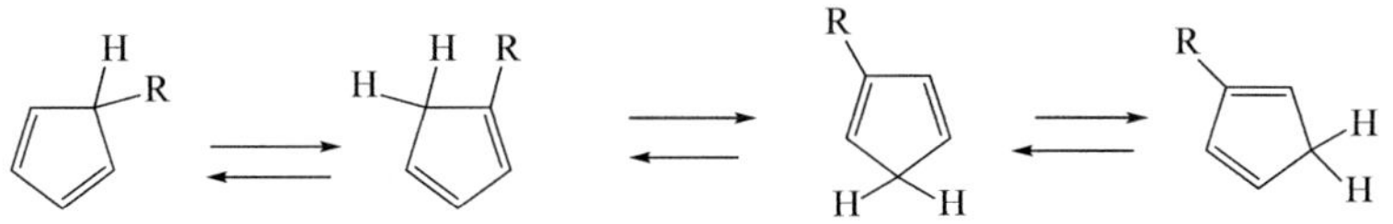

Sigmatropic rearrangements are not the only mechanism by which an alkene unit can move or the (*E*) or (*Z*) stereochemistry can change within a molecule. If an alkene moiety is present in a molecule with an electron-withdrawing group such as a carbonyl, isomerization to a conjugated system is usually facile, occurring spontaneously at ambient temperature or faster upon heating. The isomerization of cyclohex-3-en-1-one to cyclohex-2-en-1-one by exposure to an acid catalyst is one example. The geometry of an alkene that is part of a conjugated system can change with heating. An example is the molecule with a dienyl unit attached to a benzene ring. Initially, this compound was prepared and isolated as a 2:1 (*E*:*Z*) mixture, but spontaneous and complete equilibration to the (*E*)-isomer occurred after standing at room temperature for about 48 hours.[61] Heating a chloroform solution accelerated the equilibration, and complete equilibration occurred in less than 2 hours. Some reagents or compounds can initiate this isomerization, including isomerization to the less thermodynamically stable (*Z*)-isomer. Unsaturated nitriles in the presence of a riboflavin photocatalyst showed an (*E*) → (*Z*) isomerization when air or oxygen was excluded.[62]

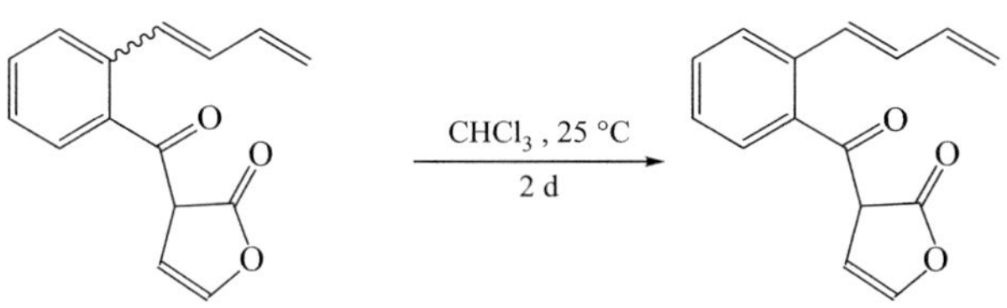

2:1 (*E*:*Z*) mixture 100% (*E*)-isomer

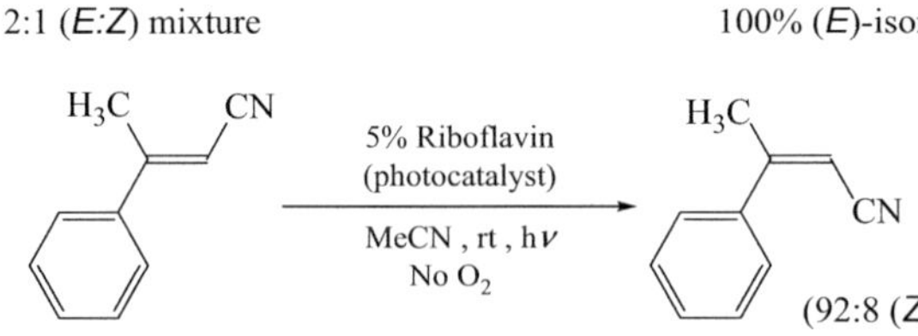

(92:8 (*Z*):(*E*)

[59] (a) Tarbell, D.S. *Org. React.* **1944**, *2*, 1; (b) Mundy, B.P.; Ellerd, M.G.; Favaloro Jr., F.G. *Name Reactions and Reagents in Organic Synthesis*, 2nd ed. Wiley-Interscience, NJ, **2005**, pp. 156–159.

[60] Alder, K.; Ache, H.-J. *Chem. Ber.* **1962**, *95*, 503.

[61] Gaenzler, F.; Guo, C.; Zhang, Y.-W.; Azab, M.E.; Salem, M.A.I.; Fan, D.P.; Smith, M.B. *Tetrahedron* **2009**, *65*, 8781.

[62] Lee, P.S.; Du, W.; Boger, D.L.; Jorgensen, W.L. *J. Org. Chem.* **2004**, *69*, 5448.

Transition metal catalysts can catalyze the isomerization of an alkene unit. A ruthenium complex catalyzed the isomerization of 3-propoxyprop-1-ene to (*E*)-1-propoxyprop-1-ene in 98% yield.[63]

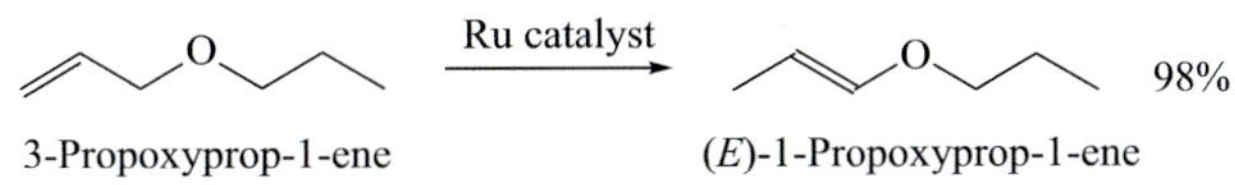

4.8 LIGASES (EC 6)

A *ligase* is an enzyme that catalyzes the coupling of two large molecules by forming a new chemical bond. Hydrolysis of a small pendant group on one of the large molecules usually accompanies the coupling, or the enzyme catalyzing the linking together of two compounds may be hydrolyzed. The ligation of DNA is also catalyzed by a *ligase*. The hydrolysis of a diphosphate bond in ATP or a similar triphosphate is an example. Naming typically takes the form *X:Y ligase (ADP-forming).*

4.8.1 Enzymatic Coupling

A carboxylase is a ligase enzyme that catalyzes the addition of a carboxyl group to a specified substrate. A *biotin carboxylase* (EC 6.3.4.14) is an enzyme that catalyzes the carboxylation of a biotin moiety. In *E. coli*, there are two monomers of the *biotin carboxylase* dimer, which have essentially the same conformation. In the biotin carboxylation reaction, the carboxyl donor is bicarbonate.[64] As shown in Figure 4.31,[64] bicarbonate forms an ATP complex, and loss of ADP (adenosine diphosphate) leads to carboxyphosphate.[64] A proximal amino acid reacts as a base during the catalysis to extract the relevant proton from bicarbonate. Loss of phosphate generates an HCO_2 moiety (coordinated CO_2) that is coordinated to a proximal amino acid. The interaction of biotin with an amino acid in the active site generates a coordinated enolate anion, which reacts with the coordinated CO_2 to give the carboxylated biotin product (carboxybiotin).

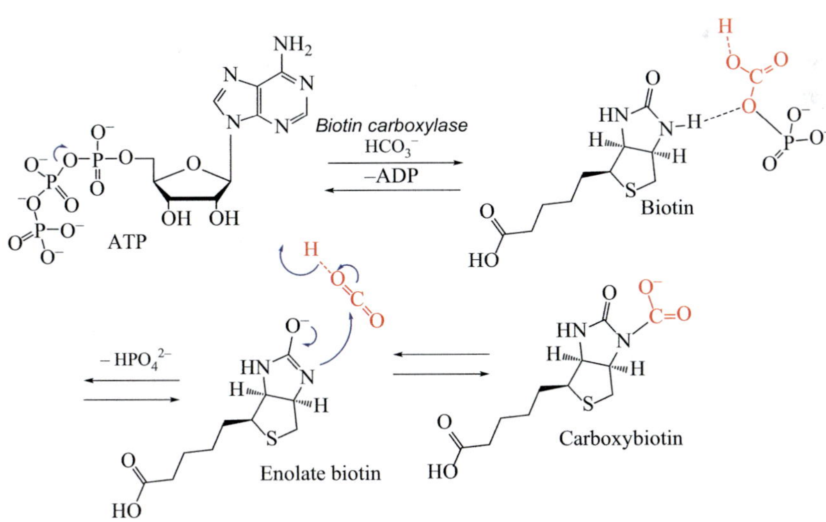

FIGURE 4.31 Proposed mechanism for ATP-dependent carboxylation of biotin with bicarbonate. (Reprinted with permission from Ogita, T.; Knowles, J.R. *Biochemistry* **1988**, *27*, 8028–8033. Copyright 1988 American Chemical Society.)

[63] Larsen, C.R.; Grotjahn, D.B. *J. Amer. Chem. Soc.* **2012**, *134*, 10357.
[64] Ogita, T.; Knowles, J.R. *Biochemistry* **1988**, *27*, 8028.

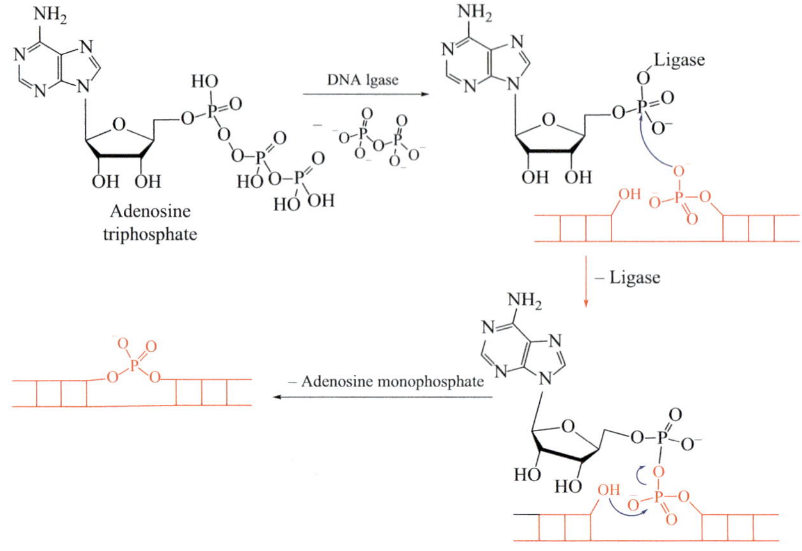

FIGURE 4.32 Repair of a Nick in DNA by *DNA ligase*. (Crut, A.; Nair, P.A.; Koster, D.A.; Shuman, S.; Dekker, N.H. *Proc. Natl. Acad. Sci.* **2008**, *105*, 6894–6899. Copyright (2008) National Academy of Sciences, U.S.A.)

DNA ligase is an enzyme (EC 6.5.1.1) that forms a phosphodiester bond in order to couple DNA strands. An important function of a *DNA liga*se is to repair nicks in a DNA strand by patching the phosphate linkage. *DNA ligases* are essential for genome integrity, and they are essential for DNA replication and repair in all organisms. *DNA ligases* form two families, the ATP-dependent *ligases* and the NAD-dependent *ligases*, and all known eukaryal cellular *DNA ligases* are ATP-dependent. All living organisms, except some bacteria, are eukaryotes.

The DNA ligation process proceeds by three nucleotidyl transfer steps, as shown in Figure 4.32.[65] The initial reaction of adenosine triphosphate (ATP) and DNA ligase results in the release of diphosphate and the formation of a covalent ligase-adenylate intermediate. In the second step, AMP is transferred to the 5′-end of the 5′-phosphate-terminated DNA strand to form a DNA-adenylate. The third step formally addresses the repair of the nick, where the enzyme catalyzes the reaction of the 3′-OH with the 5′-phosphate moiety to join the polynucleotide strands and liberate adenosine monophosphate (AMP).

4.8.2 Synthesis of Polynucleotides and Polynucleosides

Both DNA and RNA are poly(nucleosides) in which any two nucleotide fragments are connected in a 5′-3′ manner by a phosphate linkage (see Chapter 15). The disconnections are straightforward in that the 5′-*O*-phosphate unit or the 3′-*O*-phosphate unit can be disconnected. If the 5′ position has a free OH, then the 3′ unit must contain the phosphate group. Conversely, if the 3′ position has a free OH, then the 5′ unit must contain the phosphate (P indicates the phosphate, not a protecting group).

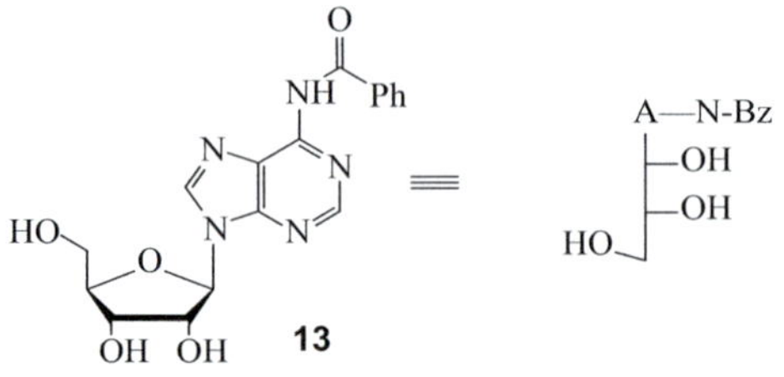

13

[65] Crut, A.; Nair, P.A.; Koster, D.A.; Shuman, S.; Dekker, N.H. *Proc. Natl. Acad. Sci.* **2008**, *105*, 6894.

A shorthand method for describing a ribonucleotide is shown for *N*-benzoyl adenosine-ribofuranose (**13**). The ribofuranose ring is represented by a modified Fischer projection (Section 5.15) that shows the stereochemistry of the three hydroxy groups. At the top of the figure is the *-ase* identifier, in this case A for adenosine, followed by the protecting group on the nitrogen. The ribofuranose units have an OH at the 2′ position, whereas the deoxyribofuranose units do not. The ribofuranose units therefore have one extra OH group that requires protection. Other protecting groups on the nitrogen or on the oxygen atoms can also be indicated.

Conversion of a thiamine deoxyribose nucleotide (**14**) into a phosphate derivative requires reaction with a derivative of phosphoric acid, but phosphoric acid itself cannot be used since there are three reactive oxygen sites. Note the use of $-CH_2-$ in **14** to indicate the 2′-deoxyribose moiety. The phosphoric acid unit must be protected. An early approach is shown for the reaction of **14** with 2-cyanoethyl hydrogen phosphate. In the presence of methanesulfonyl chloride and pyridine, coupling to the protected phosphate gives **15**. The alkoxide unit in **15** can now be coupled to a second nucleotide (the cytosine deoxyribose nucleotide), in the presence of an aromatic sulfonyl chloride, to give a dinucleotide, **16**. Removing protecting groups and adding a new phosphate-activating group allows preparation of the growing polynucleotide. Once the sugar unit, the phosphoric acid, and the heterocyclic base are protected in the proper manner, the two nucleotides must be linked together to begin the synthesis of a polynucleoside. Phosphate protecting groups are removed at the end of the synthesis.

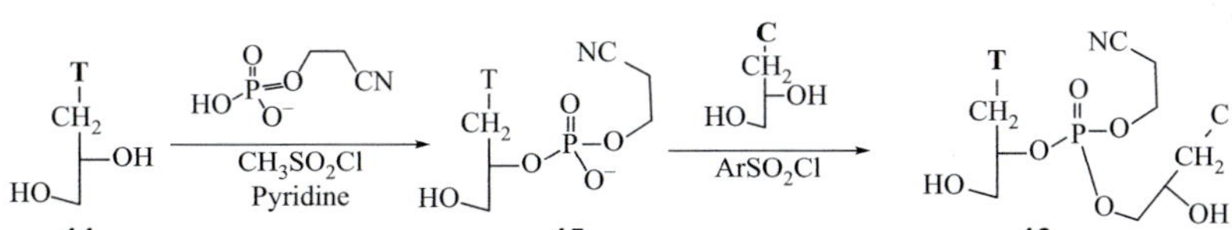

14 **15** **16**

There are several condensing agents that make a new phosphate ester bond. One of the most commonly used compounds to couple alcohols and acids to make an ester, or an amine and an acid to make an amide or a peptide bond, is dicyclohexylcarbodiimide (DCC; $c\text{-}C_6H_{11}\text{-}N{=}C{=}N\text{-}c\text{-}C_6H_{11}$). Alternative reagents used to couple nucleosides include mesitylenesulfonyl chloride (**17**) and 2,4,6-triisopropylbenzenesulfonyl chloride (**18**). These reagents are prepared by reacting either *p*-toluenesulfonyl chloride (**19**) or mesitylenesulfonyl chloride (**17**) with imidazole (**20**). Of the reagents mentioned so far, DCC is the *least* reactive. In some cases, the chloride unit of the sulfonyl chloride is not reactive enough, and that unit must be replaced with something more reactive, such as imidazole. *p*-Toluenesulfonyl imidazolide (**21**) and mesitylenesulfonyl imidazolide (**22**) are used in such cases.

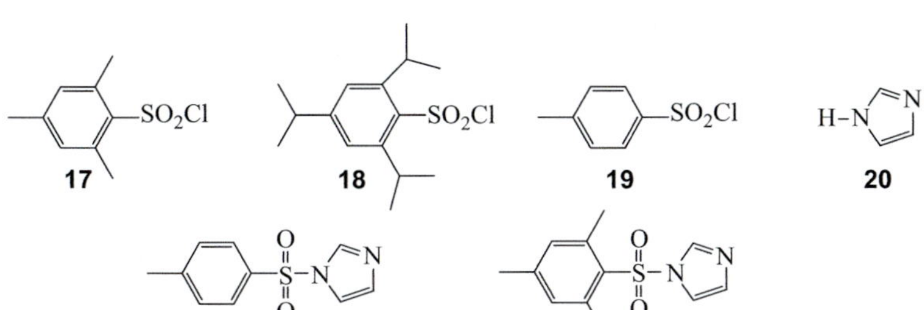

17 **18** **19** **20** **21** **22**

4.8.3 Chemical Methods for Carboxylation and Nucleotide Synthesis

The structure of carbon dioxide is O=C=O, and it is apparent that there is a carbonyl unit. It is reasonable to ask if this carbonyl will react with a nucleophile (e.g., a Grignard reagent). Based on the following experiment, the answer is yes. Chlorocyclopentane was shown to react with Mg metal in ether to give the Grignard reagent, cyclopentylmagnesium chloride. The ether solution containing the Grignard reagent was poured carefully onto crushed dry ice (CO_2), which generated the carboxylate anion, and this slurry was then poured onto crushed ice containing concentrated HCl to give cyclopentanecarboxylic acid. This reaction is reasonably general, although the yields of the reaction vary considerably (from <25% to>80% yield). *This reaction makes a carbon-carbon bond and extends the carbon chain by one carbon.* Organolithium reagents also react with CO_2, analogous to Grignard reagents, and the product is also the carboxylic acid.

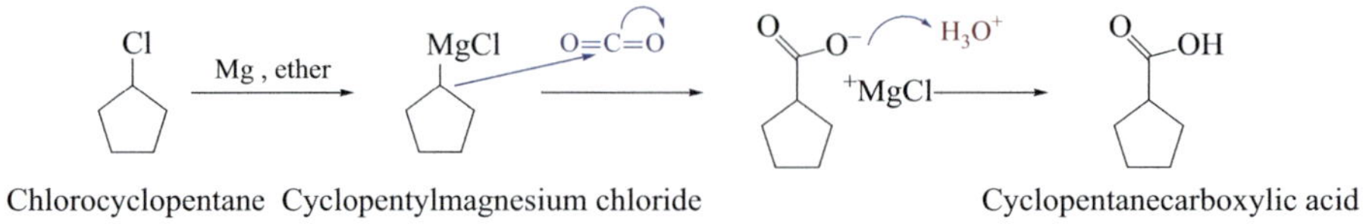

4.9 TRANSLOCASES (EC 7)

4.9.1 Enzymatic Transport Reactions

Cells have a phospholipidic membrane that tends to be insoluble, and fats are generally insoluble. In both cases, lipids are part of the structure, and the long-chain hydrophobic part of the lipid is responsible for the insolubility of these moieties.

Periplasmic enzymes of bacterial cell wall biosynthesis are common drug targets, but drugs that target the membrane-integrated catalysts are rare. The integral membrane protein MraY, which catalyzes the first step in the cytoplasmic pathway of bacterial peptidoglycan biosynthesis, is a potential target. The enzymatic activity of *MraY translocases* (phospho-N-acetylmuramoyl-pentapeptide-transferase; old EC 2.7.8.13) has been examined.[66] The integral membrane protein MraY is involved in the synthesis of lipid II. Lipid II is important for bacterial cell wall biosynthesis, and a translocase enzyme facilitates the translocation of the lipid-linked precursor from the cytoplasmic side, where lipid II is generated, to the outer side of the membrane, into the periplasmic space, which allows further elaboration.[67] The periplasm is a gel-like matrix that is found between the inner cytoplasmic membrane and the bacterial outer membrane. Note that m-DAP is meso-diaminopimelic acid, an amino acid derivative of lysine found in the cell walls of some bacteria.

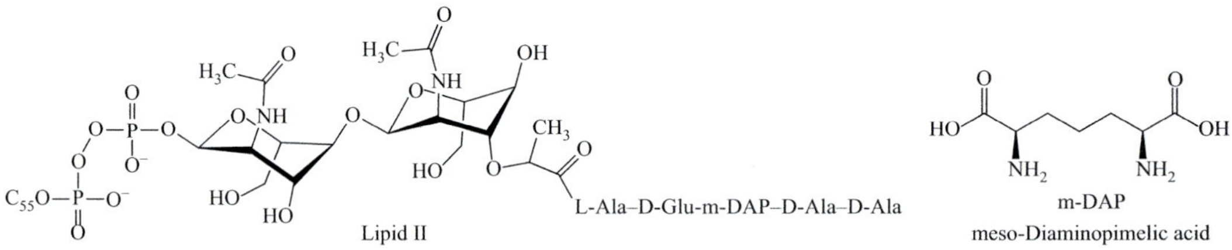

Translocases are enzymes that aid in the transport of molecules, and an *ATP/ADP translocase* has been identified that transports ATP across a lipid bilayer, which is normally impermeable to this molecule.[68] These transport proteins appear to be unique to mitochondria, plant plastids, and obligate intracellular bacteria. The *ATP translocase* is encoded by the bacterial plant pathogen *Candidatus Liberibacter asiaticus*, NttA.

---

[66] Henrich, E.; Ma, Y.; Engels, I.; Münch, D.; Otten, C.; Schneider, T.; Henrichfreise, B.; Sahl, H.-G.; Dötsch, V.; Bernhard, B. *J. Biol. Chem.* **2016**, *291*, 2535.

[67] Scheffers, D.J.; Pinho, M.G. *Microbiol. Mol. Biol. Rev.* **2005**, *69*, 585.

[68] Vahling, C.M.; Duan, Y.; Lin, H. *J. Bacteriol.* **2010**, *192*, 834.

4.9.2 Transport of Organic Materials

While organic chemistry methods are not easily amenable to a correlation with *translocases*, the transport and impact of chemicals in physiological systems are important. Humans have the blood-brain barrier and the cerebrospinal fluid barrier, for example, which are important for drug delivery. The blood-brain barrier separates the circulating blood from the brain and extracellular fluid in the central nervous system. The blood-CSF barrier is formed by the epithelial cells at the choroid plexus. The choroid plexus is a network of nerves or vessels in the body that produces the cerebrospinal fluid in the ventricles of the brain. Methods have been developed to transfer materials past these barriers based on the manipulation of lipophilic and hydrophilic groups.

Polarity and solubility are critical to the design and delivery of drugs used as medicines, an area of chemistry known as medicinal chemistry. The so-called *partition coefficient* (distribution coefficient), Log$P$, is used as a measure of a drug's ability to pass through relatively nonpolar lipid membranes from the highly polar environment of blood serum (mostly water).[69] A negative value for Log$P$ means the compound is more hydrophilic; Log$P$=0 indicates that the compound is equally partitioned between the lipid and aqueous phases; and a positive value for Log$P$ indicates that the compound is more lipophilic. This correlation has been used to predict the activity of potential drugs, but it is valid only when solubility and transport by diffusion through a membrane are important. The partition coefficient P is defined as follows:[70]

$$P = \frac{\text{Drug in the Organic Phase}}{\text{Drug in an Aqueous Phase}}$$

The values of the partition coefficient are usually measured using water or a phosphate buffer at pH 7.4 (the pH of blood) against octan-1-ol.[88] A large value of P is taken as an indication that the compound will diffuse into lipid membranes and fatty tissue, whereas a low value of P indicates that it will not diffuse easily. A large value of P is associated with compounds that are less soluble in water, typically due to a higher percentage of nonpolar organic fragments.

HOMEWORK

4.1. What is the purpose of a substrate binding to the active site of an enzyme?
4.2. What is a zero-order reaction?
4.3. What is a competitive inhibitor?
4.4. Briefly describe why penicillin G is a competitive inhibitor.
4.5. Fundamentally, what information does the Michaelis-Menten equation give about an enzyme-mediated reaction?
4.6. Give the structure of at least two vitamin B compounds that are cofactors in enzyme reactions.
4.7. What are globular proteins?
4.8. What is the purpose of a prosthetic group in an enzyme cofactor?
4.9. What is the EC number of an enzyme?
4.10. What is an oxidoreductase? A transferase? A hydrolase? An isomerase?
4.11. Identify the class of enzyme for each of the following: (a) EC2, (b) EC3, (c) EC5, and (d) EC6.
4.12. The structure of β-sitosterol is shown. Draw the product expected when this sterol reacts with *cholesterol oxidase*.

[69] (a) Leo, A.; Hansch, C.; Elkins, D. *Chem. Rev.* **1971**, *71*, 525; (b) Hansch, C.; Björkrot, J.P.; Leo, A. *J. Pharm. Sci.* **1987**, *76*, 663.

[70] Thomas, G. *Medicinal Chemistry, An Introduction*, John Wiley & Sons, Ltd., Chichester, NY, **2000**, p. 123.

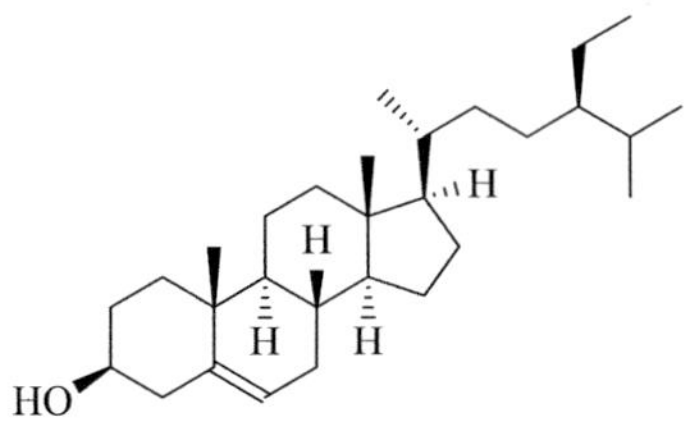

4.13. What is a catalyst?
4.14. In a chemical reaction, how many half-lives are required for the reaction to consume at least 98% of the starting material?
4.15. The S_N1 reaction is first order, and an example is the conversion of 2-methylbutan-2-ol to 2-bromo-2-methylbutane by reaction of the alcohol with KBr in aqueous medium with a catalytic amount of sulfuric acid. The first-order kinetics indicate the reaction rate depends only on the alcohol, yet a reaction with the bromide ion is required to produce the final product, a second chemical step. Explain the fact that this reaction is first order.
4.16. Look up and then draw the structure of the Hantzsch ester. Look up and then draw the structure of the vasodilator nifedipine. The 1,4-reduction of conjugated ketones with the Hantzsch ester proceeds by transfer hydrogenation, with oxidation of the Hantzsch ester. Draw this oxidation product.
4.17. Briefly discuss the similarities of the reaction of Hantzsch esters in question 4.16 with the reactions of NAD+ and NADH.
4.18. Why is the conversion of glycerol 3-phosphate to dihydroxyacetone 3-phosphate considered to be a dehydrogenase reaction?
4.19. Draw the product formed when 2,3,4,5-tetraacetyl-β-D-glucopyranose reacts with dimethyl sulfate.
4.20. What product would be expected if piperlongumine (shown) were to react with *S*-adenosyl-L-methionine? Note that piperlongumine is a constituent of the fruit of the long pepper and may have anti-cancer properties.

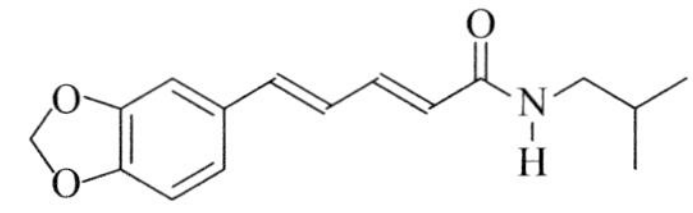

4.21. Benzene reacts with *Pseudomonas putida* to give the chiral diol, as shown. What product is expected when methylbenzene (toluene) is subjected to *P. putida*?

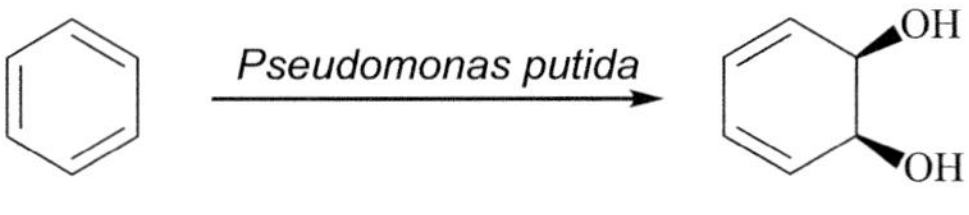

4.22. Referring to Figure 4.12, assume the lysine substrate reacts with *serine transferase*. What is the structure of the initial product formed by reaction with the imine moiety?
4.23. The structure of scopolamine is shown. Scopolamine, used to treat motion sickness and post-operative nausea, also known as hyoscine, is isolated from *Scopolia carniolica*, a poisonous plant belonging to the nightshade family. What are the products expected when scopolamine reacts with the esterase *tropinesterase*?

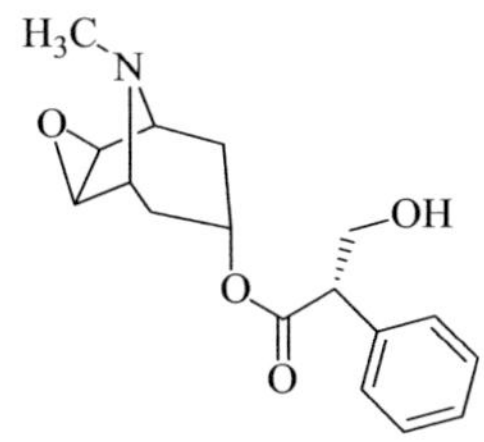

4.24. The ethyl ester of arachidonic acid, which is shown, can lead to pancreatic cell damage due to excessive consumption of alcohol. What is the product formed should this ester react with *monoacylglycerol lipase*?

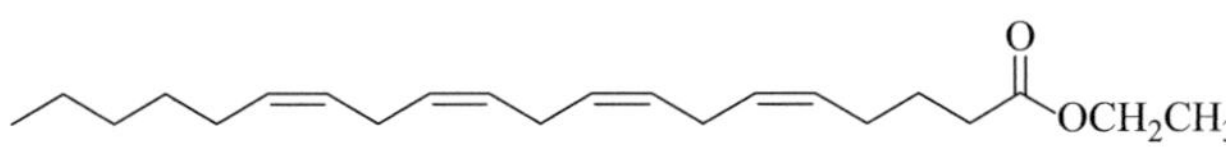

4.25. What products are expected should the ethyl ester of 1-monophosphate of α-D-ribofuranose (shown) react with phospholipase?

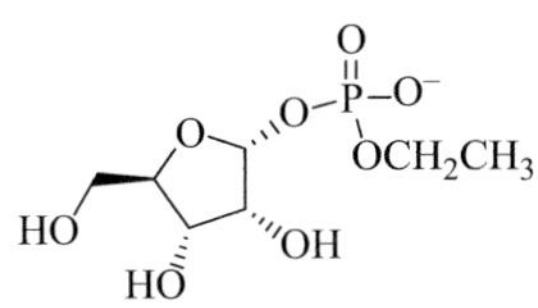

4.26. In Figure 4.21, a [2+2]-cycloaddition that gives the cis,syn cyclobutane thymine dimer T<>T1 was shown. What is the significance of this bioreaction?

4.27. In an enzyme-mediated reaction, fructose diphosphate is converted to a Schiff base by reaction with a lysine residue. Deprotonation by an aspartate residue in the presence of aldolase leads to cleavage and formation of two products. Hydrolysis of the Schiff base leads to two monophosphate products. Draw the Schiff base and the two 3-carbon products.

4.28. The equilibrium for the Cope rearrangement that occurs when 3,4-dimethylhexa-1,5-diene is heated favors the equilibrium product, octa-2,6-diene. Briefly explain why this diene is favored.

4.29. A Grignard reagent (RMgX; see Section 12.1) is formed by the reaction of an alkyl halide such as 1-bromo-2-phenylpropane with magnesium metal. When this organometallic reagent reacts with carbon dioxide and then with aqueous acid, what is the product? Also, draw the Grignard reagent that is formed.

5 A Review of Organic Chemistry

It is likely that this book will be used after, or concurrently with, a course in organic chemistry. This chapter is intended as a review of many fundamental principles of organic chemistry. It is not intended as a stand-alone treatment. The intent is to bring the reader 'up to speed' with important principles of organic chemistry that are important for the extension of those principles to biochemistry.

Arguably, the most fundamental concept in organic chemistry is the nature of the bond between two carbon atoms or between carbon and another atom. Bonding is an important concept in organic chemistry because chemical reactions involve the transfer of electrons through the making and breaking of chemical bonds. Since the molecules associated with biochemical processes are mostly organic molecules, the bonding will be similar to that of organic molecules commonly discussed in a sophomore organic chemistry course. For the most part, the bonds between carbon and another atom are covalent, so the initial focus will be on covalent bonds between two carbon atoms or covalent bonds between carbon and a different atom. The definitions of covalent bonds and polarized covalent bonds will be reviewed, as well as the concept of functional groups. The concepts of isomers, different connectivity within organic molecules, and rules for naming organic molecules will also be reviewed.

To lay the groundwork for understanding bioorganic molecules, this chapter will review relatively simple molecules, including those that have π-bonds to carbon atoms. Compounds that have carbon bonded to heteroatoms (e.g., oxygen and nitrogen) may have both σ- and π-bonds. Functional groups with both types of bonding will be reviewed. The rules of nomenclature will be extended to accommodate each new functional group and will be briefly reviewed. Relatively simple physical properties associated with polarized covalent bonds and π-bonds, especially hydrogen bonding, will be reviewed. Finally, chirality, stereochemistry, and stereoisomers will be reviewed.

5.1 BONDING AND ORBITALS

Elemental atoms are discrete entities that differ from one another by the number of protons, neutrons, and electrons that make up each atom. The motion of electrons with respect to the nucleus has some characteristics of a wave, which is expressed as a wave equation, and a solution to this equation is called a *wavefunction*. Each electron may be described by a wavefunction whose magnitude varies from point to point in space. The wavefunction is described by $\psi(x,y,z)$ by using Cartesian coordinates to define a point, which describes the position of the electron in space. *Wavefunction solutions are correlated with the space volume pictorial representations,* which are charge clouds called *orbitals*. As stated by the *Heisenberg uncertainty principle*, the position and momentum of an electron cannot be simultaneously specified, so it *is only possible to determine the probability that an electron will be found at a particular point relative to the nucleus.* Wavefunction solutions can be correlated with the position of an electron relative to the nucleus for an electron, which leads to the familiar s-, p-, and d-orbitals, as shown in Figure 5.1. The *s-orbital* is spherical, *a p-orbital* has a "dumbbell" shape, and *there are three 2p-orbitals that are degenerate (they have* the same energy).

Electrons associated with an atom are assumed to reside in *atomic orbitals*. The valence electrons are used for bonding between atoms in a molecule. The *valence electrons* are those found in the outermost orbitals (those furthest from the nucleus), and they are more weakly bound than electrons in orbitals closer to the nucleus. The number of valence electrons for an atom is calculated by subtracting the last digit of the Group number from 8. Since carbon is in Group 14, it has four valence electrons (8-4). Similarly, there are three valence electrons for N (8-5), two for O (8-6), and

DOI: 10.1201/9781003605157-5

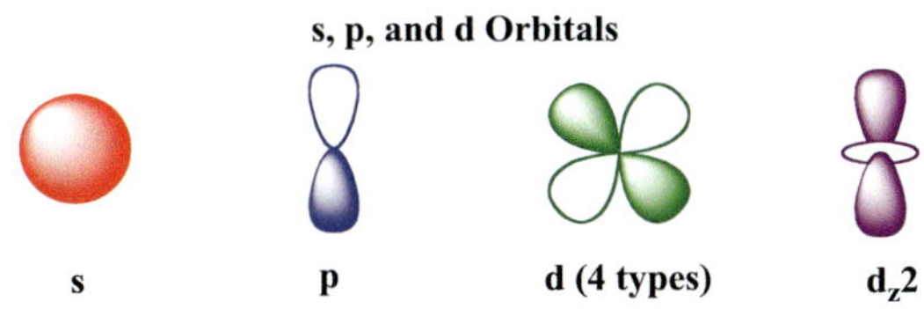

FIGURE 5.1 Common representations of the s- and p-orbitals, along with two d-orbitals.

one valence electron for F (8-7). Note that valence for an atom is the number of bonds a molecule may form using its valence electrons. Therefore, the valence of carbon is 4, that of nitrogen is 3, that of oxygen is 2, and that of fluorine is 1.

The so-called covalent chemical bond between two atoms (see Section 5.2) involves the sharing of electrons. The position of electrons in an atomic orbital of an element such as C, N, O, or F can be contrasted with the electrons in a bond between two atoms in a molecule, which are assumed to reside in *molecular orbitals*. It is important to emphasize that the positions of the atomic orbitals relative to the nucleus have different energies for the electrons found in molecular orbitals.

5.2 IONIC VERSUS COVALENT CHEMICAL BONDS

The skeleton of most molecules discussed in this book is made up of carbon, oxygen, nitrogen, phosphorus, or sulfur atoms, held together by chemical bonds between carbon and carbon atoms as well as bonds of carbon to the other atoms. Two major types of bonds will be considered for these molecules. An *ionic bond* is formed when one atom in a bond has two electrons and takes on a negative charge, while the other is electron-deficient and takes on a positive charge. A *covalent bond* is formed by the mutual sharing of electrons between two atoms, and the bond holds the atoms together.

An ionic bond holds two atoms or groups together by the electrostatic attraction of positively and negatively charged ions. Most ionic bonds in this book will be the salts of monovalent metals such as Na^+, K^+, Li^+, or a divalent metal cation such as Mg^{2+}, and the anion is a halide or the salt of a relatively strong acid: carboxylic acids, sulfonic acids, phosphoric acids, or the salt of a weak acid such as an alcohol. However, ionic bonds that involve the conjugate acid of organic bases, such as the ammonium salts formed from amines or the phosphonium salts from phosphines, are common.

When one carbon shares electrons with another carbon, a hydrogen atom, or another atom, it is a *covalent bond*, and the most common type of bonding is known as a σ-bond. There are two electrons in a σ-bond, a *single covalent bond*, or just a *single bond* between the two atoms. In other words, a covalent bond is the mutual sharing of two electrons between two atoms. The electron density of each atom is shared with the other in a covalent bond and is not localized on an individual atom. Indeed, the greatest concentration of electron density is between the nuclei. The strength of a covalent bond is related to the amount of electron density concentrated between the nuclei. When two identical atoms share electrons in a covalent bond, most of the electron density is equally distributed between the two nuclei (in the "space" between the two atoms), which leads to the strongest type of covalent bond.

If one examines the molecule methane, measurements show that all four C–H bonds are identical, and that the bond angles of each H–C–H unit are 109° 28′, the angles of a regular tetrahedron. Specifically, the four hydrogen atoms are distributed around carbon in the *shape of a regular tetrahedron*. The electrons in the C–H bonds of methane are in molecular orbitals called sp^3 hybrid orbitals. Since there are four bonds to carbon, there are four sp^3-hybrid orbitals corresponding to the four covalent bonds.

The tetrahedral array of covalent bonds about carbon, as just described, means that a three-dimensional shape is associated with each. The *Valence Shell Electron Pair Repulsion (VSEPR) model* is a useful place to begin thinking about the three-dimensional nature of organic molecules.

To use this model, a tetrahedron is imagined with C, O, or N at the center of that tetrahedron. The atoms are attached to the corners of the tetrahedron, and any unshared electrons are also attached. Examining only the atoms, carbon has a valence of four, and molecules will be tetrahedral about each carbon; nitrogen has a valence of three and will form three covalent bonds with an unshared pair of electrons that gives the molecule a pyramidal shape. Oxygen has a valence of two, with two unshared electrons, so the molecule will have an angular or bent shape. Remember, however, that this model does a *poor* job of accurately predicting bond lengths and angles. Indeed, this model *underestimates* the importance of electron pairs and does not take the *size* of the atoms or groups attached to the central atom into account.

5.3 BREAKING COVALENT BONDS

Reactions described in this book will involve making or breaking chemical bonds, which are electron transfer processes. The strength of a covalent bond is directly related to the electron density between the atoms, and that strength is usually reported as a bond energy, $H°$, as described below. Breaking a bond liberates the amount of energy that is required to keep the atoms together. This amount of energy is considered to be "stored" in a bond, and it is "released" via homolytic bond cleavage. For a covalent bond X–Y, there are two ways to break that bond. In one, another atom collides with X or Y, and energy is transferred that leads to cleavage of the covalent bond between the two atoms, with both electrons in the bond transferred to one atom. In the example in Figure 5.2 two electrons are transferred to Y as the bond breaks, generating a cation (X^+) and an anion (X^-). Breaking a bond in this manner is called *heterolytic bond cleavage*, which is a chemical process called a *chemical reaction*. Note the use of the *curved double-headed arrow* to indicate the transfer of two electrons as the bond breaks.

The alternative bond cleavage shown in Figure 5.2 breaks the covalent bond with the transfer of one electron to each of the atoms, generating a *radical* (a species with one extra electron). This process is another type of chemical reaction, and it is known as *homolytic bond cleavage*. In this case, homolytic cleavage of X–Y leads to two radicals, $X^{\bullet} + Y^{\bullet}$. Note the use of the *curved single-headed arrows* to indicate the transfer of electrons as the bond breaks.

When a bond is broken or formed, the energy required is known as the *bond dissociation enthalpy* ($D°$, or more commonly $H°$, for a bond broken or formed in a reaction). It is also called *bond dissociation energy*, and for convenience, all values listed for $H°$ will be used for both heterolytic and homolytic bond cleavages. *The bond strength of ionic bonds will be ignored for this discussion*, so only the bond dissociation energy of bonds in inorganic molecules will be considered.

If two atoms could be brought together directly to form a bond, the energy required to form that bond is the same as that required to break the bond. In organic chemistry, two elements rarely come together in a direct manner to form a bond. In most cases, two structures that contain several atoms (molecules) are involved in a chemical reaction (a bond-making and bond-breaking process), as illustrated in Figure 5.3. Much less energy is required for bond cleavage via a chemical reaction than is required for the spontaneous separation of the two atoms of the bond.

Covalent bonds are found in millions of organic molecules that participate in a multitude of organic reactions. If a covalent bond is broken in one molecule, a new covalent bond is usually formed in a new molecule. Such transformations are *chemical reactions*. In principle, each molecule will require a slightly different amount of energy for the cleavage or formation of a particular bond.

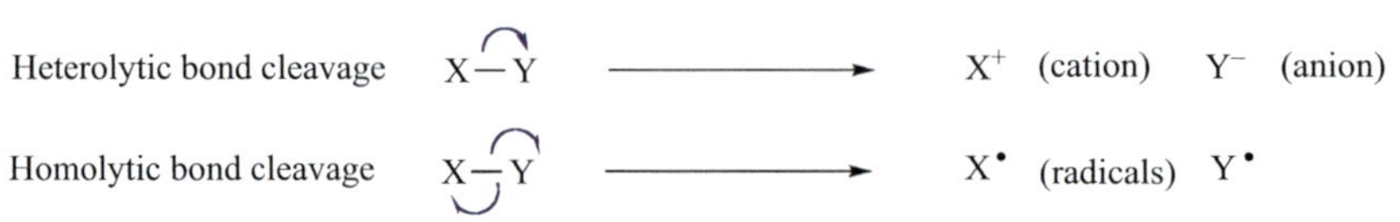

FIGURE 5.2 Heterolytic and homolytic bond cleavage.

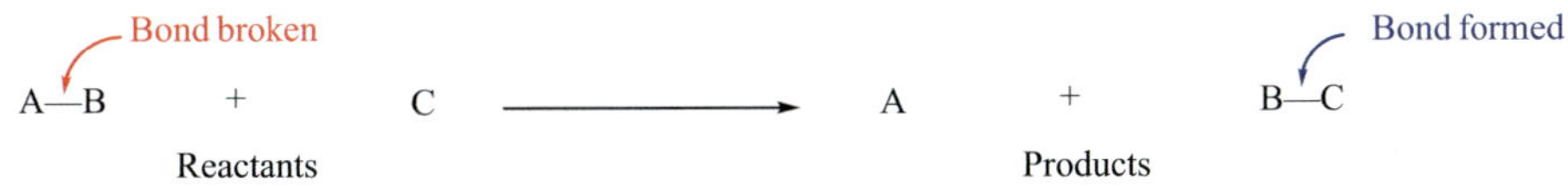

FIGURE 5.3 Fundamental components of a chemical reaction.

The molecules in which the bonds are broken are called *reactants*, whereas the molecules in which bonds are formed are called *products*. A general energy formula can be given for a reaction where reactants (A–B and C) are converted to products (A and B–C), as represented in Figure 5.3. A certain amount of energy is required to break the A–B bond (called $H°_{reactants}$) and a certain amount of energy is required to form the B–C bond (called $H°_{products}$). In this process, there is a *change* in bond dissociation energy, represented by $\Delta H°$, where the symbol Δ represents *change in*. This value is determined by subtracting the bond dissociation energy for the products ($H°_{BC}$) from the bond dissociation energy for the reactants ($H°_{AB}$). This equation leads to a general formula for $\Delta H°$.

$$\Delta H° = H°\text{products} - H°\text{reactants and for the reaction in Figure 5.3} = H°_{BC} - H°_{AB}$$

The value $H°$ is the amount of energy released when that bond is broken or the amount of energy required to form that bond. If there are no other factors, the negative sign of $\Delta H°$ indicates that this reaction is *exothermic*. In other words, this reaction is expected to produce more energy than it consumes, based solely on bond dissociation energies. If $\Delta H°$ is positive, the reaction will produce less energy than it consumes, and the reaction is *endothermic*. Note that ionic bonds have been ignored in this calculation. It is important to note that *the relative ability to break a bond* when it is involved in a chemical reaction, which is an electron-transfer process with other atoms or groups of atoms, *is not easily predicted using bond dissociation energy*. Bond cleavage depends on several factors, and it is difficult to make general comments about the relative strength of an ionic versus a covalent bond.

5.4 POLARIZED COVALENT α-BONDS

When certain atoms are collected into discrete units, they often have special physical and/or chemical properties. Such units are called *functional groups*. If the molecule is composed only of carbon and hydrogen atoms, the molecule is a *hydrocarbon*. If not all the atoms are sp^3 hybridized in the hydrocarbon, there are two possible alternatives for the carbon atoms that give two functional groups. Molecules can have one or more carbon-carbon double bonds, or they can have one or more carbon-carbon triple bonds (Section 5.9). It is also possible to have both carbon-carbon double bonds and carbon-carbon triple bonds in the same molecule. If the array of atoms includes atoms other than carbon or hydrogen (these atoms are called *heteroatoms*), several functional groups are possible, and those will be introduced later in Section 5.10.

Electronegativity will play a role in covalent bonding when one atom is bonded to a different atom with a different electronegativity. When the two atoms of the bond are identical, they have the same electronegativity, so the electron density is equally and symmetrically distributed between both nuclei. If a bond is formed between two atoms that have different electronegativities, the electron density is *not* equally distributed between the nuclei but is distorted *toward* the more electronegative atom. The net result of this electron distortion is that one atom has more electron density than the other, making one atom more negative and the other more positive. Such a bond is said to be *polarized*; *a polarized covalent bond*, is represented in Figure 5.4. The polarized bond can be represented as (+)-------(–) or by a specialized arrow (+→), where the + part of the arrow is on the more electropositive atom and the arrow (→) points to the more electronegative atom. Since

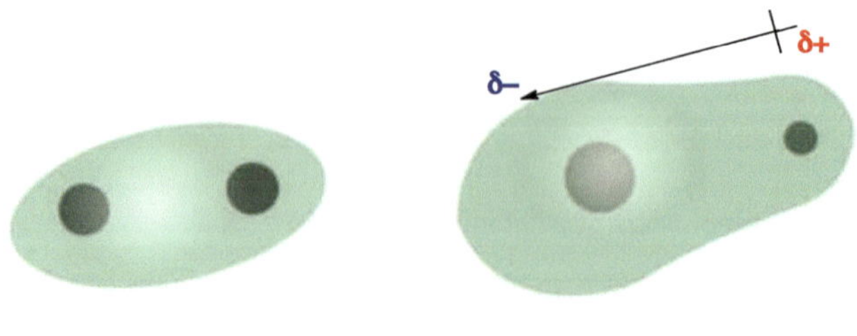

FIGURE 5.4 Nonpolar and polar covalent bonds.

these are covalent and not ionic species, *the (+) and (−) do not indicate charges, but rather bond polarization.* A molecule containing a polarized bond is usually said to be *polar.* A molecule that does not have a polarized bond is categorized as *nonpolar.*

5.5 REACTIVE INTERMEDIATES

Some reactions involve more than one sequential step. The first chemical reaction, or step, may give a *product that is unstable under the reaction conditions,* and it undergoes a subsequent chemical reaction to yield a more stable product that may be isolated. In other words, the reaction of the starting materials does not yield the final product directly; rather, a transient product is formed prior to the final product, which requires a subsequent chemical step. *Such transient products are known as intermediates,* illustrated by the generic reaction in Figure 5.5. Starting material **A** reacts with the other starting material **B** to yield a product **C**, which cannot be isolated from this reaction. When the reaction is complete, it is clear that **C** has reacted with additional **B** to yield an isolated product, **D**. In this overly generalized example, **C** is a transient species that is not isolated (an *intermediate*), but it is so high in energy (unstable) that, in this case, it reacts with **B** before it can be isolated. A *transient* and *relatively high-energy product* such as [C] is an *intermediate*, where the brackets in this case indicate a transient species. An intermediate is therefore a transient reaction product that is not isolated but reacts to give another, more stable product.

For the most part, three types of carbon intermediates will be discussed in this book. A carbocation can be viewed as an "empty" p-orbital that is localized on a carbon atom. A carbanion can be viewed as a "filled" p-orbital that is localized on a carbon atom. A carbon radical is a molecule with an orbital on a carbon atom that has only one electron.

A carbocation is formed when a covalent bond to carbon is broken in such a way that two electrons are transferred to one atom and the carbon receives no electrons during the transfer (*heterolytic cleavage*). The central carbon atom of a *carbocation* is clearly electron-deficient, with a formal charge of +1, and it is essentially an "empty" p-orbital (see **1′** in Figure 5.6). A carbon atom that bears a positive charge is an intermediate called a *carbocation* (also known as a *carbenium ion*).

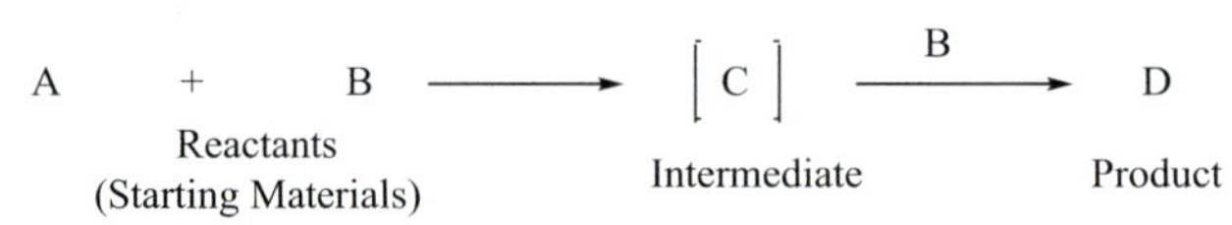

FIGURE 5.5 Formation of an intermediate in a chemical reaction.

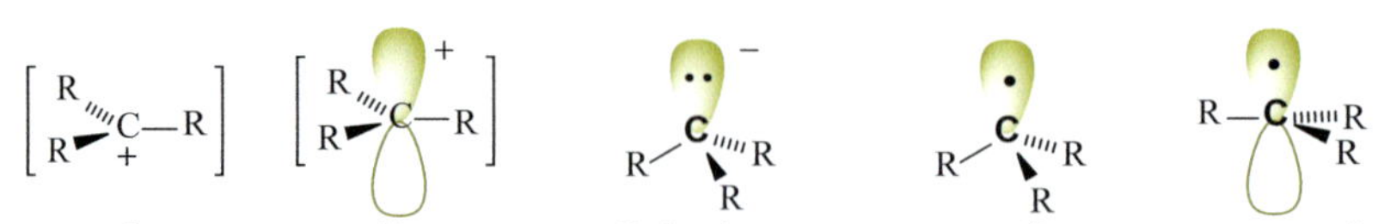

1 **1′** Carbanion **2** **3**

FIGURE 5.6 The structure of a carbocation.

Carbocations are electron deficient and will react with electron-donating species, *nucleophiles*. The positively charged carbocation has only three covalent bonds, is high in energy, unstable, highly reactive, and difficult to isolate in most cases. *The carbon of a carbocation is sp^2 hybridized and must have trigonal planar geometry* (see **1** in Figure 5.6).

A carbon cation is an electron-deficient species that has a formal charge of +1 and is attracted to, and reacts with, a species that can donate two electrons to give it eight, satisfy the valence requirements of carbon, and form the fourth bond to make carbon tetravalent. An example of such an electron-donating species is an anion with a charge of –1 or a molecule that has a heteroatom with unshared electrons, such as N or O. If the electron-donating species reacts with a carbon, that electron-donating species is called a *nucleophile*. A nucleophile is defined as an electron-rich species that donates two electrons to carbon.

An anion is a species that has an excess of electrons and bears a formal charge of –1. When the negative charge resides on carbon, this intermediate is called a *carbanion*, as shown in Figure 5.6. In general, carbanions are formed by breaking a covalent bond in such a way that two electrons are transferred to the carbon involved in that bond, and the second atom receives no electrons during the transfer (a heterolytic cleavage). A generic carbanion is shown with three covalent bonds between C and R, and *a pair of electrons in a p-orbital*. A carbanion is a high-energy intermediate, unstable and very reactive, and it will readily react with a *carbocation, a radical, or another carbanion*. A carbocation is effectively an "empty" p-orbital localized on an electron-deficient carbon atom. Carbanions are *nucleophiles*. However, if a carbanion reacts with the acidic proton of a Brønsted-Lowry acid, it is classified as a base.

When a p-orbital on any given atom has only one unshared electron, the intermediate is called a *radical*, and a carbon radical is represented as R_3C (see **2** and **3** in Figure 5.6). With three covalent bonds and one extra electron, $R_3C^{\bullet}$ is a high-energy species and a very reactive intermediate. The single electron in the orbital will slightly repel the electrons in the covalent bonds, so one might expect a squashed tetrahedron (pyramidal) shape (**2**). There is evidence, however, that a planar structure (**3**) is probably the low-energy structure rather than the pyramidal structure, at least for the methyl radical ($H_3C^{\bullet}$).

One way to form a carbon radical is by a chemical reaction between a neutral species such as methane and an existing radical (e.g., Br). In this reaction, the bromine radical donates a single electron (note the single-headed arrow, much like a fishhook) to one hydrogen atom of methane, which donates one electron from the covalent C–H bond. When this electron transfer occurs, a new H–Br bond is formed and the C–H bond is broken, with the transfer of one electron to carbon to form the methyl radical. Note that there are two electrons in the H–Br covalent bond, *one derived from the bromine radical and one from the broken C–H bond of methane*. Radicals can be formed by breaking a covalent bond in such a way that one electron is transferred to each of the two atoms involved in that bond (called *homolytic cleavage*).

Methane Bromine radical Methyl radical

5.6 ALKANES AND ISOMERS

Molecules that contain only carbon and hydrogen atoms are known as *hydrocarbons*. Since carbon forms bonds to other carbon atoms, molecules can be formed that contain chains, branches, or rings of carbon atoms. A molecule in which all carbon atoms are sp^3 hybridized, contains only single covalent bonds, and contains *only* carbon and hydrogen atoms is known as an *alkane*. When such compounds are named (see below), the suffix *-ane* in the name identifies the compound as an alk*ane*. The *general formula for alkanes* is:

$$C_nH_{2n+2}$$

where n is an integer in the series: 1,2,3,4,...

When $n=1$, the alkane formula is CH_4; when $n=3$, it is C_3H_8; when $n=100$, it is $C_{100}H_{202}$, and so on. In an alkane, there can never be more hydrogen atoms than the number obtained with this formula. There may be fewer, but never more. When generating linear carbon chains of different lengths, many different compounds can be drawn that have different empirical formulas. Different structures that have the same empirical formula are known as *isomers*. The definition of *an isomer is a molecule with the same empirical formula as another, but with a different structure.* In other words, isomers have the same formula but different connectivity; the atoms are connected in different ways. When forming alkanes with different numbers of attached carbon branches on a linear chain, many molecules with the same empirical formula can be drawn. Figure 5.7 shows seven different molecules with the empirical formula C_7H_{16}, each with a different structure. The connectivity of the atoms is different, and each is a unique molecule with unique physical properties. All of these molecules are isomers of each other.

A simple protocol is possible that allows one to draw many isomers for a given empirical formula:

1. Draw the structure with the longest possible linear chain for a given formula.
2. Remove one carbon from the chain and draw the structure with the longest possible linear chain.
3. Attach the single carbon that was removed to the new chain at as many different positions as possible.
4. Remove two carbons and draw the structure with the longest possible linear chain
5. Attach each of the individual carbons to the new chain in as many different combinations as possible.
6. Attach a two-carbon unit to the new chain in as many different ways as possible.
7. Repeat this protocol one carbon at a time, attaching all remaining carbon atoms in as many different combinations as possible.
8. Check for redundant structures.

The last item in the protocol checks for redundant structures. Indeed, following this protocol for a formula with > 4–5 carbon atoms will generate structures that are identical to some previously generated structures. This problem can be examined using Figure 5.7, where there are only *nine different structures*, although 13 structures are shown based on the protocol given. An analysis shows that compound **A** has a linear chain of seven carbon atoms, compounds **B** to **C** have a linear chain of six, compounds **D** to **H** have a linear chain of five carbon atoms, and **I** has a linear chain of four

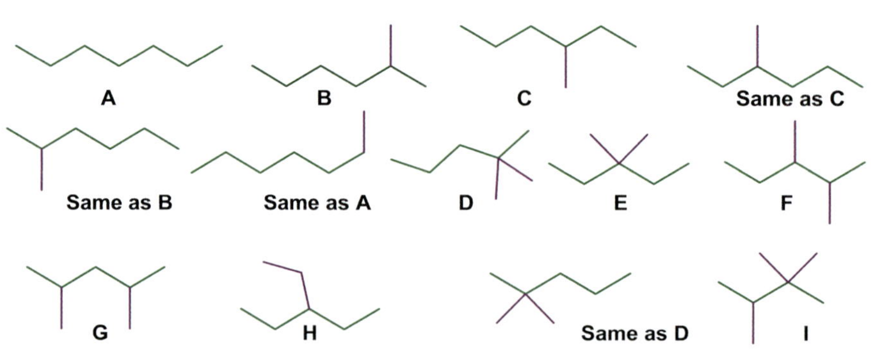

FIGURE 5.7 Structural variation in alkanes with a total of seven carbon atoms.

carbon atoms. Compounds **B** to **I** are *branched* because carbon groups are attached to the linear chain. The isomers identified in Figure 5.7 have different constitutions and different connectivity, so they are known as *constitutional isomers* (also called *structural isomers*).

5.7 THE IUPAC RULES OF NOMENCLATURE

There are a vast number of possible alkanes, certainly several million, and each unique structure requires a unique name. The nomenclature system used today is based on the number of carbon atoms in straight-chain alkanes, and it distinguishes each type of molecule or special collection of atoms known as a functional group. *Functional groups are collections of atoms that have unique physical and chemical properties.* To accommodate the structure variations, a set of "rules" has been devised that are universally used to name organic molecules. The organization that supervises these rules is the *International Union of Pure and Applied Chemistry*, I.U.P.A.C. or just IUPAC. The rules identify the number of carbon atoms in the longest continuous chain by a prefix and identify the functional group or class, or molecule by a suffix. For alkanes, the suffix is *-ane*.

Other carbon atoms or other atoms or groups can be attached to the longest continuous chain of carbon atoms, and such atoms or groups are called *substituents*. A hydrocarbon substituent that has sp^3 hybridized carbon atoms is known as an *alkyl group* or an *alkyl substituent*. If a group or atom is attached to a particular carbon, each carbon atom in the linear chain of an alkane will receive a number, and the position of each substituent is assigned the lowest possible number on that chain as part of the name.

The system for naming organic molecules begins with the first 20 straight-chain alkanes, C_1 to C_{20}.[1] The class of molecule is identified with a unique *suffix*, and for alkanes, that suffix is *-ane*. A *prefix* is added to indicate how many carbons are in the linear alkane chain. A one-carbon unit has the prefix **meth-**; two carbons are **eth-**; three carbons are **prop-**; four carbons are **but-**; and five, six, seven, eight, nine, and ten are derived from the terms: **pent-**, **hex-**; **hept-**, **oct-**, **non-**, **dec-**. These linear alkanes are followed by the equivalents of 1+10, 2+10, 3+10, and so on. The prefixes are **undec-** (11), **dodec-** (12), **tridec-** (13), **tetradec-** (14), **pentadec-** (15), **hexadec-** (16), **heptadec-** (17), **octadec-** (18), **nonadec-** (19), and **icos-** (20). By this system, a 12-carbon straight-chain alkane is called *dodecane*, the 7-carbon straight-chain alkane is *heptane*, and the 20-carbon straight-chain alkane is called *icosane*. This nomenclature system uses the so-called IUPAC selection rules[1]:

1. Determine the longest continuous chain of carbon atoms that contains the functional group of highest priority, and assign the proper prefix to indicate the number of carbon atoms.
2. Determine the class of compounds to which the molecule belongs and assign the proper suffix. For straight-chain, saturated hydrocarbons, the class name is alkane and the suffix is -ane.
3. For alkanes, determine the longest straight chain present, using a prefix for the number of carbons as in Table 5.1, and the suffix -ane.
4. Alkanes that have carbon groups attached to the longest unbranched chain (called substituents) are known as branched-chain alkanes. When that branch is an alkane fragment, it is known as an alkyl group or an alkyl substituent.

Branched-chain alkanes have a substituent or multiple substituents attached to the longest linear chain. If there is only one substituent, such molecules are named as follows:

---

[1] Flectcher, J.H.; Dermer, O.C.; Fox, R.B. *Nomenclature of Organic Compounds. Principles and Practice*, American Chemical Society, Washington, DC, **1974**, pp. 6–11.

1. Name the longest straight chain present in the compound (known as the parent chain).
2. Name the group constituting the branch (known as a substituent or substituting group) based on the number of carbon atoms.
3. The name of the side chain *precedes* the name of the parent chain.
4. For an alkane-based substituent, the -ane is dropped and replaced with -yl to indicate an alkyl substituent.
5. The nomenclature system for simple alkyl substituents is shown in Table 5.1.

For naming molecules that have other atoms or groups attached to the longest chain (substituents), the focus is on the carbon atom that serves as the point of attachment for the substituent to the main chain. The prefix for a substituent is the same as that used for naming the longest continuous chain, meth→icos. The suffix for the *alkyl substituent* is *-yl* and is combined with the prefix for the number of carbon atoms. A one-carbon substituent is methyl, a two–carbon substituent is ethyl, a three–carbon substituent is propyl, a four–carbon substituent is butyl, and so on.

The nomenclature rule states that the numbering sequence is based on the nearest locant, which is the carbon bearing the branch. In other words, the locant is the carbon bearing the substituent. *Number the longest linear chain from one end to the other using Arabic numerals and assign the lowest number to the substituent (the nearest locant).*

Common names are also listed in Table 5.1 along with the IUPAC name. One should always use the IUPAC nomenclature, but some common names appear so often that they must be recognized. The common names for straight-chain fragments (e.g., ethyl, propyl, butyl, pentyl, etc.) are straightforward. *The IUPAC rules indicate that common names should only be used for the parent alkane, not for substituents.* Substituents are named according to the IUPAC rules listed above.

Of course, hydrocarbon units are not the only type of substituent. *Halogens are not functional groups,* which means there is no suffix to indicate the presence of the halogen. When a fluorine, chlorine, bromine, or iodine substituent is attached to a carbon chain, the molecule is named just like alkyl groups. A halogen substituent is named by dropping the -ine ending of each halogen and replacing it with *-o* (i.e., fluoro, chloro, bromo, iodo).

The number assigned to a substituent on the longest continuous chain is established by determining the nearest locant that gives the lowest set of numbers, as described above. *When there is more than one identical substituent, each is given a number. The lowest sequence of numbers is assigned based on the locant closest to the end of the chain.* The rule states that another prefix

TABLE 5.1
Common Alkyl Substituents[a]

Substituent	IUPAC Name	Common Name (Abbreviation)
$\mathbf{C}H_3$–	Methyl	Methyl (Me)
$CH_3\mathbf{C}H_2$–	Ethyl	Ethyl (Et)
$CH_3CH_2\mathbf{C}H_2$–	1-Propyl	*n*-Propyl (*n*-Pr)
$CH_3CH_2CH_2\mathbf{C}H_2$–	1-Butyl	*n*-Butyl (*n*-Bu)
$(CH_3)_2\mathbf{C}H$–	2-Methylethyl	Isopropyl (iPr)
$(CH_3)_2CH\mathbf{C}H_2$–	2-Methylpropyl	Isobutyl
CH_3CH_2–$\mathbf{C}H\ CH_3$–	1-Methylpropyl	*secondary*-Butyl; *sec*-butyl (*s*-Bu)
$CH_3CH_2CH_2$–$\mathbf{C}H\ CH_3$–	1-Methylbutyl	*sec*-Pentyl
$(CH_3)_3\mathbf{C}$–	1,1-Dimethylethyl	*tertiary*-Butyl; *tert*-butyl (*t*-Bu)
$(CH_3)_3C\mathbf{C}H_2$–	2,2-Dimethylpropyl	Isoamyl (isopentyl)

[a] In each structure, the highlighted carbon (C) is the point of attachment to the longest continuous chain.

is used with the substituent: *di-* for two, *tri-* for three, *tetra-* for four, *penta-* for five, and *hexa-* for six identical substituents. First, determine the name of the substituent, and then insert the multiplying prefix. When there are multiple substituents, numbers are assigned based on the closest locant, as in all other cases. All substituents must have a number, even if they appear on the same carbon atom.

When there is more than one substituent and those substituents are different, the names for each are arranged alphabetically with their appropriate numbers. Using this rule, *i* comes before *m*, so an example is 1-iodo-3-methylheptane and *not* 3-methyl-1-iodoheptane. Similarly, in 5-bromo-2-chlorohexadecane, *b* comes before *c*, so bromine comes before chlorine, even though the chlorine atom has the smaller number.

> 6. In branched hydrocarbons having more than one identical substituent, name each substituent using rule 4 and Table 5.1, and assign an Arabic number for the position of each substituent on the longest unbranched chain.

There are structures in which the longest unbranched chain has a substituent, but the substituent also has one or more substituents. In other words, the branch has branches.

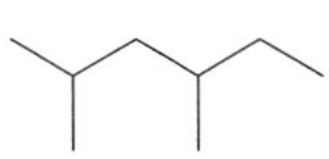

2,4-dimethylhexane

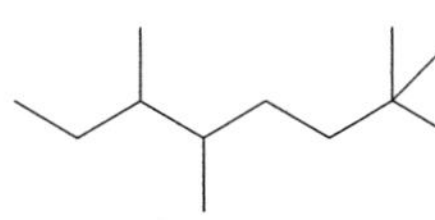

2,2,5,6-tetramethyloctane

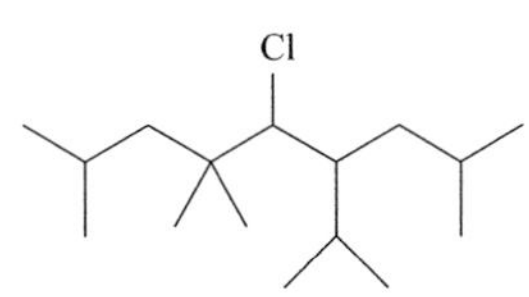

5-chloro-6-isopropyl-2,4,4,8-tetramethylnonane

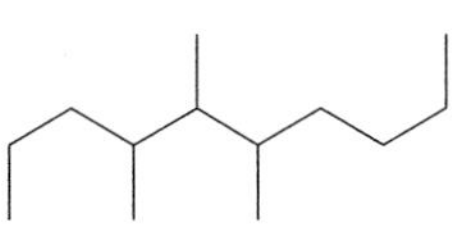

4,5,6-trimethyldecane

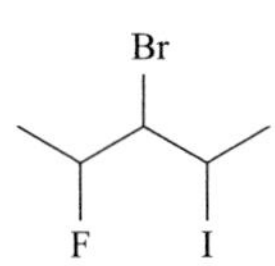

3-bromo-2-fluoro-4-iodopentane

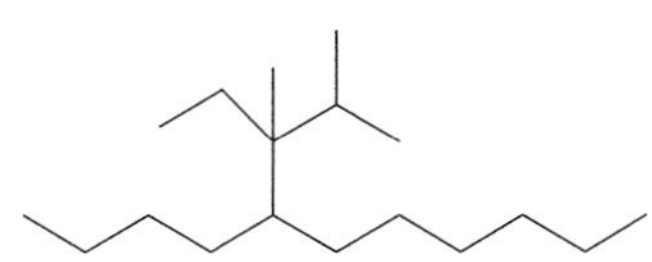

5-(2,3-dimethylpentan-3-yl)undecane

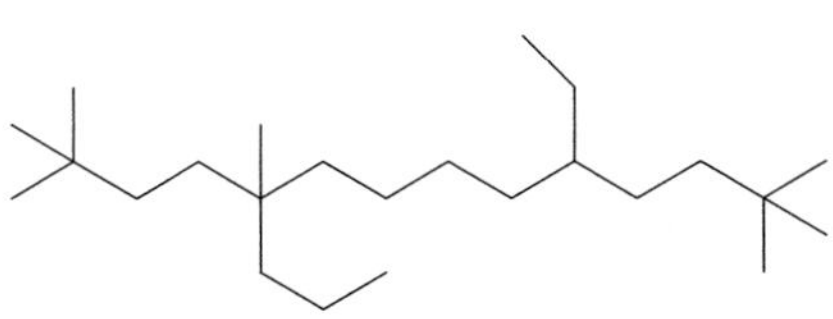

10-ethyl-2,2,5,13,13-pentamethyl-5-propyltetradecane

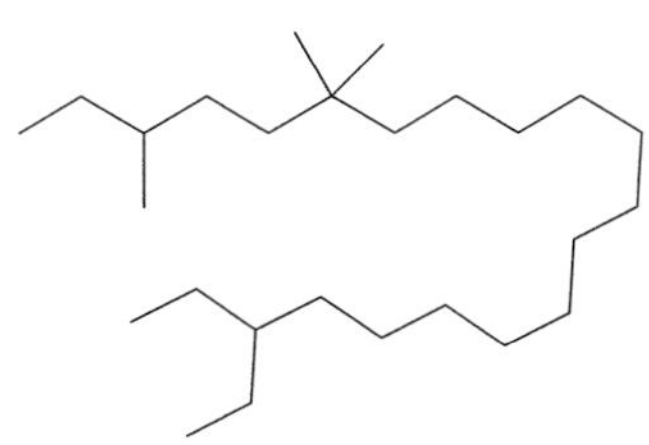

19-ethyl-3,6,6-trimethylhenicosane

> 7. If a complex substituent is present on the longest continuous chain, count the number of carbon atoms in the longest continuous part of that "side chain" and use the proper prefix. The name of a complex substituent begins with the first letter of its complete name (take the longest chain of the substituent from the point of attachment to the longest unbranched chain, and ignore di-, tri-, etc.).

Several examples illustrate these rules of nomenclature. Note that in 5-chloro-6-isopropyl-2,4,4,8-tetramethylnonane, all four methyl groups receive a number. Both methyls of the geminal dimethyl group receive a number (4,4), and the -$CNMe_2$ group is named as isopropyl. This group could also be 1-methylethyl, and the name would be 5-chloro-6-(1-methylethyl)-2,4,4,8-tetramethylnonane.

Note also that the groups are chloro, isopropyl, and tetramethyl. They are arranged alphabetically by the group: ‘c’, ‘I’, ‘m’, where these are methyl groups, so ‘m’ is used rather than ‘t’ for tetramethyl. Likewise, in 10-ethyl-2,2,5,13,13-pentamethyl-5-propyltetradecane, the groups are arranged ‘e’ for ethyl, ‘m’ for methyl, and ‘p’ for propyl. 5-(2,3-Dimethylpentan-3-yl)undecane contains a complex substituent, 2,3-dimethylpentan-3-yl. In 19-ethyl-3,6,6-trimethylhenicosane, the base name for the 21-carbon chain is henicosane.

To summarize, complex side chains are named by choosing the longest chain substituent, based on the point of attachment to the longest linear chain that gives rise to the base name; in the example shown, pentyl. The point of attachment of the substituent to a carbon of the main chain determines the number of the substituent, and numbering the side chain and the secondary substituents are numbered accordingly. To set the complex substituent apart from the other substituents, parentheses are used.

If chains of equal length are competing for selection as the main chain in a saturated branched acyclic hydrocarbon, the main chain required for naming must be chosen. The main chain can be determined in one of several ways, all of which lead to the same conclusion. If there is more than one complex substituent, but they are identical, the number of identical substituents may be indicated by the appropriate multiplying prefix: bis-, tris-, tetrakis-, pentakis-, and so on. This protocol is exactly the same as if there were two or three methyl groups (dimethyl, trimethyl, etc.). The complete expression denoting such a side chain may be enclosed in parentheses, or the carbon atoms in side chains may be indicated by primed numbers. An example is 9,9-bis(2,4-diethylhexyl)-6-methylicosane, where the term bis- is used to indicate the two identical 2,4-diethylhexyl substituents.

1. The main chain is that with the greatest number of side chains.
2. The main chain is the one whose side chains have the lowest-numbered locants.
3. The main chain will have the greatest number of carbon atoms that have smaller side chains.

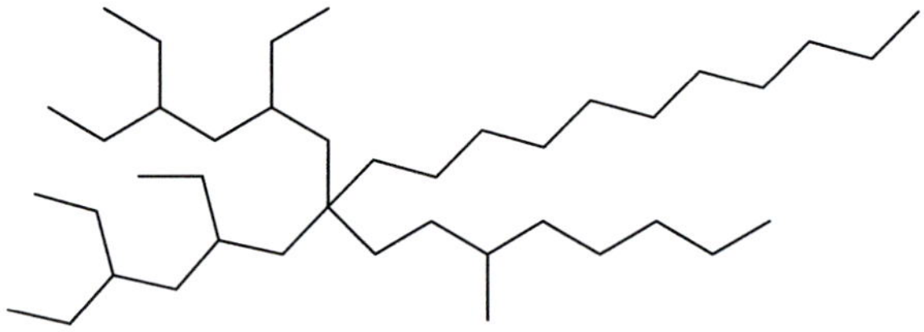

9,9-bis(2,4-diethylhexyl)-6-methylicosane

5.8 RINGS MADE OF CARBON: CYCLIC COMPOUNDS

Molecules are known that have rings of carbon atoms. If all the carbon atoms are sp^3 hybridized, they are *cyclic alkanes*. Note that a cyclic alkane has two fewer hydrogen atoms compared to the straight-chain alkane. In an acyclic alkane (acyclic means there is no ring), every carbon has the maximum number of hydrogen atoms attached to it (determined by the fact that each carbon must have four bonds). In a cyclic alkane, two carbon atoms must be joined to form a ring, resulting in *two fewer hydrogen atoms* compared to an acyclic alkane. Because of this, the *general formula for cyclic alkanes* has two hydrogens fewer than that for an alkane: C_nH_{2n}, where n is an integer: 2,3,4,.... Note that a cyclic alkane and an acyclic alkane with the same number of carbon atoms are *not* isomers because they have *different empirical formulas*. The nomenclature rule for cyclic alkanes requires that the suffix be *-ane* because they are alkanes. To distinguish between the linear 12-carbon molecule (dodecane) and the 12-membered cyclic (ring) alkane,

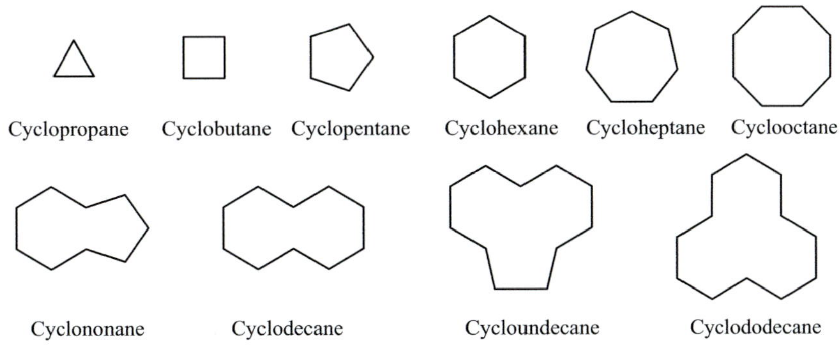

FIGURE 5.8 Cyclic hydrocarbons of 3–12 carbon atoms.

use the term *cyclo-*. If the ring is viewed as a *cycle*, it is appropriate to use the prefix *cyclo-* in front of the carbon-number prefix. In other words, place cyclo- in front of the alkane name, so the three-membered ring alkane becomes cyclopropane and the 12-membered ring alkane becomes cyclododecane (see Figure 5.8).

5.9 HYDROCARBON FUNCTIONAL GROUPS WITH π-BONDS

As noted in Section 5.4, discrete units of atoms that have special physical and/or chemical properties are known as *functional groups*. The C=C unit of alkenes and the C≡C unit of alkynes are examples of hydrocarbon functional groups. An *alkene* is a hydrocarbon that contains at least one C=C unit, which is the functional group. The general formula for an alkene is C_nH_{2n}. *Note that the generic formula for an alkene is the same as that for a cyclic alkane.* The carbon atoms are *sp*^{2} hybridized, and each carbon has three *sp*^{2} *hybrid molecular orbitals* that are used to form three σ-bonds to other atoms, as illustrated for ethene in Figure 5.9. After formation of the σ-bonds using the sp^2 hybrid orbitals, there is an "extra" molecular orbital on each carbon that is perpendicular to the plane of the atoms, and the two electrons can be shared (dispersed) over both orbitals to generate a new type of covalent bond, a *π-bond*.

The π-bond is different from the σ-bond in that the two adjacent orbitals share electron density by "sideways" overlap, leading to a covalent bond that is weaker than a covalent σ-bond. When three atoms are attached to a carbon atom that is part of a π-bond, the lowest-energy arrangement has those three atoms in a planar triangle with carbon at the center (this is called *trigonal planar geometry*). This geometry means that all four atoms are in the same plane. In general, the carbon-carbon bond distance of a double bond is shorter than that of a single bond. In other words, the internuclear distance between the two carbon atoms is shorter.

The nomenclature for alkenes uses the prefix system noted for alkanes in Section 5.7 to indicate the number of carbon atoms. For hydrocarbon molecules containing a double bond (an alkene), the suffix is taken from *the class name for an alkene (-ene)*. The name must specifically designate the position of the π-bond by *numbering the chain so that the first carbon of the π-bond receives the lowest possible number.* All substituents are numbered based on the C=C unit that is part of the longest continuous chain, which receives the smallest number.

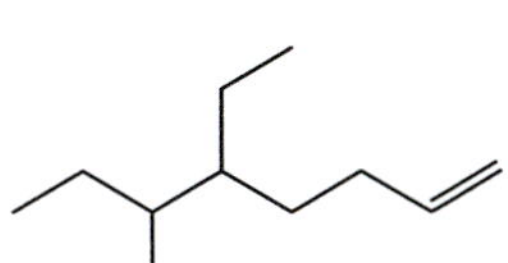

5-ethyl-6-methyloct-1-ene

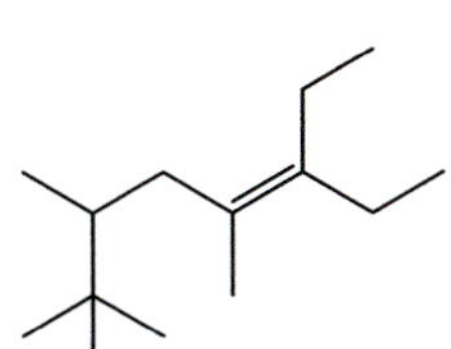

3-ethyl-4,6,7,7-tetramethyloct-3-ene

3,3-dimethylcyclopent-1-ene

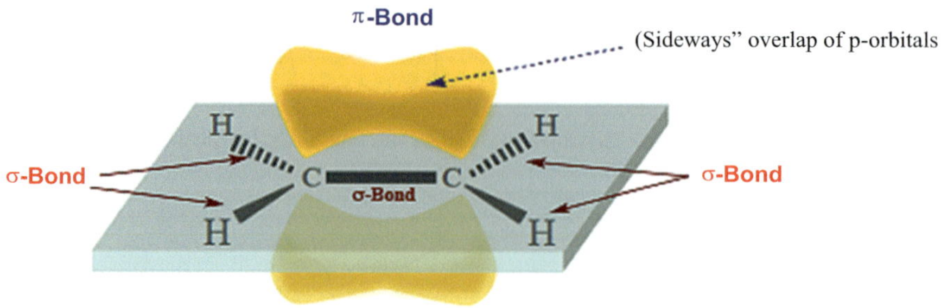

FIGURE 5.9 Interaction of two sp² hybridized carbon atoms to form three σ-bonds to each carbon and a π-bond.

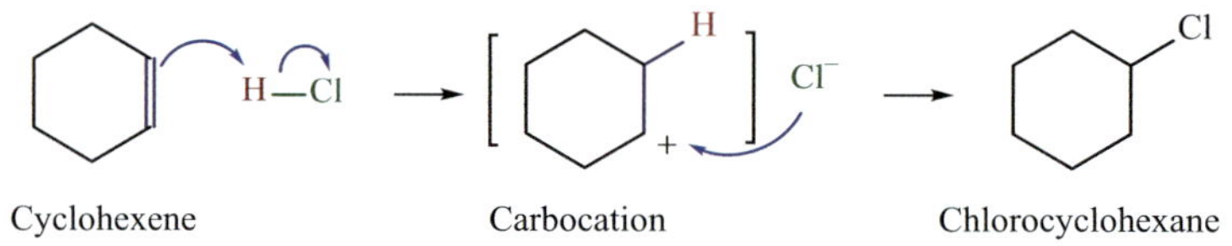

FIGURE 5.10 Acid-base reaction of ethene with HCl.

As with cyclic alkanes, the parent name for cyclic alkenes is based on the number of carbon atoms in the ring, derived from the analogous linear alkene, but the prefix cyclo- is added to the name. The six-carbon cyclic molecule that contains a C=C unit is called cyclohexene. Number the carbon atoms of the double bond in cyclic alkenes as "1" and "2" in the direction that give the substituent encountered first the *smaller* number. Note that the formula for a cyclic alkene is C_nH_{2n-2}. Examples are 5-ethyl-6-methyloct-1-ene, 3-ethyl-4,6,7,7-tetramethyloct-3-ene, and 3,3-dimethylcyclopent-1-ene.

In an alkene, the π-bond is electron-rich, and those electrons can be donated to an electrophilic center. If the electrophilic center is hydrogen, an acid-base reaction is possible. In other words, an alkene is electron-rich and reacts as an electron donor (a *Bronsted-Lowry base*). The acid (H of HCl) is electron-deficient, so it is a Bronsted-Lowry acid and accepts an electron pair from the alkene base to form a new C–H bond. The acid-base reaction of cyclohexene with HCl is shown in Figure 5.10. Donation of the electrons from the π-bond of cyclohexene (blue arrow) to the acidic proton of HCl leads to the formation of a new C–H σ-bond in a transient intermediate (a carbocation). *Note that the use of the double-headed arrow indicates the transfer of two electrons from C=C to H^+ to form a new σ-covalent bond.* In a subsequent reaction, the nucleophilic chloride ion reacts with the carbocation to give chlorocyclohexane as the final product.

Hydrocarbons that have a carbon–carbon triple bond are known as alkynes. The general formula for an alkyne is: C_nH_{2n-2} where n is an integer: 2,3,4,... *The empirical formula for a cyclic alkyne is C_nH_{2n-4}.* In alkynes, there is a σ-bond to another carbon, and a σ-bond to another atom, but there are two π-bonds between the carbon atoms as illustrated in Figure 5.11.

Focusing on the two carbons, alkynes are characterized by one σ- and two π-bonds that constitute a *carbon–carbon triple bond.* When only two other atoms are attached to a carbon atom, the sp-hybridization on each of the two carbon atoms is different from carbon atoms in an alkane or an alkene. The hybridization model leads to the formation of two *sp hybrid molecular orbitals* for each carbon atom. The simplest alkyne is ethyne (the common name is acetylene) with the formula C_2H_2. Formation of the σ-bonds leaves *two* unused p-orbitals on each carbon, and the overlap of these p-orbitals leads to two orthogonal π-bonds. The orbital used to form the σ-bond is a new type of hybrid orbital called an sp-hybrid orbital, since it is formed from one p- and one s-orbital. Using two sp-hybrid orbitals forms the two covalent σ-bonds: one to the other sp-carbon, and the other to the substituent. The molecular model in Figure 5.11 shows the concentration of electron density in a "band" between the two carbon atoms, consistent with the two orthogonal π-bonds.

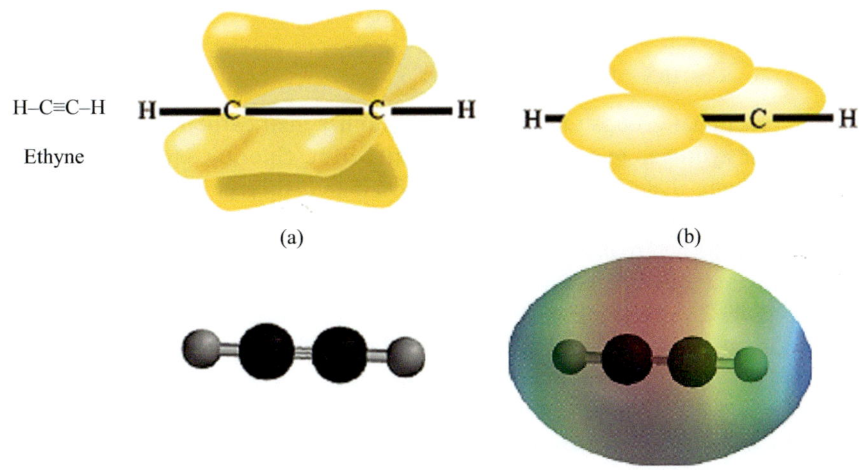

FIGURE 5.11 Two π-bonds in a sp-hybridized carbon–carbon bond.

As noted, the generic name for the class of hydrocarbons containing a triple bond (two π-bonds between adjacent carbon atoms) is an alkyne. The suffix is taken from the class name for an alkyne (*-yne*). As with alkenes, there are several possible isomers, and the position of the C≡C unit must be identified. The name specifically designates the position of the triple bond by *numbering the chain so that the first carbon of the triple bond receives the lowest possible number,* for example, oct-2-yne. In oct-1-yne, a hydrogen atom is attached to one carbon of the triple bond. In oct-2-yne, a carbon is attached to both carbons of the triple bond. Oct-1-yne is an example of a *terminal alkyne* (the triple bond occupies the terminal position and there is at least one H atom attached), whereas oct-2-yne is an *internal alkyne*. Due to the strain inherent in having the linear alkyne unit in a ring, cyclic alkynes of fewer than eight carbon atoms are not known. As with other cyclic compounds, cyclo- is part of the name, as in cyclooctyne, cyclododecyne, etc. Examples are the terminal alkyne hept-1-yne, the internal alkyne 2,7,11,11-tetramethyldodec-5-yne, and the cyclic alkyne cycloundecyne.

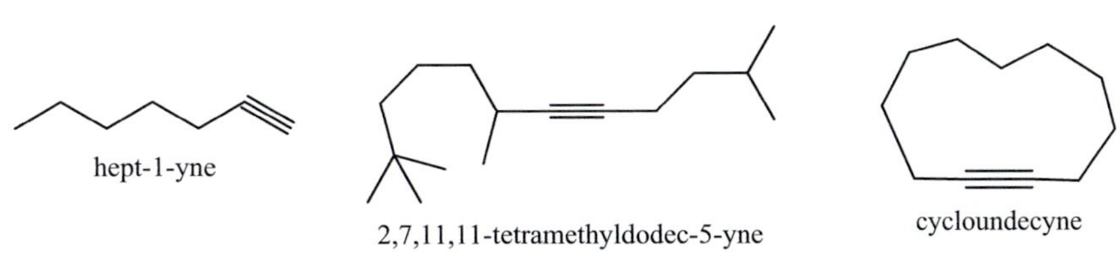

Alkynes also react as Brønsted-Lowry bases, so they react with mineral acids to form alkenes and produce the more stable carbocation. However, only one of the π-bonds reacts, so the intermediate is a *vinyl carbocation*. This reaction is illustrated by the reaction of pent-1-yne with HCl, which gives the more stable secondary vinyl carbocation. Subsequent reaction with the chloride ion leads to 2-chloropent-1-ene.

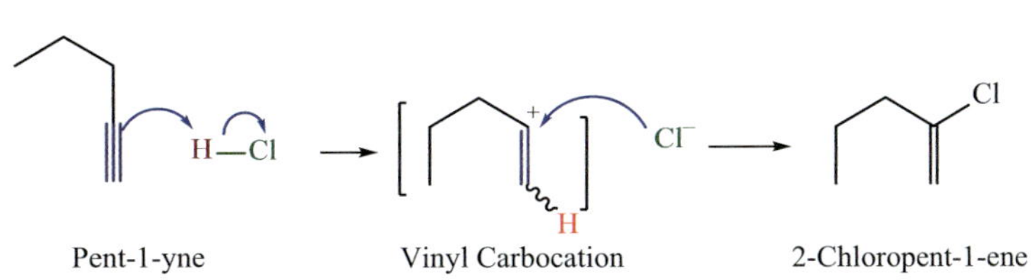

5.10 HETEROATOM FUNCTIONAL GROUPS

5.10.1 C–X Type Functional Groups

There are many functional groups that include atoms other than carbon and hydrogen. Any atom other than carbon or hydrogen (e.g., oxygen, nitrogen, sulfur, chlorine, etc.) is generically called a *heteroatom*. Heteroatoms that are more electronegative than carbon or hydrogen impart bond polarization to any single or multiple bond (C–X or C=X) in the molecule. Such molecules are considered polar. The valence of the heteroatom determines how many atoms are attached to that heteroatom. Oxygen has a valence of 2 and must form X–O–X or C=X species. When a C–O–H unit is formed, *the OH unit is called a hydroxyl group*. When OH is incorporated into a hydrocarbon molecule in place of one of the hydrogen atoms, the new molecule is called an *alcohol*. Oxygen also forms *a C–O–C unit, and molecules containing this unit are called ethers*. The OH unit, the C–O–C unit, and the amine units are *functional groups*. These functional *groups* contain polarized covalent bonds (C=X), which characterize their primary chemical reactions.

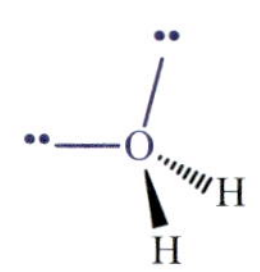

Water

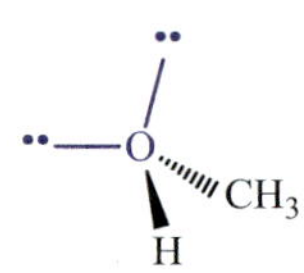

Methanol

There are several distinct structural variations of alcohols. The one-carbon molecule with a hydroxyl group is methanol. Alcohols are categorized by the number of carbon groups (known as *alkyl groups*) attached to the carbon bearing the hydroxyl group. A primary alcohol has two H and one carbon group (RCH_2–OH), a secondary alcohol has one H and two carbon groups (R_2CH–OH), and a tertiary alcohol has three carbon groups (R_3C–OH). In the IUPAC system, alcohols are named using the carbon prefix and the *suffix -ol* (taken from the generic name alcoh*ol*). The point of attachment of the oxygen of the OH to the longest linear carbon chain is used to name alcohols. *The carbon chain is numbered such that the carbon bearing the oxygen of the OH unit has the lowest possible number. Giving the oxygen-bearing carbon the lowest number supersedes the number and placement of alkyl or halogen substituents.* To name an alcohol, one identifies the hydrocarbon chain, drops the *-e* if it is an alkane backbone, and adds the suffix *-ol*. Examples are hexan-1-ol and 2-methylheptan-4-ol. Substituents on the longest carbon chain are numbered after assigning the lowest possible number to the carbon bearing the OH unit. The tertiary alcohol 5-bromo-2,3,4-trimethylheptan-2-ol has the OH unit connected to a 7-carbon chain, so it is a heptanol. Numbering the carbon chain to give the OH the lowest number results in heptan-2-ol, and the substituents are numbered accordingly, giving the name 5-bromo-2,3,4-trimethylheptan-2-ol. Cyclic alcohols are also possible; an example is 2,4,6-trimethylcyclohexan-1-ol, which has the OH group attached to a cyclohexane ring, so it is a cyclohexanol. The carbon of the ring that bears the OH is always C1, and the ring is then numbered to give the substituents the lowest numbers, resulting in the name 2,4,6-trimethylcyclohexan-1-ol.

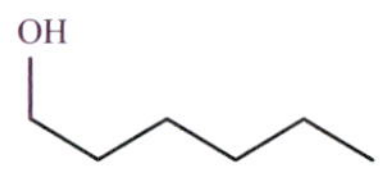

hexan-1-ol

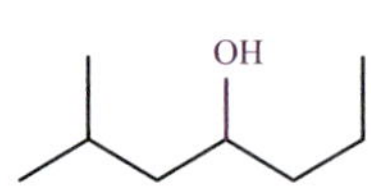

2-methylheptan-4-ol

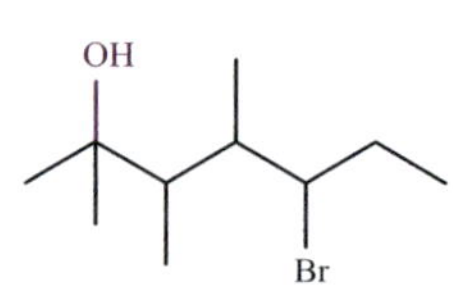

5-bromo-2,3,4-trimethylheptan-2-ol

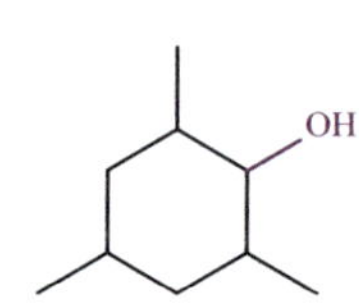

2,4,6-trimethylcyclohexan-1-ol

When two hydroxyl units are incorporated into the same molecule, it is called a *diol* (two OH units, hence diol). When there are three hydroxyl units, it is a triol, and tetraols, pentaols, and so on are known. The nomenclature for a diol identifies the longest chain that bears *both* OH units and gives all carbon atoms that bear an OH unit the lowest possible numbers. Typically, the name of the hydrocarbon chain precedes the term –diol, with numbers to identify the positions of the hydroxyl groups. For example, $HOCH_2CH_2CH_2CHOH$ is butane-1,4-diol, and $CH_3CH(OH)CH_2CH(CH_3)CH_2CH_2CH_2OH$ is 4-methylheptane-1,6-diol.

The O–H bond is polarized such that the hydrogen is partially positive. *Alcohols are therefore relatively weak Brønsted-Lowry acids*, which accounts for many of their chemical reactions of alcohols. The reaction of ethanol with a strong base (e.g., sodium amide) produces the conjugate base of the alcohol, an *alkoxide*. An alkoxide has an electron-rich oxygen atom and reacts as both a base and a nucleophile. In the example shown, the conjugate base of ethanol formed after reaction with sodium amide is known as sodium ethoxide. In this case, the base is the amide anion, which removes the proton from the O–H unit to give the conjugate acid, ammonia.

Sulfur is in group 16 of the periodic table, immediately under oxygen, and it can form molecules that are analogous to alcohols, with two covalent bonds to sulfur and two unshared electron pairs on sulfur. Sulfur has d-orbitals and can form neutral molecules that have more than two covalent bonds to sulfur. A similar neutral structure is not possible with oxygen, since it would assume a positive charge. Hydrogen forms two σ-bonds to sulfur to give hydrogen sulfide, H–S–H. Forming one hydrogen σ-bond and one carbon σ-bond to sulfur generates a thiol, R–S–H. A thiol is the direct sulfur analog of an alcohol, and they are named using the hydrocarbon portion of the alkyl unit with the suffix -thiol. Therefore, CH_4SH is methanethiol, and $CH_3CH_2CH_2CH_2SH$ is butanethiol. Mercaptan is the common name for a thiol. Therefore, another name for methanethiol is methyl mercaptan, and butanethiol is butyl mercaptan. Examples are hexane-1-thiol, 2-methylheptane-4-thiol, 5-bromo-2,3,4-trimethylheptane-2-thiol, and the cyclic compound 2,4,6-trimethylcyclohexane-1-thiol.

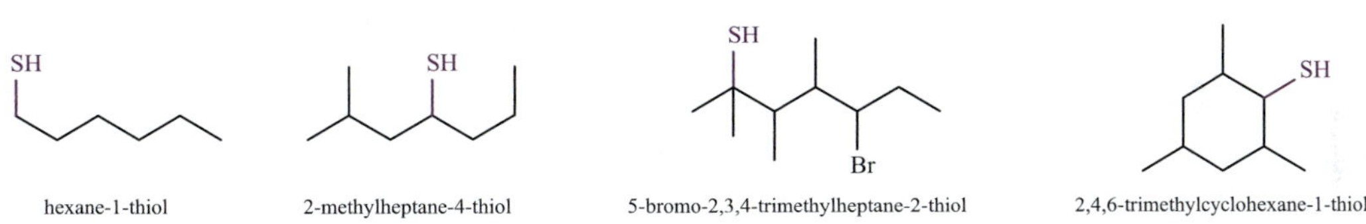

Low-molecular-weight thiols are foul-smelling compounds. Thiols react similarly to alcohols in acid-base reactions, but there are differences in reactivity because sulfur has multiple valences. Just as the hydrogen atom attached to oxygen in an alcohol is acidic, so the proton on sulfur in a thiol is also acidic. An example is the reaction of methanethiol with sodium amide. The acid-base reaction leads to the conjugate base, sodium methanethiolate, and the conjugate acid, ammonia. Thiols are generally more acidic than alcohols. The pK_a of a typical thiol is ~ 10, whereas the pK_a of a typical alcohol is ~ 15–18. Just as there are diols, there are dithiols. Ethanedithiol is $HSCH_2CH_2SH$ and butane-1,4-dithiol is $HS(CH_2)_4SH$.

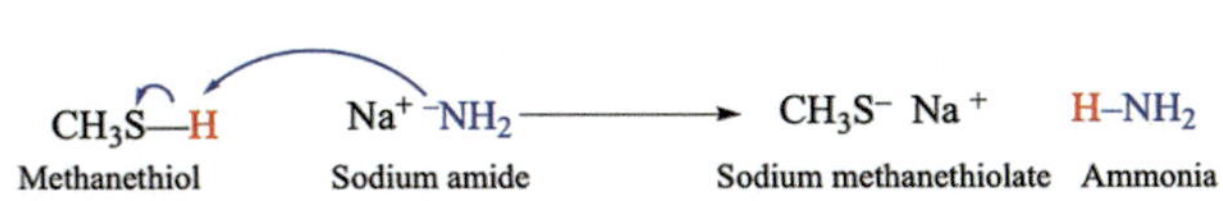

Organic molecules that contain a C–O–C unit, an oxygen atom with two alkyl groups attached to the oxygen atom, and no hydrogen atoms, are called *ethers*. Ethers are predicted by the VSEPR model to be angular, and they are characterized by their poor reactivity in a variety of reactions. Indeed, ethers are commonly used as solvents for many organic chemical reactions. The recommended method for naming identifies a long chain and a shorter linear chain attached to the oxygen. The longer chain is the parent, and the oxygen-bearing shorter chain is treated as a substituent.

If the shorter alkyl chain is a butyl, for example, the -yl used for carbon substituents is replaced with -oxy, so the shorter chain is identified as butoxy. In other words, the alkyl group becomes alkoxy: OCH_3 is methoxy, OCH_2CH_3 is ethoxy, and so on.

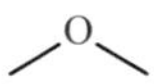

1-Methoxymethane

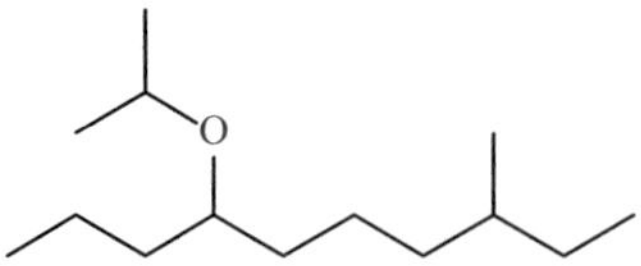

7-Isopropoxy-3-methyldecane

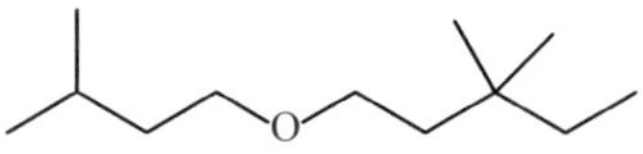

1-(Isopentyloxy)-3,3-dimethylpentane

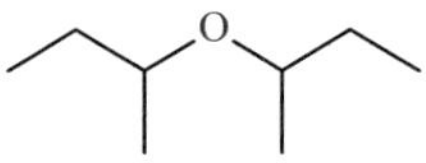

2-(*sec*-Butoxy)butane

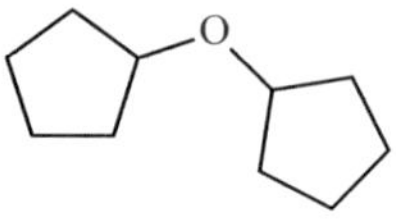

Dicyclopentyl ether

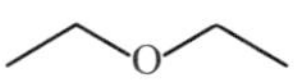

2-Ethoxyethane

Using this method, typical ethers are 1-methoxymethane and 7-isopropoxy-3-methyldecane. In 1-(isopentyloxy)-3,3-dimethylpentane, the longest chain is 5, with a branching dimethyl group. The four-carbon chain bearing the oxygen is at C1 of the five-carbon chain, and it is converted to isopentyloxy. Note that 2-(*sec*-butoxy)butane has one propyl group bearing the oxygen attached to C1 of the other propyl chain. 2-(*sec*-butoxy)butane is classified as a symmetrical ether because two sec-butoxy groups flank the oxygen. For symmetrical ethers, each alkyl group can be identified, followed by the word *ether*. The ether name for 2-(*sec*-butoxy)butane, is di-sec-butoxy ether or di(1-methylpropyl) ether. Similarly, the IUPAC name of the next example is 1-cyclopentoxycyclopentane, which is a symmetrical ether and can be called dicyclopentyl ether. Dimethyl ether is another name for 1-methoxymethane, and diethyl ether (2-ethoxyethane) is another simple example.

Replacing O of an alcohol with S leads to a thiol, R–SH. Similarly, replacing the O of an ether with S leads to a thioether (R–S–R), also known as a sulfide. Sulfides will not be discussed in detail because they are used only occasionally in this book. However, the nomenclature of simple thioethers is straightforward. The suffix used for thioethers is -sulfane. Examples are di-(1-methylpropyl)sulfane, with two 1-methylpropyl groups attached to sulfur. Unsymmetrical thioethers are named by identifying the two groups flanking the sulfur, as in ethyl(2,2-dimethyl)sulfane and (2-ethylpentyl)(propyl)sulfane. An alternative approach to naming is analogous to the nomenclature system used for ethers: identify the longest chain and attach the sulfur via the smaller group. Using this system, di-(1-methylpropyl)sulfane is 2-(1-methylpropylthio)-1-methylpropyl, ethyl(2,2-dimethyl)sulfane is (1-ethylthio)-2,2-dimethylpropane, and (2-ethylpentyl)(propyl)sulfane is 1-(propylthio)-2-ethylpentane.

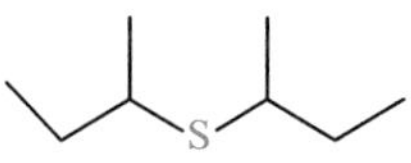

di-(1-Methylpropyl)sulfane

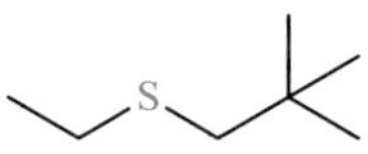

Ethyl(2,2-dimethyl)sulfane

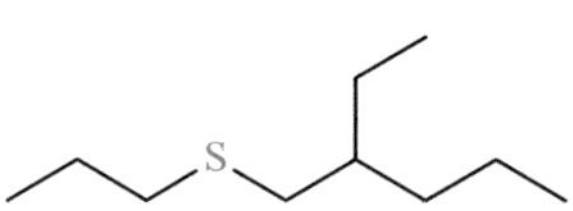

(2-Ethylpentyl)(propyl)sulfane

Nitrogen is in group 15 of the periodic table, has five electrons in its valence shell, and requires only three electrons to complete the octet, giving it a valence of three. Neutral organic compounds containing nitrogen have three covalent bonds to nitrogen and one unshared electron pair on nitrogen. When the molecule is NH_3 (ammonia), there are three N–H σ-bonds. An amine may have N–H σ-bonds and N–C σ-bonds. Indeed, amines are compounds characterized by one or more C–N bonds.

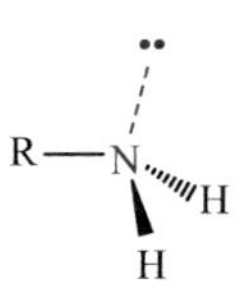

Primary amine

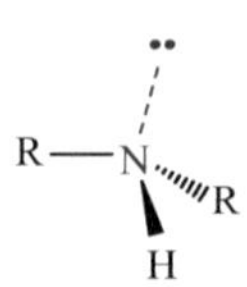

Secondary amine

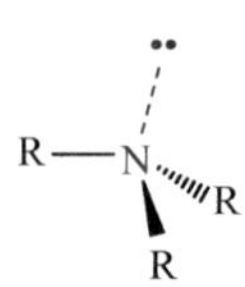

Tertiary amine

Nitrogen-containing units are known as amino groups, and, as mentioned, a molecule containing an amino group is called an amine. Nitrogen has a valence of 3 and can form three types of species containing at least one C–N bond: R–NH_2, R_2NH, and R_3N, where "R" represents a carbon group. The terms *primary, secondary, and tertiary* are used to describe the structural variations in amine structure. A primary amine has one carbon and two hydrogen atoms on nitrogen (RNH_2); a secondary amine has two carbons and one hydrogen atom on nitrogen (R_2NH); and a tertiary amine has three carbons and no hydrogen atoms on nitrogen (R_3N). The three-dimensional shape of each amine with respect to nitrogen is predicted to be similar to that of ammonia: pyramidal, with the unshared electron pair projected from the apex of the pyramid. The C–N–C or C–N–H bond angles vary with the size of the alkyl groups.

The IUPAC nomenclature system treats the amine unit as a substituent attached to the longest hydrocarbon chain. The name drops the -e ending of the hydrocarbon chain and replaces it with the suffix *-amine*, and the position of the N is indicated by a number. The one-carbon primary amine is methanamine, and the two-carbon primary amine is ethanamine, where the position of the N is obvious and the number is not required. However, the five-carbon primary amine is pentan-1-amine, since the N may be attached to more than one carbon. Pentan-2-amine and pentan-3-amine are possible, for example. Any group attached to nitrogen other than the longest hydrocarbon chain is indicated by the terms *N*-alkyl, *N,N*-dialkyl, or *N*-alkyl-*N*-alkyl. In other words, groups could be *N*-methyl, *N,N*-diethyl, or *N*-ethyl-*N*-methyl. *A substituent on the longest carbon chain is assigned a position number,* as with all other nomenclature rules encountered so far. *A substituent on the nitrogen atom, however, is assigned an N- to indicate its position.* Using this system, CH_3NHCH_3 is called *N*-methylmethanamine. In *N,N*-diethylpentan-1-amine, the 1- indicates the position of the nitrogen on the longest chain. The longest carbon chain is pentane, and the two ethyl substituents on nitrogen are indicated by *N,N*-, as shown. Similarly, *N*-ethyl-2-methylpropan-1-amine has an ethyl group attached to the nitrogen. Note that the methyl substituents on the carbon chain in *N*-ethyl-2-methylpropan-1-amine are treated the same as any substituent on the longest continuous chain that also includes the functional group (here, the amine unit). Two additional examples illustrate nomenclature for secondary and tertiary amines. In the secondary amine N-(1-methylpropyl)-3-methylhexan-2-amine, the nitrogen is attached to C2 of a six-carbon chain. A 1-methylpropyl unit is attached to the nitrogen atom as a substituent. The tertiary amine N-ethyl-N-methylheptan-3-amine has a seven-carbon chain, with the nitrogen attached to C3. Both ethyl and methyl are attached to nitrogen as substituents.

$H_3C{-}NH_2$

Methanamine

Ethanamine

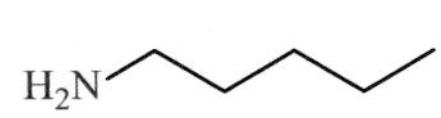

Pentan-1-amine

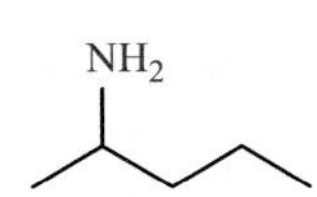

Pentan-2-amine

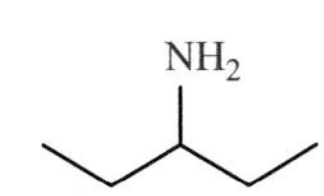

Pentan-3-amine

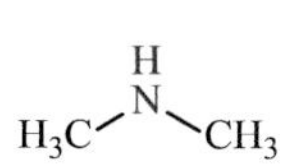

N-Methylmethanamine

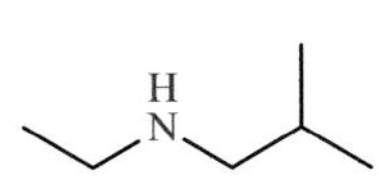

N-Ethyl-2-methyl-propan-1-amine

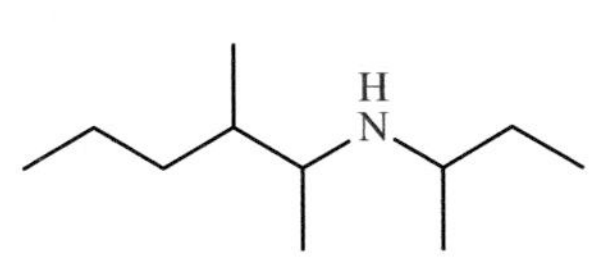

N-(1-Methylpropyl)-3-methylhexan-2-amine

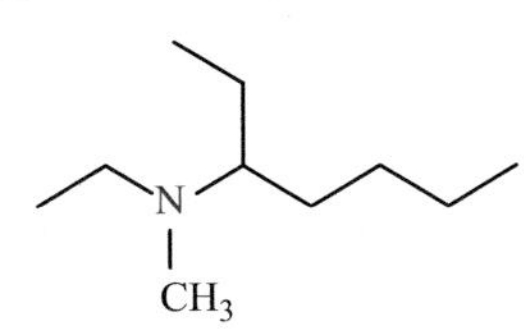

N-Ethyl-*N*-methyl-heptan-3-amine

Amines can also be named using common names, and some appear so often that the system must be noted. The system is simple in that the alkyl groups are identified, and that term is followed by the word amine. Using this system, butan-1-amine is butylamine; and *N,N*-diethylpentan-1-amine is diethylpentylamine; *N*-ethyl-2-methylpropan-1-amine is ethylisobutylamine.

5.10.2 C=X Type Functional Groups

In principle, a π-bond may form between any atom that has available p-orbitals and a valence >1. In organic chemistry and biochemistry, π-bonds are commonly formed to carbon but also to other atoms such as oxygen (C=O), sulfur (C=S), or nitrogen (C=N). In all cases, there is one strong σ-bond and one weaker π-bond. Molecules that contain N=N bonds, N=O bonds, and O=O bonds will also be encountered. It is also possible to form triple bonds between carbon and nitrogen, with one strong σ- and two weaker π-bonds.

The structural unit with one π- and one σ-bond between a carbon and oxygen, represented as C=O, is called a *carbonyl*. Both the carbon and the oxygen are sp^2 hybridized, there are two covalent bonds to oxygen, and two unshared electron pairs reside on oxygen. The molecule with only one carbon atom, which is part of a C=O unit, and two hydrogen atoms on the carbon is known as formaldehyde, $H_2C=O$. As seen in Figure 5.12, the unshared electrons are orthogonal to the π-bond and coplanar with the atoms. The H–C=O bond angle is ~ 120°, consistent with sp^2 hybridization, so the carbon, oxygen, and hydrogen atoms are all coplanar. A molecular model of formaldehyde in Figure 5.12 shows the typical trigonal planar geometry.

The C=C unit of an alkene and the C=O unit of a carbonyl are similar in that they react as Brønsted-Lowry bases in the presence of a strong acid. In the case of the carbonyl, the electronegative oxygen polarizes the C=O unit (δ^+ and δ^-). In an acid–base reaction, the oxygen is protonated to form the *oxocarbenium ion* shown in Figure 5.13a. Oxocarbenium ions are resonance-stabilized, with the two resonance contributors shown.

As pointed out above, *a carbonyl is polarized*. Because of the bond polarization, a negatively charged species (a nucleophile, X^-) reacts by donating two electrons to the carbon of the C=O unit. The reaction of the nucleophile at the carbonyl carbon forms a new σ-bond to that carbon, generating an anionic species, the alkoxide product shown in Figure 5.13b.

Analysis of the C=O unit shows that two atoms or groups must be attached to the carbonyl carbon atom. If at least one of the groups is a hydrogen atom, there is an H–C=O unit (see the box; also abbreviated as CHO), and the functional group is called an *aldehyde*. Aldehydes have one hydrogen and one carbon group attached to the carbonyl carbon. Formaldehyde is the only exception, with

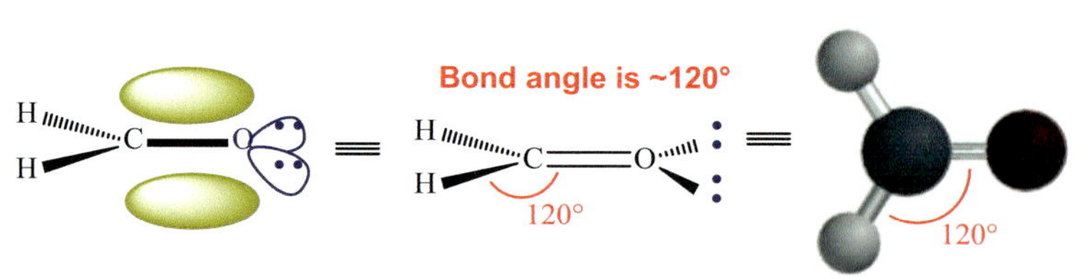

FIGURE 5.12 The carbonyl group in formaldehyde.

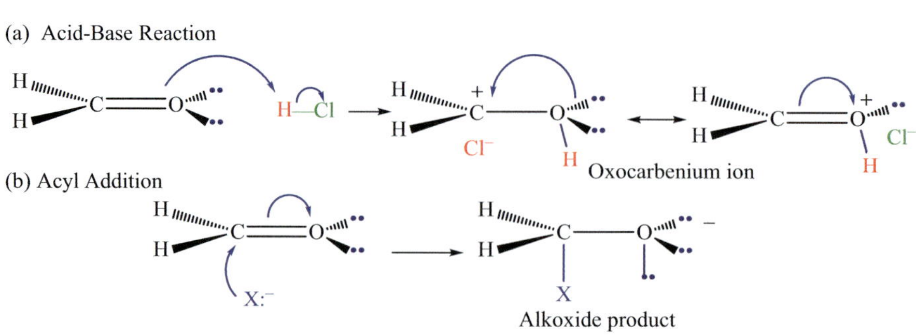

FIGURE 5.13 (a) Acid-base reaction of the carbonyl group (b) Acyl addition reaction of the carbonyl group.

two hydrogen atoms ($H_2C{=}O$). If two carbon groups (two alkyl groups) are attached to the carbonyl carbon, the generic formula is $R_2C{=}O$, and the functional group is called a *ketone.*

Aldehyde Ketone

The nomenclature system for aldehydes uses a suffix derived from the *first* two letters of *al*dehyde, *-al*. The carbonyl carbon takes priority and is given the lowest possible number, which *in all cases* will be 1, so the number is omitted. Substituents attached to the aldehyde chain are named in the usual manner, with the carbonyl carbon always numbered 1. Otherwise, the nomenclature rules concerning substituents are the same as for any other functional group. An example is 2,4-diethyl-5-methylheptanal, where the longest chain contains the aldehyde C=O (note the shorthand notation CHO for the aldehyde).

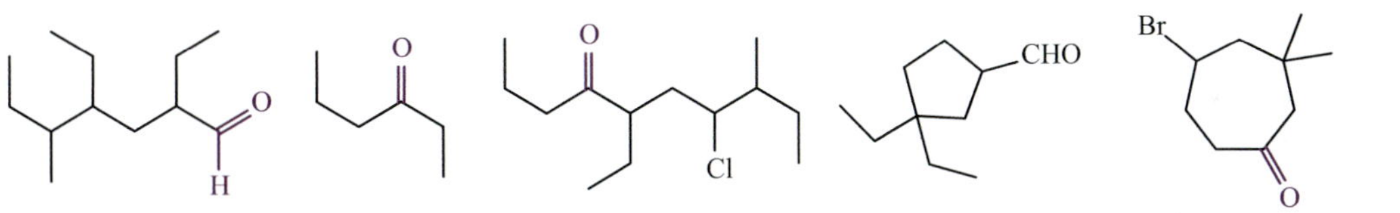

2,4-Diethyl-5-methylheptanal Hexan-3-one 7-Chloro-5-ethyl-8-methyldecan-4-one 3,3-diethylcyclopentane-1-carbaldehyde 5-Bromo-3,3-dimethylcycloheptanone

A ketone contains a carbonyl group attached to two alkyl groups (as in hexan-3-one). The suffix for ketones derives from the last three letters of ket*one*, *-one*. When numbering the longest chain, the *carbonyl carbon* receives the lowest possible number. Therefore, the six-carbon straight-chain ketone with the carbonyl carbon at C3 is hexan-3-one. Substituents are handled in the usual manner, and they are given the lowest number relative to the position of the carbonyl carbon, as in 7-chloro-5-ethyl-8-methyldecan-4-one. Note that the methyl substituent receives the closest locant in the 10-membered carbon chain, but the carbonyl carbon must receive the lowest number. Cyclic moles containing the aldehyde moiety are named carbaldehydes, with the –CHO unit attached to the ring. An example is 3,3-diethylcyclopentane-1-carbaldehyde. Cyclic ketones are also possible, but unlike aldehydes, *the carbonyl carbon is part of the ring*. In 5-bromo-3,3-dimethylcycloheptanone, the carbonyl unit is part of a seven-membered ring, and the usual protocols for rings apply. The carbonyl carbon is always C1, and the substituents are numbered accordingly.

An important functional group has a carbon atom (alkyl group) attached to a carbonyl (C=O) functional group, with a hydroxyl (OH) group also attached to the carbonyl carbon. This unit is known as a *carboxyl group*, which is the major structural feature of the class of organic molecules known as *carboxylic acids*. The structural unit of the carboxyl group is shown in the box in Figure 5.14 for ethanoic acid (the common name is acetic acid). The *carboxyl* functional group unit is also written as –COOH or $–CO_2H$.

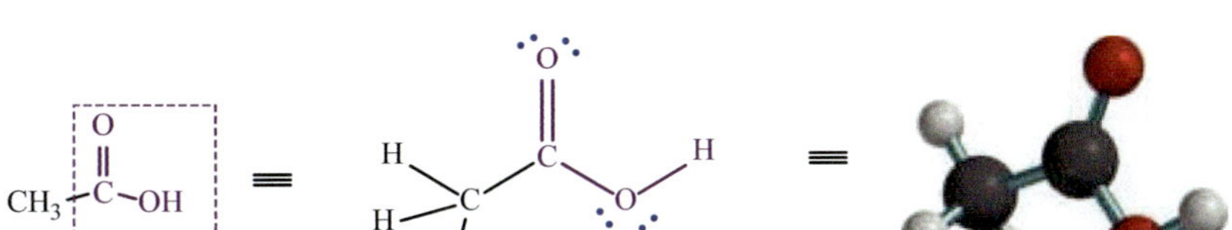

Ethanoic acid
(Acetic acid)

FIGURE 5.14 Ethanoic acid.

The carboxyl group is an O–H unit attached to a carbonyl, and the two oxygen atoms and two carbon atoms are coplanar due to the presence of an sp^2-hybridized carbonyl carbon, as shown in Figure 5.14. Ethanoic acid is also drawn as a molecular model to illustrate the spatial arrangement of the groups. The O–H proton is acidic, and a carboxylic acid is much stronger (pK_a 1–5) than an alcohol (pK_a 16–18). *The greater acidity of the carboxylic acid is largely due to the stability of the conjugate base that is formed by reaction with a suitable base.*

There are many carboxylic acids, and two simple examples are methanoic acid (CH_3COOH), which has the common name formic acid (found in some ant venoms), and ethanoic acid (CH_3CH_2COOH), which has the common name acetic acid (a dilute solution in water is called vinegar). The nomenclature rule is similar to that used with aldehydes in that the longest continuous chain for the acid *must* contain the CO_2H unit, with the carboxyl carbon as C1. The carbonyl carbon of the COOH group must receive the lowest possible number, so 1- is omitted from the name. The suffix for carboxylic acids is *-oic acid*, and the word "acid" is written separately. Note the use of COOH and CO_2H as shorthand notation for the carboxyl unit. Nomenclature is first illustrated by 2,3,3,4,5-pentamethylheptanoic acid. All substituents are assigned numbers based on C1 for the carboxyl group. A carboxylic acid with two COOH units is known as a dicarboxylic acid, or a dioic acid. The molecule $HO_2C(CH_2)_3CO_2H$, for example, should be named 1,5-pentanedioic acid, but the two carbonyl units must be at the ends of the molecule to be carboxyl units, so the 1- and 5- are redundant. The name is just pentanedioic acid. The third example is the 18-carbon acid, octadecadienoic acid, otherwise known as stearic acid. This example is also an example of a fatty acid (see Section 10.1).

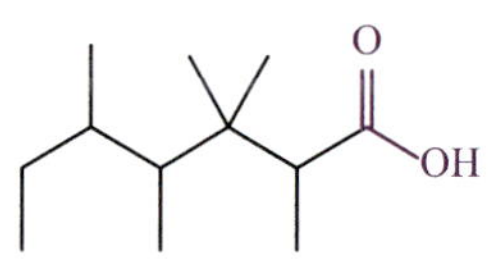

2,3,3,4,5-Pentamethylheptanoic acid

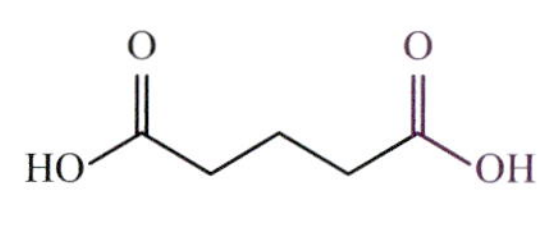

1,5-Pentanedioic acid

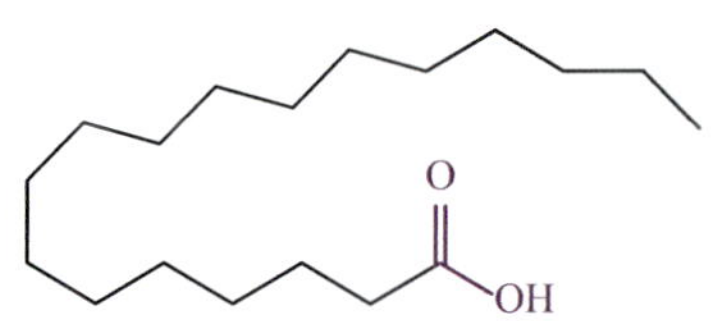

stearic acid (octadecanoic acid)

The reaction of formic acid with a base gives a carboxylate anion (the formate anion) as the conjugate base, as shown in Figure 5.15. Note that three atoms (O=C–O) are connected. One atom (O) has a negative charge, which can be viewed as a "filled" p-orbital (a p-orbital containing two electrons). In addition, there is an adjacent π-bond (C=O). However, the description of π-bonds made it clear that two adjacent and parallel orbitals overlap and share electron density to form a π-bond. The

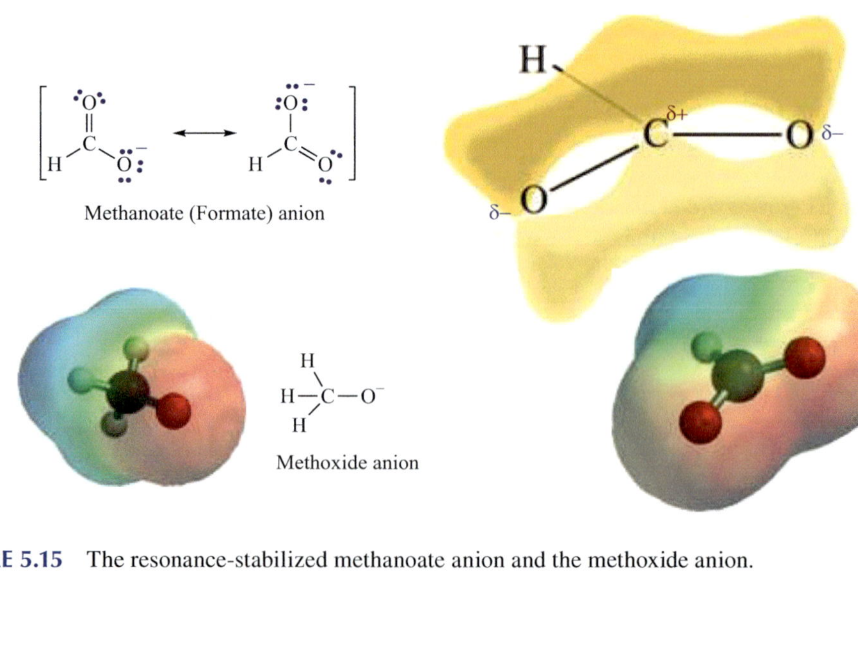

FIGURE 5.15 The resonance-stabilized methanoate anion and the methoxide anion.

concept that adjacent orbitals share electron density can be extended from two orbitals to three, four, or even more if the orbitals are on adjacent atoms and are parallel. Therefore, when three orbitals are on adjacent atoms and are parallel, *electron density is shared and dispersed (delocalized) over all three atoms.* There are a total of four electrons (two from the π-bond and two from the negative charge) dispersed over the larger surface area of three atoms, or delocalized, as represented in Figure 5.15. In other words, the electrons are not localized on a single negatively charged oxygen but are delocalized over three atoms by overlap of the two orbitals of the π-bond with the orbital on oxygen. The delocalized structure is *lower in energy* than a structure that has the charge localized on a single oxygen atom. The resonance delocalization is represented by two structures, as shown for the methanoate (formate) anion, with a *double-headed arrow to indicate resonance.* The two structures represent *one resonance-stabilized anion*, not two different molecules.

If the conjugate base of an acid–base reaction is more stable, it is less reactive and a weaker base, so the K_a is larger (the species is more acidic). The acid-base reaction of methanol makes an interesting comparison. When methanol reacts with a base, the conjugate base is the methoxide anion, so the charge is localized on a single oxygen. The electron density map of methoxide clearly shows that the highest concentration of red (electron density) is localized on the oxygen, as shown in Figure 5.15. The resonance delocalization in the methanoate anion leaves each of the two oxygen atoms with less electron density than is observed in methoxide. The methoxide anion is a stronger base than the methanoate anion, so methoxide is more reactive as an electron donor.

The carboxyl group (COOH) can be attached to a ring, but the nomenclature changes. Such molecules are named as "cycloalkane carboxylic acid." Examples are cyclobutanecarboxylic acid and 3-bromo-5-ethyl-2,4-dimethylcyclohexane-1-carboxylic acid.

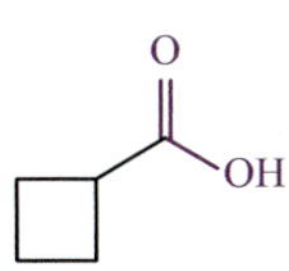

Cyclobutanecarboxylic acid

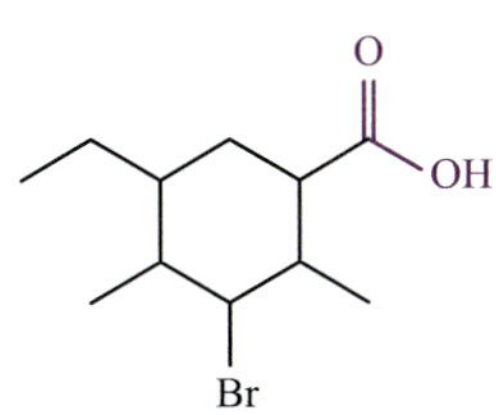

3-Bromo-5-ethyl-2,4-dimethylcyclohexane-1-carboxylic acid

5.11 NON-COVALENT BONDING AND SOLUBILITY

A *physical property* of an organic compound can be measured experimentally and is characteristic of that molecule (or a collection of like molecules). The boiling point and melting point temperatures, for example, are physical properties of the molecule. Physical properties can be used to characterize and identify a compound since the complete set of physical properties of an individual molecule is usually unique. Identifying the functional group is also important in determining the physical properties of a molecule. An alcohol has different physical properties than an amine, for example, or than a carboxylic acid.

Boiling point is defined as the temperature at which a liquid and the vapor (gas) above it are in equilibrium. At normal atmospheric pressure, the boiling point of a liquid is the temperature at which the liquid is at equilibrium with the atmosphere above it. Several factors are important in determining the boiling point of a liquid. The number of atoms in a molecule and the number and type of heteroatom functional groups play an important role. The boiling point of molecules will increase steadily as more carbon and hydrogen atoms are added. Mass is not the only factor influencing boiling point, however. Boiling point varies with the nature of the functional group. When the molecular weights of molecules are close, as in methane–methanol–formic acid–and methylamine, the functional groups lead to a vast difference in boiling point. The key reason for these differences is the intermolecular forces that keep molecules associated with each other in the liquid phase (see below). It is important to mention that as the pressure is lowered, the temperature required for the liquid to come into equilibrium is also lowered. This finding means that the boiling point is lowered as the pressure is lowered (liquids boil at a lower temperature under vacuum).

Some forces lead to attraction between molecules, even when there are no polarized bonds. Imagine that two alkane molecules come close together, and the electron density associated with the covalent bonds is attracted to the positive nuclei in the atoms of the molecules, as illustrated in Figure 5.16. In effect, subnuclear particles change polarity when they come into proximity with other particles. When two ethane molecules come close, there is no dipole, but the electrons in the covalent bonds reorganize to some extent so that there is a higher concentration of electron density on one side than on the other. This reorganization leads to a very small, induced *dipole*, illustrated in Figure 5.16a. This small, induced dipole allows a weak attraction between molecules, as indicated in Figure 5.16b.

The attractive forces just described are extremely weak and are known as *London forces*; they are also called *van der Waals forces*. Since the induced dipole results from close contact, *the larger the surface area of the molecule, the greater the van der Waals interaction*. This force is weak and can be disrupted by the application of only small amounts of energy. In other words, two molecules held close together by these forces are easily separated at low energy (low temperatures).

When a molecule has a dipole moment, it is considered to be polar. The larger the magnitude of the *dipole moment*, the greater the polarity of the molecule. Polarity in molecules leads to certain consequences that relate both to reactivity and to physical properties.

When the polarized atoms in an alkyl halide (e.g., the molecule fluoromethane) come into proximity with the dipole of another molecule of fluoromethane, as shown in Figure 5.17, the δ^+ carbon of one molecule is attracted to the δ^- fluorine of the second molecule. Remember that there are lone electron pairs on the fluorine atom, which contribute to the polarity. This electrostatic attraction is called a *dipole-dipole interaction*, and it is *stronger* than London forces, which are also present. Because of this attraction, more energy (heat) is required to separate two molecules held together

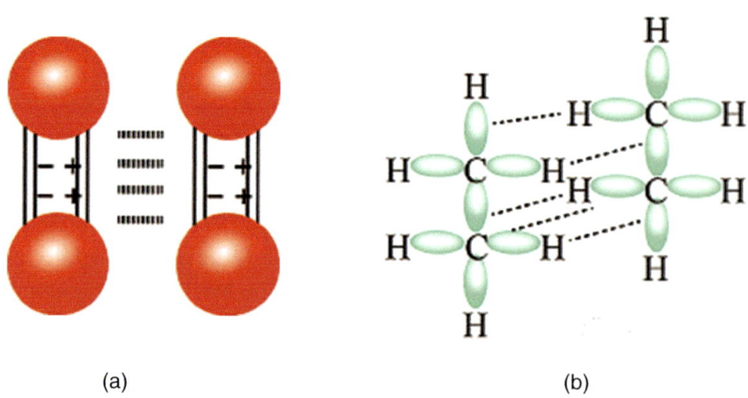

FIGURE 5.16 Attractive London Forces in ethane.

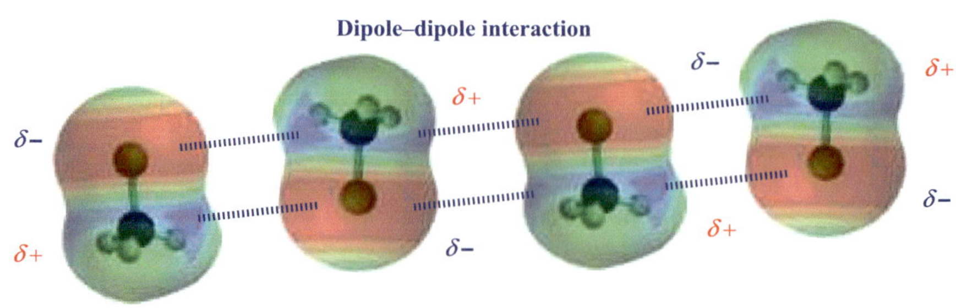

FIGURE 5.17 Attractive dipole-dipole interactions in fluoromethane.

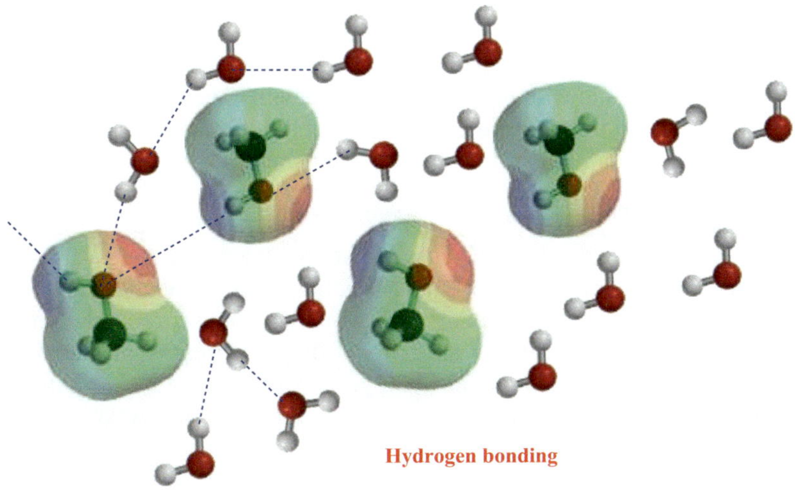

FIGURE 5.18 Hydrogen bonding in methanol.

by dipole-dipole interactions than to separate two molecules held in proximity by simple London forces. The dipole-dipole interaction does not have to be between identical molecules but can occur between two entirely different molecules. Dipole-dipole interactions can be envisioned between alkyl halides, ethers, or amines.

There is a special type of dipole-dipole interaction that occurs when one atom of the dipole is a hydrogen, as in the O–H bond of an alcohol. When the OH of one alcohol (e.g., methanol) comes into proximity with the OH unit of a second molecule of methanol (illustrated in Figure 5.18), there is a strong dipole-dipole attraction between the positively polarized hydrogen and the negatively polarized oxygen. This interaction is known as a *hydrogen bond* (Section 1.2). Figure 5.18 is meant to represent an array of *several* methanol molecules attracted to water molecules and each other by this type of hydrogen bond. Since an O–H bond is generally more polarized than a C–O bond, the hydrogen atom is rather small compared to carbon, and a positively polarized hydrogen is acidic, the O------H interaction is particularly strong. Hydrogen bonds are much stronger than common dipole-dipole interactions. Since a hydrogen bond is rather strong, it requires much more energy to disrupt it. More energy (heat) must be applied to separate associated methanol molecules than is required to separate associated fluoromethane molecules, and the least amount of energy is required to separate associated alkane molecules (e.g., ethane, shown in Figure 5.16b).

Since water (H–O–H) is such a common solvent and is frequently used in reactions with organic molecules, it is important to mention that water can form strong hydrogen bonds to itself as well as to molecules that contain a polarized X–H bond (X is usually, O, S, N, etc.), as illustrated in Figure 5.18. Examples are water with alcohols, water with amines, and water with carboxylic acids, but *not* with alkanes or alkenes, since there is no polarized bond in those molecules. A reasonably strong hydrogen bond can also be formed between water and compounds that do not contain an X–H unit but have a highly polarized functional group. These include water with ketones or aldehydes, and, to a lesser extent, ethers.

In general, a nonpolar molecule has no polarized bonds or dipole moment, and only noncovalent van der Waals interactions. Alkanes are a typical example. Only London forces are at work between two alkanes, as illustrated in Figure 5.16, leading to the association of one alkane with another. A polar molecule will have dipole–dipole interactions, or hydrogen-bonding interactions. If a polar molecule is attracted to another polar molecule, then this aggregation of polar molecules will allow them to mix. A molecule capable of hydrogen bonding or generating a dipole–dipole interaction (e.g., methanol in water, Figure 5.18) should be soluble in a molecule that is also capable of forming a hydrogen bond or a dipole interaction. The reason why nonpolar molecules are soluble in each other is more subtle.

5.12 ROTAMERS AND CONFORMATION

As molecules absorb energy from their surroundings, they "move around" as their incipient kinetic energy increases. Each molecule will undoubtedly collide with another molecule of the same type, a molecule of solvent, or the walls of the container. Each collision will dissipate some energy to another molecule, to the solvent, or to the sides of the container, and these collisions usually lead to an increase in heat for the system. Indeed, such molecular motion is one way in which molecules in solution transfer heat.

Molecules also dissipate energy by molecular vibration. One vibration mode changes the shape of the molecule by stretching covalent bonds and bending bonds, while another dissipates energy by internal rotation around single covalent σ-bonds. Rotation about carbon-carbon bonds leads to arrangements of atoms and groups in the molecule at different relative positions in space. Each different rotational arrangement of the atoms and groups attached to the carbon atoms is called a *rotamer.* Some of these arrangements of atoms lead to repulsive interactions of the electrons in the bonds, or interactions between atoms or groups induced by those units being too close together (known as *steric hindrance*). Such interactions raise the energy of that arrangement of atoms or groups relative to others. Many different arrangements are possible for most molecules, and it is possible to describe an overall "shape" for a molecule, called a *conformation.*

Rotation about the C–C bond in ethane leads to a different arrangement of bonds and atoms. If each structure could be "frozen" at a particular rotation angle, slight differences in the spatial relationships of the atoms (their relative positions in three-dimensional space) can be examined. As noted above, such imaginary "frozen" structures are called rotamers. A so-called *sawhorse diagram* is shown for two rotamers of ethane in Figure 5.19. One rotamer is shown in Figure 5.19a, with C1 in the front and C2 in the rear, as marked. Hydrogen atoms are attached to each carbon, and one hydrogen atom attached to C1 is bold. A different rotamer is shown in Figure 5.19b as a sawhorse diagram. A molecular model of each rotamer is drawn using the same perspective in order to show the relationship of the atoms and bonds. Both rotamers of ethane are shown from a different perspective in Figure 5.19. In this representation, the molecule has been turned so the view is "head on," with C1 in the front and C2 in the rear. The "front" carbon (C2) is shown as a "dot" in the center of the circle. The carbon atom in the rear (C2) is represented as a "circle," with three hydrogen atoms attached to it. This second representation is known as a *Newman projection.* A second molecular model of ethane is drawn for each rotamer from this new perspective, again to show the relationship of the bonds and atoms. Note that the arrangement of the hydrogen atoms in both representations is meant to represent a tetrahedral array.

A comparison of the hydrogen atoms in Figure 5.19a shows that the H atoms on C1, marked in violet, are "in between" the H atoms marked in green on C2. Note that the bonds are *staggered.* This arrangement is therefore identified as a *staggered rotamer.* Rotation about the C–C bond of ethane leads to a different arrangement of bonds and atoms, as in Figure 5.19b. In this second arrangement,

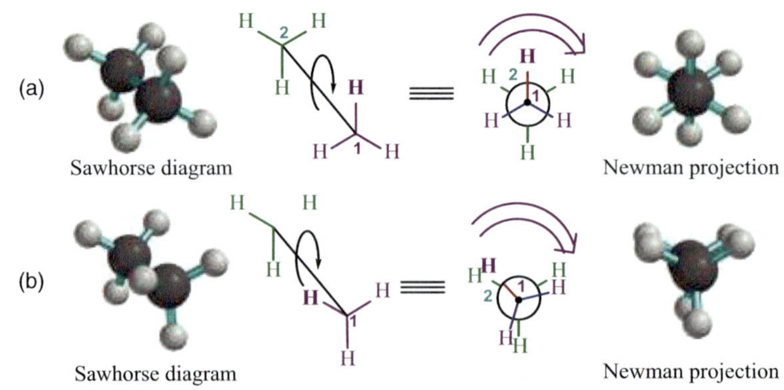

FIGURE 5.19 Sawhorse diagrams and Newman projections for rotamers of ethane.

the hydrogen atom marked in green overlaps (eclipses) another hydrogen atom. Note that the bonds *eclipse*. This arrangement is therefore identified as an *eclipsed rotamer.*

Typically, there are many conformations for a given molecule, some higher in energy and some lower. Of all the possibilities, one or, at best, a few conformations will be lower in energy than the others, and the molecule will "spend most of its time" in these low-energy conformations. Understanding the low-energy conformation(s) of a molecule is critical to understanding both its properties and its chemical reactivity.

It is possible to describe 360° rotation about the C–C bond of ethane, at 60° intervals (see Figure 5.20). Why 60°? The positions of the three hydrogen atoms on the rear or the front carbon atom are arranged at the corners of a tetrahedron, since both carbon atoms have a tetrahedral geometry. The apex of each tetrahedron is the C–C bond, and the angle between the three other bonds is 120°. Bond rotations of 120°, or one-half that amount (60°), conveniently show if groups are closer to or further away from each other. Alternating rotamers are generated, staggered and eclipsed. There are an infinite number of possible rotations, of course, since the size of the rotation angle can be any number, but the 60° rotations generate rotamers in which the hydrogen atoms are as close together or as far apart as they can be.

Figure 5.20 illustrates the "dynamic" rotation that occurs continuously, and each structure shown may be present for only a fleeting instant during the rotational vibration. Rotation about carbon-carbon single bonds therefore generates many rotamers. Every bond will have several rotamers, but each will have a different energy. It is important to identify those rotamers where interactions of the various atoms attached to the carbons will raise the energy relative to others. By examining the relative energies of such key rotamers, some conformations can be dismissed as being irrelevant because they are too high in energy. The structures in Figure 5.20 represent several "snapshots" of ethane taken while rotation occurs about the C–C bond, where the atoms would be frozen in space. Therefore, each rotamer can be thought of as a "snapshot" of rotation about the C–C bond. In a thought experiment, imagine rotation about the C–C bond of ethane at ambient temperatures, and imagine taking 100,000 "snapshots" during the rotation. Once these "snapshots' are compared, a large percentage of them would show the molecule as a staggered rotamer, and only a small percentage would show an eclipsed rotamer.

Eclipsed rotamers are higher in energy because the electron-rich bonds eclipse, which leads to electronic repulsion. In addition, the hydrogen atoms are close together in space in the eclipsed rotamer, leading to a repulsive interaction known as *steric repulsion*. In other words, steric repulsion occurs when two atoms or groups are close together in space and there is a competition for the same

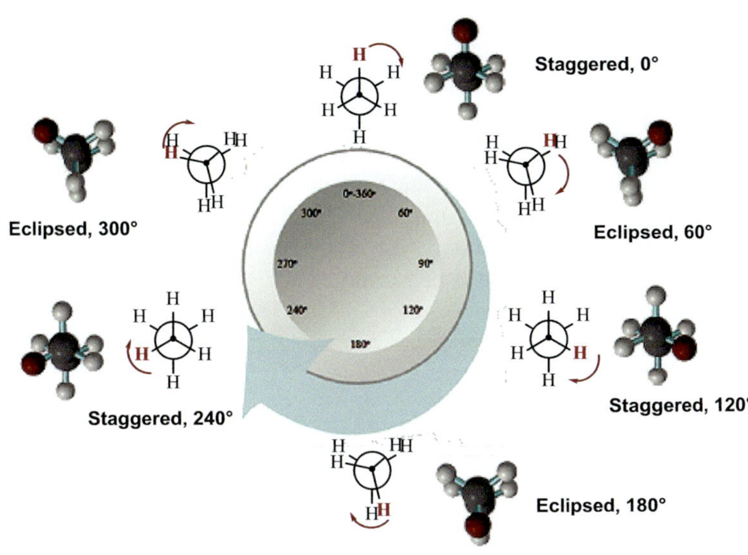

FIGURE 5.20 Rotation about the carbon–carbon bond in a "labeled" ethane using Newman projections.

space, resulting in repulsion as the atoms or groups approach one another. Both the bond repulsion and the steric hindrance are minimized in the staggered rotamer since the atoms or groups are as far apart as possible. This preference for lower-energy rotamers leads to the observation that during rotation about C–C bonds, the molecule will "spend most of its time" in the lowest-energy forms, the staggered rotamers. In a staggered rotamer of ethane, the hydrogen atoms are further apart and do not repel. In other words, the majority of the "thought snapshots" will be of staggered rotamers of ethane.

In the eclipsed rotamer, the bonds eclipse, and this repulsive energy is sometimes called *torsional energy*. Note that *torsional strain* is typically used to indicate this energy in combination with the steric strain of eclipsing atoms or groups. This increase in energy further makes the eclipsed rotamer higher in energy relative to a staggered rotamer, where the bonds are as far apart as possible. When the hydrogen atoms and the C–H bonds are staggered, torsional strain is greatly diminished.

As the C–C bond in ethane is rotated clockwise (arbitrary) through 360° degrees, at 60° intervals, six rotamers are generated, as shown in Figure 5.20. To go from one lower-energy staggered rotamer to the next lower-energy rotamer during the rotation, *a higher-energy eclipsed rotamer must form*. Therefore, *there is an experimentally measured energy barrier to rotation,* and the eclipsed rotamer will be 2.9 kcal/mol (13.95 kJ/mol) higher in energy than the staggered rotamer.[2] Rotation of 360° around the carbon-carbon bond in ethane must generate three eclipsed rotamers, each with an energy barrier of 2.9 kcal/mol (13.95 kJ/mol). The presence of this energy barrier effectively "slows down" rotation when these eclipsed rotamers are encountered, since each presents a barrier to rotation. Therefore, rotation in ethane is said to be "hindered" rather than "free." At normal temperatures, however, there is plenty of energy for rotation about carbon–carbon bonds, and organic molecules should be considered dynamic species that have a large population of different rotamers.

Rotation about a C–C bond occurs in any organic molecule with two sp^3-hybridized carbon atoms connected by a covalent σ-bond. There is rotation about all such C–C bonds. Indeed, the overall shape of a molecule results from the lowest-energy rotamer assumed by every bond. For a given arrangement of rotamers, there is an overall shape called the *conformation* of the molecule. For a molecule with several C–C bonds, there are *many* rotamers and *many* conformations. Some conformations may be lower in energy than others and therefore more abundant.

For any given molecule, one or both of the carbons of the C–C bond may have attached substituents. The presence of these groups will lead to rotamers of different energy due to increased torsional strain and/or steric hindrance. To properly understand the chemical behavior of a molecule, the energy barriers to rotation for all single covalent bonds in the molecule must be determined, and the relevant rotamers examined. Both staggered and eclipsed rotamers occur, but there are other important rotamers based on differing arrangements of atoms or groups.

There are many alkanes, and as noted, there is rotation about every C–C bond. As a practical matter, rotamers of substituted alkanes can be categorized into three fundamental types: X–C–C–H, X–C–C–X, and Y–C–C–X, as shown in Figure 5.21. The fragment X–C–C–X represents a bond with two identical substituents, whereas X–C–C–Y represents a bond with two different substituents. It is reasonable to assume that the X–H interactions are weaker than C–X or C–Y, so the X–C–C–H fragment is probably less important from an energetic perspective but cannot be ignored. The models in Figure 5.21 essentially use ethane as a reference point for all alkanes, where the X–H, X–X, and X–Y interactions must be examined.

Longer-chain alkanes or compounds with substituents are treated as substituted ethane derivatives, and the important interactions are X:H, X:X, or X:Y in X–C–C–H, X–C–C–X, and X–C–C–Y.

[2] (a) Aston, J.G.; Isserow, S.; Szasz, G.J.; Kennedy, R.M. *J. Chem. Phys.* **1944**, *12*, 336; (b) Mason, E.A.; Kreevoy, M.M. *J. Amer. Chem. Soc.* **1955**, *77*, 5808; (c) Mason, E.A.; Kreevoy, M.M. *J. Amer. Chem. Soc.* **1957**, *79*, 4851; (d) see Weinhold, F. *Nature* **2001**, *411*, 539 (Figure 1 therein).

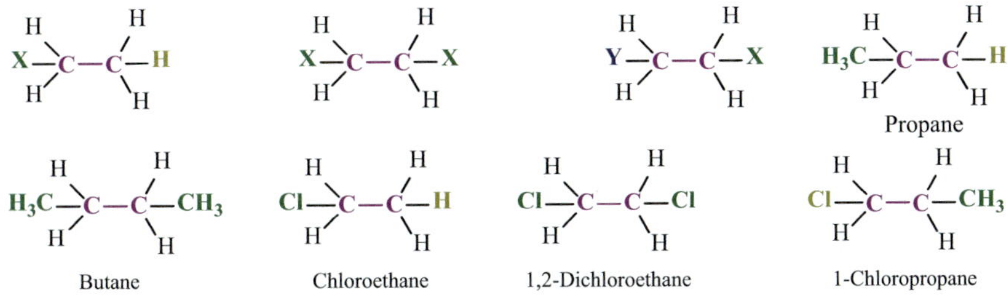

FIGURE 5.21 "Ethane" model for longer chains and substituted alkanes.

Propane is an X–C–C–H system, where X is methyl (CH_3, abbreviated Me), and the Me:H interaction should present greater steric problems than H:H interactions. Butane is an X–C–C–X system, where X=methyl, so there are Me:Me and Me:H interactions. Since the methyl groups are much larger than a hydrogen atom, the Me:Me interaction should have the highest steric interactions yet discussed. Extending this model to other systems, 1,2-chloroethane is an X–C–C–X system, where X=Cl, so there are Cl:Cl and Cl:H interactions. 1-Chloropropane is an X–C–C–Y system, where X=Cl and Y=methyl, so there are Cl:Me, Cl:H, and Me:H interactions. Clearly, as the number of atoms attached to each carbon increases, the interactions increase. The steric interactions of the larger groups lead to higher energy barriers to rotation, so the Me:Me interactions, rather than the H:H interactions, or the Cl:Me rather than Me:H or H:H interactions, are the important ones.

In butane, for example, three bonds must be considered for rotation: C1–C2, C2–C3, and C3–C4. Examination of the C2–C3 bond shows butane to be symmetrical with respect to these two carbons ($MeCH_2CH_2$–Me), making the C1–C2 and C3–C4 bonds identical. The goal is to find the highest energy barrier to rotation in butane, and to do this, C1–C2 is compared with C2–C3 (C2–C3 versus C3–C4 can also be used).

If rotation about C1–C2 is examined, it is an X–C–C–H system, and C1 has three attached hydrogen atoms, whereas C2 has an attached ethyl group and two hydrogen atoms. The C2–C3 bond is an X–C–C–X system, characterized by a methyl group on both C2 and C3. An ethyl group is larger than a methyl group, but *the steric demands of an ethyl group are not significantly greater than those of a methyl group.* Remember that there is rotation about *both* the C1–C2 bond and the C2–C3 bond. Therefore, the ethyl group (CH_3CH_2, abbreviated as Et) is attached to C2 (Et–C2), and the CH_2–C2 bond will likely assume a staggered orientation since that is the lowest-energy rotamer for that bond. Because of this staggered orientation, an Et–H interaction is not much greater than a Me–H interaction, because the Me–CH_2– unit of ethyl will have the methyl group projected "away" from the interactions involved for C1–C2.

Since the ethyl group does not give a significantly greater steric effect than methyl, the steric energy for the Et:H interaction will be close to the Me:H interaction in propane [~ 3.5 kcal (14.6 kJ)/mol], whereas the Me:Me interaction from rotation about C2–C3 is about 5.9 kcal (24.7 kJ)/mol.[3] The maximum interaction for rotation of the bonds in butane will therefore be due to the eclipsing interaction of the methyl groups (labeled the syn-rotamer of butane). The X–C–C–H interaction is less than the X–C–C–X interaction. Therefore, the focus of any discussion about energy barriers to rotation in butane will be associated with rotation about C2–C3.

Rotation about C2–C3 in butane may be tracked in the same manner as done for ethane in Figure 5.20. The C2–C3 rotation for butane is shown in Figure 5.22, which begins with the staggered rotamer of butane, where the two methyl groups are as far apart as possible, representing the energy minimum for the system. Rotation of the C2–C3 bond is shown to be clockwise through 360°, at 60° intervals, leading to a series of rotamers (follow the motion of the violet methyl group relative to the green methyl group).

[3] Kagan, H. *Organic Stereochemistry*, Halsted Press, New York, **1979**, p. 50.

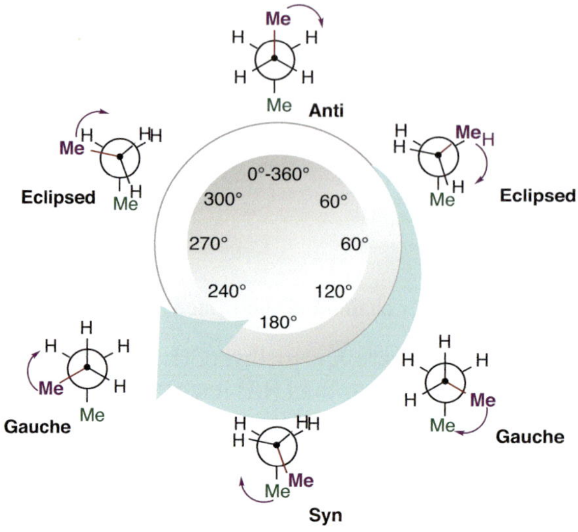

FIGURE 5.22 Rotation about C2–C3 in butane.

Rotation through 60° leads to an eclipsed rotamer where the two methyl groups *do not eclipse each other,* but one methyl eclipses a hydrogen atom, and the other methyl eclipses a different hydrogen. Rotation by another 60° leads to a staggered rotamer, but the methyl groups are closer together in space. The spatial proximity of the methyl groups leads to a steric interaction despite the fact that there are no eclipsing bonds or atoms, and it is lower in energy than any of the eclipsed rotamers. This rotamer is labeled as a *gauche-rotamer.* The rotamer where the two methyl groups are ~ 180° apart, as far apart as possible, is much lower in energy and is given a special name, the *anti-rotamer.* The gauche rotamers are higher in energy than the anti-rotamer. The next 60° rotation generates an eclipsed rotamer that is different from the first eclipsed rotamer in that the two methyl groups completely eclipse each other. The methyl groups are as close in space as they can be, and this is clearly the highest-energy rotamer. Therefore, it is given a special name, the *syn rotamer.* As rotation is continued in Figure 5.22, a second *gauche* rotamer is generated, and then a second eclipsed rotamer that is not the syn rotamer.

Recognition of the H:H, Me:H, and Me:Me interactions in Figure 5.22 allows identification of the highest barrier to rotation [measured to be 5.86 kcal (24.5 kJ)/mol] for the Me–Me interaction in the *syn*-rotamer. There are smaller energy barriers for *eclipsed*-rotamers, and the "valleys" for *gauche*-rotamers are higher in energy than the energy minimum for the *anti*-rotamer. This rotation map is relatively complex, representing a very hindered rotation about the C2–C3 bond. However, the energy required to rotate through the *syn*-rotamer constitutes the highest energy barrier to rotation, and the *anti*-rotamer is the lowest energy rotamer. These facts mean that the *anti*-rotamer should constitute a large percentage of the rotamer population of butane.

5.13 CONFORMATIONS OF MOLECULES WITH FUNCTIONAL GROUPS

In all alkanes, the carbon structure contains nothing but sp^3 carbon atoms, which are all tetrahedral. The presence of trigonal planar sp^2 carbon atoms in an alkene unit will lead to flattening about each carbon, since all four atoms connected to the planar sp^2 carbons lie in the same plane. If the C=C unit is incorporated into a chain of sp^3 carbons, as in hex-3-ene, the C=C unit "flattens" a portion of the carbon chain, as shown in Figure 5.23. Flattening of this region is apparent in the space-filling molecular model, and the sidedness of the two hydrogen atoms, as well as the two ethyl groups attached to the C=C unit, is apparent. This rigidity in structure is the result of the planar alkene unit and the inability of the groups to rotate about the C=C unit.

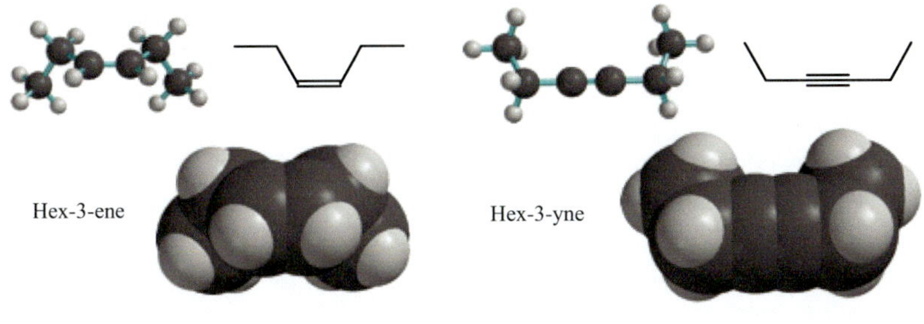

FIGURE 5.23 Influence of the π-bond on hex-3-ene and hex-3-ye.

An alkyne unit has two sp-hybridized carbon atoms, and another atom is attached to each one. Since the sp-hybridized carbons are linear, all four atoms form a linear array. In other words, the presence of the linear alkyne unit in but-3-yne (C–C≡C–C) has all four carbon atoms "in a line." The effect on the conformation is to flatten the molecule in the region of the triple bond to an even greater extent than seen with alkenes. The flattening effect is quite clear in the structure of hex-3-yne, especially in the space-filling molecular model in Figure 5.23. When the triple bond unit of an alkyne is treated as a substituent, its linear nature provides very little steric interaction when brought close to other alkyl groups.

The OH and NH units of alcohols or amines will influence rotamer populations. Both the O–H and N–H units also contain polarized hydrogen atoms that are capable of hydrogen bonding to the oxygen or nitrogen atom of another OH or NH unit, or with another heteroatom in a different molecule. Rotamers are formed in such molecules by rotation about C–C bonds, as with all other molecules. The presence of hydrogen bonding in a rotamer of an alcohol may bias the rotamer population to favor that rotamer. In effect, a rotamer that is destabilized by a steric interaction between two groups may be stabilized by a hydrogen bond elsewhere in that molecule. The two effects work at cross-purposes, and the energy "gained" or "lost" by each interaction will determine which rotamer predominates.

Amines are capable of hydrogen bonding. Introduction of an NHR or NH_2unit (from a secondary or primary amine) will have a hydrogen-bonding effect, but to a lesser degree than alcohols. The NH_2 group appears to be slightly less than a methyl in terms of its steric interactions, but the NH_2 group is capable of intermolecular hydrogen-bonding. The amine group does not hydrogen bond as effectively as the OH group. When a single amine unit is present, there can be no intramolecular hydrogen bonding, only steric interactions, but there will be intermolecular hydrogen-bonding interactions. When two NH units are present, as in 1,2-diaminoethane ($NH_2CH_2CH_2NH_2$; ethylenediamine, a *vicinal* diamine), intramolecular hydrogen bonding is possible just as in 1,2-diols. Since the N–H bond is less polarized than the O–H bond, there is weaker hydrogen bonding in diamines than in diols. The effect on the rotamer population is very similar, however.

A carbonyl group (C=O) has an sp^2-hybridized carbon and trigonal planar geometry for the groups that are attached to the carbonyl carbon. If the carbonyl is part of a ketone, as in pentan-3-one, there is localized flattening due to the sp^2 hybridization, as is apparent in the space-filling molecular model shown in Figure 5.24. The C2–C*–C4 bond angle is ~ 120°, where C* is the carbonyl carbon. This bond angle clearly influences the overall shape of the molecule, bringing the hydrogen atoms on C2 and C4 closer together than in pentane, for example. As with other heteroatoms, electron

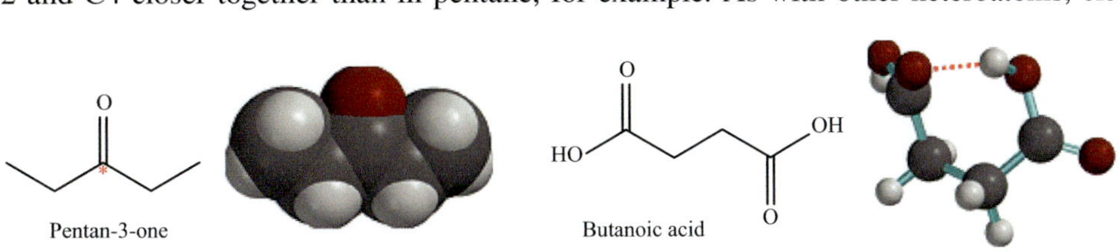

FIGURE 5.24 Confirmation of a ketone and a dicarboxylic acid.

pairs on oxygen contribute to the interaction of the oxygen with adjacent atoms or groups. An aldehyde unit (–CHO) will have a similar effect when it acts as a substituent in a given rotamer, but a hydrogen is always attached to the carbonyl, so an interaction such as that noted for C4–C6 of pentan-3-one is not possible.

A carbonyl carbon is central to the carboxyl group (COOH) in carboxylic acids. The acidic proton in the COOH unit is capable of hydrogen-bonding. When a molecule contains two carboxyl units (e.g., butane-1,4-dioic acid in Figure 5.24), it can be correlated with a X–C–C–X system, and intramolecular hydrogen-bonding can play a role, just as it does with diols. Internal hydrogen-bonding in a *gauche*-rotamer leads to greater stability for that rotamer, in the absence of hydrogen-bonding solvents.

5.14 CONFORMATION OF CYCLIC MOLECULES

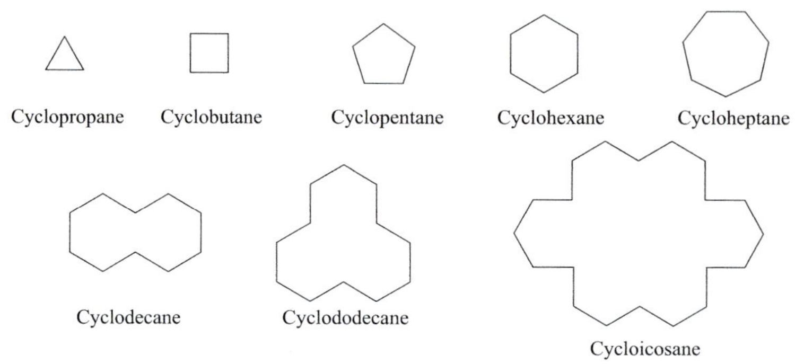

In a cyclic compound, the carbon atoms are confined to a ring, and a carbon-carbon bond in a ring cannot rotate around 360° without breaking those bonds. It is possible, however, to "partially rotate" the bond (in other words, rotate by some angle less than 360°, clockwise or counterclockwise). Such a partial rotation is called *pseudorotation*. The simplest drawings of cyclic alkanes are elementary, two-dimensional geometric figures. Using line notation, the three-membered ring cyclopropane is shaped like a triangle, the four-membered ring cyclobutane is a square, the five-membered ring cyclopentane is a pentagon, the six-membered ring cyclohexane is a hexagon, and the seven-membered ring cycloheptane is a heptagon. However, cyclic molecules are not planar but rather assume their low-energy conformation via pseudorotation.

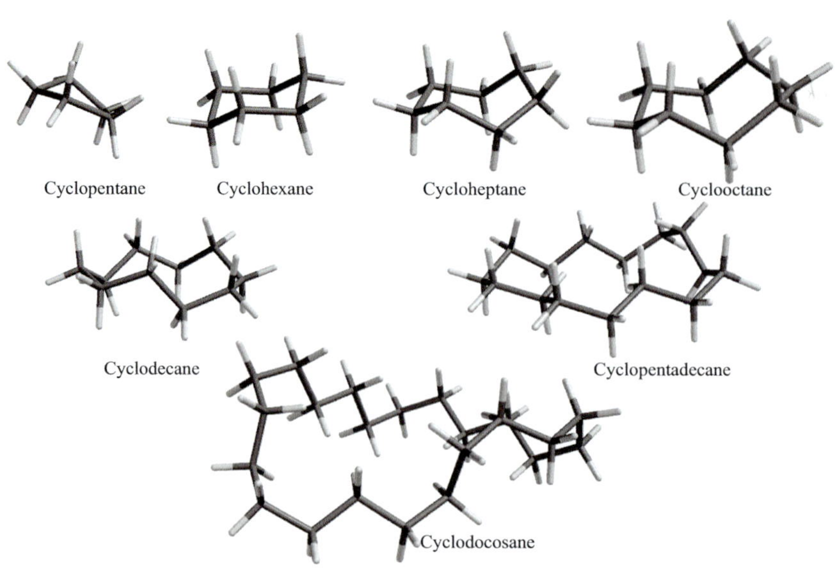

If the planar structures were correct, the C–C–C bond angle for cyclopropane would be 60°, 90° for cyclobutane, 108° for cyclopentane, 120° for cyclohexane, and 128.6° for cycloheptane. These angles are calculated from the angles of the regular polyhedral structures used to represent the structures (triangle, square, etc.), which come from the formula for a planar polygon with "n" sides;. Note that drawing the cyclic molecules as planar rings implies a shape or conformation for each molecule. Cyclic molecules do have conformations, but it must be determined whether the planar conformation is the highest or lowest in energy for a given ring. Examples of these conformations include cyclopentane, cyclohexane, and cycloheptane, which have a "crown"-type structure. Clearly, the low-energy conformation is not planar. In cyclooctane and the C10 ring cyclodecane, a more complicated conformation is shown for both the C15 ring cyclopentadecane and the C22 ring cyclodocosane. Indeed, there are myriad conformations for these rings.

The bond angle for a tetrahedral carbon is 109°28′ in methane, and if this is taken as a typical value for sp^3 hybridized carbon in a ring, most of the planar structures for three- to seven-membered rings will have significant distortion of the bond angles. The *differences in the bond angles* expected for a tetrahedral carbon in methane and the bond angles for the carbon atoms in the *planar* structures are ~49°, 19°, 1°, 11°, and 19°, respectively. The greater the distortion from the ideal tetrahedral bond angles, the greater the energy required for the molecule to assume a conformation with planar geometry, and the less stable that structure should be. The energy induced in a molecule due to distortion of the bond angles is called *Baeyer strain.*

> *The formal definition of* ***Baeyer strain*** *is the conformational energy arising in a molecule from distortion of the bond angles away from the tetrahedral ideal.*

By this analysis, planar cyclopentane should be the most stable, followed by cyclohexane, cyclobutane (which is about the same as cycloheptane), and finally cyclopropane. However, *this is not the correct order for the inherent stability of cyclic alkanes.* Energy calculations clearly show that cyclopropane is the highest in energy, but they also show that cyclohexane is lower in energy than cyclopentane.

> *The strain energy arising from eclipsing interactions in cyclic molecules of this type is called* ***Pitzer strain,*** *although it may also be called* ***torsional strain***.

Cyclic alkanes are not planar because they absorb energy, the same as acyclic alkanes. Since 360° rotation about C–C bonds is impossible in a cyclic alkane, these molecules can dissipate energy via a twisting motion about each bond. This twisting motion is called *pseudorotation*, as mentioned above, and leads to distortion away from a planar conformation of a cyclic alkane into lower-energy conformations, as shown in the conformations above. In other words, the three-dimensional structures of cyclic alkanes (their low-energy conformations) result from pseudorotation within each ring system. Pseudorotation also leads to lower-energy conformations that alleviate Baeyer strain. As the size of the ring increases, there are many possible conformations. Of all the conformations possible for a cyclic alkane, *the planar structure is one of the highest in energy* and least likely to form because of the high energy demands just described.

The lowest-energy shape or conformation is different for each cyclic alkane. Beginning with cyclopropane, the flat line drawing suggests that there is considerable strain in this molecule because the bond angles are distorted to 60° (Baeyer strain), and all the C–H bonds are eclipsed (torsional strain). Trying to "twist" this three-membered ring about any given C–C bond by pseudorotation is very difficult, however, and in fact virtually no pseudorotation is possible. A ball-and-stick model of cyclopropane is provided in Figure 5.25a, and it clearly shows that cyclopropane is flat. All the C–H bonds overlap and the hydrogen atoms are eclipsed, so there is significant torsional strain (see the space-filling molecular model, Figure 5.25b).

Since pseudorotation is so difficult, the C–C–C bond angle in cyclopropane is locked at 60°, and the shortest distance between carbon nuclei is represented by a "dashed" line in Figure 5.25c. The electron density of the σ-bonds is pushed out of linearity, however, and does not follow this line. In other words, forcing the bonds in cyclopropane close together with a 60° bond angle means there is significant "electronic distortion" of the electrons in those bonds as they push away from each other. This distortion is illustrated for the σ-bonds shown for cyclopropane in Figure 5.25c, which is said to have "banana-bonds." These are simply distorted covalent bonds, but they look a little bit like π-bonds. *They are not π-bonds!* There is less electron density between the carbon nuclei than in a normal σ-bond due to the distortion, and *the bonds are weaker than a normal alkane carbon-carbon bond*. If the bond is weaker, it is easier to break.

If one carbon is added to the ring to make cyclobutane, the ring is larger and it has more flexibility relative to cyclopropane. Therefore, cyclobutane can undergo some pseudorotation about the C–C bonds of the ring, leading to non-planar conformations. The bond angles are about 90° in planar cyclobutane, so there is less Baeyer strain than in cyclopropane, but there is a significant amount of torsional strain. Pseudorotation leads to the lowest-energy conformation, which is a "puckered" conformation called the *butterfly conformation* (see Figure 5.26). Molecular models are shown, illustrating the diminished Pitzer strain, making that conformation more stable than the planar form. Cyclopentane has one more carbon than cyclobutane, which gives the molecule even more flexibility for pseudorotation. The simple line drawing of cyclopentane is the planar structure, but

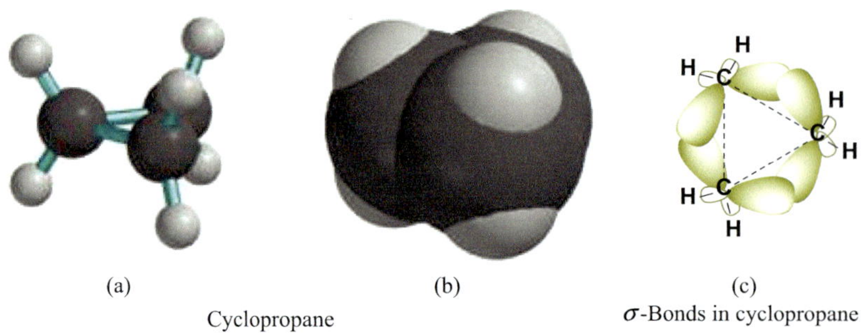

FIGURE 5.25 Bond distortion in the strained cyclopropane ring.

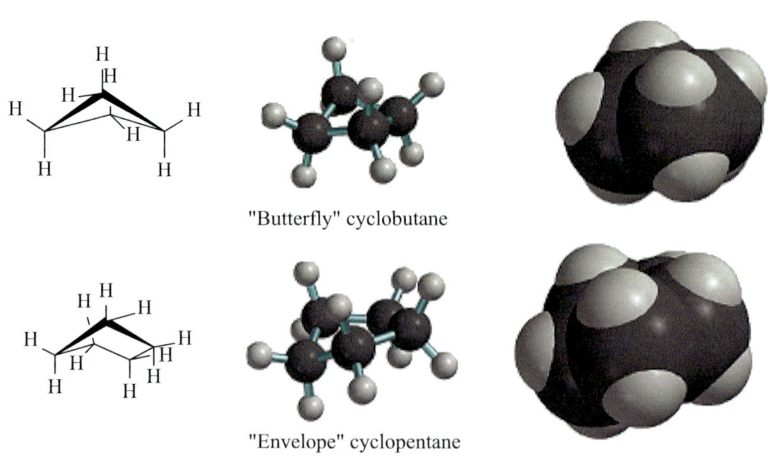

FIGURE 5.26 Low-energy conformations of cyclobutane and cyclopentane.

pseudorotation leads to significant distortion, bringing some atoms closer together while taking other atoms farther apart. The lowest-energy conformation resulting from this pseudorotation is the so-called *envelope conformation* shown in Figure 5.26. Interestingly, the bond angles for planar cyclopentane are 108°, very close to the ideal tetrahedral angle of 109°28′, so there is little Baeyer strain. Pseudorotation diminishes the torsional strain by pushing the eclipsing hydrogen atoms farther apart, but the bond angles are decreased to 105°, so there is a small increase in Baeyer strain. Relief of torsional strain competes with a slight increase in Baeyer strain. The observation that cyclopentane exists primarily in the envelope conformation is an indication that *relief of torsional strain is more important than the increase in Baeyer strain for cyclopentane.*

Cyclohexane has one more carbon atom than cyclopentane, and the greater flexibility allows for significantly more pseudorotation. Planar cyclohexane has significant torsional strain, and the bond angles would be 120°, as seen in both the line drawing and the Newman-projection-type molecular model in Figure 5.27. Of all the conformations possible due to pseudorotation, experimental data for cyclohexane indicate that one is lowest in energy. This conformation is thought to look a little like an easy chair, and it is called the *chair conformation***.** A comparison with the chair conformation, also presented as a Newman projection, shows there is virtually no Baeyer strain. The experimentally measured H–C–H bond angles of chair cyclohexane are tetrahedral, or about 109.5°, indicative of little or no torsional strain.

The chair conformation is the lowest-energy conformation of cyclohexane, but in fact there are two equilibrating chair conformations. The C1 and C4 carbon atoms are marked in both structures, and C1 is "up" in one, whereas C1 is "down" in the other, and C4 is "down" in one, but "up" in the other (see Figure 5.28). While two structures are shown in Figure 5.28, they are identical chair conformations, and they are identical in energy. Twisting the bonds (pseudorotation) in the chair on the left will interconvert it to the chair on the right. In other words, the two chair conformations in Figure 5.28 are in equilibrium, and because they are of the same energy, the equilibrium constant, K_{eq}, is unity ($K_{eq}=1$). This means that there is a 50:50 mixture of the two chair conformations.

A chair conformation of cyclohexane is shown again in Figure 5.29. Six of the 12 bonds that connect hydrogen atoms are in the vertical plane, whereas the other six bonds are in the horizontal plane. If chair cyclohexane is inserted into a "planet," the bonds in the vertical plane are aligned in the direction of the axis and are called *axial* bonds. Those bonds in an approximate horizontal plane circling the equator are called *equatorial* bonds. The hydrogen atoms attached to the axial or equatorial bonds are referred to as *axial* or *equatorial hydrogen atoms.* A molecular model of

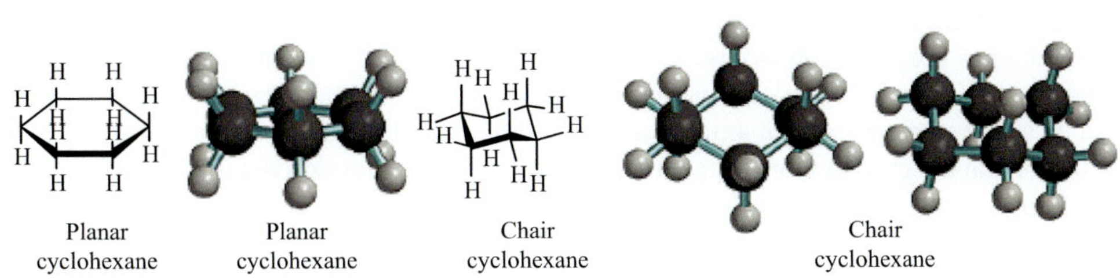

FIGURE 5.27 Planar and chair cyclohexane.

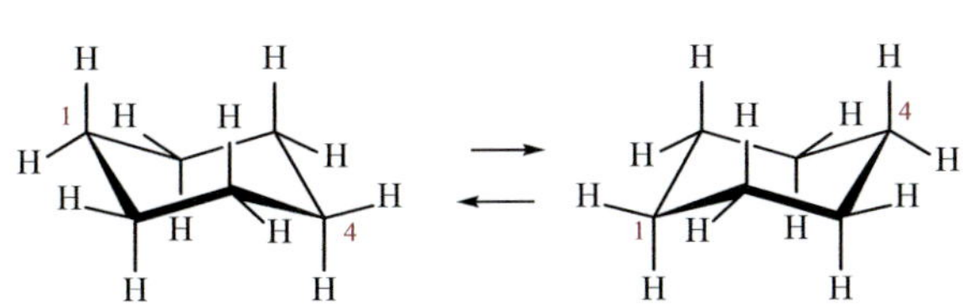

FIGURE 5.28 Equilibrating chair conformations for cyclohexane.

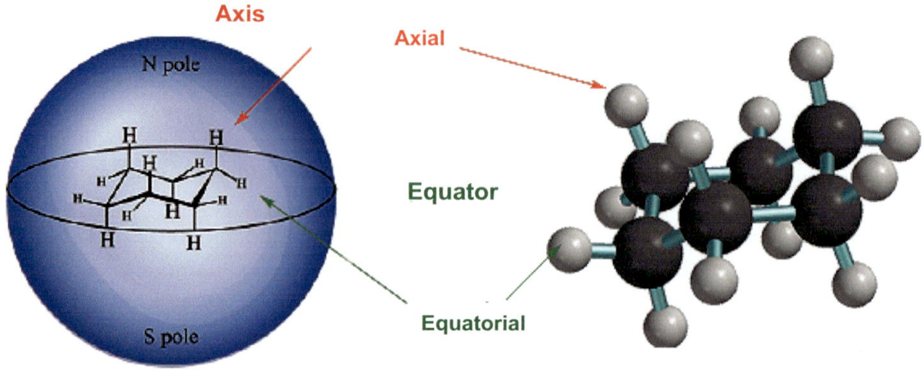

FIGURE 5.29 Axial and equatorial hydrogen atoms in chair cyclohexane.

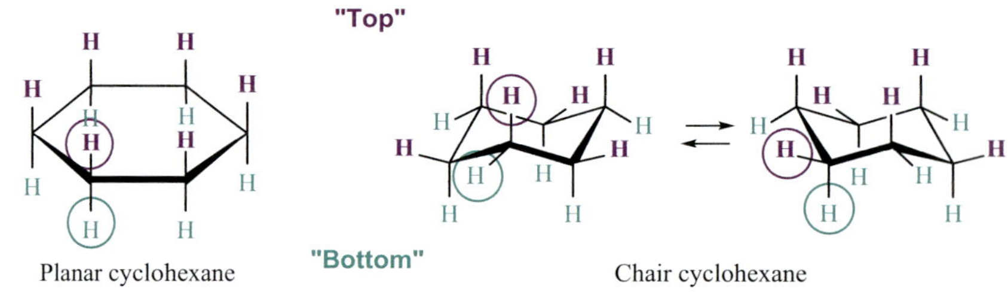

FIGURE 5.30 Interconversion of axial–equatorial hydrogen atoms in chair cyclohexane.

cyclohexane is included in Figure 5.29 to correlate the spatial relationship of axial and equatorial bonds in the actual molecule with those inside the "planet."

The equilibrating chair conformations in Figure 5.28 show an interesting phenomenon. All the axial hydrogen atoms in one chair become equatorial hydrogen atoms in the other chair, and vice versa. Examination of planar cyclohexane in Figure 5.30 reveals that the ring has two sides, "top" and "bottom." There are six hydrogen atoms on the "top" of the molecule, marked in violet, and six more on the "bottom," marked in cyan. When the planar cyclohexane is converted to the chair, three of the "top" hydrogen atoms are axial and three are equatorial. This means that all six hydrogen atoms on the "top" in the chair (in violet) show an alternating pattern: axial-equatorial-axial-equatorial-axial-equatorial (three axial and three equatorial). Likewise, the "bottom" of the chair conformation has six hydrogen atoms (in cyan) with three axial hydrogen atoms and three equatorial. The alternating axial-equatorial pattern appears on both the "bottom" and the "top" of cyclohexane, so a carbon atom with an axial hydrogen atom on the top has an equatorial hydrogen atom on the bottom, and a carbon atom with an equatorial hydrogen atom on the top has an axial hydrogen atom on the bottom.

Pseudorotation converts the chair on the left into the chair on the right in Figure 5.30. A comparison of the two structures shows that all the violet hydrogen atoms in one chair remain on the "top," but all axial hydrogen atoms in one chair are equatorial in the other. All equatorial hydrogen atoms in one are axial in the other. The same phenomenon is observed for hydrogen atoms on the bottom, in cyan. Note also that in one chair conformation the hydrogen atom on the top is axial and the hydrogen atom on the bottom, attached to the same carbon, is equatorial. This interconversion is sometimes called a "ring flip," but there is no actual flip. One chair is converted into the other by pseudorotation (twisting). Of the many conformations available to cyclohexane, the two chair conformations are lowest in energy, and cyclohexane will "spend most of its time" as equilibrating chair conformers. In other words, there is a higher percentage of the chair conformations than any of the others.

When substituents are attached to a ring carbon of cyclohexane, the position of the chair-chair equilibrium is different, and $K \neq 1$. A substituent may be axial or equatorial, and the substituent in one chair will be equatorial while it will be axial in the other. The different positions of the substituents lead to a difference in energy for the two chair conformations. The axial hydrogen atoms are not on adjacent carbons, and the interaction between the axial hydrogen atoms is not very large. If one of the axial hydrogen atoms is replaced with the larger chlorine atom, however, the steric interaction of axial Cl with the two axial H atoms is much greater, as shown in Figure 5.31. The axial hydrogen atoms on C3 and C5 in chlorocyclohexane, as well as the chlorine on C1, have a "1–3" relationship (every third carbon on the top and on the bottom has an atom in the axial position). There is a steric interaction of a substituent such as chlorine with the two axial hydrogen atoms, as seen in Figure 5.31. This steric interaction is a 1,3-diaxial interaction, commonly known as *A-strain* or *$A^{1,3}$-strain*. The axial hydrogen atoms and the axial chlorine atom arc not on adjacent atoms, so the steric interaction is literally across the six-membered ring. This type of steric strain is another type of *transannular strain.*

There is an equilibrium between the two chair conformations of chlorocyclohexane, just as in cyclohexane, but $A^{1,3}$-strain influences the equilibrium of the chair-chair interconversion. One chair conformation of chlorocyclohexane has the chlorine atom equatorial, as shown in Figure 5.31. Looking down on the molecular model with an equatorial chlorine, there is only a tiny interaction with the three axial hydrogen atoms on the top (or bottom) of cyclohexane. In any cyclohexane derivative, the interaction of these three axial hydrogen atoms is taken to be the standard interaction. This statement means that *there is no $A^{1,3}$-interaction when the chlorine atom is equatorial.* The other chair conformation of chlorocyclohexane has the chlorine atom in an axial position, and there is $A^{1,3}$-strain due to the interaction of the chlorine with the two axial hydrogen atoms. Looking down on the molecular model with the chlorine axial shows this steric interaction of the chlorine with the other axial hydrogen atoms. *This transannular steric interaction is $A^{1,3}$-strain.* To be precise, the $A^{1,3}$-strain (or just A-strain) in a monosubstituted cyclohexane is due to the interaction of the substituent with the axial hydrogen atoms on that "side" of the molecule. At equilibrium, there is a much smaller concentration of the higher-energy axial chlorine conformation and a much higher concentration of the lower-energy equatorial chlorine conformation (K_{eq} is larger, favoring the equatorial chlorine conformation).

As mentioned earlier, as the size of the ring in cyclic molecules increases, more pseudorotation is possible and there are many more conformations. Indeed, cycloheptane, cyclooctane, and larger rings have many more conformations. Since larger rings will not appear in this book, the conformational mobility of these compounds will not be discussed in this review.

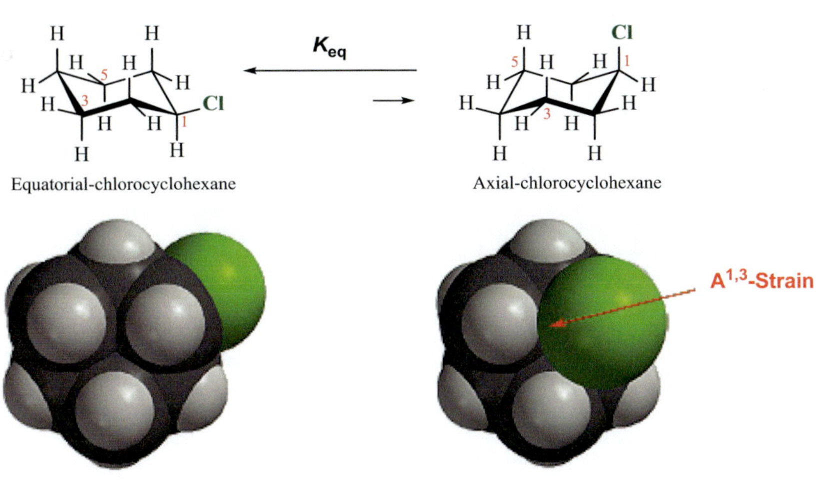

FIGURE 5.31 $A^{1,3}$-strain in chlorocyclohexane.

5.15 STEREOGENIC CARBONS AND STEREOISOMERS

When an sp^3-hybridized carbon has four different atoms or groups attached, those atoms are positioned at the corners of a regular tetrahedron around to the central sp^3-carbon. The tetrahedral arrangement results from the electron repulsion of the covalent bonds and the steric repulsion of the attached atoms, so they are as far apart as possible while all being attached to the central carbon. In other words, the attached atoms form a tetrahedral shape around that carbon atom. The relative positions of these atoms and groups to one another are fixed. Once attached to a central carbon atom, the only way to change the relative and specific positions of the four atoms or groups is to make or break bonds. If the atoms and groups attached to the sp^3 center are different from one another, a new type of stereoisomer is possible, based on generating the mirror image of the original compound.

When four different atoms are attached to a central carbon atom, the mirror image of that molecule is a completely different molecule that cannot be superimposed on the original. In other words, the *mirror image cannot be* superimposable on the original. The *molecule is asymmetric (it has no symmetry), so the molecule and its mirror image are different. The carbon in such a molecule is known as an asymmetric or chiral carbon.* The modern term is *stereogenic* carbon, and a molecule with a stereogenic carbon is said to be a chiral molecule. In the following discussion, the sp^3 atom is usually carbon, but other atoms may also be chiral.

Methane is a carbon atom with four identical hydrogen atoms attached in a tetrahedral arrangement. The attached atoms are the same, so there is a plane of symmetry, as shown in Figure 5.32, that bisects the central carbon and two of the hydrogen atoms. Of the two remaining hydrogen atoms, one projects to the left (in green) and the other to the right (in violet) of this plane. If the green H were reflected through the plane, it would superimpose on the violet H, so the atoms on either side of the plane are identical (both H). Because methane has symmetry, its mirror image of methane will superimpose. Any carbon containing groups or atoms (A,B,C,D) may be inspected in this manner to check for a plane of symmetry.

The C2 of 2-chlorobutane has four different atoms or groups attached: Cl, H, methyl, and ethyl. Therefore, C2 qualifies as a stereogenic center. Both 2-chlorobutane and its mirror image are drawn in Figure 5.33. It is not possible to define a plane through any three of the atoms (including the stereogenic center) such that the atoms or groups on either side of the plane are the same. In other words, 2-chlorobutane and its mirror image are two different molecules. A complete analysis of 2-chlorobutane shows that there is a complete lack of symmetry, which means that the central carbon is a stereogenic atom.[4]

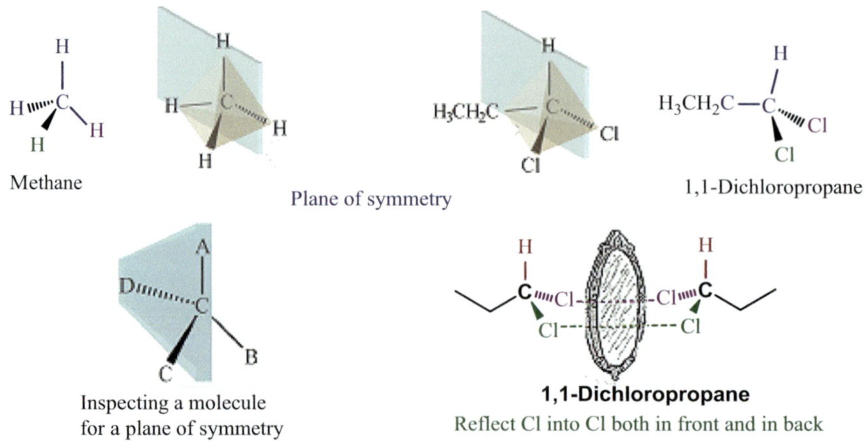

FIGURE 5.32 Planes of symmetry in methane and 1,1-dichloropropane.

[4] Eliel, E.L.; Wilen, S.H.; Mander, L.N. *Stereochemistry of Organic Compounds*, Wiley, New York, **1994**, p. 1208.

The four different groups and atoms for 2-chlorobutane in Figure 5.33 are color-coded for convenience. Using models, the two structures are easily compared to show that they are not superimposable. *It is not possible to make all the atoms and groups match up, so there are two molecules.* If molecular models of the two mirror images are superimposed, as in Figure 5.34, the two structures cannot be superimposed *atom for atom*. When 2-chlorobutane is compared with its mirror image, they have the same atoms attached to C2, so they have the same empirical formula. They differ in the spatial orientation of the atoms relative to the central carbon atom and cannot be interconverted except by breaking bonds. They are *stereoisomers* in that they have the same points of attachment but differ in their spatial orientation. They are different molecules. *Two mirror-image stereoisomers that do not superimpose are called enantiomers.* A molecule that has a stereogenic carbon and generates at least one pair of enantiomers is said to be *chiral*.[4]

The ability to determine if a molecule and its mirror image are superimposable is key to the recognition of enantiomers. The stereogenic carbon can be viewed from a different perspective, essentially from one edge of a tetrahedron. One enantiomer of 2-chlorobutane in Figure 5.35 shows the stereochemistry of the stereogenic carbon using a special line notation called a *Fischer projection*. It is easy to determine that the four groups or atoms attached to the stereogenic carbon are ethyl, methyl, chlorine, and hydrogen. The molecule is adjusted to project the methyl and the hydrogen atoms out from the front of the page (solid wedges), with the chlorine atom and the ethyl group projected behind the page (dashed lines). If the wedges and dashes are replaced with simple lines,

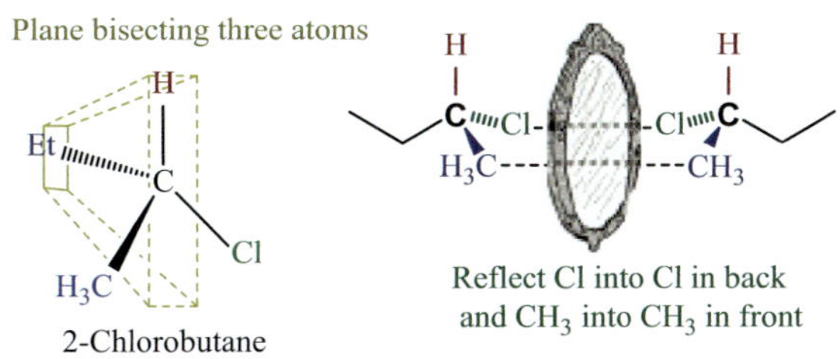

FIGURE 5.33 2-Chlorobutane and its mirror image.

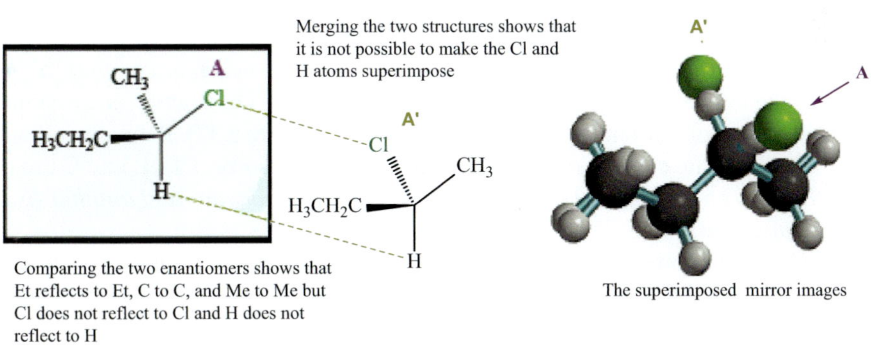

FIGURE 5.34 Attempt to superimpose enantiomers of 2-chlorobutane.

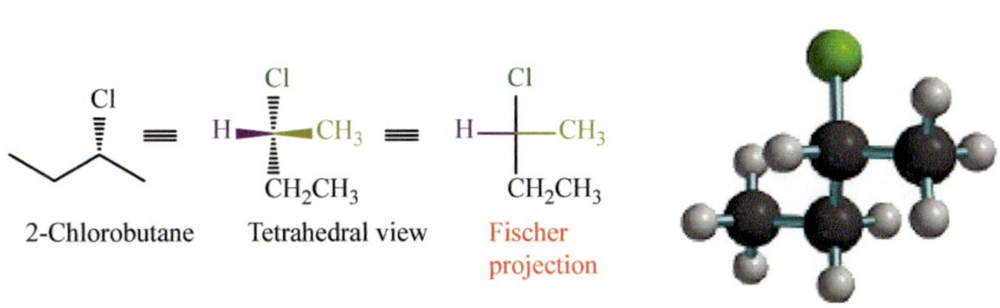

FIGURE 5.35 Fischer projection of 2-chlorobutane.

a *Fischer projection* is obtained in which two groups are attached to the vertical line and two are attached to the horizontal line. Note that the four groups appear to be attached to a "cross." *Drawing a molecule this way constitutes the definition of a Fischer projection,* in which the horizontal line is projected in front of the page and the vertical line is projected behind the page. A molecular model of the Fischer projection is also shown in Figure 5.35 to illustrate the three-dimensional arrangement of the atoms and groups.

5.16 ABSOLUTE CONFIGURATION: THE (*R*) AND (*S*) NOMENCLATURE

Enantiomers are different molecules, and each requires a unique name. Both enantiomers of 1-bromo-1-chloroethane are drawn in Figure 5.36, both in normal line notation and as a Fischer projection. In the Fischer projections, the Cl is on the left side in one enantiomer and on the right side in the other. A specific name must correlate the actual structure of a given enantiomer, and a set of standardized rules has been developed to name each enantiomer. Once the enantiomer has been named, the name can be correlated with the structure.

Any name that allows two enantiomers to be distinguished must be very specific. The two enantiomers of 1-bromo-1-chloroethane differ only in their spatial arrangement of atoms, in that they are mirror images. Each arrangement of atoms or groups constitutes the *absolute configuration* of that enantiomer. Naming uses the terms (*R*) (rectus; *Latin* for right, correct, or proper) and (*S*) (sinister; *Latin* for left or improper) to identify the absolute configuration of each enantiomer. The names of the two enantiomers in Figure 3.36 are therefore (*R*)-1-bromo-1-chloroethane and (*S*)-1-bromo-1-chloroethane. How were the names (*R*) and (*S*) determined? Which enantiomer is (*R*), and which is (*S*)?

There is a set of rules for assigning the *absolute configuration* of a given stereogenic center, and that assignment becomes part of the name for that enantiomer. All *atoms* attached to a stereogenic carbon atom are inspected with a simple goal: to assign a priority of importance to each of the atoms attached to the stereogenic carbon. The spatial arrangement of these atoms is the absolute configuration, *(R)* or *(S)*.

Initially, focus on the atomic mass of each *atom*. 1-Bromo-1-chloroethane has H, Cl, Br, and a CH_3 group attached to the stereogenic carbon. Do not use the mass of the group ($-CH_3$) but rather the mass of the atom attached to the stereogenic atom, in this case the carbon atom. Therefore, *compare the atomic masses of Br, Cl, H, and C*. The order, according to descending atomic mass, is *Br>Cl>C>H*. The priority letters a, b, c, and d are assigned to each atom, with "a" the highest priority and "d" the lowest priority; *Br*=a, *Cl*=b, *C*=c, and *H*=d. These letters are assigned to a tetrahedral representation of 1-bromo-1-chloroethane (see **A** in Figure 5.37). Remember that the *atoms* attached to the stereogenic carbon are examined, not the group, so Br, Cl, H, and C are compared. *The protocol assigns one atom as the highest priority (most important) and another as the lowest priority (least important).*

Before the (*R*) or (*S*) nomenclature can be assigned, an *assumption* must be made. The model with assigned a, b, c, and d must be viewed such that the lowest priority group (d) is projected to the rear such that a→b→c forming the base of the tetrahedron, which is projected to the front, as illustrated in Figure 5.37. The tetrahedral model of 1-bromo-1-chloroethane (**A**) is drawn as a simple tetrahedron using four different colors for the "*a→d*" atoms/groups. The same representation is shown with the tetrahedron marked. Starting with **A**, the tetrahedron is tipped back, and as the d

FIGURE 5.36 Enantiomers (non-superimposable mirror images) of 1-bromo-1-chloroethane.

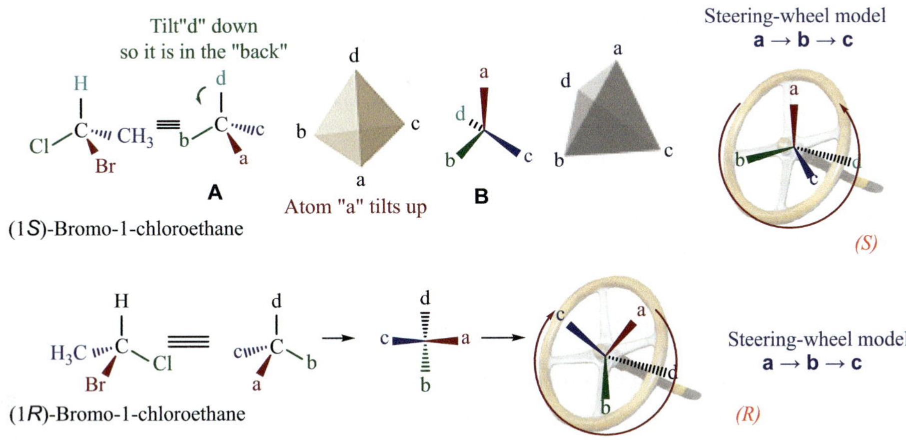

FIGURE 5.37 The steering wheel model.

group is tipped to the rear, the "*a*" tips up and the "*b*-*c*" groups remain more or less in the same position. This tilting motion leads to a different view of the tetrahedron, marked **B**. A requirement of this model is that *the lowest priority atom "d" is pointed 180° from the viewer; that is, behind the plane of the page (to the rear of the tetrahedron),* so the tetrahedron is effectively viewed from the triangular base.

With the *d* group projected to the rear, imagine that a curved arrow is drawn from the highest priority atom (a) toward the next highest priority atom (b), and finally toward (c). This imaginary arrow generates the arc of a circle, with the center being the stereogenic atom, but also connected to d (projected behind the plane of the paper). This view looks like the steering wheel of an old-time car, so this model is called the *steering-wheel model* (see Figure 5.37). In this example, the imaginary arrow from $a \rightarrow b \rightarrow c$ goes in a counterclockwise direction, and it is labeled the (*S*)-configuration. The absolute configuration, (*R*) or (*S*), becomes part of the name, so the name in this example is (1*S*)-bromo-1-chloroethane. The enantiomer is (1*R*)-bromo-1-chloroethane, and using the same protocol and priority scheme leads to an arrangement in which the arrow from $a \rightarrow b \rightarrow c$ proceeds in a clockwise rotation (also in Figure 5.37), the (*R*)-configuration.

> *The absolute configuration of an enantiomer is determined by assigning priorities a–d for atoms connected to a stereogenic center, rotating so the (d) group is projected to the rear, and, if the direction a → b → c is* ***clockwise****, the absolute configuration is* ***(R)****; if the direction is* ***counterclockwise****, the absolute configuration is* ***(S)****.*

Using the steering-wheel model, the absolute configuration of each enantiomer of 2-chlorobutane can be determined.

The two enantiomers of 2-chlorobutane are shown in Figure 5.38, where ethyl, methyl, hydrogen, and Cl are attached to the stereogenic carbon. As noted above, *only the atoms attached to the stereogenic carbon* are used for the steering-wheel model (C, C, H, and Cl). Clearly, chlorine has the highest atomic mass (it is assigned "*a*"), and hydrogen has the lowest atomic mass (it is assigned "*d*"). Two of the atoms attached to the central carbon are the same; that is, two carbon atoms. *Remember that the ethyl group is not compared with the methyl group, but rather the carbon atom of ethyl attached to the stereogenic carbon is compared with the carbon atom of the methyl attached to the stereogenic carbon.* The atomic mass rule cannot establish the priority since both atoms are the same. Another rule is required! There is a shorthand method used to help assign priorities for the carbon atoms directly attached to the stereogenic center. In Figure 5.38, the methyl group is attached to the stereogenic carbon with three hydrogen atoms, so it is represented as C^{HHH}.

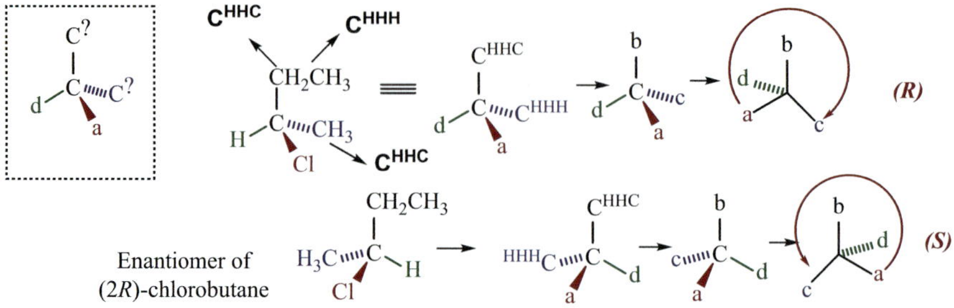

FIGURE 5.38 Determining absolute configuration for (2*R*)-chlorobutane and for (2*S*)-chlorobutane.

The ethyl group is attached by a CH_2unit to the stereogenic carbon, and that carbon has two hydrogen atoms and another carbon atom, so it is represented as C^{CHH}. This *superscript protocol* will be used in the following examples to define the rules of nomenclature for absolute configuration.

Examination of Figure 5.38 shows that analysis of the two carbon atoms attached to the stereogenic carbon reveals a structural difference based on the other atoms that are attached. Therefore, examine the other atoms that may be attached, but do *not* count the stereogenic carbon. If the methyl group is represented as C^{HHH} and the methylene group of the ethyl group is represented as C^{CHH}, there is clearly a *point of difference. At the first point of difference, the highest priority atom is determined by comparing the atomic masses of the attached atoms.*

The C^{CHH} unit has an attached carbon as the highest-mass atom, whereas the C^{HHH} unit has only an attached hydrogen as the highest-mass atom. Since carbon has a higher mass than hydrogen, C^{CHH} has a higher priority than C^{HHH} (*b* and c, respectively). This information allows the priorities to be assigned for this enantiomer as (2*R*)-chlorobutane. Applying the same protocol to the enantiomer of (2*R*)-chlorobutane leads to the opposite absolute configuration, and the name is (2*S*)-chlorobutane.

In 6-chloro-1-fluorohexan-3-ol, the same analysis used above leads to the tetrahedral representation in Figure 5.39, showing that the stereogenic center (C) is attached to H, O, C, and C. Since O is the highest priority atom (*a*) and hydrogen is the lowest priority (*d*), C and C remain unassigned. The carbon in the fluoroethyl fragment has a substitution pattern C^{CHH}, and the carbon in the chloropropyl fragment also has a C^{CHH} substitution pattern. There is no point of difference because the highest priority attached atoms are identical (C and C). The first point of difference between the groups occurs further away from the stereogenic center. A new rule is needed!

Examine each substituent atom by atom down each pertinent chain until a point of difference is found that allows priority assignment based on the atomic mass of the highest priority attached atom.

In this example, the analysis must continue further down each carbon of the two chains in question. In the chloropropyl fragment, the next carbon in the chain has the substitution pattern C^{CHH}, whereas the next carbon of the fluoroethyl fragment is C^{FHH}. At this point of difference, the priority atoms of C^{FHH} and C^{HHC} are compared. F has higher priority than C by atomic mass, and the priority scheme

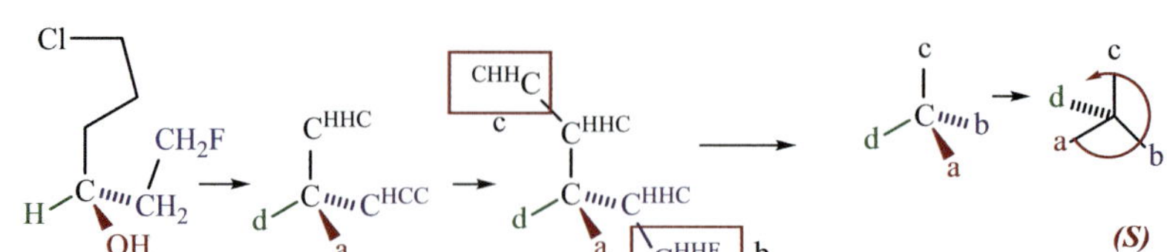

(6*S*)-Chloro-1-fluorohexan-3-ol

FIGURE 5.39 Absolute configuration of (6*S*)-chloro-1-fluorohexan-3-ol.

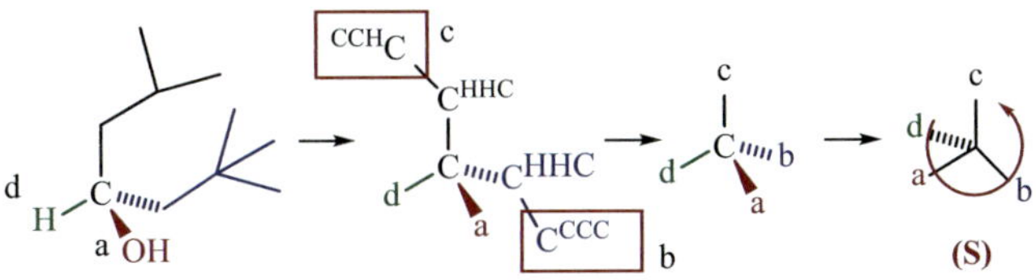

FIGURE 5.40 Absolute configuration of 2,2,6-trimethylhetpan-4-ol.

is as shown. This alcohol has an (*S*) absolute configuration, and the molecule is named 6-chloro-1-fluoro-(3*S*)-hexanol. Note that the chlorine atom in 6-chloro-1-fluorohexan-3-ol is not used for priority assignment because the point of difference is encountered before the chlorine atom in that chain.

The example in Figure 5.40 offers an interesting dilemma. For 2,2,6-trimethylheptan-4-ol, the tetrahedral representation shows that the groups attached to the stereogenic carbon are 2-methylpropyl, 2,2-dimethylpropyl, hydrogen, and OH, so the priority comparison is for C, C, H, and O. Once again, O is the highest priority (*a*) and hydrogen is the lowest priority (d). Looking at the carbon atoms attached to the stereogenic center, there is C^{CHH} for 2-methylpropyl and C^{CHH} for 2,2-dimethylpropyl. They are identical, so this is not a point of difference. As in the previous example, move down the chain to the next carbon and look for a point of difference. The next carbon in the 2-methylpropyl chain is attached to two methyl groups (labeled C^{CCH}). The next carbon in the 2,2-dimethylpropyl chain is attached to three methyl groups (labeled C^{CCC}).

While it is a point of difference, the attached atom on both carbons is C, and since C>H, the comparison is carbon with carbon for both carbon atoms. In other words, at the point of difference it is not possible to determine the priority based on atomic number. Another rule is required! In this case, C^{CCC} is compared with C^{CCH}, and since atomic mass cannot be used to determine priority, the number of priority atoms is used. The 2,2-dimethylpropyl chain has three carbons at the point of difference, whereas the 2-methylpropyl chain has two carbons at the point of difference. The assignment is (*b*) for the 2,2-dimethylpropyl group and (*c*) for the 2-methylpropyl group, making the priority assignment (*S*). The name is 2,2,6-trimethyl-(4*S*)-heptanol.

When, and ***only*** *when, the first point of difference cannot be resolved by atomic mass because the priority atoms are the same, count the number of priority atoms.*

The preceding discussion used rules to determine absolute configuration, but in narrative form. These rules are called the *Cahn-Ingold-Prelog selection rules (sometimes called the CIP rules).* The rules have been formalized and expanded by IUPAC, and they formally constitute the IUPAC rules for determining stereochemistry. The first three rules are summarized:

1. Assign a priority to the four atoms directly connected to the chiral atom based on the atomic mass of each *atom* attached to the chiral atom. The higher the atomic mass, the higher the priority. If isotopes are involved, the higher-mass isotope takes the higher priority ($^3H > ^2H > ^1H$, e.g.).
2. If any atoms directly attached to the chiral atom are the same (same atomic mass), proceed down each chain (away from the chiral atom) until a point of difference is found. At that point, use the atomic mass rule to determine the priority. If the end of a chain is reached and there is no point of difference, those groups are the same and the atom of interest is *not* chiral
3. If the first point of difference is reached and priority cannot be determined by differences in atomic mass, count the number of the highest-priority atoms at that point. The atom with the largest number of priority atoms takes the highest priority.

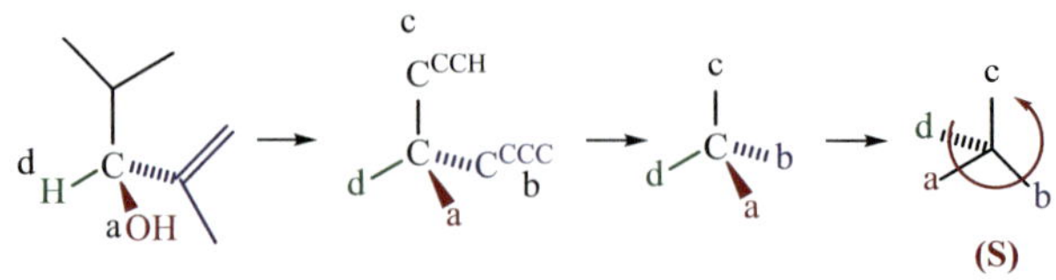

FIGURE 5.41 Absolute configuration of 2,4-dimethylpent-1-en-3-ol.

The CIP rules work well in the cases discussed, but substituents or groups that contain π-bonds have been ignored. 2,4-Dimethylpent-1-en-3-ol is an example that fits into this category, and an analysis is shown in Figure 5.41. This alcohol has the carbon of a π-bond attached directly to the stereogenic atom (the common classification of such a molecule is as an *allylic alcohol*). The stereogenic carbon is connected to H, O, C, and C, with O assigned (*a*) and H assigned (*d*). The 1-methylethyl group and the 1-methylethenyl group pose a problem, however. The 1-methylethyl (isopropyl) group shows a substitution pattern C^{CCH}, but the other carbon is part of a C=C unit. Another rule is added.

Assume that both the σ- and π-bond of the C=C carbon unit are treated as separate carbon groups, so in effect there are two carbon substituents (one for each bond).

The fourth rule is:

4. If the atom being considered is part of a π-bond, each bond is counted as being attached to a substituent (two atoms for a double bond and three atoms for a triple bond).

From the perspective of the carbon attached to the stereogenic carbon, that carbon is attached to three carbon atoms (one from each bond of the double bond+the methyl group). This analysis leads to the substitution pattern C^{CCC} shown in Figure 5.41. To determine the priority at this point of difference, C is compared with C, but application of the new rule leads to C^{CCH} and C^{CCC}. The C^{CCC} assignment arises from one carbon from the methyl group and two carbons from the σ and π-bonds of the alkene unit. Therefore, the alkene unit has the higher priority. The stereogenic carbon has an (*S*)-configuration, and the name is (3*S*)-2,4-dimethylpent-1-en-3-ol.

There is one more instance where the given rules are insufficient. For example, there are molecules where the N has four different 'groups' if one counts the lone electron pair. When one of the "groups" attached to the stereogenic atom is an electron pair rather than another atom, the electron pair is always given the *lowest* priority (*d*).

In principle, a nitrogen atom can be stereogenic if there are three different alkyl groups on nitrogen. *The fourth group is the lone pair of electrons on nitrogen.* The nitrogen is not a stereogenic center, however, because rapid inversion at nitrogen leads to the mirror image (see *N*-ethyl-*N*-methylpropan-2-amine), so both structures are present at the same time. This rapid inversion of configuration is called *fluxional inversion*. Such compounds are not optically active unless the facile *fluxional inversion* can be inhibited. Fluxional inversion is very rapid, however, and approximately 2×10^{11} inversions occur each second for ammonia.[5] The energy barrier for this inversion is somewhat higher in amines[6,7] due to the presence of alkyl groups on nitrogen that are bulkier than the hydrogen atoms in ammonia. Nonetheless, alkyl amines undergo rapid inversion.[8] The magnitude of

5 Smith, M.B. *March's Advanced Organic Chemistry*, 8th ed., John Wiley & Sons, NJ, **2020**, pp 138–139.
6 Koeppl, G.W.; Sagatys, D.S.; Krishnamurthy, G.S.; Miller, S.I. *J. Am. Chem. Soc.* **1967**, *89*, 3396.
7 Kincaid, J.F.; Henriques, Jr., F.C. *J. Am. Chem. Soc.* **1940**, *62*, 1474.
8 (a) Mislow, K. *Pure Appl. Chem.* **1968**, *25*, 549; (b) Rauk, A.; Allen, L.C.; Mislow, K. *Angew. Chem. Int. Ed.* **1970**, *9, 400*; (c) Lambert, J.B. *Top. Stereochem.* **1971**, *6*, 19.

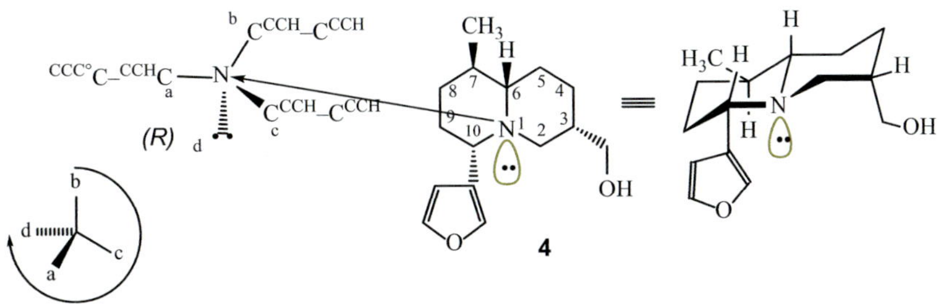

FIGURE 5.42 Determination of stereogenic centers in castoramine (**4**).

the energy barrier to inversion in amines is determined by the inter-group bond angle (α, C1–N-C2) and the corresponding bending force constant.[6] The inversion at nitrogen in acyclic amines cannot be stopped at the reaction temperatures usually employed in organic chemical reactions. Inversion can be stopped if a molecule has structural features such as those found in bicyclic and polycyclic amines. If fluxional inversion can be stopped, the nitrogen is stereogenic and enantiomers are observed.[8]

If a nitrogen atom is a stereogenic center, it must be assigned a configuration by the CIP selection rules. An example is the bridgehead nitrogen in (–)-castoramine (**4**), shown in Figure 5.42,[9] which is incapable of inversion at nitrogen because of its rigid bicyclic structure.[10] Castoramine is drawn such that the *trans*-orientation of the electron pair is indicated, which is preferred to the *cis*-orientation by ≈ 2.4 kcal (10.0 kJ)/mol.[10] The molecule is therefore effectively locked into the trans conformation. The electron pair must be considered a group, but the first five rules do not allow it to be assigned a priority. A sixth rule is necessary.

6. Lone electron pairs receive an atomic number of zero and are assigned the lowest priority. The duplicated C, O, and N atoms from rule 5 have a higher priority than an electron pair.

For (–)-castoramine, rule 6 sets the priority such that the stereogenic nitrogen center is (*R*).

Previous sections made it clear that the two enantiomers, (2*S*)-chlorobutane and (2*R*)-chlorobutane, are different molecules with different names. It is one thing to draw pictures, but it is quite another to experimentally verify the validity of the premise that the two structures are different molecules. In the case of enantiomers, there is a method for distinguishing the two enantiomers based on a difference in one physical property. The method is derived from the interaction of the chiral molecules with plane-polarized light.

5.17 SPECIFIC ROTATION

Two enantiomers differ in their spatial arrangement of atoms; i.e., they differ in their stereochemistry. However, the two enantiomers have identical physical properties, including boiling point, melting point, solubility in various solvents, refractive index, flash point, adsorptivity, and so on. There is only one physical property in which enantiomers differ. *Enantiomers differ in their interaction with plane-polarized light.* Normal light is filtered so that all the light is in a single plane (plane-polarized light). Note that virtually all methods for separating two different compounds rely on differences in physical properties, but all except one of the physical properties are identical for each enantiomer. Therefore, separation techniques based on physical properties (e.g., distillation or crystallization) cannot be used to separate a mixture of two enantiomers.

As noted, when light is passed through a polarizing filter, all the light that leaves the filter is in one plane. As the polarized light passes through a solution that contains a chiral compound, the light interacts with the chiral compound and the plane of the light is changed so that the angle of the light changes as it passes through the solution. The plane of light is rotated either to the right (clockwise) or to the left (counterclockwise) from the viewpoint of an observer. It is important to note

[9] LaLonde, R.T.; Muhammad, N.; Wong, C.F.; Sturiale, E.R. *J. Org. Chem.* **1980**, *45*, 3664.

[10] (a) Aaron, H.S.; Ferguson, C.P. *Tetrahedron Lett.* **1968**, 6191; (b) Aaron, H.S. *Chem. Ind. (London)* **1965**, 1338.

that *one enantiomer will rotate the plane of light counterclockwise, and the other enantiomer will rotate the plane of light clockwise.* If the plane of light can be detected before and after it interacts with the chiral molecule, the angle of rotation can be determined for each enantiomer. This change can be detected, and the plane-polarized light is measured in degrees. The instrument used to detect this rotation is called a *polarimeter*, and the degree of rotation is called the *observed rotation.*

A polarimeter is a device for measuring the angle of rotation of plane-polarized light for solutions containing chiral molecules. There is a light source and a polarizing filter. A *solution* of the chiral compound is placed between the polarizing light source and an eyepiece or other detection system. If the chiral compound is a solid, it must be dissolved in a solvent before it can be analyzed. Even if it is a liquid, the enantiomer is usually dissolved in a solvent. *The solvent cannot have a stereogenic center* because the rotation due to the change in the observed angle of the solvent would "swamp out" that of the molecule under examination. The concentration of the enantiomer in the solvent is measured in grams per milliliter (g/mL). This solution is added to a sample tube and placed into the polarimeter. As the plane-polarized light passes through the instrument with the solvent in the sample tube but without the chiral molecule, the plane of light is adjusted to 0° on an appropriate scale. The sample is then placed into the instrument. Sighting down the tube (through the solution), the plane of light is adjusted to determine the angle of rotation. In modern instruments, there is an instrumental detection system. Both the angle and the *direction* (clockwise or counterclockwise) are measured. These data are either recorded by the observer in arcane polarimeters or displayed digitally in modern polarimeters.

The magnitude of this angle is measured, and its direction is called the *observed rotation* and given the symbol α. Normally, (+)-α is used for a clockwise rotation and (–)-α is used for a counterclockwise rotation. A typical number read from the polarimeter will therefore be recorded as (+)-23° or (–)-56°. A molecule that rotates plane-polarized light in this manner is said to be *optically active.*

> *One enantiomer will have a (+) rotation, and its enantiomer will have a (-) rotation of exactly the same magnitude.*

The observed rotation will change with the solvent used, the concentration of the chiral compound, the length of the container used to hold the solution, and even the temperature. Therefore, the observed rotation of a chiral compound with one instrument is likely to record a *different* rotation than the analysis of the same compound using a different instrument with different parameters. The magnitude of the observed rotation (α) is measured in degrees, obtained directly from the polarimeter. As noted, this measurement is influenced (changed) by the solvent, the concentration, the temperature, the length of the polarimeter tube holding the sample, and the wavelength of the plane-polarized light. A standardized method is required that considers the differences in measurement conditions.

The standardized method converts an observed rotation to *specific rotation* (given the symbol $[\alpha]_D^{20}$), and it is a *physical property that is reported for optically active (chiral) molecules.* Specific rotation is calculated from the observed rotation α (taken directly from the polarimeter). The formula used for specific rotation is $[\alpha]_D^{20} = \dfrac{\alpha}{l \cdot c}$. In this formula, *α is the observed rotation* (the angle measured by the polarimeter). The term "*l*" is length of the sample holder (the cell that holds the sample solution), and it is measured in decimeters (dm). Most polarimeters have sample tubes that are 0.5, 1.0, 5.0, or 10.0 dm in length. The "*c*" term is the concentration of the enantiomer in solution and is measured in g/mL (*grams* of enantiomer per *mL* of solvent). If the observed rotation for a given compound is +102° at a concentration of 2.1 g/mL in ethanol, in a 5.0 dm cell, the specific rotation is $[\alpha]_D^{20} = \dfrac{+102}{5.0 \cdot 2.1} = \dfrac{+102}{10.5} = +9.71$. The D refers to the D-line of sodium when a sodium light source is used. It is the yellow line that appears in the visible spectrum with a wavelength of 589 nm. If a different wavelength of light is used from a different light source, the wavelength of light is recorded in place of D. The "20" on the bracket is the temperature (in degrees Celsius) at which the measurement was made.

The specific rotation for this example is reported as $[\alpha]_D^{20}$, +9.71 (*c* 2.1, ethanol). Reporting the specific rotation this way gives the number, the concentration, and the solvent *c*, which in this case is ethanol. With this information, anyone will be able to compare the observed rotation for the enantiomer with that reported using a different instrument but under the same conditions. Only after the physical measurement of both pure enantiomers of each named compound is made can the specific rotation of an enantiomer be correlated with the absolute configuration of that enantiomer.

There is no correlation between specific rotation and the absolute configuration (the specific location of groups attached to a stereogenic center).

If a mixture of enantiomers is prepared by mixing known amounts of each pure enantiomer, the specific rotation of the mixture can be determined because the *specific rotations of each enantiomer are additive*, using the sign of the rotation. Remember that the specific rotation of one enantiomer is identical to that of the other enantiomer but of opposite sign. In other words, if (+)-butan-2-ol has a specific rotation of +13°, then (–)-butan-2-ol has a specific rotation of –13°, and a 50:50 mixture of (+)- and (–)-butan-2-ol [labeled (±)-butan-2-ol above] will have a specific rotation of zero: $[\alpha]_D^{20}$ (mixture) $= 0.5(+13°) + 0.5(-13°) = +6.5° + -6.5° = 0$.

A 50:50 mixture of two enantiomers of a single molecule is called a ***racemic*** *mixture or a racemate.*

When this 50:50 *mixture* of enantiomers occurs, the compound is said to be *chiral, racemic*, or simply *racemic*. When butan-2-ol is labeled as (±)-butan-2-ol, it is a chiral, racemic mixture, or it can simply be said that butan-2-ol is racemic. A sample of only one enantiomer is said to be *enantiopure* [100% of (–) or 100% of (+)]. If a mixture is not a 50:50 mixture of enantiomers, as when a chemical reaction makes butan-2-ol, one enantiomer may be present to a greater extent than the other.

5.18 DIASTEREOMERS

Molecules containing more than one stereogenic center (e.g., 2,3-dichloropentane) have several stereoisomers. The presence of two stereogenic centers allows different arrangements of atoms for 2,3-diiodopentane. As shown in Figure 5.43, (2*S*,3*R*)-diiodopentane and (2*R*,3*S*)-diiodopentane are enantiomers. Likewise, (2*R*,3*R*)-diiodopentane and (2*S*,3*S*)-diiodopentane are also enantiomers. Clearly, (2*R*,3*R*)-diiodopentane and (2*S*,3*R*)-diiodopentane are stereoisomers, but they are not enantiomers. Note that changing the position of the Cl and H at C2 in (2*S*,3*R*)-diiodopentane changes the absolute configuration from (*S*) to (*R*), giving (2*R*,3*R*)-diiodopentane.

The *(2S,3S) and (2R,3R) stereoisomers are not enantiomers, and they are not mirror images of one another.* Likewise, the (2*R*,3*S*) and (2*R*,3*R*) stereoisomers are not enantiomers and are

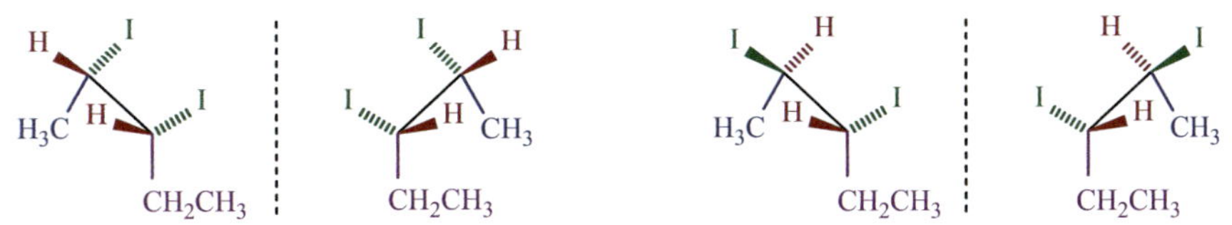

(2*S*,3*R*)-2,3-Diiodo-pentane (2*R*,3*S*)-2,3-Diiodo-pentane (2*R*,3*R*)-2,3-Diiodoo-pentane (2*S*,3*S*)-2,3-Diiodo-pentane

FIGURE 5.43 Four stereoisomers of 2,3-diiodopentane.

not mirror images. It is clear these molecules are isomers with the same connectivity, but they differ in the spatial arrangement of groups and atoms, so they are stereoisomers. Two stereoisomers that are not superimposable and not mirror images are defined as *diastereomers*. Therefore, (2*S*,3*R*)-diiodopentane is a diastereomer of (2*R*,3*R*)-diiodopentane and (2*S*,3*S*)-diiodopentane, and (2*R*,3*S*)-diiodopentane is a diastereomer of (2*R*,3*R*)-diiodopentane and (2*S*,3*S*)-diiodopentane. The definition of diastereomer may appear strange because if an apple is compared with an orange, they are nonsuperimposable, non-mirror images. However, this definition of diastereomer *applies only to*

> *For a given number of stereogenic centers (say n), there will be a* ***maximum*** *of 2n stereoisomers.*

two stereoisomers that are not the same molecule and are not mirror images of each other.

A molecule with two stereogenic centers has 2^2 or 4 stereoisomers. A molecule with four stereogenic centers will have a maximum of 2^4, or 16, stereoisomers. If a molecule has nine stereogenic centers, the maximum number of stereoisomers will be 2^9, or 512 stereoisomers. If 512 stereoisomers does not seem like a large enough number, look at a molecule with 28 stereogenic centers: 2^{28} means 2.684×10^8 stereoisomers (that's > 268 million stereoisomers) *for one constitutional isomer of a single empirical formula*. For that same empirical formula, there may be other molecules with different connectivity, and additional isomers are possible. Clearly, the presence of multiple stereogenic centers and the many stereoisomers that are possible is a concept that cannot be ignored. The 2^n rule yields the *maximum* number of stereoisomers for a molecule with more than one stereogenic center. There are never > 2^n stereoisomers, *but it is possible to have fewer stereoisomers if a molecule with two or more stereogenic centers has symmetry*. When there is more than one stereogenic carbon, similar groups may be attached that make one part of the molecule identical to another, although each individual stereogenic carbon is asymmetric with respect to that center. One such case is 2,3-dibromobutane. The line drawing of (2*R*,3*R*)-dibromobutane constitutes one stereoisomer, and the mirror image is (2*S*,3*S*)-dibromobutane, as shown in Figure 5.44.

Similar drawings are provided for (2*R*,3*S*)- and (2*S*,3*R*)-dibromobutane, which are mirror images of each other but diastereomers of the (2*R*,3*R*)- and (2*S*,3*S*)-stereoisomers. Careful inspection of (2*R*,3*S*)- and (2*S*,3*R*)-dibromobutane reveals something different from previous stereoisomers. The (2*R*,3*S*) structure is *superimposable* on the (2*S*,3*R*) structure, which means that these two structures are the same (they represent one molecule, not two). Make a model of both structures. Pick up one, rotate it by 180°, and "lay" it on top of the other. They are a perfect fit; all atoms match up. *Note that the atoms will match only in an eclipsed rotamer.*

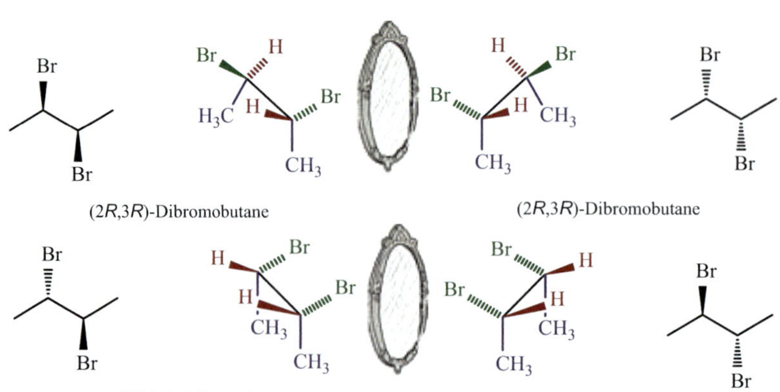

FIGURE 5.44 Stereoisomers of 2,3-dibromobutane.

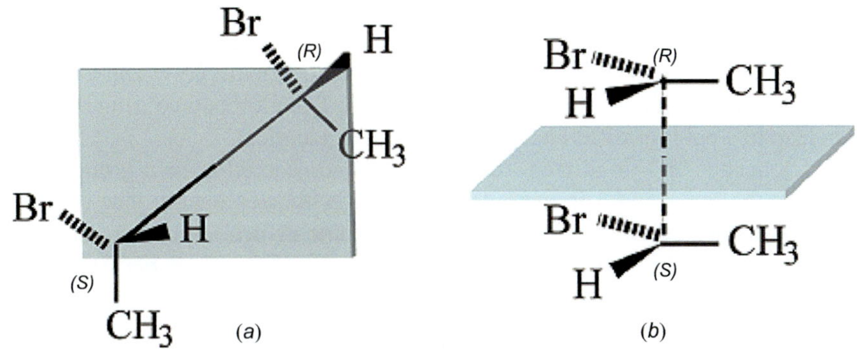

FIGURE 5.45 Plane of symmetry in (2*R*,3*S*)-dibromobutane from two perspectives, (a) and (b).

If an eclipsed rotamer of (2*R*,3*S*)-dibromobutane is examined, a slice down the middle (between the C2–C3 bond) in Figure 5.45a shows that one carbon is attached to Br, H, and Me, and the other carbon is also attached to Br, H, and Me. In other words, each stereogenic carbon atom has the same attached atoms and groups. If the eclipsed conformation of this stereoisomer is turned as in Figure 5.45b, the "top" and "bottom" are seen to be identical, but only in this eclipsed rotamer. Since the "top" and the "bottom" are identical in the eclipsed rotamer, the top reflects perfectly into the bottom and there is symmetry in the molecule (a plane of symmetry, as shown).

The mirror images are superimposable, so it is improper to identify (2R,3S)-dibromobutane and (2S,3R)-dibromobutane as separate stereoisomers because they represent the same structure. In other words, this is one molecule, not two.

When such symmetry occurs, the mirror image of one stereoisomer is superimposable on itself. Such a stereoisomer is called a *meso compound.* Since 2,3-dibromobutane has two stereogenic centers, the 2^n rule predicts a maximum of four stereoisomers, but symmetry in one stereoisomer means that it is a meso compound, so there are only three stereoisomers (the two enantiomers and the meso compound). *It is important to point out that the enantiomers and the meso compound are diastereomers.* Note that while the meso compound has two stereogenic atoms, the molecule is not optically active because it has a superimposable mirror image (it is one compound, not two).

A meso compound is an optically inactive stereoisomer of a molecule with more than one stereogenic center that arises due to symmetry in the molecule.

Cyclic molecules may have stereogenic centers, and it is possible to generate enantiomers and/or diastereomers. Problems in identifying the stereogenic center and the number of stereoisomers arise with some cyclic molecules due to pseudorotation. Similar problems do not arise in acyclic molecules.

Methylcyclohexane has one ring carbon atom connected to a methyl group, a hydrogen atom, and two carbons that are part of the six-membered ring. The two carbons in the ring that are adjacent to the methyl-bearing carbon atom (marked with a red dot) must be evaluated to determine if there is a point of difference. A plane of symmetry can be drawn along a line between C1 and C4. In effect, the carbons on one half of the six-membered ring constitute one "group," and the carbons on the other half of the ring constitute a second "group."

The CIP selection rules provide a way to compare the two red carbon atoms. The "top-left" carbon atom is C^{CHH}, and the "bottom-right" red carbon is also C^{CHH}, so there is no point of difference. Going to the next carbon atom on each side, the assignments remain C^{CHH} and C^{CHH}. Attempts to go to the next atom in each chain lead to the same carbon. Therefore, comparing each "side" of the ring does not lead to a point of difference, and each side is identical.

In methylcyclohexane and in all rings, each "side" is considered to be a group. In Figure 5.46 both "groups" are the same, so C1 in methylcyclohexane is not stereogenic. The symmetry associated with the six-membered ring makes methylcyclohexane achiral. Such symmetry is observed with many monosubstituted monocyclic compounds. Chirality differences between cyclic and acyclic molecules involve symmetry that can occur in a cyclic structure due to its rigidity, whereas free rotation is possible in an acyclic structure.

Methylcyclopentane does not contain a stereogenic center due to the inherent plane of symmetry. 1,2-Dimethylcyclopentane is a different matter, however. Two stereogenic carbons lead to two diastereomers, each with an enantiomer: the cis-diastereomer is (1*R*,2*S*)-dimethylcyclopentane, and the mirror image is (1*S*,2*R*)-dimethylcyclopentane. For the trans-diastereomer, (1*R*,2*R*)-dimethylcyclopentane, the mirror image is (1*S*,2*S*)-dimethylcyclopentane. The trans-diastereomer is drawn as the planar conformation in Figure 5.47, along with the mirror image. These planar representations emphasize the "sidedness" of the methyl groups. The two stereoisomers of *trans*-1,2-dimethylcyclopentane are *not* superimposable, so they are enantiomers. *cis*-Dimethylcyclohexane is drawn in a similar manner, along with the mirror image. If one structure is simply rotated counterclockwise by 180°, it will superimpose on the other. *cis*-1,2-Dimethylcyclopentane is a meso compound and constitutes only one stereoisomer. 1,2-Dimethylcyclopentane has only three stereoisomers because of the presence of symmetry in one stereoisomer.

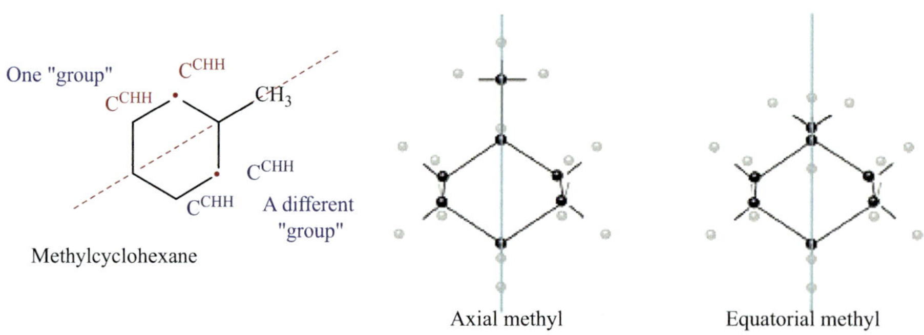

FIGURE 5.46 Symmetry in methylcyclohexane.

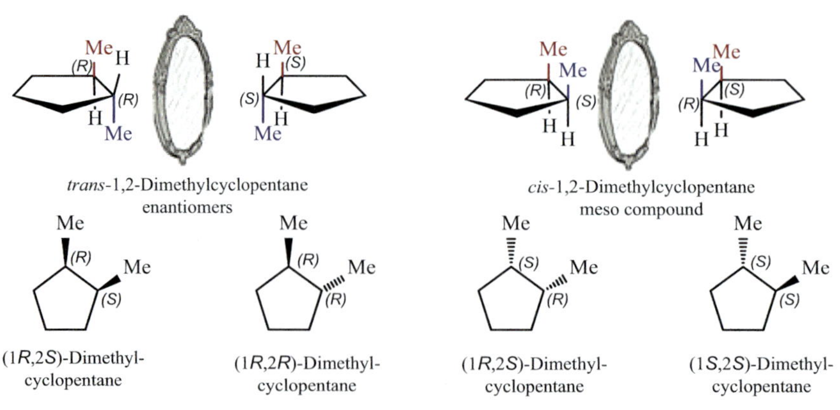

FIGURE 5.47 Dimethylcyclohexane stereoisomers.

Disubstituted cyclohexane derivatives present a more complex system, since there are two equilibrating chair conformations as well as other conformations. *For all comparisons of enantiomers and diastereomers of cyclohexane derivatives, assume that the chair conformation is the major conformer.* Even with this simplifying assumption, *both* chair conformations must be examined for *each* diastereomer. The examination is more complicated, however, because the diaxial methyl conformation is in equilibrium with the diequatorial methyl conformation. The trans-diastereomer cannot be labeled as having no symmetry until these chair conformations are inspected. Once again, there is no plane of symmetry because one chair does not superimpose on the other. Since there is no symmetry after comparing all four chair conformations, it is concluded that *trans*-1,2-dimethylcyclohexane has no symmetry, and the mirror image is an enantiomer.

5.19 ALKENE STEREOISOMERS: (*E*) AND (*Z*)-ISOMERS

Alkenes are characterized by a C=C unit to which four groups or atoms are attached. Since rotation about the C=C unit is not possible, atoms or groups attached to the C=C unit are effectively locked in space. For example, *cis*-hex-3-ene and *trans*-hex-3-ene are isomers with the same empirical formula and the same connectivity. However, they differ in the spatial arrangement of atoms and groups, so they are stereoisomers. The sp^2 carbon atoms of alkenes cannot be stereogenic, so *alkene stereoisomers are possible, but they are not enantiomers.*

But-2-ene ($CH_3CH{=}CHCH_3$) is a simple example of an alkene that exists as a stereoisomer. Structurally, but-2-ene contains a planar C=C unit with two methyl groups attached to it, as shown in Figure 5.48. Rotation around the rigid C=C unit is *impossible* (the π-bond ensures that it is locked into position), so the methyl groups can be attached on the "same side" or on "opposite" sides. The isomer with both methyl groups on the same side is marked *cis*-but-2-ene, and the isomer with the methyl groups on opposite sides of the C=C unit is marked *trans*-but-2-ene. These two alkenes are different molecules, and it is *impossible* to interconvert one into the other by rotating bonds or twisting atoms. Use a model kit to make a model of both but-2-enes and try to superimpose them! The two alkenes are isomers and stereoisomers, but they will have different physical properties. However, there is no stereogenic center, and they are not enantiomers. *Only a chemical reaction can change one into the other, and that requires making and breaking bonds.*

As noted, these isomeric alkenes are different compounds with different physical properties. Indeed, they differ slightly in boiling point (37°C for *cis*-but-2-ene and 37–38°C for *trans*-but-2-ene), melting point (–140°C and –180°C, respectively), and so on. The differences in the physical properties and structure of the alkenes can be correlated with the idea of *sidedness* in an alkene.

Another drawing of *cis*-but-2-ene is shown in Figure 5.48, in which the alkene is "tipped on its side" such that both hydrogen atoms are projected toward the front (marked in green) and both methyl groups are projected to the rear (marked in violet). In *trans*-but-2-ene, one H and one methyl are projected to the front (in green), and the other methyl group and other hydrogen atom are projected to the rear (in violet). These drawings indicate that the green atoms/groups are on the same

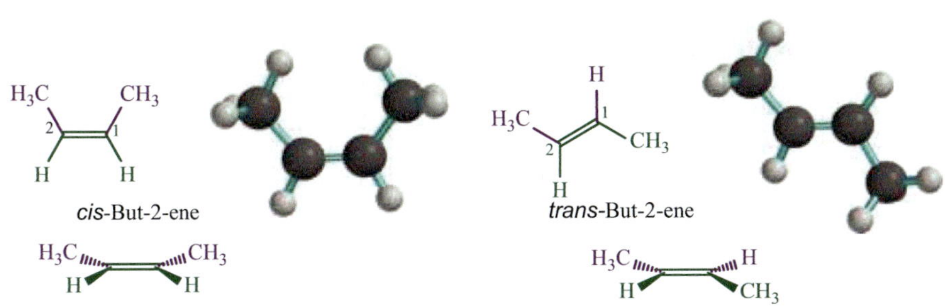

FIGURE 5.48 *cis*-But-2-ene and *trans*-but-2-ene.

side and the violet atoms/groups are on the same side. Therefore, the two methyl groups are on the same side in the cis alkene, but the methyl groups are on opposite sides in the trans alkene. The term "side" is replaced with the term "face" in many cases, such that the two methyl groups are on the same face or on opposite faces. It is important to recognize the sidedness of groups with respect to a C=C unit.

Since the two isomeric but-2-enes are different molecules, each requires a unique name. Indeed, the terms cis and trans are used, but these terms must be explained. Both molecules are but-2-ene, and they are stereoisomers, so their name must reflect the stereochemical differences of the methyl groups. There are no stereogenic centers, so the (*R*/*S*) nomenclature cannot be used. There are two methods used in nomenclature that distinguish the alkenes: the *cis–trans* nomenclature and the (*E*/*Z*) nomenclature.

In the two alkene stereoisomers, each carbon of the C=C unit (C1 and C2) has a methyl group and a hydrogen. The relative positions of those two groups cannot be changed in either. In other words, one stereoisomer cannot be transformed into the other by bond rotation. The methyl groups are on the same side of the C=C unit in the stereoisomer marked cis-, but the methyl groups are on opposite sides of the C=C unit in the stereoisomer marked trans. This observation leads to the cis-trans *definition for naming the stereoisomers.*

If two like groups are on the same side of an alkene, the molecule is a cis-alkene. If two like groups are on opposite sides of an alkene, the molecule is a trans-alkene.

A key word in these definitions is "like." The cis–trans nomenclature applies *only* when identical groups are on each carbon of the C=C unit (e.g., XYC=CXZ), where there is an X on both sp^2 carbon atoms. When the same group is on a single carbon, as in $X_2C=CYZ$, there are no stereoisomers since the two possible structures are superimposable. An example of this latter occurrence is 2-methylpent-2-ene. One carbon of the C=C unit has two identical groups, both methyl. One methyl group is cis to the ethyl group, but a methyl is also trans to the ethyl, so *2-methylpent-2-ene has no stereoisomers.* Indeed, when two identical groups are on the same carbon of the C=C unit, there is no possibility for *cis–trans* isomers.

If the groups to be compared on each carbon of the C=C unit are not the same, the cis/trans nomenclature does not apply. For example, in pent-2-ene, both a methyl group and an ethyl group are attached to the C=C unit, and two stereoisomers are possible, as shown. An alternative nomenclature method has been developed that determines the relative priority of groups attached to the C=C unit and compares those priorities. No two atoms or groups are the same in pent-2-ene, so cis or trans cannot be used. The new system for naming stereoisomers is called the *(E–Z) system.* The term (*E*) comes from the German word *entgegen*, which means "against" or "toward" or "contrary to," but it is used here to indicate opposite or apart. The term (*Z*) comes from the German word *zusammen*, which means "together." What constitutes "apart" and what constitutes "together" is determined by the Cahn-Ingold-Prelog (CIP) priority rules.

To determine the stereochemistry of the two stereoisomers of pent-2-ene, compare sidedness with groups or atoms on one side with those on the other side of the C=C unit. What constitutes a "side"? This system compares the higher priority group on each carbon (C1 vs. C2) of the C=C unit, using an atom-by-atom comparison. In other words, *the goal is to find the highest priority atom on C1 and then C2.* For one stereoisomer, carbon C1 of the C=C unit has a carbon atom and a hydrogen atom, whereas the other carbon (C2) has a carbon atom and a hydrogen atom. For C1, compare C with H, and C is clearly the higher priority. For carbon C2 of the C=C unit, a H is attached to C2 as well as a C (of the ethyl group). Using the CIP rules discussed in Section 5.16 for stereogenic atoms, carbon has a higher atomic number than hydrogen, so the methyl carbon has the higher priority at C1, and the ethyl carbon has the higher priority at C2.

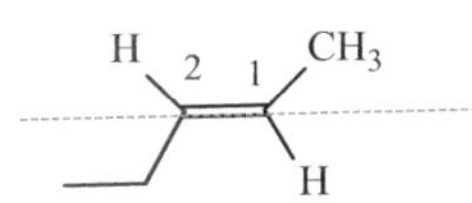

(*E*)-Pent-2-ene

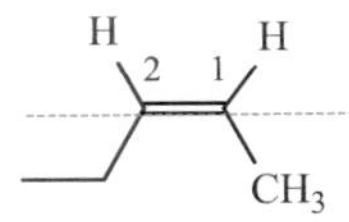

(*Z*)-Pent-2-ene

To use the *(E/Z)* nomenclature, compare the sidedness of the priority group on C1 with the priority group on C2, relative to a plane that bisects both carbon atoms of the C=C unit. This plane is shown as a yellow dashed line for the two stereoisomers of pent-2-ene. In one stereoisomer, the priority groups (ethyl and methyl) are on opposite sides, so the name is (*E*)-pent-2-ene. The stereoisomer with the two priority groups on the same side is (*Z*)-pent-2-ene. The 1-chloropent-1-enes are another example, where one chlorine, one propyl, and two hydrogen atoms are attached to the C=C units of these stereoisomers. The chlorine atom on C1 has the higher priority, and the carbon on C2 has the higher priority, and those priority groups are on opposite sides. If the chlorine and the propyl group are on *opposite* sides of the C=C unit, the stereochemistry is designated (*E*)-, and the name is 1-chloro-(1*E*)-pentene. Using a similar analysis, the stereoisomer with the chlorine and propyl groups on the *same* side is named 1-chloro-(1*Z*)-pentene.

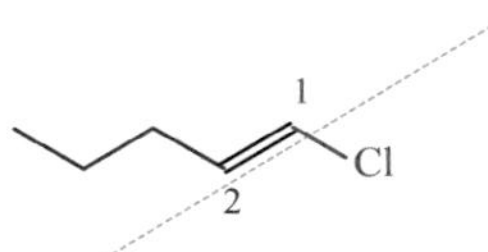

Chloropent-(1*E*)-ene

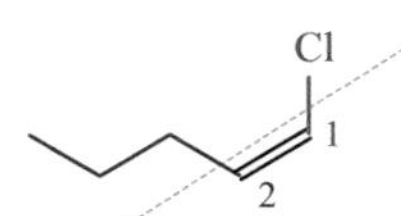

Chloropent-(1*Z*)-ene

It is important to emphasize that although (*Z*) and cis are both derived from groups being on the same side of the double bond, they arise from completely different definitions. 1-Bromo-1,2-dichlorobut-1-ene illustrates the point about cis-trans versus (*E*/*Z*). Both stereoisomers for this compound are shown. There are two identical atoms, the chlorine atoms on either carbon of the C=C unit, so the cis or trans nomenclature is appropriate. In one isomer, the two chlorine atoms are on opposite sides of the molecule, so it is named *trans*-1-bromo-1,2-dichlorobut-1-ene, whereas the chlorine atoms are on the same side in the other isomer, so it is named *cis*-1-bromo-1,2-dichlorobut-1-ene. Using (*E*/*Z*) nomenclature, the bromine and the chlorine are the two priority groups, and in the trans-alkene, the priority groups are on the same side, so it is 1-bromo-1,2-dichloro-(1*Z*)-butene. In the cis-alkene, the two priority groups are on opposite sides, so the name is 1-bromo-1,2-dichloro-(1*E*)-butene. Clearly, the *trans* alkene is the *(Z)* alkene, and the *cis* alkene is the *(E)* alkene. Choose one name or the other, but do not mix them. The formal IUPAC name should use the (*E*/*Z*) nomenclature.

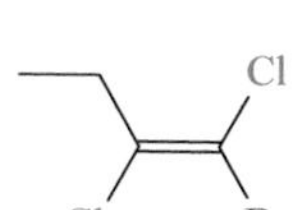

1-Bromo-1,2-dichlorobut-(1*Z*)-ene

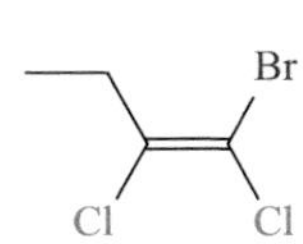

1-Bromo-1,2-dichlorobut-(1*E*)-ene

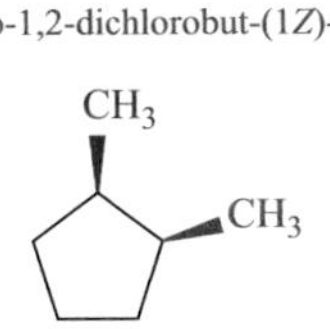

(1*R*,2*S*)-1,2-Dimethylcyclopentane

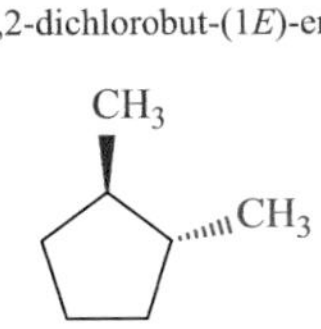

(1*R*,2*R*)-1,2-Dimethylcyclopentane

The cis–trans nomenclature can be used for substituents that are attached to rings. *The (E/Z) nomenclature cannot be used for cyclic compounds.* A carbon ring is flexible, but 360° rotation is not possible. Since complete rotation about the C–C bond is impossible, substituents on that ring are fixed (locked) onto one side of the ring or another. Therefore, two substituents can be on the same side of a ring [as in (1R,2S)-1,2-dimethylcyclopentane] or on opposite sides of that ring [as in (1*R*,2*R*)-1,2-dimethylcyclopentane]. Alternatively, the methyl groups in (1*R*,2*S*)-1,2-dimethylcyclopentane are on the same face, but the methyl groups in (1*R*,2*R*)-1,2-dimethylcyclopentane are on opposite faces. The solid wedges show the group projected out of the paper, and the dashed lines show groups projected behind the paper. If two like groups are on the same side of a ring, it is a cis-cycloalkane, and if the like groups are on opposite sides of the ring, it is a trans-cyclic alkane. Therefore, cyclic alkane (1*R*,2*S*)-1,2-dimethylcyclopentane can be called *cis*-1,2-dimethylcyclopentane, and (1*R*,2*S*)-1,2-dimethylcyclopentane can be called *trans*-1,2-dimethylcyclopentane. It is important to note that cis- and trans- are used to indicate the relationship of two groups on a ring, and the groups are not always the same.

HOMEWORK

5.1. Does the following reaction illustrate a homogeneous or a heterogeneous bond cleavage? Explain.

5.2. Give the IUPAC name for each of the following.

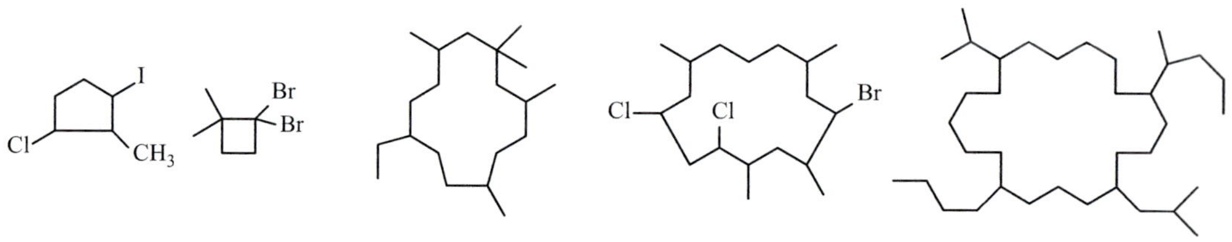

5.3. Give the structure of each '?' in the following reaction. What type of product is [?]?

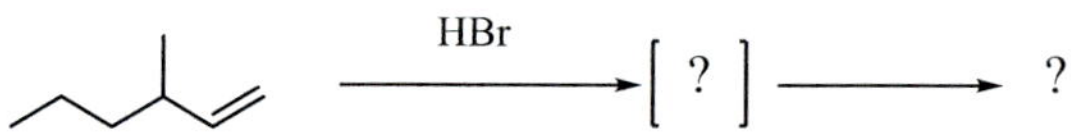

5.4. Which of the following are capable of hydrogen bonding?

CH_3OH CH_3NHCH_3 CH_3CH_2Br

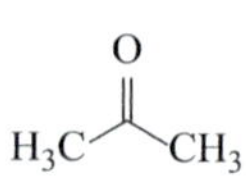

5.5. Draw both chair conformations of the following molecule. Draw both chair conformations and identify which, if either, is the major conformation. Justify your choice.

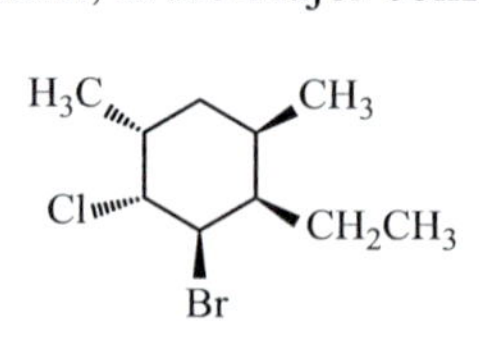

5.6. Identify the axial and the equatorial bromine atoms in the following structure. Assume the atoms are frozen in their positions and the molecule cannot undergo pseudorotation.

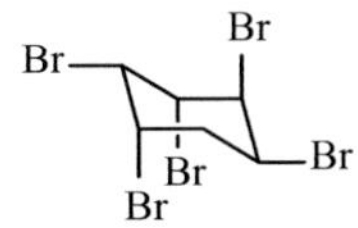

5.7. Identify each of the following as *E* or Z that is applicable. Identify as cis or trans, if applicable.

(a) 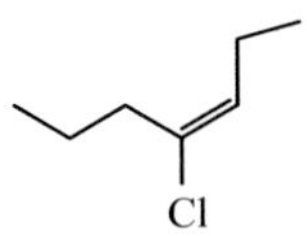 (b) 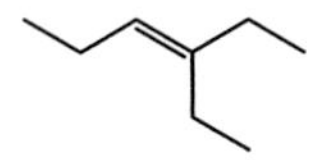 (c) 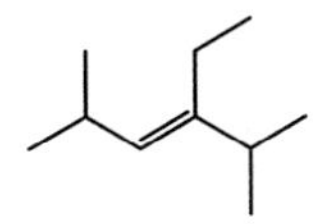

5.8. Give the electronic configuration for the following atoms: Al, He, Be, Mg, Cl, Br, Ti, Cu

5.9. Based on electron availability in the valence orbital, suggest a simple reason why potassium metal tends to be more reactive than sodium metal.

5.10. Use the VSEPR model to predict the shape for each of the following. Draw each one using line notation.

(a) CCl_4 (b) CH_3OH (c) CH_3OCH_3 (d) $(CH_3)_4N^+$ (e) $CH(CH_3)_3$ (f) $ClCH_2Cl$

5.11. Estimate the direction of the dipole moment for the molecule CH_2Cl_2.

5.12. Give the correct IUPAC name (no common names) for each of the following.

(a) 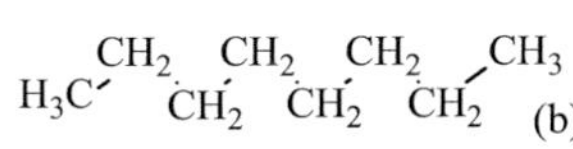 (b) 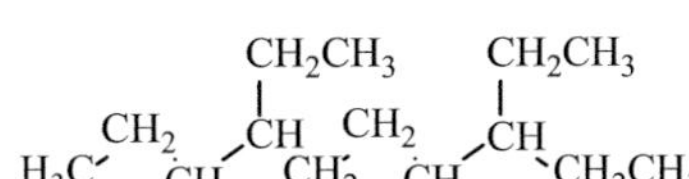

(c) 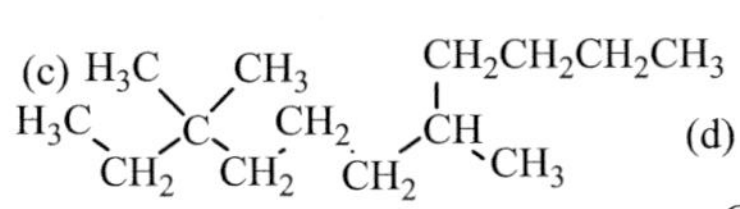 (d) 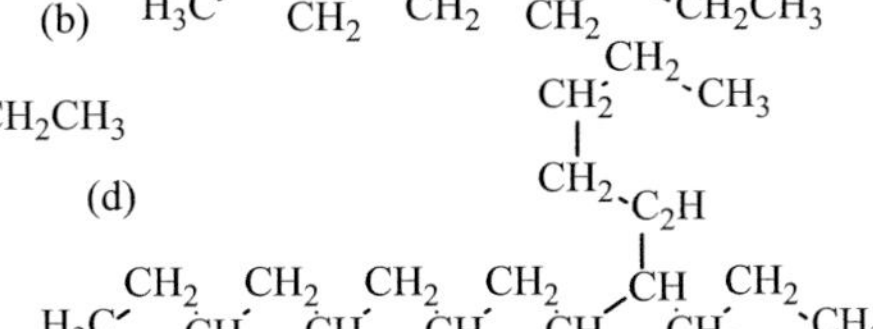

5.13. For each of the following, give the correct IUPAC name (no common names

(a) 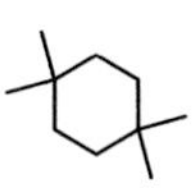 (b) 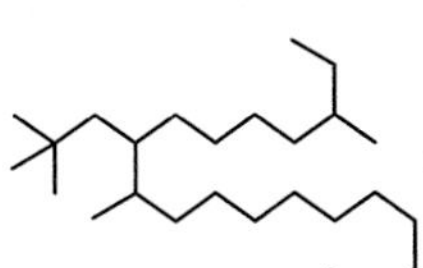 (c) 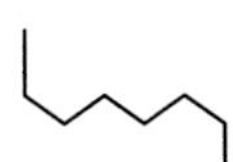 (d) 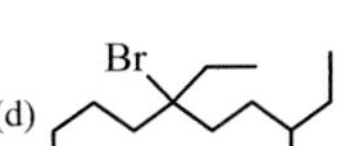

5.14. (a) Draw six *different* isomers of 1-bromooctane using line notation. Give the IUPAC name for each isomer you draw. (b) Draw the structure for five *different* isomers of the formula C_6H_{12} using line notation. Name each structure, and do *not* use common names.

5.15. (a) Identify all *isomers* of 3,5-dimethylheptane

(b) Identify all isomers of cycloheptane.

5.16. For a given reaction, explain the relationship between the energy of activation and the transition state.

5.17. When a reaction has a negative value for $\Delta H°$, it is said to be spontaneous. Does this mean that simply mixing the reactants will automatically produce the product in a reaction that generates heat? Why or why not?

5.18. Calculate the half-lives given in the following data.

(a) A first-order reaction where $k = 1.2 \times 10^{-6}$

(b) A second-order reaction where $k = 4.5$

(c) A first-order reaction where $k = 5.8 \times 103$

(d) A second-order reaction where $k = 9.25 \times 10^{-4}$

5.19. Look at the molecule drawn in the box as a chair. Note the numbering scheme for the carbons.

Draw in the bonds on the flat ring, using wedges and dashed lines, to indicate the proper arrangement of atoms for this molecule. For the flat ring, the "top" is the one facing you, as you look at the sheet of paper. The 'top' for the chair form is as indicated.

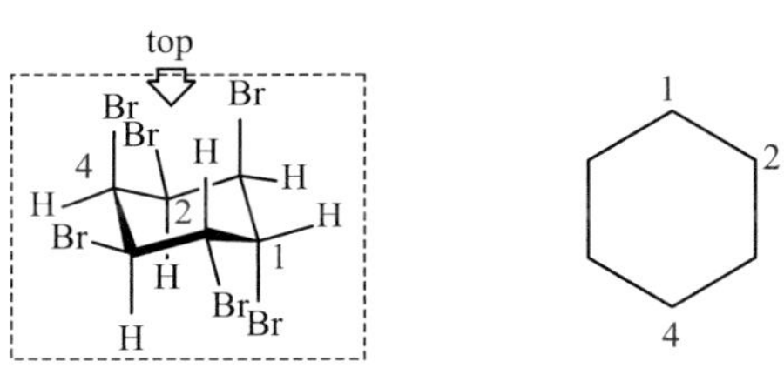

5.20. Assuming that each of the following is frozen, which of the following is the *lowest* energy conformation?

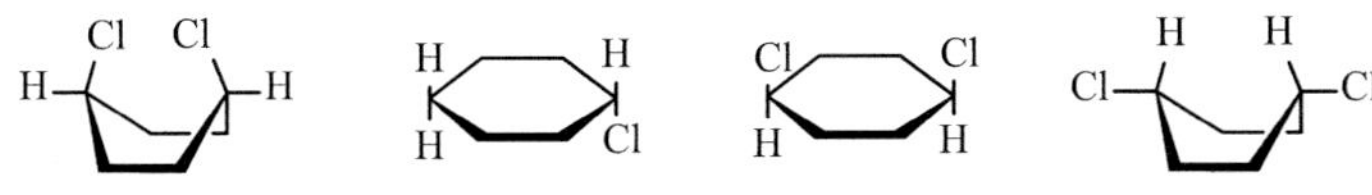

5.21. On the diagrams provided, fill in the appropriate atoms on the proper carbon and in the proper axial or equatorial position. These are two chair conformations for the same molecule, not isomers. Note the numbering scheme and assign the proper atoms to the properly numbered carbons in each chair. Which is lower in energy, A or B?

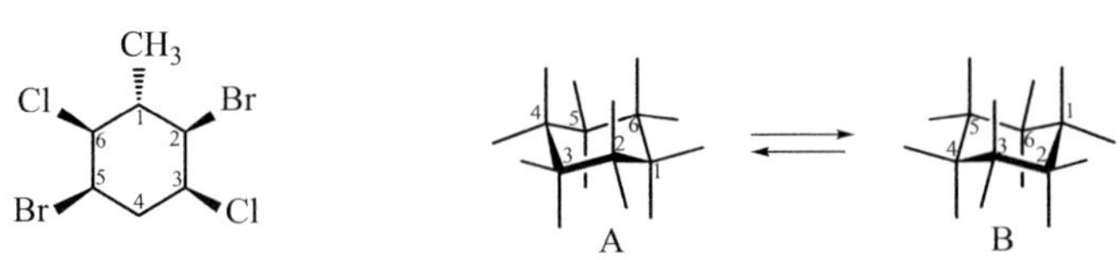

5.22. The *gauche*-rotamer of 1,2-dichloroethane [1.2 kcal (5.0 kJ)/mol] is slightly higher in energy than the *gauche*-rotamer of butane [1.0 kcal (4.18 kJ)/mol]. Explain.

5.23. Although it was stated that ethane-1,2-diol exists primarily as a *gauche*-rotamer, neat (no solvent) diol has a significant concentration of the *anti*-rotamer. In the gas phase, ethane-1,2-diol exists almost exclusively as the *gauche*-rotamer. Explain.

5.24. Which of the following solvents that cannot be used for determining [α] of a compound in a polarimeter?

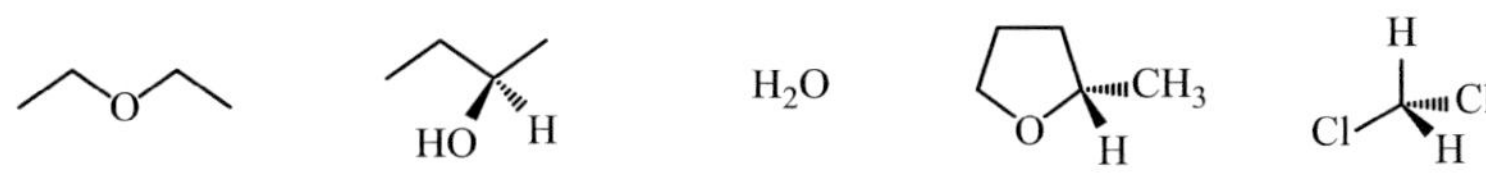

5.25. Determine the absolute configuration

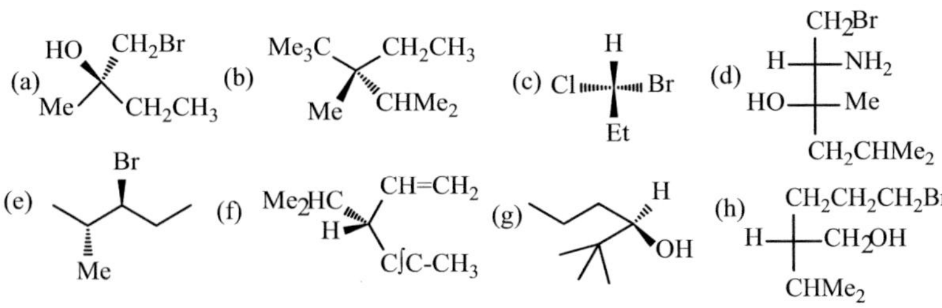

5.26. Determine the specific rotation for each of the following:
 (a) α=+18°, c=1.1 g/mL in a 25-cm tube
 (b) α=−176°, c=0.3 g/mL in a 50-cm tube
 (c) α=−1.4°, c=5.4 g/mL in a 25-cm tube
 (d) α=+94°, c=2.3 g/mL in a 30-cm tube

5.27. Draw the structure for each of the following:
 (a) 3,3-Diphenyl-(4*E*)-nonen-1-ol
 (b) 2,3,4,5-Tetrachlorohex-(2*Z*)-ene
 (c) 3-Bromo-6-fluorodeca-(3*Z*,6*E*)-dien-2-one
 (d) 3-Ethylhept-(2*E*)-ene

5.28. Determine the absolute configuration:

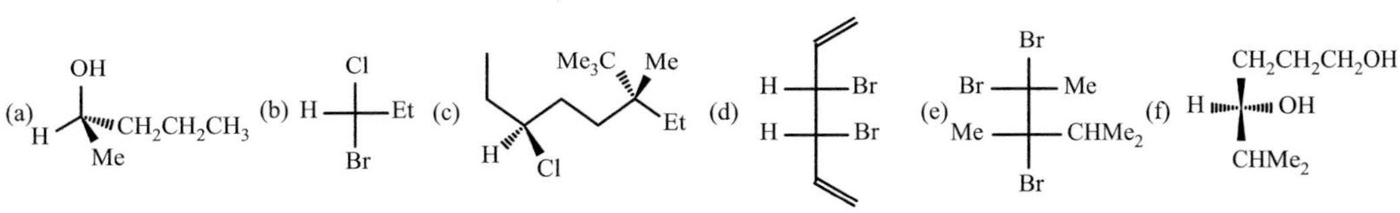

6 Nucleophilic Substitution and Elimination Reactions

Aliphatic substitution reactions are part of a typical undergraduate organic chemistry course. Such reactions involve the reaction of nucleophilic species with an electrophilic species, and in organic chemistry they usually follow first order or second order kinetics. Nucleophiles such as amines or carboxylate anions are prevalent in biochemical reactions. Substitution reactions in a typical organic chemistry course involve reactions at a carbon that is connected to a heteroatom leaving group such as a halogen or a sulfonate ester. In biochemistry, however, the leaving group is often a phosphonate ester or another biocompatible group. Nucleophilic reaction require electrophilic species. Electrophiles or electrophilic substrates are common in biochemistry, including phosphonate derivatives, carbonyl compounds and imine compounds. This chapter will discuss the S_N2 and S_N1 reactions that appear in biochemical applications and then describe these reactions as they are found in organic chemistry. Another type of nucleophilic reaction involves carbonyl compounds, including acyl addition of ketone and aldehyde moieties and acyl substitution reactions of carboxylic acid derivatives. Examples of these reactions in biochemical systems will be discussed, and then the fundamentals of both acyl addition and of acyl substitution reactions in organic chemistry will be discussed.

Elimination reactions are also presented in a typical undergraduate course. There are a biomolecular elimination reaction (E2 reactions) and unimolecular elimination reactions (E1 reactions). Both will be discussed.

6.1 ENZYMATIC S_N2 REACTIONS

Substitution reactions are well known in enzymatic pathways. A straightforward example of a biological S_N2 reaction is taken from the mechanism of action for L-DEX from *Pseudomonas* sp. YL (L-DEX YL). This Group II *haloacid dehalogenase* (EC 3.8.1.2) involves the carboxylate of an amino acid unit in an aspartate residue. As shown in Figure 6.1, the carboxylate group of an aspartic acid residue attacks the carbon atom bearing the chlorine in a haloacid in an S_N2 reaction to generate aspartate ester **1**.[1] Subsequent reaction with water gives in vivo hydrolysis to regenerate the aspartic acid residue and the hydroxy acid.

Enzymatic substitution reactions are involved in the bioremediation of contaminants in the environment. *Haloalkane dehalogenases* (EC 3.8.1.5; see Section 4.2) have been used for the conversion of chlorinated hydrocarbons into alcohols and chloride ions, allowing more facile biodegradation. "*Haloalkane dehalogenases* catalyze the conversion of chlorinated hydrocarbons into alcohols and chloride ion through nucleophilic displacement by Asp[124] in the active site. Because of its potential for bioremediation of environmental contaminants, it has been extensively studied and has been, in fact, used in applications for the treatment of contaminated soil." A computational active site mechanism is shown in Figure 6.2 for 1,2-dichloroethane, where reaction with the carboxylate moiety of an aspartic acid residue displaces a chloride ion to give an ester in a substitution reaction.[2] Subsequent reaction with water, which is coordinated to a histidine residue in the active site, leads to an acyl substitution reaction and formation of 2-chloroethanol. The overall reaction is an aliphatic substitution followed by an acyl substitution that converts 1,2-dichloroethane to 2-chloroethanol.

[1] (a) Nardi-Dei, V.; Kurihara, T.; Park, C.; Miyagi, M.; Tsunasawa, S.; Soda, K.; Esaki, N. *J. Biol. Chem.* **1999**, *274,* 20977. See Fig. 1a, p. 20978; (b) O'Hagan, D.; Schmidberger, J.W. *Nat. Prod. Rep.* **2010**, *27,* 900. Fig. 17, p. 910.

[2] Nam, K.; Prat-Resinsa, X.; Garcia-Viloca, M.; Devi-Kesavan, L.S.; Gao, J. *J. Amer. Chem. Soc.* **2004**, *126,* 1369.

DOI: 10.1201/9781003605157-6

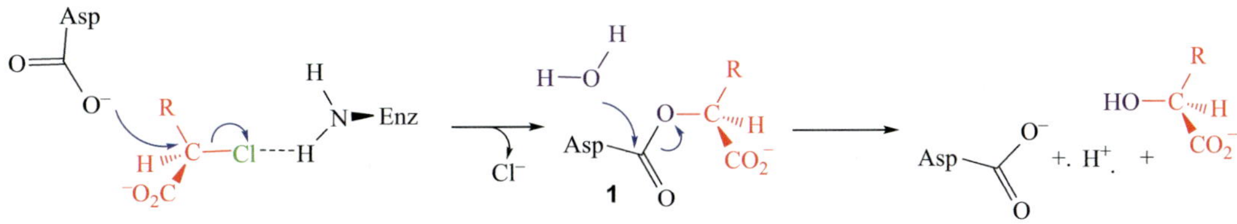

FIGURE 6.1 Enzyme-mediated nucleophilic displacement of a haloacid to give a hydroxy acid. (Nardi-Dei, V.; Kurihara, T.; Park, C.; Miyagi, M.; Tsunasawa, S.; Soda, K.; Esaki, N. *J. Biol. Chem.* **1999**, *274,* 20977. See Fig. 1a, p. 20978. Open Access.)

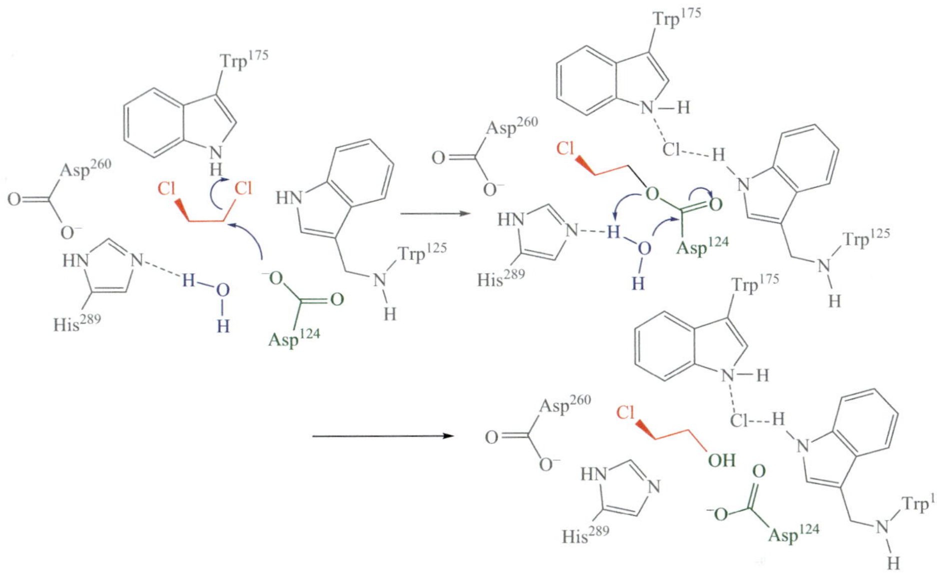

FIGURE 6.2 *Haloalkane dehalogenase*-mediated conversion of 1,2-diethoethane to 2-chloroethanol. (Reprinted with permission from Nam, K.; Prat-Resinsa, X.; Garcia-Viloca, M.; Devi-Kesavan, L.S.; Gao, J. *J. Amer. Chem. Soc.* **2004**, *126,* 1369–1376. Copyright 2004 American Chemical Society.)

Another example of a substitution reaction is the *N*-methylation reaction of norepinephrine to give epinephrine, where the nucleophilic moiety is the primary amine unit of norepinephrine and the electrophilic reagent is *S*-adenosyl methionine, as shown in Figure 6.3.[3] The *S*-methyl reagent is a common biological methylating agent and converts the amine to the *N*-methylamine as shown. It is responsible for the breakdown of glycogen in liver cells. Epinephrine is also known as adrenaline, and it is a hormone, a neurotransmitter, and a medication. Epinephrine is produced by both the adrenal glands and certain neurons, and it increases blood flow to muscles. Epinephrine injection is used for emergency treatment of severe allergic reactions (including anaphylaxis) to insect bites or stings, medicines, foods, etc. This example shows that biochemical substitution reactions can occur with amine nucleophiles; common nucleophiles include alcohols, thiols, and phosphines. Amino acids react as nucleophiles. Peptides are poly(amino acids) and are formed by the activation of amino acids with transfer RNA and coupling to form new peptide bonds on messenger RNA, as described in Sections 3.1, 3.2, and 15.2.

As mentioned, ethers are generally unreactive as bases or nucleophiles except in their reaction with strong mineral acids. The sulfur analog, thioethers, on the other hand, are much more reactive.

[3] (a) Pendleton, R.G., Snow, I.B., Gessner, G., Green, H. *Arch. Int. Pharmacodyn. Ther.* **1973**, *203*, 368; (b) See Lu, S.C. *Int. J. Biochem. Cell Biol.* **2000**, 32, 391.

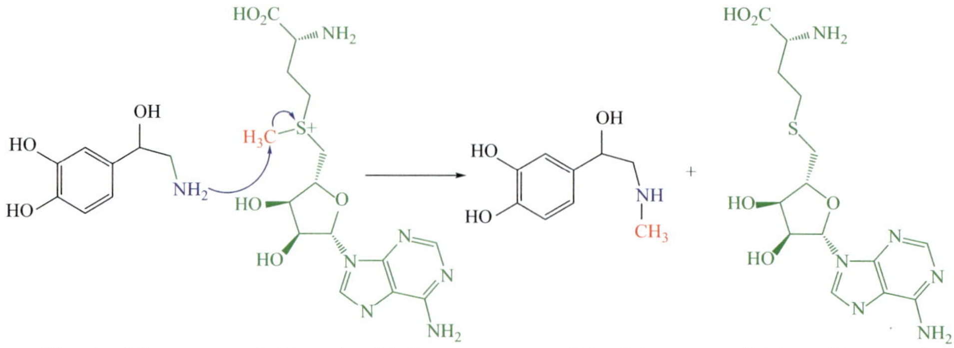

FIGURE 6.3 Methylation of norepinephrine to give epinephrine. (Reprinted from Lu, S.C. *Int. J. Biochem. Cell Biol.* **2000**, 32, 391–395. S-Adenosylmethionine, Copyright 2000, with permission from Elsevier.)

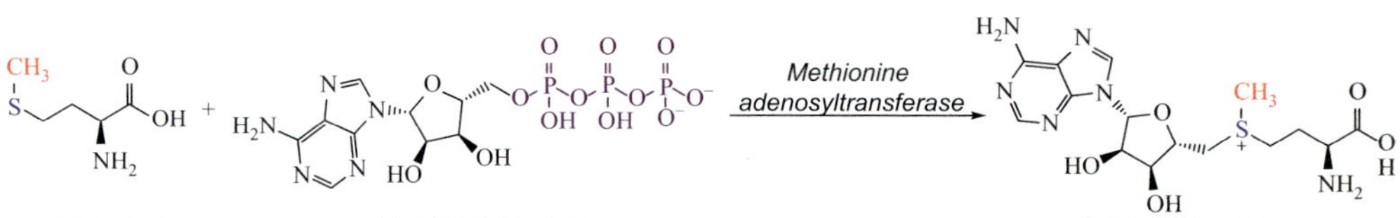

FIGURE 6.4 Formation of S-adenosyl methionine from L-methionine.

An example is the formation of the methylating agent *S*-adenosyl methionine from L-methionine and adenosine triphosphate (ATP). The reaction shown in Figure 6.4 is catalyzed by *methionine adenosyltransferase* (EC 2.5.1.6; see Section 4.2) and proceeds by the nucleophilic displacement of orthophosphate from ATP by the sulfur atom of the thioether moiety in methionine.[4]

As bis(2-chloromethyl)sulfide (mustard gas) was studied, chemical modification led to an amine derivative, 2-chloro-*N*-(2-chloroethyl)-*N*-methylethanamine, 2. This compound is one of the first clinically useful anticancer drugs. The anticancer activity arises from its reaction as a DNA intercalating agent. As shown in Figure 6.5,[5] the nitrogen mustard **2** first reacts via the N^9-nitrogen of a guanine that is part of a DNA strand (see **3**), and an S_N2-like reaction at the aziridinium salt leads to **4**. The three-membered ring is susceptible to attack by nucleophiles, similar to the reactivity of epoxides. Hence, 2-chloro-*N*-(2-chloroethyl)-*N*-methylethanamine is classified as an alkylating agent. The cross-linking (intercalating) ability arises when the tertiary amine unit in **4** reacts with the other primary alkyl chloride to give aziridinium salt **5**. A second molecule of DNA (**3**) reacts to form **6**. If two different strands of DNA react, the nitrogen mustard links the two strands, as indicated by **6**. If the DNA is double-stranded, this reaction leads to intercalation of the nitrogen mustard.

Enzymatic S_N2 reactions involving halohydrins can occur intramolecularly to give the corresponding epoxide. The bacterial *haloalcohol dehalogenase* from *Agrobacterium radiobacter* AD1,66 is one enzyme known to convert a halohydrin to an epoxide. Figure 6.6 shows an important Ser-Tyr-Arg catalytic triad.[6] It is believed that Tyr^{145} acts as a base to deprotonate the substrate alcohol and initiate epoxide ring closure. Deprotonation of the tyrosine phenol possibly occurs via an elaborate relay mechanism originating from the carboxylate of Asp^{80}, through two structural water molecules. This sequence delivers a proton to the solvent after deprotonation of Tyr^{145}. The Ser^{132} and Tyr^{145} side chains make hydrogen bonds with the oxygen atom of the epoxide product and with the hydroxyl group of a haloalcohol substrate (see Figure 6.6).[6]

[4] Cantone, G.L. *J. Amer. Chem. Soc.* **1952**, *74,* 2942.
[5] Rajski, S.R.; Williams, R.M. *Chem. Rev.* **1998**, *98*, 2723.
[6] O'Hagan, D.; Schmidberger, J.W. *Nat. Prod. Rep.* **2010**, *27,* 900.

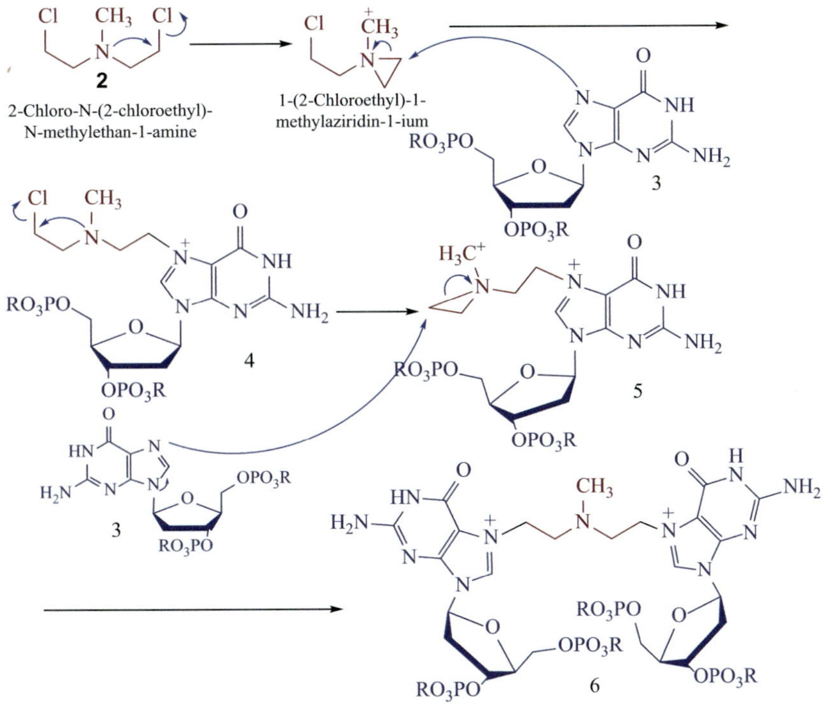

FIGURE 6.5 Intercalation of nitrogen mustard in DNA. (Reprinted with permission from Rajski, S.R.; Williams, R.M. *Chem. Rev.* **1998**, *98*, 2723. Copyright 1998 American Chemical Society.)

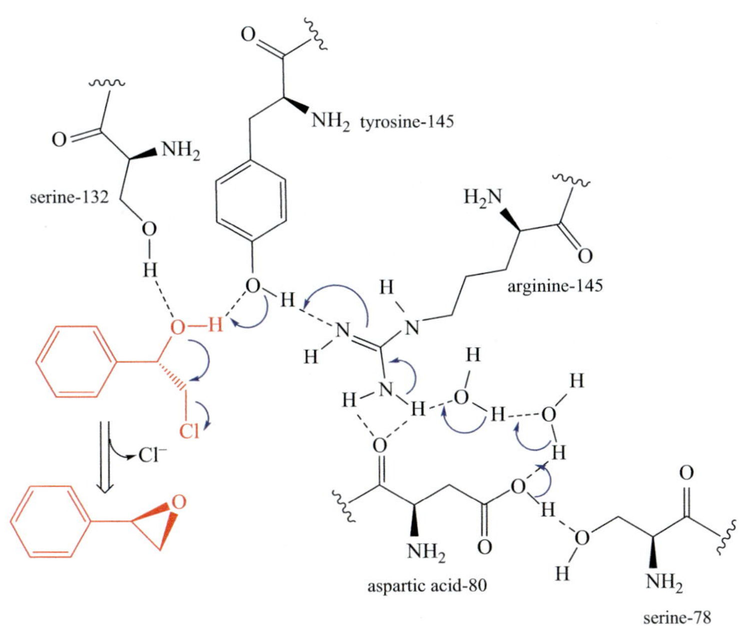

FIGURE 6.6 A schematic representation of the proposed catalytic mechanism of dehalogenation in HheC. (Taken from O'Hagan, D.; Schmidberger, J.W. *Nat. Prod. Rep.* **2010**, *27*, 900. See Fig. 21, p. 912.)

6.2 NUCLEOPHILES AND BIMOLECULAR SUBSTITUTION IN ORGANIC CHEMISTRY (THE S_N2 REACTION)

The S_N2 reaction is one of the seminal reactions in a typical undergraduate organic chemistry course. As an example, the reaction of 1-bromo-3-methylbutane with sodium iodide (NaI) using acetone as a solvent gave 1-iodo-3-methylbutane in 66% yield.[7] In terms of the structural changes, the iodide ion substitutes for the bromine, producing the bromide ion (Br^-). Iodide acts as a nucleophile in the reaction at $C^{\delta+}$ of the alkyl bromide, breaking the C–Br bond and transferring the electrons in that bond to bromine. In molecules that contain the C–Br bond, or indeed a C–X bond, where X is a heteroatom-containing group, the carbon will have a δ^+ dipole. In other words, the carbon atom is electrophilic, and the substrate that reacts with the nucleophile is called an electrophile. The reaction of a nucleophile with an aliphatic electrophile is formally called *nucleophilic aliphatic substitution*, illustrated in Figure 6.7. The displaced atom or group (e.g., chloride) departs (leaves) to become an independent ion. Displacement of chlorine leads to the chloride ion (Cl^-), but the bromide ion, iodide ion, or a sulfonate anion also correlates to X, which is referred to as a *leaving group*. In many biochemical reactions, the leaving group is a phosphate, $–O–PO_2–O–$.

A leaving group does not spontaneously "fly off" or "leave"; it is displaced by the nucleophile after collision with the electrophilic carbon atom. In other words, the incoming nucleophile effectively "kicks out" the leaving group, but only after collision with the electropositive carbon. The rate of the S_N2 reaction (the speed at which the reactants are converted into products) is proportional to the concentration of both the nucleophile and the electrophilic substrate. The rate at which this reaction occurs is proportional to the concentration of the nucleophile and the concentration of the alkyl halide, and this type of rate expression is *second order*. To change the proportionality to an equality, the expression must contain a proportionality constant, *k*.

$$\text{Rate } (S_N 2) = k\left[\text{nucleophile}\right]\left[\text{halide}\right]$$

This proportionality constant is called the *rate constant*. A large rate constant means the starting material is rapidly converted to product. For a second-order reaction, the rate constant is determined from a plot of time versus a concentration term that includes both starting materials. This reaction is a nucleophilic substitution reaction, represented by the symbol S_N, that follows second-order kinetics (also known as a bimolecular reaction). Therefore, this type of substitution is known as an *S_N2 reaction* (*nucleophilic bimolecular substitution*).

An S_N2 reaction is characterized by complete inversion of configuration. For a nucleophile to displace a leaving group, the nucleophile must *collide* with the sp^3 carbon atom, and the angle of approach must be considered because that carbon is part of a three-dimensional molecule with a tetrahedral geometry. The most reasonable approach that minimizes both electronic and steric effects is 180° (anti) relative to the iodine atom (see the blue arrow), or from the "bottom of the tetrahedron." Such an approach angle constitutes what is usually called a *backside attack*, which leads to inversion of the absolute configuration of a chiral electrophilic substrate.

No intermediate has been observed in the conversion of any of the substitution reactions discussed, so a mechanistic explanation of the reaction must involve the transition state. Using the

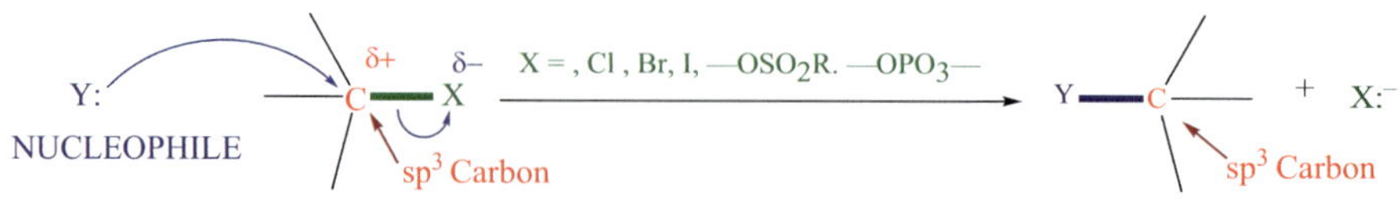

FIGURE 6.7 Nucleophilic attack at a sp^3 carbon bearing a leaving group.

[7] Furniss, B.S.; Hannaford, A.J.; Smith, P.W.G.; Tatchell, A.R. (Eds.), *Vogel's Textbook of Practical Organic Chemistry*, 5th ed., Longman, Essex, UK, **1994**, Exp. 5.62, p. 572.

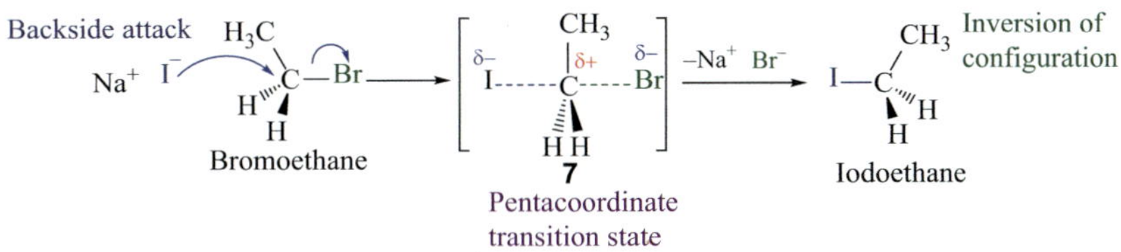

FIGURE 6.8 Pentacoordinate transition state for the reaction of sodium iodide and bromomethane.

reaction of bromoethane and NaI as an example, the backside attack of the nucleophile leads to the formation of a new C–I bond with cleavage of the C–Br bond. The transition state is the species shown in Figure 6.8 that has five atoms or groups around a central carbon. The transition state is the mid-point of the reaction and is known as a *pentacoordinate transition state* (**7**). All S_N2 reactions proceed through a similar transition state. As iodide donates two electrons to the carbon to make the I–C bond, concomitant transfer of electrons from carbon to bromine breaks the C–Br bond. As the C–I bond begins to form, the three hydrogen atoms on carbon are 'pushed away' from the incoming iodine atom, toward the departing bromine atom. This electron transfer sequence correlates with the I----C----Br motif shown in Figure 6.8. To "complete" the reaction once the pentacoordinate transition state is achieved, the iodide ion continues to approach the carbon to form a C–I bond. As the C–I bond is formed, there is simultaneous displacement of the "leaving" bromide ion. Note that the absolute configuration of the carbon atom in the iodoethane product has been inverted. *The pentacoordinate transition state (7) is necessary for the inversion of configuration.*

Differences in reactivity can be explained by the S_N2 pentacoordinate transition state. For a reaction with a primary electrophilic substrate such as bromomethane, there is minimal steric hindrance. As the hydrogen atoms on the carbon are replaced with alkyl groups (R), the energy of the pentacoordinate transition state **7** increases accordingly due to an increase in steric hindrance in the transition state. If bromoethane is used (X=Br, R^1=Me, R^2 and R^3=H), the methyl group attached to the electrophilic carbon increases the steric hindrance in **7**, and the reaction is expected to be slower because the transition state energy is higher, which means that E_{act} is higher. Such increased transition state energy makes the reaction more difficult (slower). For all practical purposes, tertiary electrophilic substrates do not react. In other words, a nucleophile may collide with a 3° carbon atom, but the activation energy for the reaction is so high that the S_N2 reaction does not occur. A relative order of reactivity for alkyl halides in the S_N2 reaction can be estimated:

$$\text{Methyl} > 1° > 2° \ggg 3°.$$

The differences in rate for primary, secondary, or tertiary alkyl halides are due to increased steric hindrance in the pentacoordinate transition state.

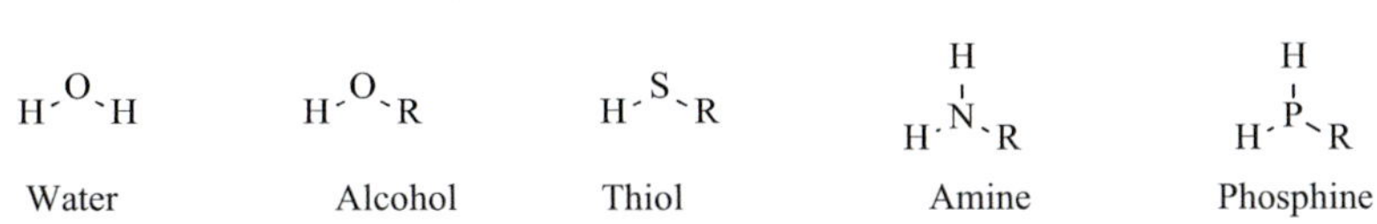

Typical nucleophiles in substitution reactions include water, alcohols, thiols, amines, and phosphines. The actual nucleophiles in such compounds are the oxygen, sulfur, nitrogen, and phosphorus. Water, H–O–H, is the most familiar covalent molecule involving oxygen. Replacing one hydrogen of the water structure with an alkyl group generates an alcohol, while replacing both hydrogen atoms with alkyl groups generates an ether. The nucleophilic potency of these atoms is enhanced when each bears a negative charge. When an alcohol (ROH) reacts with a base in an acid-base reaction, for example, the product is an alkoxide (RO^-). The enhanced charge on the oxygen makes an alkoxide a significantly better nucleophile relative to the neutral alcohol, just as hydroxide is a stronger nucleophile relative to water. The alkoxide is formally the conjugate base of the alcohol. In general, alkoxides react with primary and secondary halides in diethyl ether or THF solvents to form ether products in what is known as the *Williamson ether synthesis*.

Sulfur is immediately under oxygen in the periodic table, and C–SH compounds are called thiols (the older name is mercaptan), while R–S–R compounds are called sulfides (sulfanes). Thiols are good nucleophiles, as are the conjugate bases generated from a thiol, RS^-. The sulfur atom of sulfides also reacts as a nucleophile, but the reaction tends to be reversible. Nitrogen has a valence of three and forms three covalent bonds, as in NH_3 (ammonia). When the hydrogen atoms are replaced with alkyl groups, the resultant compounds are amines. The terms *primary, secondary, and tertiary* are used to describe the structural variations in amine structure. A primary amine has one carbon and two hydrogen atoms on nitrogen (RNH_2); a secondary amine has two carbons and one hydrogen atom on nitrogen (R_2NH); and a tertiary amine has three carbons and no hydrogen atoms on nitrogen (R_3N). Amines react as nucleophiles with alkyl halides in S_N2 reactions, as in the conversion of 1-bromopropane to propane-1-(*N,N*-dimethylammonium) bromide to *N,N*-dimethylpropan-1-amine rather than the ammonium salt. Displacement of bromide ion by the nucleophilic nitrogen in dimethylamine leads to the formation of propane-1-(*N, N*-dimethylammonium) bromide, as described above. However, this salt is a weak acid, and it is formed in the presence of dimethylamine, which is both a base and a nucleophile. A simple acid-base reaction occurs between the acidic ammonium salt and the basic secondary amine to generate the neutral amine (*N,N*-dimethylpropan-1-amine) along with dimethylammonium bromide.

Organic molecules with a trivalent phosphorus atom are known as phosphines. Indeed, alkylphosphines (PR_3) are analogs of the parent phosphine (PH_3), and it is reasonable to expect that phosphines (PR_3) should react similarly to amines (NR_3). In one example, benzyltriphenylphosphonium bromide is formed by reaction of benzyl bromide with triphenylphosphine. This S_N2 reaction with phosphine and alkyl halides is sometimes called the *Arbuzov reaction*. Primary phosphines (RPH_2) and secondary phosphines (R_2PH) also react to yield the corresponding phosphonium salts, but they are less common.

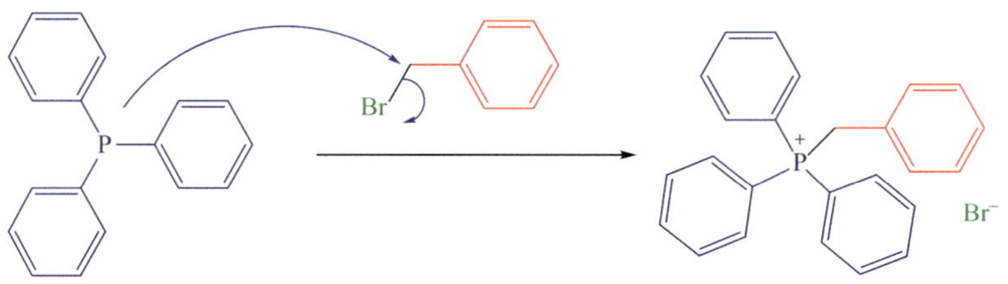

Triphenylphosphine

Benzyltriphenylphosphonium bromide

6.3 BIOLOGICAL S_N1-TYPE PROCESSES

A cation is an electron-deficient species that has a formal charge of +1. A cation is attracted to and reacts with a species that can donate two electrons (e.g., an anion). A carbon atom that bears a positive charge is called a *carbocation (a carbenium ion)*. A carbocation is formed when a covalent bond to carbon is broken in such a way that two electrons are transferred to one atom, and the carbon receives no electrons during the transfer. The structure of a generic carbocation is shown in Figure 6.9. The central carbon atom of a carbocation is clearly electron-deficient, with a formal charge of +1. This charged species has only three covalent bonds, is high in energy, unstable, highly reactive, and difficult to isolate in most cases. In other words, it is an intermediate.

The carbon of a carbocation is sp^2 hybridized and must have trigonal planar geometry (see **A** in Figure 6.9). The positive charge is localized on the carbon, and that charge is associated with an empty p orbital on that carbon, as shown in **B**. The empty p orbital in **B** is the region in space above and below the plane of the carbon and hydrogen atoms where electron density can be accepted to

FIGURE 6.9 The structure of a carbocation.

form a new bond. A carbocation will react with another species that can donate the two missing electrons to give it eight, satisfy the valence requirements of carbon, and form the fourth bond to make carbon tetravalent. In other words, a carbocation can react with a nucleophile.

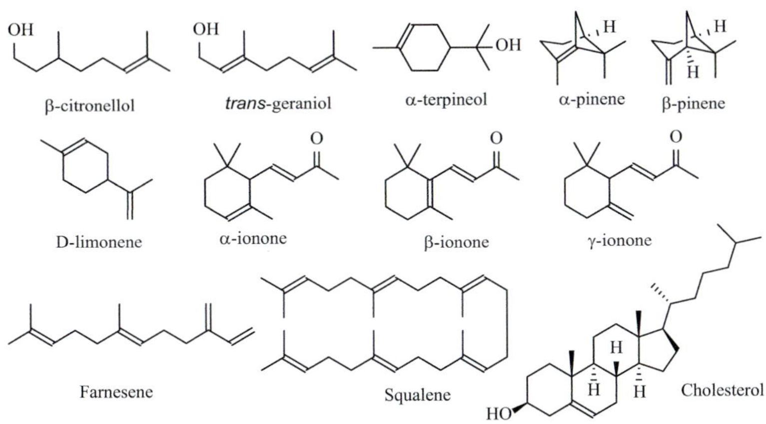

There are S_N1-type reactions found in natural processes. Terpene hydrocarbons are organic compounds that have the general formula (C_5H_8), and isoprene (2-methylbuta-1,3-diene) is an example. Terpenes, including the sesquiterpenes with three terpene units and the general formula ($C_{15}H_{24}$) or diterpenes with four terpene units and the general formula ($C_{20}H_{32}$) or are common in plants and some insects. Terpenes are constituents of essential oils and are often the main constituents of a plant's "smells," including β-citronellol and trans-geraniol, which make up about 50% of rose oil; α-terpineol is found in pine oil; and α-pinene and β-pinene, which are key constituents of pine resin and many conifers; D-limonene in lemon peel; and α-, β-, and γ-ionones in violets. Farnesene is a sesquiterpene that is found in the coating of apples and other fruits, and the characteristic odor of green apples is due to farnesene. Terpenes such as squalene, which is classified as a triterpene, are biosynthetic intermediates for the biosynthesis of steroids such as cholesterol.

Terpenes are introduced here because their biosynthesis involves the coupling of isoprene units via isopentenyl pyrophosphate. The enzyme responsible for this transformation is *farnesyl pyrophosphate synthase* (*farnesyl diphosphate synthase*; EC 2.5.1.87; see Sections 4.1.3 and 4.6).[8] Isomerization of dimethylallyl pyrophosphate is followed by ionization to the resonance-stabilized allylic carbocation shown, which reacts with more isopentenyl pyrophosphate to give geranyl pyrophosphate. This can then react with another equivalent of isopropenyl pyrophosphate to give farnesyl pyrophosphate (see Figure 6.10).[9] The biochemical transformation in Figure 6.10 invokes an allylic carbocation intermediate, which is consistent with a discussion of S_N1 reactions.

Epoxides are observed in biochemical systems, as in the conversion of arachidonic acid to 14,15-epoxyeicosatrienoic acid, catalyzed by a *cytochrome P450 epoxy oxygenase* (EC 1.14.13.69; see Section 4.1.3). Subsequent hydrolysis, catalyzed by a soluble *epoxide hydrolase* (EC 3.1.3.76),

[8] Ban Thorpe, D.V.; Bucknall, G.A.; Donna, H.J.; Donna, S.; Rowan. M.G. *Phytochemistry* **1976**, *15*, 91.
[9] Sowden, R.J.; Yasmin, S.; Rees, N.H.; Bell, S.G.; Wong, K.-L. *Org. Biomol. Chem.* **2005**, *3*, 57.

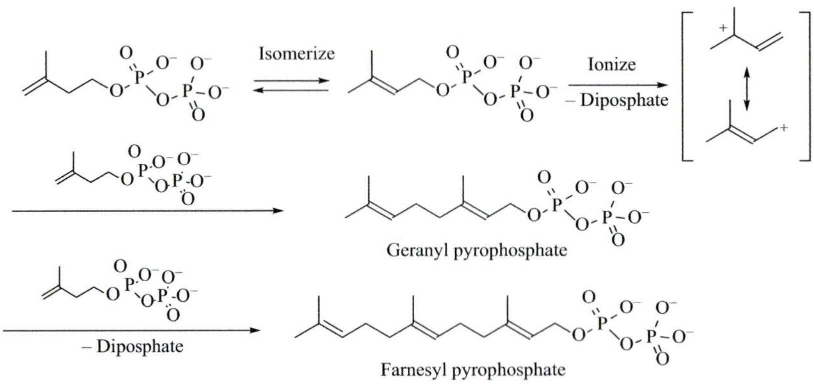

FIGURE 6.10 The biosynthesis pathway of geranyl (GPP) and farnesyl pyrophosphate (FPP) via the coupling of isoprene units. (Sowden, R.J.; Yasmin, S.; Rees, N.H.; Bell, S.G.; Wong, K.-L. *Org. Biomol. Chem.* **2005**, *3*, 57–64. Reproduced by permission of The Royal Society of Chemistry.)

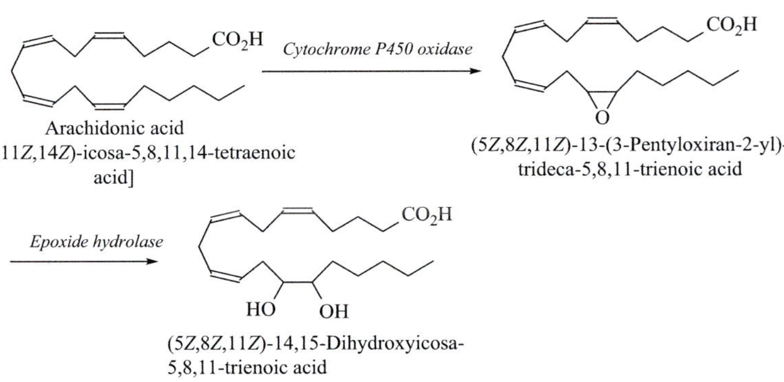

FIGURE 6.11 Epoxidation of arachidonic acid and epoxide hydrolysis. (Reprinted with permission from Wagner, K.; Inceoglu, B.; Gill, S.S.; Hammock, B.D. *J. Agric. Food Chem.* **2011**, *59*, 2816–2824. Copyright 2011 American Chemical Society.)

leads to 14,15-dihydroxyeicosatrienoic acid, as shown in Figure 6.10.[10] Soluble epoxide hydrolases are found in the kidneys, and ring opening of epoxides is important for the metabolism of epoxyeicosatrienoic acids as well as for the detoxification of drugs or their metabolites (Figure 6.11).

6.4 CARBOCATIONS AND S_N1 REACTIONS IN ORGANIC CHEMISTRY

When the tertiary halide 2-bromo-2-methylpropane was heated with potassium iodide in aqueous tetrahydrofuran (THF), the product was 2-iodo-2-methylpropane. Because the halide has a tertiary carbon, the transformation clearly cannot be an S_N2 reaction. In the presence of water, the tertiary

[10] Wagner, K.; Inceoglu, B.; Gill, S.S.; Hammock, B.D. *J. Agri. Food Chem.* **2011**, *59*, 2816. Also see Wagner, K.; Vito, S.; Inceoglu, B.; Hammock, B.D. *Prostaglandins Other Lipid Mediat.* **2014**, *113–115*, 2.

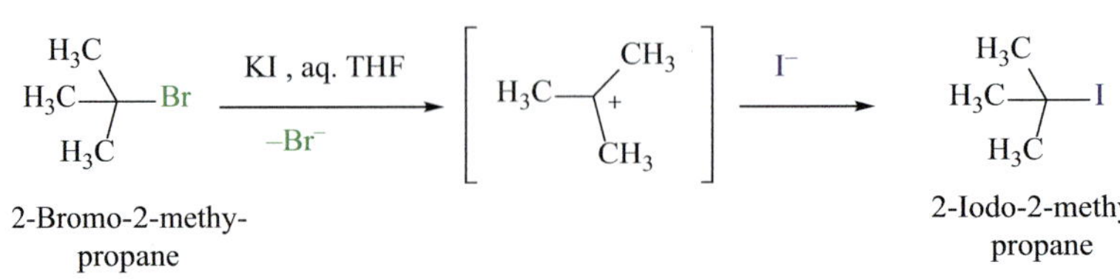

halide will ionize to an intermediate tertiary carbocation, but both the iodide ion and water can react as nucleophiles. Since the product is the iodide and not the alcohol, the carbocation clearly reacts with iodide faster than it does with water. This result suggests that iodide is a better nucleophile than water. The overall process is substitution (iodide substitutes for bromide) via a carbocation intermediate.

Obviously, 2-iodo-2-methylpropane cannot be formed if iodide does not react with the carbocation, which suggests that the collision of iodide with the cation is very fast, whereas ionization of 2-bromo-2-methylpropane is quite slow. *The rate of the overall process is determined by the rate of the slow first step (known as the rate-limiting step).* A reaction that follows this rate equation is said to be *unimolecular* (for all practical purposes, an ionization reaction), or *first order*. The example discussed here is a unimolecular nucleophilic substitution with the descriptor S_N1. *It is an* S_N1 *reaction.*

6.5 ETHERS AND THIOETHERS AS NUCLEOPHILES IN SUBSTITUTION REACTIONS

Ethers are generally unreactive except with very strong acids in organic chemistry reactions. Indeed, ethers are often used as solvents due to their limited reactivity. An exception is the reaction of ethers that react as a base with strong mineral acids (e.g., HI or HBr). When 2-isopropoxy-2-methylpropane reacts with HI, an acid-base reaction generates an oxonium ion intermediate with an iodide counterion, as shown. In this reaction, protonation of the ether oxygen is essential for the substitution because $-OCHMe_2$ and $-OCMe_3$ are not good leaving groups. The protonated ether contains $-OHCHMe_2^+$ and $-OHCMe_3^+$ units in the oxonium ion; however, these are excellent leaving groups

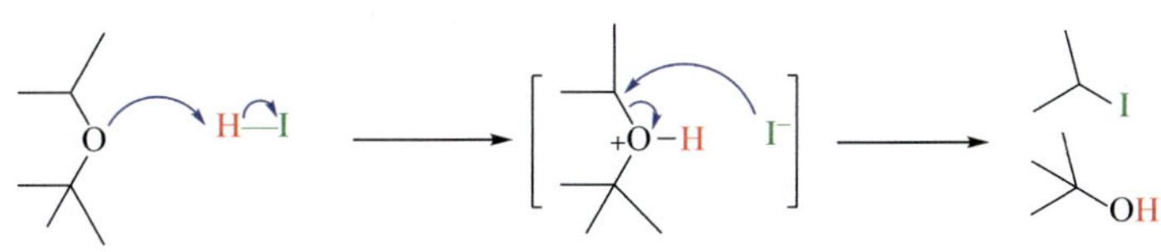

2-Isopropoxy-2-methylpropane Diethyloxonium iodide

(isopropanol and tert-butyl alcohol). The iodide ion is a good nucleophile, and an S_N2 reaction at the secondary carbon connected to oxygen rather than the tertiary carbon leads to iodoethane and ethanol as the products. In other words, iodide reacts with the less hindered carbon of the oxonium ion to yield 2-iodopropane as a product, where *tert*-butanol is the leaving group that gives the second product.

Although most ethers are rather unreactive except with HI or HBr, epoxides are a distinct exception. The strain inherent in the three-membered ring makes epoxides very reactive with a variety of reagents. 2-(3-Methylbutan-2-yl)oxirane, for example, will react with HI to form the corresponding oxonium ion. As shown in Figure 6.12, the initially formed oxonium ion (**8**) reacts with the nucleophilic iodide ion primarily at the less substituted C^a to give 1-iodo-3,4-dimethylpentan-2-ol rather than at C^b to give 2-iodo-3,4-dimethylpentan-1-ol as the major product.

Epoxides are very reactive to many nucleophiles, although they do not generally react directly with water or with alcohols in the absence of an acid catalyst. The reaction of 2,2-dimethyloxirane

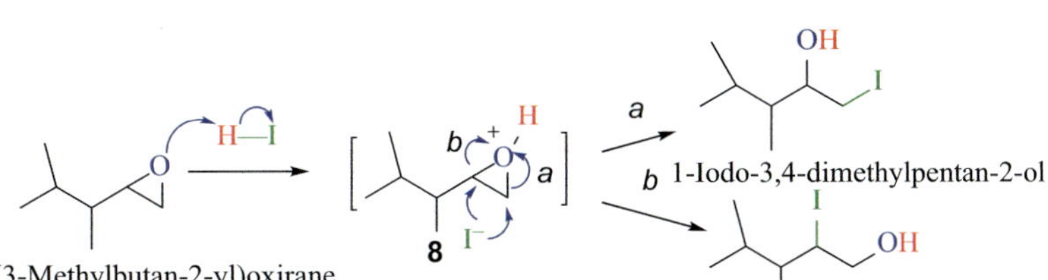

FIGURE 6.12 Reaction of an epoxide with HI.

with an acid catalyst will generate an oxonium ion, which opens in the aqueous medium to give the more stable tertiary carbocation. The oxygen atom of water attacks the positively charged carbon of this intermediate to yield an oxonium ion, and the loss of a proton gives 2-methylpropane-1,2-diol and regenerates the acid catalyst. This is a general reaction in which an epoxide reacts with a strong acid catalyst to form an oxonium ion, and then with a weak nucleophile to open the three-membered ring. Similarly, the reaction of an epoxide with NaOH will lead to a diol.

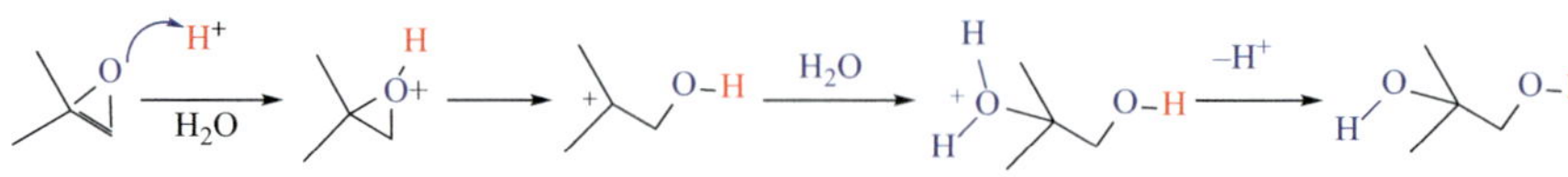

Diethyl sulfide Iodomethane Diethyl(methyl)sulfonium iodide

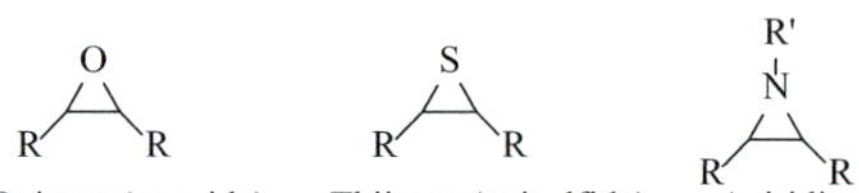

Oxirane (epoxide) Thiirane (episulfide) Aziridine

Sulfides react with alkyl halides to give the corresponding sulfonium salt. The reaction of diethyl sulfide and iodomethane, for example, gives the sulfonium iodide. Such reactivity is possible because of the extra valence available to the sulfur atom, due to the presence of d orbitals. This type of sulfonium salt is observed with *S*-adenosylmethionine (see Figure 6.4), and the stability of the sulfide accounts for the fact that *S*-adenosylmethionine is a methylating agent; i.e., a nucleophile can react with the methyl group on sulfur and displace the sulfide moiety as a leaving group. A three-membered ring thioether is called an episulfide. As expected, nucleophiles can open the strained three-membered sulfide ring to give the corresponding thiol.

A three-membered ring amine is known as an aziridine. Nucleophiles can react with this strained system to give the amine. An example of the formation of an aziridine in a biochemical pathway involves bis(2-chloromethyl)sulfide ($ClCH_2CH_2SCH_2CH_2Cl$). Substitution reactions occur in many biological processes. Bis(2-chloromethyl)sulfide, otherwise known as *mustard gas*, is a sulfide (a thioether) and a primary alkyl chloride, which makes it highly reactive (see also Figure 6.5). It was used as a poison gas in World War I since it is cytotoxic, mutagenic, and a vesicant, and exposure causes large blisters on exposed skin. These effects arise from a reaction with heterocyclic bases in DNA.

6.6 BIOCHEMICAL REACTIONS OF KETONES AND ALDEHYDES

Pyranoses are carbohydrates that exist in an equilibrium composed of two pyranose diastereomers, each of which has a stable hemiacetal unit, and a small amount of the open-chain hydroxy aldehyde. Hydrolysis of an acetal unit (C–O–C–OR) involves a lysozyme enzyme. Lysozyme functions by attacking, hydrolyzing, and breaking the mucopolysaccharide part of the peptidoglycan in the bacterial cell wall. This enzyme can also break glycosidic bonds in chitin. The carbon-oxygen bond is broken by a general-acid catalyst residue, glutamic acid (Glu), and a general-base catalyst residue, aspartic acid (Asp) or cysteine (Cys), in the active site of lysozyme. Two distinct lysozymes have been discovered in *Enterococcus hirae* ATCC 9790. Muramidase-1 is a glucoenzyme and a zymogen that requires protease activation.[11] Note that a zymogen is an inactive substance that is converted into an enzyme when activated by another enzyme. The extracellular muramidase-2 appears

[11] Shockman, G.D. *FEMS Microbiol. Lett.* **1992**, *100*, 261.

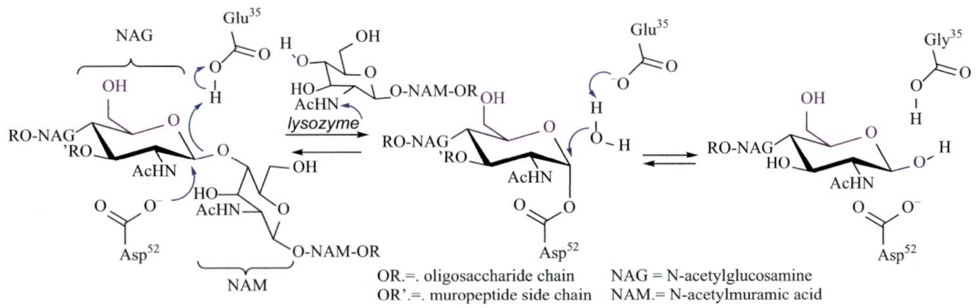

FIGURE 6.13 Lysozyme-mediated glycolysis with retention of configuration. (Reprinted from Nawaz, N.; Wen, S.; Wang, F.; Nawaz, S.; Raza, J.; Iftikhar, M. Usman, M. *Molecules* **2022**, *27*, 6305. With permission from MDPI, 2022, Open Access, Figure 1, p. 6309.)

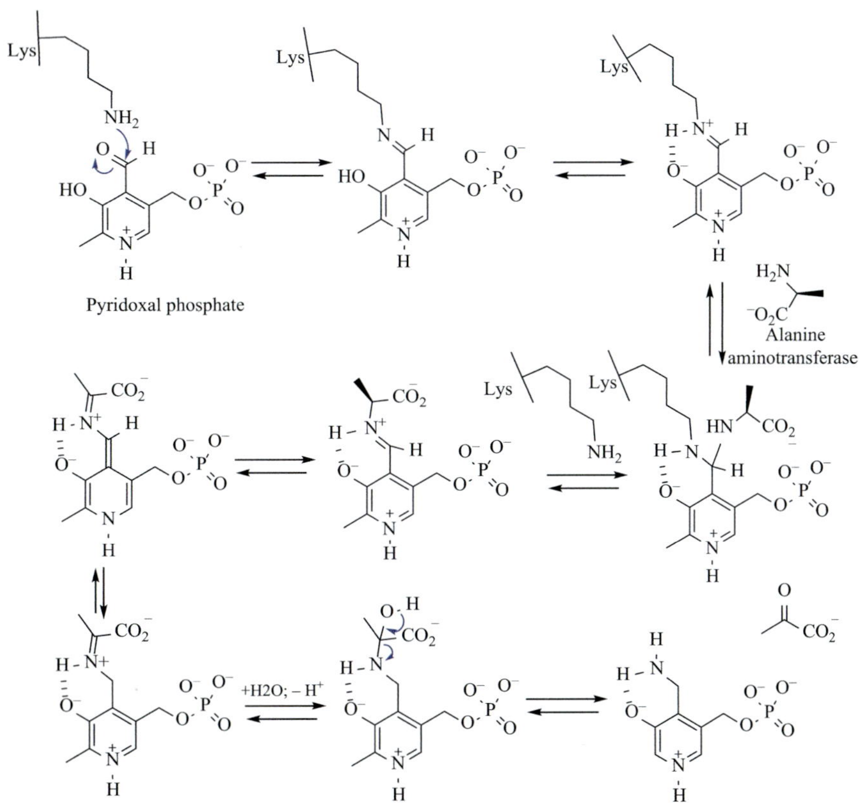

FIGURE 6.14 Transamination with alanine. (Reprinted with permission from Franco, T.M.A.; Hegde, S.; Blanchard, J.S. *Biochemistry* **2016**, *55*, 6295–6303. Copyright 2016 American Chemical Society.)

to consist of two or perhaps three functional modules that are linked in a single polypeptide chain.[12] Lysozymes are *glycosyl hydrolases* (Section 4.2), and enzymatic hydrolysis of a glycoside linkage involves two amino acid residues, aspartate and/or glutamate, in the active site of the reaction. As shown in Figure 6.13,[13] the linkage of a carbohydrate is cleaved with retention of configuration.

[12] Joris, B.; Englebert, S.; Chu, C.P.; Kariyama, R.; Daneo-Moore, L.; Shockman, G.D.; Ghuysen, J.M. *FEMS Microbiol. Lett.* **1992**, *91*, 257.

[13] Nawaz, N.; Wen, S.; Wang, F.; Nawaz, S.; Raza, J.; Iftikhar, M. Usman, M. *Molecules* **2022**, *27*, 6305.

Primary amines react with aldehydes or ketones to give an imine. When the aldehyde unit is attached to an aromatic ring, the imine is known as a Schiff base. The amine units of amino acids can be transferred to a pyridoxal unit in the presence of *aminotransferases* such as *alanine aminotransferase* (EC 2.6.1.2; see Sections 4.2 and 4.4). Pyridoxal phosphate is a coenzyme that binds to enzymes via the formation of a Schiff base, allowing an acyl addition-type reaction of the amine group of an amino acid residue, such as alanine. This reaction, shown in Figure 6.14, is taken from the biopathways used by *Mycobacterium tuberculosis*.[14] This sequence is essential for the synthesis of the branched-chain amino acids isoleucine, leucine, and valine. In this mechanism, an initial reaction of the amine unit of a lysine residue and pyridoxal pyrophosphate generates a Schiff base. Subsequent reaction with alanine leads to the formation of an imine moiety that reacts with water, allowing the loss of the amine moiety to give pyridoxamine 5′-phosphate and the formation of an α-keto acid.[14] Note that a Schiff base is an organic compound formed by the reaction of a primary amine and an aldehyde or ketone, giving an imine group.

6.7 ENZYMATIC ACYL SUBSTITUTION REACTIONS

There are important enzymatic hydrolysis reactions of carboxylic acid derivatives. The cleavage of an ester into its alcohol and carboxylic acid components requires an enzyme known as an *esterase*. A *lipase* is a type of esterase, described in detail in Section 4.5.1, and there are *pancreatic lipases* as well as mitochondrial lipases. A lipase is a water-soluble enzyme that hydrolyzes ester bonds in water-insoluble lipids, including glycerides. An *esterase* (EC 3) is a hydrolase enzyme that facilitates the hydrolysis of esters into an acid and an alcohol by reaction with water. *Esterases* (see Section 4.5.1) in humans are broadly classified as a *cholinesterase* (including *acetylcholinesterase*; EC 3.1.1.7), *carboxylesterases* (EC 3.1.1.1), and *arylesterases* (EC 3.1.1.2). *Esterases* are important enzymes for the hydrolysis of drugs. While esterase activity is typically specific for certain substrates and inhibitors, hydrolysis by more than one esterase is common. Aspirin (acetylsalicylic acid), for example, is initially hydrolyzed to salicylate by *carboxylesterases* in the liver, but later can be hydrolyzed by plasma cholinesterase and by albumin and red blood cell *arylesterases*.[15]

There are many types of *esterases*. *Acetylcholinesterase* hydrolyzes the neurotransmitter acetylcholine. *Thioesterases* specifically catalyze hydrolysis at a thiol group. A *phosphatase* hydrolyzes phosphoric acid monoesters, and *phosphodiester hydrolases* (EC 3.1.4.1) work on phosphodiesters. An *exonuclease* is an enzyme that hydrolyzes phosphodiester bonds at either the 3′ or the 5′ end of a polynucleotide (Section 15.2). An *endonuclease* cleaves phosphodiester bonds in the middle (endo) of a polynucleotide chain.

Lipases (EC 3; see Sections 4.5.1 and 4.5.2) are hydrolase enzymes that are responsible for catalyzing the hydrolysis of triglycerides to fatty acids and glycerol. A triglyceride has three ester bonds that are formed from glycerol and three fatty acid groups. Triglycerides are the main constituents of natural fats and oils. *Colipase* is a protein coenzyme required for optimal activity of *pancreatic lipase* and to prevent the inhibitory effect of bile salts on the *lipase*-catalyzed intraduodenal hydrolysis of dietary long-chain triglycerides. Esters are prevalent in humans, usually in the form of triglycerides, which are hydrolyzed by a *triacylglycerol lipase*. Triglycerides can pass through the stomach into the duodenum, where alkaline pancreatic juice raises the pH and allows hydrolysis of the ester linkages, which is mediated by an enzyme called pancreatic lipase, as well as other *esterases*.

Figure 6.15[16] shows the enzyme-mediated hydrolysis of a triglyceride to a diglyceride, and the loss of another molecule of fatty acid gives the monoglyceride, and the loss of the third fatty acid gives glycerol. In Figure 6.15,[16] *destrin/adipose triglyceride lipase* is the major triacylglycerol lipase in adipose tissue and is induced during fasting. *Hormone-sensitive lipase* (*cholesteryl ester hydrolase*; EC 3.1.1.79) is an intracellular neutral *lipase* capable of hydrolyzing a variety of esters.

---

[14] Franco, T.M.A.; Hegde, S.; Blanchard, J.S. *Biochemistry* **2016**, *55*, 6295.
[15] Williams, F.M. *Clin. Pharmacokinetics* **1985**, *10*, 392.
[16] Ahmadian, M.; Duncan, R.E.; Jaworski, K.; Sarkadi-Nagy, E.; Sul, H.S. *Fut. Lipidol.* **2007**, *2*, 229.

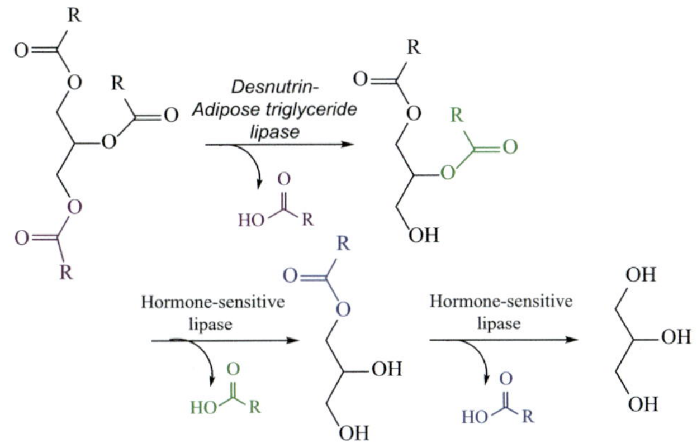

FIGURE 6.15 Enzymatic hydrolysis of a triglyceride. (Adapted from Figure 1, p. 230, Ahmadian, M.; Duncan, R.E.; Jaworski, K.; Sarkadi-Nagy, E.; Sul, H.S. *Fut. Lipidol.* **2007**, *2,* 229–237. Permission from Taylor & Francis Ltd, are acknowledged, including reprinted by permission of Informa UK Limited, trading Taylor & Francis Group, https://www.tandfonline.com.)

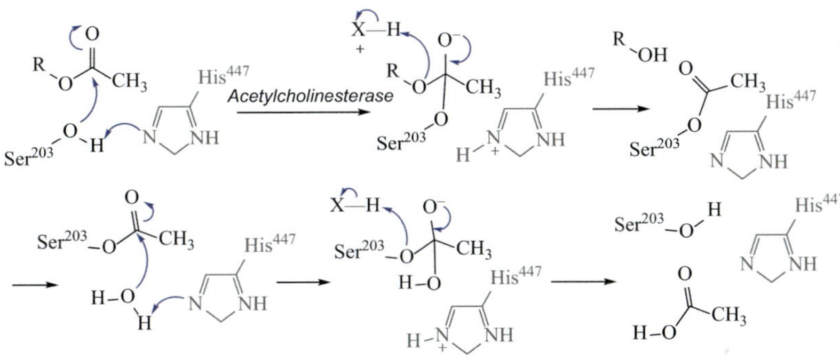

FIGURE 6.16 Acetylcholinesterase hydrolysis of an ester. (Reprinted with permission from Zhou, Y.; Wang, S.; Zhang, Y. *J. Phys. Chem. B* **2010**, *114,* 8817–8825. Copyright 2010 American Chemical Society.)

Acetylcholinesterase is a serine hydrolase and is the enzyme in the central nervous system that is responsible for the termination of impulse signaling at cholinergic brain synapses.[16] Formally, the reaction involves the hydrolysis of acetylcholine, $(CH_3)_3N^+CH_2CH_2OCOCH_3$. The mechanism of the hydrolysis of acetylcholine is acylation followed by deacylation, as shown in Figure 6.16.[17] Acetylation is simply acyl substitution by the serine residue, catalyzed by the nitrogen of the imidazole moiety of histidine. The deacylation step is the formal hydrolysis with water, again catalyzed by the histidine residue, and acyl substitution leads to hydrolysis of the methyl ester of the serine residue. The histidine acts as a base with the hydroxyl unit of serine to facilitate acyl addition to the acyl carbon of the ester. The resulting serine ester reacts with water, again facilitated by the interaction with histidine, to give a tetrahedral intermediate, and loss of serine generates the carboxylic acid, completing the acyl hydrolysis reaction. The active site is represented in Figure 6.17.[17]

Another example is the hydrolysis of a cholesterol ester with an enzyme known as *cholesterol esterase* (EC 3.1.1.13), as shown in Figure 6.18.[18] At the active site, the enzyme contains a carboxylate anion residue, an imidazole residue from the amino acid histidine and an alcohol residue. The reaction with **9**, a cholesterol ester, with an alcohol unit gives the key *tetrahedral intermediate*, **10**. The reaction of *cholesterol esterase* with cholesterol ester is initiated by an acid–base reaction of the carboxylate unit in **9** with the imidazole to generate a carboxylic acid unit on the enzyme after

[17] Zhou, Y.; Wang, S.; Zhang, Y. *J. Phys. Chem. B* **2010**, *114,* 8817.
[18] Sutton, L.D.; Froelich, S.; Hendrickson, H.S.; Quinn, D.M. *Biochemistry* **1991**, *30,* 5888.

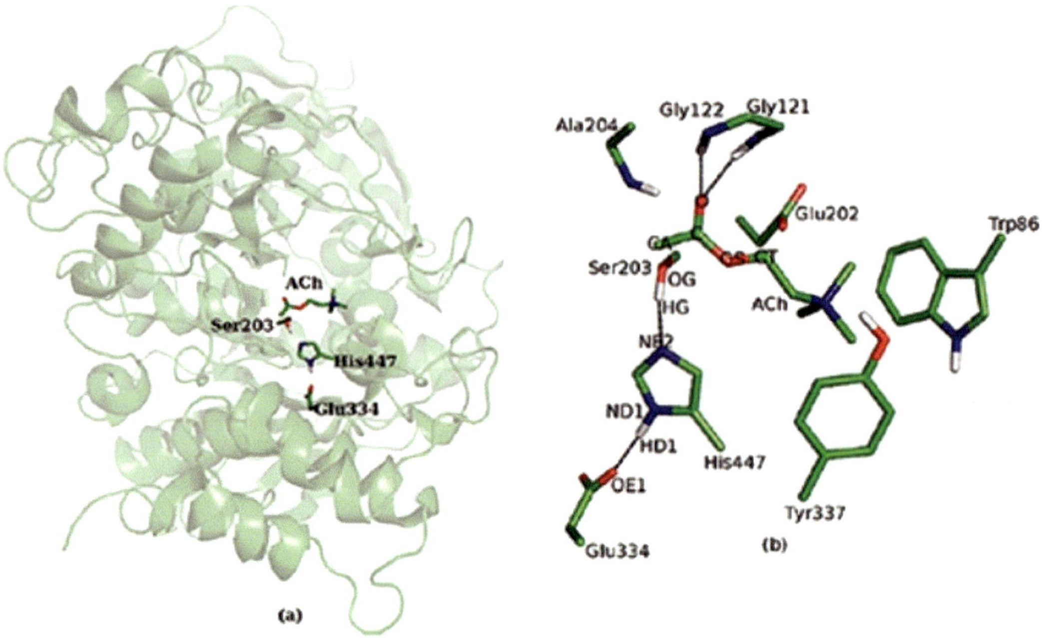

FIGURE 6.17 Acetylcholinesterase active site. (Reprinted with permission from Zhou, Y.; Wang, S.; Zhang, Y. *J. Phys. Chem. B* **2010**, *114,* 8817–8825. Copyright 2010 American Chemical Society.)

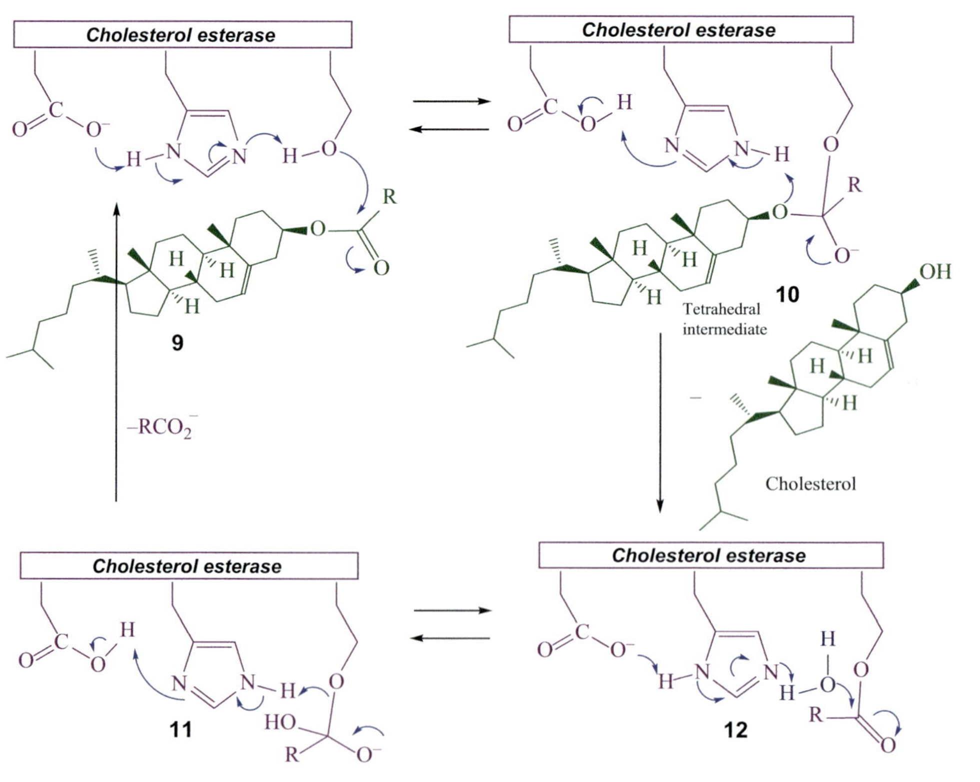

FIGURE 6.18 Hydrolysis of a cholesterol ester. (Reprinted with permission from Sutton, L.D.; Froelich, S.; Henrickson, S.; Quinn, D.M. *Biochemistry* **1991**, *30*, 5888. Copyright 1991 American Chemistry Society.)

10 is formed. When the alkoxide unit in the tetrahedral intermediate regenerates the carbonyl unit to complete the acyl substitution reaction, it yields cholesterol and the ester unit in **11**. Note that the

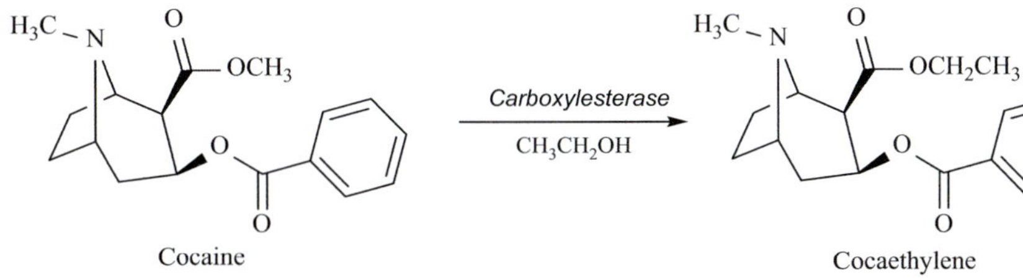

alcohol unit in cholesterol is converted to an ester in **11,** and the ester unit in the cholesterol ester is converted to an alcohol in cholesterol. This transformation is called a *transesterification* reaction (the transformation of one ester into a different ester). Another acid-base reaction of **11** with water in the cellular medium generates **12**, which regenerates *cholesterol esterase* by loss of a carboxylate unit, RCO_2^-. Cholesterol is a part of mammalian cell membranes and is very important for membrane permeability and fluidity. Cholesterol is also an important precursor for the biosynthesis of bile acids and some fat-soluble vitamins.

People who use the illicit drug cocaine often consume alcohol as well. This dual use results in the formation of the ethanol transesterification product of cocaine, in which the 2-carboxymethyl is transesterified to the carboxyethyl derivative. This reaction is probably catalyzed by hepatic and possibly renal carboxylesterases localized in the endoplasmic reticulum. This metabolite, known as "cocaethylene," is found in urine, and post-mortem studies have found this metabolite in the liver tissue of victims.[19]

One of the principal pancreatic *lipases* is known as *steapsin* (EC 3.1.1.3). Cleavage occurs at the C1 and C3 positions, and "the hydrolysis depends on bile salts (bile salts are carboxylic acid salts linked to a steroid), which act as detergents to emulsify the triglycerides and facilitate the

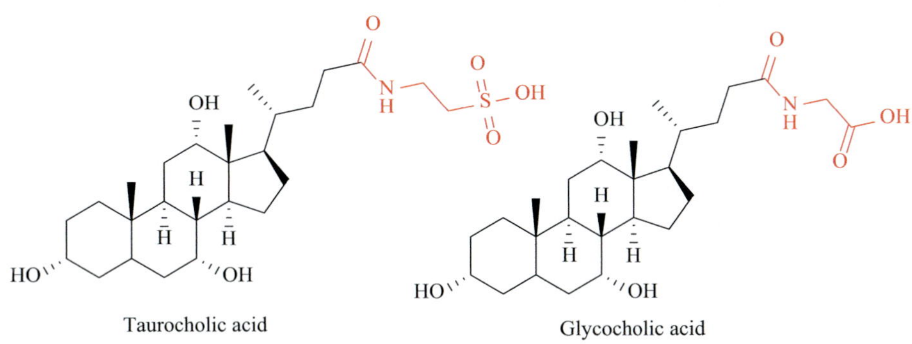

ester cleavage reaction. The long-chain fatty acid products are converted to micelles via interaction with the bile salts, and carried to the surface of epithelial cells, where they react with glycerol to form new triglycerides, which aggregate with lipoproteins to form particles called chylomicrons."[20] Chylomicrons are lipoprotein particles that transport dietary lipids from the intestines to other parts of the body. In the basic environment of the duodenum, the fatty acid is converted to the salt, which is emulsified by the bile salts.[20] Bile salts are commonly amide derivatives of steroids. Two bile acids are taurocholic acid and glycocholic acid.

[19] Boyer, C.S.; Petersen, D.R. *J. Pharmacol. Exp. Ther.* **1992**, *260,* 939.

[20] Garrett, R.H.; Grisham, C.M. *Biochemistry*, Saunders, Fort Worth, TX, **1995**, pp. 734 and 735.

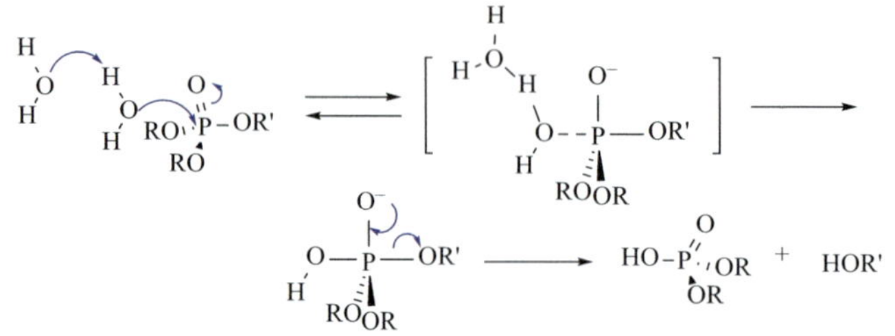

FIGURE 6.19 Hydrolysis of a phosphorus triester with inversion at phosphorus. (Reprinted with permission from Kirby, A.J.; Nome, F. *Acc. Chem. Res.* **2015**, *48,* 1806–1814. Copyright 2015 American Chemical Society. https://pubs.acs.org/doi/10.1021/acs.accounts.5b00072.)

Phospholipids (Section 4.5.1) are lipids that constitute a major component of cell membranes, forming lipid bilayers. Most contain a diglyceride unit (a diester of glycerol and two fatty acids), a phosphate ester unit, and usually an organic molecule. One example is dipalmitoyl phosphatidylcholine, which is a major component of lecithin. Lecithin is found in egg yolk and soybeans. Phosphatidylcholines are a major component of biological membranes and are components of pulmonary surfactant used to calculate fetal lung maturity.

Enzymes that catalyze the transfer of a phosphate group from adenosine triphosphate (ATP) to an alcohol acceptor are called kinases (EC 2; see Section 4.4.1). Nucleophilic attack by the alcohol oxygen on the γ-phosphorus of ATP, which is electrophilic, expels adenosine diphosphate (ADP). While carbon is the most familiar atom, phosphorus is a common electrophilic center in biological substitution reactions. A phosphate transfer reaction at phosphorus involves an alcohol nucleophile, which approaches opposite the leaving group. As the leaving group is displaced, the geometry at the phosphorus atom changes from tetrahedral to trigonal bipyramidal at the pentavalent (5-bond) transition state, and inversion of the absolute configuration occurs. The hydrolysis of a phosphate triester is shown in Figure 6.19,[21] with formation of the pentacoordinate phosphorus intermediate and inversion at the phosphorus atom.

This reaction is important for the enzyme *glucose kinase* (EC 2.7.1.1; see Section 4.4.1). Glucose is phosphorylated in the first step of the glycolysis pathway, forming glucose-6-phosphate, which is the first step in glucose metabolism and activates glucose for oxidation through glycolysis.[22]

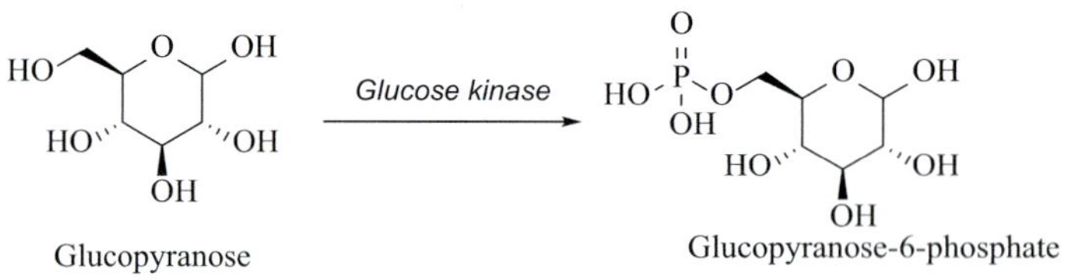

6.8 CHEMICAL REACTIONS OF CARBONYL GROUPS IN ORGANIC CHEMISTRY

The structural unit with one π- and one σ-bond between a carbon and oxygen (C=O) is called a *carbonyl.* Both the carbon and the oxygen are sp^2 hybridized, there are two covalent bonds to oxygen, and two unshared electron pairs reside on oxygen. If at least one of the groups is a hydrogen atom, there is an H–C=O unit (also abbreviated as CHO), and the functional group is called an *aldehyde.* If two carbon groups (two alkyl groups) are attached to the carbonyl carbon, the generic formula is $R_2C=O$, and the functional group is called a *ketone.*

[21] Kirby, A.J.; Nome, F. *Acc. Chem. Res.* **2015**, *48,* 1806.
[22] Marrero, J.; Trujillo, C.; Rhee, K.Y.; Ehrt, S. *PLoS Pathogens* **2013**, *9*, e1003116.

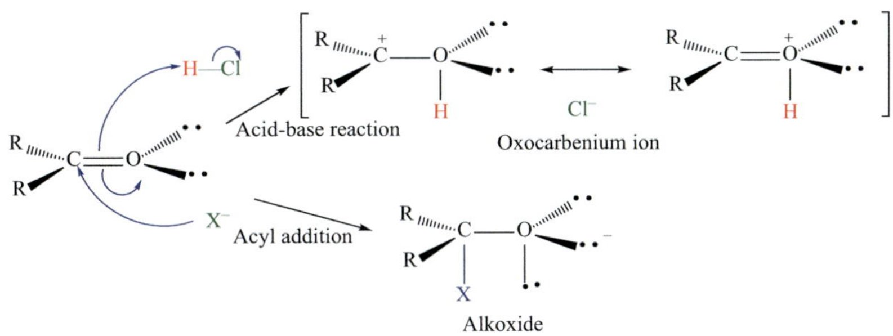

FIGURE 6.20 Donation of electrons to a carbonyl with cleavage of the π-bond.

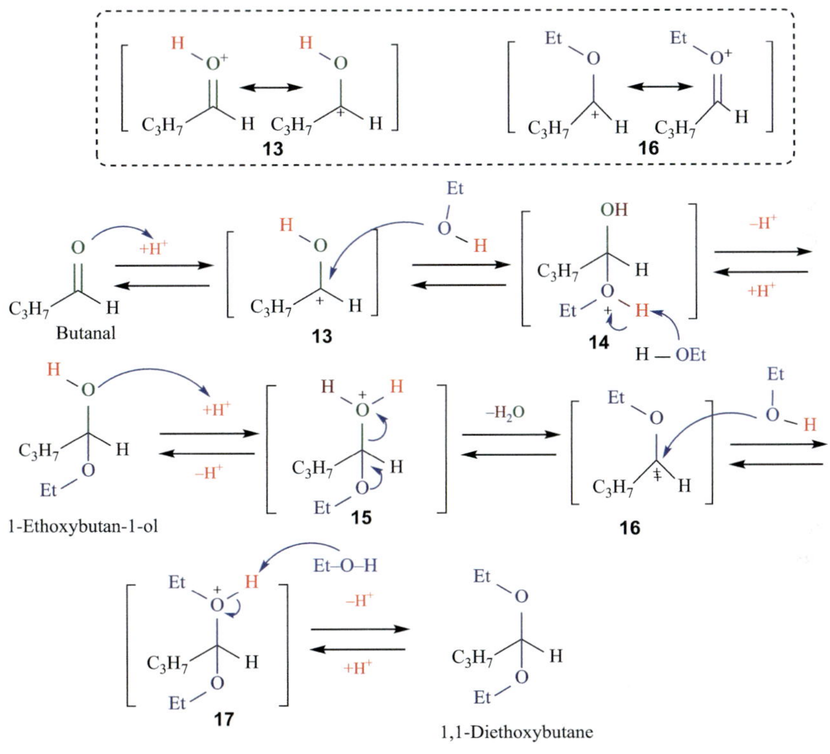

FIGURE 6.21 Mechanism of acetal formation for the reaction of butanal with ethanol.

The electronegative oxygen atom polarizes the C=O unit (δ^+ and δ^-). An aldehyde or a ketone will react as a Brønsted-Lowry base with a Brønsted-Lowry acid (H_2SO_4, HCl, nitric acid, methanesulfonic acid) to form a resonance-stabilized oxocarbenium ion, as shown in Figure 6.20. Two resonance contributors are formed for oxocarbenium ions. In addition, because of bond polarization, aldehydes and ketones react with nucleophiles to form a new σ-bond to the carbonyl carbon, generating an anionic species (the alkoxide product in Figure 6.20).

When a nucleophile reacts with the carbonyl carbon (the acyl carbon), the overall process is called *nucleophilic acyl addition*, since a nucleophile "adds" to the carbonyl carbon (an *acyl* carbon) and breaks the π-bond. In biochemical transformations, the nucleophiles involved in the reaction are water, alcohol moieties, or amine moieties. These three types of nucleophiles are classified as weak nucleophiles, and the initial acyl addition is followed by the formation of intermediate products, such as a hemiacetal that reacts further to give the final product, an acetal. The mechanism is shown for the reaction of ethanol with butanal in Figure 6.21. In this reaction, butanal reacts with an excess of ethanol in the presence of an acid catalyst to give the acetal product, 1,1-diethoxy-butane, along with water as a second product. The initially formed oxocarbenium ion (**13**) reacts with ethanol to give **14**, which is deprotonated to form the hemiacetal. Protonation of the OH unit generates the water leaving group **15,** and loss of water gives oxocarbenium ion **16**. Reaction with the second equivalent of ethanol gives **17,** and deprotonation yields the final product, 1,1-diethoxy-butane. This reaction is completely reversible, so an acetal reacts with water and an acid catalyst in an identical manner to give the corresponding aldehyde and two equivalents of the alcohol. As with similar reversible reactions, removal of the product water from the reaction will shift the equilibrium toward the acetal. Note that a key intermediate in this mechanism is 1-ethoxybutan-1-ol, a hemiacetal. In this and most cases of relatively simple aldehydes and ketones, the hemiacetal intermediate is very unstable under the reaction conditions and loses water to give the acetal product.

Although hemiacetals are unstable (highly reactive under the reaction conditions) for the most part, carbohydrates such as glucose exist primarily in the pyranose form, which is a hemiacetal. In fact, glucose and other carbohydrates (see Sections 13.1 and 13.4) are in equilibrium with the open-chain or aldehyde form and two isomeric hemiacetal forms, as shown in Figure 6.22. An equilibrium such as this, generated by carbohydrates, is known as *mutarotation* (Section 13.3). The two hemiacetal stereoisomers of D-glucose undergo mutarotation to form the α- and β-pyranose forms, which are the two diastereomers with different stereochemistry at the hydroxyl group of the acetal carbon. The α-stereoisomer has the hydroxyl group axial, or on the "bottom" side of the molecule, as shown. The β-pyranose stereoisomer has an equatorial hydroxyl group on the "top" side of the molecule, as shown.

When the nitrogen of an amine reacts with the electrophilic carbon of a carbonyl, a new C–N bond is formed as the π-bond is broken, as in any other nucleophilic acyl addition reaction. As shown in Figure 6.23, acetophenone reacts with the acid catalyst to form an oxocarbenium ion, **18** (only one resonance contributor is shown). The nucleophilic nitrogen atom of the amine reacts with **18** to yield the ammonium salt, **19**. The ammonium salt is a weak acid (cf. ammonium chloride, NH_4Cl, pK_a=9.25), and in an equilibrium acid-base reaction, this acid can react with unreacted

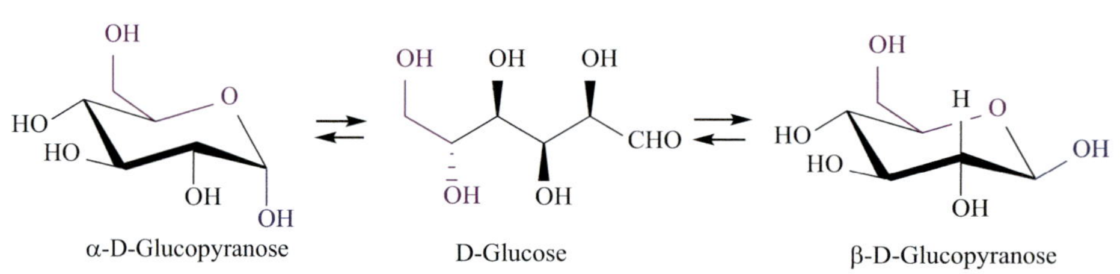

FIGURE 6.22 Mutarotation in D-glucose.

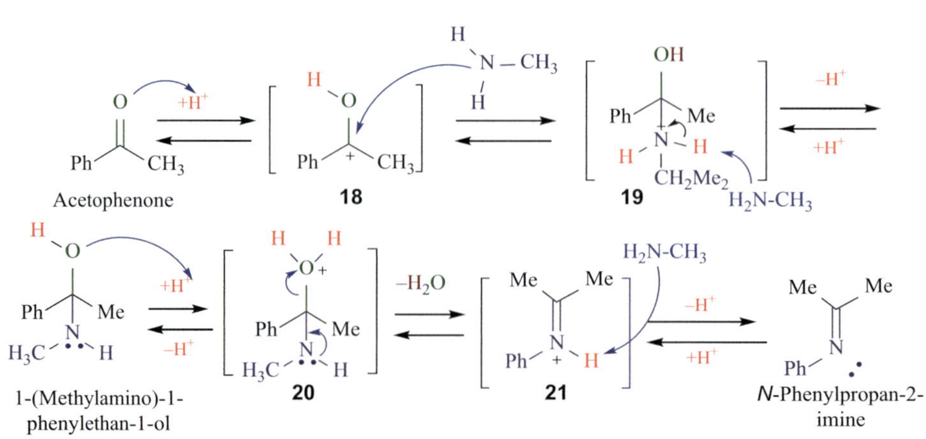

FIGURE 6.23 Mechanism of imine formation in the reaction of acetophenone and methanamine.

amine. The amine reacts as a base with **19** to remove the acidic proton from that species, giving the neutral amino alcohol, 1-(methylamino)-1-phenylethan-1-ol. Under these reaction conditions, this amino alcohol (known as a *hemiaminal*) is unstable and reacts to form a new product. Note that a hemiaminal is formally the nitrogen analog of a hemiacetal.

Since water is a product of this reaction, the OH unit of the hemiaminal (1-(methylamino)-1-phenylethan-1-ol) must be protonated to yield **20**, which then contains the $-OH_2^+$ leaving group. The electron pair on nitrogen in **20** can displace the leaving group (water) to form a new π-bond between carbon and nitrogen (a C=N bond) in **21**, an *iminium salt*. The proton on the nitrogen of the iminium salt is a weak acid, and reaction with a suitable base (e.g., the amine) gives a neutral product, the *imine* (N-phenylpropan-2-imine). Note that ammonia reacts with aldehydes or ketones in a manner similar to a primary amine. However, the imine products formed from ammonia are much less stable, as a rule. Imine-forming reactions therefore focus on reactions of primary amines rather than ammonia. Schiff bases are imines with an aryl group attached to the nitrogen atom, as previously mentioned.

6.9 CARBOXYLIC ACID DERIVATIVES AND ACYL SUBSTITUTION

Carboxylic acids contain an important functional group in which a carbon atom (alkyl group) is attached to a carbonyl (C=O) functional group, with a hydroxyl (OH) group also attached to the carbonyl carbon. This unit is known as a *carboxyl group*, which is the major structural feature of *carboxylic acids*. An important structural feature of the carboxyl functional group is the presence of the highly polarized O–H unit with the hydrogen δ^+, as shown in Figure 6.24 (same as Figure 5.14). The polarization induced by the carboxyl oxygen makes the carboxyl carbon atom positive, which leads to the oxygen of the OH unit being negatively polarized and the hydrogen positively polarized, as shown. In other words, the proton is acidic. *The greater acidity of the carboxylic acid is largely due to the stability of the conjugate base that is formed* (see Section 5.10.2).

There are four important derivatives of carboxylic acids in which the OH unit in RCOOH is replaced by a halogen, $-O_2CR$, –OR, or $-NR_2$, all attached to a carbonyl unit. The first type is an *acid halide*, generated when the OH unit in RCOOH is replaced with a halogen atom (e.g., chlorine, an *acid chloride*). An *acid anhydride* is formed when the OH group is replaced by another acid unit (O_2CR). As the name suggests (anhydride=without water), anhydrides are essentially two carboxylic acid units joined together with the loss of a water molecule. If the OH group in RCOOH is replaced by an OR′ group (from an alcohol), it is called an *ester* or a carboxylic ester. An ester is essentially a combination of a carboxylic acid and an alcohol. Finally, if the OH in RCOOH is replaced with an amine group (NH_2, NHR^1, or NR^1R^2), the derivative is called an *amide*. An amide is essentially a combination of a carboxylic acid and an amine. For each of these carboxylic acid derivatives, the unit that has replaced the OH unit is shown in violet. The structures of several acid derivatives and their names are shown in Table 6.1, along with the parent acid. The nomenclature rules for each type of molecule are described below. Most of the biologically important derivatives are ester or amide derivatives.

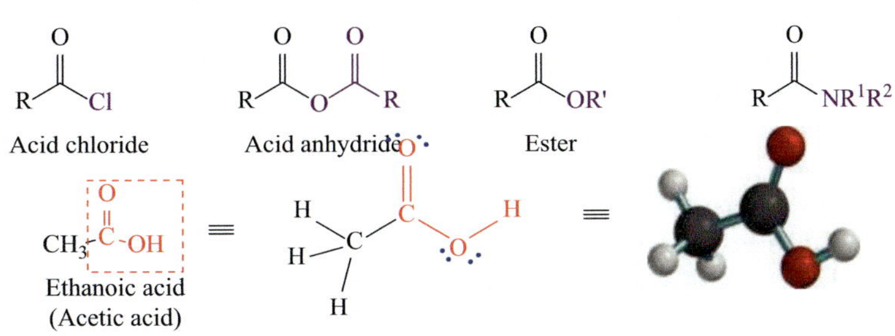

FIGURE 6.24 Ethanoic acid.

TABLE 6.1
Typical Acid Derivatives

Structure	Parent Acid	Name
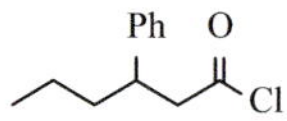	3-Phenylhexanoic	3-Phenylhexan**oyl chloride**
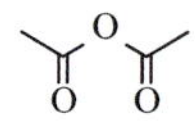	Ethanoic (acetic)	Ethanoic **anhydride** (acetic **anhydride**)
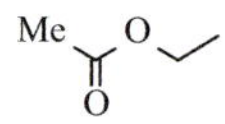	Ethanoic (acetic)	**Ethyl** ethan**oate** (**ethyl acetate**)
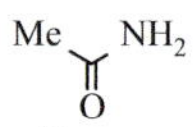	Ethanoic (acetic)	Ethan**amide** (acet**amide)**
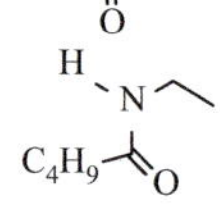	Pentanoic	*N*-Ethyl pentan**amide**
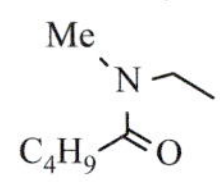	Pentanoic	*N*-Ethyl-*N*-methylpentan**amide**

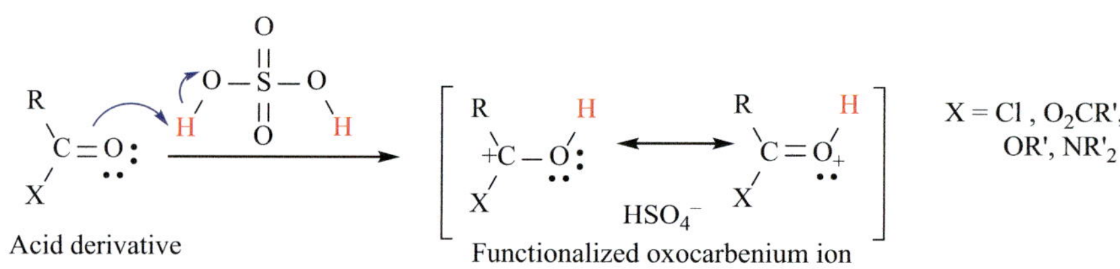

FIGURE 6.25 Formation of an oxocarbenium ion from an acid derivative.

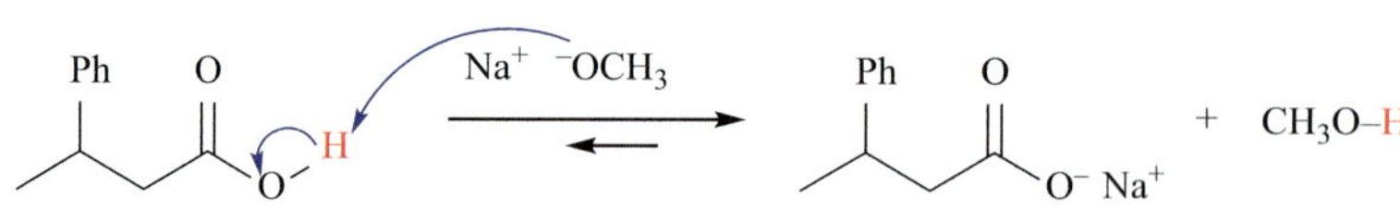

3-phenylbutanoic acid sodium 3-phenylbutanoate

The acid-base reaction dominates the chemistry of a carboxylic acid because *the acid-base reaction is much faster than the acyl substitution reaction.* Therefore, sodium methoxide reacts with 3-phenylbutanoic acid to yield the sodium salt, sodium 3-phenylbutanoate (the conjugate base), and methanol (the conjugate acid). Nucleophilic acyl substitution reactions dominate with acid derivatives.

When nucleophiles react with acid derivatives, there is a significant difference compared with aldehydes and ketones. Acid derivatives have a *leaving group* attached to the carbonyl, whereas aldehydes and ketones do not. This difference leads to a new type of reaction known as *acyl substitution.*

When a generic acid derivative reacts with an acid catalyst such as sulfuric acid, the acid-base reaction of the carbonyl gives the functionalized oxocarbenium ion shown in Figure 6.25, which is resonance-stabilized. In this reaction, the oxygen of the carboxylic acid derivative is the Brønsted-Lowry *base* (the proton acceptor), and sulfuric acid is the acid (the proton donor). The reaction occurs for carboxylic acid derivatives, where the X group is Cl for an acid chloride, O_2CR

for an anhydride, OR for an ester, or NR_2 for an amide. The presence of a leaving group in acid derivatives (e.g., alkoxy in an ester) leads to a new type of substitution reaction in which the leaving group is replaced with a nucleophile (Y). This process is called nucleophilic *acyl substitution*, and the intermediate is known as a *tetrahedral intermediate*. In general, *acid derivatives undergo acyl substitution, whereas aldehydes and ketones undergo acyl addition.*

If a nucleophile reacts with an acid derivative such as an ester, the nucleophile reacts with the acyl carbon to form a functionalized alkoxide, a *tetrahedral intermediate*, as shown in Figure 6.26. Remember that "Y" is the nucleophile and "CH_3O" is the leaving group. In the tetrahedral intermediate, there is one C–C σ-covalent bond to an alkyl group, but there is also a much weaker C–OR bond. The alkoxy group is a leaving group in the tetrahedral intermediate, and expulsion leads to the formation of a new acid derivative.

Acid chlorides, anhydrides, esters, and amides all undergo this reaction, but acid chlorides are more reactive than esters, which are more reactive than amides because of the greater reactivity of the tetrahedral intermediate. An important interpretation of this fact is that Cl is a better leaving group when compared to OR, which in turn is a better leaving group than NR_2.

Acid derivatives such as esters are hydrolyzed with aqueous acid under acidic conditions. When ethyl butanoate is treated with an acid catalyst in water, the products are butanoic acid and ethanol, as shown in Figure 6.27. The OH unit in butanoic acid is clearly derived from the water, and the OEt unit in the ethanol product is derived from the OEt unit of the ester. The carbonyl oxygen of ethyl butanoate reacts with the acid catalyst to give the resonance-stabilized oxocarbenium ion **22** (only

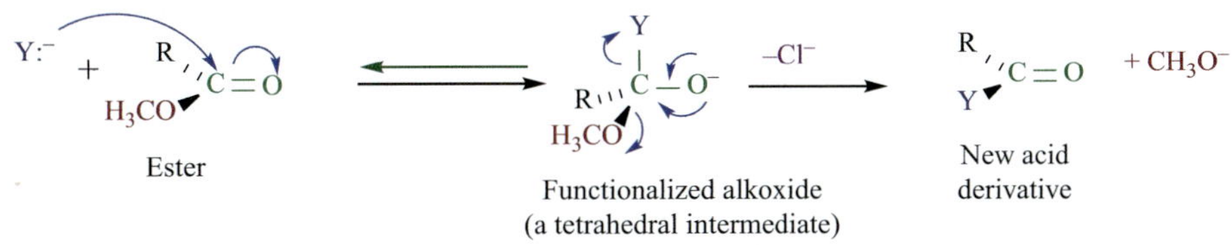

FIGURE 6.26 Acyl substitution.

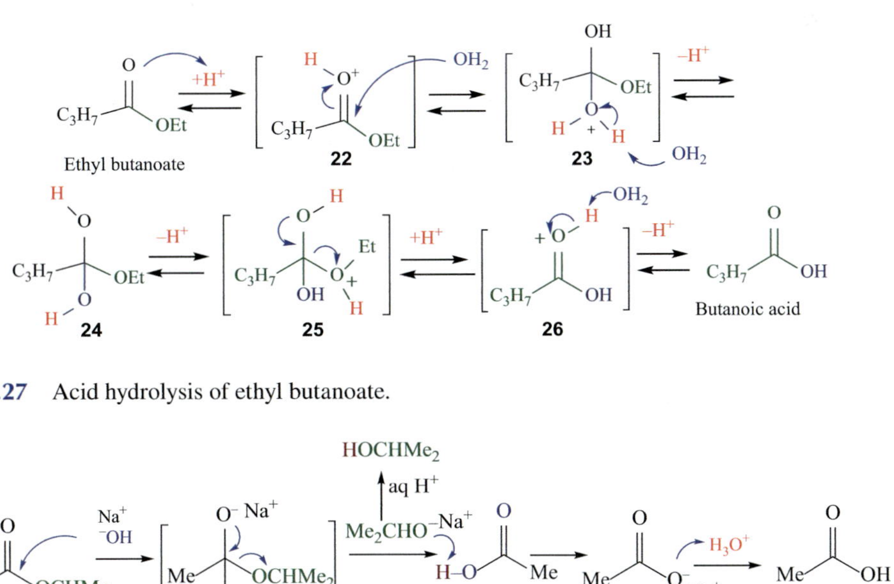

FIGURE 6.27 Acid hydrolysis of ethyl butanoate.

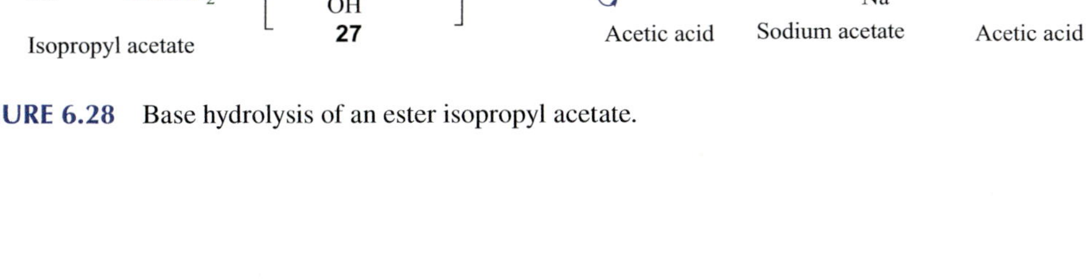

FIGURE 6.28 Base hydrolysis of an ester isopropyl acetate.

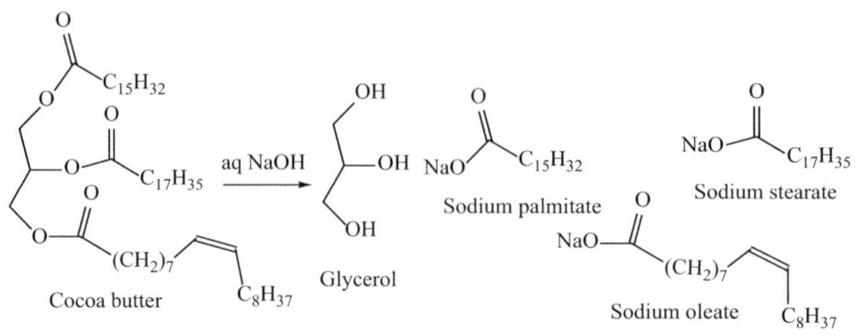

FIGURE 6.29 Saponification of cocoa butter.

one of the resonance contributors is shown). Subsequent reaction with water (the nucleophile) gives the oxonium ion (**23**). Loss of the acidic proton leads to the "hydrate-like" intermediate (**24**), which is the tetrahedral intermediate. Further reaction of the OEt unit with an acid catalyst is necessary so that ethanol can be lost as a leaving group. As shown in Figure 6.28, oxonium ion **25** is formed, and loss of ethanol gives oxocarbenium ion **26**, which is the protonated form of the carboxylic acid. Loss of the acidic proton yields ethanol and butanoic acid. Note that the acid-base reactions in the mechanism are completely reversible.

Esters also react with water under basic conditions to yield the acid, which is essentially the reverse of esterification, but the leaving group is now RO^-. Basic hydrolysis conditions require two steps to convert the ester to the acid. Isopropyl acetate (1-methylethyl ethanoate), for example, is heated to reflux in aq NaOH and then neutralized with aq H_2SO_4 in a second step to give a quantitative yield of two products: acetic acid and propan-2-ol (commonly known as isopropyl alcohol).[23] This reaction is explained by the addition of hydroxide (in blue) to the acyl carbon of the ester to yield tetrahedral intermediate **27**, as shown in Figure 6.28. The alkoxide unit ($C–O^-$) can transfer electrons back to the electrophilic carbon, but there are two potential leaving groups in **27**, ^-OH and $^-O\textit{i}Pr$. While alkoxy groups are not exceptional leaving groups, they are generally better than OH, so in tetrahedral intermediate **27**, O*i*Pr is lost rather than OH. The isolated products are acetic acid and propan-2-ol, but because acetic acid is formed in the presence of hydroxide (with isopropoxide also present), there is a rapid acid-base reaction that leads to the final product of the first chemical step: a carboxylate ion, here sodium acetate. To isolate acetic acid, the acetate ion (the conjugate base of acetic acid) is treated with aqueous acid in a second chemical step.

This two-step process, shown in reaction form for isopropyl acetate to give acetic acid + propan-2-ol, is known as *saponification*. Saponification means "to make soap," and the term comes from the ancient practice of using wood ashes (rich in potassium hydroxide) to convert animal fat into soap. Soap making evolved from the basic hydrolysis of fats or oils derived from animals or plants. Animal fat, as well as vegetable oils, is usually a mixture of triglycerides, the triester derivatives of fatty acids and glycerol. Fats are esters of glycerol and long-chain carboxylic acids (fatty acids). Many different fatty acids are constituents of fats, but octadecanoic acid (common name stearic acid) and hexadecanoic acid (common name palmitic acid) are common. Oils, typically obtained from plant sources, usually have alkenoic acids as part of the triglyceride rather than alkanoic acids. In other words, they contain one or more C=C units in the carbon skeleton, and this structural feature usually lowers the melting point of the compound, making it a liquid. Under saponification conditions, basic hydrolysis of all three ester units leads to the formation of glycerol and the salts of the fatty acids. An example is the saponification of cocoa butter, which is extracted from cocoa beans. Cocoa butter, shown in Figure 6.29, is 3-(palmitoyloxy)-2-(stearoyloxy)propyl oleate. Basic hydrolysis leads to glycerol (propane-1,2,3-triol) along with the sodium salts of the three carboxylic acid components: sodium palmitate, sodium stearate, and sodium oleate.

[23] Hurd, C.D.; Blunck, F.H. *J. Amer. Chem. Soc.* **1938**, *60*, 2419.

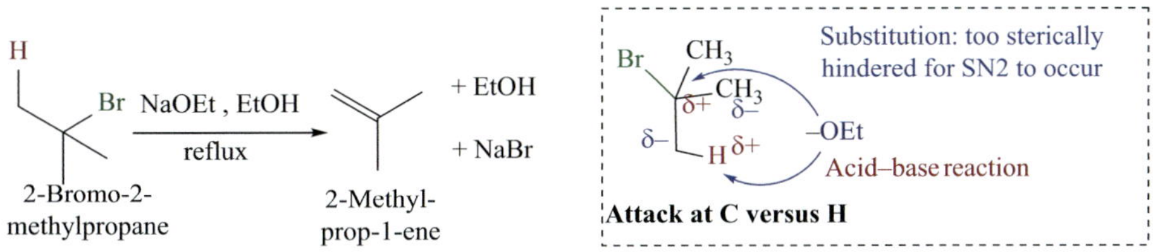

FIGURE 6.30 Elimination of 2-bromo-2-methylpropane.

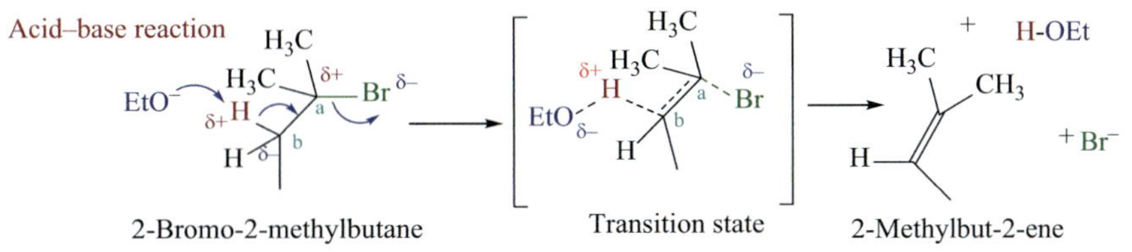

FIGURE 6.31 Conversion of 2-bromo-2-methylbutane to 2-methylbut-2-ene.

6.10 ELIMINATION REACTIONS OF ALKYL HALIDES (E2 AND E1 REACTIONS)

When 2-bromo-2-methylpropane was heated with sodium ethoxide (the conjugate base of ethanol) at reflux in ethanol, the isolated product was an alkene, 2-methylprop-2-ene, along with sodium bromide and ethanol. A comparison of the starting material and the product in Figure 6.30 shows that both the bromine and a hydrogen atom are lost from 2-bromo-2-methylpropane, and an alkene is formed, 2-methylprop-1-ene. Structurally, C–Br and C–H bonds are broken, and a π-bond is formed. The bromine atom is converted to the bromide ion (as NaBr), and ethoxide reacts with the acidic hydrogen atom (on the β-carbon relative to bromine, so it is called a β-hydrogen) to form ethanol. In this reaction, the ethoxide ion reacts as a base to form a new O–H bond in the conjugate acid (ethanol), and bromine is the leaving group that gives the bromide ion as the conjugate base.

In the reaction of sodium ethoxide with a tertiary halide, the activation barrier for an S_N2 reaction is too high due to steric hindrance in the transition state. Polarization of the bromine atom on the α-carbon is δ^-, and the induced dipole on the second carbon away from the bromine is also δ^-, which leads to a δ^+ hydrogen (see Figures 6.30 and 6.31). This small positive dipole makes the hydrogen on the β-carbon slightly acidic. The bond polarization of the C–Br bond in 2-bromo-2-methylbutane leads to a δ^+ β-hydrogen. While the ethoxide ion is attracted to the δ^+ carbon atom, an S_N2 reaction does not occur; instead, the ethoxide ion is attracted to the β-hydrogen. The collision of hydroxide with the β hydrogen initiates an acid–base reaction that removes the β hydrogen with concomitant formation of 2-methylbut-2-ene, as shown in Figure 6.31. Therefore, the overall transformation is called an *elimination reaction* (the elements of H and Br are eliminated to form a π-bond).

No intermediate has been detected under the conditions under which the reaction is normally done. If there is no intermediate, then the characteristics of the reaction are described by the transition state shown in Figure 6.31. If there is no intermediate, all the bond making and breaking occur simultaneously (this is a *synchronous* reaction). The initial reaction of ethoxide and the β-hydrogen is a collision process, and experimental kinetic data show it to be a *second-order reaction*. Since the overall transformation is an elimination reaction, with loss of H and Br, the symbol E for elimination is used. It is a bimolecular elimination reaction, so the symbol 2 is used. *This is an E2 reaction.* Note that it is the electrons in the C^β–βH bond that expel the bromine leaving group as the π-bond is formed. Therefore, the leaving group and the βH must be *anti in the E2 transition state,* as shown in Figure 6.31.

Many alkenes have more than one β-hydrogen atom, and removal of those different β-hydrogen atoms via an E2 reaction can lead to different isomeric alkenes. When 3-bromo-3-methylpentane is heated with KOH in ethanol, for example, the major product is 3-methylpent-2-ene. However, as shown in Figure 6.32, 3-bromo-3-methylpentane has two β-hydrogen atoms (H_a and H_b). In principle, there should be two products, 3-methylpent-2-ene and 2-ethylbut-1-ene, formed by loss of each of the two β-hydrogen atoms. Experimentally, there are a major and a minor product. The observed major product is 3-methylpent-2-ene, formed by removal of H_b, and the observed minor product is 2-ethylbut-1-ene, formed by removal of H_a.

3-Methylpent-2-ene is a trisubstituted alkene with three carbon substituents attached to the C=C unit, whereas 2-ethylbut-1-ene is a disubstituted alkene with two carbon substituents. This preference is explained by the thermodynamic stability of the alkene product. A carbon substituent (an alkyl group) is electron-releasing relative to the adjacent carbon atom, and a C=C unit with more alkyl groups has more electron density released into the π-bond, making the bond stronger and the alkene more stable. Based on the number of substituents attached to the C=C unit, a trisubstituted alkene is more stable than a disubstituted alkene, and a tetrasubstituted alkene is more stable than a trisubstituted alkene. In an E2 reaction, the more thermodynamically stable (more substituted) alkene is formed as the major product (the *Zaitsev rule*). This observation is consistent with the formation of 3-methylpent-2-ene as the major product of the E2 reaction shown in Figure 6.32.

There is an alternative reaction pathway to the E2 reaction that is most prevalent in an aqueous medium. If elimination occurs by removal of the β hydrogen by the base in a second step after initial ionization to a carbocation, the reaction is termed a *unimolecular elimination, E1.*

If 2-bromo-2-methylbutane is heated with KOH in aq THF, an S_N2 reaction is not possible due to steric hindrance, but an E2 reaction is certainly possible. However, it is known that water facilitates ionization by "pulling" the leaving group via hydrogen bonding and solvates and helps stabilize positive and negative ions, which assists the separation of the ions. In this case, ionization will give the bromide ion and the tertiary carbocation, as shown in Figure 6.33. Therefore, in an aqueous solvent, formation of a carbocation is possible and often favored. There are two reactions: ionization to a carbocation and subsequent removal of the β-hydrogen atom by a base. In this reaction, loss of bromide to yield the carbocation via ionization has a much higher activation energy and is *a significantly slower step, whereas removal of the β hydrogen by the base has a low activation energy and is a much faster step.* Therefore, ionization is the rate-determining step, and this reaction is termed a unimolecular elimination, or E1.

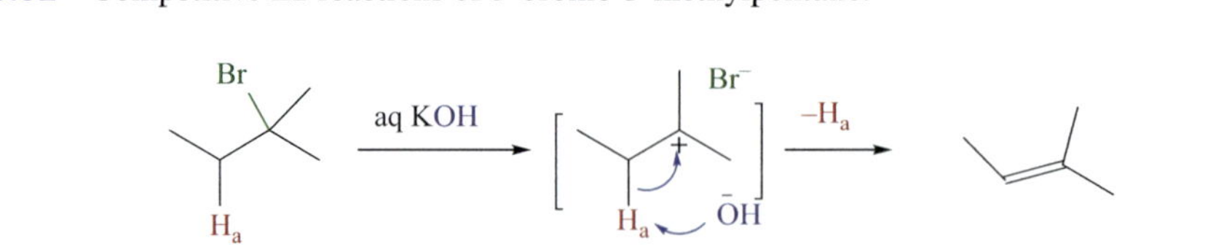

FIGURE 6.32 Competitive E2 reactions of 3-bromo-3-methylpentane.

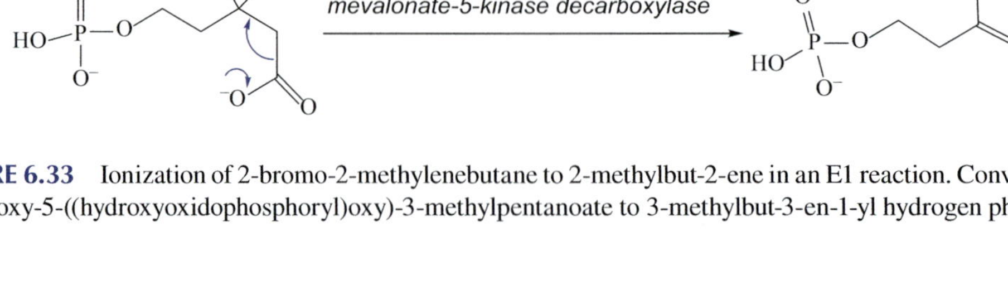

FIGURE 6.33 Ionization of 2-bromo-2-methylenebutane to 2-methylbut-2-ene in an E1 reaction. Conversion of 3-hydroxy-5-((hydroxyoxidophosphoryl)oxy)-3-methylpentanoate to 3-methylbut-3-en-1-yl hydrogen phosphate.

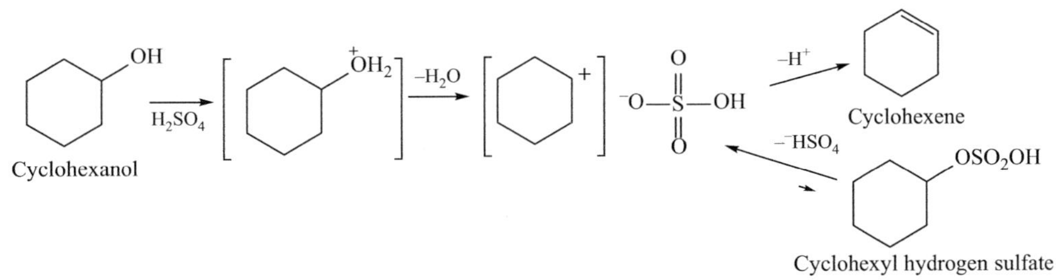

FIGURE 6.34 Dehydration of cyclohexanol.

In the carbocation intermediate, the β-hydrogen of the carbocation is more polarized and more acidic when compared to the β-hydrogen atom (H_a) in 2-bromo-2-methylbutane (Figure 6.33). It is analogous to S_N1, and the product will be 2-methylbutan-2-ol. Water may also react with the carbocation, and in the basic medium, the product is also the alcohol. In fact, the attraction of hydroxide or water to C^+ is largely a function of the solvent, and in a protic solvent (e.g., water), substitution is usually faster. If a base is rather nucleophilic and the reaction is done in an aprotic solvent, the S_N1 reaction also competes with E1 if there is a carbocation intermediate. If the reaction is done in a protic solvent, elimination is usually faster than the S_N1 reaction, but this obviously depends on the nucleophile.

Elimination also occurs in biological systems. A simple example is the conversion of mevolonate-5-phosphate to isopentenyl diphosphate, as shown in Figure 6.33.[24] The building block for isoprenoid compounds is the decarboxylation-elimination step shown for the conversion of 3-hydroxy-5-((hydroxyoxidophosphoryl)oxy)-3-methylpentanoate to 3-methylbut-3-en-1-yl hydrogen phosphate using *mevalonate-5-kinase decarboxylase*.[24] This step is part of the mevalonate pathway.[25]

Since the reaction conditions used for an E1 reaction often favor S_N1 rather than E1, it is difficult to find a "pure" E1 reaction. In general, this is true, but there are exceptions when the base used in the reaction is a poor nucleophile or when the S_N1 product is unstable and leads to a reversible reaction. If cyclohexanol is treated with concentrated sulfuric acid, for example, a rapid elimination reaction leads to the observed product, cyclohexene. The experimentally determined mechanism involves an acid-base reaction of the oxygen from the OH unit (the base) with sulfuric acid to form an oxonium ion, the conjugate acid.

Loss of water from the oxonium ion yields a secondary carbocation and the hydrogen sulfate anion, as shown in Figure 6.34. The hydrogen sulfate anion is highly stabilized due to resonance and is not very nucleophilic. If the S_N1 reaction were faster, the product would be cyclohexyl hydrogen sulfate, but this product is very unstable under these reaction conditions, so the reaction is reversible and unfavorable. In other words, the instability of cyclohexyl hydrogen sulfate leads to an equilibrium that favors the carbocation. On the other hand, the hydrogen sulfate anion is sufficiently basic that it can remove a β-hydrogen from the carbocation intermediate, yielding the cyclohexene product. Cyclohexene is the E1 product.

In most E1 reactions, an aqueous medium or a highly protic medium such as sulfuric acid or perchloric acid is used to generate the carbocation. The E1 product can be formed when secondary and tertiary alcohols react with concentrated sulfuric acid or concentrated perchloric acid. When a good nucleophile is present, as in a dilute aqueous solution of sulfuric acid, S_N1 products usually dominate (e.g., the alcohol product when a carbocation reacts with water).

[24] Goldstein, J.L.; Brown, S.B. *Nature* **1990**, *343*, 425.
[25] Buhaescu, I.; Izzedine, H. *Clin. Biochem.* **2007**, *40,* 575.

HOMEWORK

6.1. Draw the S_N2 transition state for the reaction of norepinephrine and *S*-adenosyl methionine (see Figure 6.3)

6.2. Draw the transition state and the final product that is formed when triethylphosphine reacts with (2*R*)-iodobutane in tetrahydrofuran solvent.

6.3. What is the structure of the product formed when the monosodium salt of glutamic acid methyl ester reacts with 1-bromopent-2*E*-ene?

6.4. What is the structure of the product formed with S-adenosyl methionine reacts with butan-1-amine?

6.5. Draw gulopyranose-3-phosphate. See Section 13.2 for the structure of gulose.

6.6. What is the structure of the carbocation formed by ionization of bromotriphenylmethane?

6.7. What is the structure of farnesyl monophosphate?

6.8. What are the intermediate and the final products formed when *N*-acetyl-*O*-methyl serine reacts with HI?

6.9. What is the final product formed by a glycolysis reaction of β-D-iodopyranose residue, where the anomeric group is OR?

6.10. What product is expected if cyclooctene were to react with cytochrome P450 oxidase?

6.11. Draw the structure of α-D-altropyranose and of β-D-talopyranose. See Sections 13.2 and 14.1.

6.12. What is the product if the amino group of glutamine were to react with the open-chain form of α-D-altropyranose? See Sections 13.2 and 14.2.

6.13. Draw the products formed by saponification of tripentanoyl glycerol.

7 Dienes and Conjugated Carbonyl Compounds in Biochemistry

This chapter discusses dienes, alkene-ketones, alkene-aldehydes, or alkene-esters with a particular emphasis on those molecules that are conjugated. In conjugated molecules, two or more π-bonds are directly connected with no intervening sp^3 atoms. Conjugated dienes react similarly to other alkenes, but due to conjugation the carbocation or radical intermediates formed from conjugated dienes are resonance stabilized. Therefore, there are two sites of reactivity: the carbonyl as well as the C=C unit and so there are differences in product distribution when compared to simple alkenes.

7.1 ENZYME-MEDIATED CONJUGATE ADDITIONS

Enzyme-mediated addition to conjugated carboxylic acids is known. *Argininosuccinate lyase* produced by *Saccharomyces cerevisiae* catalyzes the Michael addition reaction of L-arginine to fumaric acid to give L-argininosuccinate, as shown in the reaction.[1] L-argininosuccinate plays a role in the biosynthesis of secondary metabolites as well as other amino acids.

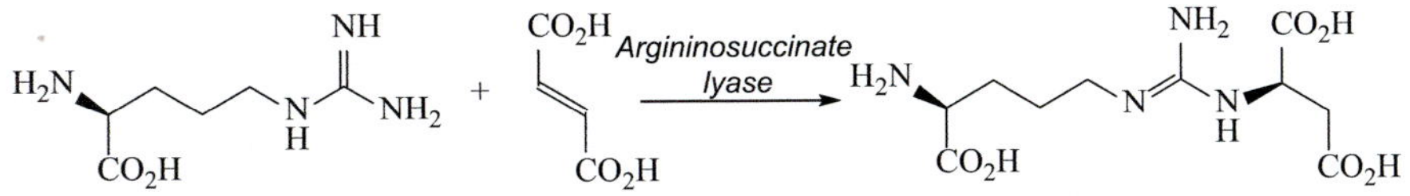

L-Arginine   Fumaric acid   L-Argininosuccinate

A ribozyme has been discovered that mediates a Michael reaction similar to the rate-limiting step of the enzyme *thymidylate synthase* (EC 2.1.1.45; see Section 4.1.3). This enzyme catalyzes a key (Reprinted from Sengle, G.; Eisenführ, A.; Arora, P.S.; Nowick, J.S.; Famulok, M. *Chem. Biol.* **2001**, *8*, 459–473, Novel RNA catalysts for the Michael reaction, Copyright 2001, with permission from Elsevier.) step in the biosynthetic pathway for the reductive methylation of dUMP, using tetrahydrofolate [4-(4-(((2-amino-4-oxo-3,4,5,6,7,8-hexahydropteridin-6-yl)methyl)amino) benzamido)-4-carboxybutanoate] as the methyl donor. In the reaction shown, dUMP is activated via a 1,4-Michael addition of the thiol group of cysteine[198] residue, mediated by *thymidylate synthase*.[2]

---

[1] Schoenenberger, B.; Wszolek, A.; Meier, R.; Brundiek, H.; Obkirchera, M.; Wohlgemuth, R. *RSC Adv.* **2017**, *7*, 48952.
[2] Sengle, G.; Eisenführ, A.; Arora, P.S.; Nowick, J.S.; Famulok, M. *Chem. Biol.* **2001**, *8*, 459.

DOI: 10.1201/9781003605157-7

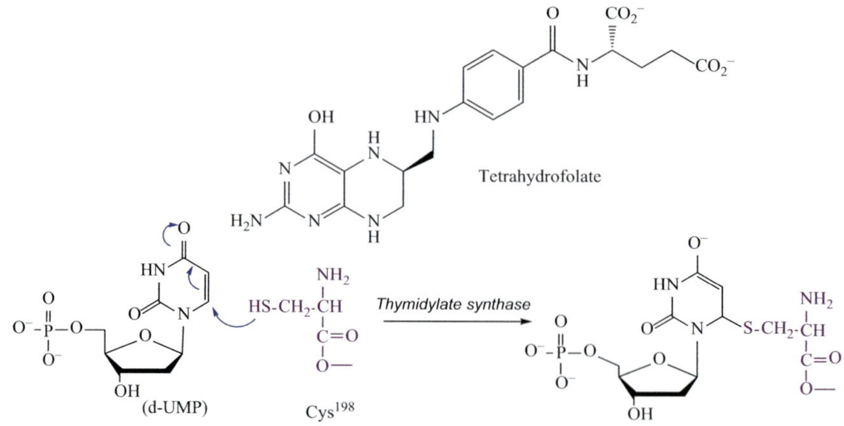

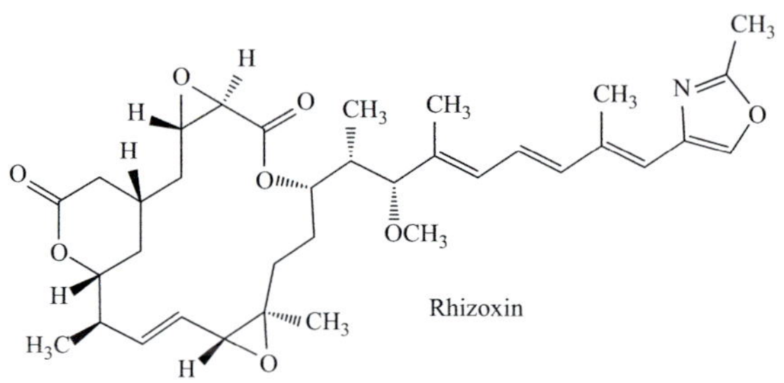

A *polyketide synthase* (EC 2.3.1.260)-mediated Michael addition has been found, as shown in Figure 7.1,[3] as part of polyketide-chain branching in the rhizoxin biosynthetic pathway. The Michael addition involves a malonate-type addition to the conjugated ester, and the resulting product is subsequently converted to the lactone moiety shown. Rhizoxin is an antimitotic agent (it blocks cell growth by stopping mitosis) with anti-tumor activity, isolated from the pathogenic plant fungus

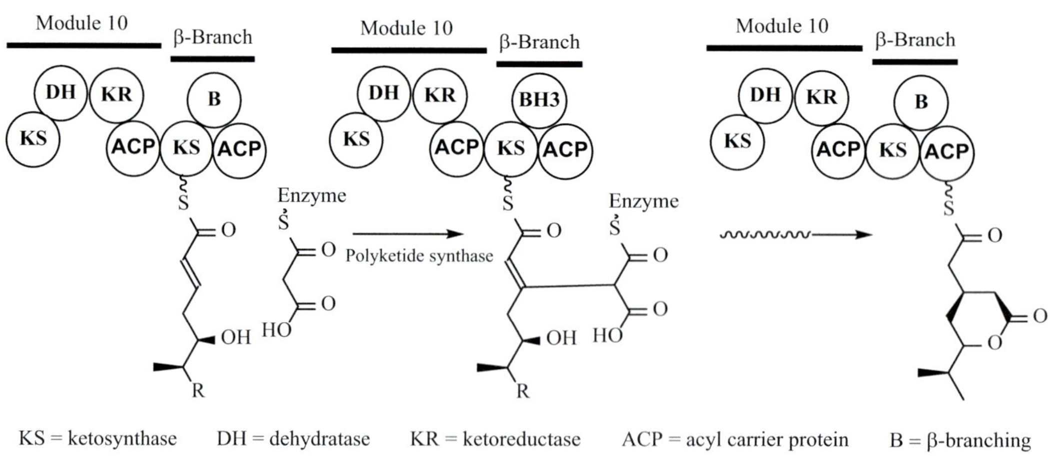

FIGURE 7.1 A *polyketide synthase*-mediated Michael addition. (Kusebauch, B.; Busch, B.; Scherlach, K.; Roth, M.; Hertweck, C. *Angew. Chem. Int. Ed.* **2009**, *48*, 5001–5004. Copyright 2009 Wiley-VCH Verlag GmbH & C KGaA. Scheme 2, p. 5002 and Scheme 4, p. 5003 therein. Copyright 2009 Wiley-VCH Verlag GmbH & C KGaA. Scheme 2, p. 5002 and Scheme 4, p. 5003 therein. Reproduced with permission of John Wiley & Sons. Ltd.)

[3] Kusebauch, B.; Busch, B.; Scherlach, K.; Roth, M.; Hertweck, C. *Angew. Chem. Int. Ed. Engl.* **2009**, *48*, 5001.

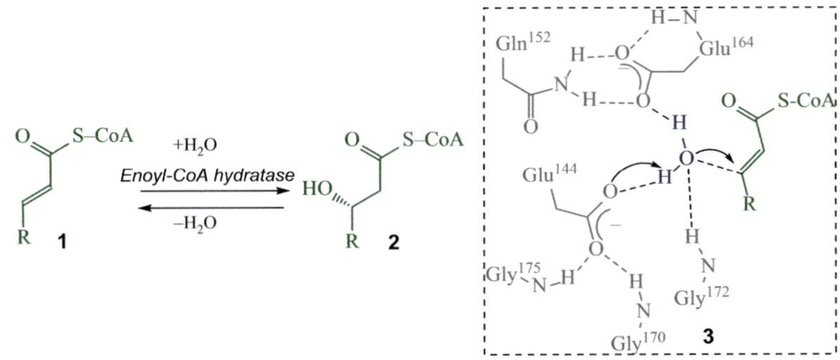

FIGURE 7.2 Enzyme-mediated conjugate addition of water to an acyl–CoA molecule. (Reprinted in part with permission from Bahnson, B.J.; Anderson, V.E.; Petsko, G.A. *Biochemistry* **2002**, *41,* 2621–2629. Copyright 2002 American Chemical Society.)

(*Rhizopus microsporus*). However, rhizoxin is not biosynthesized by the blight fungus but by bacteria that live within the fungal cytosol.[3] The KS, DH, KR, ACP, and B modules are enzymes that are part of the biosynthetic gene cluster for rhizoxin. A gene cluster is a group of two or more related genes found in DNA that encode similar products.

There is an enzyme known as *enoyl-CoA hydratas*e (EC 4.2.1.150; also known as *crotonase*; see Section 4.6.1) that facilitates the conjugate addition of water to the C=C unit of an acyl–CoA molecule (e.g., **1**) to yield the 3-hydroxy thioester (**2**), as shown in Figure 7.2.[4] The fragment CoA is coenzyme A, which forms a thioester unit as seen in **1** and **2**. This process is essential for the metabolism of fatty acids and the production of energy,[4] in which *enoyl-CoA hydratase* catalyzes the second step in the breakdown of fatty acids, or the second step of β-oxidation in fatty acid metabolism. The active site of the enzyme involves water, coordinated to two glutamic acid residues (E) (E^{144} and E^{164}; see **3**), stabilized with a glutamine residue (Q^{152}), and glycine residues (G^{175} and G^{170}) are also essential. The water, stabilized by G^{172} and E^{164}, is a hydroxide surrogate by means of the interaction with the carboxylate unit of glutamic acid[144], allowing conjugate addition to C3 of the enzyme-bound conjugated thioester for conversion to the β-hydroxyl product.[4]

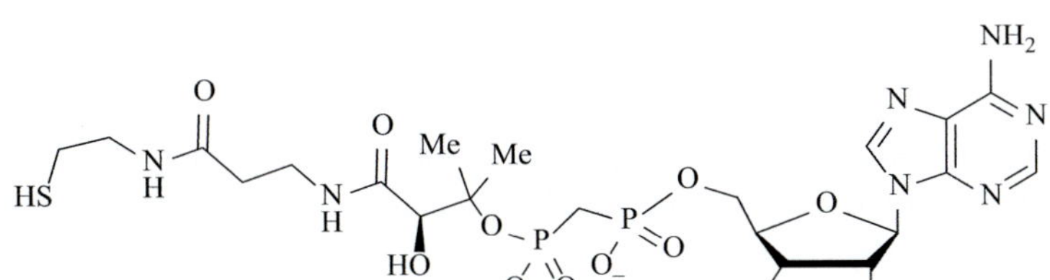

CoenzymeA (CoA)

7.2 CONJUGATED DIENES AND CONJUGATED CARBONYL COMPOUNDS

A molecule containing one π-bond as part of a C=C unit is called an alkene, and when a molecule contains two π-bonds in two C=C units, it is called a diene. There are three fundamental structural categories for a diene: (i) those where the C=C units are separated by sp^3-hybridized atoms, (ii) those where the C=C units are connected to form a C=C–C=C unit, and (iii) those that contain two π-bonds that share an sp-hybridized atom. Dienes in category (1) are called *nonconjugated dienes,* and an example is hexa-1,5-diene, using the standard nomenclature rules for alkenes introduced in Chapter 5. Dienes in category (2) are called *conjugated dienes*, illustrated by hexa-1,3-diene, which can exist as a mixture of (*E,E*), (*E,Z*), or (*Z,Z*)-isomers. Dienes in category (3) are 1,2-dienes, and an

[4] Bahnson, B.J.; Anderson, V.E.; Petsko, G.A. *Biochemistry* **2002**, *41,* 2621.

example is buta-1,2-diene, although they have the common name of allenes. An allene is an example of a *cumulative* π-system, specifically a *cumulative diene.* In benzene, the conjugated π-bonds in a six-membered ring lead to the aromatic stability of *benzene* (Section 11.2), and *it is not a triene.*

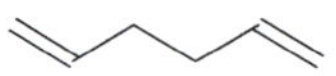

Hexa-1,5-diene

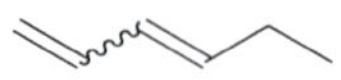

Hexa-1,3-diene

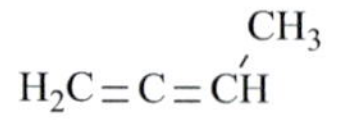

Buta-1,2-diene

The nomenclature and identification of conjugated and nonconjugated dienes are straightforward. The chemical reactions of such compounds are more complicated. In a nonconjugated diene, each π-bond behaves more or less independently, so hexa-1,5-diene is expected to react with one molar equivalent of HBr to form an alkyl bromide that has a C=C unit elsewhere in the molecule. This example shows that a reagent (e.g., HCl or HBr) will react with one C=C unit without affecting the other, and the second C=C unit does not necessarily influence the reaction. The situation is quite different for a conjugated diene such as hexa-1,3-diene. Since the π-bonds are linked together, a reaction at one C=C unit is influenced by the presence of the other C=C unit.

Before discussing the formal reactions of conjugated dienes, it is useful to analyze the diene unit (C=C–C=C) in a typical conjugated diene (e.g., buta-1,3-diene). The first observation is that a carbon-carbon single bond connects the two C=C units. The bond length of the C=C unit is 1.48 Å (148 pm).[5] In a typical alkane, the C–C bond distance is about 1.54 Å (154 pm),[5a] so the C—C unit in butadiene is shorter. The C=C bond distance in buta-1,3-diene is 1.34 Å (134 pm),[5b] which compares favorably with the 1.32 Å (132 pm) bond distance of the C=C unit in ethene[5a] (a typical C=C unit has a bond distance of ~ 1.337 Å, 133.7 pm).[5a]

The shorter bond distance of the C–C unit is explained by the presence of p-orbitals on the "middle carbons," which are relatively close. While the π-electrons are localized in the π-bonds of the C=C units, and buta-1,3-diene has distinct single and double bonds, there is a small amount of overlap of the p-orbitals that makes the bond slightly shorter than expected. *Butadiene is not resonance-stabilized*, but there is an interaction of the π-bonds, which is consistent with a slightly shorter bond distance. The electron potential map for the molecular model of butadiene is shown in Figure 7.3, where the red areas indicate high concentrations of electron density. The highest electron density is localized on the π-bonds, between C1–C2 and C3–C4, rather than between the single bond C3–C4, consistent with no resonance for this molecule.

The C3–C4 bond is a single covalent bond, and there is rotation about that bond, with two rotamers in equilibrium that result from the high- and low-energy interactions of the C=C units. When the C=C units have an eclipsed relationship (as shown in Figure 7.3), buta-1,3-diene is said to be in a *cisoid or s-cis conformation.* When the two C=C units have a staggered relationship, buta-1,3-diene is said to be in a *transoid or s-trans conformation.* The steric interaction of the hydrogen atoms makes the s-cis rotamer higher in energy, which means there is a higher percentage of *s-trans*-buta-1,3-diene at equilibrium. In other words, buta-1,3-diene spends most of its time in the s-trans conformation.

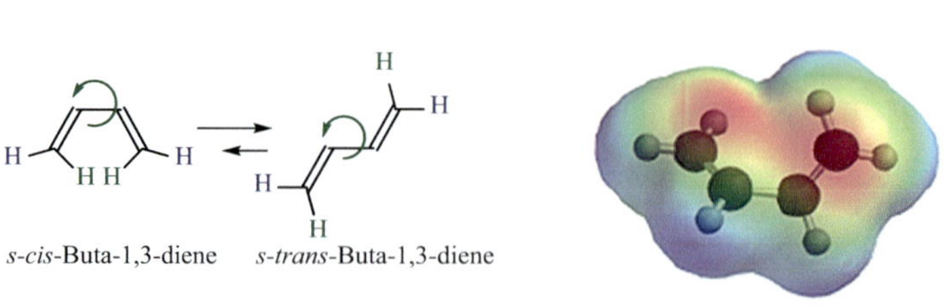

FIGURE 7.3 Rotamers in buta-1,3-diene.

[5] (a) Dean, J.A. *Handbook of Organic Chemistry*, McGraw-Hill, New York, **1987**, pp. 3–12 to 3–15; (b) Smith, M. *March's Advanced Organic Chemistry*, 8th ed., John Wiley and Sons, Inc., Hoboken, NJ, **2019**, pp. 25 and 40.

Just as it is possible to have two C=C units conjugated to form a diene, two C≡C units may be conjugated to form a diyne. It is fair to say, however, that conjugated diynes are not as common as conjugated dienes. It is also possible to have an ene-yne, with a C=C unit as well as a C≡C unit in the same molecule. For the most part, diynes and ene-ynes will be ignored, and the focus will be on the chemistry of dienes.

Cyclopenta-1,3-diene

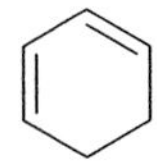
Cyclohexa-1,3-diene

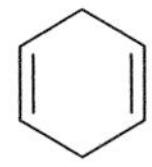
Cyclohexa-1,4-diene

In addition to the acyclic compounds just discussed, it is common to see cyclic molecules with conjugated π-bonds. An important cyclic diene is cyclopenta-1,3-diene. For cyclohexa-1,3-diene derivatives, there is the conjugated cyclohexa-1,3-diene and the nonconjugated cyclohexa-1,4-diene. With all dienes that have larger rings, there is one conjugated isomer, but there may be several possible nonconjugated isomers.

Dienes are a well-known class of compounds, but diynes and ene-ynes are less common, especially in biochemical systems. There are, by contrast, many molecules that have a C=C unit as well as a C=O unit (a carbonyl). Conjugated carbonyl compounds are characterized and identified by *the presence of a C=C–C=O unit,* and nonconjugated compounds have *sp^3-hybridized carbon atoms that separate the C=C and C=O units.* Nonconjugated molecules of this type include the ketone hex-5-ene-2-one and the ester ethyl pent-4-enoate. For all practical purposes, the C=C unit and the C=O units in nonconjugated compounds behave independently of each other. This observation is not always true, but it is a good working assumption in most cases. Hydrogenation of hex-5-ene-2-one using a palladium catalyst, for example, generally yields hexan-2-one by reduction of the C=C unit, whereas switching to a PtO_2 catalyst leads to reduction of the C=O unit to yield hex-1-ene. In other words, the two functional groups react nearly independently.

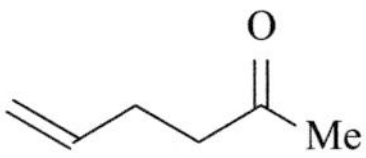

Hex-5-en-2-one

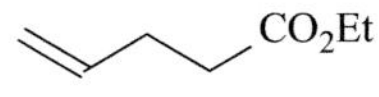

Ethyl pent-4-enoate

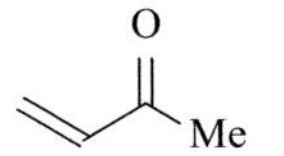

But-3-en-2-one

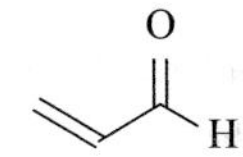

Prop-2-enal
(Acrolein)

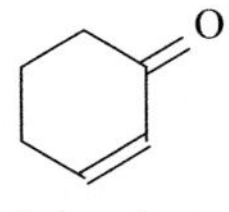

Cyclohex-2-enone

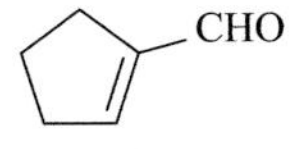

Cyclopent-1-enecarbaldehyde

Conjugated carbonyl compounds are also known as α,β-unsaturated carbonyl compounds (α,β-unsaturated ketones, α,β-unsaturated aldehydes, α,β-unsaturated esters, etc.), and are characterized by a C=C–C=O unit. Typical examples include but-3-en-2-one (the common name is methyl vinyl ketone) and prop-2-enal (the common name is acrolein). In these molecules, the presence of the carbonyl will influence reactions at the C=C unit, and the presence of the alkene will influence reactions at the C=O unit. In other words, it is difficult to do a chemical reaction at one functional group without the other either reacting or influencing the course of the reaction. As in buta-1,3-diene, *there is some overlap of the p-orbitals of C1 (the carbonyl carbon) and C2, but as with buta-1,3-diene, there is no resonance.*

Cyclic ketones can have a conjugated carbonyl unit within the ring. An example is cyclohex-2-enone. In an aldehyde, the CHO unit must be attached to a ring, and the conjugating C=C unit is in the ring. An example of this type of aldehyde is cyclopent-1-enecarbaldehyde. Several difunctional compounds involve a C=C unit and the carbonyl unit of a carboxylic acid (or an acid chloride), ester, or amide. Typical conjugated carboxylic acids are prop-2-enoic acid (acrylic acid), but-(2*E*)-enoic acid (also known as crotonic acid), and 2-methylpropenoic acid (also known as methacrylic acid). There are cyclic derivatives of the carboxylic acids in which the carboxyl unit is attached to the ring. An example is cyclobut-1-enecarboxylic acid. A C=C unit or a C=O unit is conjugated if it is connected to an aromatic ring. Ethenylbenzene (styrene) is a conjugated alkene, and both benzaldehyde and acetophenone are conjugated carbonyl compounds.

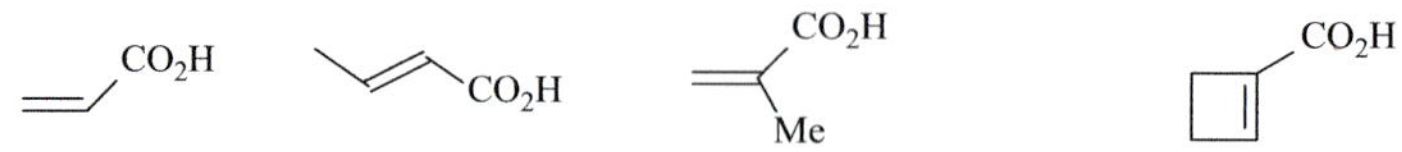

Prop-2-enoic acid (Acrylic acid) But-(2E)-enoic acid 2-Methylprop-2-enoic acid (Methacrylic acid) Cyclobut-1-enecarboxylic acid

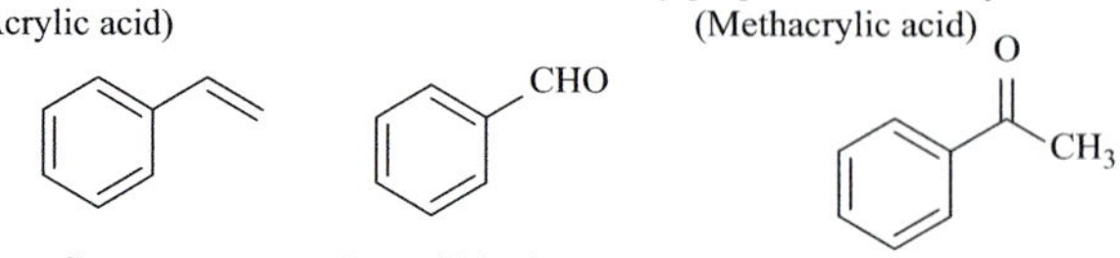

Styrene Benzaldehyde Acetophenone

Poly(alkene) molecules, or polyenes, are structural features of many compounds. Some high-molecular-weight compounds, such as natural rubber, are polymers that have the polyene structure shown.

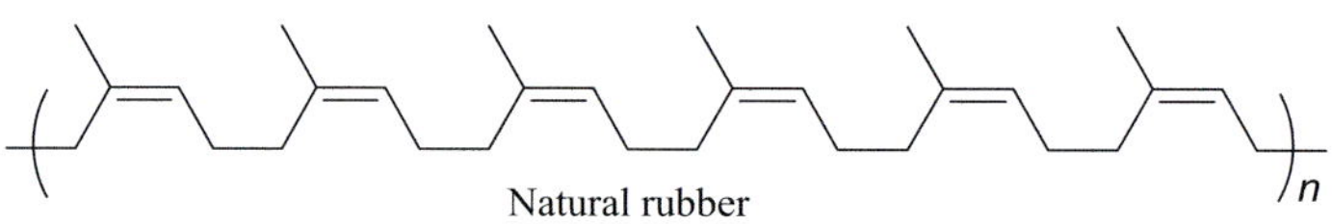

Natural rubber

7.3 REACTIONS OF CONJUGATED COMPOUNDS

A conjugated diene contains two C=C units connected by a covalent single bond, as in buta-1,3-diene. The C=C unit of an alkene reacts as a Brønsted-Lowry base in the presence of an acid (HX) to form a new C–H bond and generate a carbocation. That is exactly what happens when an acid (e.g., HCl) reacts with buta-1,3-diene, except that the influence of the second C=C unit must be taken into consideration.

Experimentally, when buta-1,3-diene is treated with HCl under various conditions, a mixture of (*E*)-1-chlorobut-2-ene and (Z)-1-chlorobut-2-ene is obtained in ~ 80% yield, along with ~ 20% of 3-chlorobut-1-ene.[6] The chlorine atom appears at different positions on the four carbon atoms of buta-1,3-diene, as shown in Figure 7.4. The reaction of an alkene proceeds via a carbocation intermediate, but there must be more than one electrophilic site to account for the products. In other words, the positive charge is not localized on a single carbon, but rather delocalized.

These results are explained by recognizing that one of the C=C units reacts with the acidic proton to generate a carbocation with two resonance contributors, **4A** and **4B**, as shown in Figure 7.4. The resonance-stabilized carbocation **4** is known as an *allylic carbocation*. The presence of the

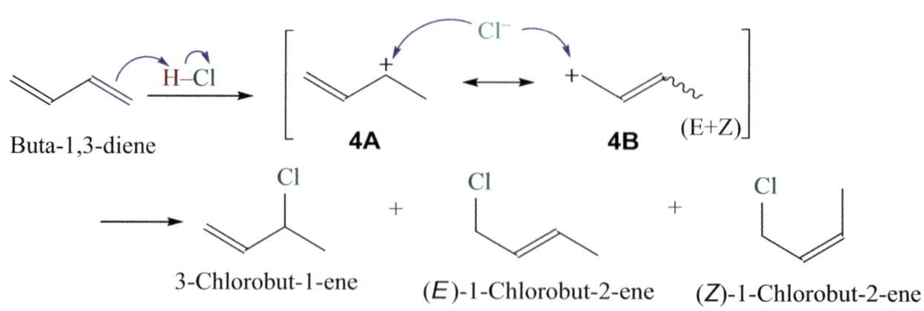

FIGURE 7.4 Mechanism of the reaction of buta-1,3-diene with HCl.

[6] Fieser, L.F.; Fieser, M. *Advanced Organic Chemistry*, Reinhold Pub., New York, **1961**, p. 199.

conjugating C=C unit in an allylic carbocation allows *resonance delocalization of the positive charge*. This intermediate is *more stable* than a non-allylic carbocation.

To explain the products, a new question can be posed about resonance contributors: *Are the resonance contributors equal in energy?* There are two electrophilic carbon atoms in buta-1,3-diene, and the reaction of the nucleophilic chloride ion with each positive carbon will lead to two regioisomeric products, 3-chlorobut-1-ene (from **4A**) and 1-chlorobut-2-ene (from **4B**). This latter alkene is a mixture of both the (*E*) and (*Z*) stereoisomers since the geometrical identity of the C=C unit is lost in the resonance-stabilized cation **4** because there is free rotation about the C2–C3 bond, which explains the two 1-chlorobut-2-ene products. The major products are 1-chlorobut-(2*E*)-ene and 1-chlorobut-(2*Z*)-ene, and these products result from the reaction of the chloride ion with resonance contributor **4B**, which suggests that the electron dispersal in the resonance contributors is not equal. The (*E*)-alkene is generally more stable since the C=C unit has more electron-releasing alkyl groups, and a higher percentage is expected.

Another factor that should be considered is what helps explain the product distribution in Figure 7.4. Assume that the reaction of buta-1,3-diene with HCl has a *late transition state*, so *the major product will be determined by the relative stability of the products.* It is known that a disubstituted alkene is more stable than a monosubstituted alkene. Based on the *Hammond postulate* and the invocation of a late transition state, the 1-chlorobut-2-enes will be the major products of this reaction. Therefore, the predominance of 1-chlorobut-(2*E*)-ene and 1-chlorobut-(2*Z*)-ene can be explained in terms of a late transition state that favors formation of the more stable product.

Special terminology is used for the products formed when buta-1,3-diene reacts with HCl. A simplistic comparison of buta-1,3-diene shows that 3-chlorobut-1-ene is formed by adding the elements of H and Cl to one C=C unit of C=C–C=C. Therefore, 3-chlorobut-1-ene is called a *1,2-addition product.* A similar comparison of buta-1,3-diene with 1-chlorobut-2-ene shows that the elements of H and Cl add to C1 and C4 of C=C–C=C, so it is called a *1,4-addition product.*

There is another category of conjugated alkene compounds that must be discussed. When the C=C unit is conjugated to a benzene ring, as in styrene, the reaction with HCl proceeds as if it were a simple alkene to yield the more stable cation intermediate, **5**. This carbocation is more stable because the positive charge can be delocalized into the benzene ring, leading to the four resonance contributors shown. This carbocation is known as a *benzylic cation.* This intermediate traps the nucleophilic chloride ion to yield (1-chloroethyl)benzene (benzyl chloride). Note that styrene reacts with diatomic chlorine the same as any other alkene to generate the expected dichloride product, 1,2-dichloro-1-phenylethane.

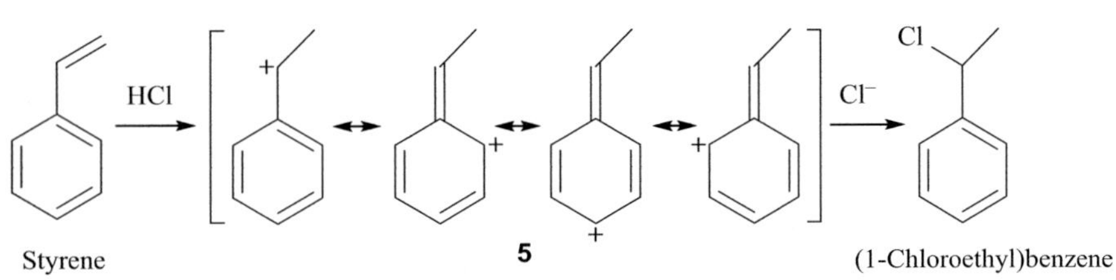

Another common reaction of alkenes is with diatomic halogens (e.g., bromine, Br_2) to form 1,2-dibromides. In this reaction, the alkene reacts as a Lewis base with the bromine atom to form a bromonium ion. When buta-1,3-diene reacts with bromine, a bromonium ion is formed by the reaction of one C=C unit, but the proximity of the second C=C unit leads to alternative reaction pathways. Indeed, both 1,2- and 1,4-addition products are formed, 3,4-dibromobut-1-ene and a mixture of (*E*) and (*Z*)-1,4-dibromobut-2-ene.

Initial reaction with bromine generates bromonium ion **6**, and as shown in Figure 7.5, the bromide ion will attack the less substituted carbon to form a new C–Br bond, leading to 3,3-dibromobut-1-ene. The conjugating C=C unit in **6** provides another site for reaction with the bromide ion.

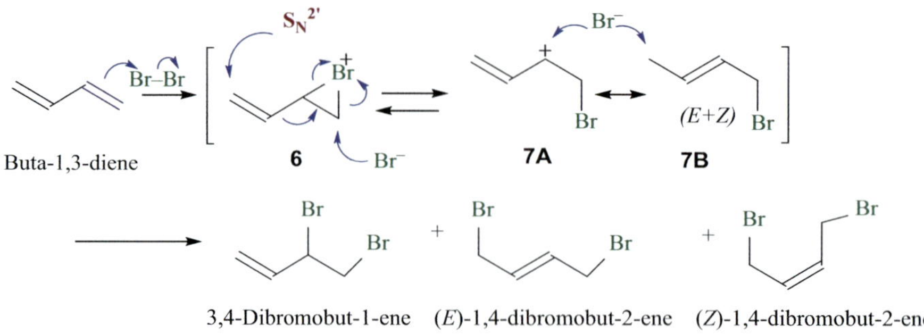

FIGURE 7.5 Mechanism of the reaction of buta-1,3-diene with diatomic bromine.

Both carbon atoms of the three-membered ring have a δ^+ dipole due to the electronegative bromine, and the nucleophilic bromide ion can react with either carbon.

As noted, reaction usually occurs at the less substituted carbon. The attached C=C unit effectively extends this reactivity, and inductive effects lead to a δ^+ dipole at the terminal carbon of the alkene unit. Formation of 1,4-dibromobut-2-ene is an example of a reaction at the second site of reactivity in the molecule, the C=C unit. The extension of reactivity due to conjugating π-bonds is known as *vinylogy*. Formation of the 1,4-dibromobut-2-ene products is due to vinylogy. Reaction of the bromide ion at the C=C unit yields the two stereoisomeric products, 1,4-dibromobut-(2*E*)-ene and 1,4-dibromobut-(2*Z*)-ene. When the bromide ion attacks the C=C unit, the π-bond is broken and those electrons are transferred to the adjacent carbon, breaking the C–Br bond of the three-membered ring, forming a new π-bond and yielding a dibromide product. Nucleophilic attack of this type is called an S_N2' *reaction* (nucleophilic substitution at an allylic carbon with displacement of the leaving group).

An alternative mechanism is the ring opening of the three-membered ring in **6** to generate the stable allyl cation, **7**, which reacts with the bromide ion to yield the observed products (see Figure 7.5). By and large, ionization of this type only occurs when water is present, so this ring-opening mechanism is more commonly observed in aqueous media. In terms of product distribution, buta-1,3-diene reacts with Br_2 *at low temperatures* to yield 3,4-dibromobut-1-ene as the major product, but *at higher temperatures,* the major products are 1,4-dibromobut-(2*E*)-ene and 1,4-dibromobut-(2*Z*)-ene. At higher temperatures, the thermodynamically more stable products, 1,4-dibromobut-(2*E*)-ene and 1,4-dibromobut-(2*Z*)-ene, are the major products, again consistent with a late transition state.

The stereochemistry of alkenes (*E*) or (*Z*) is also important in many biological processes. The polyene molecules retinol (Vitamin A_1) and dehydroretinol (Vitamin A_2) are found in animal products (e.g., eggs, dairy products, animal livers and kidneys, as well as fish liver oils). These retinoids are known to act as signaling molecules that regulate aspects of cell differentiation, embryonic development, growth, and vision.[7] In mammals, retinol is obtained by cleavage of the tetraterpenoid β-carotene at the highlighted bond in Figure 7.6,[7] usually in the intestine, to give the aldehyde retinal, which is enzymatically reduced to retinol.

The biological process known as vision involves the conversion of retinol to retinal via an oxidation reaction that uses the cofactor $NADP^+$ (nicotinamide adenine dinucleotide phosphate).[7] Note the (*E*)-geometry of the C11–C12 double bond in retinol (the numbering is based on the numbering that begins in the cyclohexene unit, as shown). The key process here is the change of an (*E*)-alkene to a (*Z*)-alkene using an enzyme. Coordination to an enzyme and interaction with light changes the alkene stereochemistry back to (*E*)-. Enzymatic conversion of the C11-(*E*)-double bond to the C11-(*Z*)-double bond (11-*cis*-retinal) allows reaction with an amine unit on a protein

[7] Dewick, P.M. *Medicinal Natural Products*, 2nd ed., Wiley, West Sussex, UK, **2002**, pp. 230–231 and Figure 6.73.

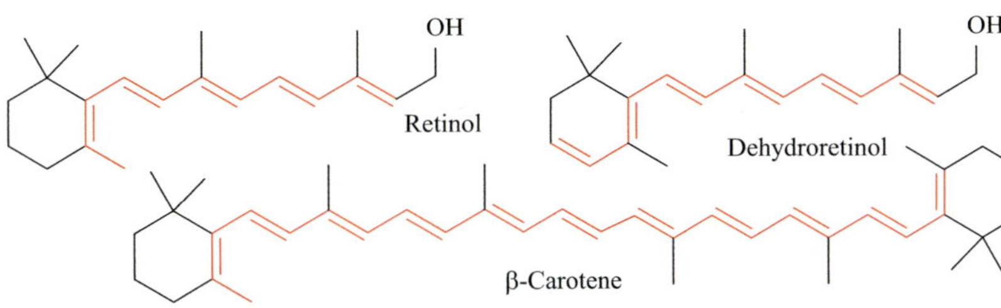

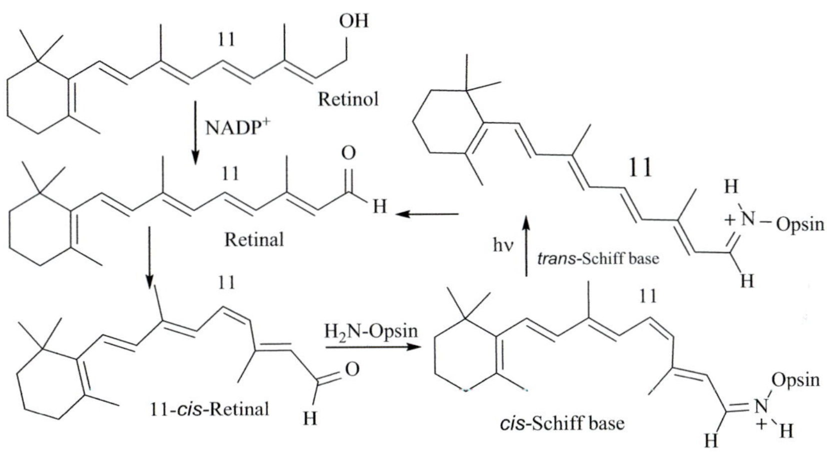

FIGURE 7.6 cis-trans-Isomerization in retinal. (Dewick, P.M. *Medicinal Natural Products*, 2nd ed., Wiley, West Sussex, UK, **2002**, pp. 230–231 and Figure 6.73. Copyright 2002, reproduced with permission of John Wiley & Sons. Ltd.)

called opsin to form an imine (a *cis*-Schiff base). Retinal is the red visual pigment, and exposure to light (*hν*) converts the C11-(*Z*)-double bond back to the C11-(*E*)-double bond (a trans-Schiff base), and this triggers a nerve impulse to the brain.[3] Conversion of the trans-Schiff base back to retinal allows the process to cycle again.

7.4 CONJUGATE (MICHAEL) ADDITION

There is a well-known reaction of conjugated carbonyl compounds in which a nucleophile adds to the C=C unit. The reaction of the C=C unit of the conjugated system as a Brønsted-Lowry base involves the donation of electrons to the acidic proton. In but-3-en-2-one, a C=C unit is directly attached to a C=O unit. The carbonyl carbon is electrophilic due to polarization of the oxygen atom, and the π-system effectively extends the reactivity of the carbonyl to the C=C unit (*vinylogy*). When a nucleophile donates electrons to the C=C unit of the conjugated system, it is commonly called *Michael addition* (or the *Michael reaction*).

In its initial form, Michael addition was the reaction of amines with conjugated carbonyl compounds. However, this reaction is often reversible, and while reaction with amines is useful, the Michael reaction has come to include the 1,4-addition of any nucleophile to a conjugated carbonyl system. A generalized form of this reaction shows the bond polarization induced by the electronegative carbonyl oxygen atom. The δ^+ dipole of the C=C unit reacts with a nucleophile (Y^-) to form a new bond (C–Y), breaking the π-bond with formation of an enolate anion, **8**, which is resonance stabilized. In a second step, treatment with aqueous acid gives the isolated product in which the nucleophile has added to C3 relative to the carbonyl carbon atom. There are many examples of this reaction, and some of them generate new carbon–carbon bonds when Y^- is a carbon nucleophile. *Reactions of this type are formally a 1,4-addition, or conjugate addition, but the common term is Michael addition.*

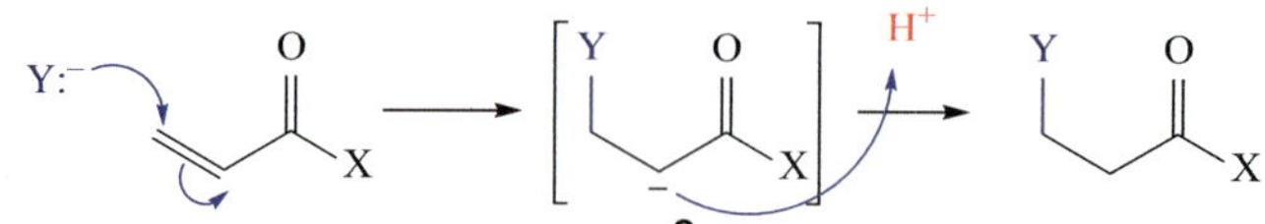

8

The *Candida antarctica* lipase B (Novozym 435) catalyzes a novel decarboxylative Michael addition in vinylogous carbamate systems to synthesize 1,4-benzoxazinone derivatives. The reaction involves Michael addition, ester hydrolysis, and decarboxylation. The reaction is catalyzed by lipase and involves the amino acids serine, histidine, and aspartate in the enzyme active site. The carbonyl group in chalcone is initially activated by hydrogen bonding to serine. The 1,4-benzoxazinone is subsequently deprotonated by the His-Asp system, followed by nucleophilic attack of the vinylogous carbamate on the activated chalcone Michael acceptor. In other words, the ester is first activated by the lipase, and subsequently carbon dioxide is released prior to conversion of the enamine to the final product. An example is the reaction of (*E*)-3-(2-chlorophenyl)-1-(thiophen-2-yl)prop-2-en-1-one with **9**, catalyzed by Novozym 435 to give **10** in 78% yield.[8]

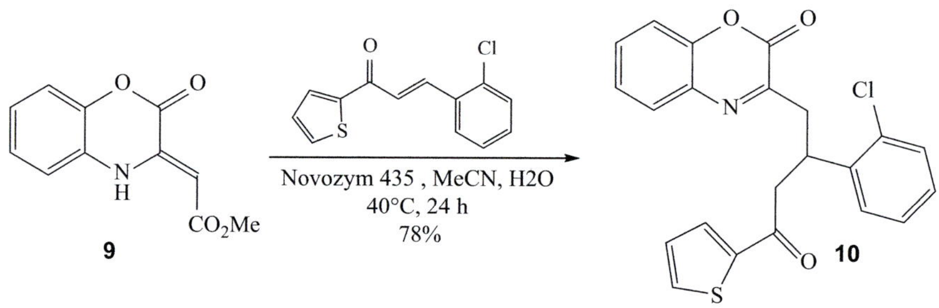

9 **10**

7.5 CYCLOADDITION REACTIONS

7.5.1 Enzymatic Cycloaddition Reactions

Cycloaddition reactions are catalyzed by certain enzymes to give the cycloadduct. Natural enzymes can catalyze the cycloaddition, but enzymes have also been designed for use in organic synthesis. *Diels-Alderase* enzymes have been discovered that facilitate [4+2]-cycloaddition, commonly known as Diels-Alder reactions.[9] There are two categories of naturally evolved [4+2]-cyclases.[10] The first comprises those enzymes that appear to have evolved to catalyze oxidation reactions but have a protein scaffold capable of facilitating cyclization of the primary reaction product.[11] The second

---

[8] Bavandi, H.; Shahedi1, M.; Habibi, Z.; Yousefi, M.; Brask, J.; Mehdi Mohammadi, M. *Nature* **2022**, *12,* 12713. https://doi.org/10.1038/s41598-022-16291-3.

[9] (a) Watanabe, K. *J. Nat. Med.* **2021**, *75,* 434; (b) Klas, K.; Tsukamoto, S.; Shermanc, D.H.; Williams, R.M. *J. Org. Chem.* **2015**, *80,* 11672; (c) Zhang, B.; Ge, H.M. *Natl. Sci. Rev.* **2022**, *9,* nwac229. https://doi.org/10.1093/nsr/nwac229.

[10] Maschio, L.; Back, C.R.; Alnawah, J.; Bowen, J.I.; Johns, S.T.; Mbatha, S.Z.; Han, L.-C.; Lees, N.R.; Zorn, K.; Stach, J.E.M.; Hayes, M.A.; van der Kamp, M.W.; Pudney, C.R.; Burston, S.G.; Willis, C.L.; Race, P.R. *Chem. Sci.* **2024**, *15,* 1172.

[11] Oikawa, H.; Katayama, K.; Suzuki, Y.; Ichihara, A. *J. Chem. Soc., Chem. Commun.* **1995**, 1321; Serafimov, J.M.; Gillingham, D.; Kuster, S.; Hilvert, D. *J. Am. Chem. Soc.* **2008**, *130,* 7798.

category includes enzymes that function as explicit [4+2] cyclases,[12] including the *Diels-Alderase AbyU*.[13] The proposed abyssomicin C biosynthetic pathway includes a Diels-Alder reaction that converts **11** to **12**. The enzyme *spirotetronate cyclase AbyU* catalyzes the key [4+2] cycloaddition reaction between the exo-alkene and the terminal diene, with complete stereocontrol via an endo cyclization that is a key step in the biosynthesis of abyssomicin C.[14]

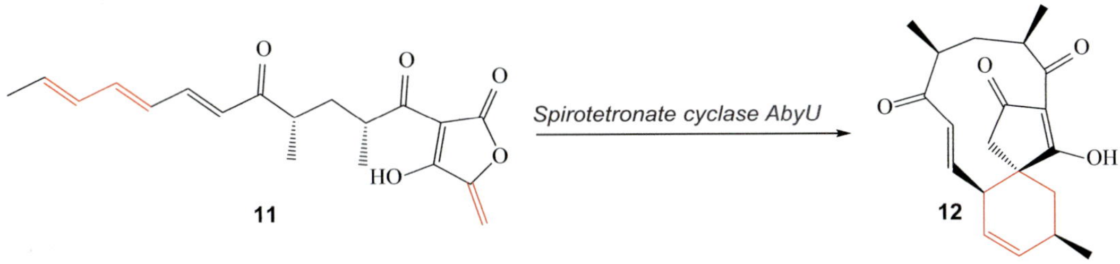

11 **12**

Diels-Alderases catalyze concerted [4+2] cycloadditions, but biosynthetic enzymes with [2+2]-cyclase activity have not been known. However, PloI4, a β-barrel-fold protein, catalyzes competitive [2+2] and [4+2]-cycloaddition reactions.[15] PloI4 is believed to catalyze an *endo*-[4+2] cycloaddition in the biosynthesis of pyrrolosporin A; however, when the substrate precursor of pyrroindomycins was treated with PloI4, an *exo*-[2+2]-adduct was produced in addition to the *exo*- and *endo*-[4+2]-cycloadducts. This work illustrates an enzymatic thermal [2+2]-cycloaddition and provides evidence of a process through which an enzyme evolves along with its substrate for specialization and activity improvement.[15]

As noted, [2+2] cycloadditions are thermally forbidden, but a photoenzyme has been developed that operates through a triplet EnT mechanism, one that has been inaccessible to biocatalysis.[16] A genetically encoded photosensitizer was installed into the beta-propeller scaffold of DA_20_00.[17] This change converted a de novo Diels-Alderase into a photoenzyme for [2+2] cycloadditions (EnT1.0). This engineered enzyme was further developed to give an enantioselective enzyme (EnT1.3). Specifically, an engineered *Methanococcus jannaschii tyrosyl-tRNA synthetase*/tyrosyl-tRNA (MjTyrRS/MjtRNATyr) pair was used to incorporate 4-benzoylphenylalanine at several positions around the active site pocket. Benzophenone proved to be a suitable photosensitizer for intramolecular [2+2] cycloaddition reactions. The photocatalytic activity of EnT1.3 toward [2+2]-cycloadditions is illustrated by the conversion of **13** to the optically enriched product **14**.[17]

---

[12] (a) Gao, L.; Su, C.; Du, X.; Wang, R.; Chen, S.; Zhou, Y.; Liu, C.; Liu, X.; Tian, R.; Zhang, L.; Xie, K.; Chen, S.; Guo, Q.; Guo, L.; Hano, Y.; Shimazaki, M.; Minami, A.; Oikawa, H.; Huang, N.; Houk, K.N.; Huang, L.; Dai, J.; Lei, X. *Nat. Chem.* **2020**, *12*, 620; (b) Zheng, Q.; Guo, Y.; Yang, L.; Zhao, Z.; Wu, H.; Zhang, H.; Liu, J.; Cheng, X.; Wu, J.; Yang, H.; Jiang, H.; Pan, L.; Liu, W. *Cell Chem. Biol.* **2016**, *23*, 352. See references therein.

[13] Byrne, M.J.; Lees, N.R.; Han, L.-C.; van der Kamp, M.W.; Mulholland, A.J.; Stach, J.E.M.; Willis, C.L.; Race, P.R. *J. Am. Chem. Soc.* **2016**, *138*, 6095.

[14] Devine, A.J.; Parnell, A.E.; Back, C.R.; Lees, N.R.; Johns, S.T.; Zulkepli, A.Z.; Barringer, R.; Zorn, K.; Stach, J.E.M.; Crump, M.P.; Hayes, M.A.; van der Kamp, M.W.; Race, P.R.; Willis, C.L. *Angew. Chem., Int. Ed.* **2023**, *135*, e202300382.

[15] Wang, H.; Zou, Y.; Li, M.; Tang, Z.; Wang, J.; Tian, Z.; Strassner, N.; Yang, Q.; Zheng, Q.; Guo, Y.; Liu, W.; Pan, L.; Houk, K.N. *Nat. Chem.* **2023**, *15*, 177.

[16] Trimble, J.S.; Crawshaw, R.; Hardy, F.J.; Colin W. Levy, C.W.; Brown, M.J.B.; Fuerst, D.E.; Heyes, D.J.; Obexer, R.; Green, A.P. *Nature* **2022**, *611*, 709.

[17] Siegel, J.B.; Zanghellini, A.; Lovock, H.M.; Kiss, G.; Lambert, A.R.; St. Clair, J.L.; Gallaher, J.L.; Hilvert, D.; Gelb, M.H.; Baker, D. *Science* **2010**, *329*, 309.

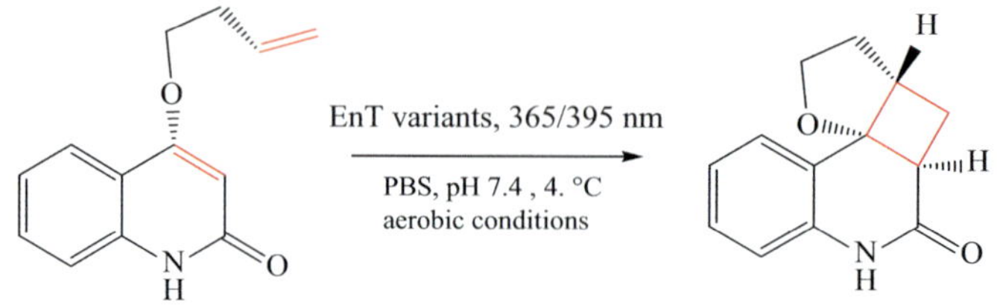

13 **14**

7.5.2 [2+2]-Photocycloaddition

The [2+2]-cycloaddition reaction is common in organic chemistry. A simple example of this type of reaction is the photocycloaddition of two molecules of ethene to give cyclobutane. This reaction is called a [2+2]-cycloaddition because a 2π-system (one alkene) reacts with another 2π-system (the other alkene). The reaction is explained by frontier molecular orbital theory, although the explanation of this process is beyond the scope of this discussion. The reader is referred to other sources to understand the reaction.[18] There are many examples of photochemical [2+2]-cycloadditions involving enones and alkenes, or alkenes with alkenes. The relative stereochemistry observed in many [2+2]-photocycloaddition reactions generally corresponds to a suprafacial addition for both alkene fragments.[19] For intermolecular cycloaddition reactions, free rotation around the original carbon-carbon double bond can take place in the excited state of the molecule. Because of such rotation, photocyclization of but-2-ene leads to four tetramethylcyclobutane isomers.[18a] Similar results are observed when unsymmetrical alkenes are photocyclized. Selectivity is improved with stereochemically pure alkenes. Reaction of pure *cis*-but-2-ene gave only two cyclobutanes with complete retention of configuration of each of the two *cis*-but-2-ene reactants. Similarly, pure *trans*-but-2-ene gave the other two cyclobutanes with complete retention of configuration of each of the two *trans*-but-2-ene reactants. Despite the selectivity shown in the reaction of but-2-ene, the relative proportions of cis and trans cycloadducts may depend on the reaction conditions. In general, reaction of an alkene and a conjugated carbonyl derivative gives a derivative such as **15**, and coupling of two nonconjugated alkenes gives **16**, as shown in Figure 7.7.

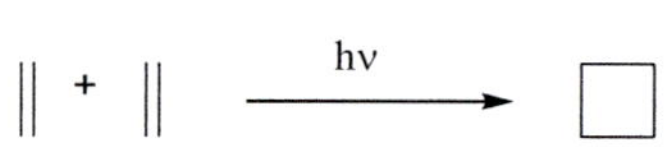

7.6 [4+2]-CYCLOADDITION REACTIONS IN ORGANIC CHEMISTRY

Conjugated dienes react with alkenes to give cyclohexene products. An example is the reaction of buta-1,3-diene and furan-2,5-dione (maleic anhydride) to give a bicyclic anhydride, 3a,4,7,7a-tetrahydroisobenzofuran-1,3-dione, in 90% yield[20] when heated in an autoclave. This reaction generates a ring, and it is classified as a *[4+2]-cycloaddition*. There are *no* intermediates in a

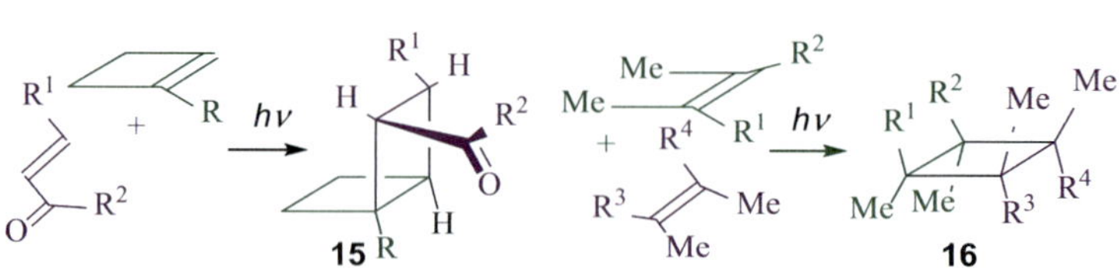

FIGURE 7.7 The [2+2]-photocycloaddition of alkenes.

[18] (a) Fukui, K.; Yonezawa, T.; Nagata, C.; Shingu, H. *J. Chem. Phys.* **1954**, *22*, 1433; (b) Fukui, K. in *Molecular Orbitals in Chemistry, Physics and Biology*, Löwdin, P.-O.; Pullman, B. (Eds.), Academic Press, New York, **1964**, p. 513.

[19] Yamazaki, H.; Cvetanovi, R.J. *J. Amer. Chem. Soc.* **1969,** *91*, 520.

[20] Fieser, L.F.; Novello, F.C. *J. Amer. Chem. Soc.* **1942**, *64*, 802.

[4+2]-cycloaddition reaction, and it is *synchronous*, or *concerted*. The [4+2]-cycloaddition is known as the *Diels-Alder reaction*.

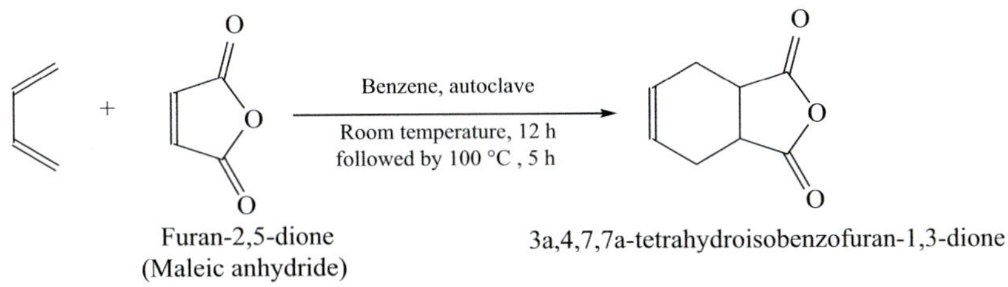

Another Diels-Alder reaction mixes buta-1,3-diene and ethene to give cyclohexene. As mentioned, the reaction is *concerted* and described by a *six-centered transition state*, as illustrated in Figure 7.8. Only the π-orbitals of buta-1,3-diene and ethene are involved in the Diels-Alder reaction. The four molecular orbitals on butadiene and the two on ethene each have a p-orbital on each carbon atom. Both the diene and the alkene have a HOMO and a LUMO. The *HOMO* is the highest energy molecular orbital that contains electrons. The *LUMO* is the lowest energy molecular orbital that does not contain electrons. The reaction proceeds by transfer of π-electrons from a filled molecular orbital of the diene to an empty molecular orbital of the alkene. For a Diels-Alder reaction, the $HOMO_{diene}$ interacts with the $LUMO_{alkene}$, and transfers two electrons to initiate the reaction. Six π-electrons are transferred in the cyclic transition state to generate cyclohexene.

The energy required to lose an electron from the diene or the alkene is the first *ionization potential (IP)*, and these values are taken to be the energy of the HOMO. Similarly, *electron affinity (EA)* is the energy associated with gaining one electron, and EA is taken to be the energy of each LUMO. The electron-rich HOMO is associated with electron donation, and the electron-deficient LUMO is associated with electron capture. The HOMO and LUMO for the diene and the alkene are the *frontier molecular orbitals*. Figure 7.9 is a simplified MO diagram for the reaction of buta-1,3-diene with ethene, and the energy of each HOMO and LUMO is provided. Frontier Molecular Orbital (FMO) theory describes electron transfer from the electron-rich $HOMO_{diene}$, with four electrons, to the electron-poor $LUMO_{alkene}$ in the Diels-Alder reaction. Figure 7.9 shows the molecular orbitals as lines for a simple Diels-Alder reaction between buta-1,3-diene and ethene. The electrons for each molecular orbital are shown, and the HOMO and LUMO for each molecule are indicated. The difference in energy between the $HOMO_{diene}$ and the $LUMO_{alkene}$ (ΔE) is taken to be the *activation energy* (E_{act}). A larger value of E_{act} (larger ΔE) indicates the reaction is more difficult to initiate, requiring higher heat and/or pressure for the reaction to proceed. Conversely, a lower E_{act} (smaller ΔE) indicates the reaction is more facile and requires less heat for the reaction to proceed. Based on a knowledge of the E_{act} for each reaction, a diene is predicted to react faster with some alkenes and slower with others.

If an alkene has an electron-withdrawing group attached, both the $HOMO_{alkene}$ and the $LUMO_{alkene}$ are lower in energy because there is less electron density in the π-bond. Since the $LUMO_{alkene}$ is lower, in reactions with a diene there is a smaller E_{act} and a faster reaction under milder reaction

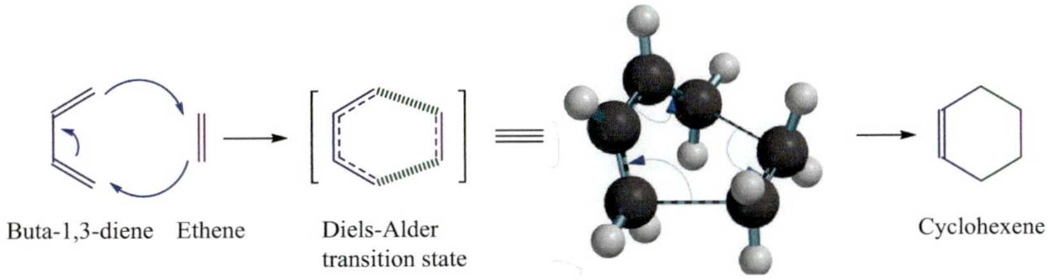

FIGURE 7.8 The Diels-Alder reaction of buta-1,3-diene and ethene.

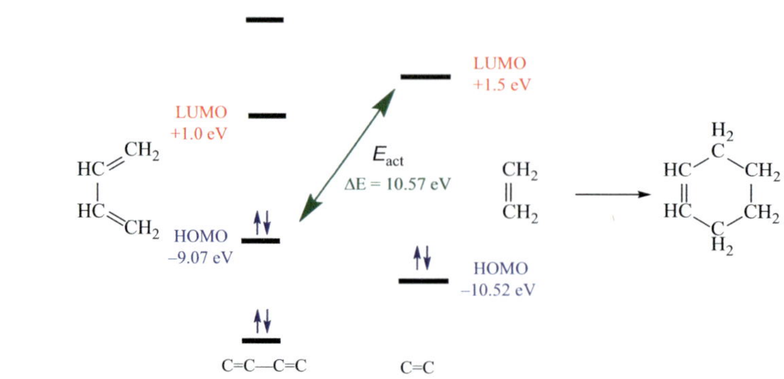

FIGURE 7.9 The molecular orbital diagram for the $HOMO_{diene}$ of buta-1,3-diene and the $LUMO_{alkene}$ of ethene showing the ΔE, which is the E_{act} for the reaction.

conditions. Therefore, the reaction of methyl acrylate proceeds faster. When the alkene has an electron-releasing group attached to it (e.g., OR), both the $HOMO_{alkene}$ and the $LUMO_{alkene}$ are higher in energy than ethene. An electron-releasing group increases the electron density in the π-bond. In reactions with a diene, there is a larger E_{act} and a slower reaction requiring more vigorous reaction conditions. Therefore, methoxyethene reacts more slowly than methyl acrylate. In other words, the Diels-Alder reaction of methyl acrylate requires milder reaction conditions.

7.7 ENZYME-MEDIATED SIGMATROPIC REARRANGEMENTS

Sigmatropic rearrangements are known in biological systems. One report shows that the last step in the aerobic biosynthesis of the corrin macrocycle of vitamin B12 in *Pseudomonas denitrificans* is an enzyme-catalyzed reaction, as shown in Figure 7.10.[21] The enzyme is *precorrin-8x methyl mutase* (EC 5.4.99.61, abbreviated as CobH; see Section 4.7.2), in which the methyl group attached to C11 of the substrate, precorrin-8x, that is bound to a protein, migrates from C11 to C12 to give the product, hydrogenobyrinic acid.[21] This transformation is known as a 1,5-sigmatropic methyl shift.

The enzyme *chorismate mutase* (EC 5.4.99.5; see Section 4.7.1) accelerates the Claisen rearrangement of chorismate into prephenate, a key step in the shikimate biosynthetic pathway that leads to aromatic amino acids.[22] Chorismate is an important biochemical intermediate in plants and microorganisms and a precursor for phenylalanine, tryptophan, and tyrosine. Prephenate is an

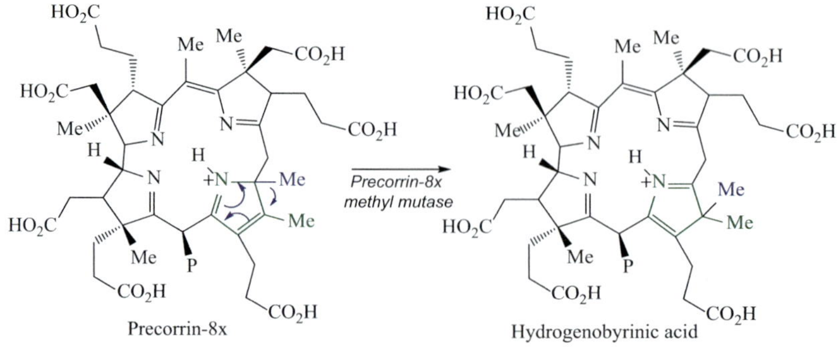

FIGURE 7.10 [1,5]-Sigmatropic Migration of methyl from C11 to C12 to give hydrogenobyrinic acid. (Shipman, L.W.; Li, D.; Roessner, C.A.; Scott, A.I.; Sacchettini, J.C. *Structure* **2001**, *9*, 587–596. Crystal Structure of Precorrin-8x Methyl Mutase. Copyright 2001, with permission from Elsevier.)

[21] Shipman, L.W.; Li, D.; Roessner, C.A.; Scott, A.I.; Sacchettini, J.C. *Structure* **2001**, *9*, 587.
[22] Khanjin, N.A.; Snyder, J.P.; Menger, F.M. *J. Amer. Chem. Soc.* **1999**, *121*, 11831.

intermediate in the biosynthesis of the aromatic amino acids, phenylalanine and tyrosine. The shikimate pathway (see Section 11.1) is a metabolic route used by bacteria, fungi, algae, some protozoa, and plants for the biosynthesis of folates and aromatic amino acids. The shikimate pathway is found only in microorganisms and plants, never in animals.

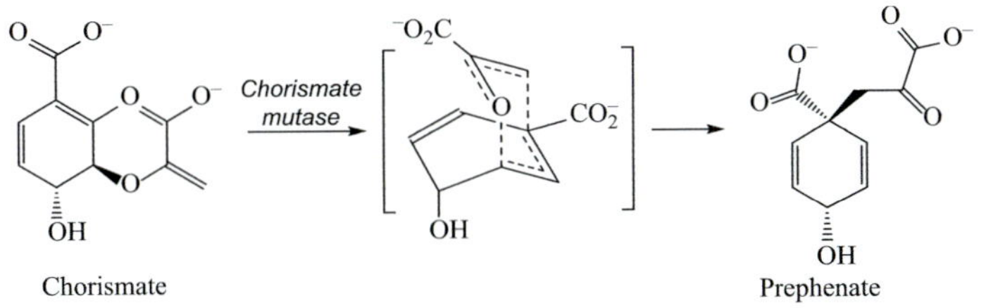

7.8 SIGMATROPIC REARRANGEMENT REACTIONS IN ORGANIC CHEMISTRY

1,5-Dienes undergo a rearrangement to a different 1,5-diene in what is known as a sigmatropic rearrangement. Similarly, allyl vinyl ethers rearrange to form alkenyl aldehydes or ketones in a closely related sigmatropic rearrangement. Many years of work led to the understanding of a new mechanism that explains reactions of this type, based on what is known as *frontier molecular orbital (FMO) theory*. This mechanism is based on the interaction of π-electrons in the diene with those of the alkene.

Several important reactions involve the transfer of electrons and the migration of atoms or groups across a π-system, but that do not form rings. These reactions are categorized as *sigmatropic rearrangements. Sigmatropic rearrangements are defined as reactions in which a σ-bond, meaning an atom or a substituent, moves across a conjugated system to a new site.*[23]

There are numbers attached to the length of each fragment associated with the move, and they are placed in brackets. In Figure 7.11, a [1,3]-shift involves migration of one fragment across another. In this case, one fragment is a single atom and the other fragment is three carbon atoms: a 1,3-shift of a proton, where the hydrogen atom migrates from C3 to C1 to yield a new propene. A [1,5]-shift is shown that involves the migration of two fragments. One fragment is a single atom, and the other fragment is five atoms in length. An example is the migration of a hydrogen atom from C5 in penta-1,3-diene to C1. The [3,3]-shift involves migration of two fragments, each three atoms in length. In the hexa-1,5-diene derivative shown in Figure 7.11, there is a three-carbon allylic unit (C1–C3) connected to another three-carbon allylic unit (C4–C6). The rearrangement involves breaking the bond between C3–C4 in the starting diene and forming a new bond between C1–C6 in the product diene. It is important to understand that these sigmatropic rearrangements are thermally reversible, as shown.

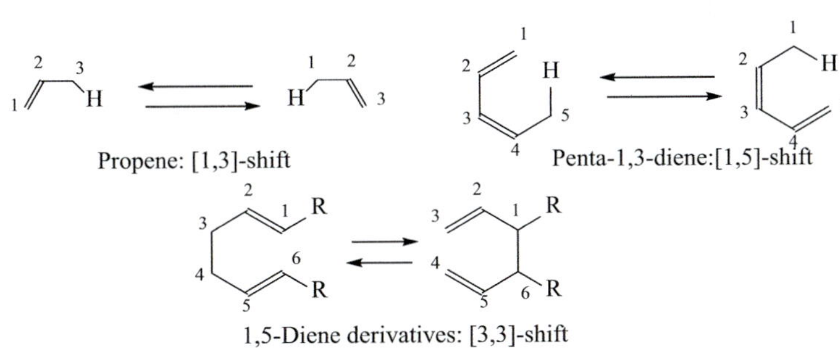

FIGURE 7.11 1,3-, 1,5-, and 1,7-Sigmatropic rearrangements.

---

[23] Fleming, I. *Frontier Orbitals and Organic Chemical Reactions*, Wiley, London, UK, **1976**, p. 98.

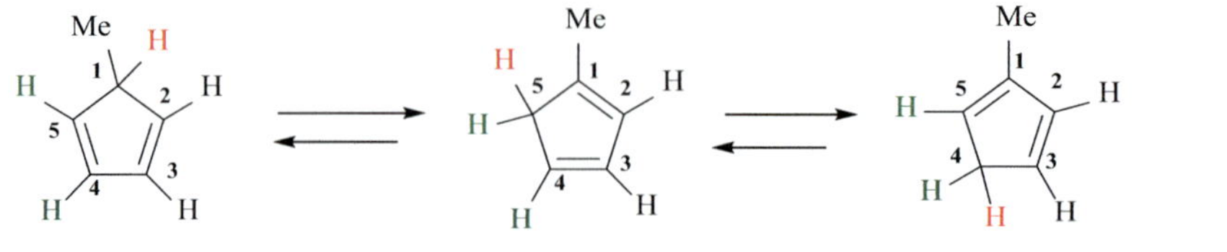

FIGURE 7.12 [1,5]-Sigmatropic hydrogen shift of methylcyclopenta-1,3-diene.

Methylcyclopenta-1,3-diene provides an example of a [1,5]-sigmatropic shift of a hydrogen atom. When 5-methylcyclopenta-1,3-diene is heated, the hydrogen marked in red moves as shown to give 1-methylcyclopenta-1,3-diene via a [1,5]-sigmatropic shift (see Figure 7.12). The hydrogen atom is a one-atom fragment, and it moves across a five-carbon fragment (C1–C5). Similarly, the hydrogen in 1-methylcyclopenta-1,3-diene moves to yield 2-methylcyclopenta-1,3-diene (from C5 to C4, as marked), and all these isomers are in equilibrium with each other. The hydrogen atoms in green and red are marked to show the positional movement of the hydrogen atom and the π-bonds during the [1,5]-sigmatropic rearrangement.

Rearrangement of this type is called a *suprafacial shift* because *the σ-bond to the hydrogen atom is made and broken on the same side of the conjugated system.* [1,7]-Sigmatropic shifts are observed in some systems. Although [1,5]-sigmatropic rearrangements are common, [1,3]-sigmatropic shifts are not, because moving a hydrogen atom in a [1,3]-shift requires that the σ-bond be *broken on one side* of the conjugated system but *made on the opposite side* of that system. This movement is known as an *antarafacial shift.* Antarafacial shifts are generally not observed in this system.

[3,3]-Sigmatropic rearrangements are another type of sigmatropic shift that involves the migration of alkyl groups rather than hydrogen atoms. There is no intermediate, and the prototype [3,3]-sigmatropic rearrangement involves heating a 1,5-diene (e.g., hexa-1,5-diene), as shown. The sigmatropic rearrangement proceeds by a concerted six-centered transition state that is represented by **17** and gives hexa-1,5-diene back again. The product is identical to the starting material, but if the C=C units of the diene are "labeled" with substituents, as with 3,4-dimethylhexa-1,5-diene, it is clear that the [3,3]-sigmatropic rearrangement has occurred because it leads to a different 1,5-diene, octa-2,6-diene. The "squiggle" lines indicate a mixture of isomers at those stereocenters.

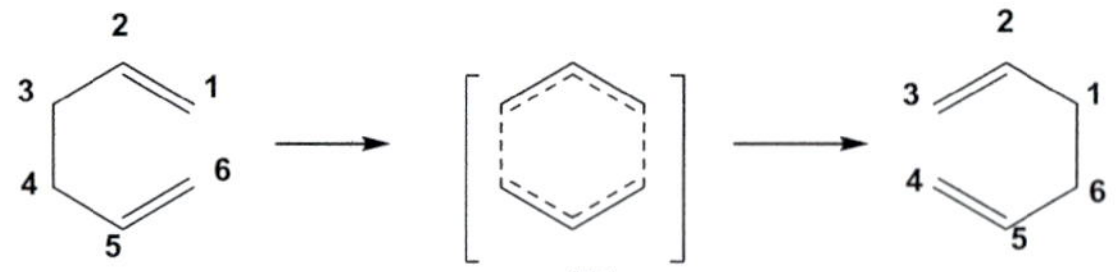

17 Hexa-1,5-diene

These two dienes are in equilibrium, and the equilibrium favors the *thermodynamically more stable diene*. In this case, the more stable diene is octa-2,6-diene, where each C=C unit is disubstituted. This rearrangement is called the *Cope rearrangement.* The temperatures required for a Cope rearrangement are sometimes quite high, and the fact that an equilibrium is established between the two dienes is a problem. If the Cope rearrangement leads to a different 1,5-diene but the substitution pattern of the dienes is identical, the mixture of products may be difficult to separate. Another example is the Cope rearrangement of the 1,5-diene 4-ethylhepta-(1,5*E*)-diene, which upon heating yields 4-methylocta-(1,5*E*)-diene. There is one disubstituted C=C unit in (*E*)-4-methylocta-1,5-diene and one disubstituted C=C unit in 4-methylocta-(1,5*E*)-diene, so the equilibrium is unlikely to favor one over the other, and a close to 1:1 mixture of these two dienes is expected. In other words, this reaction will produce a mixture, and it is probably difficult to separate one from the other.

While the Cope rearrangement can be problematic, another variation of a [3,3]-sigmatropic rearrangement is particularly useful. If one $-CH_2-$ unit of a 1,5-diene is replaced with an oxygen atom, the resulting structure is an allylic vinyl ether such as 3-(vinyloxy)prop-1-ene. This compound has two C=C units that constitute a 1,5-diene, and heating leads to a [3,3]-sigmatropic rearrangement reaction that proceeds via transition state **18** to yield the aldehyde pent-4-enal. The equilibrium in this reaction favors pent-4-enal because the C=O bond is favored over a C=C bond, so the equilibrium favors formation of the aldehyde. The presence of oxygen leads to an acceleration of the reaction rate, so the reaction temperature is relatively low compared to the Cope rearrangement. This particular [3,3]-sigmatropic rearrangement is called the *Claisen rearrangement*, as shown.

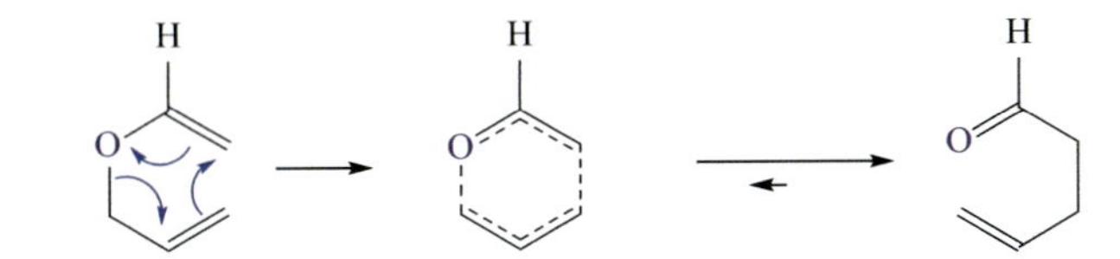

HOMEWORK

7.1. In the structure shown for a lanthipeptide, draw the product if a LanC enzyme is present. The LanC enzyme is a lanthionine-containing peptide (lanthipeptide synthetases) and acts as cyclase.

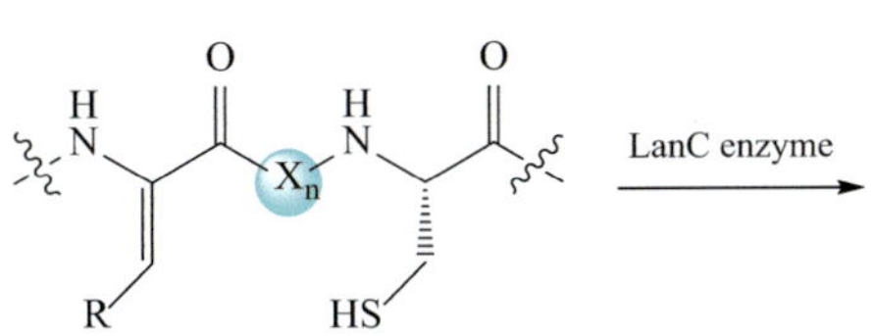

7.2. Suggest the product for the following reaction catalyzed by the enzyme AbyU.

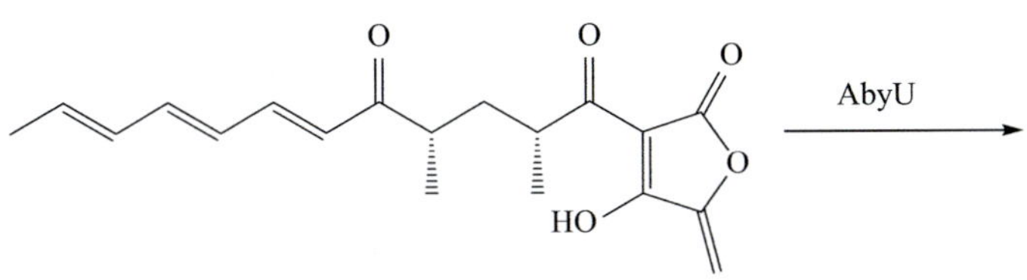

7.3. What is the product of this conversion in the presence of the enzyme *chorismate mutase*?

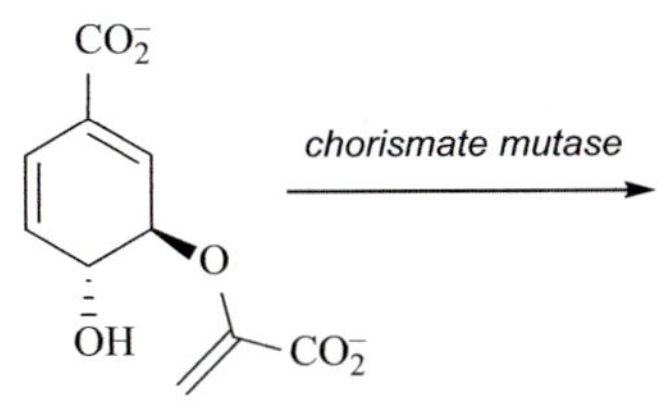

7.4. Give the major product of the following reactions.

(a) 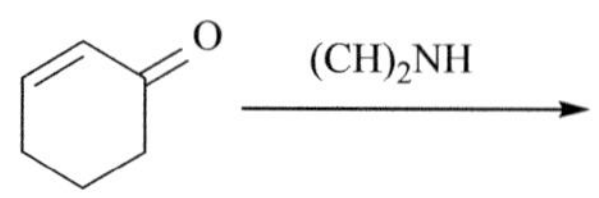

(b) 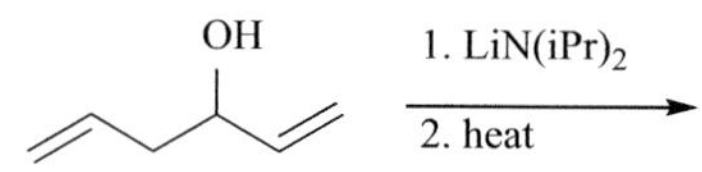

7.5. Assume that cysteine reacts with cyclohex-2-ene-1-one. Draw the expected product.
7.6. Look up and draw the structure of folic acid.
7.7. What is the structure of precorrin-8x?

8 Radical Reactions

Radicals are highly reactive intermediates. Radicals play an important role in many biological processes. There are several methods for generating radicals, and many involve diatomic oxygen, O=O•. Once radicals are generated in a biological environment they tend to react quickly in a localized environment. This chapter will describe radical reactions in biological systems and radical reactions in organic chemistry.

8.1 STRUCTURE OF RADICALS

A radical is a molecule that contains at least one unpaired electron, which makes radicals highly reactive. In organic chemistry, radicals are usually transient intermediates. An example in nature is that the ground state (triplet) of O_2 is a radical: O:O•, whereas the excited (singlet) state of oxygen is usually shown as O=O. Radicals form when a single covalent bond breaks to give two atoms or molecules, but an unpaired electron remains on one or both species. Such a process is called *homolytic cleavage*. Examples generated by a chemical reaction include halogen-free radicals such as Cl• or Br•. Radicals with a single electron on oxygen include hydroxyl radicals, HO•; alkoxide radicals, RO•; and superoxide, which is an anion radical (i.e., $^{\bullet}O_2^{-}$).

Carbon-based radicals are well-known in organic chemistry. A carbocation can be described as a carbon atom bearing an "empty" p-orbital (no electrons), and a carbanion can be described as a carbon atom with a "filled" p-orbital (two electrons). When a p-orbital on any given atom has only one unshared electron, it is called a *radical*, and a carbon radical is represented as R_3C (see **1** and **2**). With three covalent bonds and one extra electron, $R_3C^{\bullet}$ is a high-energy species and a very reactive intermediate. The single electron in an orbital will slightly repel the electrons in the covalent bonds, so one might expect a squashed tetrahedron (pyramidal) shape (**1**). There is evidence, however, that a planar structure (**2**) is probably the low-energy structure rather than the pyramidal structure, at least for the methyl radical ($H_3C^{\bullet}$).

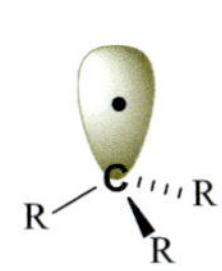

1

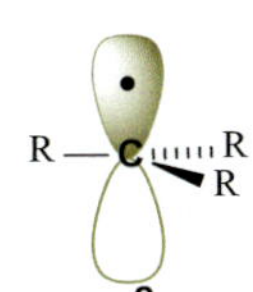

2

A carbon radical can be viewed as a trivalent species containing a single electron in a p-orbital (i.e., it is an atom that has one electron in an orbital). Such a species can theoretically be tetrahedral, planar, or "in between." It is generally conceded that carbon radicals without significant steric encumbrance are planar, as represented by **2**. In terms of its reactivity, this radical may be considered electron-rich or electron-poor. In most of its reactions, the electron-deficient characterization is useful for predicting products. Another way to state this characterization is to say that *radicals, such as this, are not considered to be nucleophilic.*

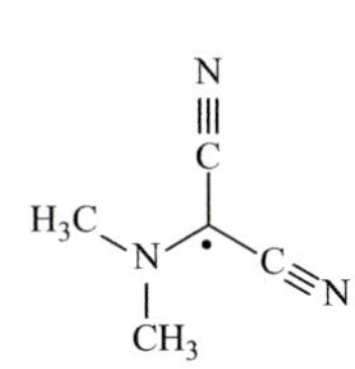

dicyano (dimethylamino) methyl
free radical

DOI: 10.1201/9781003605157-8

Another type of radical is the so-called *captodative radical*, which contains both an electron-withdrawing group and an electron-releasing group. The so-called captodative effect stabilizes radicals by a synergistic effect of an electron-withdrawing substituent and an electron-donating substituent.[1] Such stabilization leads to an increased rate of reaction for free radicals. An example is the dicyano(dimethylamino)methyl free radical.[2]

8.2 RADICALS IN BIOLOGICAL SYSTEMS

Radicals are highly reactive intermediates in myriad biochemical reactions. Some proteins catalyze chemically challenging reactions due to the high reactivity of radicals. Such proteins contain or form a radical and are called "radical enzymes."[3] Radicals formed in radical enzymes are not "free radicals" but "protein-bound radicals," so their reactivity is clearly influenced by the proteins. At least one of the roles of proteins in radical enzymes is to serve as a negative catalyst towards side reactions.[3] In living organisms, reactive oxygen species and other free radicals may react with nearby biomolecules, including lipids, proteins, and DNA. These radical reactions cause serious tissue injury, DNA damage, and cell death. Therefore, radical reactions may contribute to the progression of various diseases, cancer, and aging. Radical species in radical enzymes can be classified into two major groups: one derived from cofactors, such as adenosylcobalamin (AdoCbl) (coenzyme B12) or *S*-adenosylmethionine (SAM), and the other derived from amino acid residues of proteins.[3] In addition to radicals that have deleterious effects, there are radical reactions that involve NO and superoxide radicals that are important in biological defense.

AdoCbl-dependent ethanolamine ammonia-lyase (see **3**)[4] is a naturally occurring organometallic compound. A 5′-deoxy-5′-adenosyl (adenosyl) radical is generated and activated upon binding to apoproteins.[4] The adenosyl radical generated functions as a catalytic radical and abstracts a hydrogen atom from a substrate such as propane-1,3-diol to form radicals and 5′-deoxyadenosine. The substrate radical rearranges to the corresponding product-derived radicals, which back-abstract a hydrogen atom from 5′-deoxyadenosine to form the propanal product and the adenosyl radical.[4] Elimination of water occurs to form the final aldehyde products, as shown in Figure 8.1. The adenosyl radical finally recombines with Co(II) of cob(II)alamin to regenerate the coenzyme.[4]

S-Adenosylmethionine (SAM; see Figure 6.3 in Section 6.1.1) is a sulfonium compound (**4**) and an electrophilic methyl-group donor in many biological methylation reactions.[4] Another use of SAM involves the enzyme lysine 2,3-aminomutase, which catalyzes the conversion between L-α-lysine and L-β-lysine, the first step of L-lysine fermentation in Clostridia.[5] The enzyme contains pyridoxal 50-phosphate, high-spin Co(II), and an iron-sulfur cluster with SAM. It has also been reported[6] that in the presence of a substrate, α-lysine, an adenosyl radical (see above) is formed from SAM by reductive cleavage of its C–S bond. The one-electron donor for this cleavage is the reduced form of an iron-sulfur cluster $[4Fe\text{-}4S]^+$ in the same enzyme molecule. SAM is a cofactor for many biological radical reactions.[7]

[1] Viehe, H.G.; Janousek, Z.; Merenyi, R.; Stella, L. *Acc. Chem. Res.* **1985**, *18*, 148.
[2] De Vries, L. *J. Am. Chem. Soc.* **1978**, *100*, 926.
[3] Shibata, N.; Toraya, T. *J. Biochem.* **2015**, *158*, 271.
[4] Babior, B.; Gould, D.C. *Biochem. Biophys. Res. Commun.* **1969**, *34*, 441.
[5] Chirpich, T.P.; Zappia, V.; Costilow, R.N.; Barker, H.A. *J. Biol. Chem.* **1970**, *245*, 1778.
[6] Lieder, K.W.; Booker, S.; Ruzicka, F.J.; Beinert, H.; Reed, G.H.; Frey, P.A. *Biochemistry* **1998**, 37, 2578.
[7] Broderick, J.B.; Broderick, W.E.; Brian, M.; Hoffman, B.M. *FEBS Lett.* **2023**, *597*, 92.

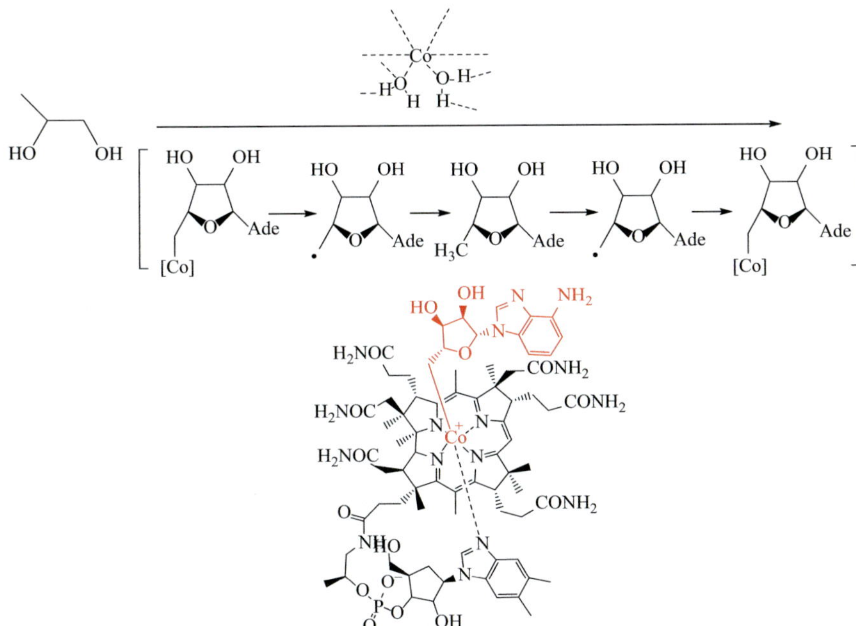

FIGURE 8.1 Structures and a catalytic mechanism of radical B12 enzyme: Adenosylcobalamin (AdoCbl). The adenosyl group is shown in red. AdoCbl-bound models of *diol dehydratase* for the Co–C bond cleavage and for the radical access to a substrate. The proposed mechanism for diol dehydratase. Co is cobalamin, and Ade is the 9-adeninyl group. (Shibata, N.; Toraya, T. *J. Biochem.* **2015**, *158,* 271. See Figure 2A,B,F, p. 274. Copyright © 2015, © The Authors 2015. With permission from Oxford University Press on behalf of the Japanese Biochemical Society.)

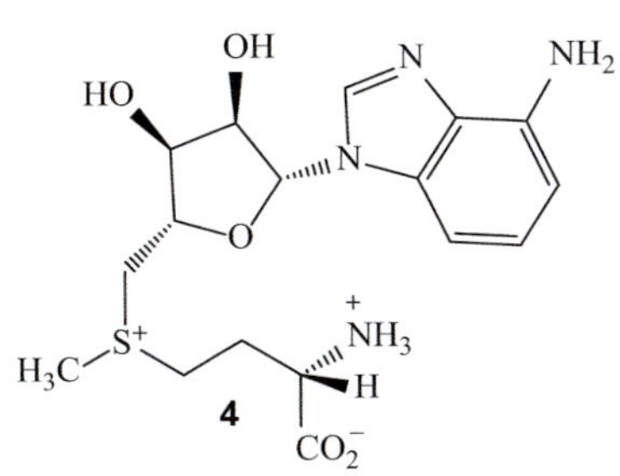

4

Other radicals in biological systems are well known. Both superoxide and hydroperoxyl are generated by the immune system to kill invading microorganisms. In phagocytes, superoxide is produced in large quantities by the enzyme *NADPH oxidase* (EC 1.6.3.1; see Section 4.3.1) for use in oxygen-dependent killing mechanisms of invading pathogens. A phagocyte is a type of cell within the body capable of engulfing and absorbing bacteria and other small cells and particles. Superoxide is toxic at high concentrations, so nearly all organisms that exist in the presence of oxygen contain superoxide-scavenging enzyme such as *superoxide dismutase* (EC 1.15.1.1), which catalyzes the disproportionation of superoxide to give O_2 and H_2O_2. Disproportionation is a reaction in which a compound is simultaneously oxidized and reduced, giving two different products.

Reactive free radicals formed within cells can lead to cell death and tissue injury. Cells have developed enzymes to decompose peroxides, proteins to sequester transition metals, and many antioxidants to "scavenge" free radicals, to prevent free radical formation or limit their deleterious effects. The most important source of radicals in aerobic cells is oxygen and the production of superoxide, hydroxyl radical, and hydrogen peroxide. Radicals are usually accidental by-products of metabolism, and they are produced continuously in cells or during phagocytosis. Phagocytosis is the process of a cell engulfing a solid particle to form a phagosome (a vacuole in the cytoplasm of a cell containing an engulfed particle within a part of the cell membrane). This process is a major mechanism used to remove pathogens and cell debris. Enzymes and toxic peroxides digest the pathogen within the phagolysosome (also called an endolysosome; it is a cytoplasmic body formed by the fusion of a phagosome with a lysosome). Important biological radicals are produced by several pathways.

Hydroxyl radical (HO) is highly reactive and is one of the more important biological radicals. Hydroxyl radicals, along with hydrogen peroxide, are produced from oxyhemoglobin and methemoglobin. Oxyhemoglobin reduces Fe(III) NTA [ferric nitrilotriacetate; $Fe^{+3}N(CH_2CO_2{}^{3-})$] aerobically to become methemoglobin and Fe(II)NTA. Oxyhemoglobin is a bright red substance formed by the attachment of a molecule of oxygen irreversibly attached to the iron atom in the Fe^{3+} state in the hemoglobin molecule, and its function is the transportation of oxygen from the lungs to the tissues. Methemoglobin is a bluish chocolate-brown form of hemoglobin in which the iron in the heme group is in the Fe^{3+} state, not the $Fe^{2+,}$ and cannot bind oxygen, which means it cannot carry oxygen to tissues. The reversible binding of molecular oxygen to form oxyhemoglobin is accompanied by side reactions in which methemoglobin is formed with the attendant formation of hydroxyl radicals. It is known that hydroxyl free radicals are necessary for the conversion of lysine and proline to hydroxylysine and hydroxyproline, respectively, which are necessary for collagen biosynthesis.[8]

Fosfomycin is a broad-range antibiotic.[9] An iron(IV)-oxo intermediate is recognized as the principal oxidant in nonheme enzymes for C–H bond activations and functionalizations. The ferryl species is bypassed in (*S*)-2-hydroxypropylphosphonic (S-HPP) acid epoxidase (HppE). The actual oxidant is an HO radical hydrogen-bonded to a ferric-hydroxo complex, which leads to the homolytic dissociation of a hydrogen peroxide complex, Fe(II)–H_2O_2. The HO• radical is highly selective, and H-abstraction gives *cis*-fosfomycin. Either FeIII–(HO•) and/or FeIII–OH (HO•) is formed in the biological Fenton system,[10] and hydrogen abstraction leads to intermediate **5** (Figure 8.2). An intramolecular reaction of oxygen generates the epoxide moiety in fosfomycin.[11]

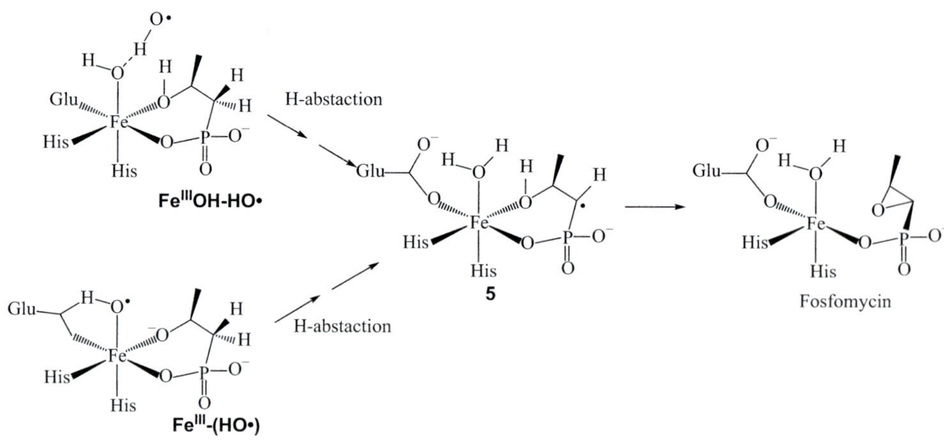

FIGURE 8.2 Role of hydroxyl radical in the biosynthesis of Fosfomycin. (Wang, B.; Lu, J.; Dubey, K.D.; Dong, G.; Lai, W.; Shaik, S. J. *Am. Chem. Soc.* **2016**, *138,* 8489. Copyright 2016 American Chemical Society.)

[8] Wright, D.T.; Cohn, L.A.; Li, H.; Fischer, B.; Li, C.M.; Adler, K.B. *Environ. Health Perspect.* **1994**, *102* (Suppl 10), 85.

[9] (a) Wang, C.; Chang, W.C.; Guo, Y.S.; Huang, H.; Peck, S.C.; Pandelia, M.E.; Lin, G.M.; Liu, H.W.; Krebs, C.; Bollinger, J. M. *Science* **2013**, *342,* 991; (b) Huang, H.; Chang, W.C.; Lin, G.M.; Romo, A.; Pai, P. J.; Russell, W.K.; Russell, D.H.; Liu, H.W. *J. Am. Chem. Soc.* **2014**, *136,* 2944; (c) Liu, P.H.; Liu, AM.; Yan, F.; Wolfe, M.D.; Lipscomb, J.D.; Liu, H.W. *Biochemistry* **2003**, *42*, 11577.

[10] See (a) Prousek, J. *Pure Appl. Chem.* **2007**, *79*, 2325; (b) Soltani, F.; Nidjouy, N.; Rahimneja, M. *RSC Adv.* **2022**, *12*, 5184.

[11] Wang, B.; Lu, J.; Dubey, K.D.; Dong, G.; Lai, W.; Shaik, S. J. *Am. Chem. Soc.* **2016**, *138,* 8489.

The ferric/perferryl peroxidase couple is a biochemical catalyst that produces ′OH from $O_2^{\bullet-}$ and H_2O_2.[12] This is the iron-catalyzed Haber-Weiss reaction. This reaction can be measured by the degradation of deoxyribose. Both $O_2^{\bullet-}$ and H_2O_2 can be generated by the oxidase reaction of horseradish peroxidase in the presence of NADH. The •OH-producing activity of horseradish peroxidase can be inhibited by inactivators of heme iron or by various $O_2^{\bullet-}$ and •OH scavengers.[8] Heme iron is an iron-porphyrin complex.

Superoxide, $O_2^{\bullet-}$, is generated from two major sources: the mitochondrial respiratory chain and phagocytic *nicotinamide adenine dinucleotide phosphate oxidase* (*NADPH oxidase*; EC 1.6.3.1; see Section 4.3), so it is a byproduct of respiration. Superoxide is generated by the immune system to kill invading microorganisms, especially in phagocytes. Superoxide is implicated in harmful biological processes, including protein denaturation and lipid peroxidation. During mitochondrial respiration, oxidative phosphorylation creates a flow of electrons across the inner mitochondrial membrane, which is used as an energy source for the synthesis of ATP, as shown in Figure 8.3. Electrons are transferred to molecular oxygen (O_2) through a chain of enzymatic complexes (I–IV); partial reduction of O_2 can result in the formation of superoxide.

Superoxide released from the mitochondrial electron transport chain can target enzymes containing 4Fe-4S. In aconitase, 4Fe-4S is the most sensitive to superoxide and other reactive oxygen species. The reaction of superoxide with aconitase converts the 4Fe-4S cluster to a 3Fe-4S cluster and releases ferrous iron and hydrogen peroxide. This sequence leads to the formation of hydroxyl radicals via Fenton-type chemistry.[13] *Fenton's reagent* is a solution of hydrogen peroxide (H_2O_2) and an iron catalyst. It is used to oxidize contaminants or wastewater. Inactivation of aconitase may disrupt the mitochondrial tricarboxylic acid cycle and alter carbon metabolism. Radical electrons leaked from the mitochondrial electron transport chain assist the release of free iron from an iron cluster, leading to lipid peroxidation reactions and ferroptosis (Figure 8.3).[14] A consequence of this sequence is the peroxidation of polyunsaturated fatty acids, leading to mitochondrial impairment. As shown in Figure 8.3, superoxide is released from the electron transport chain under ferroptotic stimuli and

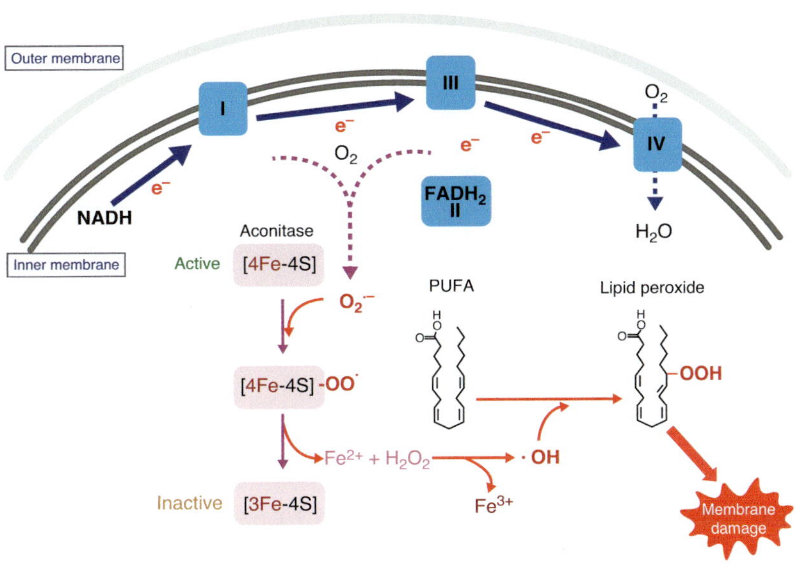

FIGURE 8.3 Proposed role of superoxide in mitochondrial membrane damage. (Figure 3, p. 516 in Fujii, J.; Homma, T.; Osaki, T. *Antioxidants* **2022**, *11*, 501. Open Access via MDPI.)

[12] Chen, S.-X.; Schopfer, P. *Eur. J. Biochem.* **1999**, *260*, 726.
[13] Vasquez-Vivar, J.; Kalyanaraman, B.; Kennedy, M.C. *J. Biol. Chem.* **2000**, *275*, 14064.
[14] Fujii, J.; Homma, T.; Osaki, T. *Antioxidants* **2022**, *11*, 501.

may become attached to 4Fe-4S clusters in proteins, notably aconitase. Subsequently, ferrous iron and hydrogen peroxide are released. Polyunsaturated fatty acids, fatty acids that possess two or more double bonds in their chain, in mitochondrial membranes undergo peroxidation by reactive oxygen species, and mitochondrial membrane integrity is destroyed. Note that ferroptosis is initiated by inhibition of cystine uptake or the inactivation of the lipid repair enzyme *glutathione peroxidase*.

Hydrogen peroxide is generated by mitochondrial respiration and is a toxic by-product of metabolism in organisms. When the immune system is activated in response to bacteria, large amounts of hydrogen peroxide are produced to fight the infection. Hydrogen peroxide must be decomposed into its products, water and oxygen, to mitigate any harmful effects on the cells. *Superoxide dismutase* (EC 1.15.1.1; see Section 4.2) is an important antioxidant agent in nearly all living cells, promoting the disproportionation of superoxide into oxygen and hydrogen peroxide. Hydrogen peroxide is rapidly decomposed by the enzyme *catalase* into oxygen and water.

Nitric oxide (NO) is synthesized by *nitric oxide synthase* (NOS, EC 1.14.13.39; see Section 4.2) from arginine. Oxidation of the terminal guanidine nitrogen atoms of L-arginine occurs in the vascular endothelium, by cytotoxic macrophages, by neutrophils, and by brain *synaptosomes* (an isolated synaptic terminal from a neuron). Nitric oxide synthase utilizes tetrahydrobiopterin, nicotinamide adenine dinucleotide phosphate (NADPH), and molecular oxygen to convert L-arginine to L-citrulline and NO (see Figure 8.4).[15] These enzymes are soluble and require Ca^{2+} in brain synaptosomes. Nitric oxide is a gas and a highly toxic compound with a short half-life and is diffusible. Nitric oxide is part of the mammalian immune system. It kills foreign intruders and cancer cells. It also functions as a cellular signal by stimulating cGMP synthesis. Nitric oxide affects O_2 supply and demand by regulating vascular tone and blood flow. It activates soluble guanylate cyclase in the vascular smooth muscle and controls mitochondrial O_2 consumption by inhibiting cytochrome C oxidase.[16] Within the vasculature, NO induces vasodilation, inhibits platelet aggregation, prevents neutrophil/platelet adhesion to endothelial cells, inhibits smooth muscle cell proliferation and migration, regulates programmed cell death (apoptosis), and maintains endothelial cell barrier function.[16] When NO is generated by neurons, it acts as a neurotransmitter, whereas NO generated by macrophages in response to invading microbes acts as an antimicrobial agent. It is likely that NO is an important regulator of the biology and physiology of the reproductive system.[17]

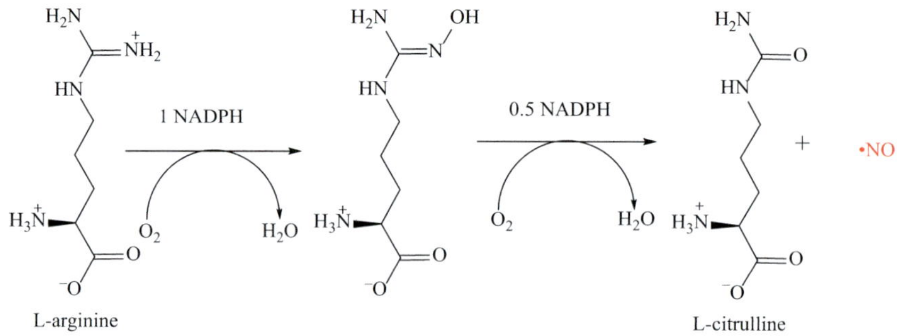

FIGURE 8.4 Reaction catalyzed by NOS. (Tejero, J.; Stehr, D. *UBMB Life* **2013**, *65*, 28. Copyright © 2013 International Union of Biochemistry and Molecular Biology, Inc. With permission from: John Wiley and Sons.)

[15] (a) Tejero, J.; Stehr, D. *UBMB Life* **2013**, *65*, 28; (b) See Król, M.; Kepinska, M. *Int J Mol Sci.* **2020**, *22*, 56.
[16] Chen, K.; Pittman, R.N.; Popel, A.S. *Antioxid. Redox Signal.* **2008**, *10*, 1185.
[17] Rosselli, M.; Keller, P.J.; Dubey, R.K. *Human Reprod. Update.* **1998**, *4*, 3. Also see Wink, D.A.; Kasprzak, K.S.; Maragos, C.; Elespuru, R.K.; Misra, M.; Dunams, T.M.; Cebula, T.A.; Koch, W.H.; Andrews, A.W.; Allen, J.S.; Keefer, L.K. *Science* **1991**, *254*, 1001.

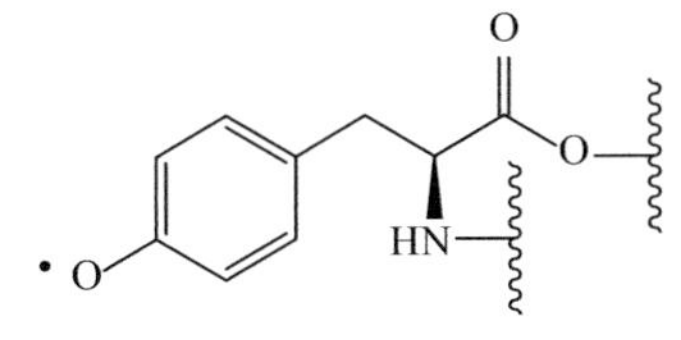

Tyrosine free radical

Ribonucleotide reductase (EC 1.17.4.1; see Section 4.2) is an enzyme with a stable free radical that is essential to its activity.[18] It is an essential enzyme that reduces ribonucleotides to their corresponding deoxyribonucleotides, which is the first step of DNA synthesis.[19] A tyrosine free radical is found in iron-containing ribonucleotide reductases. The enzymatic reaction is the reduction of the 2′-C alcohol in the ribose ring to a $-CH_2-$ unit to give the deoxyribonucleotide. An example is the reduction of adenine monophosphate.[18] In *Escherichia coli*, *ribonucleotide reductase* is a homodimer of two different protein subunits in an overall tetramer. One subunit functions as the binding site for the substrate and catalyzes the dehydroxylation of the 2′-hydroxyl group of the ribose ring, whereas the other subunit contains one binuclear iron cluster, which generates and stabilizes the tyrosine radical at Tyr[122].[20] A diiron cluster (see Section 12.2) is known to be involved in the transformation.[19]

Adenine monophosphate 2'-Deoxyadenine monophosphate

Radicals produced in living organisms are highly reactive, reacting with lipids, sulfur-containing proteins, and DNA. Membranes are largely made of unsaturated lipids and proteins and are particularly susceptible to reaction with radicals. Such damage has been referred to as "oxidative stress." Polyunsaturated fatty acid derivatives (see Section 10.4), for example, react with radicals to form conjugated allylic radicals that react with oxygen to give alkyl hydroperoxides via hydrogen abstraction from another lipid. Such reactivity propagates a radical chain reaction process. Note that oxidative stress is defined as an imbalance between the production of free radicals and reactive metabolites (oxidants) and their removal by antioxidants, which causes damage to important biomolecules and cells. Oxidative stress is associated with many degenerative diseases.

The sulfur moieties of thiol-containing proteins are particularly vulnerable to oxidation by reaction with oxygen, free radicals, or peroxides. Such oxidation leads to conformational changes in the protein, which leads to diminished bioactivity. Histidine residues can also be oxidized to 2-oxohistidine. Radicals react with nucleic acids and nucleotides. The hydroxyl radical, for example, oxidizes guanosine or thymine to 8-hydroxy-2-deoxyguanosine and thymine glycol, respectively. Changes in DNA can lead to mutagenesis and carcinogenesis.

Living organisms usually evolve antioxidant mechanisms to deal with damage by endogenous radicals. The two major antioxidant mechanisms are primary antioxidants, which react directly with the radicals (radical scavengers), and secondary antioxidants that trap chain-propagating radicals. There are several common radical scavengers, including ascorbic acid, glutathione (which contains

[18] Graslund, A.; Sahlin, M.; Sjobergt, B.-M. *Environ. Health Perspect.* **1985**, *64*, 139–149.
[19] See Sjoberg, B.-M., Reichard, P., Graslund, A.; Ehrenberg, A. *J. Biol. Chem.* **1978**, *253*, 6863.
[20] Han, W.-G.; Liu, T.; Lovell, T.; Noodleman L. *J. Amer. Chem. Soc.* **2005**, *127*, 15778.

the SH unit), thiol-containing proteins, α-tocopherol and β-carotene, and enzymes such as *superoxide dismutase,* catalase (EC 1.15.1.1), and *glutathione peroxidase* (EC 1.11.1.9) (see Section 4.1.3).

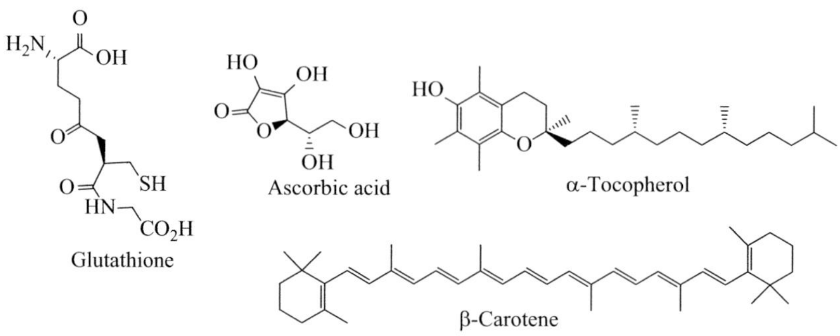

During the normal use of oxygen by humans, radicals are produced continuously. The organelles that produce the energy required for the endergonic processes of living cells are the mitochondria, which are the most important cellular source of free radicals. Endothermic, as defined in metabolism, is an endergonic process that is anabolic (energy is stored). Radicals produced in the mitochondria include the superoxide anion ($O_2^{\bullet -}$), nitric oxide (NO), and H_2O_2 (hydrogen peroxide).[21] When OH is generated in vivo, it is extremely reactive and reacts close to its site of formation. The nitric oxide radical, $NO^{\bullet}$, is generated in biological tissues by specific *nitric oxide synthases*, which metabolize arginine to citrulline.[22] Hydrogen peroxide can be generated from the superoxide anion by reaction with superoxide dismutase, amino acid oxidase, or xanthine oxidase.[21] Singlet oxygen is a non-radical but exists in an excited state when O_2 receives energy, and it is very reactive toward biomolecules since there is no spin restriction. Singlet oxygen can also be formed by the reaction of the superoxide anion with H_2O_2 or with chloroamines in cells and tissues.

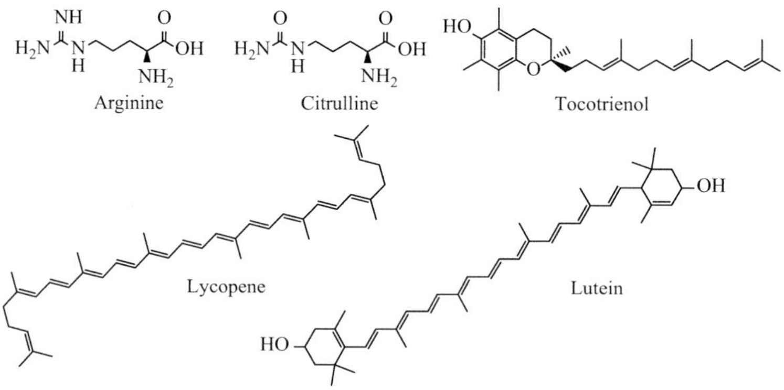

An "antioxidant" is any molecule that can stabilize or deactivate free radicals before they attack cells. Important antioxidants include *superoxide dismutase* (EC 1.15.1.1), *catalase* (EC 1.11.1.6), *glutathione peroxidase* (EC 1.11.1.9), and coenzyme Q (which has a quinone head attached to 9 or 10 isoprene units). Non-enzymatic antioxidants include tocopherol and tocotrienol, L-ascorbic acid (vitamin C; in citrus), carotenes such as lycopene and β-carotene, xanthophylls such as lutein (from green leaves) and zeaxanthin (from corn), and phenolic compounds (ArOH) that are common in plants, as well as lipoic acid, biliverdin, and bilirubin.

---

[21] (a) Sisein, E.A. *Sch. Acad. J. Biosci.* **2014**, *2*, 110; (b) Halliwell, B.; Murcia, M.A.; Chirico, S.; Aruoma, O.I. *Crit. Rev. Food Sci. Nutr.* **1995**, *35*, 7.

[22] Ghafourifar, P.; Cadenas, E. *Trends Pharmacol. Sci.* **2005**, *26*,190.

The hydroxyl radical is a main initiator of lipoperoxidation. Peroxidation in polyunsaturated fatty acids is initiated by the attack of a reactive species such as the hydroxyl radical, giving a resonance-stabilized allylic radical, as shown in Figure 8.5.[23] Subsequent reaction with oxygen generates a peroxy radical, and a H atom transfer reaction, presumably from another lipid, generates the hydroperoxide. The peroxidative process proceeds by a chain reaction. Hydroperoxides can be further oxidized to ketones, and the end product of further oxidation is malonaldehyde. Two specific oxidized products are 4-hydroxynon-2-enal and 4-hydroxyhex-2-enal. 4-Hydroxynon-2-enal has been found in many diseases, including atherosclerosis, neurodegenerative diseases, and cancer.[24] The formation of aldehydes and ketones from lipids also leads to poor food flavor and a reduction in nutritional value due to the destruction of essential fatty acids and lipid-soluble vitamins.

Free radical damage to nucleic acids and nucleotides likely occurs by radical attack at the anomeric carbon, where the pyrimidine unit forms a bond with the deoxyribose unit. Subsequent cleavage of the sugar-phosphate bond liberates the pyrimidine base, thereby cleaving a base pair in the DNA strand, as shown in Figure 8.6.[25]

Proteins, especially thiol-containing proteins, are damaged by free radical reactions. Thiol-containing moieties in a protein react with a radical source to give thiyl radicals (R–S), and subsequent coupling with another thiyl radical leads to a disulfide linkage, (R-S-S-R). Formation of the disulfide induces conformational changes in the protein, and in an enzyme, the bioactivity may be diminished or disappear entirely. Disulfide bonds formed under oxidative stress conditions are often nonspecific and can cause irreversible damage to proteins. An example is the

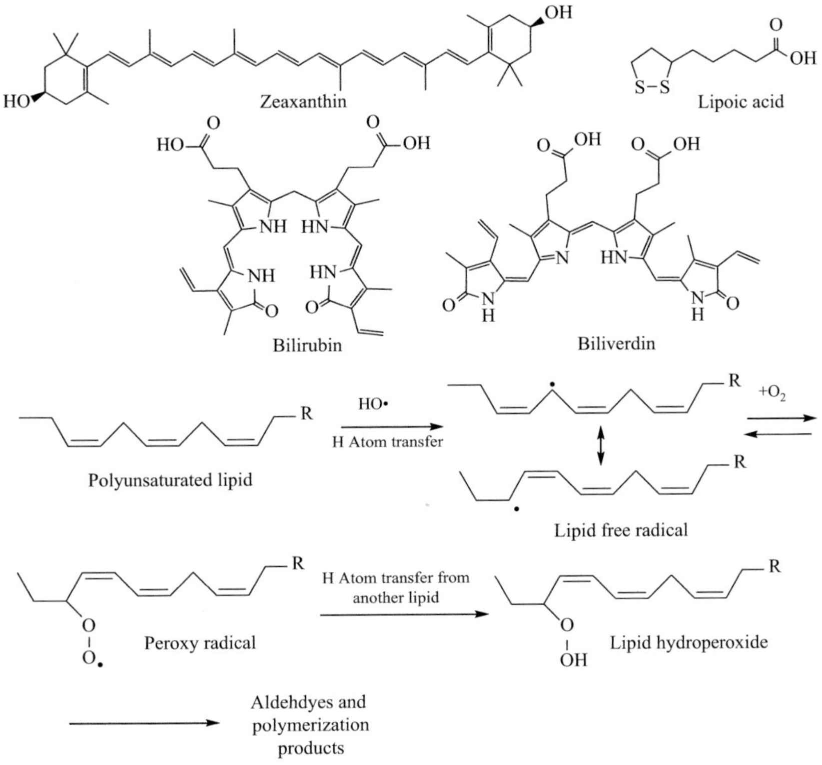

FIGURE 8.5 Lipid peroxidation.

[23] Webster, N.R.; Nunn, J.F. *Brit. J. Anaesth.* **1988**, *60*, 98.
[24] Tao, L. *Adv. Food Technol. Nutrit. Sci. Open J.* **2015**, *1*, 135.
[25] Dizdaroglu, M.; Jaruga, P. *Free Radic. Res.* **2012**, *46*, 382.

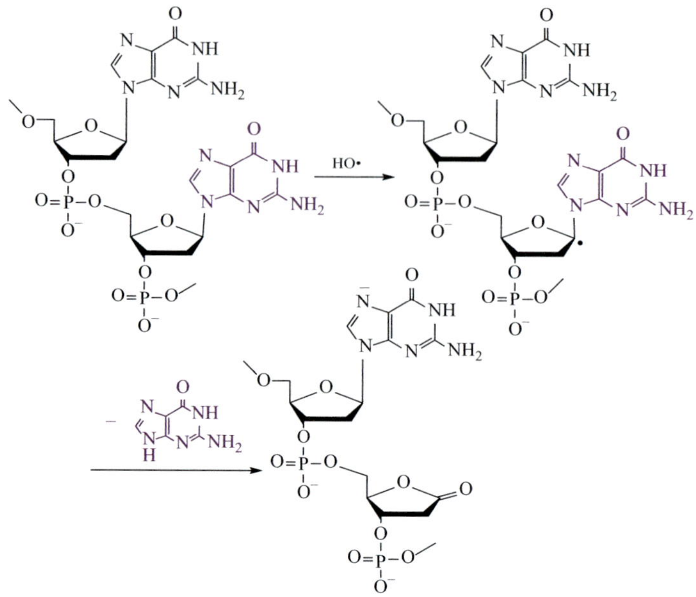

FIGURE 8.6 Hydroxyl radical-induced single-strand cleavage of DNA. (Dizdaroglu, M.; Jaruga, P. *Free Radic. Res.* **2012**, *46*, 382–419, reprinted by permission of the publisher (Taylor & Francis Ltd, http://www.tandfonline.com). Figure 24, p. 404. Permission from Taylor & Francis Ltd, is acknowledged, including reprinted by permission of Informa UK Limited, trading as Taylor & Francis Group, https://www.tandfonline.com.)

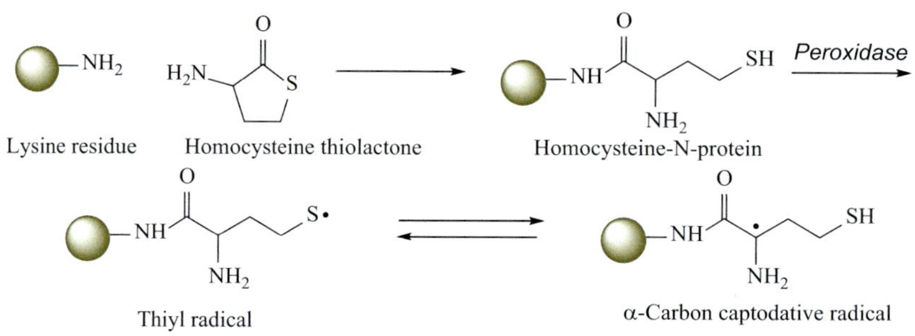

FIGURE 8.7 A captodative radical.

transcription factor Yaplp [a transcription factor, a 650-amino-acid-long bZIP DNA-binding protein (the Basic Leucine Zipper Domain is found in many DNA-binding eukaryotic proteins) of the AP-1 family (proteins that generally bind DNA as heterodimers composed of one member each of two different families of related bZIP proteins)] in *Saccharomyces cerevisiae*. Upon oxidative stress, one or more disulfide bonds form in the redox-regulated transcription factor Yaplp. While both the free thiol form and the disulfide form migrate from the cytoplasm into the nucleus, the disulfide form of Yaplp cannot form a complex with the export receptor Crmlp to be transported back into the cytoplasm.[26]

The preceding discussion of thiyl radicals leads to an example involving captodative radicals (Section 8.1), which are known in biochemistry. The homocysteine side chain can exist as a stable five-membered ring thiolactone [homocysteine thiolactone], which reacts with the ε-amine of lysine residues to form an isopeptide bond in the Hcy-N-protein (see Figure 8.7).[27] An isopeptide bond is an

[26] Linke, K.; Jakob, U. *Antioxid. Redox Signal.* **2003**, *5*, 425.

[27] Sibrian-Vazquez, M.; Escobedo, J.O.; Lim, S.; Samoei, G.K.; Strongin, R.M. *Proceed. Nat. Acad. Sci.* **2010**, *107*, 551.

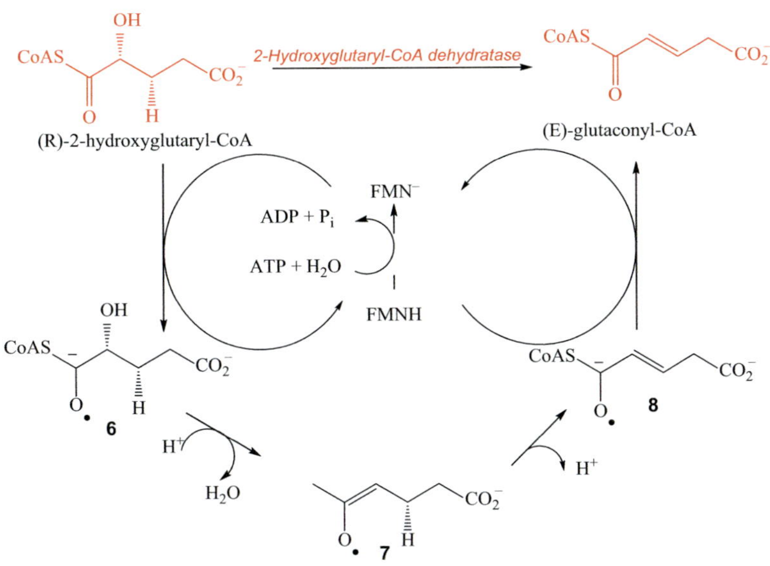

FIGURE 8.8 Proposed mechanism of the dehydration of (*R*)-2-hydroxyglutaryl-CoA. FMN is flavin mononucleotide (FMN) or riboflavin-5′-phosphate. (Buckel, W.; Golding, B.T. *FEMS Microbiol. Rev.* **1999**, *22*, 523. Copyright © 1998, © 1999 Federation of European Microbiological Societies. With permission from: Oxford University Press.)

amide bond in which at least one of the carboxyl or amino units is part of the side chain of an amino acid residue. A thiyl radical can form, and a hydrogen transfer leads to an α-amino carbon-centered radical, as shown in Figure 8.7, which is a captodative radical. The increased stabilization of this captodative radical favors the hydrogen transfer from the thiyl radical. This radical can abstract αC-H protons from proximal amino acid residues of proteins, and subsequent reaction with oxygen can form superoxide and carbonyl compounds, which lead to fragmentation of the peptide.[28]

Radical-induced elimination reactions are known. *2-Hydroxyglutaryl-CoA dehydratase* was obtained from *Acidaminococcus fermentans*.[29] The heterodimeric enzyme (K, 54 kDa; L, 42 kDa) facilitates the reversible syn-dehydration of (*R*)-2-hydroxyglutaryl-CoA to (*E*)-glutaconyl-CoA, as shown in Figure 8.8.[29] This elimination reaction is initiated by sub-stoichiometric amounts of ATP and requires high concentrations of Mg^{2+} and a reducing agent such as Ti(III) citrate. The mechanism postulated for the dehydration of 2-hydroxyglutaryl-CoA involves the initial reduction of the thiol ester carbonyl to a ketyl radical anion (**6**, Figure 8.8). Elimination of the adjacent hydroxyl group gives an enoxy radical (**7**), which is deprotonated to give ketyl (**8**). Subsequent oxidation gives glutaconyl-CoA.[30] After oxidation, the dehydratase activity is recovered by the residual ATP and excess reductant.[30] Note that FMN is flavin mononucleotide, also known as riboflavin-5′-phosphate, which is produced from riboflavin (vitamin B2).

8.3 RADICALS AND CANCER

Radicals are important in cancer, from carcinogenesis to the progression of malignant tumor cells. Radicals are also important in the application of radiation therapy for treating cancerous tumors. An in-depth discussion of this topic is beyond the scope of this work, but a few comments are relevant to this chapter. Free radical-induced oxidative stress can be linked to inflammatory processes that

---

[28] Schöenich, C. *Chem. Res. Toxicol.* **2008**, *21*, 1175.
[29] Buckel, W.; Golding, B.T. *FEMS Microbiol. Rev.* **1999**, *22*, 523.
[30] Muëller, U.; Buckel, W. *Eur. J. Biochem.* **1995**, *230*, 698.

lead to cancer. The application of chemotherapy and radiotherapy in cancer treatments can also favor oxidative stress. Nitric oxide, NO, is a ubiquitous radical in biological systems. It is known that endothelial *nitric oxide synthetase* modulates various cancer-related processes, such as apoptosis, angiogenesis, cell cycle, invasion, and metastasis, and so it is an important enzyme in tumor development. At high concentrations, $NO^{\bullet}$ has cytotoxic effects that inhibit mitochondrial enzymes, including succinate and *ubiquinone oxidoreductase* (EC 1.5.5.1; see Section 4.3) and *aconitase* (EC 4.2.1.3; see Section 4.3), which are important in cell metabolism. The cleavage and damage to DNA due to reactions with free radicals were discussed above, and such damage can cause mutations that may lead to cancer.[31]

8.4 FORMATION OF RADICALS IN ORGANIC CHEMISTRY

Radicals can be formed by breaking a covalent bond in such a way that one electron is transferred to each of the two atoms involved in that bond (called *homolytic cleavage*). The bonds in some molecules easily undergo homolytic cleavage. One example is the O–O bond of a peroxide, which has the structure R–O–O–R. An important route to radical intermediates involves the reaction of a radical (X) with a neutral molecule (X–Y), producing a new radical ($Y^{\bullet}$) and a new neutral molecule (X–X), as shown in Figure 8.9. The equilibrium constant will determine if there is a significant concentration of radicals, and the position of the equilibrium depends on both the relative bond strength of X–Y and the relative stabilities of $X^{\bullet}$ and $Y^{\bullet}$. Raising the temperature of the reaction usually shifts the equilibrium toward a higher concentration of radicals in homolytic cleavage reactions.

In organic chemistry, a carbon radical can be formed by a chemical reaction between a neutral species (e.g., methane) and an existing radical (e.g., Br). In this reaction, the bromine radical donates a single electron (note the single-headed arrow; much like a fishhook) to one hydrogen atom of methane, which donates one electron from the covalent C–H bond. When this electron transfer occurs, a new H–Br bond is formed and the C–H bond is broken, with transfer of one electron to carbon to form the methyl radical. Note that there are two electrons in the newly formed H–Br covalent bond, *one derived from the bromine radical and one from the broken C–H bond of methane.*

Several reagents generate radicals when heated or exposed to light (the symbol for a photon of light is $h\nu$), and several examples are shown in Figure 8.10. When heated, many peroxides (e.g., alkyl hydroperoxides, ROOH, and dialkyl peroxides, ROOR) undergo homolytic cleavage to generate radicals RO + $^{\bullet}$OH or two molar equivalents of $RO^{\bullet}$, respectively.

An example with the structure ROOH is *tert*-butyl hydroperoxide, and heating yields the *tert*-butoxy radical and the hydroxyl radical. Another example is dibenzoyl peroxide (or just benzoyl peroxide), which yields two molar equivalents of the acyl radical shown when heated. Benzoyl peroxide is classified as a diacyl peroxide. A useful reagent that generates radicals is an *azo compound*

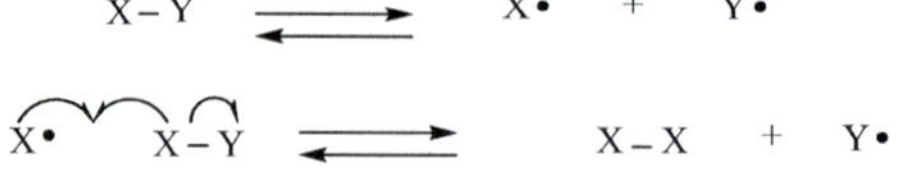

FIGURE 8.9 Homolytic cleavage reactions.

[31] (a) Coussens, L.M.; Werb, Z. *Nature* **2002**, *420*, 860; (b) Klaunig, J.E.; Kamendulis, L.M.; Hocevar, B.A. *Toxicolog. Pathol.* **2010**, *38*, 96.

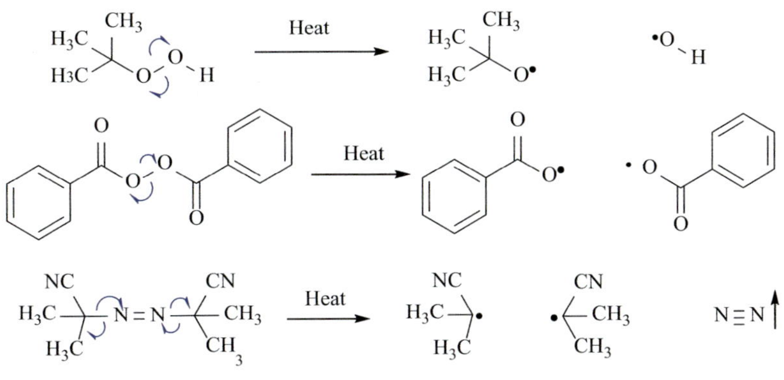

FIGURE 8.10 Radical initiators.

called *azobis*isobutyronitrile (AIBN). When heated, AIBN easily decomposes to produce two molar equivalents of the radical shown, along with nitrogen gas, which escapes from the reaction.

8.5 REACTIONS OF RADICALS

A radical is a highly reactive and unstable intermediate, and once formed, it can react with another molecule in the reaction medium. Several reactions are typical of radicals. In *coupling reactions*, two radicals combine to form a new carbon bond, as in the coupling of two ethyl radicals (CH_3CH_2) to give butane.

In *addition, reactions* of a radical can add to a π-bond to generate a new alkyl radical species, which can undergo coupling or other reactions. *Atom transfer reactions* occur when a radical is generated in the presence of certain reducing agents such as tributyltin hydride, Bu_2SnH. The reaction of the radical (R) with the hydrogen leads to the reduction of the radical (R–H) and the formation of $Bu_3Sn^•$. *Atom transfer* can occur with other reactive species such as Br_2, where $R^•$ and Br_2 give R–Br and $Br^•$. Such *atom transfer reactions* lead to *substitution reactions*, as when a bromine radical abstracts a hydrogen atom from the α-carbon of toluene to give a resonance-delocalized benzylic radical and HBr. The benzylic radical then reacts with the neutral bromine in an *atom transfer reaction* to give the substitution product, benzyl bromide, and another carrier bromine radical. A *fragmentation reaction* can occur, and a radical can fragment to form a new radical and a neutral molecule, typically an alkene, so the overall process is an elimination.

Radical polymerization reactions are possible. Heating benzoyl peroxide generates the radical, which reacts with an alkene to form a new radical and begin a *chain radical reaction*, as shown in Figure 8.11. The initial cleavage of the peroxide first generates the radical, and since products cannot form until a radical is formed, this reaction is called the chain *initiation step*. Once the radical is formed, reactions continue to form products until there are no longer reactive radicals. Each step in the sequence that produces a reactive radical, which goes on to react with another molecule of alkene, is called a chain *propagation step,* since the production of product is propagated. When there are no more radicals, the reaction sequence stops. Therefore, a step where two radicals react to produce a neutral molecule is called a chain *termination step.*

If the radical polymerization process continues, the final product will be a large molecule known as a *polymer*, $–(C–C)_n–$, where n is a large number (e.g., 500, 2000, or higher) that represents the number of times that unit is repeated. In other words, there are 500 repeating C–C units or 2000 repeating C–C units. *A polymer is simply a high-molecular weight substance composed of repeating small-molecule units called monomers.* A *low polymer* has a molecular weight below ~ 10,000–20,000, whereas a *high polymer* has a molecular weight between 20,000 and several million. *A monomer is defined as the single-molecule precursor to a polymer.* An alkene monomer may be polymerized to form a polymer.

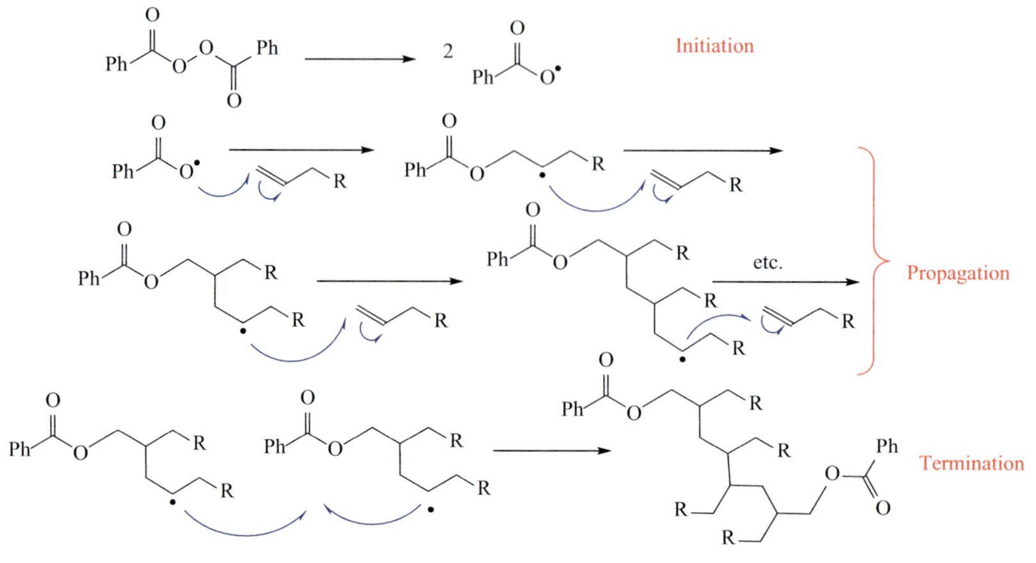

FIGURE 8.11 Mechanism of radical polymerization of alkenes.

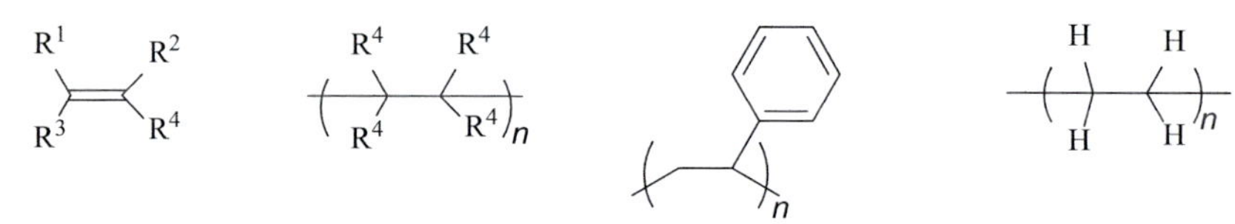

Alkene Polymer Poly(styrene) Poly(ethlene)

A specific example uses purified styrene (phenylethene) heated in a special receptacle at 125°C for several days. The product is poly(styrene). The *n* in poly(styrene) indicates the number of times the styrene monomer unit is repeated, which varies with the way the polymer is formed. Under the reaction conditions just described, the polymer has a molecular weight of ~ 150,000. If the monomeric C_8H_8 unit (mass = 104) repeats, a polymer weight of 150,000 would correspond to $n \approx \left(\frac{150000}{104}\right)$, which is ≈ 1440. When benzoyl peroxide is heated in the presence of ethene (ethylene), radical addition will continue until the ethene is consumed, and the final product is polyethylene, which is an example of a linear polymer.

Poly(styrene) is widely used to manufacture plastic dinnerware, including plastic knives and spoons. Plastic model kits and smoke detector housings are also manufactured using poly(styrene). It is used to make polystyrene foam, marketed under the name Styrofoam by Dow Chemical Co. Poly(ethylene) is prepared in several forms, including ultra-high molecular weight poly(ethylene) (UHMWPE), high-density polyethylene (HDPE), and medium-density poly(ethylene) (MDPE). Ultra-high molecular weight poly(ethylene) is used to manufacture machine joints and gears and is used as a component of artificial hip and knee replacements. High-density polyethylene is used in milk jugs, detergent bottles, "plastic" toys, and water pipes, while MDPE is used in gas pipes and packaging film.

HOMEWORK

8.1. Draw the structures of oxygen, 5-deoxy-5′-adenosyl radical, hydroxy radical, and superoxide.

8.2. Draw all resonance forms of the following radical.

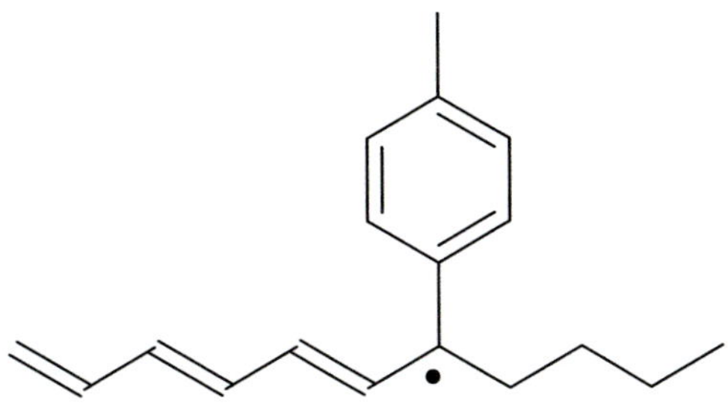

8.3. With respect to the C–H bond, is the following reaction an example of heterolytic cleavage or homolytic cleavage?

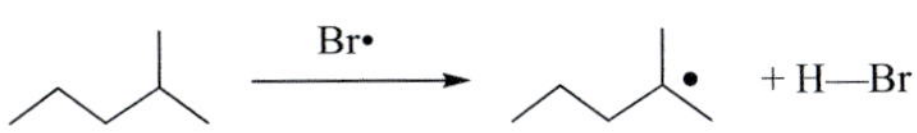

8.4. Draw the structure of *S*-Adenosylmethionine (SAM).

8.5. Look up and draw the structure of methemoglobin.

8.6. What are the products formed by the enzymatic decomposition of superoxide? What type of reaction is this?

8.7. Give the structure of three radical scavengers.

8.8. Give the structure of five antioxidants.

8.9. What is a captodative radical?

8.10. What is the structure of heme iron?

9 Enols and Enolate Anions

Ketones and aldehydes react with nucleophiles primarily by acyl addition, whereas carboxylic acid derivatives react by acyl substitution. In these reactions the carbonyl carbon is an electrophilic center. The oxygen of a carbonyl compound reacts as a Bronsted-Lowry base in the presence of a suitable acid, generating a protonated carbonyl intermediate known as an oxocarbenium ion. An oxocarbenium ion rapidly reacts with nucleophiles. In the presence of a protic solvent, tautomerization leads to an equilibrium concentration of an enol form that can react as a nucleophile in some cases. Aldehydes, ketones, or esters that have a proton on the α-carbon directly connected to the carbonyl carbon, the α-carbon are weak acids. With a suitable base, these carbonyl compounds generate a carbon nucleophile called an enolate anion. This chapter will discuss the formation and nucleophilic reactions of enols and enolate anions.

9.1 ENZYME-MEDIATED ENOLASE CONDENSATIONS

Enolase enzymes catalyze a wide variety of reactions and have diverse roles in metabolism. The catalyzed reactions all involve the abstraction of a proton from a carbon adjacent to the carbonyl carbon of a carboxylic acid. The reaction usually requires a divalent metal ion. *Enolases* are structurally related proteins that share the common ability to catalyze the abstraction of the (*R*)-protons of carboxylic acids. *Phosphopyruvate hydratase*, commonly called enolase, catalyzes the conversion of 2-phospho-D-glycerate to phosphoenolpyruvate,[1] the next-to-last step of glycolysis[2] (the conversion of glucose to pyruvate or lactate). Note that *phosphopyruvate hydratase* belongs to the family of lyases.

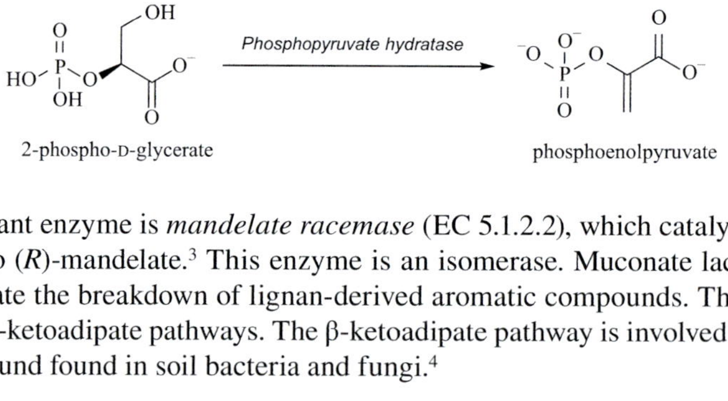

Another important enzyme is *mandelate racemase* (EC 5.1.2.2), which catalyzes the conversion of (*S*)-mandelate to (*R*)-mandelate.[3] This enzyme is an isomerase. Muconate lactonizing enzymes (EC 5.5.1.1) facilitate the breakdown of lignan-derived aromatic compounds. These enzymes catalyze the bacterial β-ketoadipate pathways. The β-ketoadipate pathway is involved in the degradation of aromatic compound found in soil bacteria and fungi.[4]

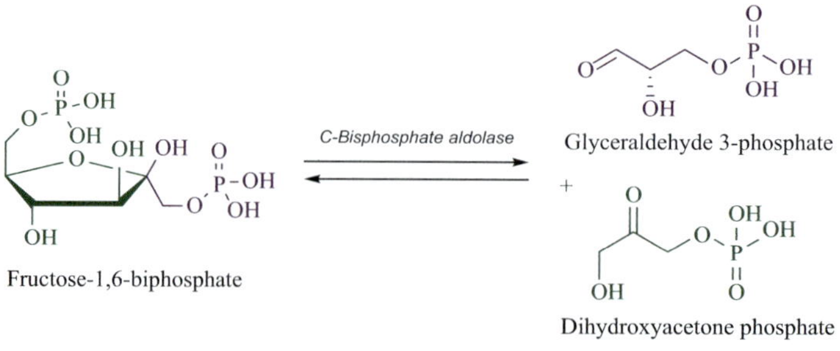

[1] Cui, J.; Maloney, M.I.; Olson, D.G.; Lynd, L.R. *Metab. Eng. Commun.* **2020**, *10*, e00122.
[2] (a) Akram, M.J. *Cancer Educ.* **2013**, *28*, 454; (b) Kim, B.H.; Gadd, G.M. *Bacterial Physiology and Metabolism*, 3rd ed., **2011**.
[3] Gunsalus, C.F.; Stanier, R.Y.; Gunsalus, I.C. *J. Bacteriol.* **1953**, *66*, 548.
[4] Harwood, C.S.; Parales, R.E. *Annu. Rev. Microbiol.* **1996**, *50*, 553.

DOI: 10.1201/9781003605157-9

Enzymes known as aldolases can catalyze both the forward and reverse aldol reactions. *Aldolase A* (EC 4.1.2.13; *bisphosphate aldolase C*; see Section 4.6.1) is a human enzyme that catalyzes the reversible reaction of fructose-1,6-bisphosphate to glyceraldehyde 3-phosphate and dihydroxyacetone phosphate. *Aldolase A* is found in the developing embryo and is produced in even greater amounts in adult muscle, and deficiency of this enzyme has been associated with myopathy (a muscle tissue disease) and hemolytic anemia (the abnormal breakdown of red blood cells). An *aldolase* is an enzyme that facilitates an aldol reaction or the reverse reaction initiated by enzymes such as *sialic acid aldolase* (EC 4.1.3.3; also found in *N-acetylneuraminate pyruvate lyase*, occurring mainly in pathogens), which forms sialic acid (*N*-acetylneuraminic acid) and breaks apart an aldol. An example is the cleavage of *N*-acetylneuraminic acid to give pyruvate and *N*-acetyl-D-mannosamine.[5]

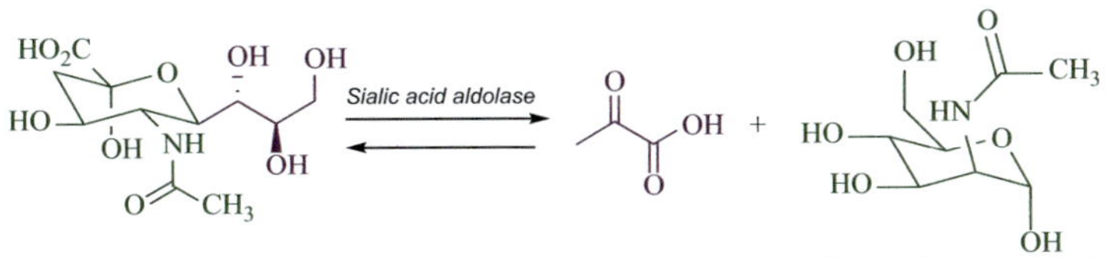

N-Acetylneuraminic acid (sialic acid) Pyruvate *N*-Acetyl-D-mannosamine

An example is the retro-aldol reaction catalyzed by the enzyme *L-ribulose-5-phosphate-4-epimerase* (EC 5.1.3.4; see Section 4.7.2), found in both prokaryotes and eukaryotes, to invert the stereochemistry of the hydroxyl-bearing carbon in **1** to that in **4**. In other words, L-ribulose-5-phosphate is epimerized (the stereogenic center is inverted to the opposite absolute configuration) to give D-xylulose-5-phosphate, as shown in Figure 9.1.[6] The first step is a retro-aldol of the Zn^{2+} coordinated L-ribulose-5-phosphate (**1**), induced by the enzyme to yield the aldehyde 2-oxoethyl

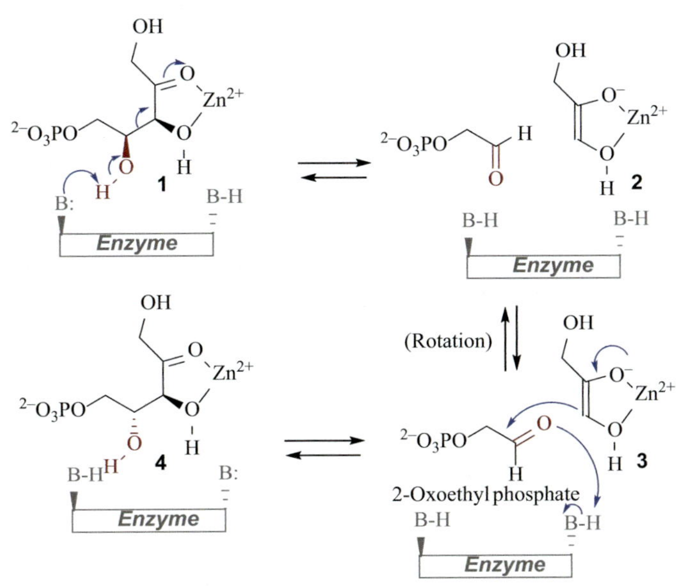

FIGURE 9.1 Enzyme-mediated epimerization of L-ribulose-5-phosphate to D-xylulose-5-phosphate. (Reprinted with permission from Tanner, M.E. *Acc. Chem. Res.* **2002**, *35*, 237. Copyright 2002 American Chemical Society.)

[5] Sánchez-Carrón, G.; García-García, M.I.; López-Rodríguez, A.B.; Jiménez-García, S.; Sola-Carvajal, A.; García-Carmona, F.; Sánchez-Ferrer, A. *Appl. Environ. Microbiol.* **2011**, *77*, 2471.

[6] Tanner, M.E. *Acc. Chem. Res.* **2002**, *35*, 237.

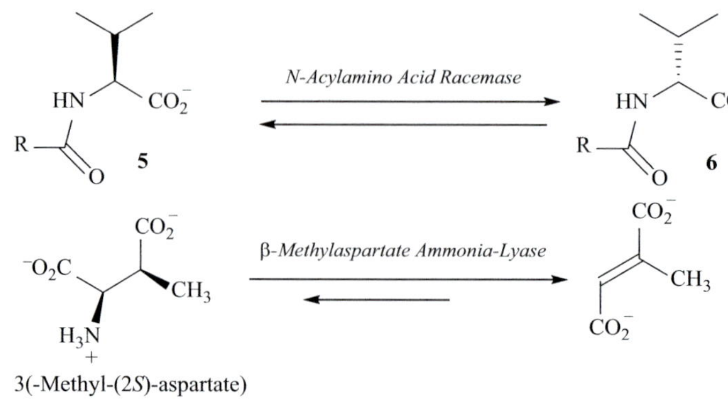

FIGURE 9.2 Epimerase reactions.

phosphate (glycoaldehyde phosphate) and the zinc-coordinated enolate anion **2** (dihydroxyacetone enolate).[6] Before the aldol reaction occurs, there is a bond rotation to generate a different rotamer of 2-oxoethyl phosphate (**3**). The aldehyde unit is now positioned differently, such that an aldol reaction will give the aldolate **4**, but with a different absolute stereochemistry. The overall enzyme process leads to epimerization of the hydroxyl-bearing carbon. Note that prokaryotes are organisms whose cells lack a nucleus and other organelles. All animals, plants, fungi, seaweeds, and many unicellular organisms are eukaryotes.

Epimerization, or changing the absolute stereochemistry of a given stereogenic center, is an important part of many enzymatic reactions. *Enolase* enzymes can catalyze the thermodynamically difficult step of proton abstraction, which often leads to epimerization. These proteins include enolase as well as more specialized enzymes such as *mandelate racemase* (EC 5.1.2.2; see Section 4.7), *galactonate dehydratase* (EC 4.2.1.6), *glucarate dehydratase* (EC 4.2.1.40), muconate-lactonizing enzymes (EC 5.1.1.X), *N-acylamino acid racemase* (EC 5.1.1.4), *α-methylaspartate ammonia-lyase* (EC 4.3.1.2), and o-*succinylbenzoate synthase* (EC 4.2.1.113).[7]

Two examples are shown in Figure 9.2 involving *N-acylamino acid racemase* and *β-methylaspartate ammonia-lyase. N-Acylamino acid racemase* is a racemase and catalyzes a 1,1-proton transfer reaction that is stereorandom with respect to the abstraction of the (*R*) proton of the amino acid residue. It appears to contain two bases in its active site positioned for proton abstraction from either the (*R*)- or (*S*)-enantiomers of the amino acid. A simple example is the epimerization of acyl valine derivative **5–6**, which proceeds by removal of the α-proton to form an enol or an enolate intermediate. *β-Methylaspartate ammonia-lyase* initiates epimerization, but it involves the presence of oxygen, leaving groups. Either diastereomer of 3-methyl-(2*S*)-aspartate ((2*R*,3*S*)-2-ammonio-3-methylsuccinate) is a substrate for this enzyme. An example is the conversion of (2*R*,3*S*)-2-ammonio-3-methylsuccinate to 2-methylfumarate via loss of ammonia, as shown in Figure 9.2. Abstraction of the (*R*)-proton by the enzyme leads to loss of ammonia and formation of the C=C unit. Epimerization is completed by the delivery of ammonia from another nitrogen source to the conjugate acid unit to regenerate the epimeric methyl aspartate.

9.2 ALDEHYDES AND KETONES ARE WEAK ACIDS

The compound with an OH unit attached to a sp^2-hybridized carbon of a C=C unit, rather than a sp^3-hybridized carbon, is known as an *enol*. The carbonyl form of a ketone or an aldehyde is known as the *keto* form and is in equilibrium with the alkene form known as the *enol*, as shown in the reaction. *Keto-enol tautomerism* is an equilibrium reaction that interconverts an enol form and a keto

[7] Babbitt, P.C.; Hasson, M.S.; Wedekind, J.E.; Palmer, D.R.; Barrett, W.C.; Reed, G.H.; Rayment, I.; Ringe, D.; Kenyon, G.L.; Gerlt, J.A. *Biochemistry* **1996**, *35*, 16489.

form. This equilibrium greatly favors the keto form. The carbonyl form is favored over the enol unless there is some special structural feature, such as the presence of a second electron-withdrawing group on an α-carbon. Keto-enol tautomerism is possible because the hydrogen atom of the O–H group in the enol is acidic and subject to intramolecular attack by the π-bond of the C=C unit, which is a weak base. Such a reaction is therefore an internal acid-base reaction.

Enol form Keto form

A ketone or an aldehyde exists with small amounts of the enol form in most cases. Using acetone (propan-2-one) as an example, the ketone predominates, and experiments indicate that only 1.5×10^{-4} % of the enol (prop-1-en-2-ol) is present at equilibrium.[8] In other words, acetone exists almost exclusively in the ketone form. As mentioned, the proton of an enol is acidic, and the α-proton on the α-carbon is also acidic. Therefore, the hydrogen atom on the α-carbon, the α-proton, makes ketones and aldehydes weak acids. The molar concentration of enol is small and present in an equilibrium.[9] Either the O–H proton of the enol or the α-proton of the carbonyl form can be removed by an appropriate base. Since the keto form usually predominates in a keto-enol tautomerism, the focus is typically on the α-proton.

It is known that the acid-catalyzed reaction of acetone begins with an acid-base reaction to give an oxocarbenium ion, in which the carbonyl oxygen reacts as a base. The presence of an acid catalyst is also known to increase the percentage of the enol. The presence of heteroatoms or electron-withdrawing functional groups at the α-carbon will stabilize the enol form via internal hydrogen bonding. This observation is important because an increase in the enol content is associated with an increase in acidity (lower pK_a). There is an increase in the enol content of 1-chloropropan-2-one and 1,1-dichloropropan-2-one relative to acetone.

An ester (e.g., methyl ethanoate) has an α-carbon directly connected to a carbonyl unit, and the attached α-proton is polarized and acidic. It is known that the α-proton of an ester is a weaker acid when compared to a ketone or an aldehyde. The pK_a of the α-proton of an ester is ~ 24–25, whereas the pK_a of the α-proton of a ketone is typically 19–21. If an α-carbon has more than one electron-withdrawing group, that proton becomes increasingly acidic as the number of groups attached to the carbonyl increases. The pK_a of propan-2-one is ~ 20, whereas the pK_a of protons on the CH_2 unit of pentane-2,4-dione is ~ 9, for example. Since an increase in enol content is associated with increased acidity, pentane-2,4-dione should have an increased enol content due to the two carbonyl units. Titration of pentane-2,4-dione (known as acetylacetone) with bromine yields 3-bromopentane-2,4-dione, and this experiment showed that pentane-2,4-dione has 79.7% enol, consistent with the concept of enol stabilization by an adjacent carbonyl.

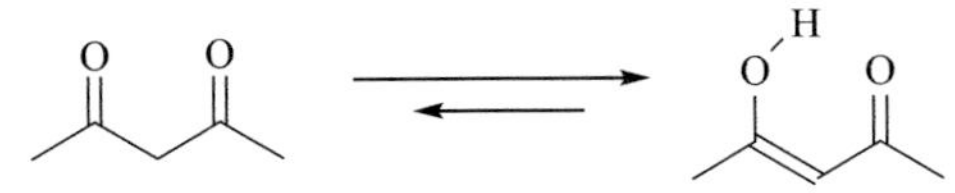

Pentane-2,4-dione (*Z*)-4-Hydroxypent-3-en-2-one

In general, dicarbonyl compounds with two carbonyl units, where one carbonyl has a 1,3-relationship with a second carbonyl (O=C–CH_2–C=O), will be more acidic. All such compounds will be more acidic than simple aldehydes (e.g., acetaldehyde) or simple ketones (e.g., acetone).

[8] Gero, A. *J. Org. Chem.* **1954**, *19*, 469.
[9] Chiang, Y.; Kresge, A.J.; Tan, Y.S.; Wirz, J. *J. Amer. Chem. Soc.* **1984**, *106*, 460.

Examples include 1,3-propanedial (malonaldehyde), the diketone pentane-2,4-dione, 3-oxoesters (β-ketoesters) (e.g., methyl 3-oxobutanoate), and finally 1,3-dioic esters (malonates). The pK_a of malonaldehyde is 5.9,[10] the pK_a of pentane-2,4-dione is 9.0,[10] the pK_a of ethyl 3-oxobutanoate is 10.68, and the pK_a of diethyl malonate is 13.3.[10]

9.3 FORMATION OF ENOLATE ANIONS

If the α-proton of a ketone or aldehyde is a weak acid, treatment with a strong base should initiate an acid-base reaction, forming the conjugate base of the aldehyde or ketone, known as an *enolate anion*. The reaction of acetone (propan-2-one) with an unspecified base, for example, yields the conjugate acid (H:base) and the enolate anion (2-oxopropan-1-ide), which is the conjugate base of acetone (see Figure 9.3). Bases used for deprotonation include sodium methoxide or sodium hydroxide in protic solvents such as an alcohol or water, or lithium diisopropylamide (LDA) in aprotic solvents such as diethyl ether or THF.

Apart from inductive effects that make the α-proton more acidic in a ketone or aldehyde, another major contributing factor to K_a is the charge delocalization (resonance stability) of the conjugate base. A more stable conjugate base shifts the equilibrium of the acid-base reaction to the right (more product), leading to a larger K_a (smaller pK_a), which is equated with greater acid strength. Therefore, to properly assess the acidity of acetone or any other carbonyl compound, the conjugate base of the acid-base reaction must be examined. As shown in Figure 9.3, the conjugate base is the resonance-stabilized *enolate anion*, 2-oxopropan-1-ide.

As noted in the previous section, the keto form of acetone is the dominant species in keto-enol tautomerism, but both the α-proton in the keto form and the acidic proton of the enol form can be deprotonated to generate an enolate anion. The acid-base reactions for both the keto and the enol forms of acetone are shown in Figure 9.3. Deprotonation of the keto form leads to one resonance contributor of the enolate anion, with the negative charge on carbon. Deprotonation of the enol form leads to the other resonance contributor, with the negative charge on oxygen. Both resonance contributors represent 2-oxopropan-1-ide, so it is possible to view enolate anion formation by deprotonation of the enol rather than the keto form. Since the keto form dominates, deprotonation of the α-proton is commonly shown, and the resonance contributor with the charge on carbon is generated directly. Examination of the two resonance contributors for the enolate anion in Figure 9.3 shows that the negative charge is delocalized on oxygen in one resonance

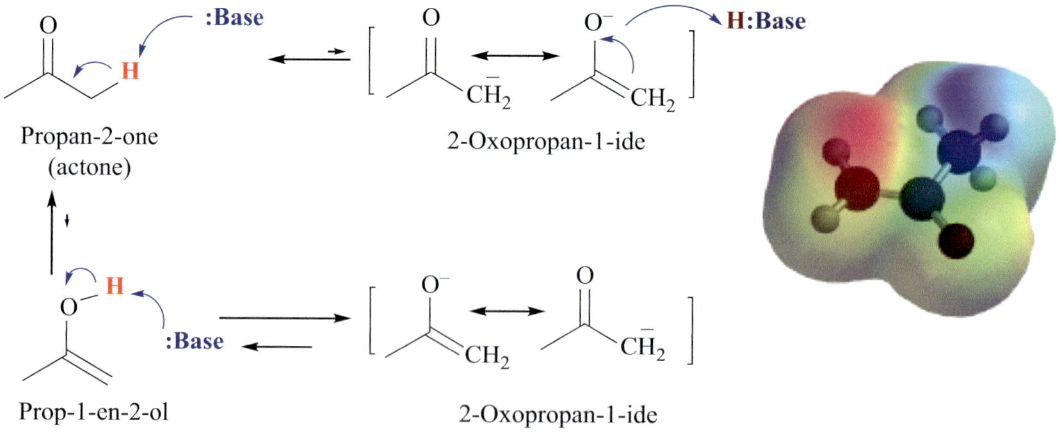

FIGURE 9.3 Enolate anion formation from acetone.

[10] Pearson, R.G.; Dillon, R.C. *J. Amer. Chem. Soc.* **1953**, *75*, 2439.

contributor and on carbon in the other. Enolate anions are carbanions because the resonance-stabilized enolate anion has a larger concentration of electron density on the carbon relative to oxygen. The electron potential map shows the charge distribution for this enolate anion, where more red (higher electron density) is concentrated on the carbon than on the oxygen, although this is influenced by the counterion.

9.4 ENZYME-MEDIATED CLAISEN CONDENSATION

Enzyme-catalyzed condensation reactions are used in fatty acid biosynthesis.[11] The enzymatic Claisen condensation is facilitated by the decarboxylating condensing enzymes of fatty acid and polyketide synthesis, including *β-ketoacyl-ACP synthase III* (*FabH*; EC 2.3.1.180; see Section 4.2) and the *β-ketoacyl-ACP synthases I and II*, *FabB* (EC 2.3.1.41) and *FabF* (EC 2.3.1.179), respectively. Fatty acid synthesis can be imitated by the condensation of acetyl-CoA and malonyl-ACP by the enzyme FabB or FabH. An example is the *ketosynthase FabB*-catalyzed condensation of butyryl-ACP (ACP = acyl carrier protein) with malonyl-ACP in what is essentially an overall Claisen condensation to give a fatty acid-ACP after 5–7 rounds of coupling with malonyl-CoA and ketosynthase enzyme, as shown in Figure 9.4.[12] The correlation with enol-condensation reactions and with the Claisen condensation is apparent in the proposed mechanism for the conversion of acetyl-CoA to malonyl-CoA in a Claisen-type process.[13]

The intramolecular Claisen condensation, known as the Dieckmann condensation, occurs in biosynthetic transformations. Menaquinone (vitamin K_2) functions as a redox-active cofactor in the electron transport chain of many bacteria.[14] In the biosynthesis of menaquinone from chorismate in *Escherichia coli*, a key step is the reaction catalyzed by *1,4-dihydroxynaphthoyl-CoA synthase* (EC 4.1.3.36; *MenB*; see Section 4.2), which is essentially an intramolecular Claisen (or Dieckmann) condensation involving *O*-succinylbenzoate.[15] Formation of an enol intermediate leads to attack of the distal acyl carbon to form the six-membered ring, and tautomerization of the product leads to the product, 1,2-dihydroxyl-2-napthyl-CoA (see Figure 9.5).

The proposed mechanism of this transformation is shown in Figure 9.6.[15] It is clear that the interaction with the proximal amino acid residues facilitates enol formation and subsequent acyl addition to give the tetrahedral intermediate that leads to the cyclized product. The carboxylate moiety is positioned to abstract the pro-2*S* proton, generating a carbanion/enolate that reacts with the protonated carboxylic acid to give a tetrahedral intermediate. Elimination of water yields a β-ketothioester, which tautomerizes to the enol form, which constitutes the aromatic ring.

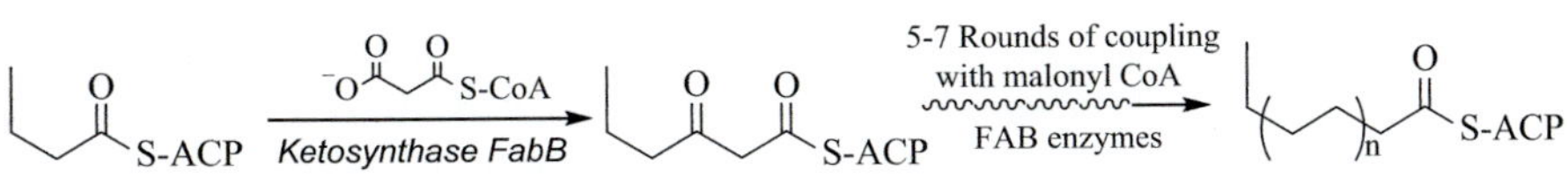

FIGURE 9.4 Biosynthesis of fatty acid-ACP from Butyryl-ACP. (Yu, X.; Liu, T.; Zhu, F.; Khosla, C. *Proc. Natl. Acad. Sci.* **2011**, *108*, 18643–18648. Copyright (2011) National Academy of Sciences, U.S.A.)

[11] Heath, R.J.; Rock, C.O. *Nat. Prod. Rep.* **2002**, *19*, 581.
[12] (a) Yu, X.; Liu, T.; Zhu, F.; Khosla, C. *Proc. Nat. Acad. Sci.* **2011**, *108*, 18643; (b) Also see Chapter 21, Tabas, I. in *Biochemistry of Lipids, Lipoproteins and Membranes*, 5th ed., Vance, J.E.; Vance, D.E. (Eds), Elsevier, The Netherlands, **2002**.
[13] Qiu, X.; Janson, C.A.; Smith, W.W.; Head, M.; Lonsdale, J.; Konstantinidis, A.K. *J. Mol. Biol.* **2001**, *307*, 341.
[14] Bishop, D.H.L.; Pandya, K.P.; King, H.K. *Biochem. J.* **1962**, *83*, 606.
[15] Li, H.-J.; Li, X.; Liu, N.; Zhang, H.; Truglio, J.J.; Mishra, S.; Kisker, C.; Garcia-Diaz, M.; Tonge, P.J. *Biochemistry* **2011**, *50*, 9532.

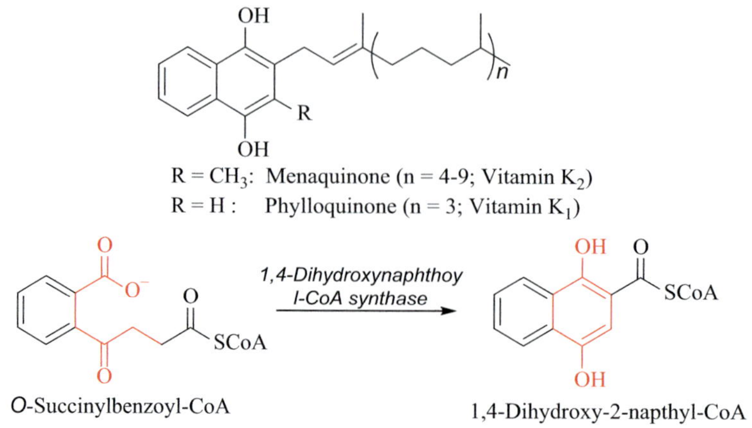

FIGURE 9.5 Enzymatic catalysis of a Dieckmann-like condensation.

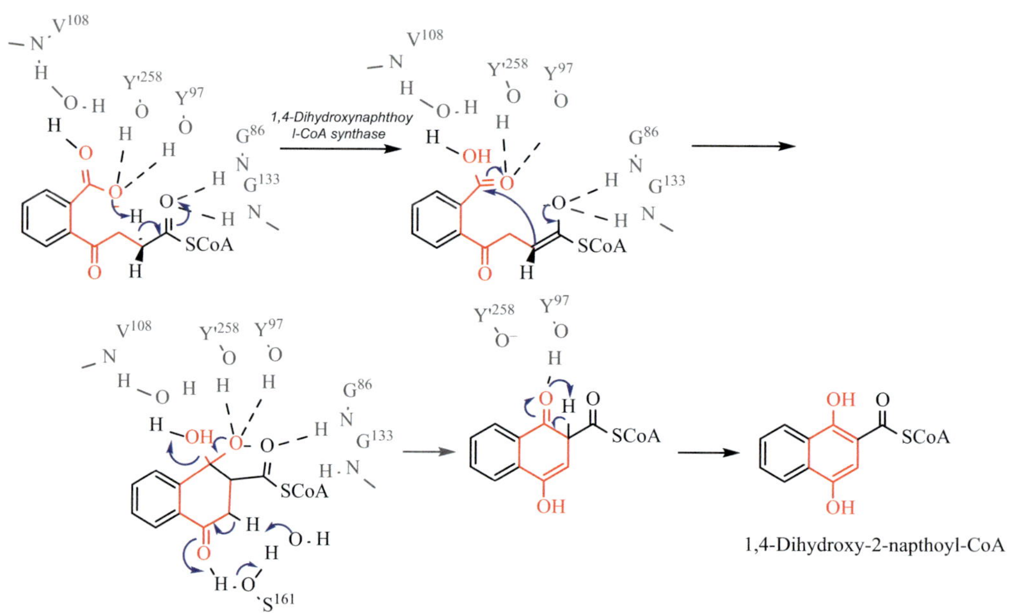

FIGURE 9.6 Mechanistic rationale for the synthesis of 1,4-dihydroxy-2-napthoyl-CoA. (Reprinted with permission from Li, H.-J.; Li, X.; Liu, N.; Zhang, H.; Truglio, J.J.; Mishra, S.; Kisker, C.; Garcia-Diaz, M.; Tonge, P.J. *Biochemistry* **2011**, *50*, 9532–9544. Copyright 2011 American Chemical Society.)

9.5 THE ALDOL CONDENSATION

In acyl addition reactions, nucleophiles react with aldehydes or ketones to form a new C–C bond in the alkoxide product. Enolate anions react with aldehydes and ketones via nucleophilic acyl addition. The enolate anion of acetophenone is a carbanion nucleophile and reacts with acetaldehyde, as shown.

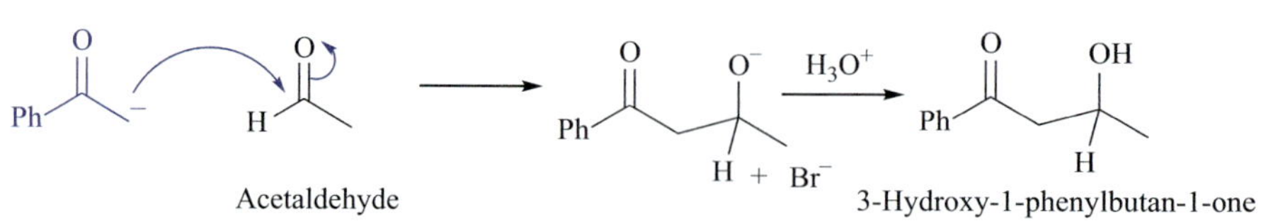

A new C–C bond is formed to generate an alkoxide, and the carbonyl of the ketone enolate anion remains, so the product is the keto alkoxide. In other words, the product is the alkoxide of 3-hydroxy-1-phenylbutan-1-one. An aqueous workup converts the keto alkoxide to hydroxy ketones or hydroxy aldehydes, in this case 3-hydroxy-1-phenylbutan-1-one. Such β-hydroxy ketones or aldehydes are known as *aldols*, and this transformation has come to be called the *aldol condensation*.

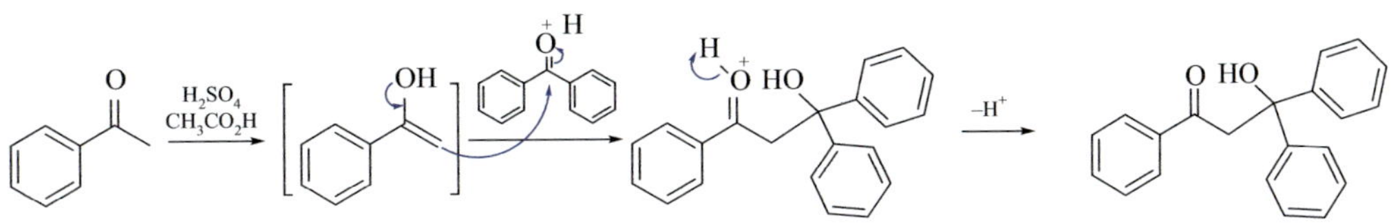

3-Hydroxy-1,3,3-triphenylpropan-1-one

Direct asymmetric catalytic aldol reactions have been successfully performed using aldehydes and unmodified ketones together with commercially available chiral cyclic secondary amines as a base. The reaction of acetone and 4-nitrobenzaldehyde in dimethyl sulfoxide (DMSO) was reported using proline, for example.[16] This aldol product is 4-hydroxy-4-(4-nitrophenyl)butan-2-one.

Acid-catalyzed aldol condensation reactions are also known, which may be more pertinent to biological aldol condensation reactions. A simple example is the condensation reaction of acetophenone and benzaldehyde with sulfuric acid in acetic acid, as shown. In this example, acetophenone is the only carbonyl partner with α-hydrogen atoms, and in the presence of the acid, the enol is formed. In the acidic medium, the α-carbon of the enol reacts with the protonated carbonyl of benzaldehyde, as shown, to give the protonated aldol coupling product. Loss of a proton gives the aldol product 3-hydroxy-1,3-diphenylpropan-1-one. The lesson of this example is that the acidic medium facilitates formation of the enol, and reaction with the other carbonyl partner leads to the aldol product.

9.6 THE CLAISEN CONDENSATION

In the late 19th century, an experiment was reported in which an ester was heated with a base in ethanol, and the isolated product was a β-keto ester. This product results from the enolate anion of one molecule of the ester condensing with a second molecule via an acyl substitution reaction. In a typical experiment, ethyl 2-methylpropanoate was heated with sodium ethoxide in ethanol. This reaction mixture was then acidified with glacial acetic acid (i.e., 100% acetic acid), and the final isolated product was ethyl 2,2,4-trimethyl-3-oxopentanone.[17] It is clear that this reaction is a self-condensation of an ester enolate reacting with another molecule of the same ester. The product is a β-keto ester, and the reaction is now called the *Claisen condensation*.

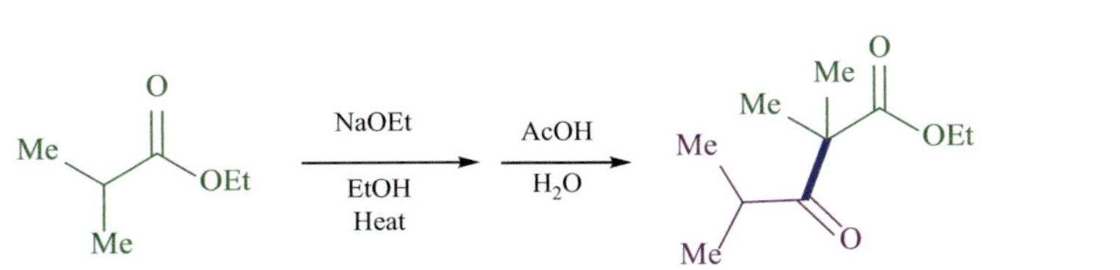

Ethyl 2-methylpropanoate

Ethyl 2,2,4-trimethyl-3-oxopentanoate

To explain the reaction with ethyl 2-methylpropanoate and the formation of the product ethyl 2,2,4-trimethyl-3-oxopentanoate, the mechanism shown in Figure 9.7 requires deprotonation of the acidic α-proton by the base to form a resonance-stabilized enolate anion. Under equilibrium

---

[16] Sakthivel, K.; Notz, W.; Bui, T.; Barbas III, C.F. *J. Amer. Chem. Soc.* **2001**, *123*, 5260.

[17] For a similar experiment, see Furniss, B.S.; Hannaford, A.J.; Smith, P.W.G.; Tatchell, A.R. (Eds.), *Vogel's Textbook of Practical Organic Chemistry*, 5th ed., Longman, Essex, UK, **1994**, Exp. 5.176, pp. 741–742.

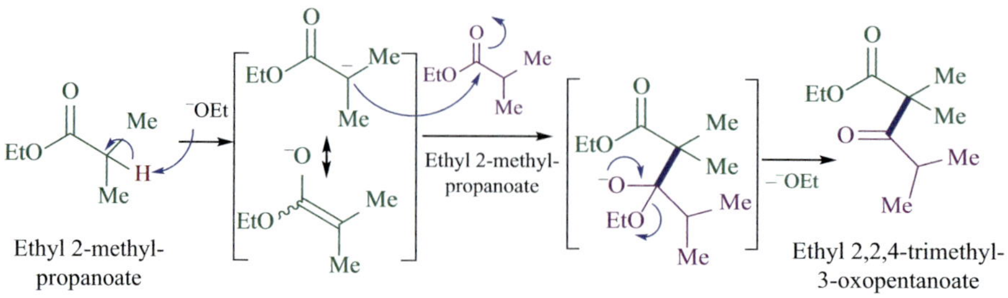

FIGURE 9.7 Claisen self-condensation of ethyl 2-methylpropanoate.

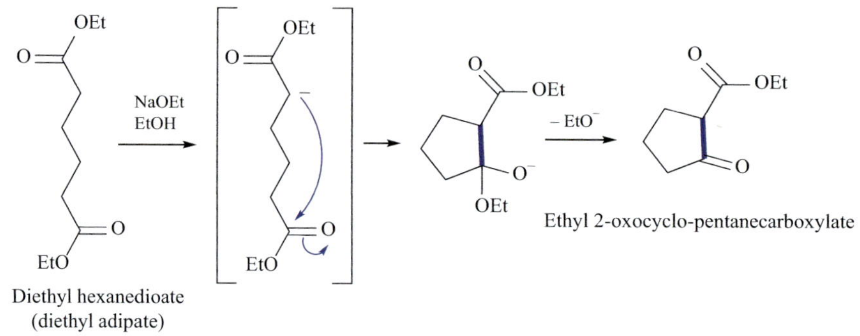

FIGURE 9.8 Dieckmann condensation.

conditions, both the enolate anion and ethyl 2-methylpropanoate are present, and condensation leads to a tetrahedral intermediate. Loss of ethoxide leads to the formation of the ketone unit in ethyl 2,2,4-trimethyl-3-oxopentanoate, the Claisen condensation product. Note that ester enolates usually exist as mixtures of (*E*)- and (*Z*)-isomers, and the Claisen condensation can lead to isomers. This aspect of the Claisen condensation, however, will not be discussed further in this work.

There is an intramolecular version of the Claisen condensation, the Dieckmann Condensation, illustrated by the treatment of diethyl 1,6-hexanedioate (diethyl adipate) with sodium ethoxide in ethanol, and the isolated final product is the intramolecular Claisen condensation product, ethyl 2-oxocyclopentanecarboxylate, as shown in Figure 9.8. In this reaction, the enolate anion reacts with the carbonyl (C6) at the other end of the molecule, and the intramolecular Claisen condensation leads to the tetrahedral intermediate in Figure 9.8. Loss of ethoxide ($^{-}$OEt) leads to a cyclic β-ketone ester (ethyl 2-oxocyclopentanecarboxylate). This intramolecular Claisen condensation is formally known as the *Dieckmann condensation*.

9.7 ENZYMATIC DECARBOXYLATION

Decarboxylation is important in many biopathways. Parkinson's disease is the second most common neurodegenerative disorder, affecting 1% of the global population over the age of 60, and has recently been correlated with alterations in microbial gut composition.[18] The primary treatment of Parkinson's disease is levo-Dopa (L-3,4-dihydroxyphenylalanine or L-DOPA) in combination with an aromatic amino acid decarboxylase inhibitor (primarily carbidopa).[19] However, the bioavailability of L-DOPA/decarboxylase inhibitor required to ensure sufficient amounts of dopamine reach the brain[20] varies significantly among Parkinson's disease patients depending on the time of treatment,

[18] Pereira, P.A.B.; Aho, V.T.E.; Paulin, L.; Pekkonen, E.; Auvinen, P.; Scheperjans, F. *Relat. Disord.* **2017**, *38*, 61 and references cited therein.

[19] Deleu, D.; Northway, M.G.; Hanssens, Y. *Clin. Pharmacokinet.* **2002**, *41*, 261.

[20] Pinder, R.M. *Nature* **1970**, *228*, 358.

necessitating more frequent drug doses.. Because of this, the L-Dopa/decarboxylase inhibitor is ineffective in a subset of patients, and its efficacy decreases due to decarboxylation with bacterial *tyrosine decarboxylase*, which converts L-DOPA to dopamine as shown.[21] Several amino acid decarboxylases have been identified in bacteria. Tyrosine decarboxylase genes have been encoded in the genomes of several bacterial species in the genera Lactobacillus and Enterococcus.[22] Species of the genera Lactobacillus and Enterococcus have been reported to harbor this enzyme.[22a,23]

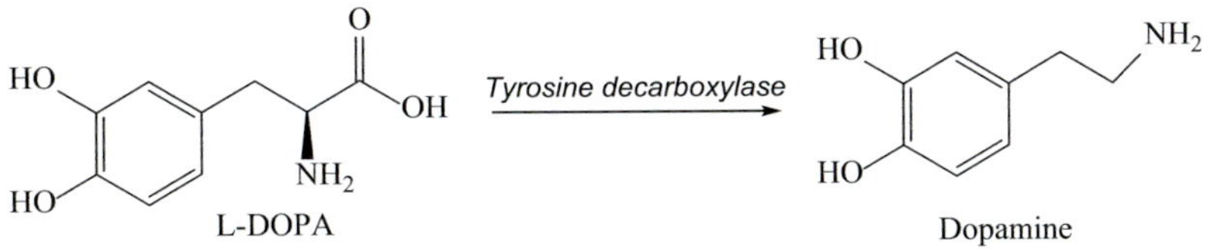

Another example of an enzymatic decarboxylation is taken from the biosynthesis of microcin C.[24] This decarboxylation of a pyridoxal 5′-phosphate (PLP)-dependent decarboxylation is found in microcin C biosynthesis. Microcin C is an antibiotic found in *E. coli* and closely related bacteria. Decarboxylation of 3-amino-3-carboxypropylated MccA (**7**) is catalyzed by the N-terminal domain of the MccE enzyme (MccENTD), a PLP-dependent decarboxylase, as shown in Figure 9.9.[25] The product is microcin C.

In bacteria, the biosynthesis of complex polyketides utilizes assembly-line-like megaenzymes that are part of type 1 modular polyketide synthase (PKS) classes. Figure 9.10 illustrates the active site of a decarboxylation in polyketide β-methyl branch Incorporation, in which complex **8** is predicted to be deprotonated by His[230].[26] Loss of CO_2 generates enol **9**, which gives the conjugated system, **10**. In addition to His[230], Gly[66], and Gly[108], the active site of PksI comprises the residues Phe[81], Phe[136], Phe[234], and His[235].

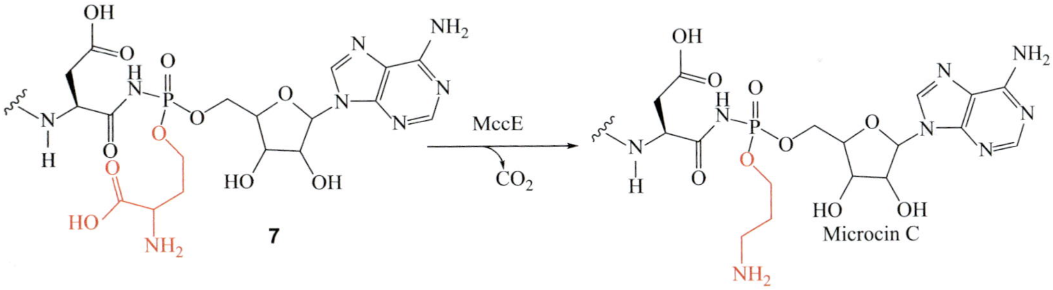

FIGURE 9.9 Microcin C maturation pathway and the formation of the bioactive processed Microcin C, and the structure of Asp-AMP, the native substrate of the aspartyl-tRNA synthetase. (Kulikovsky, A.; Serebryakova, M.; Bantysh, O.; Metlitskaya, A.; Borukhov, S.; Severinov, K.; Dubiley, S. *J. Am. Chem. Soc.* **2014**, *136*, 11168. Copyright 2014 American Chemical Society.)

---

[21] van Kessel, S.P.; Frye, A.K.; El-Gendy, A.O.; Castejon, M.; Keshavarzian, A.; van Dijk, G.; El Aidy, S. *Nat. Commun.* **2019**, *10*, 310. https://doi.org/10.1038/s41467-019-08294-y. Also see Williams, B.B.; Van Benschoten, A.H.; Cimermancic, P.; Donia, M.S.; Zimmermann, M.; Taketani, M.; Ishihara, A.; Kashyap, P.C.; Fraser, J.S.; Fischbach, M.A. *Cell Host Microb.* **2014**, *16*, 495.

[22] (a) Pereira, P.A.B.; Aho, V.T.E.; Paulin, L.; Pekkonen, E.; Auvinen, P.; Scheperjans, F. *Appl. Microbiol. Biotechnol.* **2015**, *99*, 3547; (b) Zhu, H.; Xu, G.; Zhang, K.; Kong, X.; Han, R.; Zhou, R.; Ni, Y. *Sci. Rep.* **2016**, *6*, 27779.

[23] Zhang, K.; Ni, Y. *Protein Expr. Purif.* **2014**, *94*, 33.

[24] Nguyen, N.A.; Forstater, J.H.; McIntosh. J.A. *JACS Au* **2024**, *4*, 2715.

[25] Kulikovsky, A.; Serebryakova, M.; Bantysh, O.; Metlitskaya, A.; Borukhov, S.; Severinov, K.; Dubiley, S. *J. Am. Chem. Soc.* **2014**, *136*, 11168.

[26] Nair, A.V.; Alice Robson, A.; Ackrill, T.D.; Till, M.; Matthew J. Byrne, M.J.; Back, C.R.; Tiwari, K.; Davies, J.A.; Willis, C.L.; Race, P.R. *Nat. Rep. Sci. Rep.* **2020**, *10*, 15323. https://doi.org/10.1038/s41598-020-71850-w.

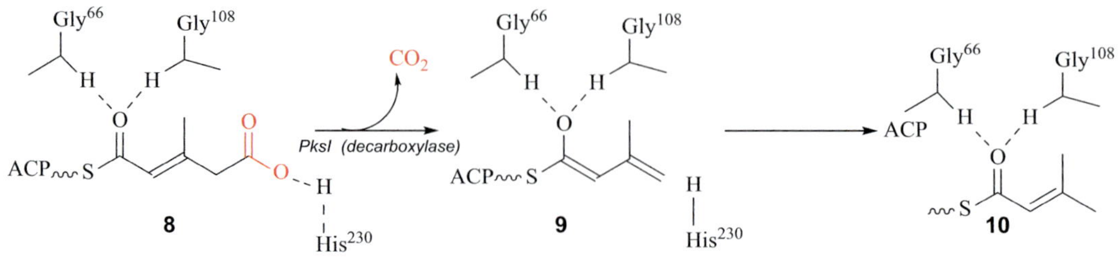

FIGURE 9.10 Proposed catalytic mechanism of PksI (decarboxylase). (Figure 5, p.15330 in Nair, A.V.; Alice Robson, A.; Ackrill, T.D.; Till, M.; Matthew J. Byrne, M.J.; Back, C.R.; Tiwari, K.; Davies, J.A.; Willis, C.L.; Race, P.R. *Nat. Rep. Sci. Rep.* **2020**, *10,* 15323. Open Access.)

9.8 DECARBOXYLATION IN ORGANIC CHEMISTRY

There is a specialized elimination reaction known as *decarboxylation* that is unique to 1,3-dicarboxylic acids, as well as β-keto acids. In general, decarboxylation requires the presence of a carboxyl group (COOH) attached to a carbon that has another carbonyl substituent, or, in some cases, a substituent with a π-bond. With this structural motif, an internal acid–base reaction leads to cleavage of a C–C bond and loss of the neutral molecule carbon dioxide (CO_2; the structure is O=C=O). Malonic acid has the requisite structural features, and when heated to 200°C, carbon dioxide (CO_2) is lost from the molecule. In other words, it undergoes *decarboxylation.*

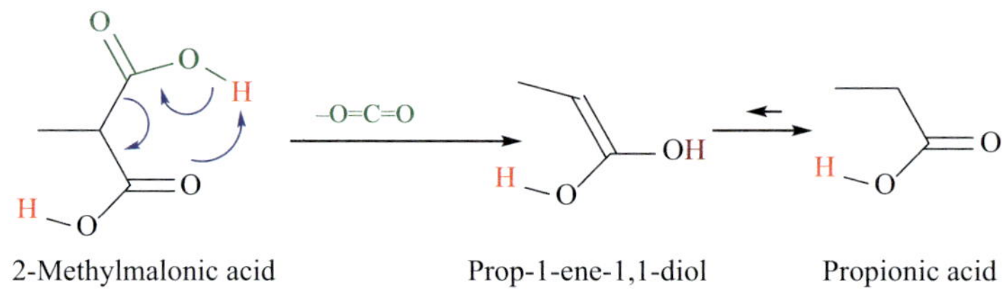

A specific example is 2-methyl-1,3-propanedioic acid (2-methylmalonic acid), which yields propanoic acid upon heating. The oxygen atom of one carboxyl unit acts as a base, donating electrons to the proton of the other carboxyl unit, as shown, via the conformation that generates a six-centered transition state. The product is an enol, prop-1-ene-1,1-diol, the enol of a carboxylic acid, and keto–enol tautomerization of an enol favors the keto form. In this case, tautomerization of prop-1-ene-1,1-diol leads to the carboxyl unit in propanoic acid. Decarboxylation occurs only when the dicarboxylic acid is heated to a relatively high temperature.

HOMEWORK

9.1. Why is the pK_a of cyclohexa-1,3-dione lower than that of cyclohexanone?
9.2. How would the *Phosphopyruvate hydratase* catalyzed conversion of 2-phosphoglycerate to phosphoenolpyruvate be characterized in organic chemistry?
9.3. What is a glycolysis transformation in biochemistry?
9.4. How are *enolase* enzymes important in biological transformation that leads to epimerization?
9.5. Briefly describe the purpose of the interaction of a histidine residue with an aspartic acid residue in an aldolase enzyme active site.
9.6. Briefly describe the purpose of a racemase enzyme.

9.7. In Figure 7.2, the acetyl-CoA enol reacts with an acyl carbon to give an intermediate. What is the name of such an intermediate in organic chemistry?

9.8. What is the product formed when methyl 3-methylcyclopentane-1-carboxylate reacts with sodium methoxide in methanol at reflux, followed by neutralization with dilute acid?

9.9. What is the chemical function of acetyl-Co-A in enzymatic Claisen condensation reactions?

9.10. Give the major product of the following reactions.

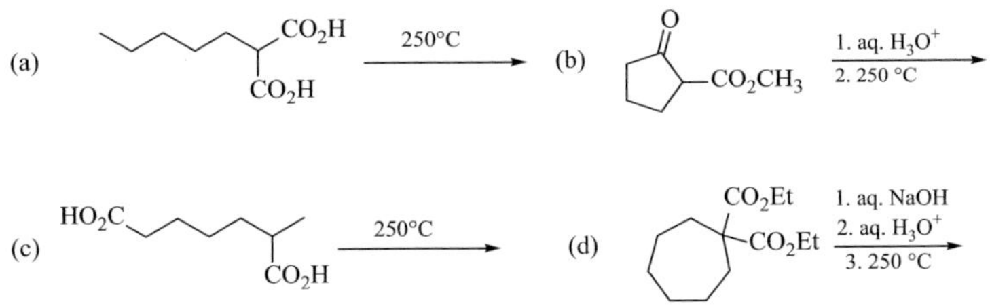

9.11. The initial product of decarboxylation of 2-ethyl-1, 3-propandioic acid is not the carboxylic acid. Draw the initially formed product.

10 Lipids

Lipids are organic compounds such as fatty acids or natural oils, waxes, and steroids that are insoluble in water but soluble in organic solvents. Fatty acids are components of triglycerides and the hydrolysis of triglycerides is important for energy conversion in living systems. Lipids are important for the formation of lipid bilayers and the chemistry of esters is important to the discussion of lipids.

10.1 LIPID CLASSES

Lipids are a class of organic molecules that are lipophilic and hydrophobic. In other words, lipids have a relatively large hydrocarbon contribution to their structure that makes them poorly soluble or insoluble in aqueous media and generally soluble in organic solvents. In biological systems, their insolubility in aqueous media is important. Triglycerides are an important class of lipids that provide energy reserves in biological systems. The hydrophobic nature of lipids makes them important building blocks for biological membranes. Lipids have important regulatory functions, such as bile acids, which are important in the emulsification of lipids during the digestion of fats. There are several biologically important classes of lipids, including fatty acids such as stearic acid and triglycerides, which are fatty acid esters of glycerol. A fatty acid is a carboxylic acid with a long aliphatic chain, as described in Section 5.10.2, which is either saturated or unsaturated.

Phospholipids are like triglycerides, but one or more of the hydroxyl units are functionalized as a phosphoric acid ester, while the other hydroxyl units are fatty acid esters [see 2,3-bis(stearoyloxy)propyl phosphate in Figure 10.1]. Phospholipids are major components of biological membranes. There are several variations in the structure of phospholipids, largely determined by the nature of the ester unit. Several important examples include phosphatidylcholine, which has a choline moiety (the structure of choline is marked in red in Figure 10.1) attached to the phosphatidic acid unit. Choline is the *N,N,N*-trimethylethanolammonium cation, and attachment to the phosphatidyl unit gives phosphatidylcholine, which is also known as lecithin. Choline is a water-soluble, vitamin-like essential nutrient. Lecithins are a group of yellow-brownish fatty substances that occur in animal and plant tissues and are amphiphilic, meaning they attract both water and fatty substances; i.e., they are both hydrophilic and lipophilic. This phospholipid is a major component of lecithin, and it is classified as a plasmalogen, which is a complex membrane lipid. It is also found in egg yolk and soybeans.

Sphingolipids are glycerol derivatives that have an attached sphingosine, as shown in Figure 10.2. When the sphingosine is *N*-acetylated with a fatty acid, the result is a ceramide. It is noted that sphingolipids are important components of the myelin sheath of nerve fibers. Sphingomyelin is an

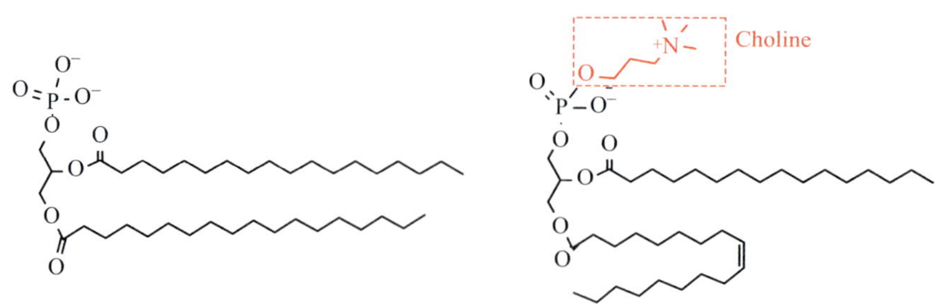

Phospholipid (2,3-bis(stearoyloxy)propyl phosphate) Phosphatidylcholine

FIGURE 10.1 Phospholipids.

DOI: 10.1201/9781003605157-10

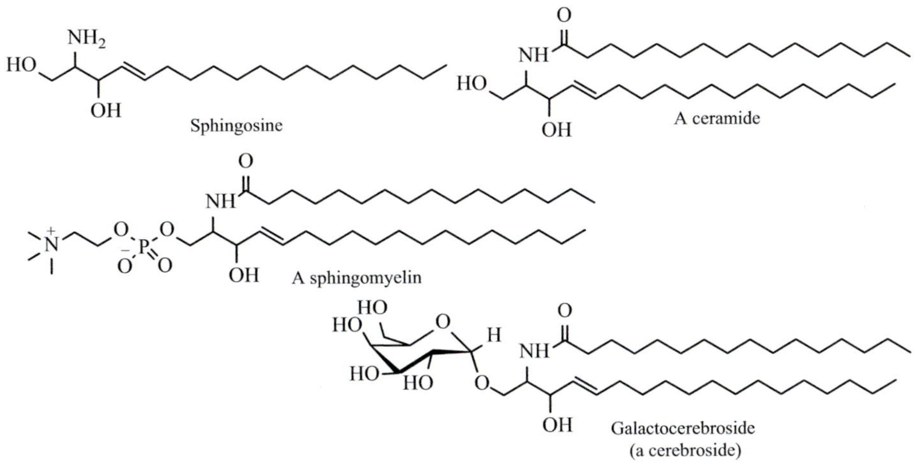

FIGURE 10.2 Sphingolipids and ceramides.

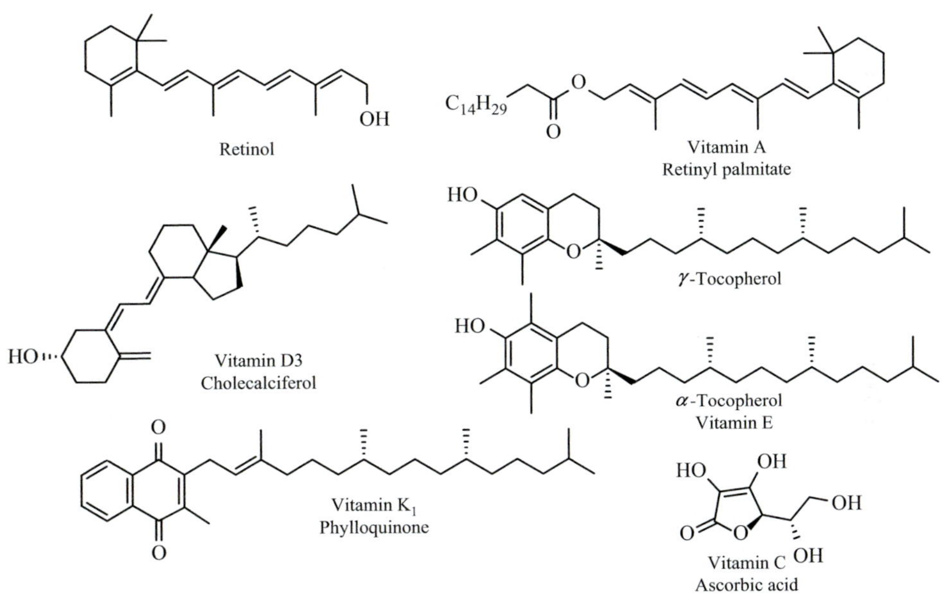

FIGURE 10.3 Vitamins.

important sphingolipid found in the myelin sheath, and it is formed by the attachment of phosphatidylcholine to a ceramide. Glycosphingolipids are found in the cell membrane, and they are generated by attaching a carbohydrate to the sphingosine backbone of a ceramide. The example shown in Figure 10.2 is galactocerebroside, a cerebroside that contains a galactose unit.

An important class of lipophilic compounds is the lipophilic vitamins (A, D, E, and K), which are isoprene-based lipids (see Figure 10.3). There are several vitamin A compounds, but the most prevalent is retinyl palmitate, which is converted to retinol in the small intestine. There are several vitamin D compounds, including cholecalciferol or vitamin D_3. The major natural source of this vitamin is the synthesis of cholecalciferol in the skin from cholesterol via a chemical reaction induced by exposure to UVB radiation in sunlight. Vitamin E exists in eight chemical forms: the tocopherols and the tocotrienols. The most common in a natural diet is γ-tocopherol, followed by α-tocopherol. Phylloquinone (vitamin K1) is essential for blood coagulation, and uncontrolled bleeding occurs without vitamin K. Vitamin K_1 is found in the highest amounts in green leafy vegetables. These vitamins are essential nutrients stored in the liver and fatty tissues, with a diverse range of functions.

Ascorbic acid, vitamin C, is an essential nutrient for humans as well as many other species of living organisms and is required for many essential metabolic reactions in all animals and plants. Humans obtain vitamin C from citrus, or as a food additive or supplement. A deficiency of vitamin C causes scurvy in humans. Note that vitamin C exists in the form of an enol (Section 9.1) rather than in the keto form.

Waxes are a diverse class of organic compounds that are mixtures of long-chain nonpolar lipids. Waxes form a protective coating called cutin (in the cuticle) on plant leaves and fruits and are also found in honeybee wax, the cuticular lipids of insects, the spermaceti of the sperm whale, and skin lipids. Waxes usually include higher alkanes and lipids, typically with melting points above about 40°C, melting to give low-viscosity liquids. Typically, waxes are esters of an alcohol other than glycerol and a long-chain acid (wax esters). Triacontanyl palmitate, a typical wax ester, is derived from triacontanyl alcohol and palmitic acid. The fatty acids in wax esters derived from plants typically range from C12 to C24, and the alcohols in plant waxes tend to be very long, typically C24 to C30.[1]

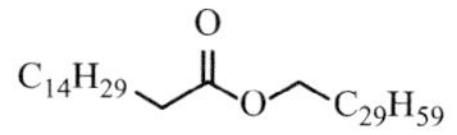

Triacontanyl palmitate

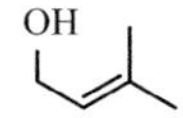

Prenol

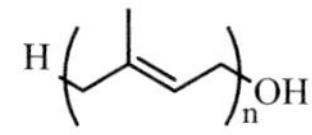

Polyprenol

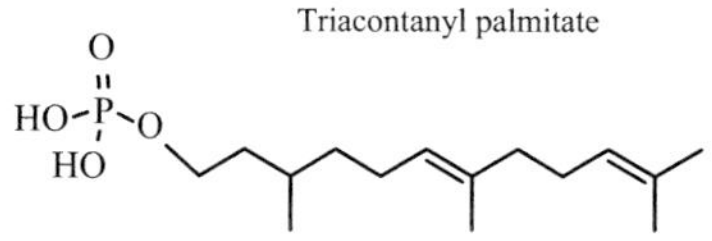

Dolichol phosphate

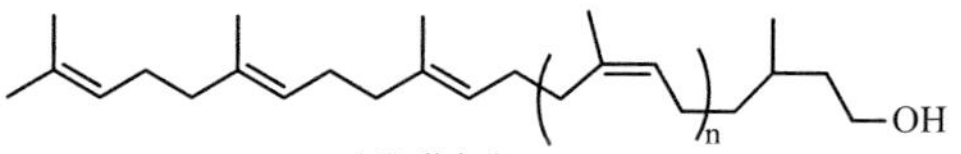

A Dolichol

The transport of oligosaccharides across membranes is an important biological function, and polyprenols and their phosphorylated derivatives are important in this process. Polyprenols are natural bioregulators. Prenol is simply 3-methylbut-2-en-1-ol, and a polyprenol is a polymer with a repeating isoprene unit, where technically, *n* is greater than 4, but *n* >45 is more common. An example of an important polyprenol is the dolichols, which are membrane anchors for the formation of an oligosaccharide. Dolichol phosphate is an important intermediate to produce polyprenol monophosphate and diphosphate sugars that are essential in glycosylation reactions.[2] An enzyme called *dolichol kinase* (EC 2.7.1.108) catalyzes the reaction of cytidine triphosphate (CTP) and dolichol to give dolichyl phosphate.

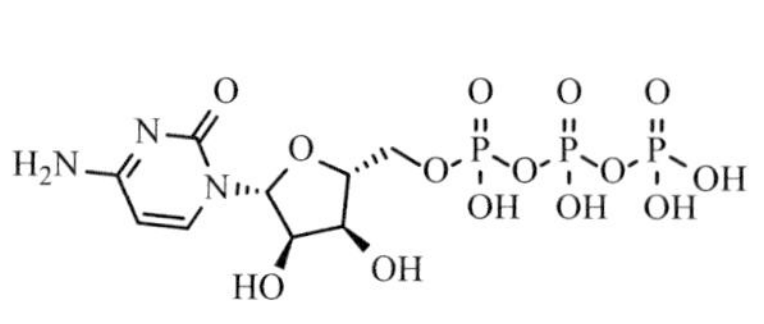

Cytidine triphosphate

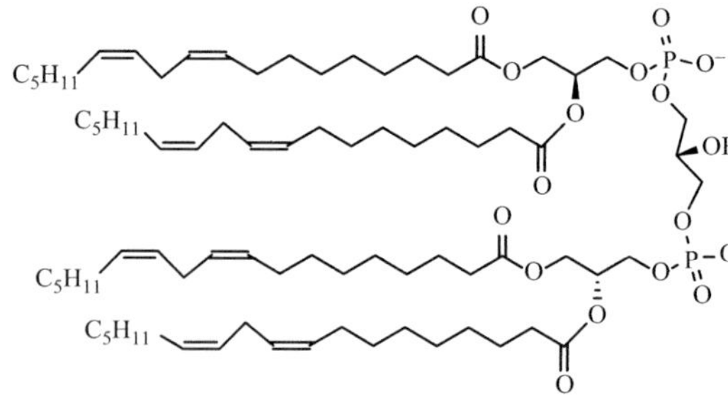

Cardiolipin (a 1,3-*bis*(*sn*-3'phosphatidyl)-*sn*-glycerol

Cardiolipin is a phospholipid located in mitochondria that consists of two phosphate residues and four kinds of fatty acyl chains.[3] Cardiolipin is important for the regulation of mitochondrial

[1] Hargrove, J.L. *Exper. Biol. Med.* **2004**, *229*, 215.
[2] See Jankowski, W.; Mańkowski, T.; Chojnacki, T. *Biochim. Biophys. Acta* **1974**, *337*, 153.
[3] Nakagawa, Y. *Yakugaku Zasshi*. **2013**, *133*, 561.

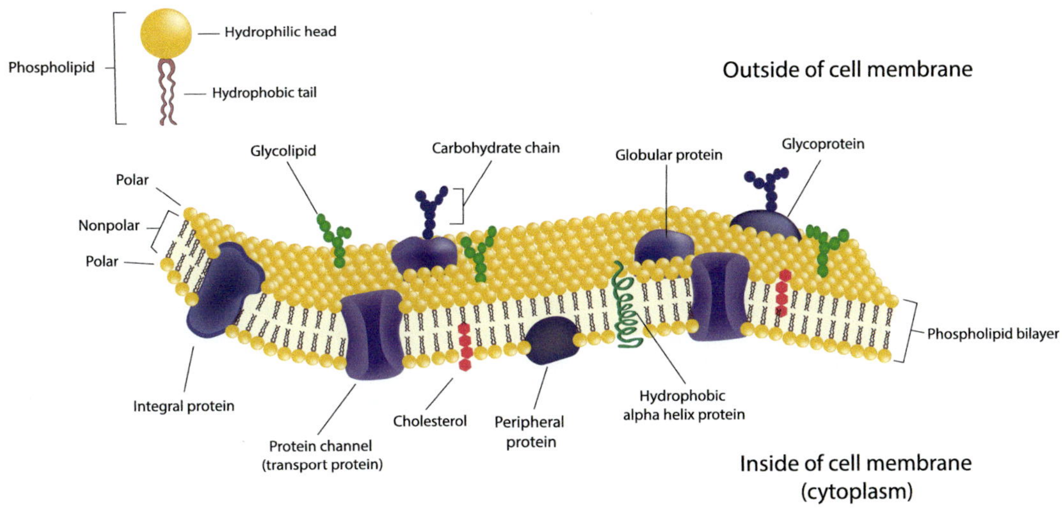

FIGURE 10.4 Structure of phospholipid bilayers cell membrane or cytoplasmic membrane. Stock Vector ID: 1163460010.

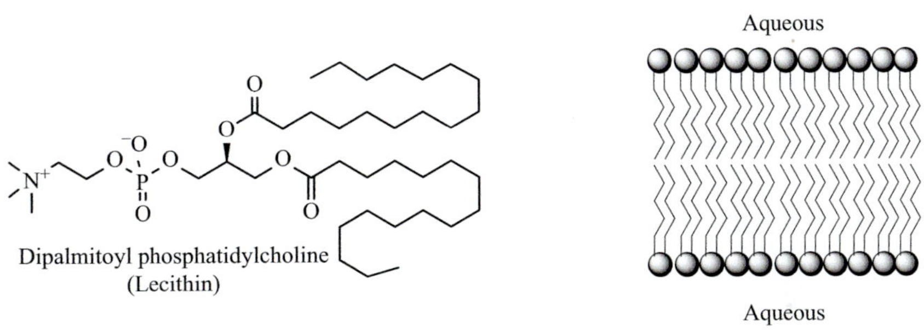

FIGURE 10.5 Lecithin and a lipid bilayer.

proteins, including electron transport complexes, carrier proteins, and *phosphate kinases*. It is essential for the organization of mitochondrial structures.[3]

Phospholipids form lipid bilayers that constitute a major component of cell membranes (Figure 10.4). Most have a diglyceride unit (a diester of glycerol and two fatty acids), a phosphate ester unit, and usually an organic molecule. One example is dipalmitoyl phosphatidylcholine, which is a major component of lecithin.

Lecithin is found in egg yolk and soybeans. Phosphatidylcholines are a major component of biological membranes and are components of pulmonary surfactant used to calculate fetal lung maturity. The lipid bilayer (or phospholipid bilayer) is a thin polar membrane made of two layers of lipid molecules, forming flat sheets that create a continuous barrier around all cells. In the model, the polar and hydrophilic moieties are on the outside, with the lipophilic or hydrophobic moieties on the inside (see Figure 10.5).

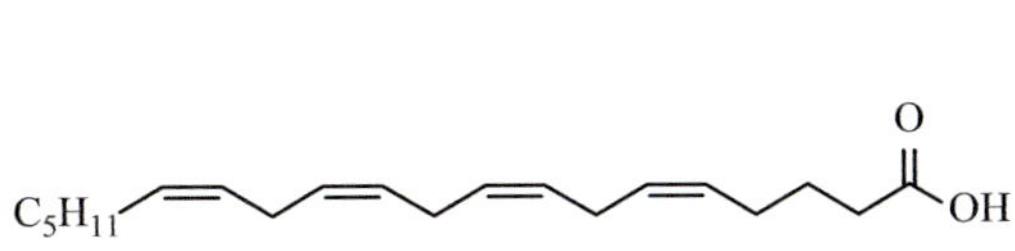

Arachidonic acid

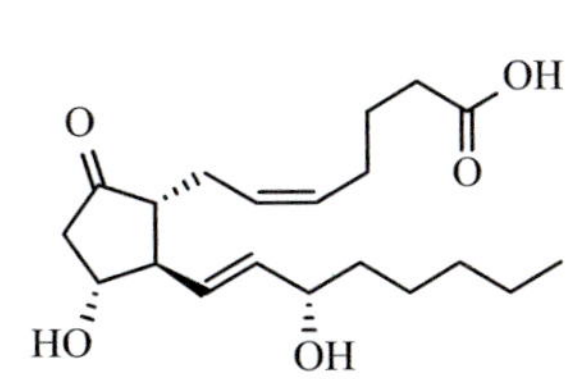

Prostaglandin E_2

Eicosanoids are signaling molecules made by the enzymatic or non-enzymatic oxidation of arachidonic acid or other polyunsaturated fatty acids, such as arachidonic acid. Arachidonic acid is found in membranes and is abundant in the brain, muscles, and liver. Another example of an eicosanoid is prostaglandin E2. Prostaglandins are physiologically active lipids that have diverse hormone-like effects in animals.

10.2 BIOSYNTHESIS AND BIODEGRADATION OF ESTERS

Fatty acid biosynthesis has been studied extensively for nearly 60 years. An accepted mechanism is shown in Figure 10.6.[4a] and begins with acetyl-CoA, which reacts with malonyl-CoA in the presence of β-ketoacyl-ACP synthase to give the corresponding acyl carrier proteins (ACP) derivative the acyl carrier protein. Acetyl-CoA participates in many biochemical reactions in protein, carbohydrate, and lipid metabolism, where it delivers the acetyl group to the citric acid cycle that can be oxidized for energy production. This protein is an important component in fatty acid biosynthesis, with the growing chain bound during synthesis as a thioester at the distal thiol of a 4'-phosphopantetheine moiety, an essential prosthetic group of several ACP involved in pathways of primary and secondary metabolism, including the ACP of *fatty acid synthases. Fatty acid synthase* (EC 2.3.1.85; Sections 4.2 and 4.6) is a multi-enzyme protein that catalyzes fatty acid synthesis. It is not a single enzyme but a whole enzymatic system. Its main function is to catalyze the synthesis of long-chain saturated fatty acids from acetyl-CoA and malonyl-CoA in the presence of NADPH. Indeed, acetyl-S-ACP and malonyl-S-ACP are converted to acetylacetonate-S-ACP in the presence of fatty acid synthase. The stereoselective reduction to the β-hydroxy compound occurs with NADPH. Nicotinamide adenine dinucleotide phosphate is a cofactor used in anabolic reactions that require NADPH. NADPH is the reduced form of $NADP^+$. Elimination of the alcohol moiety gives the conjugated derivative, and further reaction with NADPH gives the butanoyl-S-ACP derivative. This sequence is repeated to generate palmitoleoyl-ACP (Figure 10.6), which is hydrolyzed to give the C16 fatty acid, palmitic acid.

Fatty acids are oxidized by most tissues in the body. Beta oxidation is the metabolic process by which fatty acids are broken down in the cytosol of prokaryotes and in the mitochondria of eukaryotes to generate acetyl-CoA. Fatty acids are oxidized by the sequential removal of two-carbon fragments from the carboxyl end of the acid, following steps of dehydrogenation, hydration, and oxidation to form a β-keto acid, like a reversal of fatty acid synthesis, with a final thiolysis step. This process generates acetyl-CoA, which is converted into ATP, CO_2, and H_2O through the citric acid

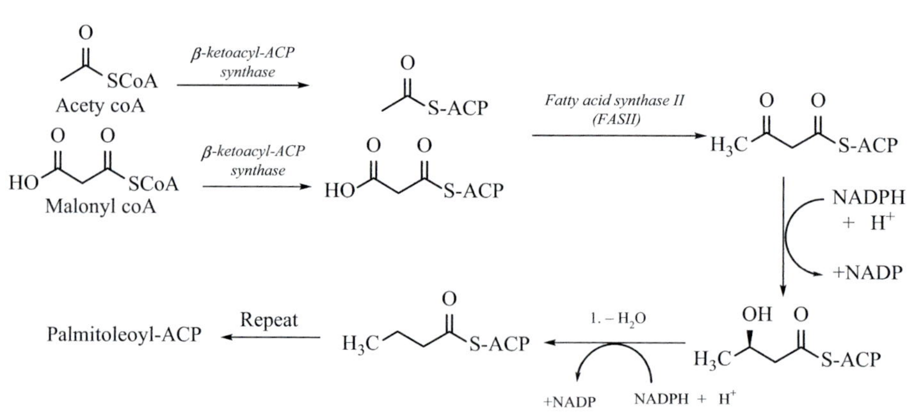

FIGURE 10.6 Biosynthesis of palmitoleoyl-ACP.

[4] (a) Wakil, S.J. *Lipid Res.* **1961**, *2*, 1; Also see (b) Dijkstra, A.J.; Hamilton, R.J.; Hamm, W. *Fatty Acid Biosynthesis. Trans Fatty Acids*, **2008**, Blackwell Pub., Oxford, p. 12; (c) "MetaCyc pathway: superpathway of fatty acids biosynthesis (E. coli)", see https://biocyc.org/META/new-image?type=PATHWAY&object=PWY-6285.

cycle. Therefore, energy is made available by the complete oxidation of palmitic acid when fat is being broken down for energy if there is little or no glucose available.

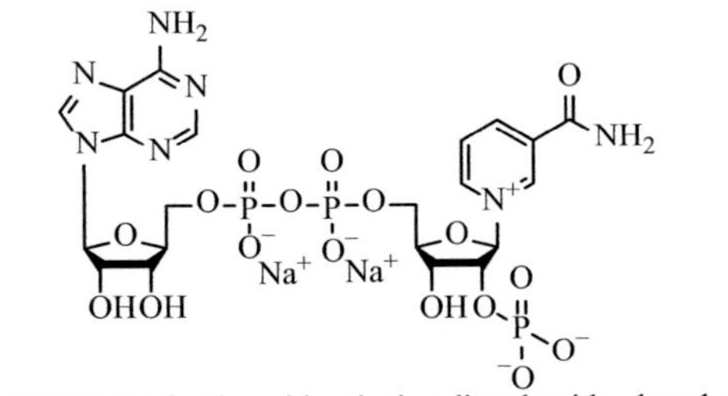

NADPH: Nicotinamide adenine dinucleotide phosphate

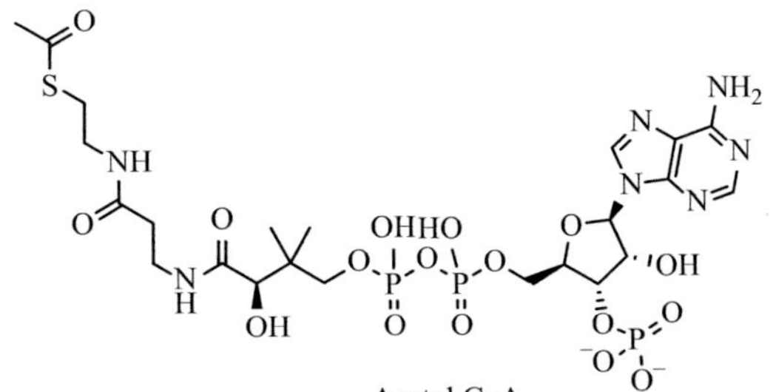

Acetyl CoA

A biopathway for the synthesis of glycerolipids has been reported, as shown in Figure 10.7.[5] The acylation of glycerol-3-phosphate is initiated by *sn-1-glycerol-3-phosphate acyltransferase* (GPAT; EC 2.3.1.n4) in the presence of acetyl-CoA, and the product is lysophosphatidic acid [2-hydroxy-3-(phosphonooxy)propyl palmitate]. The enzymatic reaction to attach a second fatty acid also uses acetyl-CoA but involves *1-acylglycerol-3-phosphate acyltransferase* [AGPAT, also called

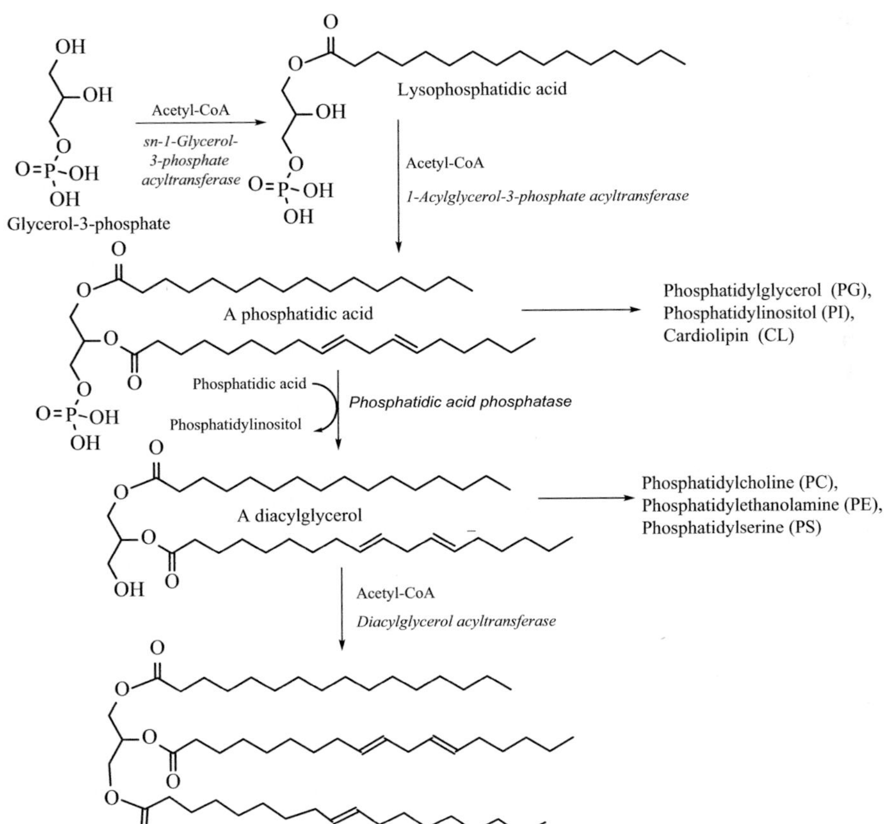

FIGURE 10.7 Glycerolipid synthesis pathway. (Reprinted with permission from Coleman, R.A.; Mashe, D.G. *Chem. Rev.* **2011**, *111*, 6359–6386. Copyright 2011 American Chemical Society.)

[5] Coleman, R.A.; Mashe, D.G. *Chem. Rev.* **2011**, *111*, 6359.

LPA acyltransferase; EC 2.3.1.n3], providing the diacylglyceride, a phosphatidic acid [1-(palmitoyl oxy)-3-(phosphonooxy)propan-2-yl (9*E*,12*E*)-octadeca-9,12-dienoate]. As the precursor of cytidine diphosphate diacylglycerol, phosphatidic acid is the precursor for the synthesis of phosphatidylglycerol (PG), phosphatidylinositol (PI), and cardiolipin (CL). Phosphatidic acid is hydrolyzed by *phosphatidic acid phosphatase* (EC 3.1.3.4; also called *lipin*) to remove the phosphatidyl group to give the diacylglycerol [1-hydroxy-3-(palmitoyloxy)propan-2-yl (9*E*,12*E*)-octadeca-9,12-dienoate], which is the precursor for the synthesis of the major phospholipids in mammalian cells: phosphatidylcholine (PC), phosphatidylethanolamine (PE), and phosphatidylserine (PS). The final esterification step is catalyzed by the enzyme diacylglycerol acyltransferase (DGAT; EC 2.3.1.20) and acetyl-CoA, producing the triacylglycerol 1-(((*E*)-octadec-9-enoyl)oxy)-3-(palmitoyloxy)propan-2-yl (9*E*,12*E*)-octadeca-9,12-dienoate.[5]

The process of lipid metabolism synthesizes and degrades lipid stores and produces the structural and functional lipids characteristic of individual tissues. The cleavage of an ester into its alcohol and carboxylic acid components requires an enzyme known as an *esterase*. An *esterase* is a *hydrolase* enzyme that splits esters into an acid and an alcohol in a chemical reaction with water called hydrolysis. There are myriad esterases, and many were introduced in Chapter 4. A *lipase* is a type of *esterase*, and there are pancreatic lipases as well as mitochondrial lipases. A *lipase* is a water-soluble enzyme that hydrolyzes the ester bonds in water-insoluble lipids, including glycerides. A *triacylglycerol lipase* will cleave a triglyceride, for example. Triglycerides can pass through the stomach into the duodenum, where alkaline pancreatic juice raises the pH and allows hydrolysis of the ester linkages mediated by an enzyme called pancreatic lipase, as well as other esterases.

Triglyceride lipases (EC 3.1.1.3) are lipolytic enzymes that hydrolyze ester linkages, which are required for fat metabolism, including the digestion of dietary fats, the uptake of fats into various tissues, and the mobilization of fats inside cells.[6] The cleavage of dietary triglycerides to free fatty acids and monoacylglycerols must occur before they are absorbed because triglycerides are not absorbed by intestinal enterocytes. In humans, the digestion of dietary triglycerides begins in the stomach, and lipases, secreted by the pancreas, complete fat digestion. Pancreatic *triglyceride lipase* is essential for the efficient digestion of dietary triglycerides. One of the principal *pancreatic lipases* that catalyzes the hydrolysis of fats to fatty acids and glycerol is known as *steapsin*. Cleavage occurs at the C1 and C3 positions of a triglyceride, and "the hydrolysis depends on bile salts (bile salts are carboxylic acid salts linked to a steroid). In the basic environment of the duodenum, the fatty acid is converted to the salt, which is emulsified by the bile salts".[6] In other words, the bile salts act as detergents to emulsify the triglycerides and form micelles, which facilitate the cleavage of an ester linkage. Bile salts are commonly amide derivatives of steroids, and examples are taurocholic acid and glycocholic acid. Cholesterol is a steroid alcohol, a sterol, and an example of another type of lipid that is found in all animal cells. It is an essential structural component of all animal cell membranes. Aldosterone is a steroid hormone made by the adrenal cortex (the outer layer of the adrenal gland) that helps control the balance of water and salts in the kidneys. It helps retain sodium and release potassium from the body. The release of too much aldosterone leads to high blood pressure and a buildup of fluid in body tissues. Cortisone helps the body respond to stress or danger. It increases the body's metabolism of glucose and controls blood pressure. Cortisol is the primary stress hormone, increasing glucose in the bloodstream. It enhances the brain's use of glucose and increases the availability of substances in the body that repair tissues. Boldenone undecylenate is an androgen and anabolic steroid used in veterinary medicine. It is used to improve the physique and performance of horses. It has strong anabolic effects and moderate androgenic effects. Prednisone is a glucocorticoid used to treat rashes, inflammatory bowel disease, asthma, and other conditions. It suppresses the immune system and decreases inflammation in conditions such as asthma, COPD, and rheumatologic diseases. It is also used to treat high blood calcium associated with cancer and adrenal insufficiency.

[6] Lowe, M.E. *Ann. Rev. Nutr.* **1997**, *17*, 141.

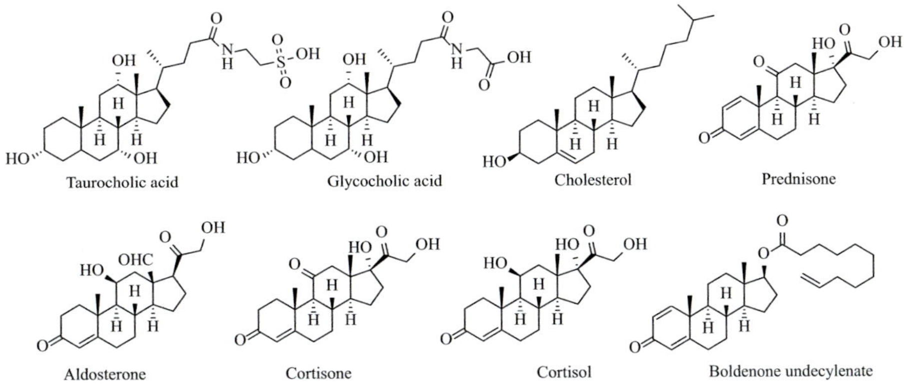

The long-chain fatty acid micelles are carried to the surface of epithelial cells, and reaction with glycerol forms new triglycerides that aggregate with lipoproteins to form particles called chylomicrons.[7] Chylomicrons are lipoprotein particles that transport dietary lipids from the intestines to other parts of the body. It is noted that *pancreatic lipase* is secreted in its final form, but it becomes efficient only in the presence of colipase in the duodenum. Colipase is a protein and a cofactor for *pancreatic lipase*. Colipase forms a stoichiometric complex with the lipase and binds to the bile salt-covered triacylglycerol interface, allowing the enzyme to anchor itself to the water-lipid interface.

The enzyme-mediated hydrolysis of an ester linkage, such as those found in triglycerides, can be illustrated by the model in Figure 10.8,[8a] using methyl acetate as a typical ester. The *canonical*

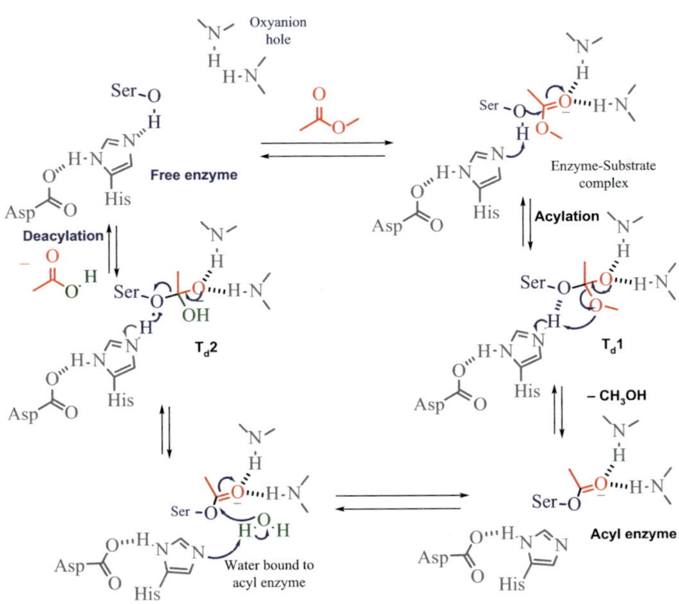

FIGURE 10.8 Canonical esterase mechanism for the hydrolysis of methyl acetate, a typical ester. (Reprinted with permission from Rauwerdink, A.; Kazlauskas, R.J. *ACS Catal.* **2015**, *5*, 6153–6176. Copyright 2015 American Chemical Society; Reprinted with permission from Smith, A.J.T.; Müller, R.; Toscano, M.D.; Kast, P.; Hellinga, H.W.; Hilvert, D.; Houk, K.N. *J. Amer. Chem. Soc.* **2008**, *130,* 15361–15373. Copyright 2008 American Chemical Society.)

[7] Garrett, R.H.; Grisham, C.M. *Biochemistry*, Saunders, Fort Worth, TX, **1995**, pp. 734 and 735.

[8] (a) Rauwerdink, A.; Kazlauskas, R.J. *ACS Catal.* **2015**, *5*, 6153. Also see (b) Smith, A.J.T.; Müller, R.; Toscano, M.D.; Kast, P.; Hellinga, H.W.; Hilvert, D.; Houk, K.N. *J. Amer. Chem. Soc.* **2008**, *130,* 15361.

esterase mechanism uses the core Ser-His-Asp catalytic machinery, and this mechanism uses an acyl-serine enzyme intermediate. The free acyl enzyme coordinates with the Asp-His-Ser residues and the two main-chain amide N–H's oxyanion to bind the substrate, methyl acetate, in the active site. Attack of serine on the ester carbonyl carbon yields the Td1 tetrahedral intermediate as the histidine residues react as a base to deprotonate the serine. Reformation of the carbon-oxygen double bond leads to the loss of methanol via reaction with the histidine residue and the formation of the acyl enzyme intermediate, which is a serine ester. In the next step, water binds to the acyl enzyme intermediate, and the histidine residue again reacts as a base to deprotonate the water, with concomitant attack at the carbonyl carbon of the acyl enzyme to form tetrahedral intermediate, Td2. As the carbon–oxygen double bond is regenerated, acetic acid is lost and the free enzyme is restored.[8a]

10.3 BIOSYNTHESIS OF LIPIDS

It is clear from the previous sections that lipids represent many different structural types. Therefore, only a few of these structures will be discussed to give an idea of the diversity in this area. This section will focus on the biopathways associated with fatty acids, triglycerides, phospholipids, sphingolipids, and steroids. The biopathways of glycerophospholipids, glycerolipids, sphingolipids, and cholesterol have been discussed in detail.[9]

An enzyme is involved in the synthesis of triglycerides: RCO-S-CoA + α,β-diglyceride -+ triglyceride + CoASH.[10] This enzyme and phosphorylcholine-glyceride transferase are important for many glycerides as acceptors. It has been shown that L-α-glycerophosphate, phosphatidic acids, and α,β-diglycerides are intermediates in the synthesis of triglycerides.

Dietary bioactive lipids are one of the three primary nutrients. *Desaturase* and *elongase* are actively involved in the fatty acids' synthesis metabolic pathway.[11] Δ9-Desaturase is a central enzyme in the biosynthesis of long-chain monounsaturated fatty acids from long-chain saturated fatty acids, catalyzing stearic acid and palmitic acid to oleic acid and palmitoleic acid (Figure 10.9).[11] Several Δ9-desaturases have been identified that share three histidine-conserved boxes that would be the catalytic position of Δ9 fatty acid desaturase.[11]

The glycerophospholipids phosphatidylcholine (PC) and phosphatidylethanolamine (PE) account for greater than 50% of the total phospholipid species in eukaryotic membranes and thus play major roles in the structure and function of those membranes.[12] In most eukaryotic cells, PC and PE are synthesized by an aminoalcoholphosphotransferase reaction, which uses *sn*-1,2-diradylglycerol and either CDP-choline (cytidine 5′-diphosphocholine or citicoline) or CDP-ethanolamine (cytidine 5′-phosphoethanolamine), respectively. This is the last step in a biosynthetic pathway known as the Kennedy pathway, shown in Figure 10.10.[13] In many eukaryotes, the CDP-ethanolamine and CDP-choline branches of the Kennedy pathway[13] produce phosphatidylcholine and phosphatidylethanolamine, with ethanolamine and choline/ethanolamine phosphotransferases (EPT and CEPT, respectively) catalyzing the final reactions in the respective pathways. Although the site of PE and PC synthesis is commonly known to be the endoplasmic reticulum, detailed information on the localization of the different phosphotransferases is lacking.

Sphingolipids are a class of lipids with 18-carbon amino-alcohol backbones, synthesized in the endoplasmic reticulum from nonsphingolipid precursors.[14] They play an essential role in mammalian systems. Sphingosine, phytosphingosine, and dihydrosphingosine are the backbones that lead to further structural complexity. For example, phosphorylation of the C1 hydroxyl group leads

---

[9] Jacquemyn, J.; Cascalho, A.; Ro Goodchild, R.E. *EMBO Rep.* **2017**, *18*, 1905. https://doi.org/10.15252/embr.201643426.

[10] Weiss, S.B.; Kennedy, E.P.; Kiyasu, J.B. *J. Biol. Chem.* **1960**, *235*, 40.

[11] Zhuang, X.-Y.; Zhang, Y.-H.; Xiao, A.-F.; Zhang, A.-H.; Fang, B.-S. *Front. Nutr.* **2022**, *9* |Article 851402. https://doi.org/10.3389/fnut.2022.851402.

[12] Farine, L.; Niemann, M.; Schneider, A.; Bütikofer, P. *Sci. Rep.* **2015**, *5*, 16787. https://doi.org/10.1038/srep16787.

[13] Gibellini, F.; Smith, T.K. *IUBMB Life* **2010**, *62*, 414.

[14] Gault, C.R.; Obeid, L.M.; Hannun, Y.A. *Adv. Exp. Med. Biol.* **2010**, *688*, 1.

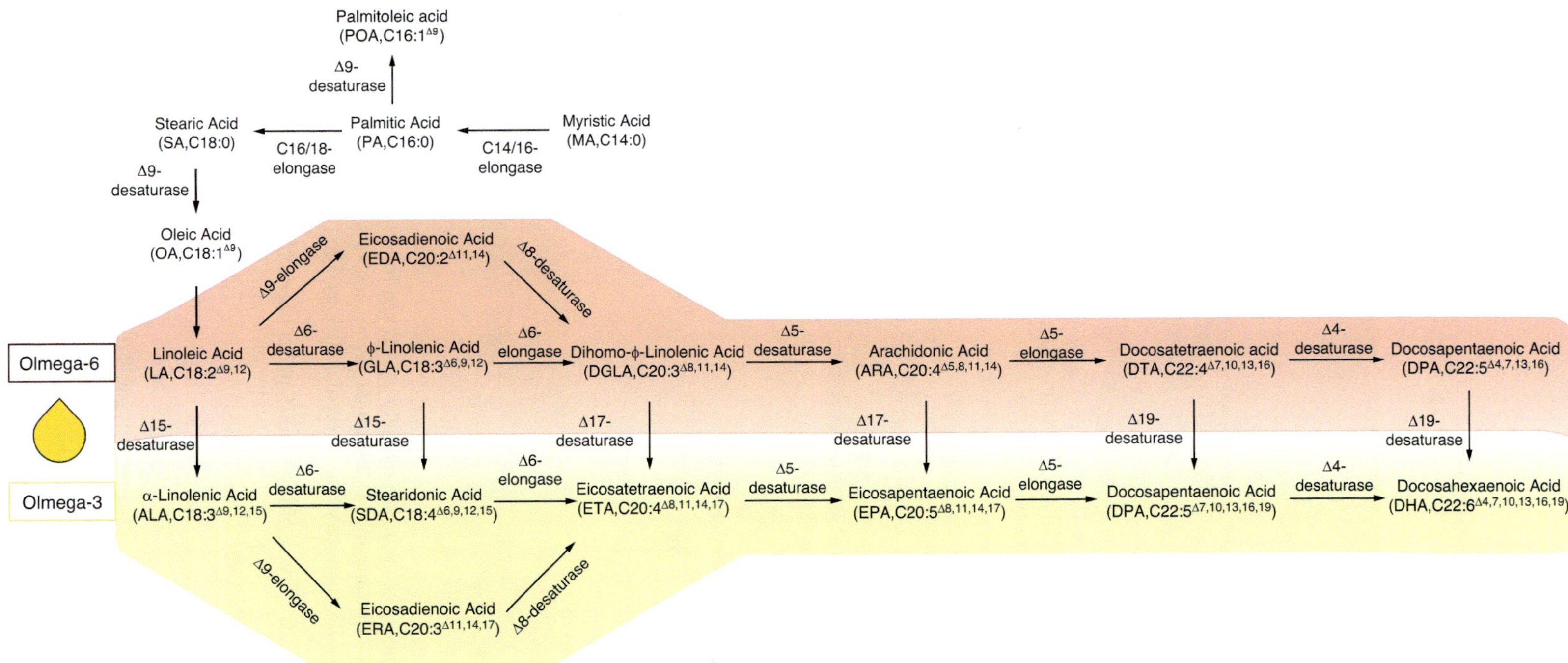

FIGURE 10.9 Key enzymes in the fatty acids' synthesis pathway. (Taken from *Front. Nutr.* **2022**, *9* |Article 851402. Fig. 1, p.3, doi:10.3389/fnut.2022.851402. Open Access.)

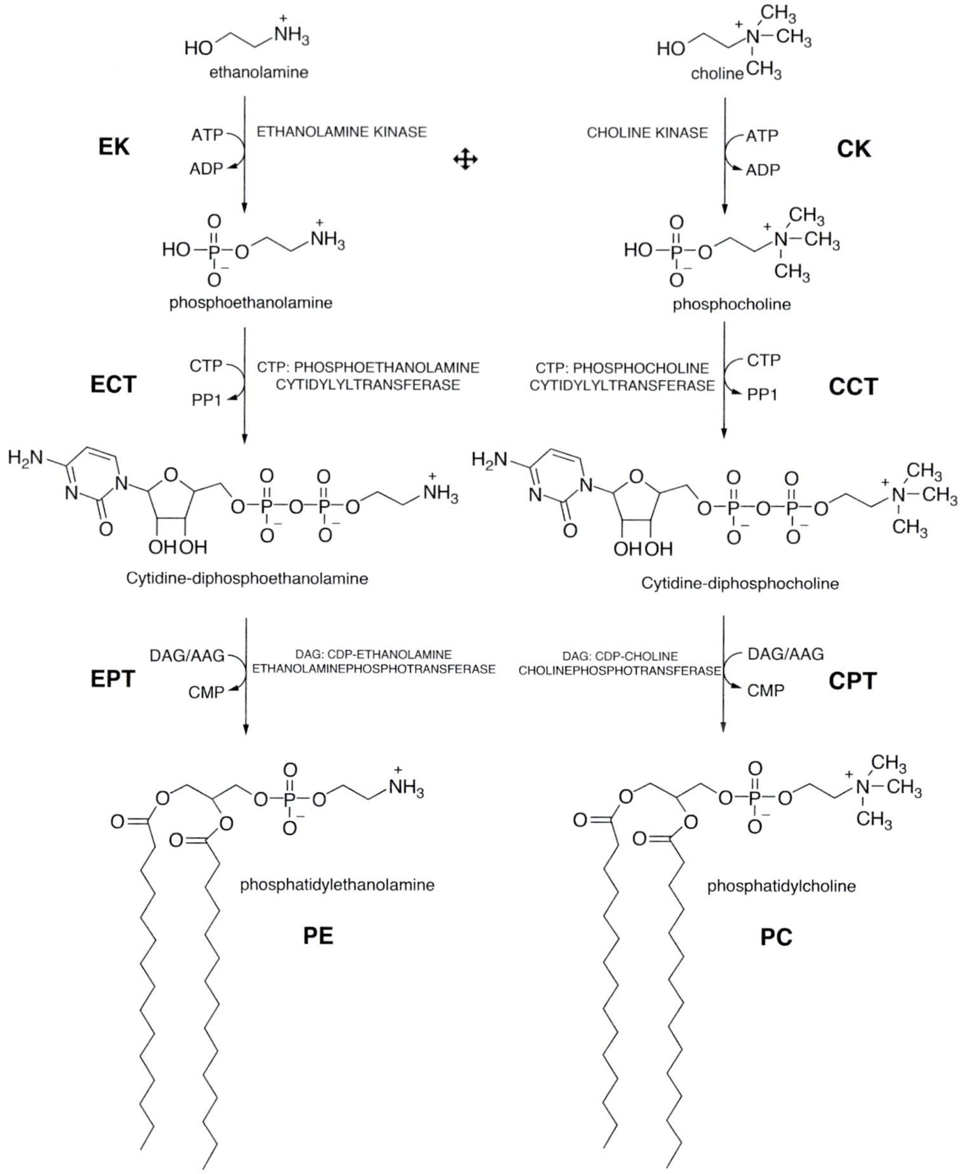

FIGURE 10.10 Two branches of the Kennedy pathway. (Taken from Gibellini, F.; Smith, T.K. *IUBMB Life* **2010**, *62*, 414. Figure 1, p. 415 Copyright © 2010 with permission from John Wiley and Sons. Copyright © 2010 International Union of Biochemistry and Molecular Biology, Inc.)

to sphingosine-1-phosphate, phytosphingosine-1-phosphate, and dihydrosphingosine-1-phosphate. Acylation of sphingosine, phytosphingosine, or dihydrosphingosine with the appropriate acyl-CoA molecules via the appropriate ceramide synthases gives ceramide, phytoceramide, or dihydroceramide.[14] The sphingolipid metabolic network is illustrated in Figure 10.11.[14]

The biosynthesis of the steroids lanosterol and cycloartenol arises from the cyclization of squalene. Steroid biosynthesis is an anabolic pathway that produces a variety of steroids from dimethylallyl and isopentenyl pyrophosphate precursors by the mevalonate pathway, outlined in Figure 10.12.[15] Squalene is the biosynthetic precursor to lanosterol, which is the precursor to

[15] Goldstein, J.L.; Brown, S.B. *Nature* **1990**, 343, 425. See https://en.wikipedia.org/wiki/Steroid#Biosynthesis_and_metabolism

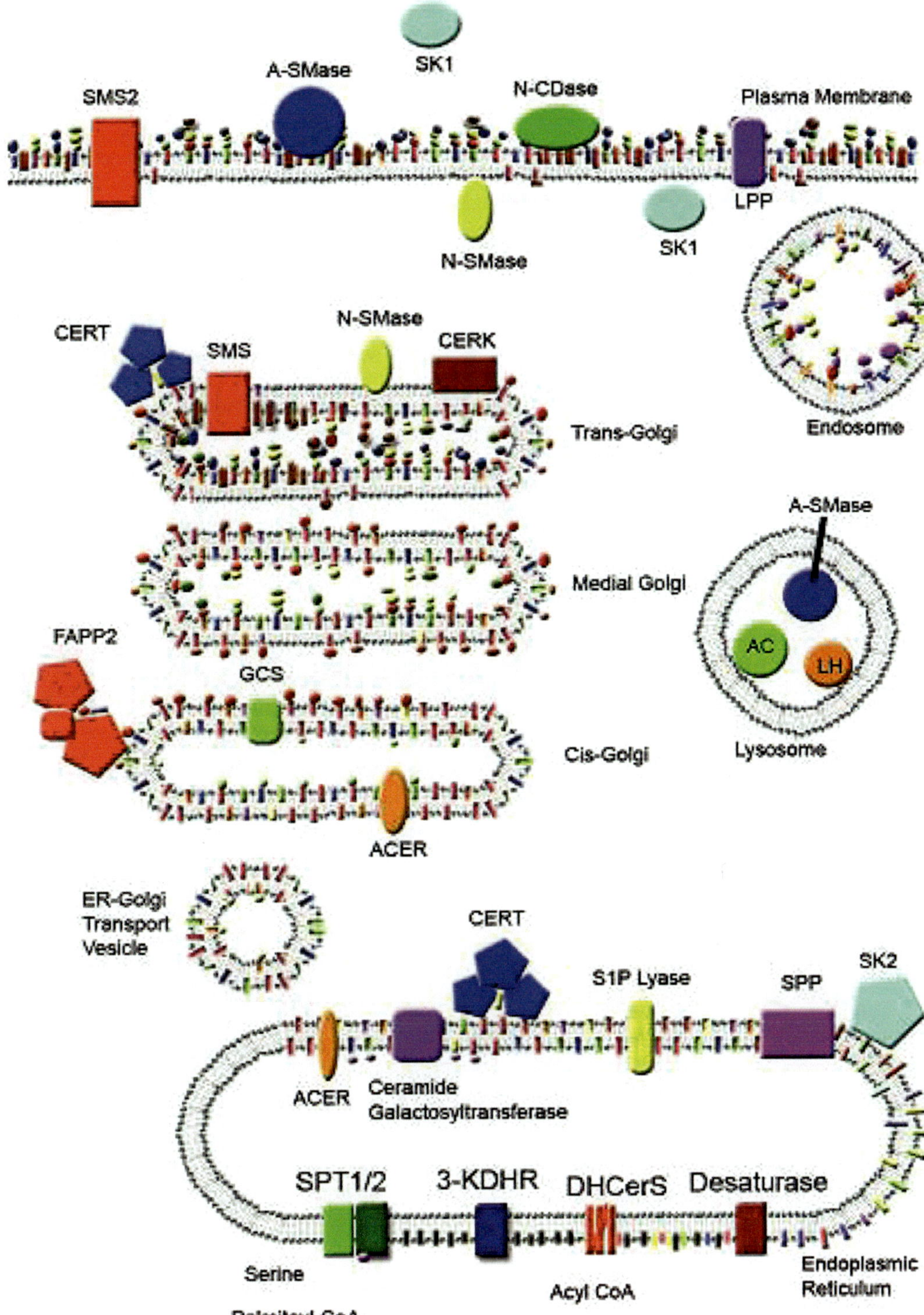

FIGURE 10.11 The sphingolipid metabolic network. (Taken from Gault, C.R.; Obeid, L.M.; Hannun, Y.A. *Adv. Exp. Med. Biol.* **2010**, *688*, 1. Figure 2, p. 28. Copyright © 2010 with permission from Springer Nature. Copyright © 2010, Landes Bioscience and Springer Science Business Media.)

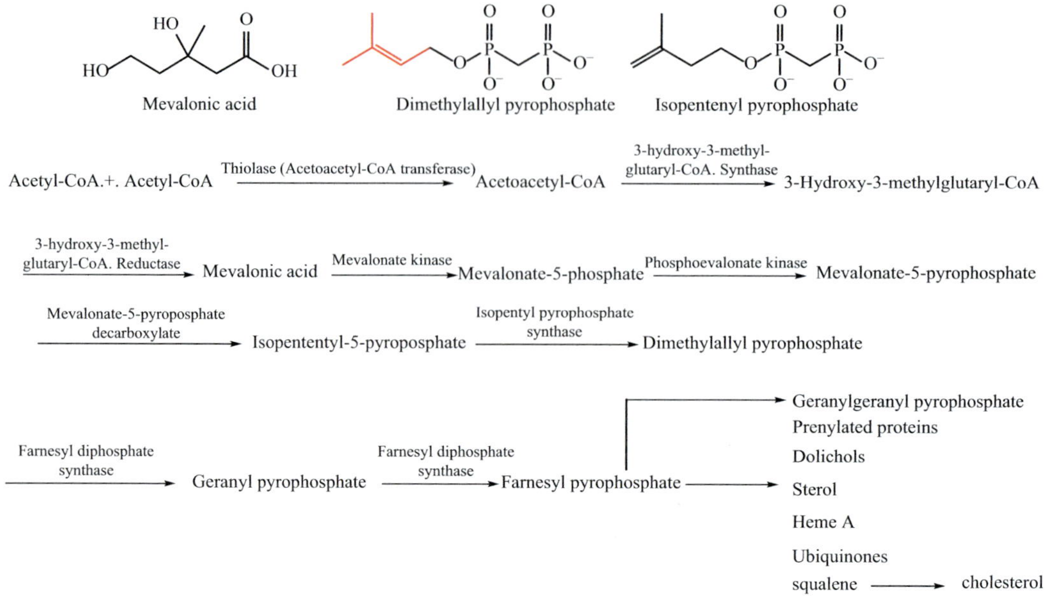

FIGURE 10.12 The Mevalonate pathway.

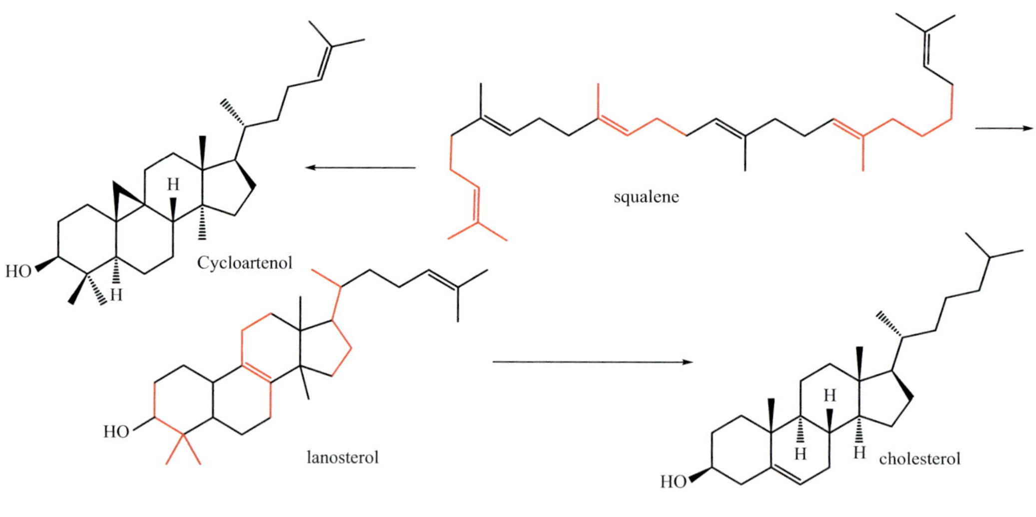

FIGURE 10.13 Lanosterol is the biosynthetic precursor to cholesterol, both derived from squalene.

cholesterol, as shown in Figure 10.13.[16] Cycloartenol is a steroid that is the starting point for the synthesis of almost all plant steroids.[17]

The biochemical pathways for the biosynthesis and metabolism of steroid hormones are well defined. Steroid hormone biosynthesis and metabolism by the liver and peripheral tissues include the serum and urine steroid metabolomes.[18] This work presents a schematic overview of adrenal

[16] See (a) Schroepfer G.J. *Ann. Rev. Biochem.* **1981**, *50*, 585–621; (b) Chatuphonprasert, W.; Jarukamjorn, K.; Ellinger, I. *Front. Pharmacol.* **2018**, *9*, *1027*. https://doi.org/10.3389/fphar.2018.01027.

[17] Schaller, H. *Prog. Lipid Res.* **2003**, *42*, 163.

[18] Schiffer, L.; Barnard, L.; Baranowski, E.S.; Gilligan, L.C.; Taylor, A.E.; Arlt, W.; Shackleton, C.H.L.; Storbeck, K.-H. *J. Steroid Biochem. Mol. Biol.* **2019**, *194*, 105439.

steroidogenesis and peripheral modulation of steroid bioactivity[18] that shows the enzymatic evolution of several steroids from cholesterol.[18]

Enzymatically oxidized lipids are biomolecules that function as key signaling mediators and hormones.[19] Examples include polyunsaturated fatty acid oxygenated derivatives such as "oxylipins," endocannabinoids, oxidized phospholipids, or cholesterol derivatives such as oxysterols, steroid hormones, and bile acids. These compounds regulate various cellular and physiological processes, ranging from metabolism and cell death to inflammation and the immune response. Enzymes that prepare these compounds include lipoxygenases, cyclooxygenases, cytochrome P450s, and aldo-keto reductases. Many of the biopathways to these oxidized derivatives are known.[19]

10.4 CARBOXYLIC ACIDS AND ESTERS

Carboxylic acids are functional groups with a carbonyl (C=O) and a hydroxyl (OH) group attached to the carbonyl carbon. This unit is known as a *carboxyl group*, also written as –COOH or $-CO_2H$, and it is the major structural feature of the class of organic molecules known as *carboxylic acids*. This structural arrangement is shown in the box on ethanoic acid (commonly known as acetic acid; Figure 10.14). The polarization induced by the electronegative carboxyl oxygen makes the carboxyl carbon atom positive, which leads to the oxygen of the O–H unit being negatively polarized, and the hydrogen positively polarized; i.e., the proton is acidic. The O–H proton of a carboxylic acid is much stronger (pK_a 1–5) than the O–H proton of an alcohol (pK_a 16–18).

Fatty acids are long-chain carboxylic acids, and there are several classes. Certain terminology is used, particularly for fatty acids that are found in nature. The carboxylic acid end is polar, associated with water solubility, and is labeled the delta (Δ) end of the fatty acid, as shown in Figure 10.15. The methyl end is nonpolar, more soluble in oils, and is labeled the omega (Ω) end of the fatty acid.

The so-called essential fatty acids are those required in the diet because they cannot be synthesized by human cells. Common naturally occurring fatty acids include the C_{14} acid (myristic acid), the C_{16} acid (palmitic acid), the C_{18} acid (stearic acid), and the C_{20} acid (arachidic acid). Myristic acid (tetradecanoic acid) is found in nutmeg. Palmitic acid (hexadecanoic acid) is found in butter, cheese, and meat. It is also a component of palm oil. Other examples are α-linolenic acid, eicosapentaenoic acid, and docosahexaenoic acid. All these fatty acids are examples of Ω3-fatty acids, where the first double bond encountered from the methyl group distal to the COOH moiety is found at the third carbon from that methyl group.

Unsaturated fatty acids (Figure 10.15) are characterized by having an alkene, diene, or polyene unit in the long carbon (lipophilic) chain. Additional nomenclature for unsaturated fatty acids is used to identify the site of C=C units relative to the terminal methyl group (the Ω end). Because the closest double-bond to the methyl group in oleic acid is nine carbon atoms away, oleic acid is called an omega-9 (Ω-9 or *n*-9) fatty acid. In α-linolenic acid, the double bond closest to the methyl group is only three carbons away, so it is an omega-3 (Ω-3 or *n*-3) fatty acid. Several other acids are suitably labeled in Figure 10.15. Omega-3 fatty acids are known to lower triglycerides and blood pressure. Omega-9 fatty acids reduce the risk of cardiovascular disease and stroke. These fatty acids may help eliminate plaque buildup in arteries by increasing high-density lipoproteins (HDL or good cholesterol) and lowering low-density lipoproteins (LDL or bad cholesterol). Common unsaturated fatty acids are palmitoleic acid (C16), oleic acid (C18), linolenic acid (C18), α-linolenic acid (C_{18}), and γ-linolenic acid (C_{18}). Other examples include arachidonic acid (C20), erucic acid (C22; found in mustard seed), and nervonic acid (C24; important for the biosynthesis of nerve cell myelin).

[19] Hajeyah, A.A.; Griffiths, W.J.; Wang, Y.; Finch, A.J.; O'Donnell, V.B. *Front. Endocrinol.* **2020**, *11*, 591819.

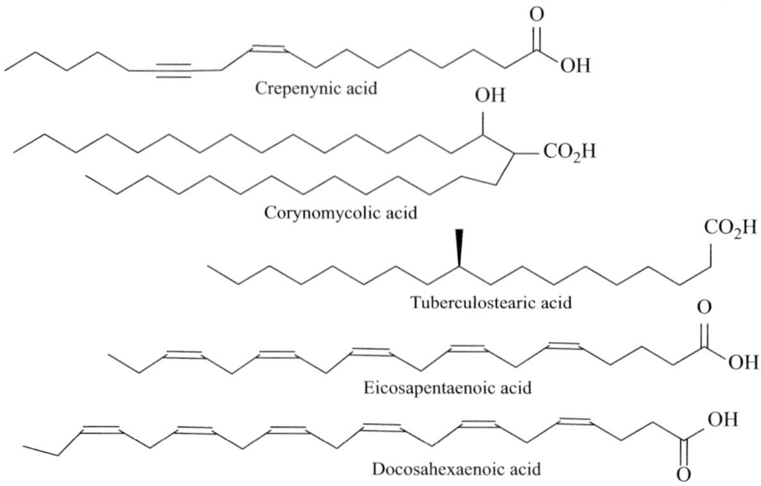

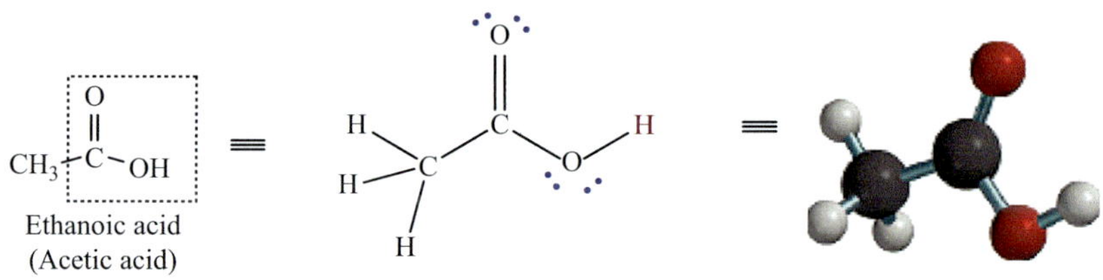

FIGURE 10.14 Ethanoic acid.

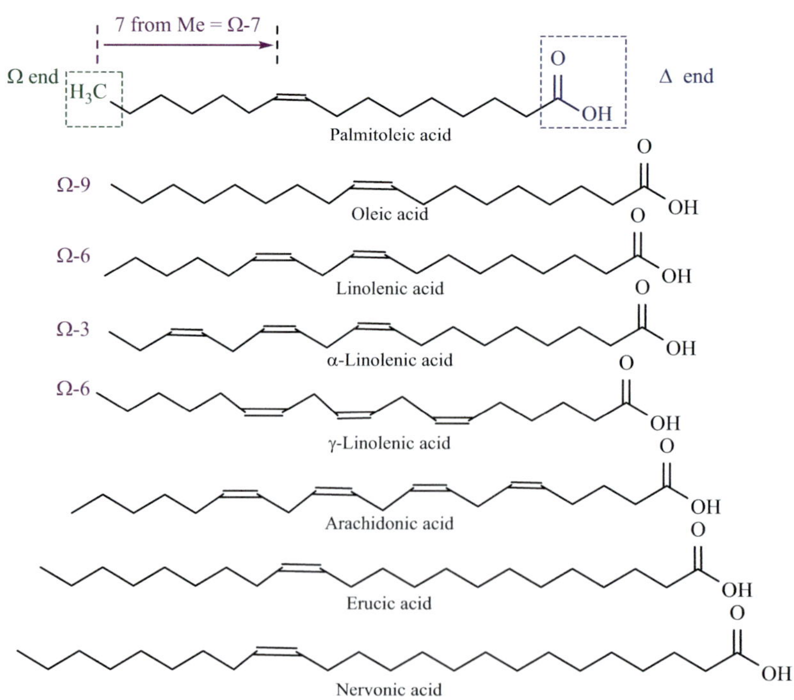

FIGURE 10.15 Unsaturated fatty acids.

The essential fatty acids are unsaturated fatty acids that cannot be made in the body but are required for normal, healthy functioning. These fatty acids must be obtained from foods (e.g., nuts, sunflower oil or other vegetable oils, and oil-rich fish). Wild salmon, for example, is rich in Ω-3 fatty acids. A deficiency of these essential fatty acids may result in hyperactivity, reduced growth, or, in extreme cases, death. There are two families of essential fatty acids: the Ω-3 and Ω-6 fatty acids. Nutritionally essential Ω-3 fatty acids are α-linolenic acid, eicosapentaenoic acid, and docosahexaenoic acid. The human body cannot synthesize ω-fatty acids using its biochemical machinery, but it can form eicosapentaenoic acid and docosahexaenoic acid from α-linolenic acid. Stearic acid is used in cosmetics, as a dietary supplement, and in candle making. Arachidic acid (eicosanoic acid) is found in peanut oil as well as in fish oil.

Linoleic acid is an essential fatty acid, and it is used by the body to produce arachidonic acid, which is physiologically significant because it is the precursor for prostaglandins. Prostaglandins activate the inflammatory response, the production of pain, and fever. They are implicated in other biological functions, and thromboxane A2 is one of the prostaglandins that stimulates constriction and clotting of platelets. Note that Ω-9 fatty acids are not classified as essential fatty acids because they can be created by the human body from unsaturated fat (see below) and are therefore not essential in the diet.

There are four important derivatives of carboxylic acids in which the OH unit in RCOOH is replaced by a halogen, $-O_2CR$, –OR, or $-NR_2$, all attached to a carbonyl unit. If the OH group in RCOOH is replaced by an OR′ group (from an alcohol), it is called an *ester* or a carboxylic ester. An ester is essentially a combination of a carboxylic acid and an alcohol.

Acid chloride   Acid anhydride   Ester   Amide

Esters are essentially a combination of a carboxylic acid and an alcohol. The "acid part" is named after the parent carboxylic acid, and the "alcohol part" is named after the alcohol. For example, an ester with a methanol component is a methyl ester, and an ester with a propanol component is a propyl ester. The alcohol part of the name is followed by the acid part, where the -oic acid part of the name is replaced with -*oate*. Ethyl hexanoate is one example [from ethyl alcohol (ethanol or EtOH) and hexanoic acid], and ethyl ethanoate ($MeCO_2Et$) is another example. Ethyl ethanoate is a special case in the sense that it is derived from ethanoic acid, which has the common name acetic acid. Therefore, this ester has a common name that is used most often: ethyl acetate. Note the "shorthand" way of writing ethyl acetate (EtOAc), where Et is ethyl and OAc is the acetate unit, O_2CCH_3.

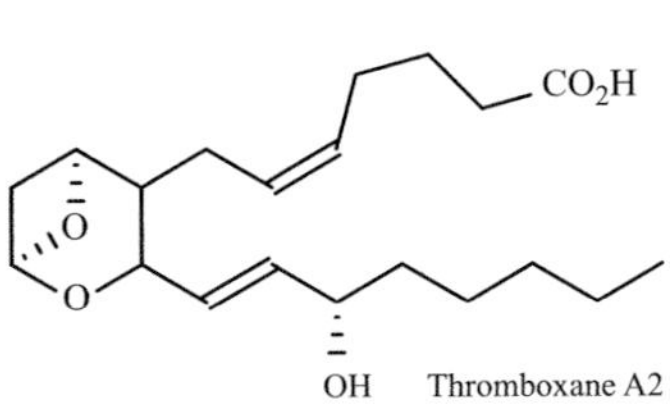

Thromboxane A2

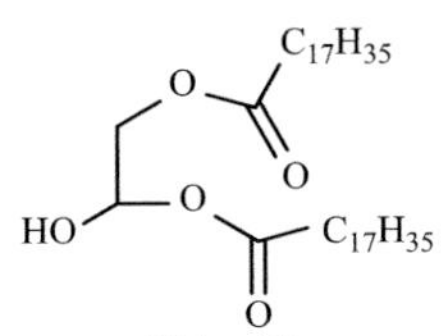

Diglyceride

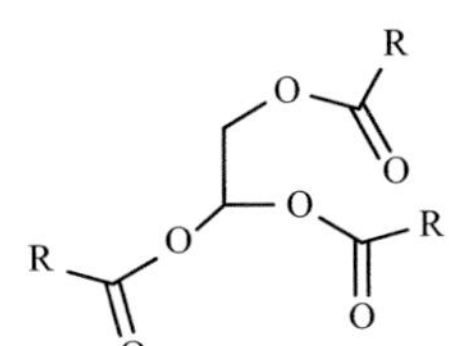

Triglyceride

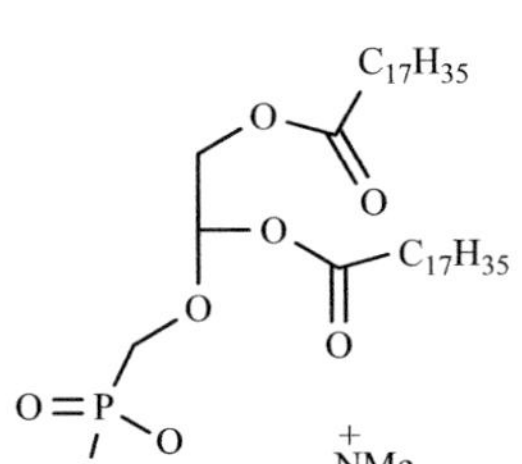

Phosphatidylcholine (lecithin)

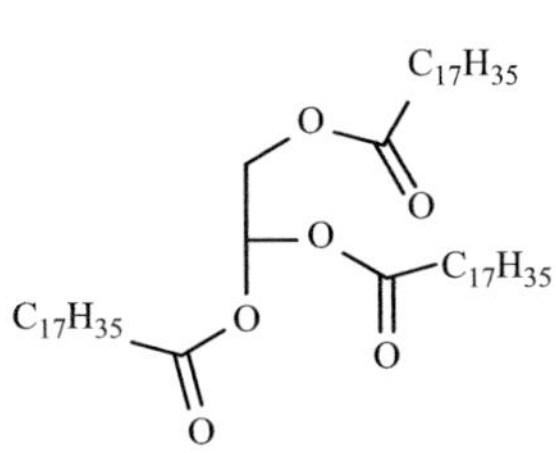

Stearate triglyceride

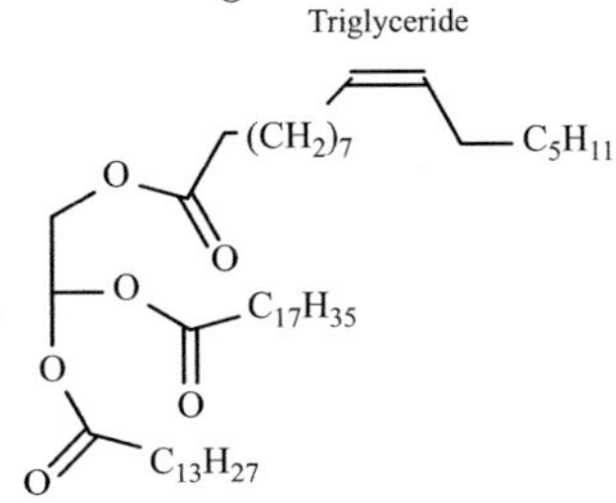

Myristic acid, stearic acid and palmitoleic acid triglyceride

Esters are prevalent in the body, often in the form of triglycerides. Polyketides, with the generic formula $-[CH_2CO]_n-$, are an important class of naturally occurring compounds that include fatty acids and prostaglandins. Esters of 1,2,3-propanetriol (glycerol) are generated by the formation of mono-, di-, or triesters from fatty acids. Diesters of glycerol are known as diglycerides or triglycerides. If these compounds (known as lipids) are solids, they are called *fats*, whereas if they are liquids, they are known as *oils*. An important diglyceride is phosphatidic acid (with stearic acid ester units called stearates), which is the parent compound for glycerol-based phospholipids. One example is phosphatidylcholine (also known as lecithin), which contains the trimethylammonium ethanolamine unit. Lecithin is used as an emulsifier, and it keeps cocoa and cocoa butter in a candy bar from separating, for example. It is produced by the liver and is a building block of cell membranes (see Figure 10.4). Lecithin is also found in the protective sheaths surrounding brain cells.

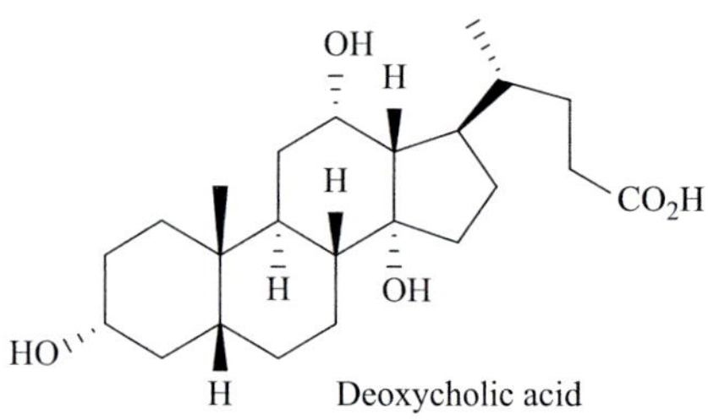

Deoxycholic acid

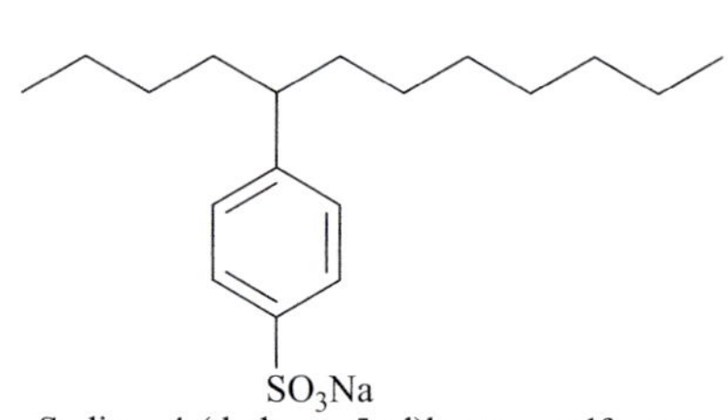

Sodium 4-(dodecan-5-yl)benzenesulfonate

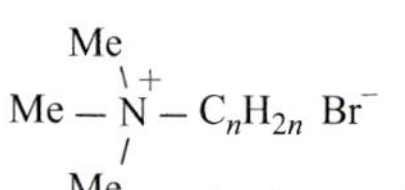

A mixture of n = 12, 14, 16
cetrimide

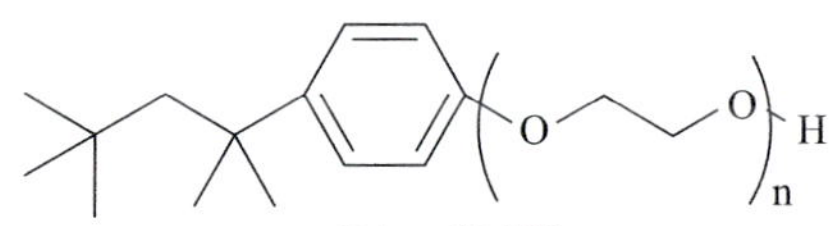

Triton X-100

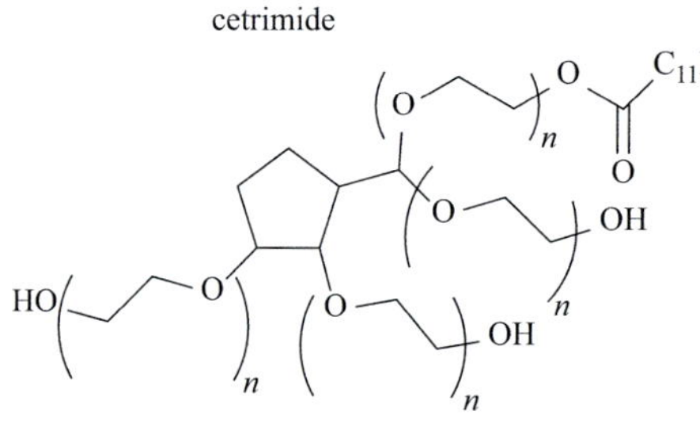

Tween-20 (polysorbate-20)

Most natural fats and oils are mixed triglycerides (triesters of glycerol), in which the fatty acid constituents are different (in the example, the fatty acids are myristic acid, stearic acid, and palmitoleic acid).[20] A simple triglyceride usually has three identical fatty acids, such as three stearic acid units.[20] Naturally occurring triglycerides are usually composed of fatty acids with 16, 18, and 20 carbons. The fats obtained from the food humans eat are the usual source of the triglycerides found in our blood plasma, and an excess of triglycerides in blood plasma is called hypertriglyceridemia. An elevated level of triglycerides is linked to the occurrence of coronary artery disease.

10.5 NITRATE ESTERS, SULFATE ESTERS, AND PHOSPHATE ESTERS

All discussions of acid derivatives in the preceding sections and chapters focus on "organic acids" (e.g., carboxylic acids and sulfonic acids). Notably missing are derivatives of the mineral acids nitric, sulfuric, and phosphoric acid. For the most part, such acid derivatives are used sparingly in organic chemical transformations. Phosphorus derivatives, however, are widely observed in

[20] Garrett, R.H.; Grisham, C.M. *Biochemistry,* Saunders, Fort Worth, TX, ***1995***, pp. 279–284

biological systems. This section offers a brief introduction to some of the organic chemistry of the mineral acids nitric, sulfuric, and phosphoric acid.

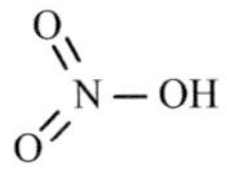

Nitric acid

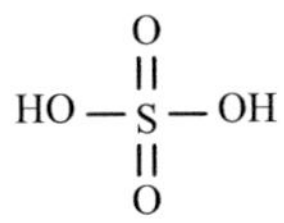

Sulfuric acid

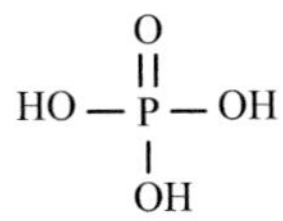

Phosphoric acid

The structures of the mineral acids suggest that replacing one or more OH units will lead to acid derivatives. Esters are possible for all three mineral acids. Nitric acid yields a nitrate ester, methyl nitrate. Nitrate esters are often unstable, and the trinitrate ester of glycerol (1,2,3-trinitroxypropane) is the well-known explosive nitroglycerin. Apart from their explosive properties, highly specialized nitrate esters (e.g., nitroglycerin) are known to relax vascular smooth muscle, which leads to vasodilation (relaxation of the muscle wall of blood vessels that leads to widening of those vessels). Indeed, nitroglycerin and related nitrate esters are important compounds used to treat episodes of angina (chest pain) in people who have coronary artery disease (narrowing of the blood vessels that supply blood to the heart).[21] Amide derivatives, $O_3N–NH_2$, are known as nitramides, but they will not be discussed.

Acid derivatives of sulfuric acid are more complicated because there are two acidic protons and two OH units. If one OH unit is replaced with Cl, this is the acid chloride equivalent, sulfurochloridic acid. However, this compound is more commonly known as chlorosulfonic acid, and its chemistry is related more to the sulfonic acids (RSO_2OH). The amide derivative is known as sulfamic acid, which is used in the manufacture of some sweetening agents and in some pharmaceutical preparations. The chemistry of sulfamic acid derivatives will not be discussed further. The ester derivatives of sulfuric acid include the monoesters and diesters, represented by methyl hydrogen sulfate and dimethyl sulfate. Dimethyl sulfate is an alkylating agent that reacts with alcohols to give ethers. Organic sulfates are used in some chemical transformations and in pharmaceutical preparations, but this chemistry will not be discussed further.

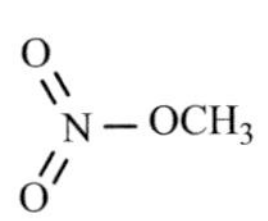

Methyl nitrate

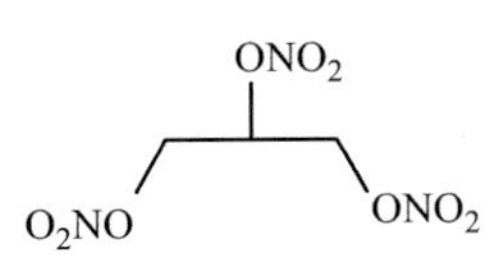

Propane-1,2,3-triyl trinitrate
(nitroglycerine)

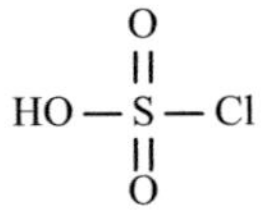

Sulfurochloridic acid
(Chlorosulfonic acid)

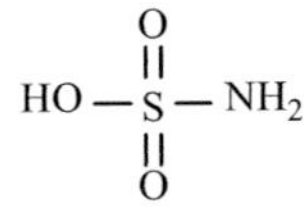

Sulfamic acid

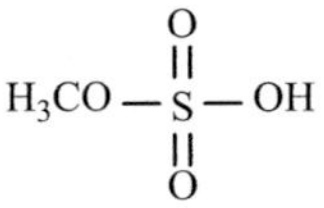

Methyl hydrogen sulfate

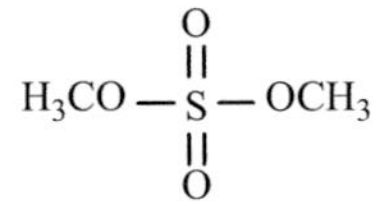

Dimethyl sulfate

Apart from their use as chemical reagents, sulfates are important biological compounds. One example is chondroitin sulfate, a biopolymer that is an important structural component of cartilage. The sulfate unit is marked in red. Chondroitin sulfate has been used as a dietary supplement for the treatment of osteoarthritis. Another important biomolecule is heparin, which has both sulfate units

[21] Katzung, B.G.; Chatterjee, K. in *Basic and Clinical Pharmacology* (Katzung B.G., Ed), **1989**, Appleton and Lange, Norwalk, CT, p. 1017.

(in red) and sulfamic acid units (in green). It is used as an injectable anticoagulant, and it is effective in the prevention of deep vein thrombosis and pulmonary emboli.[22]

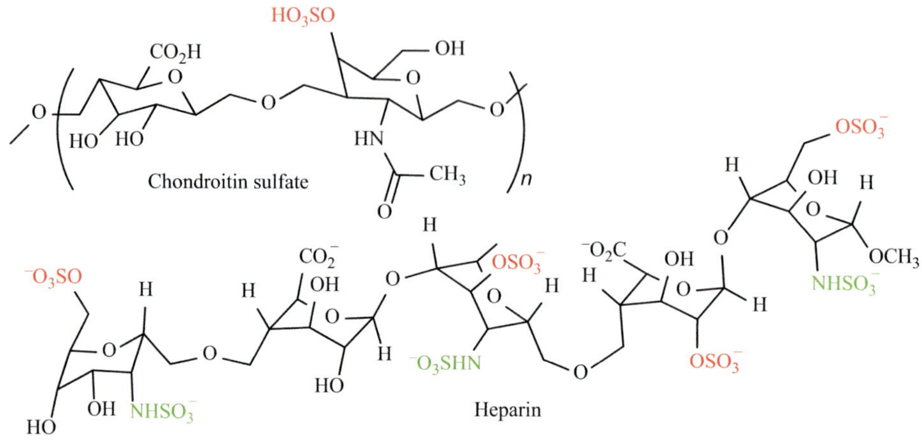

An important reason for this discussion is to introduce derivatives of phosphoric acid. An acid chloride derivative is $ClPO(OH)_2$, chlorophosphoric acid, and an amide derivative is phosphoramic acid, $H_2N–PO(OH)_3$. These particular acid derivatives will not be discussed, but phosphoric acid esters are especially important, particularly in biological systems. There are three ester derivatives of phosphoric acid: a monoester, a diester, and a triester. Using the methyl esters as examples, the monoester is methyl dihydrogen phosphate, the diester is dimethyl hydrogen phosphate, and the triester is trimethyl phosphate. There are also derivatives of pyrophosphoric acid (diphosphoric acid), which may have various ester derivatives by replacing any or all of the OH units.

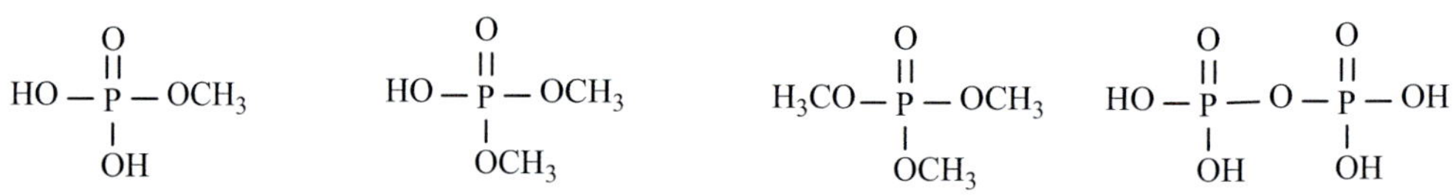

Methyl dihydrogen phosphate — Dimethyl hydrogen phosphate — Trimethyl phosphate — Diphosphoric acid

Phosphate esters have been seen in myriad compounds throughout this book. In principle, the preparation of phosphate esters or other acid derivatives should be like the chemistry of carboxylic acids and sulfonic acids, but it is more complex because several derivatives are possible, and there are synthetic routes that are specific to phosphorus compounds. Some of these routes are rather specific.

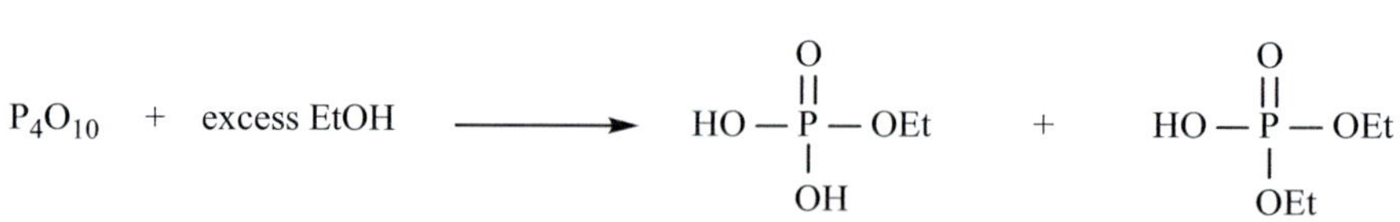

Ethyl dihydrogen phosphate — Diethyl hydrogen phosphate

[22] (a) Agnelli, G.; Piovella, F.; Buoncristiani, P.; Severi, P.; Pini, M.; D'Angelo, A.; Beltrametti, C.; Damiani, M.; Andrioli, G.C.; Pugliese, R.; Iorio, A.; Brambilla, G. *New England J. Med.* **1998**, *339*, 80; (b) Bergqvist, D.; Agnelli, G.; Cohen, A.T.; Eldor, A.; Nilsson, P.E.; Le Moigne-Amrani, A.; Dietrich-Neto, F. *New England J. Med.* **2002**, *346*, 975.

For example, the reaction of phosphoric acid and an alcohol may give the ester, but the reaction is slow and not a good preparative method. An alternative is the reaction of phosphorus pentoxide with an excess of an alcohol to yield a mixture of phosphate esters. The reaction of ethanol and P_4O_{10} (the dimeric form of phosphorus pentoxide) yields a mixture of ethyl dihydrogen phosphate and diethyl hydrogen phosphate. These two phosphates are separable based on differences in solubility with alkaline earth salts.[23]

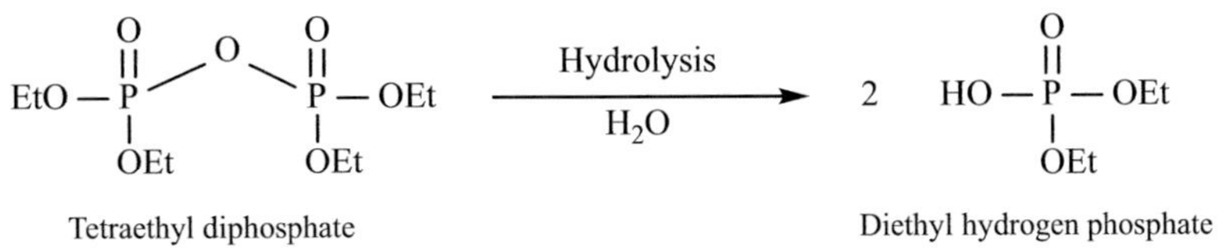

Hydrolysis of a phosphate ester may yield the parent phosphoric acid, but it is also possible to observe "partial hydrolysis." Derivatives of diphosphoric acid often yield phosphates. The hydrolysis of pyrophosphate esters, such as tetraethyl diphosphate, for example, yields the phosphate ester diethyl hydrogen phosphate.

The chemistry of phosphate esters is rich and varied. Phosphate esters are important in biological systems. The phosphate ester of a nucleoside (a nucleobase attached to a ribose derivative) is called a *nucleotide*. These are structural components of deoxyribonucleic acid (DNA) and ribonucleic acid (RNA). Using adenosine as an example, there are three possible monophosphate esters shown in Figure 10.16. The symbol A is used to designate an adenosine derivative in biology, so adenosine 5'-monophosphate is abbreviated 5'-AMP, adenosine 3'-monophosphate is 3′-AMP, and adenosine 2′-monophosphate is 2′-AMP. The pyrophosphate (diphosphate) derivative is adenosine 5′-diphosphate, or 5′-ADP.

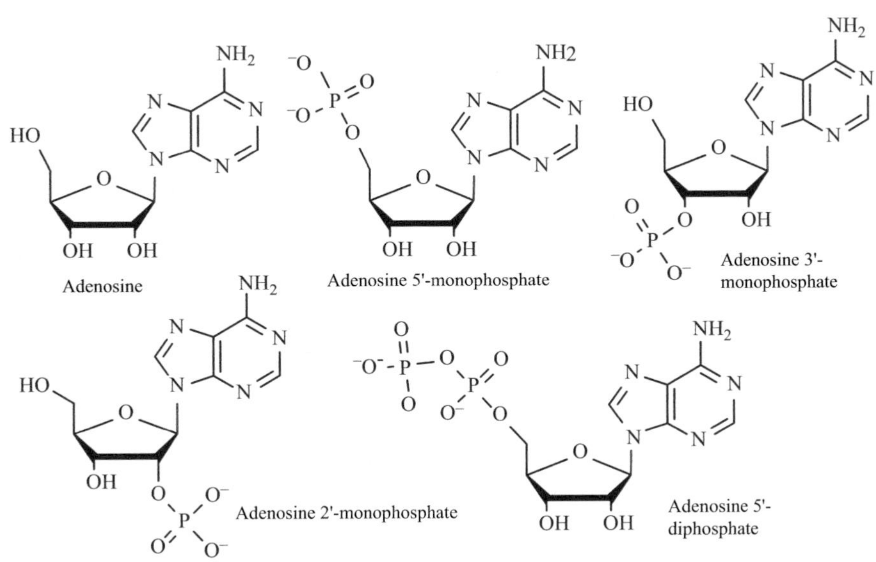

FIGURE 10.16 Adenosine, adenine monophosphates, and an adenine diphosphate.

[23]Corbridge, D.E.C. *Phosphorus 2000: Chemistry, Biochemistry & Technology*, Elsevier, Amsterdam, The Netherlands, **2000**, pp. 248–249.

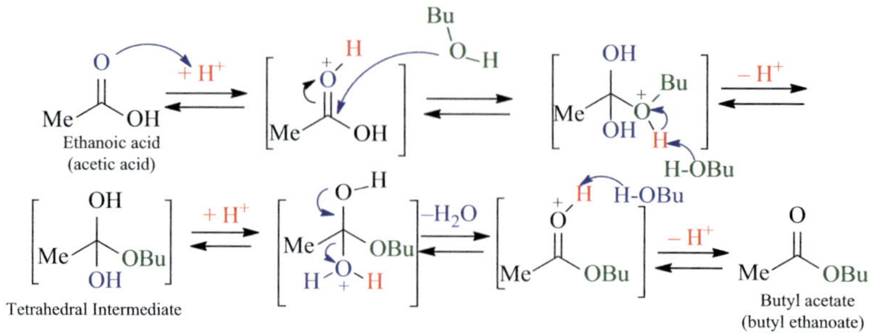

FIGURE 10.17 Acid-catalyzed esterification of acetic acid.

10.6 CHEMICAL SYNTHESIS OF ESTERS

A common method for preparing esters simply reacts a carboxylic acid with an alcohol in the presence of an acid catalyst. This reaction is reversible, and using butanol as the solvent (a large molar excess of the alcohol) helps shift the equilibrium toward the ester (recall Le Chatelier's principle). In the absence of the sulfuric acid catalyst, virtually no ester is formed.

The first step in the mechanism for the conversion of ethanoic acid to butyl ethanoate is shown in Figure 10.17, where protonation of ethanoic acid yields an oxocarbenium ion. The oxygen atom of butanol reacts with the oxocarbenium ion to yield an oxonium ion, which reacts as an acid with either butanol or ethanoic acid to yield a tetrahedral intermediate. To convert butanoic acid to butyl acetate, water must be lost from the molecule, so the OH unit in the tetrahedral intermediate is protonated to yield an oxonium ion, which loses water to yield an oxocarbenium ion (the protonated form of the acid). Loss of a proton completes the mechanism to yield butyl acetate. This reaction is shown again using the "stacked reaction" protocol rather than a mechanism, and the ester is formed.

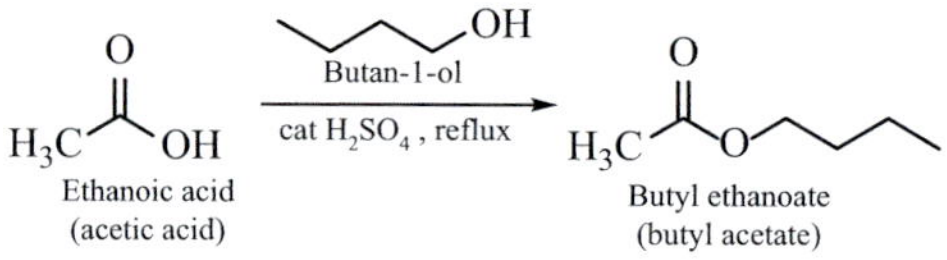

The mechanism for acid-catalyzed esterification of carboxylic acids is completely reversible, but it is driven to the right (toward the ester product) by application of Le Chatelier's principle and removal of water. A different application of Le Chatelier's principle uses a large excess of the alcohol (butanol for the formation of butyl acetate) to shift the equilibrium toward the ester by increasing the probability that it will react with alcohol rather than with water. Note that if butyl acetate is treated with an acid catalyst and water rather than butanol, the reverse of the mechanism shown will convert the ester back to the acid.

The esterification reaction is an acid-catalyzed acyl substitution. Acyl substitution of an ester to give a different ester is possible, where one alcohol unit replaces another. In other words, replacing one OR group in an ester with a new group (OR') will yield a different ester. This reaction is known as *transesterification*. If ethyl 3-phenylpropanoate is heated with a large excess of propanol and an acid catalyst, the product is propyl 3-phenylpropanoate. Analysis of the reaction shows that the OPr unit of propanol attacks the acyl carbon with loss of ethanol. This also requires a proton transfer, and the acid catalysts suggest formation of an oxocarbenium ion. This acyl substitution must proceed by a tetrahedral intermediate that contains both the OPr (propoxy) and the OEt (ethoxy) units. This sequence is driven toward the ethyl ester because a large excess of ethanol is used (ethanol is the solvent), in accord with Le Chatelier's principle. The overall sequence interchanges OEt for OPr. This is the reason the reaction is called transesterification.

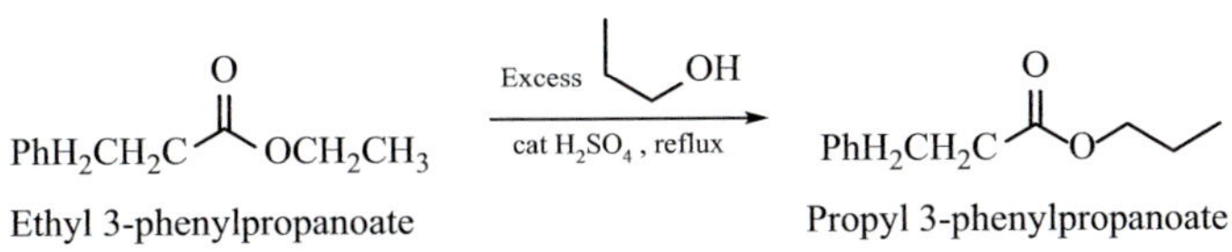

In any acid-catalyzed transesterification reaction, there is a potential problem if the OR unit in the alcohol solvent is *different* from the OR unit of the ester. If, for example, methanol is used as a solvent for the reaction of an ethyl ester, the final product may be a mixture of methyl and ethyl esters. If those esters must be separated, this mixture may be a problem. If the alcohol is the same as the "alcohol part" of the ester, transesterification generates the same ester. In other words, if methanol is used with a methyl ester and ethanol is used with an ethyl ester, there will be no structural changes in the ester unit of the final product.

HOMEWORK

10.1. Give an example of an omega-3 fatty acid. Of an omega-9 fatty acid.
10.2. What is an essential fatty acid?
10.3. What is the structure of choline?
10.4. Draw an example of a phospholipid.
10.5. Using 3'-deoxyribofuranose as a template, what is the structure of guanosine 5'-triphosphate? Of thymidine 3'-monophosphate? See Section 15.1.
10.6. What is the structure of a ceramide diphosphate?
10.7. What is the typical biological molecule that functions as a two-carbon unit used in the biosynthesis of fatty acids?

11 Aromatic Compounds and Heterocyclic Compounds

Benzene is a particularly stable hydrocarbon. Derivatives are known in which the hydrogen atoms of benzene have been replaced with substituents and/or functional groups. There are hydrocarbons related to benzene that have two, three, or more rings fused together (polycyclic compounds). The unifying concept of all these molecules is that they are aromatic, which means that they are especially stable with respect to their bonding and structure.

11.1 BIOSYNTHETIC ROUTES TO A BENZENE RING

The biosynthesis of simple aromatic compounds is well known in bacteria, fungi, and plants. These organisms all share an exclusive biosynthetic route, the *shikimate pathway*, also known as the *shikimic acid pathway*. Aromatic compounds are important in many biosynthetic pathways for myriad organisms. The shikimate pathway, which includes enzymatic steps for the production of chorismate via shikimate from *phosphoenolpyruvate* and *erythrose-4-phosphate*, is found in various organisms for the biosynthesis of aromatic amino acids and most biogenic benzene derivatives. The shikimate pathway is shown in Figure 11.1[1] for the biosynthesis of chorismate, which can be converted to tyrosine, phenylalanine, and tryptophan.

Another example is the proposed biopathway that prepares 2-hydroxy-6-methyl-benzaldehyde in the defense secretions of the oribatid mite *Archegozetes longisetosus* starting with D-glucose and phenylalanine.[2] The oribatid mite *Archegozetes longisetosus* produces small aromatic compounds using a horizontally acquired putative polyketide synthase-like mechanism. Experimental data point to the biochemical pathway shown in Figure 11.2[2] that appears to explain the data. Malonyl-CoA and/or acetyl-CoA produced via glycolysis and gluconeogenesis from sugars such as D-glucose and amino acids such as phenylalanine lead to a C8-polyketide-like intermediate via a head-to-tail condensation. Subsequent ring-closing cyclization via an aldol reaction yields a second intermediate, which eventually undergoes several H_2O enolization and H_2O reduction reactions,[3] leading to aromatization of the ring and the formation of the final product, 2-hydroxy-6-methyl-benzaldehyde.

A mixture of yellow pigments, grixazone A and grixazone B, are secondary metabolites of *Streptomyces griseus*. 3-Amino-4-hydroxybenzoic acid (3,4-AHBA) is an intermediate and a precursor for several secondary metabolites produced by *Streptomyces*. However, it has been proposed that 3-amino-4-hydroxybenzoic acid is derived from a pathway other than a shikimate-type pathway. A study on the grixazone biosynthesis gene cluster led to the biosynthesis pathway of 3,4-AHBA, where only two enzymes are needed.

3-Amino-4-hydroxybenzoic acid is a precursor for several secondary metabolites, including grixazone, produced by *Streptomyces griseus*.[4] The biosynthesis pathway to grixazone A and grixazone B utilizes two primary metabolites (2-amino-4-oxobutanoic acid and 3-hydroxy-2-oxo-propyl

---

[1] Almeida, A.M.; Abrahão, J.; Seixas, F.A.V.; Bueno, P.S.A.; de Oliveira, M.A.S.; Tomazini, L.F.; Constantin,R.P.; dos Santos, W.D.; Marchiosi, R.; Osvaldo Ferrarese-Filho, O. *Agronomy* **2024**, *14*, 930. Figure 1, p. 932. Also see Herrmann, K.M.; Weaver, L.M. *Ann. Rev. Plant Physiol. Plant Mol. Biol.* **1999**, *50*, 473; Maeda, H.; Dudareva, N. *Annu. Rev. Plant Biol.* **2012**, *63*, 73.

[2] Brückner, A.; Kaltenpoth, M.; Michael Heethoff, M. *Proc. R. Soc. B* **2020**, *287*, 20201429.

[3] Rocha, D.F.; Wouters, F.C.; Zampieri, D.S.; Brocksom, T.J.; Machado, G.; Marsaioli, A.J. *Molecules* **2013**, *18*, 429. https://doi.org/10.3390/molecules180911429.

[4] (a) Ohnishi, Y.; Furusho, Y.; Higashi, T.; Chun, H.-K.; Furihata, K.; Sakuda, S.; Horinouchi, S. *J. Antibiot.* **2004**, *57*, 218; (b) Suzuki, H.; Furusho, Y.; Higashi, T.; Ohnishi, Y.; Horinouchi, S. *J. Biol. Chem.* **2006**, *281*, 824.

DOI: 10.1201/9781003605157-11

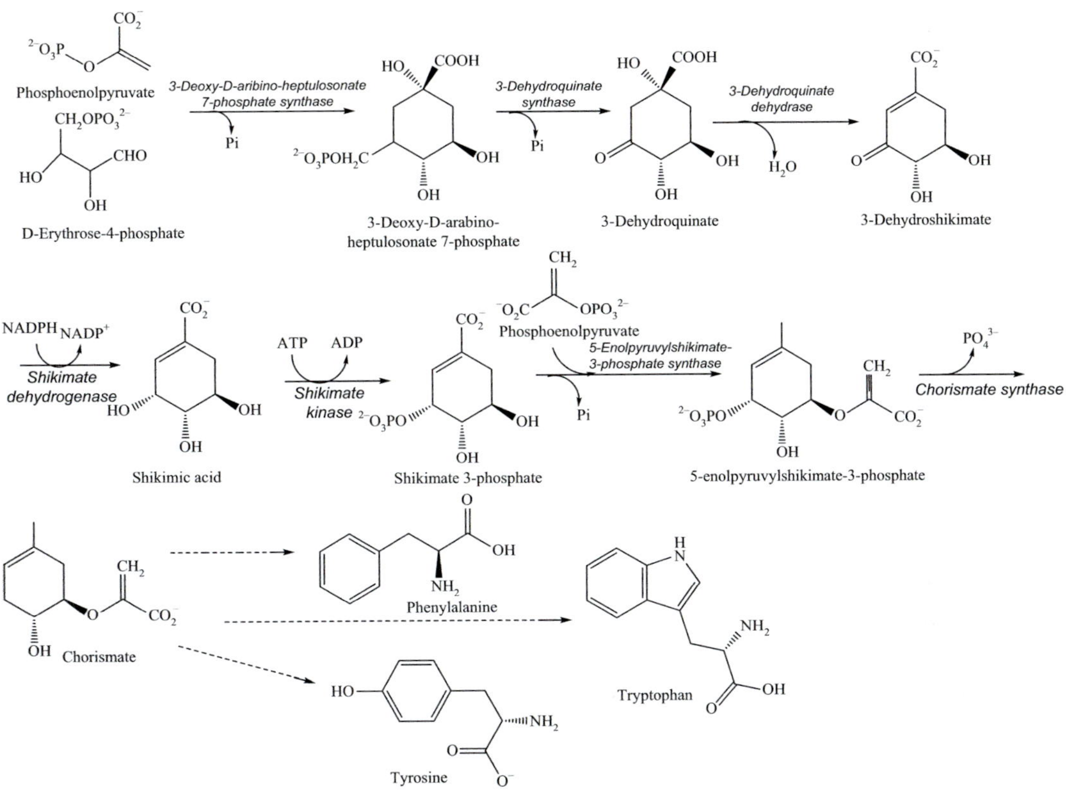

FIGURE 11.1 The Shikimate pathway. (Taken from Almeida, A.M.; Abrahão, J.; Seixas, F.A.V.; Bueno, P.S.A.; de Oliveira, M.A.S.; Tomazini, L.F.; Constantin,R.P.; dos Santos, W.D.; Marchiosi, R.; Osvaldo Ferrarese-Filho, O. *Agronomy* **2024**, *14*, 930. Figure 1, p. 932. Open Access.)

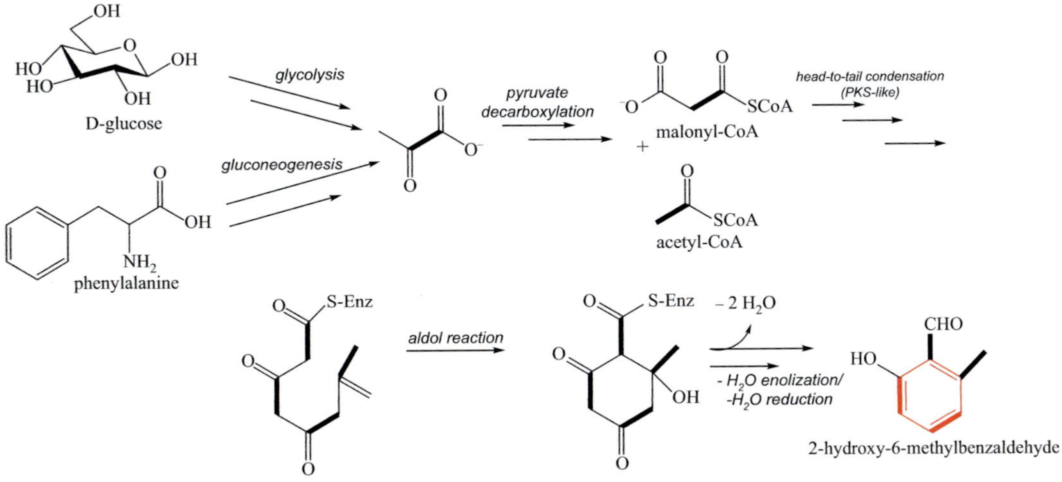

FIGURE 11.2 Proposed biosynthetic pathway leading to 2-hydroxy-6-methyl-benzaldehyde in the oribatid mite *Archegozetes longisetosus,* starting with D-glucose and phenylalanine as primary substrates. (Taken from Brückner, A.; Kaltenpoth, M.; Michael Heethoff, M. *Proc. R. Soc. B* **2020**, *287,* 20201429. Open Access.)

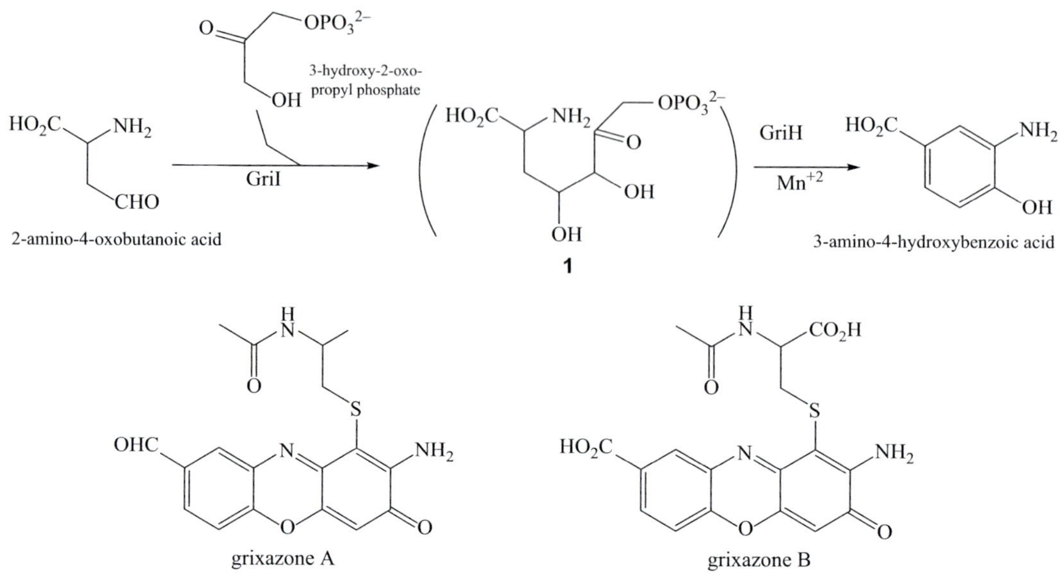

FIGURE 11.3 Biosynthesis pathway to form a benzene ring in 3-amino-4-hydroxybenzoic acid and its derivatives in *S. griseus*. (Taken from Suzuki, H.; Ohnishi, Y.; Furusho, Y.; Sakuda, S.; Horinouchi, S. *J. Biol. Chem.* **2006**, *281,* 36944. Open Access.)

phosphate). Two genes, GriI and GriH, within the biosynthesis gene cluster are responsible for the biosynthesis of 3-amino-4-hydroxybenzoic acid.[5] 3-Amino-4-hydroxybenzoic acid is formed via two reactions catalyzed by GriI and GriH. As shown in Figure 11.3,[5] GriI catalyzed an aldol condensation to form a 7-carbon product, **1** (2-amino-4,5-dihydroxy-6-one-heptanoic acid-7-phosphate), which was subsequently converted to 3-amino-4-hydroxybenzoic acid by GriH. The latter reaction required Mn^{2+} ions but no cofactors involved in reduction or oxidation. This pathway is independent of the shikimate pathway, representing a novel, simple enzyme system responsible for the synthesis of a benzene ring from the C3 and C4 primary metabolites.

11.2 BENZENE AND AROMATICITY

To understand biosynthetic pathways to aromatic compounds, one must first understand benzene itself and aromaticity. Benzene is a hydrocarbon with the formula C_6H_6. It is the parent of a large class of compounds known as aromatic hydrocarbons. The structure and chemical reactivity of aromatic hydrocarbons are so unique that benzene derivatives are given their own nomenclature system. This discussion begins with the unique structure of benzene.

The structure of benzene is shown in Figure 11.4. It is known that the C–C bond length in ethane is 1.53 Å (153 pm), and the C–C bond length in cyclohexane is 1.536 Å (153.6 pm). The bond distance for the C=C bond in ethene is 1.34 Å (134 pm). These data indicate that a C=C unit has a shorter bond distance than a C–C unit. If benzene has a structure with both single and double bonds, it should have three longer C–C single bonds and three shorter C=C units. It would then be called cyclohexatriene. It has been experimentally determined that all six carbon–carbon bonds have a measured bond distance of 1.397 Å (139.7 pm), a value that lies between those for the C–C bond in an alkane and the C=C unit of an alkene. This observation means that the C–C bonds in benzene are *not* single bonds, nor are they C=C double bonds where the π-electrons are localized between two carbons in a π-bond. *This molecule is not cyclohexatriene; it is benzene*. Each carbon in benzene

[5] Suzuki, H.; Ohnishi, Y.; Furusho, Y.; Sakuda, S.; Horinouchi, S. *J. Biol. Chem.* **2006**, *281,* 36944.

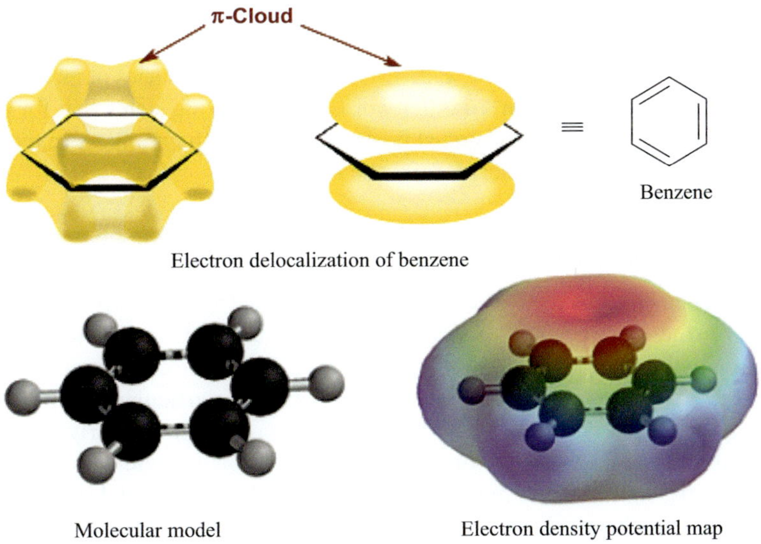

FIGURE 11.4 Structure and delocalization of electron density in benzene.

is sp^2 hybridized, which means each has a trigonal planar geometry, leading to a planar geometry for the molecule. The planar geometry is seen more clearly in the molecular model of benzene in Figure 11.4.

Each carbon atom has a p-orbital that is perpendicular to the plane of the carbon and hydrogen atoms, and there is one π-electron associated with each p-orbital. The six electrons are delocalized over six parallel p-orbitals on six adjacent sp^2 carbon atoms in a ring. Figure 11.4 shows six delocalized p-orbitals as a π-cloud of electrons above and below the plane of the atoms. The electron density map in Figure 11.4 shows a red area in the center of the ring, which indicates an area of high electron density, consistent with the π-cloud.

The bonding in benzene is not properly described by the single structure shown in Figure 11.4, because that structure shows localized single and double bonds. As just described, the delocalized electron density leads to a structure in which all the bonds have the same bond length and strength. In other words, the bonding is "in between" C–C and C=C, which is consistent with resonance stabilization. To accommodate these structural features, the two structures drawn for benzene are resonance contributors. Neither of the two individual resonance contributors accurately represents the structure of benzene, so *both* structures are required. Therefore, the proper way to represent benzene is with two resonance structures and a double-headed arrow to indicate they are resonance contributors, as shown. Each of the two resonance contributors is known as a Kekulé structure. Taken *together*, the two Kekulé structures represent the resonance delocalization and the bonding in benzene. Benzene is sometimes drawn as a single six-membered ring with a circle in it, but it is more common for one of the Kekulé structures to be used to represent benzene.

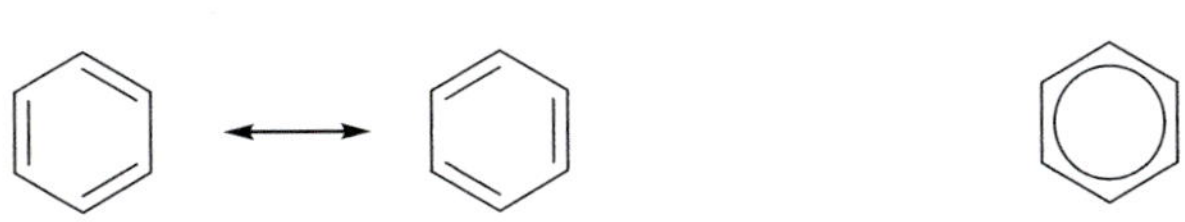

BENZENE (two resonance contributors)

Alternative representation of benzene

In benzene, six π-electrons are confined to a ring, and every carbon atom in that ring is sp^2 hybridized with no intervening sp^3 atom, and a p-orbital is attached to each carbon. Further, the p-orbitals are contiguous and continuous. These features lead to special stability for benzene, and benzene is said to be *aromatic*. To explain the special stability associated with these molecules, some important rules have been developed to predict aromaticity. Perhaps the most important is called *Hückel's rule*.

As the value of n changes in the series 0, 1, 2, 3, 4, and so on, a new series of numbers is generated by $4n+2$: 2, 6, 10, 14, 18, 22, and so on. In other words, Hückel's rule states that for a hydrocarbon to be aromatic, the number of π-electrons must be equal to one of the numbers in the $4n+2$ series. Using *Hückel's rule* generates a $4n+2$ series from the n series:

n series:	0, 1, 2, 3, 4, 5, 6, …
$4n+1$ series:	2, 6, 10, 14, 18, 22, 26, …

Using *Hückel's rule*, benzene has six π-electrons that are confined to a planar ring, and six π-electrons correspond to $n = 1$ in the $4n+2$ series. This means that six is a number in the Hückel series: 2, 6, 10, 14, and so on. When a hydrocarbon fits the criteria described in this discussion and has a number of π-electrons equal to a value in the $4n+2$ series, it is aromatic and particularly stable.

An aromatic hydrocarbon is resonance-stabilized and therefore particularly stable. The lack of reactivity is a result of the fact that electrons are not localized on a single atom but rather delocalized over several atoms, so the structure tends to be more stable. The special stability of benzene is easily shown by a simple chemical reaction. The π-bond of an alkene reacts as a Brønsted–Lowry base with an acid HX (HCl, HBr, HI, etc.). The π-bonds of benzene do not react with HX acids. This fact is evidence that the more tightly a π-bond holds its electrons, the more stable the molecule will be and the less likely it is to donate electrons. In other words, the molecule is a weaker base.

The effect of aromaticity can be examined with a comparison of the relative reactivity of HBr with cyclohexene, cyclohexadiene, and benzene (remember that using one Kekulé structure implies both resonance forms). The reaction of cyclohexene with HBr is shown in Figure 11.5. This reaction is rapid, and the product is a carbocation, which traps bromide to give bromocyclohexane. Cyclohexa-1,3-diene also rapidly reacts with HBr to yield an allylic carbocation (a $C{=}C{-}C^+$ unit). The allylic carbocation is resonance-stabilized, easily formed, and reacts with the bromide ion to yield an allylic bromide, 3-bromocyclohex-1-ene. While both cyclohexene and cyclohexa-1,3-diene react quickly and easily with HBr, *benzene does not react with HBr,* even with heating**.** Benzene is therefore *a weaker base* than an alkene due to aromatic stabilization.

11.3 BENZENE IS A CARCINOGEN

"Benzene is one of the best-studied of the known human carcinogens. It causes leukemia in humans and a variety of solid tumors in rats and mice."[6] "To be carcinogenic, benzene must first be metabolized in the liver, mainly via cytochrome P4502E1. The major product is phenol, shown in Figure 11.6,

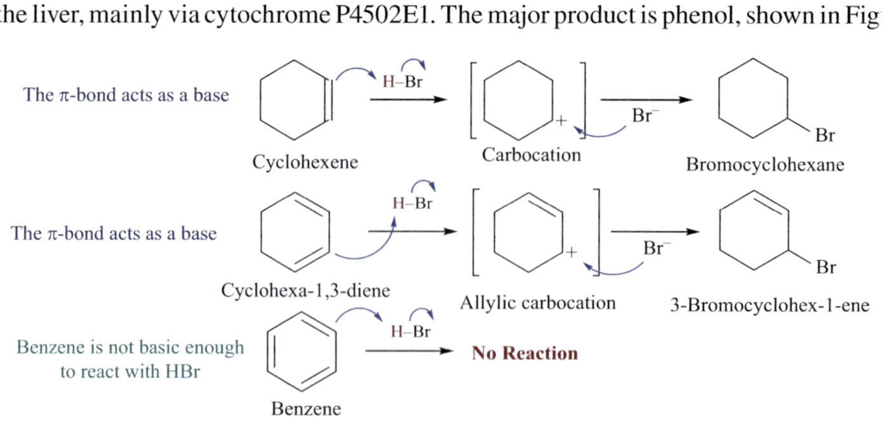

FIGURE 11.5 Comparative reactivity of HBr with cyclohexene, cyclohexadiene, and benzene.

[6] Cox Jr., L.A. *Risk Anal.* **1991**, *11*, 453.

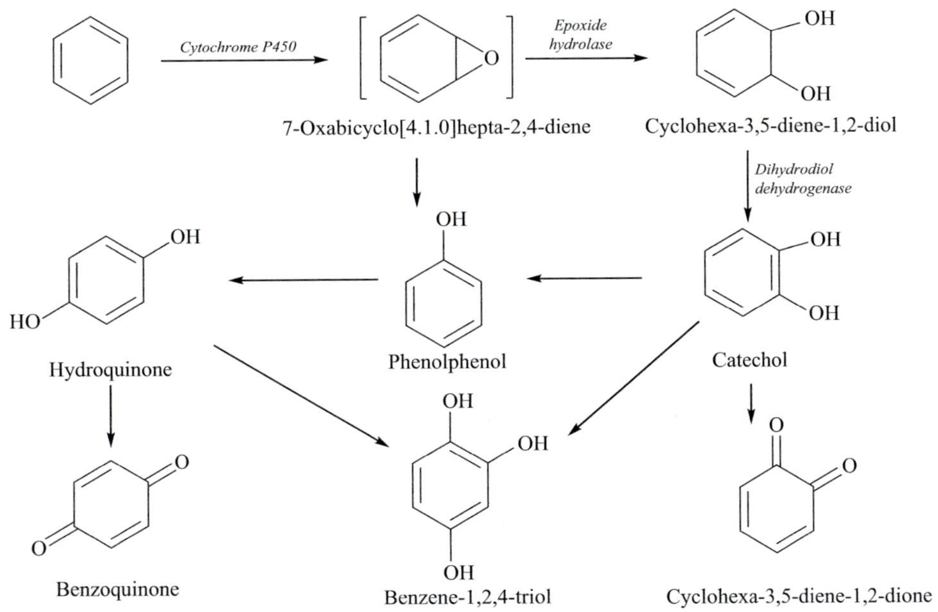

FIGURE 11.6 Metabolic pathway for the conversion of benzene to cancer-inducing metabolites. (Smith, M.T. *Environ. Health Perspect.* **1996**, *104,* 1219–1225; Snyder, R.; Hedli, C.C. *Environ. Health Perspect.* **1996**, *104* (Supplement 6), 1165–1171. *Environmental Health Perspectives* is a publication of the U.S. government, and lies in the public domain and is therefore without copyright.)

which is either conjugated (primarily to phenyl sulfate in humans) or further hydroxylated by P4502E1 to hydroquinone. Other major metabolites include catechol and *trans-trans*-muconic acid [(2*E*,4*E*)-hexa-2,4-dienedioic acid]. The latter is presumed to be formed from the ring opening of benzene epoxide (7-oxabicyclo[4.1.0]hepta-2,4-diene or benzene oxepin), or perhaps benzene dihydrodiol (cyclohexa-3,5-diene-1,2-diol)."[7]

A series of reactions has been reported that is taken to be the metabolic pathway for the conversion of benzene to cancer-inducing metabolites.[8] Benzene is epoxidized by *cytochrome P*-450 (EC 1.6.2.4) to yield hydroquinone, and *epoxide hydrolase* (EC 3.3.2.10) catalyzes the opening of the epoxide to give the diol. This intermediate is converted to catechol, catalyzed by *dihydrodiol dehydrogenase* (EC 1.3.1.87), which in turn can be converted to phenol. Phenol can be converted to hydroquinone, which gives rise to benzoquinone and 1,2,4-trihydroxybenzene (benzene-1,2,4-triol). Catechol can be oxidized enzymatically to the 1,2-quinone (*o*-quinone; cyclohexa-3,5-diene-1,2-dione).

11.4 ENZYMATIC S_EAr REACTIONS

The enzyme *2,6-DHBA decarboxylase* (EC 4.1.1.X) from *Rhizobium sp.* catalyzes the reversible and regioselective carboxylation of catechol. Both hydroxybenzoic (EC 4.1.1.X) and salicylic acid decarboxylases show high ortho selectivity.[9] In this reaction, catechol is converted to 2,3-dihydroxybenzoic acid.[10] The source of carbon dioxide is bicarbonate, and Zn^{2+} facilitates the reaction. In the mechanism shown in Figure 11.7,[9] bicarbonate is deprotonated by catechol, facilitated by Zn^{2+} and elimination of water regenerates the aromaticity and concomitant tautomerization in a similar fashion as the chemical variant.

[7] Smith, M.T. *Environ. Health Perspect.* **1996**, *104,* 1219.
[8] Snyder, R.; Hedli, C.C. *Environ. Health Perspect.* **1996**, *104* (Supplement 6), 1165.
[9] Matsui, T.; Yoshida, T.; Yoshimura, T.; Nagasawa, T. *Appl. Microbiol. Biotechnol.* **2006**, *73,* 95.
[10] Pesci, L.; Glueck, S.M.; Gurikov, P.; Smirnova, I.; Faber, K.; Liese, A. *FEBS J.* **2015**, *282,* 1334.

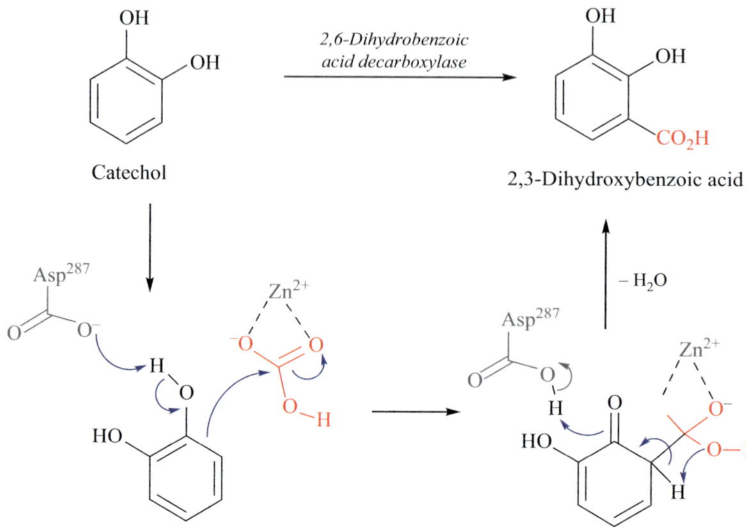

FIGURE 11.7 Enzymatic carboxylation of catechol. (Reprinted with permission from Pesci, L.; Glueck, S.M.; Gurikov, P.; Smirnova, I.; Faber, K.; Liese, A. *FEBS J.* **2015**, *282*, 1334–1345. Scheme 2, p. 1342. Copyright © 2015, reproduced with permission of John Wiley & Sons. Ltd.)

Thyroxine (3,5,3′,5′-tetraiodothyronine, T4) is the major hormone secreted by the follicular cells of the thyroid gland. In the hypothalamus, T4 is converted to triiodothyronine, T3, via an enzyme called a *deiodinase* (*tetraiodothyronine 5′ deiodinase;* EC 1.21.99.4). Here, T3 is the main inhibitor of thyroid-stimulating hormone. "Thyroxine (also known as Levothyroxine or T4) is the ultimate metabolism regulator. Its reactions and products influence carbohydrate metabolism, protein synthesis and breakdown, and cardiovascular, renal, and brain function. Thyroxine is essential to an animal's functions, and it is essential for development in the young. Tadpoles will not develop into frogs, for example. Untreated human babies will develop cretinism, a condition marked by severe mental and physical retardation. Adult humans with low thyroxine levels (hypothyroidism) suffer mental slowness, weight gain, depression, and fatigue."[11] It is believed that thyroxine is formed in nature from the amino acid tyrosine via diiodotyrosine. When iodine is taken into the body, it is covalently bound to tyrosine residues in thyroglobulin molecules, and the enzyme *thyroperoxidase* (EC 1.11.1.8) converts the bound tyrosine to monoiodotyrosine (MIT) and diiodotyrosine (Figure 11.8).[11] This transformation is an enzyme-mediated electrophilic aromatic substitution reaction.

There are important compounds produced in nature, or derived from natural products, that contain chlorine, including vancomycin and cryptophycin A. Vancomycin is used to treat infections caused by Gram-positive bacteria, and it has traditionally been reserved as a "drug of last resort," used only after treatment with other antibiotics has failed. Cryptophycin A has anti-cancer activity.

The biosynthesis of vancomycin involves regioselective chlorination by flavin-dependent halogenases, and one of these enzymes is *tryptophan 7-halogenase* (PrnA; EC 1.14.19.9), which regioselectively chlorinates tryptophan using hypochlorous acid (HOCl), which is produced biosynthetically from chloride ion. The tryptophan-mediated transformation is shown in Figure 11.9.[12]

Using amino acid residue K^{79} (a lysine residue), binding HOCl allows electrophilic substitution on the indole ring, as shown in tryptophan, to yield an arenium ion, **2**. Loss of the proton by the acid–base reaction with the carboxylate anion unit of glutamic acid residue E^{346} completes the electrophilic chlorination reaction, giving the chlorinated indole unit in **3**.

[11] The Top Pharmaceuticals That Changed the World. *Chem. Eng. News 83* (25) (6/20/05).
[12] Karabencheva-Christova, T.G.; Torras, J.; Mulholland, A.J.; Lodola, S.; Christov, C.Z. *Sci Rep.* **2017**, *7*, 17395.

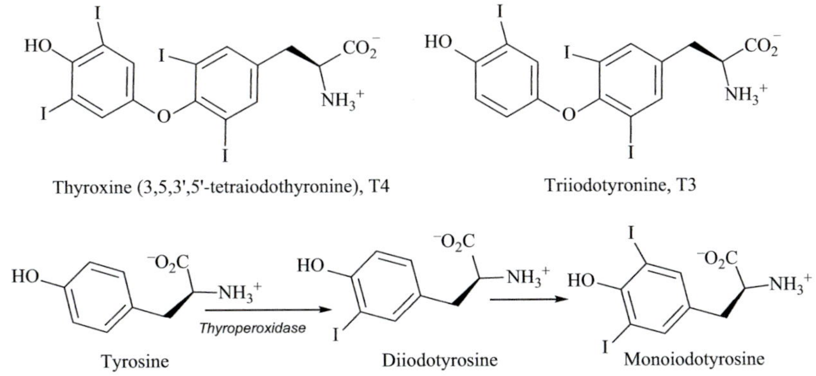

FIGURE 11.8 Conversion of thyroxine (3,5,3′,5′-tetraiodothyronine, T4) to triiodotyronine, T3.

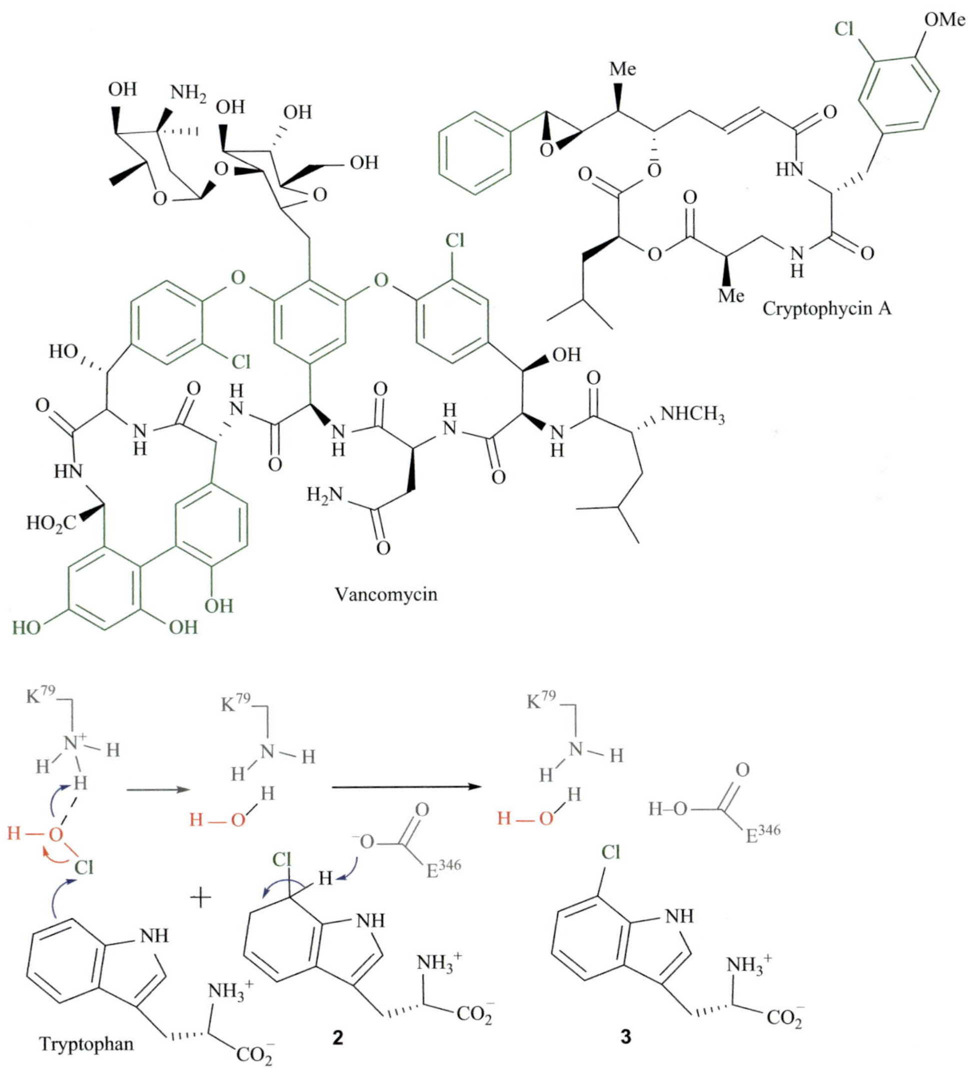

FIGURE 11.9 Tryptophan-mediated transformation. (From Karabencheva-Christova, T.G.; Torras, J.; Mulholland, A.J.; Lodola, S.; Christov, C.Z. *Sci Rep.* **2017**, *7*, 17395. Open Access.)

Aromatic substitution with alkyl substituents is also known in biochemistry. *Dimethylallyltryptophan (DMAT) synthase* (EC 2.5.1.80) is a *prenyltransferase* that catalyzes the alkylation of L-tryptophan at C4 by dimethylallyl diphosphate, as shown.[13] This step is the first in the biosynthesis pathway of the ergot alkaloids. Reported work suggests that the enzyme catalyzes a stepwise reaction via carbocation intermediates.

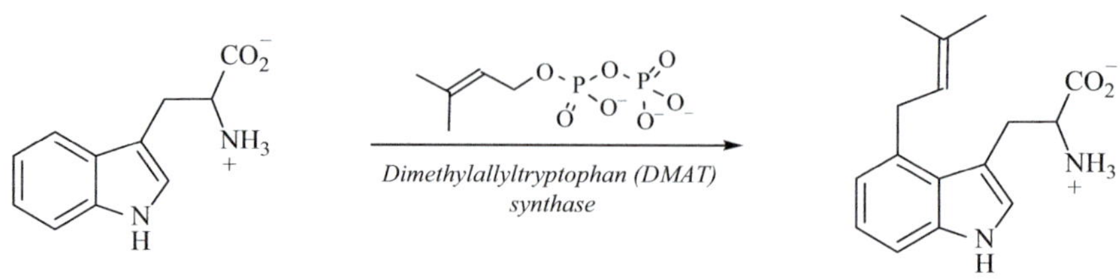

L-Tryptophan 4-Prenyl-L-tryptophan

11.5 FUNCTIONALIZED BENZENE DERIVATIVES

Benzene is the parent of a large class of compounds known as aromatic hydrocarbons. The abbreviation "Ph" is used when benzene is a substituent, called a phenyl group. An example is methylbenzene, or $PhCH_3$ (see methylbenzene; toluene discussed below). Benzene derivatives may be viewed as molecules in which one or more hydrogen atoms have been replaced with alkyl groups, heteroatom substituents, or functional groups. When a hydrogen atom on the benzene ring is replaced with an alkyl group (methyl, ethyl, etc.), the resulting molecule is called an *arene*. The benzene derivative with one methyl group is named *methylbenzene*, but the *common* name is *toluene*. The benzene derivative with an ethyl group is *ethylbenzene*. The alkyl substituent of an arene may have longer alkyl chains, which poses a problem with respect to nomenclature. In general, if a benzene ring is attached to an alkane carbon chain of more than six carbon atoms, the benzene ring is treated as a substituent, so it is a phenyl alkane. If the carbon chain is six carbons or fewer, it is named as an alkylbenzene (like ethylbenzene). Using this protocol, a benzene ring attached to C4 of a nine-carbon chain is named 4-phenylnonane. There are many arenes that have two or more substituents on the ring: disubstituted, trisubstituted, and so on. Three dimethylbenzene derivatives are known: 1,2-dimethylbenzene, 1,3-dimethylbenzene, and 1,4-dimethylbenzene, which are the IUPAC names, but the common name for a dimethylbenzene is *xylene*.

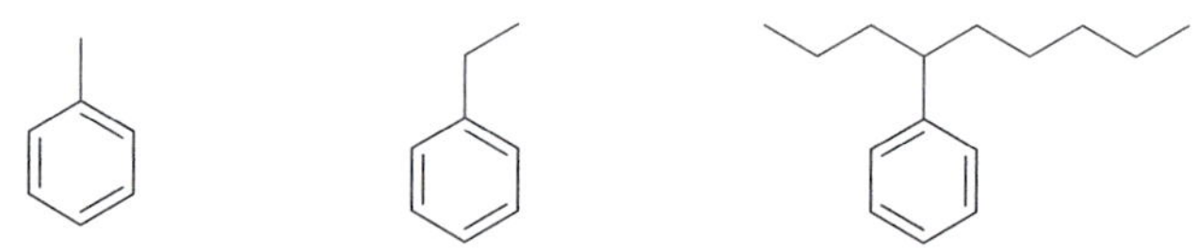

Methylbenzene (Toluene) Ethylbenzene 4-Phenylnonane

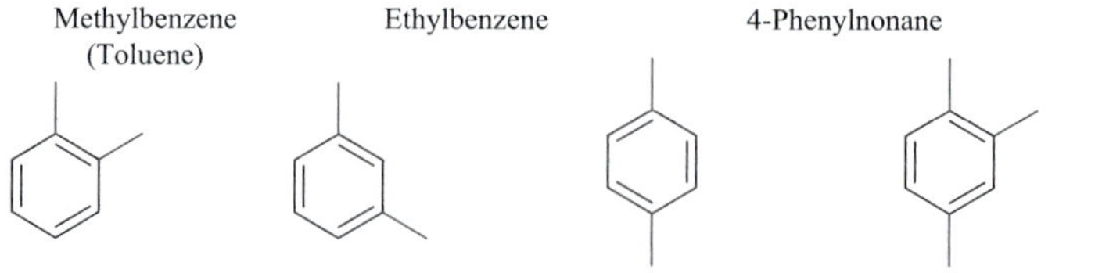

1,2-Dimethylbenzene (o-Xylene) 1,3-Dimethylbenzene (m-Xylene) 1,4-Dimethylbenzene (p-Xylene) 1,2,4-Trimethylbenzene

For disubstituted benzene derivatives, a specialized *common* naming system has been used for many years, other than numbers, to designate on which carbon atoms the methyl groups appear. In

[13] Gebler, J.C.; Woodside, A.B.; Poulter, C.D. *J. Amer. Chem. Soc.* **1992**, *114*, 7354.

this common nomenclature, two substituents with a 1,2-relationship are said to be *ortho*; those with a 1,3-relationship are said to be *meta*, and those with a 1,4-relationship are said to be *para*. Using the common names, 1,2-dimethylbenzene is called *ortho*-xylene (or *o*-xylene), 1,3-dimethylbenzene is called *meta*-xylene (or *m*-xylene), and 1,4-dimethylbenzene is called *para*-xylene (or *p*-xylene). The IUPAC system will be used most often, but ortho, meta, and para are so common that one must be familiar with this system. It is also possible to use ortho-, meta-, and para- in a comparative sense, and two groups can be said to be ortho or para to one another. A chlorine and a bromine in a molecule may have an ortho relationship, for example. It is important to remember *not* to mix the IUPAC and common nomenclature systems. The arene shown with three methyl substituents is 1,2,4-trimethylbenzene.

There are hydrocarbon substituents other than the simple alkyl fragments just discussed. Alkene and alkyne groups may be attached to a benzene ring. The benzene derivative with an attached ethenyl group (a vinyl group) is known as ethenylbenzene ($PhCH{=}CH_2$), but the common name is *styrene*. The hydrocarbon shown with two benzene rings attached to a C=C unit is (*E*)-1,2-diphenylethene (PhCH=CHPh), but it is commonly called *trans-stilbene* (there is also *cis*-stilbene). The halobenzenes are named fluorobenzene, chlorobenzene, bromobenzene, and iodobenzene.

Naming a molecule with an OH group on a benzene ring requires a variation of the usual alcohol nomenclature. If the OH is the group attached to the benzene ring, the -ol ending for an alcohol is used rather than hydroxybenzene. Therefore, the prefix is modified to accommodate benzene by dropping the *-e* and replacing it with *-ol*, so the name is *benzenol* using the IUPAC system. The common name of benzenol is *phenol*, which has such widespread use that this name is often preferred. The derivative with one methoxy group is called *methoxybenzene* (*anisole*), and the derivative with one ethoxy group is *ethoxybenzene* (*phenetole*).

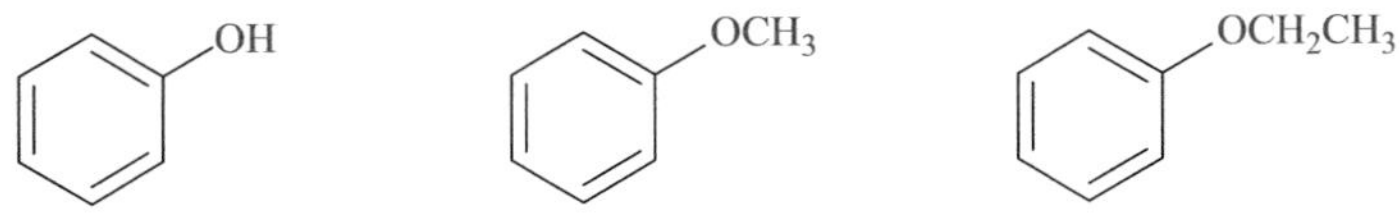

Phenol (hydroxybenzene) — Anisole (methoxybenzene) — Ethoxybenzene (ethoxybenzene)

The molecule with a single NH_2 group is named *benzeneamine*, but the common name is *aniline*. Substituents may appear on the benzene ring, as in *4-methylbenzeneamine* (*p*-methylaniline; a common name is *p-toluidine*), or on the nitrogen, as in *N*-methylbenzeneamine (*N*-methylaniline). Conversion of benzeneamine to the corresponding amide gives *N*-acetylbenzeneamine (*N*-acetylaniline or *N*-phenylacetamide). Note that a common name is acetanilide. A molecule that has an attached COOH (carboxyl) group is a carboxylic acid, and the name is *benzoic acid*. Benzoic acid is the parent structure for this class, and the usual acid derivatives, such as esters (*benzoates*), acid chlorides (*benzoyl chlorides*), and amides (*benzamides*), are all known.

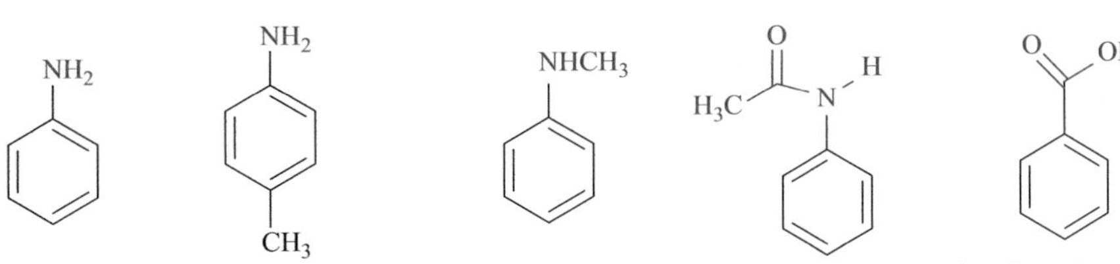

Benzeneamine (Aniline) — 4-Methylbenzeneamine (*p*-Toluidine) (*p*-Methylaniline) — *N*-Methylbenzeneamine (*N*-Methylaniline) — *N*-Acetylbenzeneamine (*N*-Phenylacetamide) (*N*-Acetylaniline) — Benzoic acid

The compound with a CHO group attached directly to the benzene ring is named *benzenal*, but the common name is *benzaldehyde*. *Acetophenone* is the common name for the ketone with a methyl and a phenyl group attached to a carbonyl, but the IUPAC system for ketones identifies the longest chain as two carbons with the C=O unit at C1. Therefore, the IUPAC name is *1-phenylethanone*. Another example is 1-phenylbutan-1-one. When two phenyl rings are attached to a carbonyl, the name is *benzophenone*.

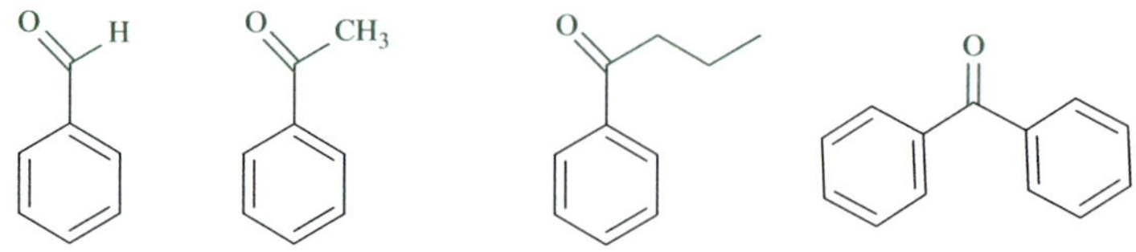

Benzeneal (Benzaldehyde) 1-Phenylethanone (Acetophenone) 1-Phenylbutan-1-one Diphenylmethanone (Benzophenone)

More than one functional group may be attached to a benzene ring. In some cases, the presence of two or more functional groups leads to a special name for that compound. The ortho dihydroxy derivative is shown, and it is named *1,2-benzenediol*, but its common name is *catechol*. The meta compound is named *1,3-benzenediol* (*resorcinol*), and the para compound is named *1,4-benzenediol* (*hydroquinone*). There are also three compounds that have two carboxyl groups, the dicarboxylic acids.

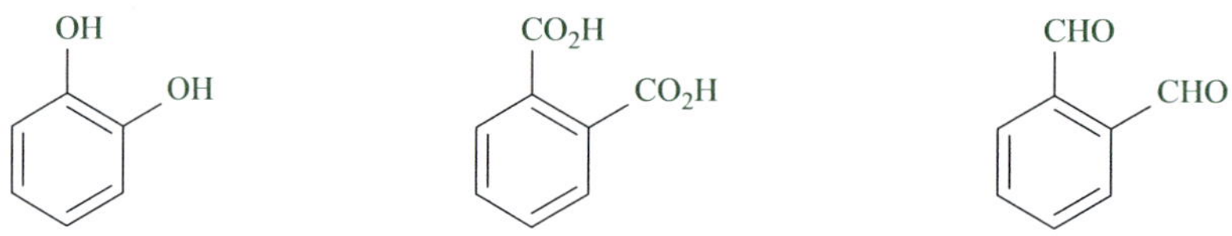

1,2-Benzenediol (Catechol) 1,2-Benzenedioic acid (Phthalic acid) 1,2-Benzenedial (Phthalaldehyde)

1,2-Benzenedioc acid (*phthalic acid*) is shown, and it has two carboxyl groups with an ortho relationship. In *1,3-benzenedioc acid* (*isophthalic acid*), the carboxyl groups are meta, and in *1,4-benzenedioc acid* (*terephthalic acid*), the carboxyl groups have a para relationship. Dialdehyde compounds have related common names: *1,2-benzenedial* (*phthaladehyde*; also phthalic dicarbaldehyde) is shown, and the other isomers are *1,3-benzenedial* (*isophthalaldehyde*; also isophthalic dicarbaldehyde) and *1,4-benzenedial* (*terephthalaldehyde*; also terephthalic dicarbaldehyde).

11.6 ELECTROPHILIC AROMATIC SUBSTITUTION: THE S_EAr REACTION

As noted above, benzene does not react with bromine. If, however, benzene is mixed with Br_2, in the presence of ferric bromide ($FeBr_3$), a reaction occurs, and bromobenzene is formed in good yield, along with HBr. Diatomic bromine reacts with a Lewis acid (e.g., ferric bromide) to form an 'ate' complex; Br^+ $FeBr_4^-$ is formed. Benzene is a sufficiently strong Lewis base to react with the cationic Br^+, but it is not a strong enough Lewis base to react with diatomic bromine.

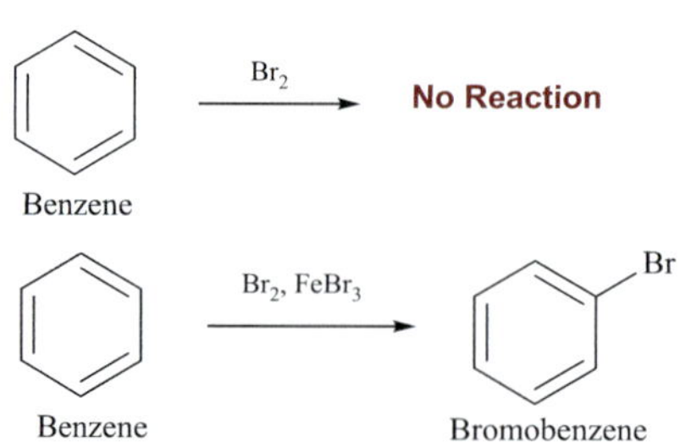

Analysis of the overall conversion of benzene to bromobenzene shows that a bromine atom has replaced one H on the benzene ring. Therefore, this reaction is a *substitution* (S). Experiments have shown that the reaction proceeds via a *cation intermediate*, so it is an *electrophilic* (E) reaction,

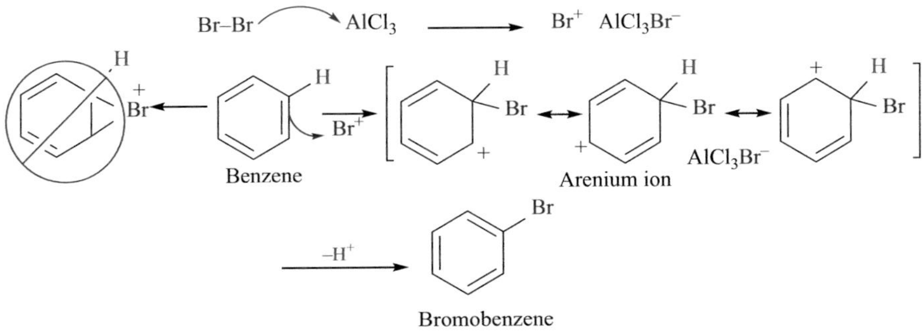

FIGURE 11.10 Mechanism of electrophilic aromatic substitution with benzene and bromine.

and it clearly involves an *aromatic* species (benzene or aryl, Ar). This type of reaction is labeled *electrophilic aromatic substitution*, and it accounts for most of the chemistry of benzene and its derivatives. This reaction is labeled S_EAr. A carbocation intermediate formed in an S_EAr reaction is called a *Wheland intermediate*. Such intermediates have also been called *Meisenheimer adducts* and even σ-adducts. The modern term, which will be used exclusively, is an *arenium ion*.

The aromatic stability of benzene is disrupted to form a resonance-stabilized arenium ion, as shown in Figure 11.10, and the aluminate anion ($AlCl_3Br^-$) can remove a proton from the carbon bearing the bromine to generate a new C=C unit, regenerating the aromatic benzene ring to yield the final product, bromobenzene. Conversion of an arenium ion to a benzene derivative is f*ormally an El reaction*, and the energetic driving force for losing this proton (for the El reaction) is the regeneration of the aromatic ring. From a mechanistic viewpoint, an electrophilic aromatic substitution is in reality two reactions. The first is the Lewis acid–base reaction of benzene with Br^+, which is generated by the reaction of bromine with the Lewis acid. The second is an El-type reaction to yield the substitution product, and it is an acid–base reaction. The reaction of benzene and bromine in the presence of aluminum chloride to yield bromobenzene is a *S_EAr reaction*.

Many Lewis acids other than ferric bromide can be used, including aluminum bromide ($AlBr_3$), aluminum chloride ($AlCl_3$), boron trifluoride (BF_3), and ferric oxide (Fe_2O_3). Indeed, old lab experiments for this reaction suggest that adding a "rusty nail" (rust is ferric oxide) to benzene and then slowly adding bromine will yield bromobenzene. As long as a suitable Lewis acid is present to generate Br^+ from Br_2, the subsequent reaction with benzene is straightforward.

Electrophilic aromatic substitution is not limited to bromine, and if it is possible to form a cationic species X^+, electrophilic aromatic substitution should occur. The nature of X^+ may be varied, including extending the reaction observed with diatomic bromine to diatomic chlorine. The reaction of chlorine and benzene, in the presence of $AlCl_3$, generates Cl^+, and the subsequent reaction with benzene yields chlorobenzene. A nitrogen electrophile (X=N) is the nitronium ion, NO_2^+, formed when nitric acid is mixed with the stronger sulfuric acid, where the nitric acid functions as a base. Once the nitronium ion is generated, benzene reacts as a Lewis base to yield an arenium ion and the product, nitrobenzene. Sulfur electrophiles in this S_EAr reaction are sulfuric acid and sulfur trioxide (SO_3). The sulfonation of benzene by concentrated sulfuric acid involves the reaction of benzene with the electrophilic sulfur atom. When benzene reacts with sulfuric acid, it donates two electrons to sulfur to yield the sulfonium arenium ion intermediate, and loss of water gives benzenesulfonic acid ($PhSO_3H$).

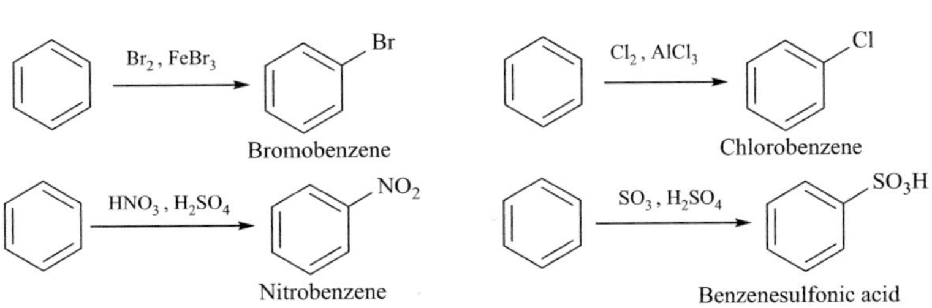

When a substituent is attached to the benzene ring, adding one substituent can lead to three different isomers. Using a generic benzene derivative in Figure 11.11, with a substituent X, selectivity for the isomeric products can be explained. This carbon atom is defined as the *ipso carbon* (.) in Figure 11.11. In a reaction with the bromonium ion, the resulting arenium ion from reaction with Br^+ at the ortho position (**4**, **5**) is formed, from reaction at the para position, and **6** is formed by reaction at the meta position. If X is an electron-releasing group (e.g., methyl), a positive charge on the ipso carbon is stabilized in the arenium ion. Releasing electrons to a positive center will stabilize the charge, as shown in Figure 11.11, and lead to a lower activation energy for the formation of that intermediate. Therefore, that intermediate is easier to form, and the reaction is faster. The positive charge resides on the carbon attached to the X group (the ipso carbon) in arenium ions **4** and **5**, so the electron-releasing effect of a methyl group will stabilize the charge, making those intermediates more stable and lower in energy than arenium ion **6**. Arenium ions **4** and **5** should be about equal in energy since there are the same number of resonance contributors, so the activation energy for formation of 5 and 5 is lower than the activation energy for formation of **6**, which has no charge on the ipso carbon. If intermediates **4** and **5** are more stable and have a lower activation energy, they will form faster.

Because the activation energy for their formation is lower, the rate of reaction for generating the ortho and para products is faster than the rate for generating the meta product when X is electron-releasing. If the ortho and para arenium ions are about equal in energy, in the absence of other mitigating factors, the ortho and para products should be formed as the major products in roughly equal amounts. A substituent (X) with unshared electrons (O, N, halogen) attached to a benzene ring is classified as electron-releasing relative to a positively charged ipso carbon. When the charge is on the ipso carbon, delocalization on X leads to a fourth resonance contributor. As shown in Figure 11.11, this fourth resonance contributor is an *indication of greater charge delocalization and greater stability* for those arenium ions. This extra charge delocalization contributes to the fact that the ortho and para positions are favored over the meta position, and the overall

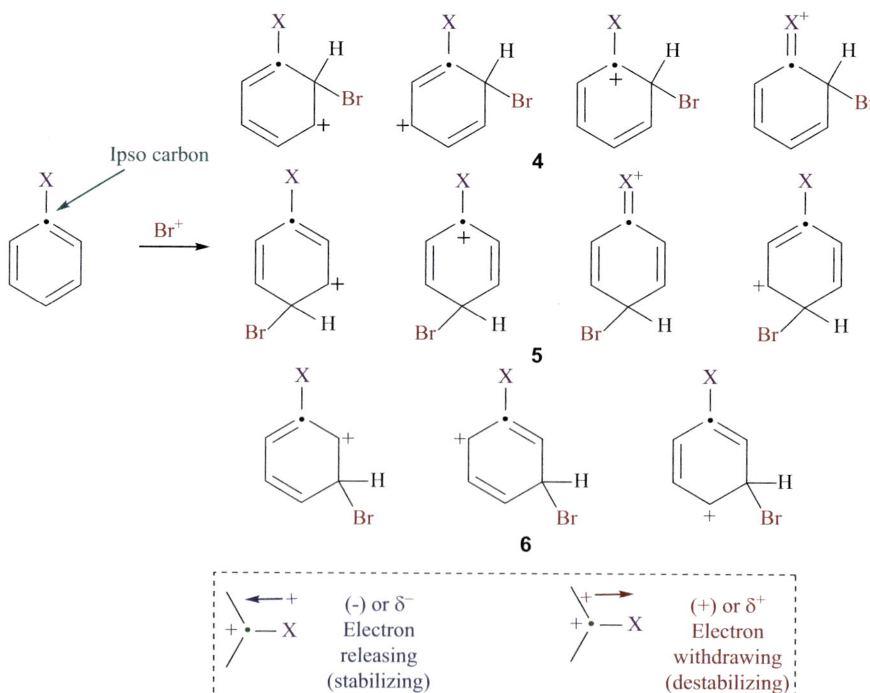

FIGURE 11.11 Arenium ion intermediates for electrophilic aromatic substitutions for ortho, meta, and para substitution.

reaction is faster than benzene because the arenium ion intermediates are more stable than the arenium ions derived from benzene.

If the X group in Figure 11.11 is electron-withdrawing (a group with an atom that has a positive charge or one that is polarized δ^+), then reaction at the ortho or para position places a positive charge on the ipso carbon of **4** or **5**. When a positive charge resides on an atom that is adjacent to another positive charge (see Figure 11.11), the repulsive interaction of two like charges will destabilize the arenium ion. The net result of the electron-withdrawing effect of a group (e.g., nitro) is to make the arenium ion less stable and more difficult to form (a higher activation barrier). Destabilization of the arenium ions from nitrobenzene explains why that reaction is slower than the reaction with benzene.

For reaction at the meta position, no resonance contributor has the positive charge in the ring on the ipso carbon. Therefore, the arenium ion formed when Br attaches to the meta carbon **3** is less destabilized when compared to **4** or **5**. The activation energy for reaction at the *meta carbon* is lower than the energy of the arenium ions from reaction at the ortho or para carbon atoms**.** When the X group in Figure 11.11 is electron-withdrawing, electrophilic aromatic substitution at the meta position will be faster than the reaction at the ortho and para positions. This observation is consistent with the formation of the meta product because it is less destabilized when compared to the ortho and para products.

11.7 BIOLOGICAL REDUCTION OF AROMATIC RINGS

Benzoyl-CoA is an important intermediate in the metabolism of many aromatic compounds in anaerobic bacteria. The enzyme catalyst is *benzoyl-CoA reductase* for the reduction of the benzene ring to the cyclohexa-1,5-diene-1-carboxyl-CoA derivative.[14] The reduction is mediated by the conversion of two adenosine triphosphate units to two adenosine diphosphate units, along with two equivalents of phosphate. The reaction also uses two ferredoxins (reduced), which are converted to the oxidized form. This reduction is effectively a *Birch reduction* of the aromatic ring (Section 11.8).

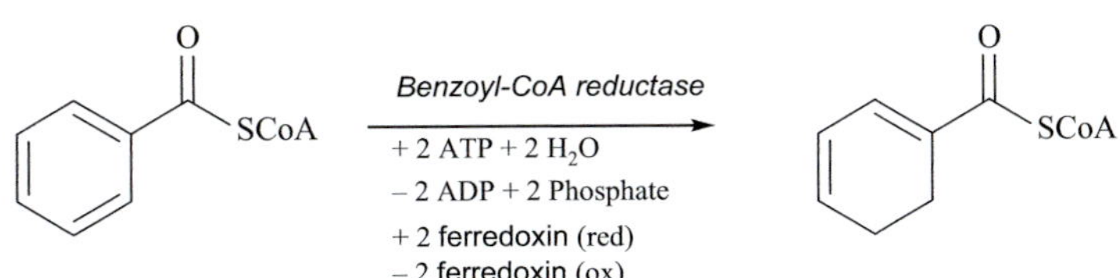

Benzoyl-CoA 1,5-Cyclohexadiene-carboxyl-CoA

Another example is the hydroxylation of phenylalanine to give tyrosine by reaction with oxygen, which is formally an oxidation. The oxidation requires the cofactor tetrahydrobiopterin, which is reduced to dihydrobiopterin during the reaction, as shown in Figure 11.12.[15] The oxygen is converted to water, and the reaction is mediated by NADP+, which is converted to NADPH. From the standpoint of tetrahydrobipterin, the transformation is effectively a dehydration reaction, catalyzed by *phenylalanine-4-monooxygenase* (*phenylalanine hydroxylase*; EC 1.14.16.1). The focus in this example is the nicotinamide adenine dinucleotide phosphate (NADP).

11.8 REDUCTION OF AROMATIC COMPOUNDS IN ORGANIC CHEMISTRY

There are several reactions of benzene derivatives that involve the benzene ring itself. Reduction reactions are particularly important for the preparation of many useful derivatives. When benzene is treated with one molar equivalent of hydrogen gas, the product is expected to be cyclohexa-1,3-diene, which reacts with hydrogen faster than benzene. It is, therefore, often difficult to isolate cyclohexa-1,3-diene in good yield, and a mixture of products is common (cyclohexadiene,

[14] Unciuleac, M.; Boll, M. *Proc. Nat. Acad. Sci.* **2001**, *98*, 13619.
[15] Daubner, S.C.; Le, T.; Wang, S. *Arch. Biochem. Biophys.* **2011**, *508*, 1.

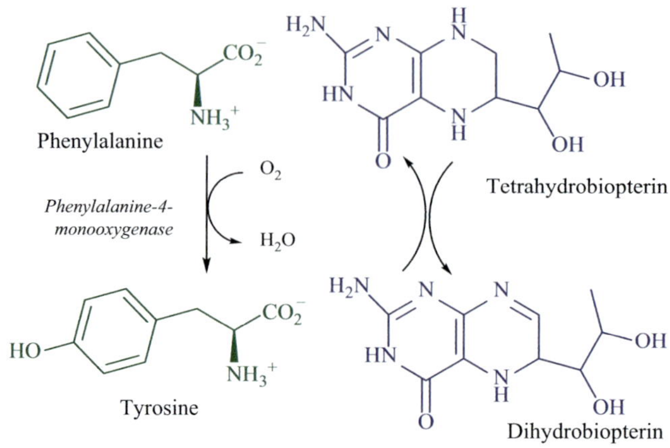

FIGURE 11.12 Biosynthesis of tyrosine. (Daubner, S.C.; Le, T.; Wang, S. *Arch. Biochem. Biophys.* **2011**, *508*, 1–12. Tyrosine hydroxylase and regulation of dopamine synthesis, Copyright (2011), with permission from Elsevier.)

cyclohexene, and cyclohexane). Control of the amount of hydrogen gas, the catalyst, and the reaction temperature allows isolation of the cyclohexene product. *With an excess of hydrogen gas (three or more molar equivalents), benzene is cleanly converted to cyclohexane.* When benzene is heated with three molar equivalents of hydrogen gas in the presence of a Raney nickel catalyst, abbreviated Ni(R), reduction yields cyclohexane as the product. Raney nickel refers to nickel prepared by a specified procedure. Hydrogenation of benzene is also possible using a palladium (Pd) or rhodium (Rh) catalyst.

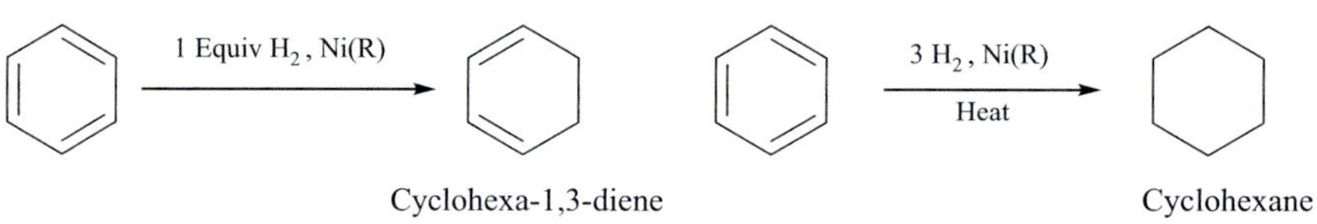

An alternative method for the reduction of benzene rings uses alkali metals (group 1 or 2), such as sodium or lithium, in liquid ammonia, often in the presence of ethanol. This method is used for the reduction of alkynes to (*E*)-alkenes, and when benzene reacts with sodium and ethanol in liquid ammonia, the product is cyclohexa-1,4-diene, as shown. This reaction is known as the *Birch reduction.*

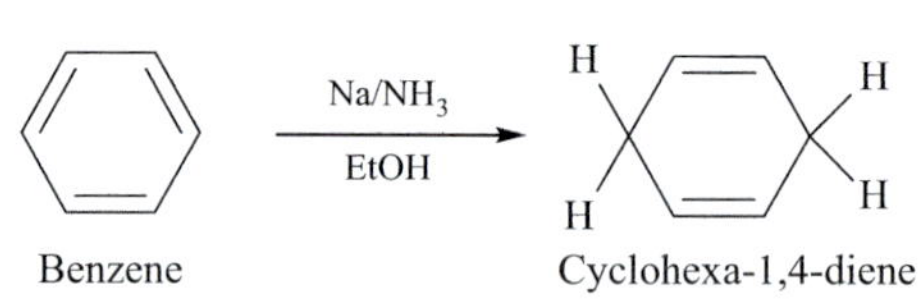

In this reaction, two hydrogen atoms are incorporated into the benzene ring with a net reduction of one C=C unit. Note that the two hydrogen atoms are incorporated from the solvent (ethanol) and that the remaining C=C units are not conjugated. The mechanism shown in Figure 11.13 accounts for the nonconjugated nature of the product and the incorporation of hydrogen atoms from the solvent.

Group 1 and 2 metals, such as Na, Li, or K, initially transfer a single electron to benzene to give a radical anion, **7**. Although there are three resonance contributors to **7**, the lowest-energy resonance contributor will arise by separation of the two electrons of the carbanion and the single electron of the radical to diminish electronic repulsion. Since the (C^-) resonance contributor in **7** is a very powerful base, the proton of the N—H unit in liquid ammonia acts a strong acid in this system. Therefore, **7** reacts with ammonia to yield radical **7** and the conjugate base, which is the amide anion, $^-NH_2$.

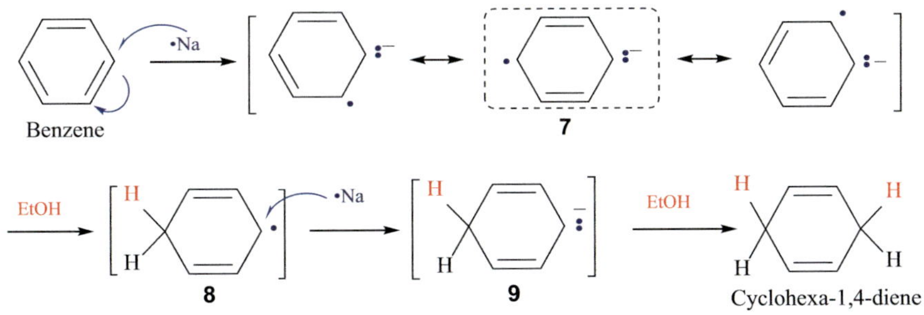

FIGURE 11.13 Mechanism of the Birch reduction of Benzene.

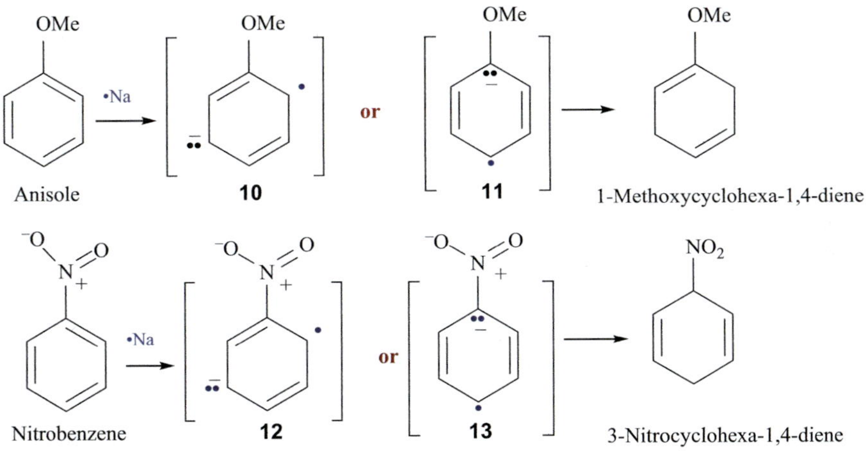

FIGURE 11.14 Birch reduction of anisole and nitrobenzene.

Ethanol (pK_a 15.7) is a stronger acid than ammonia (pK_a 36), so it is often added to the reaction so that the protonation reaction is much faster. If **7** reacts with ethanol rather than ammonia, the product is also **7**, a radical, and reaction with a second molar equivalent of sodium metal transfers one electron to generate a new carbanion (**9**). Subsequent reaction with ethanol via another acid-base reaction yields the product, cyclohexa-1,4-diene. Note that the nonconjugated diene motif is set by the formation of **7**.

The electron-transfer mechanism just described also applies to substituted benzene derivatives, but different products are formed when there is an electron-withdrawing substituent versus an electron–releasing substituent, as shown in Figure 11.14. If anisole is treated with sodium in ethanol and liquid ammonia, initial electron transfer can yield resonance-stabilized intermediates **10** or **11**. Intermediate **11** has a resonance contributor with the negative charge on the ipso carbon, but the resonance contributor in **10** does not. Electronic repulsion of the negative charge on the ipso carbon with the proximal OMe unit will destabilize **11,** so **10** is more stable and leads to the major product. Subsequent reaction with ethanol, electron transfer from a second equivalent of sodium metal, and protonation with ethanol will yield 1-methoxycyclohexa-1,4-diene.

Birch reduction of nitrobenzene differs in that the substituent is the electron-withdrawing nitro group, with a positively charged nitrogen atom. As shown in Figure 11.14, initial electron transfer from sodium can generate radical anion **12** or **13**, and both are resonance-stabilized. The negative charge is on the ipso carbon adjacent to the positively charged nitrogen atom of the nitro group in **13**, and the attraction of the positive and negative charges is stabilizing, making this intermediate more stable. Therefore, **13** will be the intermediate that leads to the major final product, 3-nitrocyclohexa-1,4-diene.

11.9 ENZYMATIC S_NAr REACTIONS

The aromatic hydroxylation reaction in Figure 11.12 shows the conversion of phenylalanine to tyrosine. The emphasis in that discussion was on the reduction of the pyridine ring, but the hydroxylation of the aromatic ring is an example of an S_NAr type reaction (nucleophilic aromatic substitution).

Enzyme-mediated aromatic substitution has several interesting variations. In one, an amine group is replaced by a hydroxy group. In other words, an aniline derivative is converted to a phenolic derivative. Thaxtomin is the main phytotoxin from *Streptomyces scabies*, but an important metabolite is 5-nitroanthranilic acid (5NAA), the nitrated analog of a key tryptophan biosynthesis intermediate. This metabolite is degraded by *Bradyrhizobium sp.* JS329, a bacterial strain isolated from potato field soil. *5-Nitroanthranilate aminohydrolase* (EC 3.5.99.8, *5NAA deaminase*) catalyzes a deamination reaction in which the para-amino group reacts with water to give a hydroxyl group, facilitated by the proximal glutamic acid residue, as shown in Figure 11.15.[16] Therefore, the product is the para-hydroxyl group in 5-nitrosalicylic acid. This aromatic substitution reaction is mediated by divalent metals (see Chapter 12), including Mn^{2+} and Zn^{2+}.[16]

The replacement of a halogen atom on an aromatic ring with a nucleophilic species is a common example of aromatic nucleophilic substitution. Chlorinated aromatic compounds are important environmental pollutants that are resistant to biodegradation, including polychlorinated biphenyls (PCBs), dioxins, and pesticides such as dichlorodiphenyltrichloroethane (DDT). There are at least three enzymatic pathways for removing chlorine from aromatic rings: hydrolytic *dehalogenases* replace chlorine substituents with hydroxyl groups that are derived from water; reductive *dehalogenases*

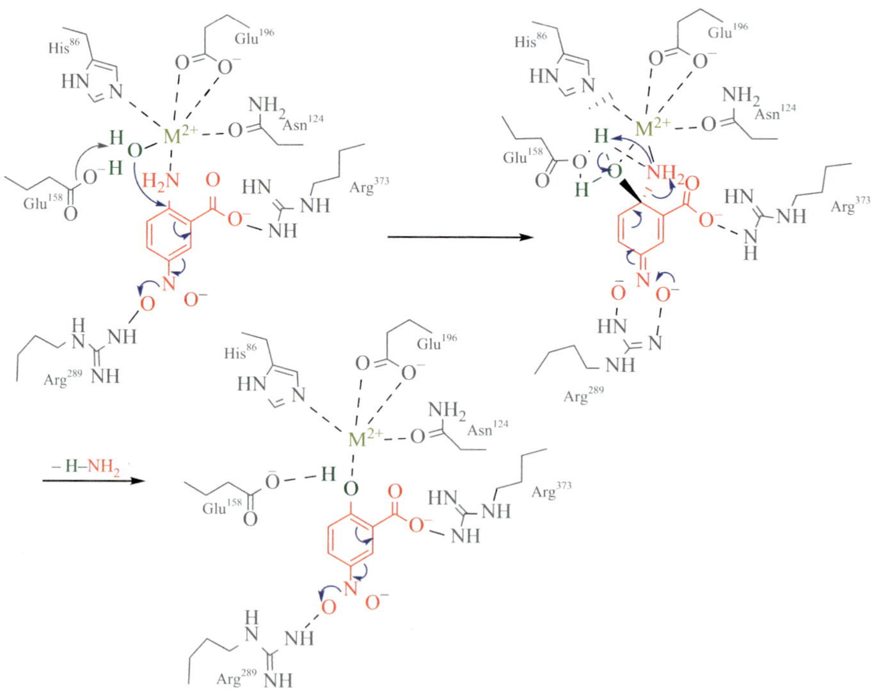

FIGURE 11.15 Enzymatic conversion of 5-nitroanthranilic acid to 5-nitrosalicylic acid. (Reprinted by permission from Springer Nature Customer Service Centre GmbH: Springer Nature, Kalyoncu, S.; Heaner Jr., D.P.; Kurt, Z.; Bethel, C.M.; Ukachukwu, C,U.; Chakravarthy, S.; Spain, J.C.; Lieberman, R.L. *Nat. Chem. Biol.* **2016**, *12*, 1031–1036. Copyright, 2016.)

[16] Kalyoncu, S.; Heaner Jr., D.P.; Kurt, Z.; Bethel, C.M.; Ukachukwu, C,U.; Chakravarthy, S.; Spain, J.C.; Lieberman, R.L. *Nat. Chem. Biol.* **2016**, *12* 1031.

replace chlorine substituents with hydrogen atoms; and oxygen-dependent *dehalogenases* introduce hydroxyl groups from sources other than water. An important hydrolytic aromatic *dehalogenase* is *4-chlorobenzoyl coenzyme A (CoA) dehalogenase* (EC 3.8.1.7), found in bacteria that degrade 4-chlorobenzoate, a metabolite of some PCBs. In the initial step, Asp[145] is the source of the OH moiety, as shown in Figure 11.16,[17] and the nucleophilic aromatic substitution reaction leads to expulsion of chloride to form an acyl-enzyme intermediate. Attack by water gives a tetrahedral intermediate via acyl substitution that leads to 4-hydroxybenzoyl CoA and regenerates the free enzyme.

The glutathione transferases (EC 2.5.1.18) are detoxification enzymes that react with xenobiotics or potentially harmful metabolic products by conjugating a glutathione molecule. The resulting complex improves water solubility, which allows excretion of the harmful compounds.[18] The enzymatic reaction replaces an aromatic halogen with glutathione, as shown in Figure 11.16,[19] with the reaction of 2,4-dinitrochlorobenzene and glutathione, catalyzed by *glutathione transferase*. The initial reaction with glutathione generates a resonance-stabilized complex (**14**) analogous to those formed by an S_NAr. Loss of chlorine from complex **14** leads to the substitution product, **15** (Figure 11.17).

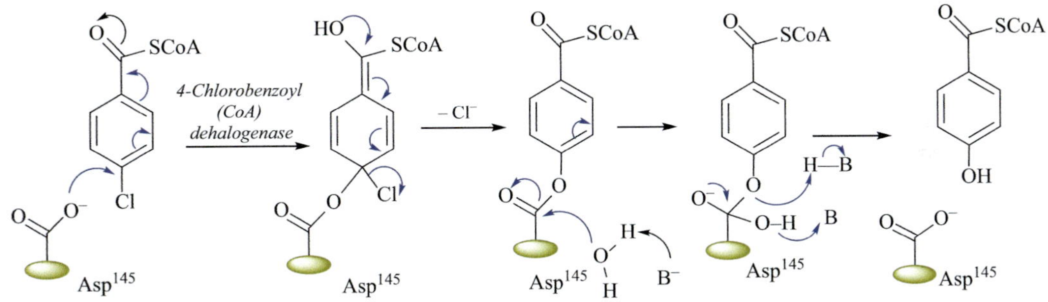

FIGURE 11.16 Enzyme-mediated replacement of chlorine with hydroxyl. (Copley, S.D. *Chem. Biol.* **1997**, *4*, 169–174, Diverse mechanistic approaches to difficult chemical transformations: microbial dehalogenation of chlorinated aromatic compounds. Copyright (1997), with permission from Elsevier.)

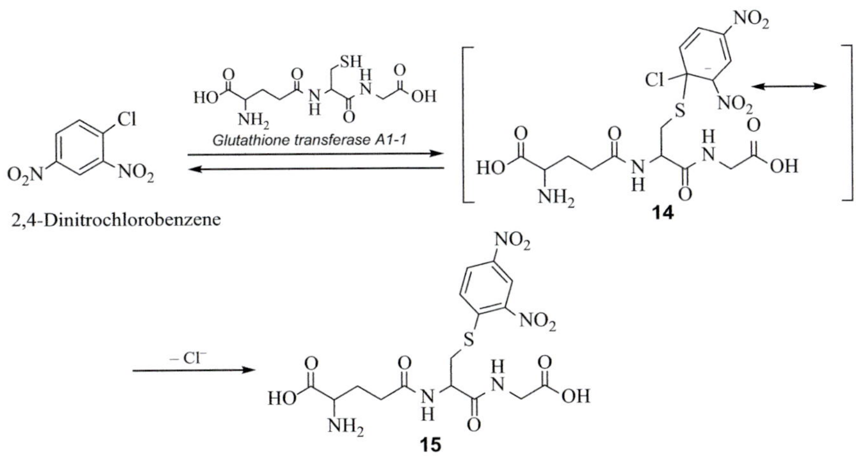

FIGURE 11.17 Enzyme-mediated reaction of glutathione with aromatic nitro compounds. (Reprinted with permission from Widersten, M.; Björnestedt, R.; Mannervik, B. *Biochemistry* **1996**, *35*, 7731–7742. Copyright 1996 American Chemical Society.)

[17] Copley, S.D. *Chem. Biol.* **1997**, *4*, 169.
[18] Lien, S.; Gustafsson, A.; Andersson, A.-K.; Mannervik, B. *J. Biol. Chem.* **2001**, *276*, 35599.
[19] Widersten, M.; Björnestedt, R.; Mannervik, B. *Biochemistry* **1996**, *35*, 7731.

11.10 NUCLEOPHILIC AROMATIC SUBSTITUTION: THE S_NAr REACTION

Electrophilic aromatic substitution reactions (S_EAr) were discussed in Section 11.6. An alternative reaction involves a nucleophile donating electrons *to* a benzene ring. Such a reaction is known as *nucleophilic aromatic substitution* and is given the symbol *S_NAr*. When chlorobenzene was heated to 50°C in aqueous hydroxide, there was no reaction, and clearly, attack of the negatively charged hydroxide on the electron-rich aromatic ring is difficult. However, the reaction can be "forced" if the reaction is done in a sealed reaction vessel and heated to 350°C so that it is pressurized, and under these conditions, chlorobenzene is converted to phenol. In general, nucleophilic aromatic substitution with chlorobenzene requires high reaction temperatures because the benzene ring is electron-rich, and hydroxide is an electron-rich species.

The reaction works as shown in Figure 11.18, with attack of hydroxide as a nucleophile on the sp^2 carbon that bears the chlorine (the *ipso carbon*) to generate carbanionic intermediate **16**. Although this intermediate is resonance-stabilized, there is a high activation energy for this reaction, and once formed, it is very reactive. The chloride ion is a good leaving group in **16**, however, and it is rapidly expelled to initially form phenol. Since phenol is formed in a basic solution of aqueous hydroxide and phenol is a moderately strong acid ($pK_a \sim 10$), it reacts with the hydroxide to form phenoxide. When the basic phenoxide ion is neutralized with dilute acid, the acid–base reaction will regenerate phenol. Therefore, the conversion of chlorobenzene to phenol is a two-step process: reaction of chlorobenzene with aqueous NaOH followed by treatment with aqueous acid. This reaction is an example of an aromatic substitution process and is usually called *nucleophilic aromatic substitution*. This reaction is sometimes referred to as an *aromatic S_N2 reaction*, but this term is incorrect since it implies a concerted reaction. The correct term for this two-step process is *an S_NAr reaction*.

Another example of an S_NAr reaction heats benzenesulfonic acid with aqueous NaOH (to 300°C) in a reaction bomb. The fastest reaction is the deprotonation of the sulfonic acid by hydroxide ion to yield sodium benzenesulfonate, which is an acid–base reaction. This sulfonate anion reacts with hydroxide to give the resonance-stabilized carbanionic intermediate by an S_NAr pathway. Loss of sodium sulfite (Na_2SO_3) yields phenoxide in the aqueous NaOH medium. Cooling the reaction and neutralization with acid yields phenol. The SO_3 unit is the leaving group in this reaction, and it is a straightforward S_NAr reaction. In fact, it is the basis for an important industrial preparation of phenol.

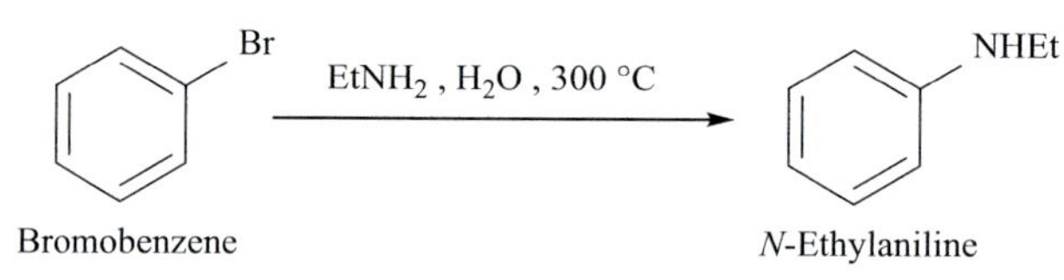

The S_NAr reaction can be used to make aniline derivatives. When bromobenzene is heated with ethylamine ($EtNH_2$) in aqueous solution (to ~ 300°C), the S_NAr reaction is slow, but the product is *N*-ethylaniline. In general, aromatic S_N2 reactions are slow in water or alcohol solvents, but they are much faster when tertiary amide solvents [e.g., *N,N*-dimethylformamide (DMF) or *N,N*-dimethylacetamide] are used. The reaction of bromobenzene with ethylamine, for example, produces a good yield of *N*-ethylaniline at a reaction temperature of only 160°C–200°C when DMF is used as the solvent.

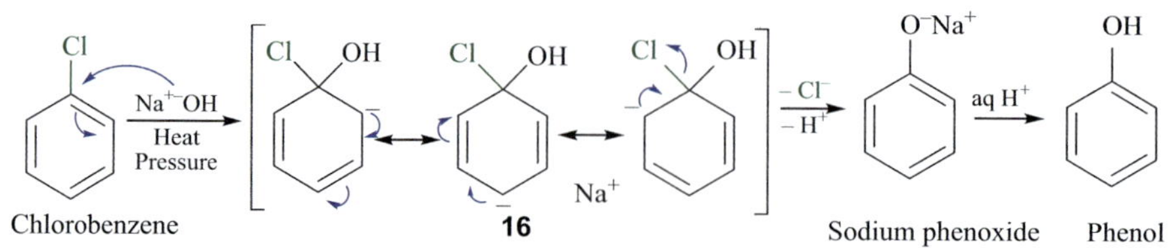

FIGURE 11.18 Reaction of chlorobenzene and hydroxide to yield phenol.

11.11 POLYNUCLEAR AROMATIC HYDROCARBONS

There are many aromatic compounds that meet the $4n+2$ rule, with structures where the π-electrons are not confined to one ring, but rather contained in several rings that are fused together. They are called *polynuclear (or polycyclic) aromatic molecules*. Three simple examples are shown: naphthalene, anthracene, and phenanthrene. Naphthalene is a bicyclic aromatic compound with the formula $C_{10}H_8$. It was the main constituent of "mothballs" for many years, but toxicity led to its replacement with 1,4-dichlorobenzene (characterized as an insecticidal fumigant). Anthracene is a polycyclic aromatic compound (14 π-electrons) with three rings fused together and the formula $C_{14}H_{10}$. Anthracene is used to prepare other important compounds and is used in wood preservatives and insecticides. Phenanthrene is an isomer of anthracene in which the point of attachment of the "third ring" on the "middle ring" is different. Like anthracene, phenanthrene is derived from coal tar and is used in the synthesis of dyes, explosives, and drugs.

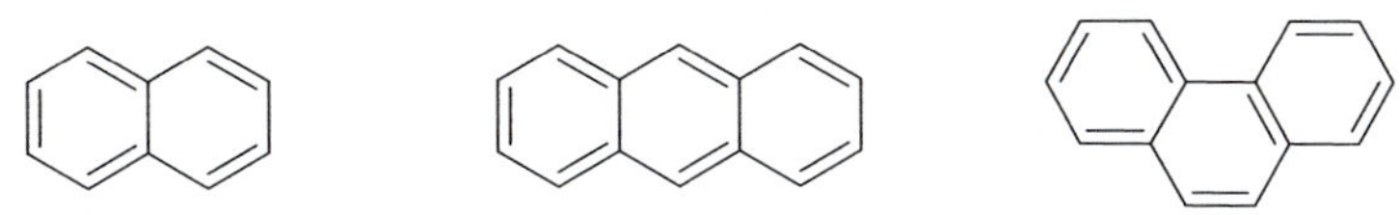

Naphthalene Anthracene Phenanthrene

Polynuclear aromatic hydrocarbons (e.g., naphthalene, anthracene, and phenanthrene) undergo electrophilic aromatic substitution reactions in the same manner as benzene. A significant difference is that there are more carbon atoms, more potential sites for substitution, and more resonance structures to consider in the arenium ion intermediates. The reaction of naphthalene with Cl_2/$AlCl_3$ gives 1-chloronaphthalene as the major product, one of only two *different* positions for substitution, C1 and C2. Chlorination of naphthalene at C1 leads to the arenium ion shown in Figure 11.19, with five resonance structures.

A similar reaction at C2 yields the arenium ion in Figure 11.19, which has the five resonance structures shown. There is a subtle difference in these two intermediates that makes substitution at C1 preferred to that at C2. Note that the arenium ions in Figure 11.19 each have what appears to be one or two Kekulé structures in some of the resonance contributors. These Kekulé structures represent fully aromatic benzene rings (an "*intact benzene ring*"), and there is an additional Kekulé structure for each intact ring. Therefore, the arenium ion from reaction at C1 has four Kekulé structures and a total of seven resonance structures, but the arenium ion from reaction at C2 has only two Kekulé structures and a total of six resonance contributors. The extra resonance contributor makes

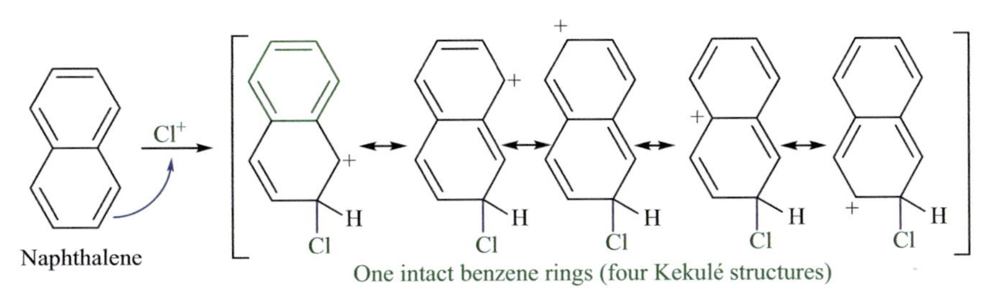

FIGURE 11.19 Arenium ion for chlorination at C1 of naphthalene.

the arenium ion for substitution at C1 more stable than the arenium ion for substitution at C2. The more stable arenium ion will form faster and lead to the major product, 1-chloronaphthalene.

Anthracene has three different positions (C1, C2, and C9), and phenanthrene has five different positions (C1, C2, C4, C5, and C9). Electrophilic aromatic substitution in both compounds can give more than one product. However, substitution of anthracene leads to substitution primarily at C9 since that arenium ion has the most resonance forms and the most "intact benzene rings." A comparison of attack at C1 and C2 in anthracene shows that there are more resonance forms for attack at C1 and more fully aromatic rings. Attack at C9 leads to an intermediate with even more resonance contributors, but electrophilic substitution usually leads to a mixture of C9 and C1 products, with little reaction at C2.

Electrophilic substitution of phenanthrene is more complicated since there is less difference in the stability of the intermediate cations formed from S_EAr reactions at each substitution position. Nitration of phenanthrene leads to five aromatic substitution products, and reactions of this type will not be discussed further for phenanthrene. The "middle" ring in phenanthrene is not stabilized by aromaticity to the same extent as the others, and C9 and C10 are susceptible to reactions not usually observed with aromatic systems; those positions are commonly referred to as the *bay region* of phenanthrene. Phenanthrene reacts with diatomic bromine in the *absence* of a Lewis acid, much like a simple alkene, to yield 9,10-dibromo-9,10-dihydrophenanthrene. This addition-type reaction does not occur with benzene or naphthalene.

There are many polycyclic aromatic hydrocarbons. A handful of examples are shown here. These include chrysene, pyrene, coronene, tetraphene, benzo[a]pyrene, and ovalene. Pyrene was first isolated from coal tar. Animal studies have shown pyrene to be toxic to the kidneys and liver. Pyrene also affects several living functions in fish and algae. Chrysene is a natural part of coal tar, is also present in creosote, and is a constituent of tobacco smoke. Coronene is of theoretical interest because of its aromaticity and naturally as the very rare mineral carpathite. It is believed that coronene was formed from contact of magma with fossil fuel deposits. Tetraphene (benz[a]anthracene) is produced during the incomplete combustion of organic matter and is one of the carcinogenic constituents of tobacco smoke. Benzo[a]pyrene is the result of incomplete combustion of organic matter and can be found in coal tar, tobacco smoke, and many foods, especially grilled meats. Ovalene forms in deep-sea hydrothermal vents and in the hydrocracking process of petroleum refining.

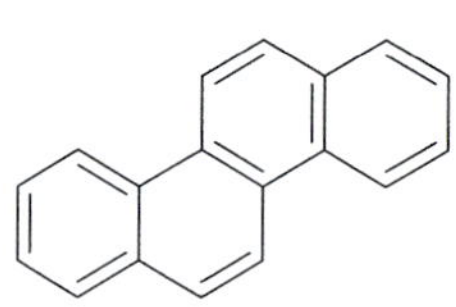
Chrysene

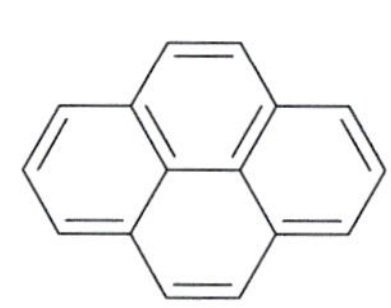
Pyrene

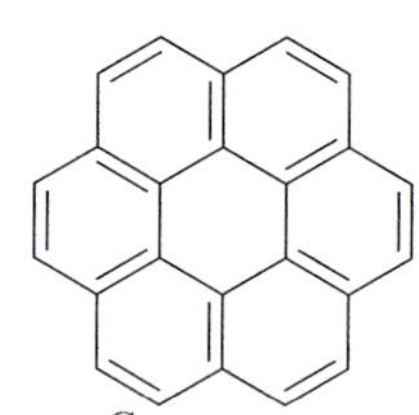
Coronene

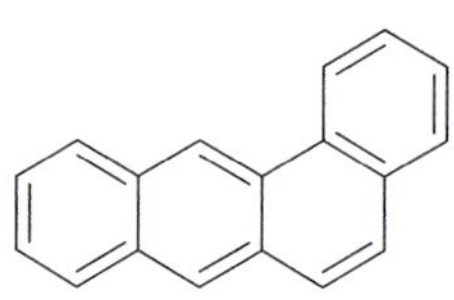
Tetraphene

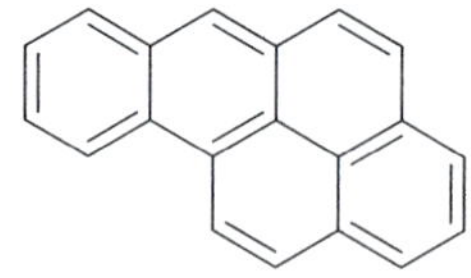
Benzo[a]pyrene

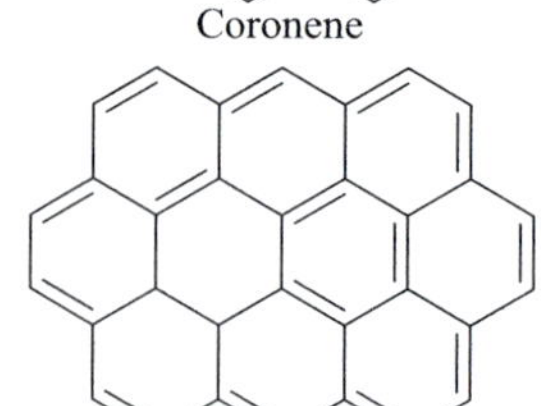
Ovalene

11.12 ENZYMATIC REACTIONS THAT GENERATE HETEROCYCLIC COMPOUNDS

Aromatic compounds are important components of many bioactive molecules, and the biosynthesis of aromatic compounds is important. The purine or pyrimidine units in DNA and RNA have been discussed, including biochemical pathways that produce these compounds (Section

15.3). Other aromatic compounds, some with heterocyclic rings, appear in biological systems that include common amino acids. Perhaps the most common biopathway to produce aromatic compounds is the Shikimate pathway, which was first introduced in Figure 11.1.[20] The Shikimate pathway is not used to synthesize all aromatic compounds, but many amino acids that have an aromatic unit are biosynthesized by this route, including phenylalanine, tyrosine, and tryptophan. As shown in Figure 11.20,[20a] the pathway begins with the *3-deoxy-*D*-arabino-heptulosonate 7-phosphate synthase* (EC 2.5.1.54)-catalyzed reaction of erythrose 4-phosphate and phosphoenolpyruvate to give 3-deoxy-D-aribino-heptulosonate 7-phosphate. The subsequent reaction, catalyzed by *3-dehydroquinate synthase* (EC 4.2.3.4), gives 3-dehydroquinate. The reaction with *3-dehydroquinate dehydratase* (EC 4.2.1.10) and *shikimate dehydrogenase* (EC 1.1.1.25), in the presence of NADPH, gives 3-dehydroshikimate as a transient product via loss of water, which is reduced by the *dehydrogenase* to give shikimate. *Shikimate kinase* (EC 2.7.1.71) generates shikimate 3-phosphate, and *5-enolpyruvylshikimate 3-phosphate synthase* (EC 2.5.1.19) leads to loss of phosphate and

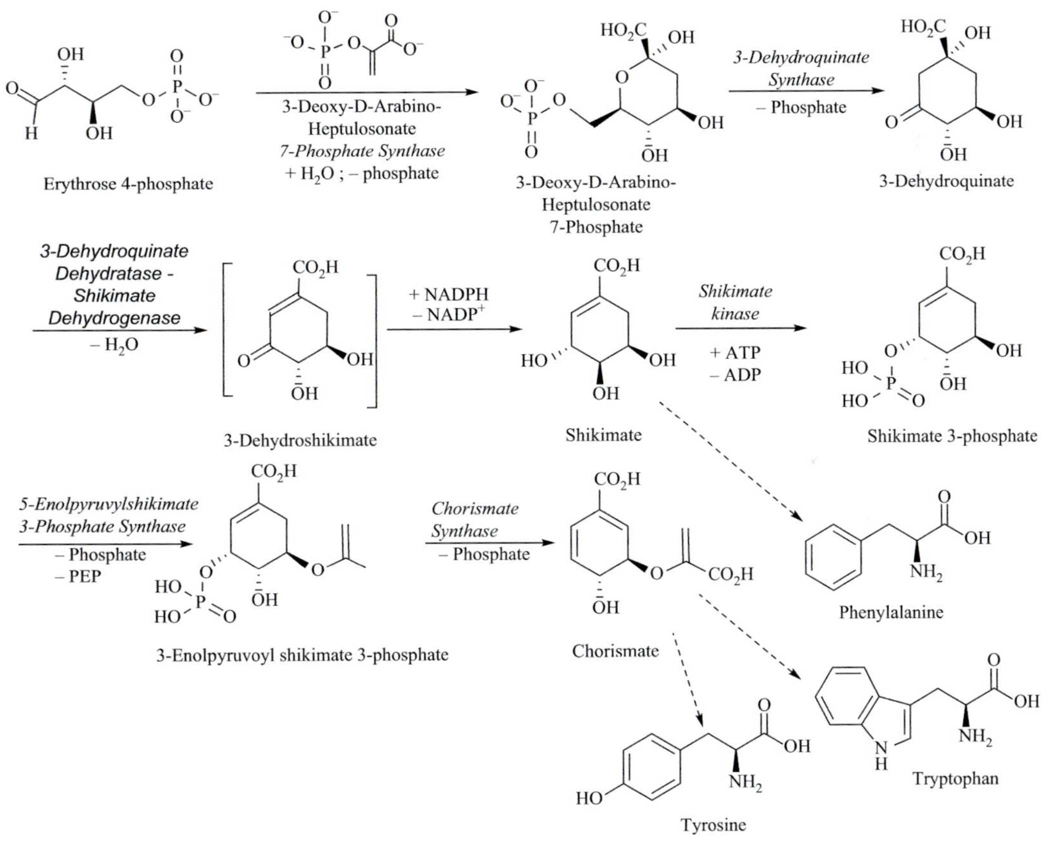

FIGURE 11.20 Shikimate pathway. (Adapted from: Reprinted with permission from Ran, N.; Frost, J.W. *J. Amer. Chem. Soc.* **2007**, *129*, 6130–61392. Copyright 2007 American Chemical Society; Reprinted with permission from He, Z.; Stigers, K.D.; Lavoie, S.; Bartlett, P.A.; Toney, M.D. *J. Amer. Chem. Soc.* **2004**, *126*, 2378–2385. Copyright 2004 American Chemical Society; Reprinted with permission from Choo, H.J.; Ahn, J.-H. *J. Agric. Food Chem.* **2019**, *67*, 8581–8589. Copyright 2019 American Chemical Society; Reprinted with permission from Yang, S.; Cao, Y.; Sun, L.; Li, C.; Lin, X.; Cai, Z.; Zhang, G.; Song, H. *ACS Synth. Biol.* **2019**, *8*, 70–81. Copyright 2019 American Chemical Society.)

[20](a) Ran, N.; Frost, J.W. *J. Amer. Chem. Soc.* **2007**, *129*, 6130; (b) He, Z.; Stigers, K.D.; Lavoie, S.; Bartlett, P.A.; Toney, M.D. *J. Amer. Chem. Soc.* **2004**, *126*, 2378; (c) Choo, H.J.; Ahn, J.-H. *J. Agric. Food Chem.* **2019**, *67*, 8581; (d) Yang, S; Cao, Y.; Sun, L.; Li, C.; Lin, X.; Cai, Z.; Zhang, G. Song, H. *ACS Synth. Biol.* **2019**, *8*, 70.

phosphoenolpyruvate to give 3-enolpyruvyl shikimate 3-phosphate. In the final step, chorismate is formed by loss of phosphate, catalyzed by *chorismate synthase* (EC 4.2.3.5). Once chorismate is formed, reaction with *anthranilate synthase α/anthranilate synthase β* (EC 4.2.3.5) leads to anthranilic acid. Chorismate is also the precursor to the aromatic substituted amino acids phenylalanine, tryptophan, and tyrosine.

Pyridine derivatives are also important biomolecules. Quinolinate is the biosynthetic precursor of nicotinamide adenine dinucleotide (NAD), which is an important coenzyme for oxidation–reduction reactions. In prokaryotes, quinolinate is formed from aspartate and dihydroxyacetone phosphate in a two-step pathway using *NadB* (*L-aspartate oxidase*, EC 1.4.3.16) to oxidize the amino group of L-aspartate to the imino group, giving α-iminoaspartate, as shown in Figure 11.21.[21] In the second step, *NadA* (*quinolinate synthase*; EC 2.5.1.72) converts α-iminoaspartate and dihydroxyacetone phosphate to quinolinic acid.

Pyrroles are heterocyclic aromatic compounds, and they are important structural units in many bioactive compounds. A well-studied mechanism for pyrrole biosynthesis is the generation of porphobilinogen, an early intermediate in porphyrin and corrin ring synthesis. The synthesis of the pyrrole moiety is catalyzed by δ-aminolevulinate dehydratase, whereby the condensation of two molecules of δ-aminolevulinate gives porphobilinogen, as shown in Figure 11.22.[22a]

Other heterocycles have been identified as key components of biomolecules. Thiamin contains a thiazole ring as well as a pyrimidine ring, and it is an essential component of the human diet. Thiamin is a water-soluble vitamin of the B complex (vitamin B_1). While humans cannot synthesize thiamin, many bacteria synthesize it and use it in the form of thiamin pyrophosphate. *Bacillus subtilis* produces thiamin by a biosynthetic route, and it synthesizes the thiazole unit found in thiamin from the amino acid glycine and 1-deoxy-D-xylulose-5-phosphate, **17**.[23] As shown in Figure 11.23,[23] the enzyme *DXP synthase* (EC 2.2.1.7) couples **17** and glycine to give iminium salt **18**. Note that this

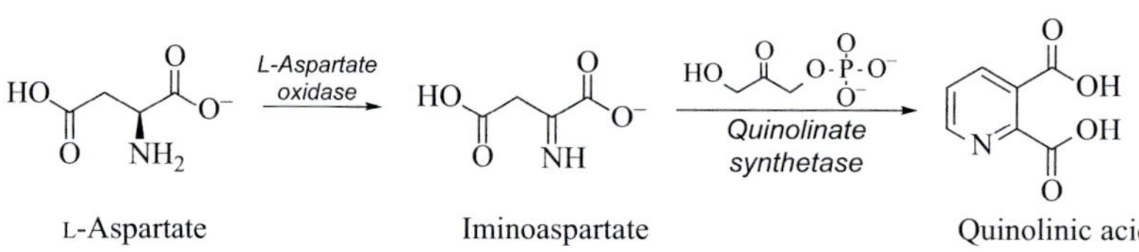

FIGURE 11.21 Biosynthesis of quinolinic acid. (Lima, W.C.; Varani, A.M.; Menck, C.F.M. *Mol. Biol. Evol.* **2009**, *26*, 399–406. Reproduced with permission from Oxford University Press.)

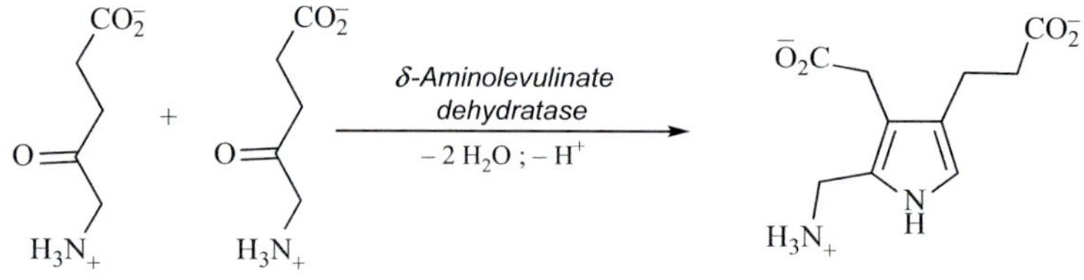

FIGURE 11.22 Enzymatic synthesis of Porphobilinogen. (Thomas, M.G.; Burkart, M.D.; Walsh, C.T. *Chem. Biol.* **2002**, *9*, 171–184, Conversion of L-Proline to Pyrrolyl-2-Carboxyl-S-PCP during Undecylprodigiosin and Pyoluteorin Biosynthesis. Copyright (2002), with permission from Elsevier.)

[21] Lima, W.C.; Varani, A.M.; Menck, C.F.M. *Mol. Biol. Evol.* **2009**, *26*, 399.

[22] (a) Thomas, M.G.; Burkart, M.D.; Walsh, C.T. *Chem. Biol.* **2002**, *9*, 171. Also see (b) Jordan, P.M. *Curr. Opin. Struct. Biol.* **1994**, *4*, 902.

[23] Begley, T.P.; Downs, D.M.; Ealick, S.E.; McLafferty, F.W.W.; Van Loon, A.G.M.; Taylor, S.; Campobasso, N.; Chiu, H.-J.; Kinsland, C.; Reddick, J.J.; Xi, J. *Arch. Microbiol.* **1999**, *171*, 293–300. https://doi.org/10.1007/s00203005071.

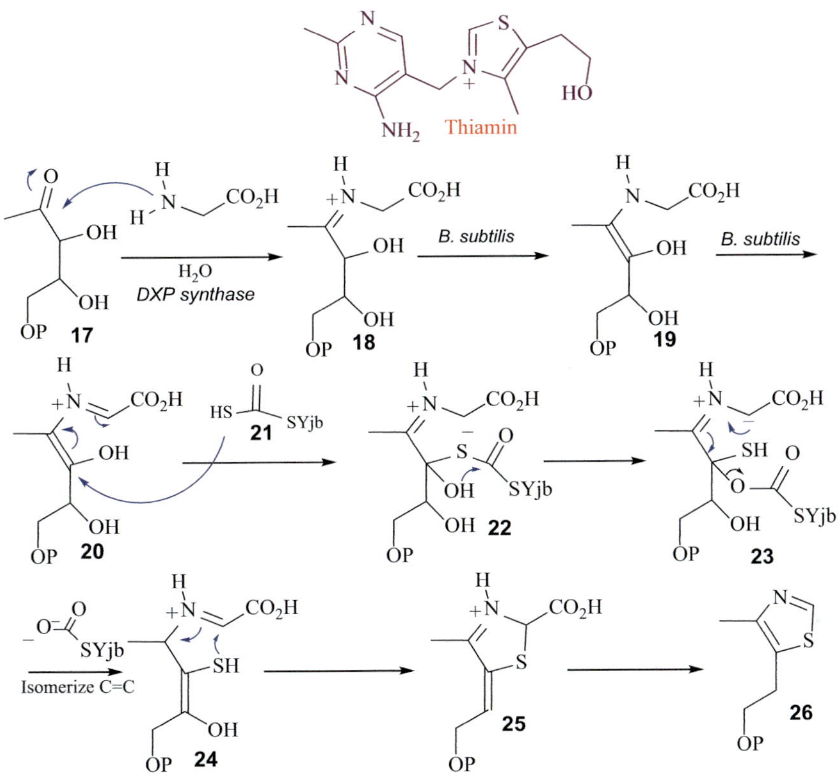

FIGURE 11.23 Biosynthesis of a thiazole unit. (Springer and Begley, T.P.; Downs, D.M.; Ealick, S.E.; McLafferty, F.W.W.; Van Loon, A..G.M.; Taylor, S.; Campobasso, N.; Chiu, H.-J.; Kinsland, C.; Reddick, J.J.; Xi, J. *Arch. Microbiol.* **1999**, *171*, 293. With kind permission from Springer Science and Business Media.)

reaction is effectively an amine–ketone reaction to yield an iminium salt, and an enzymatically catalyzed shift of the double bond generates the enamine form (**19**). The enzyme also removes a hydrogen atom α to the carboxyl carbonyl to generate enol **20**. Subsequent reaction with thiocarbonyl derivative **21** yields thioester **22**. Note that Yjb in Figure 11.23 is a gene cluster that contains the COSH unit.

An internal transesterification reaction converts **22** to **23**, and the enolate anion provides an electronic driving force to expel the acetate unit containing the gene cluster, and isomerization of the resulting double bond yields **24**. Internal addition of the thiol unit to the iminium salt generates the five-membered ring in **25**, and aromatization leads to the thiazole unit in **26**. Other enzyme reactions use **26** to prepare thiamine.

11.13 HETEROAROMATIC COMPOUNDS: NITROGEN, OXYGEN, OR SULFUR

Another class of aromatic compounds is important in chemistry and biology, and they are characterized by the replacement of one or more ring carbons with a heteroatom. These compounds are collectively known as *heterocycles* or *heterocyclic aromatic compounds*, and they comprise a class of compounds so large that an entire course could easily be built around their chemistry. Common heterocycles include five- and six-membered monocyclic derivatives that contain nitrogen, oxygen, or sulfur. In addition, there are several important bicyclic derivatives that contain nitrogen, oxygen, or sulfur.

In a "thought experiment," replace the CH moiety of cyclopentadiene with an N–H unit, and the result is the aromatic compound pyrrole, a constituent of coal tar that is also found in bone oil. Pyrrole is an aromatic compound because the nitrogen atom has an unshared electron pair in an

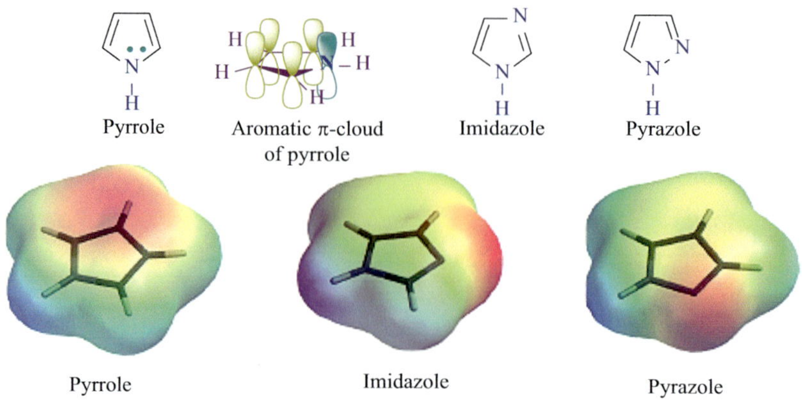

FIGURE 11.24 Nitrogen-containing five-membered ring aromatic heterocycles.

orbital that is parallel with the four π-electrons in the C=C units, forming a six π-electron system in a π-framework (see Figure 11.24). Pyrrole is a planar molecule, and the hydrogen atom is coplanar with the carbon atoms and nitrogen, as shown in Figure 11.24. Since the lone electron pair is part of the aromatic sextet, the electrons cannot be donated without disrupting the aromaticity, and pyrrole is not very basic. There are other aromatic five-membered ring amines, but they have two nitrogen atoms in the ring. The two nitrogen atoms in *imidazole* have a 1,3-relationship (note that an older term for imidazole is *azole*). The two nitrogen atoms in *pyrazole* have a 1,2-relationship. The convention for naming these compounds makes the nitrogen bearing the hydrogen atom 1, and the numbering around the ring is such that the second nitrogen receives the lowest possible number (2 or 3).

Both imidazole and pyrazole are important units in many pharmaceutical preparations, as well as in naturally occurring compounds. Histidine is an important amino acid, and histamine is a neurotransmitter that is important in cells during antigen–antibody reactions. Important physiological reactions attributed to histamine include anaphylaxis (a significant drop in blood pressure that can result in shock) and allergy. Pilocarpine is an imidazole derivative that is used to treat glaucoma. Note the presence of a lactone ring in pilocarpine. The imidazole derivatives clotrimazole (Lotrimin) and miconazole (Monistat, Micatin) are antifungal agents that are applied topically in preparations to treat Athlete's foot.

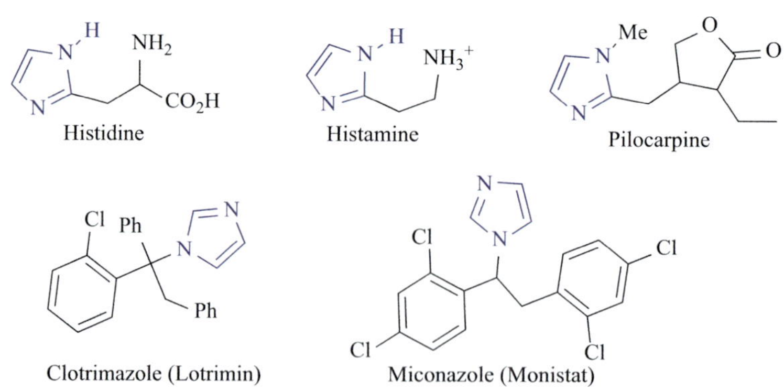

Pyridine is an aromatic six-membered ring compound that contains one nitrogen atom, first isolated in 1846 from coal tar. Pyridine is aromatic, so it is a planar compound with an aromatic π-cloud above and below the plane of the ring. The nitrogen atom is sp^2 hybridized, but the lone electron pair on nitrogen is *not* part of the aromatic π-cloud. That electron pair is perpendicular to the π-cloud (Figure 11.25), and pyridine is a good Brønsted–Lowry and Lewis base. The electron

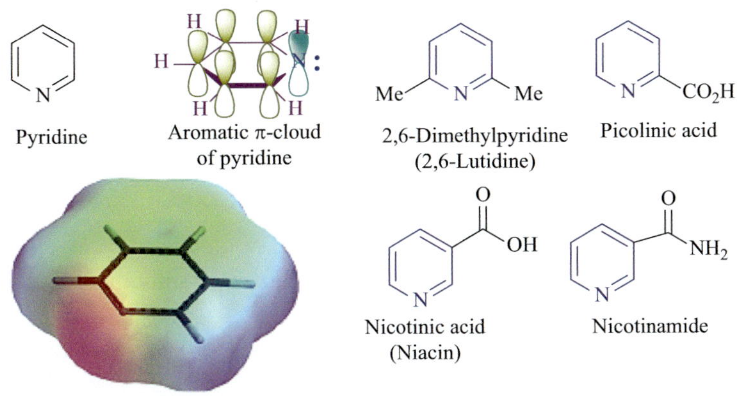

FIGURE 11.25 Pyridine and pyridine derivatives.

density map of pyridine clearly shows a high concentration of electron density (more red) on the nitrogen. Important substituted pyridine derivatives include 2,6-lutidine (2,6-dimethylpyridine) and picolinic acid. Many pyridine derivatives are found in pharmaceutically active compounds. One is nicotinic acid (niacin, vitamin B3), found in liver, yeast, and meat. A deficiency in this vitamin can lead to pellagra (a wasting disease). Nicotinamide (niacinamide) is one of the two principal forms of the B-complex vitamin niacin. Nicotinamide may be useful for individuals with type 1 (insulin-dependent) diabetes.

Three different aromatic six-membered ring compounds contain two nitrogen atoms in the ring: *pyrazine*, *pyrimidine*, and *pyridazine*. Pyrazine has two nitrogen atoms in a 1,4-relationship, pyrimidine has two nitrogen atoms in a 1,3-relationship, and pyridazine has two nitrogen atoms in a 1,2-relationship. In each case, both nitrogen atoms are sp^2 hybridized, and *both* nitrogen atoms are basic. As with all aromatic compounds, these are planar compounds with an aromatic π-cloud above and below the plane of the ring.

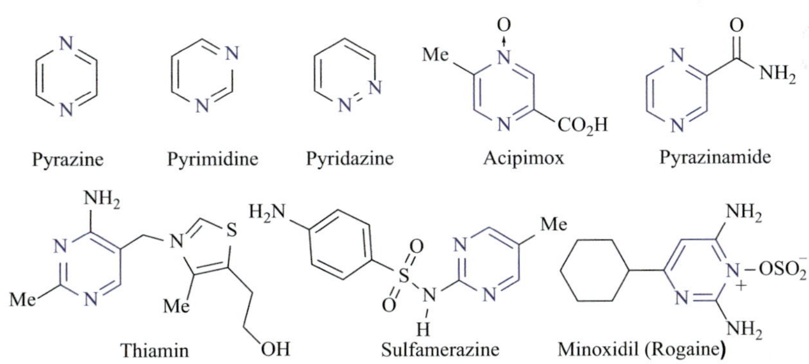

Pyrazine, pyrimidine, and pyridazine derivatives are found in pharmaceutically important compounds. The pyrazine derivative (5-carboxy-2-methylpyrazine 1-oxide) is called Acipimox, and it is used to lower levels of cholesterol and triglycerides. Note that the pyrazine ring is an oxidized form of the amine, known as an *N*-oxide. Pyrazinamide (pyrazine-2-carboxamide) is an antibacterial agent.

The pyrimidine ring system is found in thiamin (known as vitamin B1). A deficiency of thiamin is associated with beriberi. Symptoms of beriberi include loss of appetite, fatigue, digestive irregularities, and a feeling of numbness and weakness in the limbs and extremities that may lead to nerve degeneration in extreme cases. Another pyrimidine, sulfamerazine, is a broad-spectrum antibacterial agent. Minoxidil (Rogaine) is classified as a vasodilator used to treat high blood pressure, but it

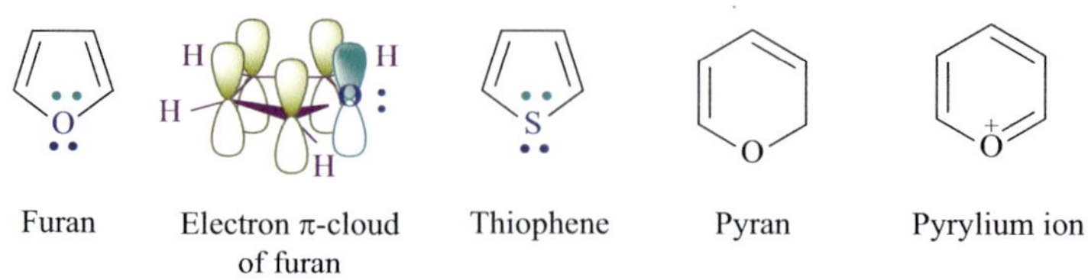

FIGURE 11.26 Furan, thiophene, and pyran.

is also used in hair-restoring preparations. A vasodilator relaxes the smooth muscle in blood vessels, causing the vessels to dilate.

There are five-membered ring aromatic compounds that contain an oxygen or a sulfur, but the analogous six-membered ring compounds are *not* aromatic. Furan is an aromatic compound that is distilled from pine wood rosin, and its vapors are narcotic. As shown in Figure 11.26, one of the two electron pairs on oxygen is involved in the aromatic π-cloud (those two electrons are needed to make a total of six), but the other lone electron pair is perpendicular to the π-cloud. Therefore, furan can function as a base, and it should be a *stronger* base than pyrrole because of the availability of those electrons. While furan is a somewhat stronger base than pyrrole, it is generally categorized as a weak base. Furan is an aromatic ether, and the furan ring plays a prominent role in many naturally occurring and synthetic compounds.

If the oxygen atom is replaced with sulfur, the resulting compound is the aromatic compound thiophene, which is an aromatic thioether. As with furan, one electron pair is involved in the six π-electron aromatic cloud, and the other is perpendicular to that π-cloud. Thiophene is a weaker base than furan. Oxygen is divalent, so incorporation into a neutral six-membered ring does not allow an aromatic system to be generated, but the non-aromatic pyran is well known. Indeed, pyran is the six-membered ring analog, but for oxygen to remain neutral, one carbon must be sp^3 hybridized, as shown. An aromatic system can be generated by adding a charge to the oxygen to give the pyrylium ion, which is a transient intermediate in some reactions. In order to focus on neutral compounds that are aromatic, the following discussions of aromatic oxygen and sulfur compounds are restricted to furan and thiophene.

11.14 REACTIONS OF HETEROAROMATIC COMPOUNDS

Aromatic heterocyclic compounds undergo electrophilic aromatic substitution reactions like reactions of aromatic hydrocarbons. Five-membered heterocyclic rings undergo electrophilic aromatic substitution reactions (S_EAr) faster than benzene, so they are activated aromatic rings. Six-membered heterocyclic rings react slower than benzene in S_EAr reactions, so they are deactivated. In other words, the nature of the heteroatom and the size of the ring have a profound influence on the rate of reaction, as well as the site of reaction. The basic principles of reactivity and regioselectivity in these cases are governed by the same fundamental principles discussed for benzene derivatives in Section 11.5. For S_EAr reactions of heterocycles, an arenium ion intermediate is formed, but the influence of the electron-rich heteroatom must be taken into account. The major site of substitution in this reaction is the one that yields the more stable arenium ion intermediate.

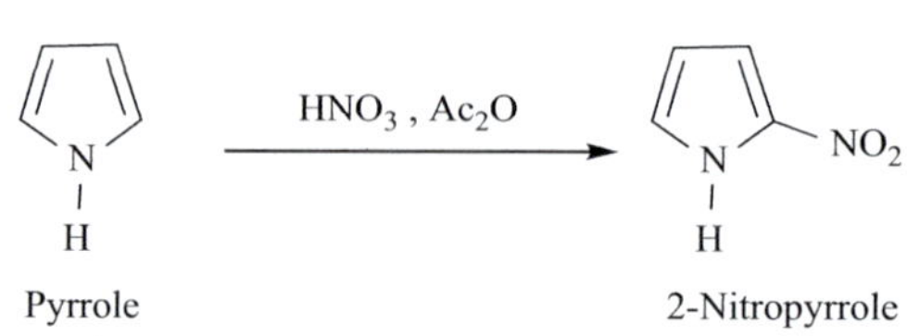

Pyrrole reacts with nitric acid in the presence of acetic anhydride by the S_EAr mechanism to yield 2-nitropyrrole, with high selectivity for substitution at C2. Pyrrole is a very electron-rich aromatic system and, as noted, it is much more reactive than benzene in S_EAr reactions, so *it is an activated aromatic ring.* There are two potential sites for substitution, C2 and C3, when an electrophile reacts with pyrrole. If the pyrrole ring attacks the nitronium ion, NO_2^+ attaches to C2 to yield intermediate **27**, as shown in Figure 11.27, whereas intermediate **28** is formed by attachment at C3. It is apparent that reaction at C3 can generate only two resonance forms, but reaction at C2 generates three. In both cases, a positive charge resides at the carbon adjacent to an sp^2 nitrogen, leading to a stable iminium ion structure. Therefore, nitration of pyrrole yields 2-nitropyrrole as the major product. The same preference for C2 substitution is true if the heteroatom is oxygen or sulfur. Indeed, both furan and thiophene yield predominantly C2 substitution in S_EAr reactions.

Pyridine is a tertiary amine and a good base, and many of the electrophilic reagents used for aromatic substitution coordinate with the electron pair on nitrogen (an acid-base reaction). Specifically, the Lewis acids used for electrophilic aromatic substitution coordinate with the electron pair on nitrogen, so they cannot be used in substitution reactions.

Electrophilic aromatic substitution does occur with pyridine, but the reaction is slow, and harsh reaction conditions are required. Pyridine reacts with potassium nitrate, but the reaction requires heating to 330°C to yield 3-nitropyridine, as shown in Figure 11.28. Relative to nitrogen, C3 and C5 have the greatest π-electron density, and those are the major sites for reaction. The intermediates generated from pyridine in electrophilic aromatic substitution reactions are relatively unstable, so the reaction is slow. Indeed, *the pyridine ring is deactivated for electrophilic aromatic substitution.*

The reaction of pyridine with potassium nitrate (KNO_3) to yield 3-nitropyridine reinforces the statement that pyridine is resistant to electrophilic aromatic substitution and reacts much slower than benzene, in contrast to the five-membered ring aromatic compounds. It also shows that substitution occurs at C3. If the NO_2^+ reagent is used in the reaction, reaction at C2 leads to intermediate **29**, reaction at C3 leads to **30**, and reaction at C4 leads to **31** (see Figure 11.28). Reaction at C2 and C5 (**29** and **31**) leads to three resonance forms that are particularly destabilized since the positive charge is directly on nitrogen, so the rate of their formation is very slow. Reaction at C3 (**30**) yields a relatively unstable arenium ion, but the positive charge is never on

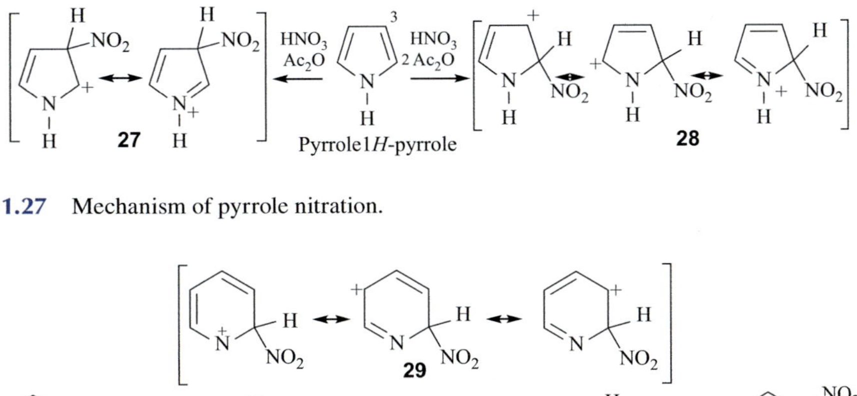

FIGURE 11.27 Mechanism of pyrrole nitration.

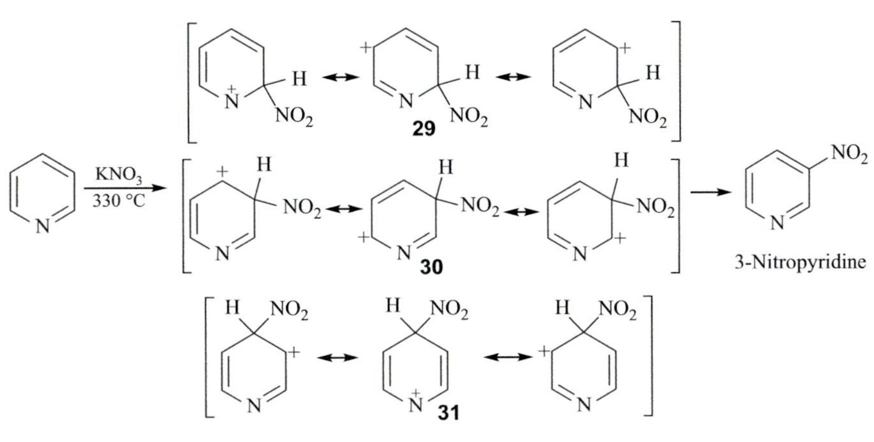

FIGURE 11.28 Electrophilic aromatic substitution of pyridine.

nitrogen, so it is *less* destabilized relative to **29** and **31**, and the rate of formation of **31** is significantly faster. Therefore, *pyridine reacts with electrophilic reagents to yield the 3-substituted derivative, 3-nitropyridine*. For the same reasons given for pyridine, S_EAr reactions of simple pyrimidines are difficult.

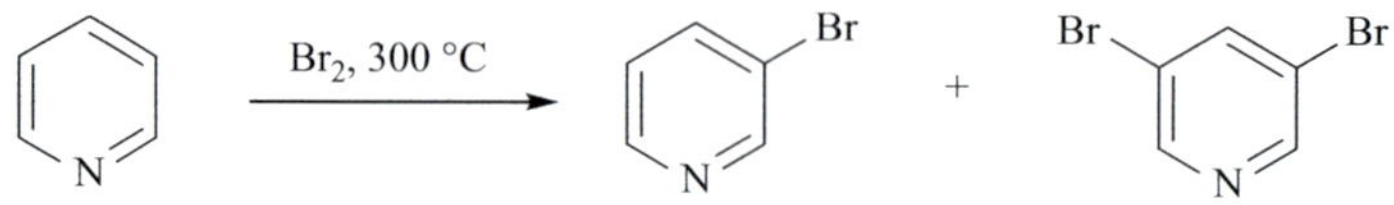

Pyridine 3-Bromopyridine 3,5-Dibromopyridine

11.15 REDUCED FORMS OF NITROGEN, OXYGEN, AND SULFUR HETEROCYCLES

There are several important compounds that result from the reduction of heterocycles, and they are often also identified as heterocyclic compounds. Many reduced heterocycles are commonly used in medicine, and they also have industrial uses. Nitrogen-containing ring systems are classified by the nature and number of heteroatoms, as well as by the size of the ring. For a monocyclic system with one nitrogen, the three-membered ring derivative is called aziridine, the four-membered ring is azetidine, the five-membered ring is pyrrolidine, and the six-membered ring is named piperidine. Pyrrolidine is prepared by catalytic hydrogenation of pyrrole, which can also be named tetrahydropyrrole. Similarly, catalytic hydrogenation of pyridine leads to piperidine, which can also be named hexahydropyridine. These compounds are key units in important pharmaceuticals. All four of these cyclic amines are secondary amines and react as bases.

One important compound that contains reduced heterocyclic rings is mitomycin C, an important drug used to treat cancer. It contains an aziridine ring, a pyrrolidine ring, and a dihydropyrrole unit. Azetidine-2-carboxylic acid shows growth-inhibitory activity on cultures of *Escherichia coli*. Many important compounds contain a pyrrolidine ring, including proline, which is a common amino acid. Nicotine is found in tobacco, and the carboxylate salt of kainic acid is a neurotoxin. A piperidine derivative, β-eucaine, is used as a local anesthetic.

Aziridine Azetidine Pyrrolidine Piperidine

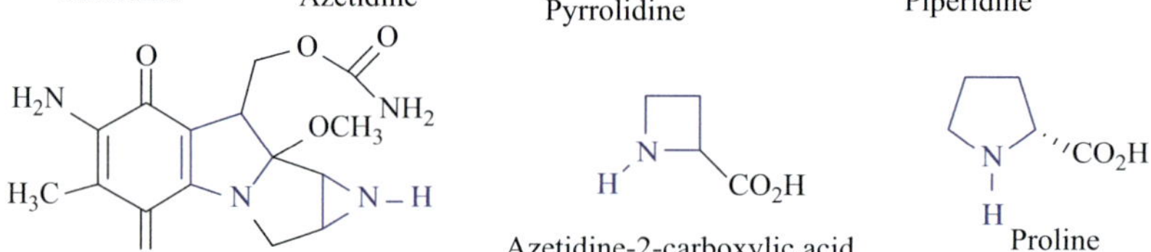

Mitomycin C Azetidine-2-carboxylic acid Proline

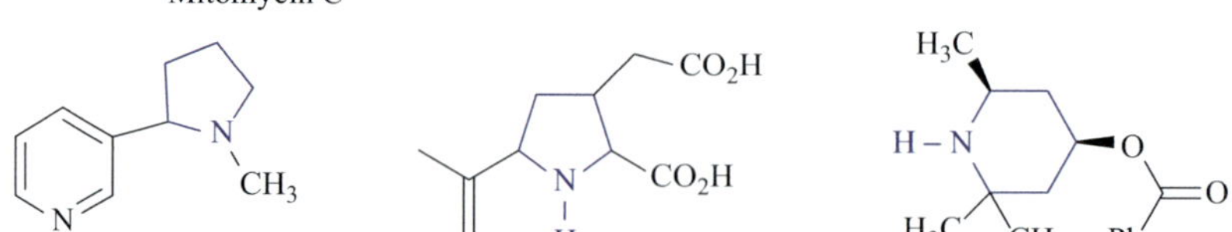

Nicotine Kainic acid β-Eucaine

Oxygen-containing ring systems are important, and all are cyclic ethers. The three-membered ring ether is oxirane, and this is the "parent" of the epoxides; the four-membered ring ether is oxetane, and the five-membered ring ether is oxolane, or tetrahydrofuran (THF). Tetrahydrofuran is an important cyclic ether prepared by catalytic hydrogenation of furan and is used many times in this book as a polar, aprotic solvent. The six-membered ring ether is called oxane, or tetrahydropyran, and is obtained by catalytic hydrogenation of pyran. There are two six-membered ring derivatives, 1,3-dioxane and 1,4-dioxane. Cyclic ethers are components of many natural products as well as synthetic compounds.

Oxirane Oxetane Tetrahydrofuran Tetrahydropyran

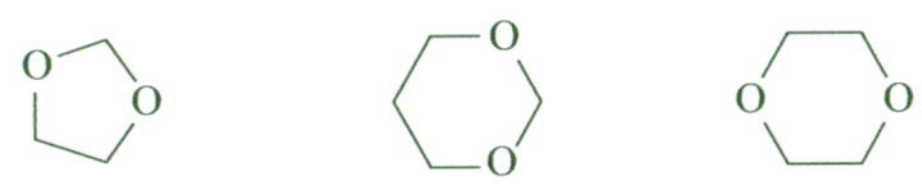

1,3-Dioxolane 1,3-Dioxane 1,4-Dioxane

11.16 BIOSYNTHESIS OF BICYCLIC HETEROCYCLES

A hypothetical enzymatic biosynthesis of quinine is shown in Figure 11.29.[24] Strictosidine is a known biochemical intermediate from secologanin and tryptamine,[25] and strictosidine glucosidase (CrSGD) mediates the conversion to strictosidine aglycone. Secologanin is synthesized from geranyl pyrophosphate in the mevalonate pathway.[26] Three enzymes were discovered as part of the early

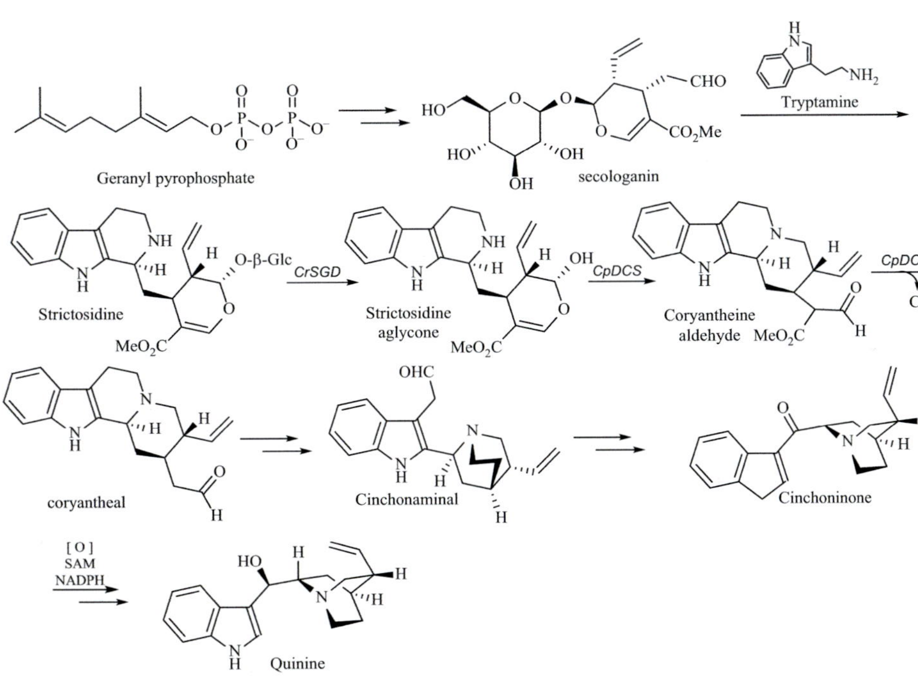

FIGURE 11.29 Hypothetical pathway of quinine biosynthesis. (Trenti, F.; Yamamoto, K.; Hong, B.; Paetz, C.; Nakamura, Y.; O'Connor, S.E. *Org. Lett.* **2021**, *23*, 1793. Taken from Scheme 1, p. 1794.© 2015 The Author(s). Licensee InTech. This chapter is distributed under the terms of the Creative Commons Attribution License (http://creativecommons.org/licenses/by/3.0), which permits unrestricted use, distribution, and reproduction in any medium, provided the original work is properly cited.)

[24] Trenti, F.; Yamamoto, K.; Hong, B.; Paetz, C.; Nakamura, Y.; O'Connor, S.E. *Org. Lett.* **2021**, *23*, 1793.

[25] Hay, C.A.; Anderson, L.A.; Phillipson, J.D.; Curless, D.; Brown, R.T. *Plant Cell Tissue Org. Cult.* **1987**, *9*, 197; Guirimand, G.; Courdavault, V.; Lanoue, A.; Mahroug, S.; Guihur, A.; Blanc, N.; Giglioli-Guivarc'h, N.; St-Pierre, B.; Burlat, V. *BMC Plant Biol.* **2010**, *10*, 182.

[26] Miettinen, K; Dong, L.; Navrot, N.; Schneider, T.; Burlat, V.; Pollier, J.; Woittiez, L.; van der Krol, S.; Lugan, R.; Ilc, T.; Verpoorte, R.; Oksman-Caldentev, K.-M.; Martinoia, E.; Bouwmeester, H.; Goossens, A.; Memelink, J.; Werck-Reichhart, D. *Nat. Commun.* **2014**, *5*, 3606.

and late stages of the biosynthesis.[24] In the putative biosynthesis, strictosidine aglycone is converted to coryantheine aldehyde by a dehydrogenase (CpDCS). An esterase then mediates decarboxylation, and an esterase (CpDCE) yields the biosynthetic intermediate corynantheal from coryantheine aldehyde. Additionally, the discovery of an *O*-methyltransferase specific for 6′-hydroxycinchoninone suggested that the final step order is cinchoninone hydroxylation, methylation, and keto-reduction.[24]

Quinoline and isoquinoline are commonly found in the structures of many alkaloids that are common in plants. An example in Figure 11.30 is the quinoline biosynthesis via the enzyme-mediated synthesis of quinolin-2,4-diol.[27] Quinoline alkaloids are generally derived from 3-hydroxyanthranilic acid. 3-Hydroxyanthranilic acid is a metabolite formed, mediated by oxygen and *indoleamine 2,3-dioxygenase*, to give *N*-formylquinurenine. *Formamidase* leads to quinurenine, and *quinureninase* gives 3-hydroxyanthranilic acid. The condensation of malonyl-SCoA and 3-hydroxyanthranilic acid, followed by cyclization, produces the quinoline alkaloid quinoline-2,4-diol.[27,28] The naturally occurring quinolone alkaloids, buchapine and 3-(3-methylbut-2-en-1-yl)-4-((3-methylbut-2-en-1-yl)oxy)quinolin-2-ol,[29] were evaluated for anti-HIV potential in the human CD4+ T cell line CEM-GFP, infected with HIV-1NL4.3 virus, by p24 antigen capture ELISA assay. Several analogs were prepared and found to have some activity against HIV.[29]

The biosynthesis of isoquinoline is also derived from the cyclization of the Schiff base formed between dopamine and 2-(4-hydroxyphenyl)acetaldehyde.[28a] The biosynthesis of (*S*)-reticuline in Figure 11.31 illustrates a typical pathway.[30] The reaction of tryptophan with *tyrosine decarboxylase* leads to hydroxylation and decarboxylation to give dopamine. The reaction with transaminase and then decarboxylase gave by transamination, decarboxylation, and hydroxylation to form 2-(4-hydroxyphenyl)acetaldehyde. The condensation of these two intermediates with *norcoclaurine synthase* gives (*S*)-norcoclaurine, which is converted to (*S*)-1-(4-hydroxybenzyl)-6-methoxy-1,2,3,4-tetrahydroisoquinolin-7-ol by *norcoclaurine6-O-methyltransferase*. The amine is N-methylated by

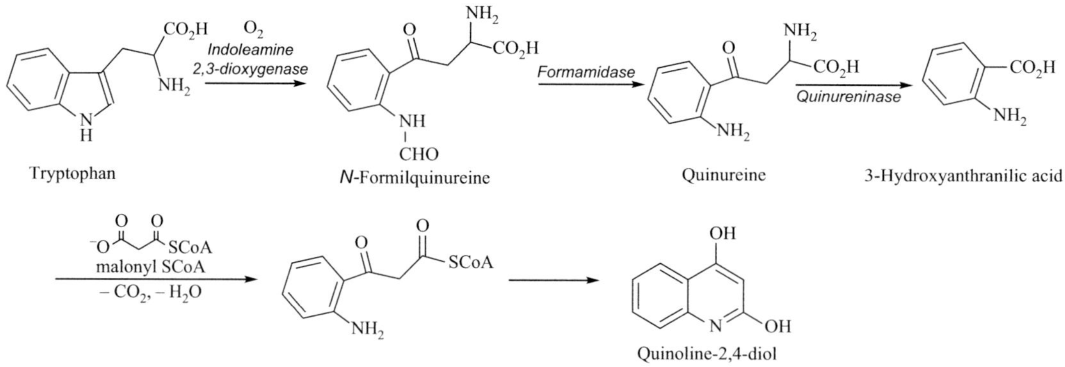

FIGURE 11.30 Biosynthesis of Quinoline-2,4-diol. (Diaz, G.; Miranda, I.L.; Diaz, M.A.N. Chapter 6, Quinolines, Isoquinolines, Angustureine, and Congeneric Alkaloids–Occurrence, Chemistry, and Biological Activity in Phytochemicals - Isolation, Characterization and Role in Human Health, Rao, L.G.; Rao, A.V. (Eds.), IntechOpen, 2015. ISBN: 13-978-9535121701. Figure 4, p. 149. This chapter is distributed under the terms of the Creative Commons Attribution License (http://creativecommons.org/licenses/by/3.0), which permits unrestricted use, distribution, and reproduction in any medium, provided the original work is properly cited.)

[27] Diaz, G.; Miranda, I.L.; Diaz, M.A.N. Chapter 6, *Quinolines, Isoquinolines, Angustureine, and Congeneric Alkaloids–Occurrence, Chemistry, and Biological Activity in Phytochemicals - Isolation, Characterization and Role in Human Health*, Rao, L.G.; Rao, A.V. (Eds.), IntechOpen, **2015**.

[28] (a) Mann, J. *Secondary Metabolism*, Oxford, Clarendon Press, **2005**; (b) Souza, M.V.N.; Almeida, M.V.; Hyaric, M.; Cardoso, S.H.; Amarante, W.G. *Química Nova* **2003**, *26*, 694; (c) Sultana, N. *Pharm. Sci.* **2013**, *2*, 13.

[29] Ahmed, N.; Brahmbhatt, K.G.; Sabde, S.; Debashis Mitra, D.; Singh, I.P.; Bhutani, K.K. *Bioorg. Med. Chem.* **2010**, *18*, 2872.

[30] Civjan, N. *Natural Products in Chemical Biology*, John Wiley & Sons, Hoboken, NJ, **2012**.

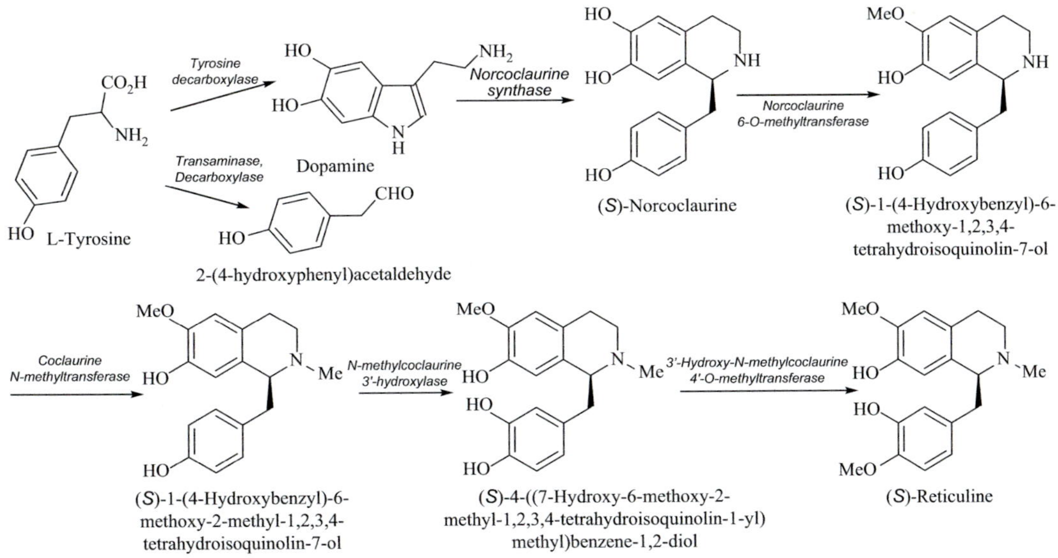

FIGURE 11.31 Biosynthesis of isoquinoline alkaloids. (Diaz, G.; Miranda, I.L.; Diaz, M.A.N. Chapter 6, Quinolines, Isoquinolines, Angustureine, and Congeneric Alkaloids–Occurrence, Chemistry, and Biological Activity in Phytochemicals - Isolation, Characterization and Role in Human Health, Rao, L.G.; Rao, A.V. (Eds.), IntechOpen, 2015. ISBN: 13-978-9535121701. Figure 4, p. 150. This chapter is distributed under the terms of the Creative Commons Attribution License (http://creativecommons.org/licenses/by/3.0), which permits unrestricted use, distribution, and reproduction in any medium, provided the original work is properly cited.)

coclaurine N-methyltransferase, and then (*S*)-4-((7-hydroxy-6-methoxy-2-methyl-1,2,3,4-tetrahydroisoquinolin-1-yl)methyl)benzene-1,2-diol is formed, mediated by *N-methylcoclaurine 3'-hydroxylase*. This intermediate is converted to (*SS*)-reticuline by *3'-hydroxy-N-methylcoclaurine-4'-O-methyltransferase*. (*S*)-Reticuline is a biosynthetic intermediate of all isoquinoline alkaloids, including morphine. It also depresses the central nervous system.[31]

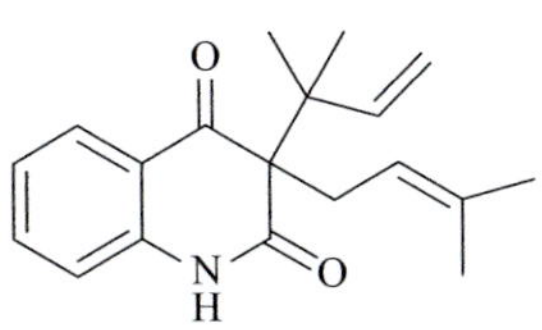

Buchapine

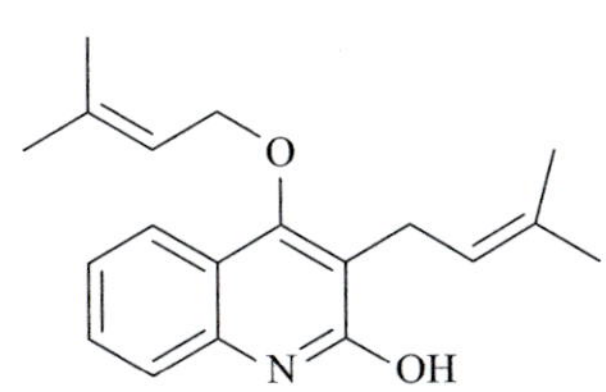

3-(3-Methylbut-2-en-1-yl)-4-((3-methylbut-2-en-1-yl)oxy)quinolin-2-ol

11.17 BICYCLIC HETEROAROMATIC COMPOUNDS

Polycyclic aromatic hydrocarbons were discussed previously, including naphthalene. There are important polycyclic systems that contain heteroatoms, and they are found in many biological systems as well as in many medicines. If the nitrogen in the bicyclic system is at position "1," the amine is named *quinoline*. When the nitrogen is in position "2," the amine is named *isoquinoline*. Many *alkaloids* (natural products containing nitrogen) have significant biological activity and contain either the quinoline or isoquinoline ring systems, or hydrogenated forms of these rings. Quinine is a common quinoline derivative, and it is an important antimalarial isolated from *Cinchona* bark. Another antimalarial compound is primaquine. Camptothecin is an important

[31] Dholvitayakhun, A.; Trachoo, N.; Sakee, U.; Cushnie, T.P.T. *Nat. Prod. Commun.* **2013**, *8*, 385.

anticancer drug, and it contains the quinoline unit. A typical isoquinoline derivative is papaverine, which is related to morphine, also isolated from the opium poppy. Papaverine relaxes smooth muscle tissue in blood vessels.

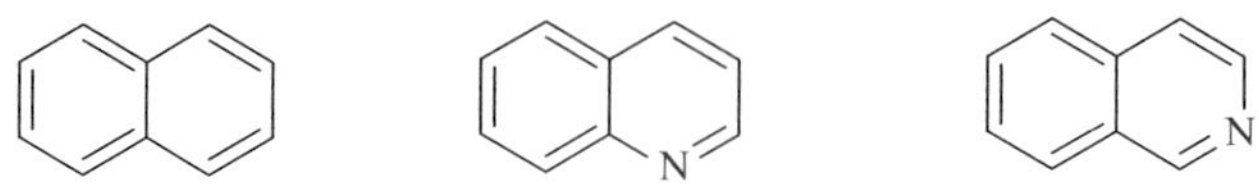

Naphthalene Quinoline Isoquinoline

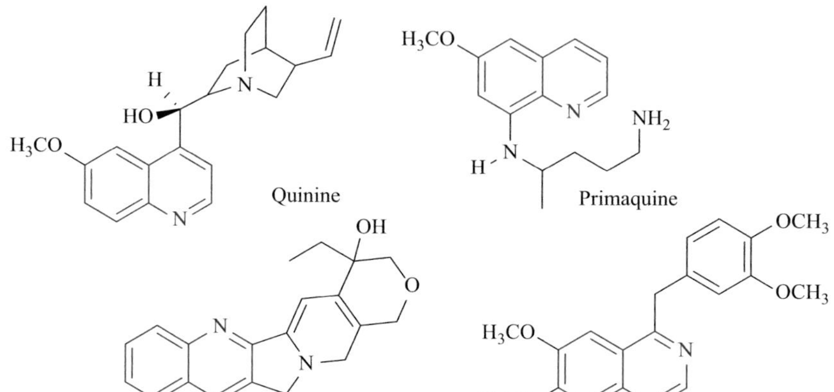

Quinine Primaquine Camptothecin Papaverine

Indole is a very important bicyclic aromatic heterocycle with a six-membered ring fused to a five-membered ring. A common compound containing an indole ring is the important amino acid tryptophan. The closely related compound serotonin is a neurotransmitter that causes smooth muscle effects in the cardiovascular and gastrointestinal systems. Some indole-containing compounds have hallucinogenic effects, as seen with lysergic acid, which is found in a fungal parasite common to rye and wheat. Lysergic acid, in combination with several related compounds found in this fungus, is responsible for "ergot poisoning," which killed>40,000 people in Europe in the 10th and 12th centuries. Note that lysergic acid also includes a reduced quinoline unit.

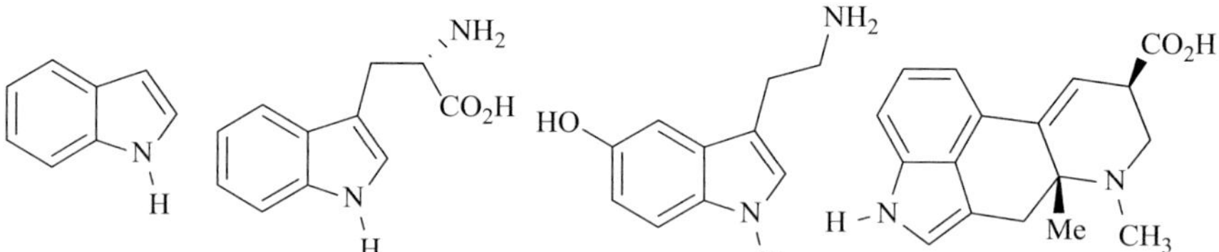

Indole Tryptophan Serotonin Lysergic acid

There are derivatives that have one or two nitrogen atoms at different positions in bicyclic six-six, six-five, or five-five fused aromatic rings. It is also possible to incorporate three, four, or even more nitrogen atoms into these rings. The purines adenine and guanine are components of deoxyribonucleic acid (DNA) and ribonucleic acid (RNA), as discussed in Section 15.2. In addition, both uric acid (a component of urine) and caffeine (found in coffee and tea) have a purine skeleton. A buildup of uric acid in the blood leads to gout, which is an inflammatory form of arthritis. Caffeine is found in many plants, and it has insecticidal or insect-repellent properties for some insects.

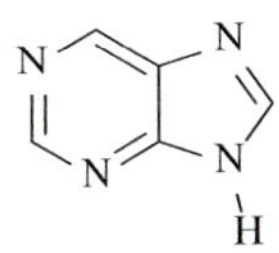

9H-Purine

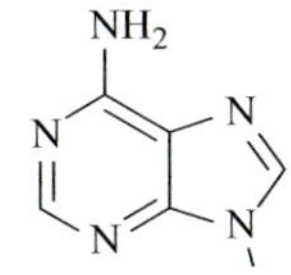

Adenine

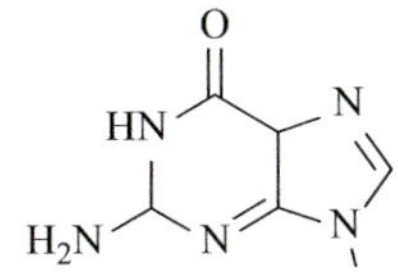

Guanine

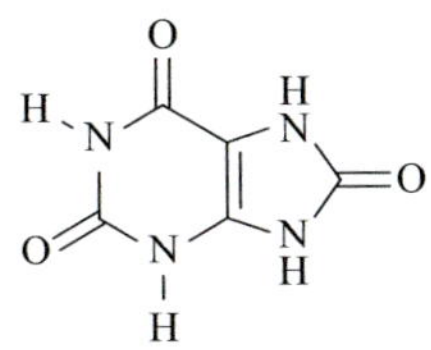

Uric acid

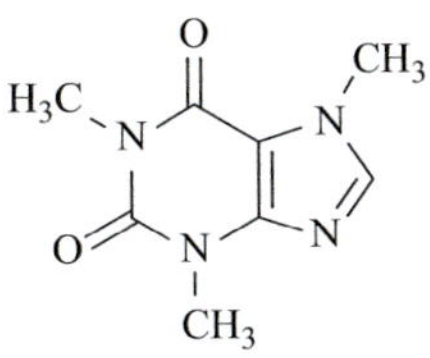

Caffeine

Both quinoline and isoquinoline contain a pyridine unit as well as a benzene ring unit. One ring (a benzene ring) has a nitrogen atom *attached*, and it is expected to be activated for S_EAr reactions. Therefore, electrophilic aromatic substitution will occur exclusively in the benzene ring unit. The indole system contains a pyrrole unit and a benzene ring unit, and pyrrole is highly activated to electrophilic aromatic substitution reactions. In S_EAr reactions, indole reacts at the activated pyrrole unit rather than the benzene ring unit.

HOMEWORK

11.1. Figure 11.7 shows the biosynthesis pathway for the carboxylation of catechol. In one step, the benzene ring attacks a carbonyl group. What well-known reaction in organic chemistry is similar?

11.2. Briefly explain why benzene does not react with HBr.

11.3. Why does the transformation of cyclohexa-3,5-diene-1,2-diol to catechol require a dehydrogenase enzyme? What EC number is used for a generic dehydrogenase?

11.4. Suggest a reason why benzoyl-CoA reductase converts benzoyl-CoA to the 1,5-cyclohexadiene rather than the 1,4-cyclohexadiene.

11.5. What is an ipso carbon?

11.6. What is the biological function of a halogenase enzyme?

11.7. If it is known that a reaction at a heteroatom is more likely, what product would likely be formed if norepinephrine (shown) reacted with a methyltransferase enzyme?

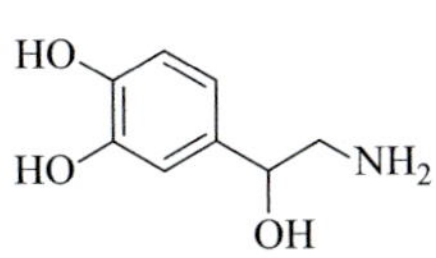

11.8. If an allyldiphosphate were mixed with an activated aromatic substrate, what can be predicted about the product after a reaction with an allyltransferase?

11.9. In Figure 11.16, the enzyme-mediated replacement of chlorine with hydroxyl begins with an S_NAr-type reaction. What type of organic transformation is required to complete the aromatic substitution reaction?

11.10. Draw all resonance forms for the intermediate formed when 2,4,6-trinitrobromobenzene reacts with NaOH in hot water.

11.11. Classify the following enzymatic transformation shown in Figure 11.9 as a known type of organic reaction.

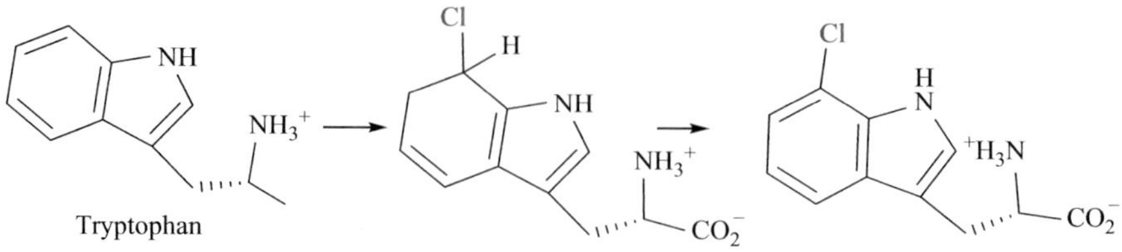

11.12. Draw all resonance forms for the reaction of Cl^+ at C-1 of anthracene.
11.13. Indole is known to react with Cl^+ at the five-membered ring rather than the six-membered ring. Briefly explain why.
11.14. Quinoline is known to react with Cl+ preferentially at either the nitrogen-containing ring or at the ring that does not contain nitrogen. Predict which ring and briefly explain the choice.
11.15. If a proline intermediate were to react with an enzymatic decarboxylase, what would the product be?

12 Metals
Carbon-Metal Bonds, Chelating Agents, and Coordination Complexes

When a metal is incorporated into a molecule with a metal-carbon bond, the resulting molecule is called an *organometallic compound.* In organic chemistry, organometallic reagents are reactive species and often used as reagents. In biology, metals are important at cofactors for the efficient activity of enzymes, and their ability to form multivalent compounds leads to the use of metals as chelating agents and their importance for the formation of coordination complexes.

12.1 ORGANOMETALLICS

Carbon-metal bonds are important in many areas of organic chemistry. This section will briefly review simple examples. A carbon-boron bond and a carbon-mercury bond are important examples of a class of compounds known as *organometallics* (organic molecules that incorporate one or more metal atoms). As a practical matter, the type of metal can be divided into two categories, alkali metals (Li, Na, Mg) or transition metals (Cu, Pd, etc.). While some definitions of an organometallic compound are limited to compounds that contain a transition metal, the broader definition includes both types of metal. For convenience, the term organometallics in this book will include both alkali and transition metals.

Carbon forms bonds to many metals to yield molecules that contain a C–M unit, where M is the metal. The more common metals include lithium, magnesium, and copper, and organometallic compounds based on those metals are introduced in this chapter. Other organometallic compounds will be introduced as they are required.

12.2 BIOLOGICALLY RELEVANT METALS

Metals are essential to many biological processes.[1] Indeed, certain metals are important to human life, and they serve as important cofactors for many enzymes. The two major classes are metal-activated enzymes and metalloenzymes. For metal-catalyzed enzymes, the catalytic activity is enhanced by the presence of a mono- or divalent metal ion. The metal can activate the substrate, coordinate directly with the enzyme, or lead to more favorable binding in an equilibrium reaction. The metal is usually found in large excess, perhaps a 2–10-fold excess relative to the concentration of the enzyme. Examples of metal-activated enzymes include *pyruvate kinase* (EC 2.7.1.40), catalyzed by K^{+}, and *DNase* (EC 3.1.21.1), *RNase* (EC 3.1.27.5), and *ATPase* (EC 3.6.1.3), catalyzed by Mg^{2+}. Metalloenzymes have one or more metals that are part of a specific region on the protein surface. The most common metals are Fe, Zn, Cu, and Mn. Metalloenzymes contain a metal ion that is key to the activity of the enzyme, and tightly bound metal ions include Fe^{2+}, Cu^{2+}, Zn^{2+}, Mn^{2++}, Co^{3+}, Ni^{3+},

[1] Sigel, A.; Sigl, H.; Sigel, R.K.O. (Eds), *Metal-Carbon Bonds in Enzymes and Cofactors (Metal Ions in Life Sciences, 6)*, 1st ed., Walter de Gruyter, **2015**. ISBN-13: 978-3110442793; Siegel, A.; Siegel, H.; Freisinger, E.; Siegel, R.K.O. (Eds), *Metallo-Drugs: Development and Action of Anticancer Agents (Metal Ions in Life Sciences, 18)*, 1st ed., De Gruyter, **2018**. ISBN-13: 978-3110469844.

DOI: 10.1201/9781003605157-12

and Mo^{6+}.[2] Examples are *alcohol dehydrogenase* (zinc; EC 1.1.1.1), *ascorbic acid oxidase* (copper; EC 1.11.1.11), *cytochrome* (iron), *cytochrome oxidase* (copper; EC 1.9.3.1), *glutamate mutase* (cobalt; EC 5.4.99.1), *glutathione peroxidase* (selenium; EC 1.11.1.9), *urease* (nickel; EC 3.5.1.5), *xanthine oxidase* (molybdenum; EC 1.17.3.2), and cobalt in vitamin B12. Other enzymes include *superoxide dismutase* (EC 1.15.1.1), containing Cu^{2+} and Zn^{2+}; catalase (EC 1.11.1.6), which contains Fe; *alcohol dehydrogenase* (EC 1.1.1.1) or *DNA polymerase* (EC 2.7.7.6), containing Zn; *pyruvate carboxylase* (EC 6.4.1.1), which contains Mn; and *alpha-amylase* (EC 3.2.1.1), which contains Ca.

The biological impact of metals is wide-ranging and important. For example, Na^+ and K^+ are necessary to regulate cell wall osmotic pressures and, along with Ca^{2+} and Mg^{2+}, are required to maintain nerve sensitivity and help control muscle function.[3] It is known that Ca^{2+} helps maintain the correct rhythm of the heartbeat, the conversion of fibrinogen to fibrin, and the formation of shells, bones, and cellulose structures.[3] The Mg^{2+} ion complexes with nucleic acids inside cells and is necessary for nerve impulse transmission, and all phosphate transfer enzymes usually require Mg^{2+} or Mn^{2+}.[3]

There are a few general principles that govern the metal-containing active sites in enzymes. If a metal reacts as a Lewis acid, there must be an appropriate charge and coordination geometry. A metal used in a redox reaction must have an appropriate reduction potential, and there should be minimal structural rearrangements upon charge transfer. Metals such as Mg^{2+} and Ca^{2+}, which are quite similar, may lead to very different catalytic properties for an enzyme due to large-scale structural effects induced by the metal. In some enzymes, differences in the electronic effects of the metal may lead to differences in catalysis.

Tyrosinase (EC 1.10.3.1) is an enzyme that catalyzes the production of melanin and other pigments from tyrosine by oxidation, as in the blackening of a peeled or sliced potato exposed to air. Cytochrome oxidase is one of a family of proteins that act as the terminal enzymes of respiratory chains. Both of these enzymes require Cu^{2+} or Cu^+ as a cofactor.[4] *Phosphohydrolases* (EC 3.1) are a class of enzymes that cleave phosphoric acid from phosphate ester linkages to give an orthophosphate. Phosphotransferases (EC 2.7) are a class of enzymes that include the *kinases* and catalyze the transfer of phosphorus-containing groups from one compound to another. These latter two classes of enzymes require Mg^{2+} as a cofactor.[3]

The metal zinc has been shown to be important in many enzymes, including *dehydrogenases*, *aldolases*, *peptidases*, and *phosphatases*. The portion of LPS (lipopolysaccharides) identified as lipid A anchors LPS to the cell surface. The enzyme UDP-3-*O*-[(*R*)-3-hydroxymyristoyl]-*N*-acetylglucosamine deacetylase (LpxC) catalyzes the second overall step in the biosynthesis of lipid A: the deacetylation of UDP-3-*O*-[(*R*)-3-hydroxymyristoyl]-*N*-acetylglucosamine to form UDP-3-*O*-[(*R*)-3-hydroxymyristoyl] glucosamine and acetate.[5] The crystal structure of the LpxC active site contains two metal ion binding sites (Figure 12.1).[5] The zinc cofactor marked ZnA is catalytic, but ZnB is an inhibitory site.

This model is shown to illustrate the involvement of a metal in the active site of the enzyme, and how coordination or binding of the metal to the substrate may differ with different substrates and from metal to metal. Many enzymes use zinc. *Thermolysin* (EC 3.4.24.27), for example, is a metalloproteinase produced by various *Bacillus* species. This enzyme requires one zinc ion for enzyme activity and four calcium ions for structural stability. *Thermolysin* specifically catalyzes the hydrolysis of peptide bonds that contain hydrophobic amino acids.

Magnesium is required by some *phosphatases* (EC 3.1; enzymes that remove a phosphate group), *carboxylases* (enzymes that catalyze decarboxylation or carboxylation), and some proteolytic enzymes (enzymes that cleave peptide bonds). Magnesium is important in several biologically important compounds. One is chlorophyll A, one of the green pigments found in photosynthetic cells, which forms a magnesium complex with the porphyrin nitrogen atoms, as shown in Figure 12.2.

---

[2] Riordan, J.F. *Ann. Clin. Lab. Sci.* **1977**, *7*, 119.
[3] Ainscough, E.W.; Brodie, A.M. *J. Chem. Ed.* **1976**, *53*, 156.
[4] Lehninger, A.I. *Biochemistry*, Worth Pub., NewYork, **1970**, p. 149.
[5] Hernick, M.; Gattis, S.G.; Penner-Hahn, J.E.; Fierke, C.A. *Biochemistry* **2010**, *49*, 2246.

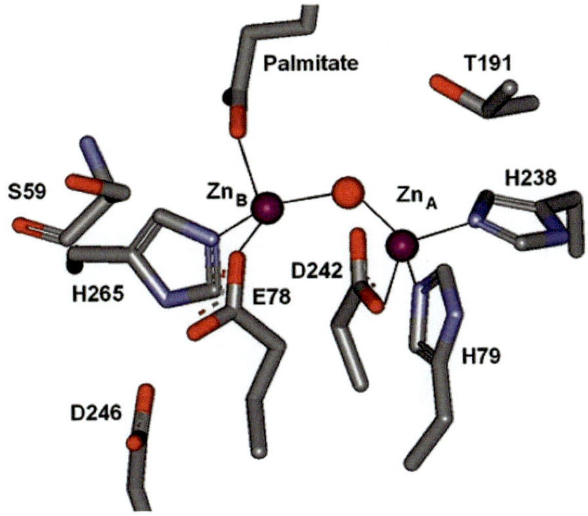

FIGURE 12.1 Active site of LpxC from *Aquifex aeolicus* containing two zinc ions: ZnA (catalytic) and ZnB (inhibitory). (Reprinted with permission from Hernick, M.; Gattis, S.G.; Penner-Hahn, J.E.; Fierke, C.A. *Biochemistry* **2010**, *49*, 2246–2255. Copyright 2010 American Chemical Society.)

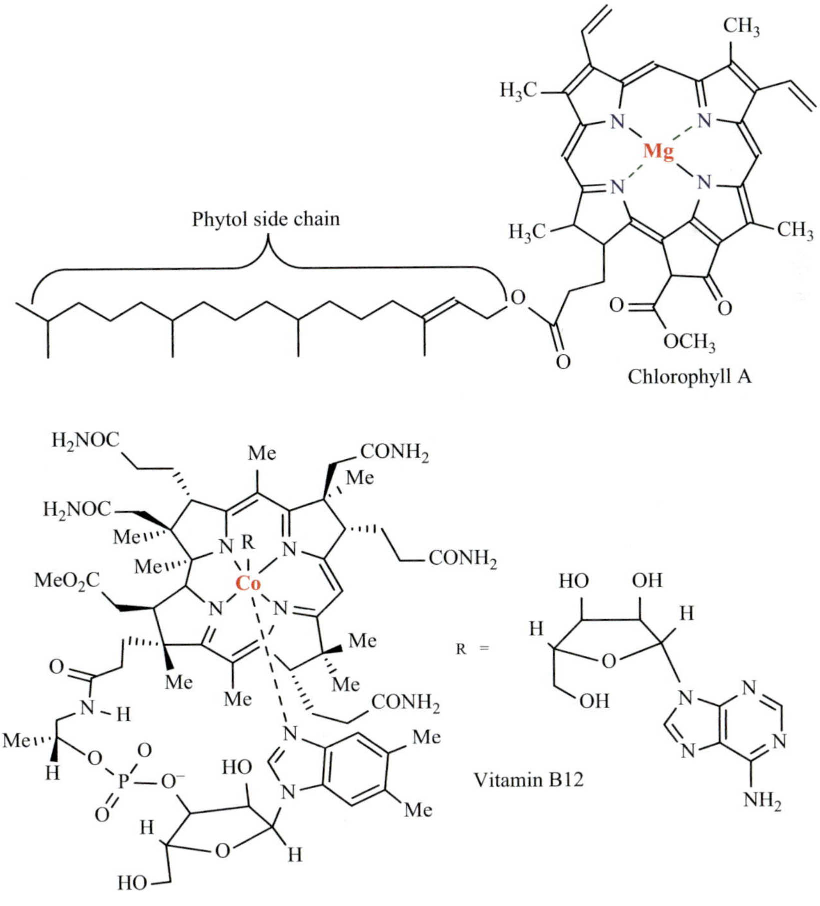

FIGURE 12.2 Chlorophyll A.

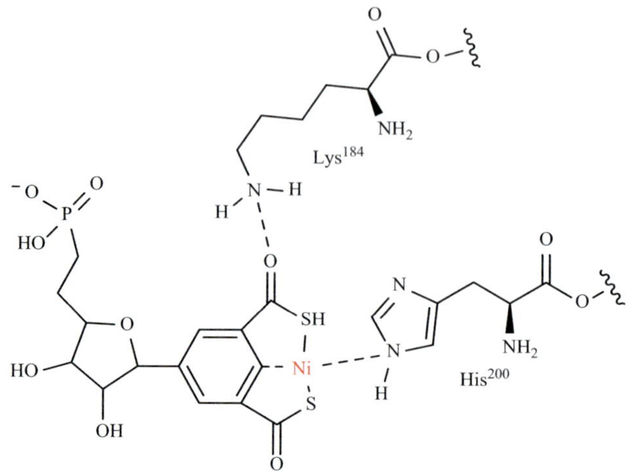

FIGURE 12.3 The nickel-pincer nucleotide (NPN) cofactor in Lactate racemase. (Hausinger, R.P. *Curr. Opinion Struct. Biol.* **2019**, *59*, 1. Figure 1b, p. 2. Copyright (2019), with permission from Elsevier.)

Chlorophyll is typically extracted from the leaves of trees and plants. The active form of vitamin B12 is a cobalt complex,[6] and the metal is again coordinated to the nitrogen atoms of a porphyrin system.

Myriad examples of metal cofactors that are part of an active enzyme site have been reported.[7] A metal cofactor was discovered in lactate racemase (EC 5.1.2.1) from *Lactobacillus plantarum*, a nickel-pincer nucleotide (NPN,[8] see Figure 12.3).[7] The organic portion of this cofactor, pyridinium-3-thioamide-5-thiocarboxylic acid mononucleotide, is covalently tethered to Lys^{184} of the enzyme, and it is a tridentate ligand to nickel. This molecule is bound to the pyridinium carbon atom, the two sulfur atoms, and His^{200}. This is an example of the highly studied pincer complexes in inorganic chemistry.[9] The enzymatic function of the NPN cofactor is to participate in a proton-coupled hydride-transfer reaction.[10] Pyruvate was identified as the catalytic intermediate, and the enzyme chromophore was perturbed in the presence of lactate or sulfite. The crystal structure of the enzyme in the presence of sulfite revealed an adduct of the reagent bound to C4 of the pyridinium group.[7]

Studies of *methanol dehydrogenases from Methylacidiphilum fumariolicum* SolV were shown to require a lanthanide metal ion (lanthanum, cerium, neodymium, praseodymium, samarium, europium, or gadolinium) for growth. For example, cerium (see Figure 12.4)[7] was incorporated into its *methanol dehydrogenase* (EC.1.1.244).[11] Structural studies of the enzyme from this organism in the presence of different metals revealed a cerium-PQQ cofactor,[11,12] where PQQ is pyrrolo-quinoline quinone. A *methanol dehydrogenase* is an oxidoreductase enzyme (Section 2.3, EC 1.1) that catalyzes the reversible interconversion of formaldehyde and methanol. Note that the PQQ-methanol dehydrogenase enzyme structure has PQQ in the center.

---

[6] Halpern, J. *Pure Appl. Chem.* **2001**, *73*, 209.

[7] Hausinger, R.P. *Curr. Opinion Struct. Biol.* **2019**, *59*, 1.

[8] Desguin, B.; Zhang, T.; Soumillion, P.; Hols, P.; Hu, J.; Hausinger, R.P. *Science* **2015**, *349*, 66.

[9] Peris, E.; Crabtree, R.H. *Chem. Soc. Rev.* **2018**, *47*, 1959.

[10] Rankin, J.A.; Mauban, R.C.; Fellner, M.; Desguin, B.; McCracken, J.; Hu, J.; Varganov, S.A.; Hausinger, R.P. *Biochemistry* **2018**, *57*, 3244.

[11] Pol, A.; Barends, T.R.M.; Dietl, A.; Khadem, A.F.; Eygensteyn, J.; Jetten, M.S.M.; Op den Camp, H.J.M. *Environ. Microbiol.* **2014**, *16*, 255.

[12] Jahn, B.; Pol, A.; Lumpe, H.; Barends, T.R.M.; Dietl, A.; Hogendoorn, C.; Op den Camp, H.J.M.; Daumann, L.J.: *ChemBioChem.* **2018**, *19*, 1147.

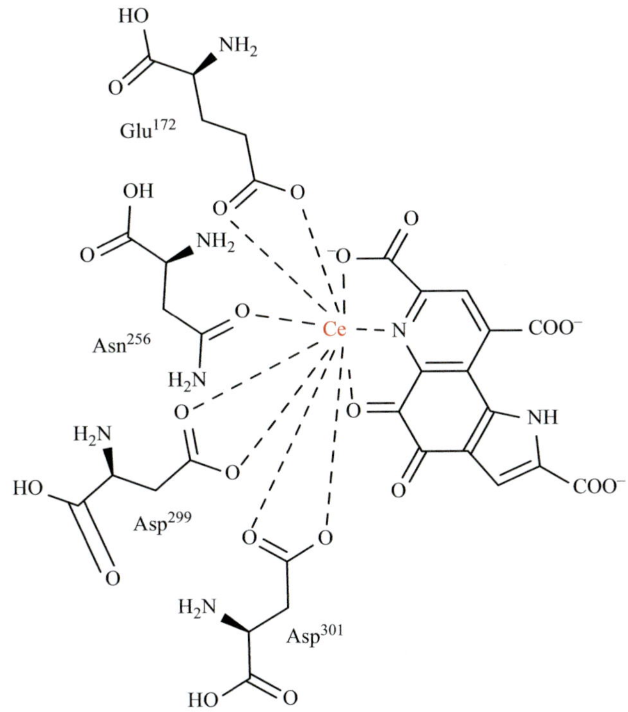

FIGURE 12.4 Active site of methanol dehydrogenase (MDH) with its lanthanide-PQQ cofactor [PDB ID: 4MAE (cerium). (Hausinger, R.P. *Curr. Opinion Struct. Biol.* **2019**, *59*, 1. Figure 2a, p. 3. Copyright (2019), with permission from Elsevier.)

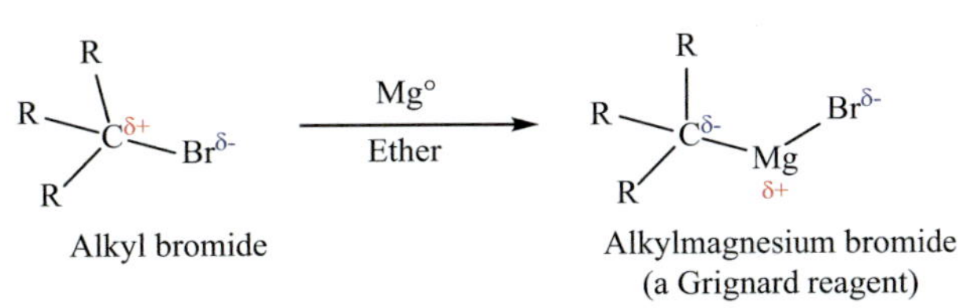

FIGURE 12.5 Formation of a Grignard reagent.

12.3 ORGANOMETALLICS IN ORGANIC CHEMISTRY

An important class of carbon nucleophiles is formed by the reaction of alkyl halides (R–X, where X=halogen) with alkali metals. The alkali metals (groups 1 and 2) most used include lithium and magnesium, and the product of this reaction contains a carbon–metal bond (C–Na, C–Li, C–Mg). As noted above, such compounds are generically called *organometallics.*

Magnesium reacts directly with an alkyl halide R_3C–X (X=Cl, Br, or I) to yield an organometallic product, as shown in Figure 12.5. Primary, secondary, and tertiary alkyl halides react, as well as vinyl halides (C=C–X) and aromatic halides (Ar–X). For the reaction shown in Figure 12.5, the organometallic is characterized by a C–Mg bond and is formally an alkylmagnesium bromide (an organomagnesium bromide). However, such compounds are typically known as *Grignard reagents*, RMgX.

In bromoethane, and in all common alkyl halides, the dipole induced by bromine leads to $^{\delta+}$C–Br$^{\delta-}$. This bond polarization is the result of the heteroatom being more electronegative than carbon. However, *carbon is more electronegative than magnesium*, so the electron density in a C–Mg bond

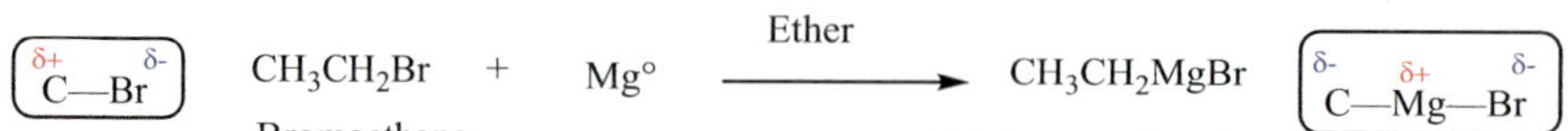

Bromoethane Ethylmagnesium bromide

FIGURE 12.6 Bond polarization in ethylmagnesium bromide.

of a Grignard reagent is distorted toward carbon, making carbon δ^- and magnesium δ^+, as shown in Figure 12.6. Therefore, the carbon atom of this organometallic is nucleophilic. In most cases, an ether solvent such as diethyl ether or tetrahydrofuran (THF) is used in these reactions.

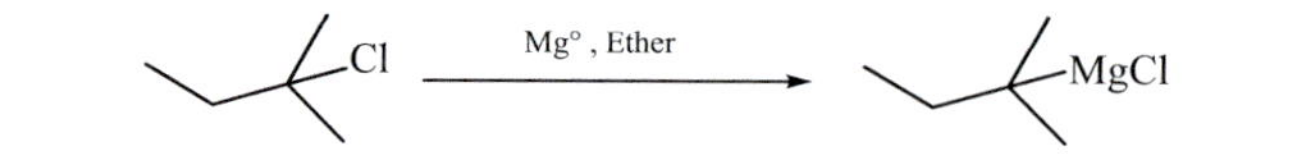

2-Chloro-2-methylbutane (2-Methylbut-2-yl)magnesium chloride

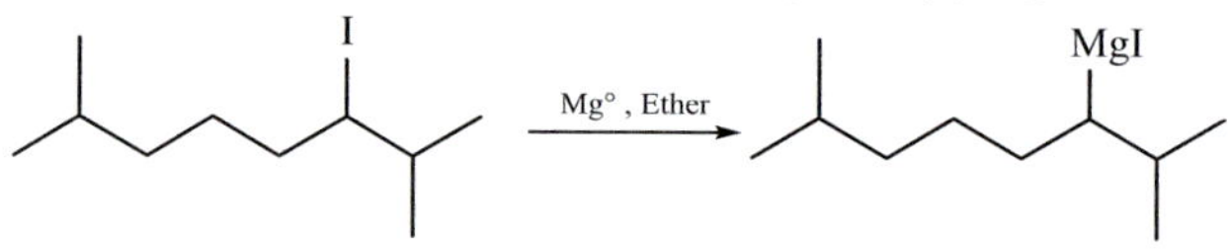

3-Iodo-2,7-dimethyloctane 2,7-Dimethyloctyl-3-magnesium iodide

The overall reaction that converts an alkyl halide to a Grignard reagent is called a *metal insertion reaction,* and it occurs with most alkyl halides. 2-Chloro-2-methylbutane reacts with magnesium to form (2-methylbut-2-yl)magnesium chloride, for example, and 3-iodo-2,7-dimethyloctane reacts to form 2,7-dimethyloctyl-3-magnesium iodide. Primary, secondary, and tertiary halides react with Mg metal to form the corresponding Grignard reagent.

The carbon of the C–Mg–X unit is polarized δ^- and can, therefore, donate electrons to an electrophilic atom. Therefore, Grignard reagents are nucleophiles if they react with an electrophilic carbon, but a Grignard reagent can also donate electrons to a proton, so it is a Brønsted-Lowry base. A Grignard reagent is a powerful base that reacts with very weak acids. Indeed, water, alcohols, terminal alkynes, and even amines react with a Grignard reagent to form the corresponding alkane (the conjugate acid) and the conjugate base of each of the weak acids mentioned. An example is methylmagnesium bromide, formed by reaction of iodomethane and magnesium, and this Grignard reagent will react with water to form methane (*the conjugate acid*) and hydroxide ion (BrMgOH, the conjugate base). This reaction effectively destroys the Grignard reagent (the C–Mg bond is broken to form C–H). Similarly, methylmagnesium bromide reacts with ethanol or acetic acid to give methane and ethoxide or the acetate anion, respectively. Note that the conjugate acid of this reaction is methane, an extremely weak acid, so the acid-base equilibrium lies far to the right, and the organometallic is a very powerful base, as noted.

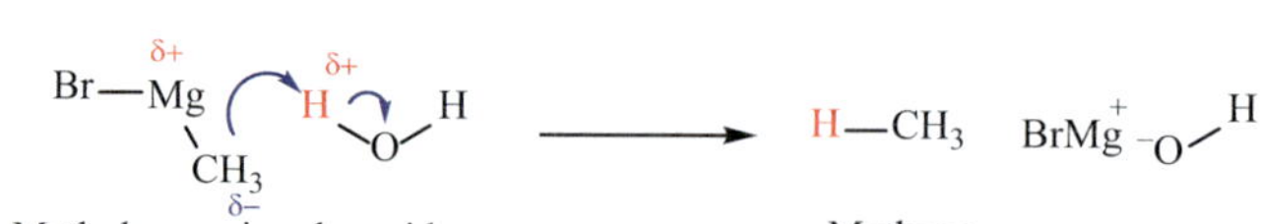

Methylmgnesium bromide Methane

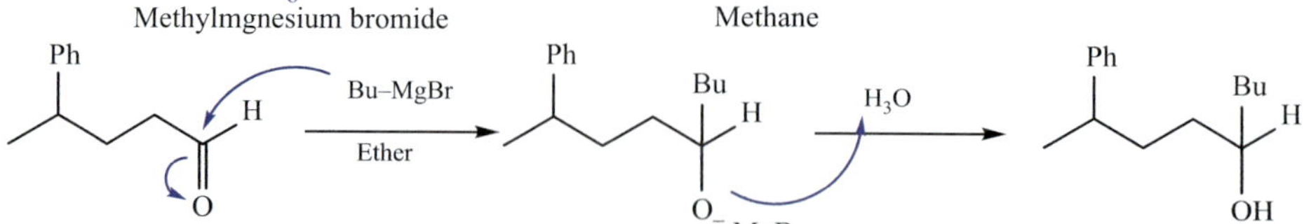

4-Phenylpentanal 2-Phenylnonan-5-ol

As noted, the presence of the δ^- carbon in a Grignard reagent makes that carbon a nucleophile. In a subsequent reaction, a Grignard reagent reacts with a ketone or an aldehyde in what is now called a *Grignard reaction.* For example, 4-phenylpentanal reacts with butylmagnesium bromide

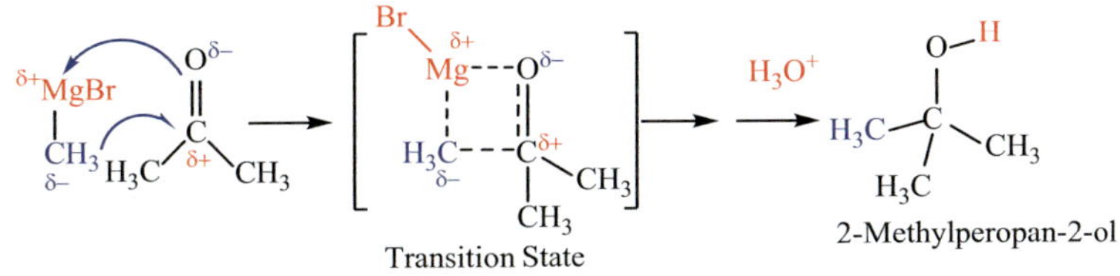

FIGURE 12.7 Transition state for acyl addition of a Grignard reagent to a ketone.

in diethyl ether to yield an alkoxide product via acyl addition. Treatment of the alkoxide with aqueous acid gives 2-phenylnonan-5-ol. The reaction of Grignard reagents and aldehydes (or ketones) to yield an alcohol product is a two-step process known as a *Grignard reaction.* Ketones also react with Grignard reagents to give a tertiary alcohol.

The acyl addition reactions of Grignard reagents with aldehydes or ketones just introduced have no intermediate but proceed by a *four-center transition* state, as shown in Figure 12.7. In this example, methylmagnesium bromide reacts with acetone to give 2-methylpropan-2-ol via the four-center transition state (Figure 12.7). A four-center transition state is taken to represent the transition state for all Grignard reactions with aldehydes or ketones.

The reaction of an alkyl halide with the group 1 metal lithium (Li) yields an organometallic called an *organolithium reagent* (RLi). For example, 1-bromo-3-methylbutane and lithium metal form the organometallic known as isopentyllithium, where lithium has replaced bromine. This organolithium reagent is characterized by a C–Li bond. Virtually any alkyl halide, primary, secondary, or tertiary, will react with lithium metal to form the corresponding organolithium reagent. However, tertiary organolithium reagents are significantly less stable than primary or secondary organolithium reagents.

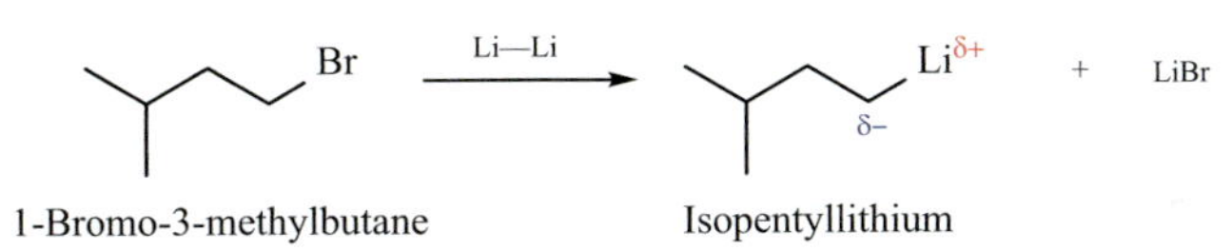

Like the bond polarization observed in Grignard reagents, organolithium reagents have a δ^- carbon connected to a Li atom. Therefore, organolithium reagents react as carbanion nucleophiles with aldehydes or ketones, giving the secondary or tertiary alcohols after reaction with an aldehyde or a ketone, respectively, followed by a workup with aqueous acid. Organolithium reagents are very powerful bases, analogous to the reactivity noted for Grignard reagents. In fact, an organolithium reagent is a significantly stronger base than a comparable Grignard reagent. It is such a strong base that it reacts with the protons of very weak acids, some with a $pK_a > 30$. Methyllithium reacts with water to yield the conjugate base, LiOH (lithium hydroxide), and the conjugate acid, methane, which is a very weak acid as pointed out earlier. Indeed, methyllithium and all organolithium reagents are very strong bases.

Many compounds that are not thought of as acids react with organolithium reagents to yield interesting conjugate bases that may be used for other purposes. As shown in Figure 12.8, alkynes (e.g., pent-1-yne) and amines (e.g., diethylamine) are weak acids with pK_a values close to 25 and 36, respectively. In both cases, *n*-butyllithium (1-lithiobutane) reacts to form the anion and butane as the conjugate acid. Pent-1-yne yields an alkyne anion, 1-lithiopent-1-yne. An alkyne anion is a good nucleophile in S_N2 reactions. Similarly, diethylamine reacts with butyllithium to yield the conjugate base, lithium diethylamide, and the conjugate acid, butane.

Experimentally, both Grignard and organolithium reagents are generally unreactive with simple alkyl halides, but the addition of CuBr or $FeBr_3$ to the organometallic enhances reactivity with alkyl halides to give a coupling product. Indeed, a new organometallic species is formed. When

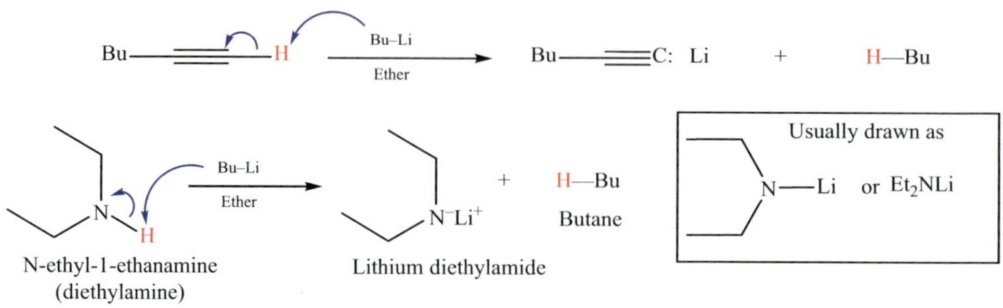

FIGURE 12.8 Terminal alkyne and secondary amines react as acids with butyllithium.

an organolithium reagent (RLi) reacts with cuprous iodide (CuI) in ether at −10°C, for example, a reaction takes place to generate what is called an *organocuprate*, R_2CuLi. If two molar equivalents of *n*-butyllithium react with CuI, the product is lithium dibutylcuprate. Note the nomenclature that is used, where the alkyl unit of the organolithium reagent is combined with the term *cuprate*, which describes the oxidation state of the copper. The carbon attached to copper in the organocuprate is δ^- (it is carbanionic), and it is very reactive with most alkyl halides, leading to the coupling product.

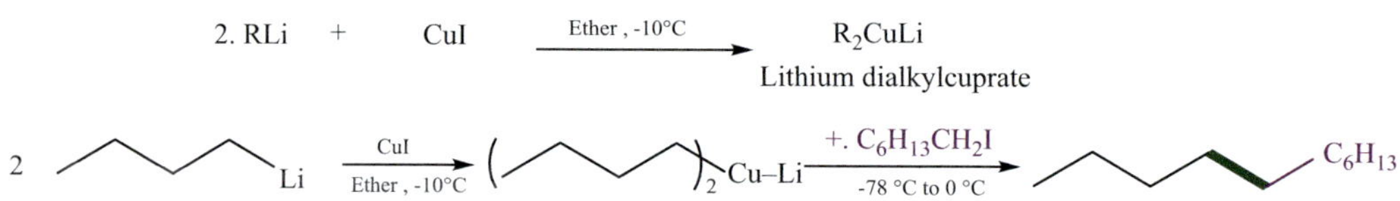

12.4 CHELATING AGENTS

Some metals have multiple oxidation states and can coordinate or bind multiple atoms or groups. Iron has a range of oxidation states, from −2 to +7, although the ferrous ion (+2) and the ferric ion (+3) are the most common. Ferrous chloride, or iron(II) chloride, $FeCl_2$, and ferric chloride, or iron (III) chloride, $FeCl_3$, are therefore common compounds. Note that the chlorine atoms in such compounds are attached to iron by covalent rather than ionic bonds.

Metals, particularly transition metals, can form bonds to atoms, ions, or molecules such as carbon monoxide (CO), amines (RNH_2), or phosphines (PR_3). An example is cobalt tetracarbonyl hydride, $HCo(CO)_4$. As shown in the structure, the four CO units are bound to the metal. The CO units are known as ligands. A *ligand* is an atom, ion, or molecule attached to a metal atom by coordinate bonding. A second example is chloridotris(triphenylphosphane)rhodium(I), which has three triphenylphosphine (PPh_3) units as ligands and one chlorine atom. Note the different geometry of the ligands about the metal in these two examples, due to differences in the metal oxidation state as well as the number and type of ligands.

Cobalt tetracarbonyl hydride Chloridotris(triphenylphosphane)rhodium(I)

With this brief introduction in mind, *chelation* can be introduced as a type of bonding that involves the formation or presence of two or more separate coordinate bonds between a ligand with multiple bonds and a single central atom. A *chelating agent* is a molecule that can form multiple bonds to a metal ion. The chelating agent is said to be a multidentate ligand. Many enzymes form chelation complexes that are essential for their bioactivity. Indeed, chelation is important for the transport of metals in biological systems, and essential transition metals such as Cu, Zn, Cr, Fe, Mn, and Co exist as binary and ternary chelates of amino acids, carboxylic acids, and proteins.

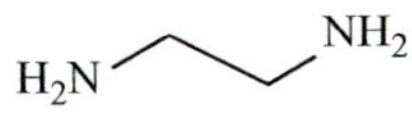

Ethylene diamine

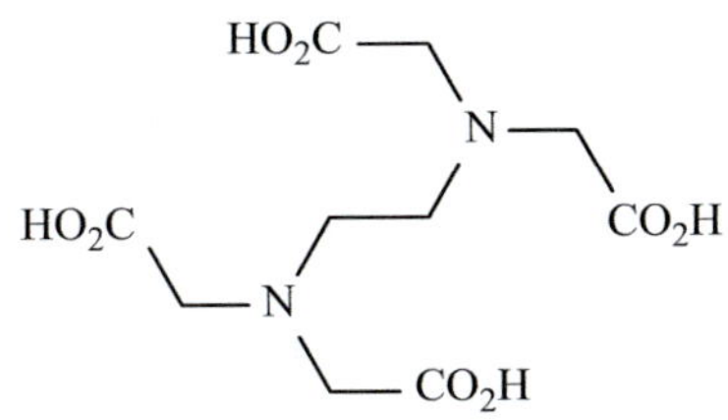

Ethylenediaminetetraacetic acid

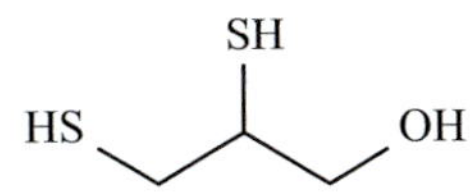

2,3-Dimercapto-1-propanol (dimercaprol)

Typical chelating ligands used in organic chemistry applications include ethylenediamine, ethylenediaminetetraacetic acid, and 2,3-dimercaptopropan-1-ol (known as dimercaprol). Note the presence of multiple chelating atoms or groups in each of these ligands.

Several factors are important for metal chelates, including the stability of the complex, pH, competition with other metals and ligands, lipophilicity, and the rates of hydrolysis and formation of the chelate.[13] An example of a metal-EDTA complex is shown in Figure 12.9,[14] where EDTA is ethylenediaminetetraacetic acid.

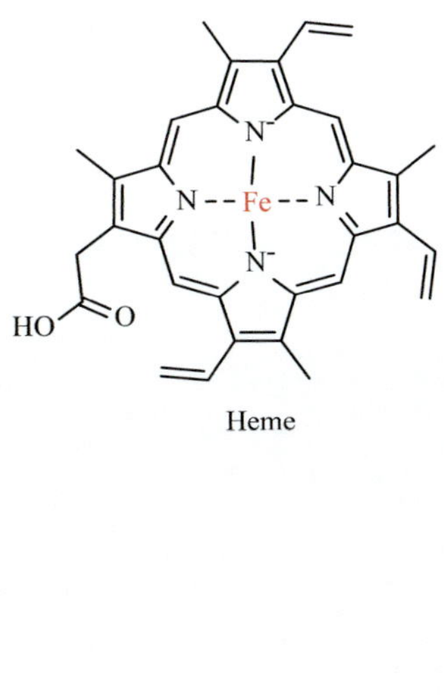

Heme

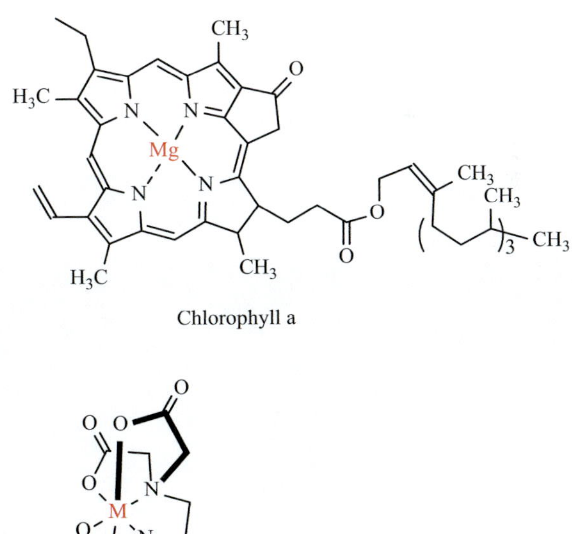

Chlorophyll a

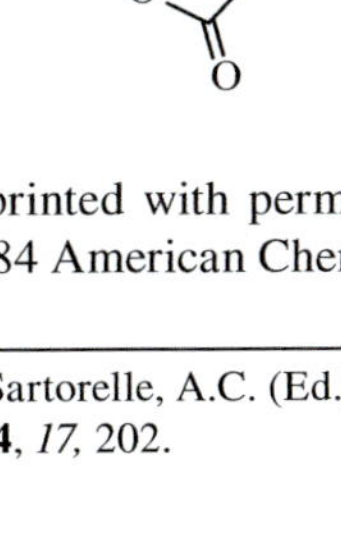

FIGURE 12.9 Metal-EDTA complex. (Reprinted with permission from Meares, C.F.; Wensel, T.G. *Acc. Chem. Res.* **1984**, *17*, 202–209. Copyright 1984 American Chemical Society.)

[13] Mellor, D.P. in *The Chelation of Heavy Metals*, Sartorelle, A.C. (Ed.), Pergamon Press, Oxford, **1975**.

[14] Meares, C.F.; Wensel, T.G. *Acc. Chem. Res.* **1984**, *17*, 202.

Many essential biological chemicals exist as chelates, and many enzymes are chelated. Heme is an important example. Heme is a coordination complex of Fe^{2+} coordinated to a porphyrin. Hemes are components of hemoglobin, the red pigment in blood. Another example is chlorophyll, the green pigment in plants. A porphyrin unit is chelated to a central magnesium. Chlorophyll is responsible for the absorption of light that is converted to chemical energy for photosynthesis in plants. The structure of one form, chlorophyll *a*, is shown. In photosynthesis, the chemical reaction is the conversion of carbon dioxide and water into carbohydrates and oxygen.

Metallothioneins and phytochelatins are cysteine-rich metal-binding proteins that act as chelators.[15] Specifically, metallothioneins are proteins found in living systems with a high content of cysteine residues that bind various heavy metals. They are important for the regulation of cell growth and for DNA damage repair.[8] Phytochelatins are oligomers of glutathione, produced by the enzyme *phytochelatin synthase* (EC 2.3.2.15), which act as a chelator for heavy metal detoxification. These compounds are found in plants, fungi, nematodes, and algae.

A biological application of nuclear magnetic resonance (NMR) spectroscopy is known as magnetic resonance imaging (MRI). When an MRI is taken of a person, or a part thereof, the result is a high-quality image of the inside of the human body. The MRI scan of a human brain is shown in Figure 12.10. This noninvasive technique is a great help to medicine. In many cases, examination of human tissue is complicated because it may be difficult to distinguish one tissue type from another or to bring out detail for one organ or tissue type.

As noted, there are some problems associated with MRI scans. The difference between a cancerous tumor and normal brain tissue may be so small that the detection of the tumor by MRI is difficult. Metal-chelated chemicals can be injected into a patient to make it easier to distinguish tissue types. These are known as *contrasting agents*. They typically make one or more tissue types appear brighter, leading to greater contrast. Alternatively, a tissue may appear darker, providing a significant contrast. Some typical contrast agents[16] are Gd-EDTA, Gd-DTPA, and Gd-DOTA. *Gadodiamide* is useful for imaging the central nervous system. *Mangafodipir trisodium* is used for lesions of the liver. Note that the organic fragments of the contrast agents are amino acid derivatives (Chapter 2), phosphates, or amides. These compounds include ferric ammonium citrate (sold commercially as Geritol), which is used to image the stomach and upper small intestine.

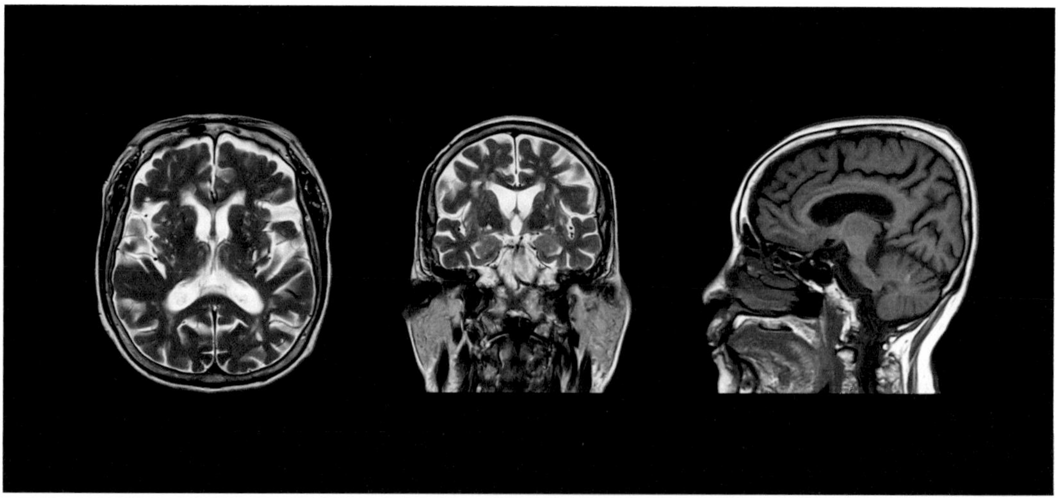

FIGURE 12.10 Magnetic resonance image (MRI) of a human brain: Comparison MRI of the brain, Axial, Coronal. (Reprinted from Shutterstock Asset id: 1390007888.)

[15] Grennan, A.K. *Plant Physiol.* **2011**, *155*, 1750.
[16] Hornak, J.P. *The Basics of MRI*, an online book, **2008**. In Chapter 12.

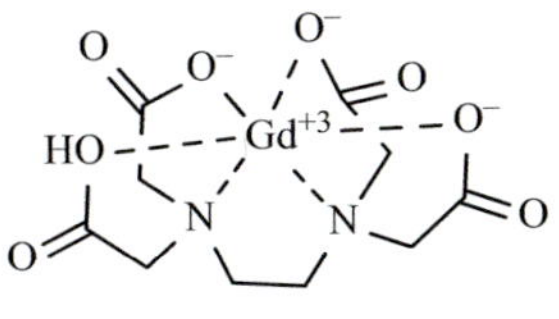
Gd-EDTA

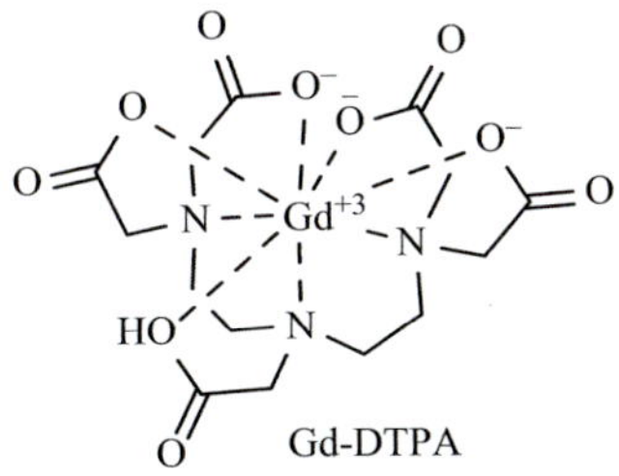
Gd-DTPA

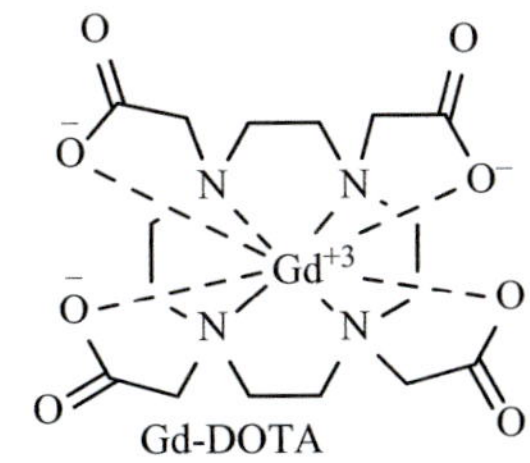
Gd-DOTA

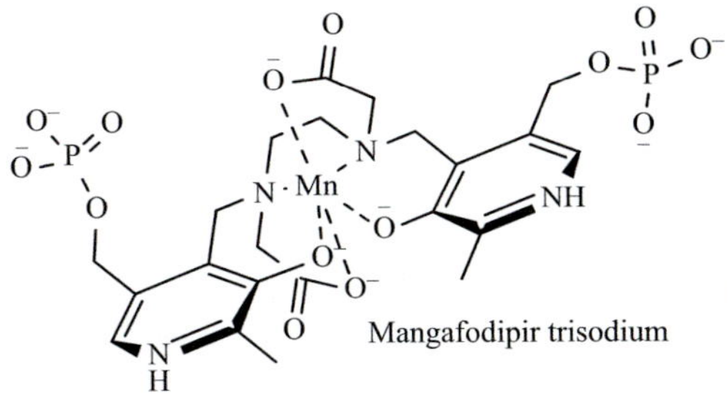
Mangafodipir trisodium

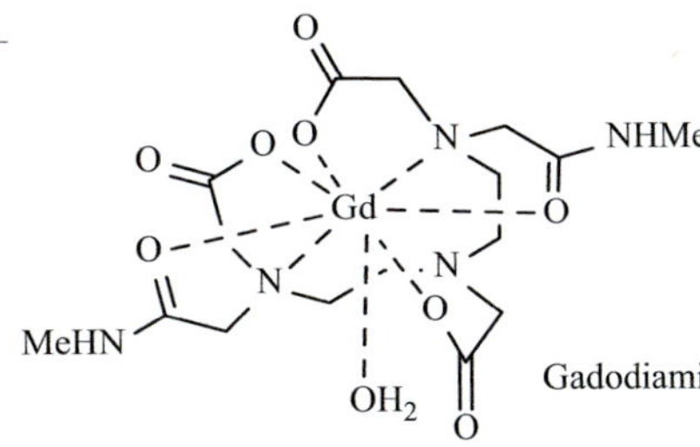
Gadodiamide

Most contrasting agents are organic complexes of paramagnetic metal ions (e.g., gadolinium, Gd and Gd^{3+} compounds, ferric compounds, Fe^{3+}, or manganese compounds, Mn^{2+}).[17] Heavy metals are often toxic, but complexation with other molecules or ions often diminishes the toxicity and facilitates the clearance of those metals from the body. If one tissue type complexes the metal more than another, that tissue type (or organ) "stands out" from the background. Most tumors, for example, have a greater Gd uptake than the surrounding tissues, leading to a stronger signal as the tumor stands out from the background.[18]

HOMEWORK

12.1. The biologically important metals, Mg^{2+} and Ca^{2+}, have different ionic radii. This can lead to very different catalytic properties. Why?
12.2. Draw the transition state for the reaction of phenylmagnesium bromide with pentan-3-one.
12.3. What is the function of the phytol side chain in chlorophyll A?
12.4. Where are metals usually found in metalloenzymes?
12.5. How does cobalt interact with heme?
12.6. List at least six ligands that coordinate with transition metals.
12.7. Suggest a structure formed when one molecule of ethylenediamine reacts with $CoCl_6$ in an octahedral geometry.

[17] Rajan, S.S. *MRI. A Conceptual Overview*, Springer, New York, **1998**, p. 66.
[18] Hornak, J.P. *The Basics of MRI*, an online book, **2008**. In Chapter 1.

13 Carbohydrates

An important class of molecules that are critical to an understanding of biology processes are carbohydrates, commonly known as sugars. Carbohydrates are key components of glycosides and cells, and they comprise the backbone of nucleotides, including DNA and RNA. This chapter will introduce the fundamentals of carbohydrate structure and nomenclature.

13.1 (POLY)HYDROXY CARBONYL COMPOUNDS

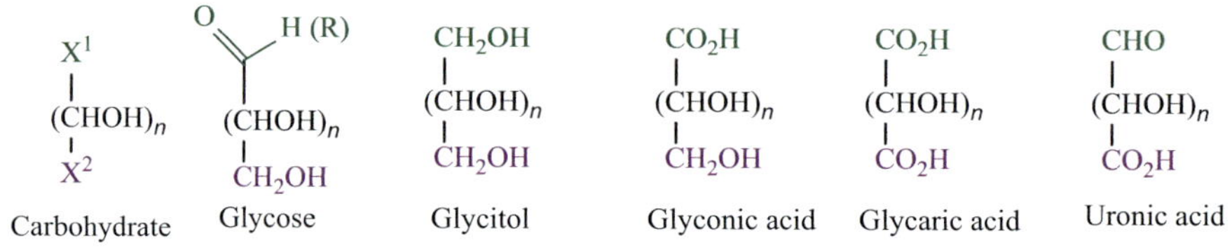

Most carbohydrates are (poly)hydroxylated aldehydes or ketones, although there are other (poly) hydroxylated derivatives. There are two fundamental ways to classify carbohydrates. The first is based on the type of functional group that accompanies the hydroxyl units. The second is based on the number of individual units that make up the carbohydrate**.** Carbohydrates are classified by the nature of the functional groups X^1 and X^2. In a carbohydrate, there are several repeating CHOH units defined by the integer "*n*," where $n = 3$, 5, or 6, and so on. For example, when $n = 3$ in a carbohydrate, the structure is X^1–CHOH-CHOH-CHOH–X^2. Most carbohydrates are defined by making X^1 and/or X^2 = CH_2OH, CHO, COR (a ketone), or COOH. A *glycose* has a CH_2OH and an aldehyde or ketone unit, and a *glycitol* has two CH_2OH units (sometimes called an *alditol*). When one group is a carboxylic acid and the other is CH_2OH, it is a *glyconic acid* (sometimes called an *aldonic acid*). A *glycaric acid* has two carboxyl units and is a hydroxy-dioic acid (sometimes called an *aldaric* acid), and a *uronic acid* has a carboxyl group and an aldehyde group.

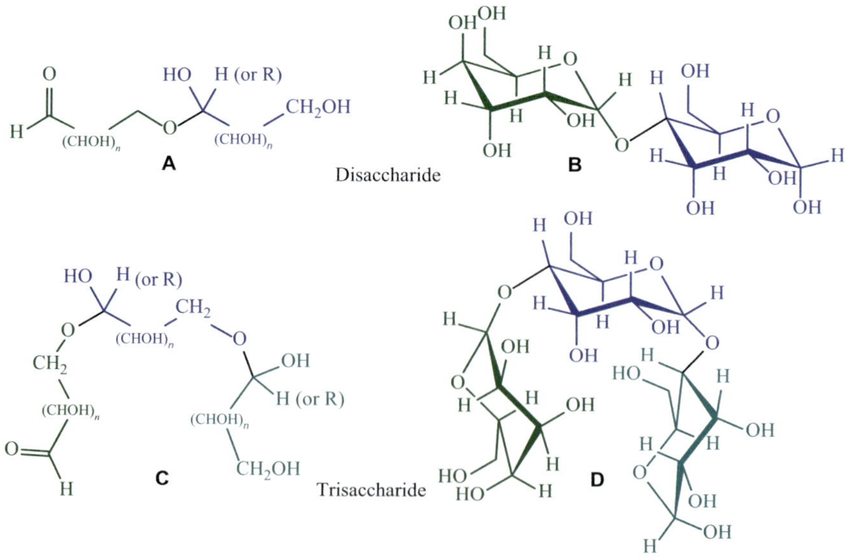

DOI: 10.1201/9781003605157-13

The second way to categorize carbohydrates is by the number of sugar units the carbohydrate contains. The general carbohydrate structure shown above contains one carbohydrate unit, and it is categorized as a *monosaccharide*. If two monosaccharides are coupled together, the resulting molecule is a *disaccharide,* and a molecule with three monosaccharide units is a *trisaccharide*. The structures shown for the disaccharide include both the open-chain form (**A**) and the pyranose form (**B**; see Section 13.2 and Figure 13.3). Likewise, the trisaccharide is shown as the open-chain form (**C**) and the pyranose form (**D**). Linking 5–15 monosaccharides yields an *oligosaccharide*. A *polysaccharide* has > 15 monosaccharide units linked together. Note that in both disaccharides and trisaccharides, the monosaccharides are linked together by an acetal linkage (Section 6.8).

13.2 MONOSACCHARIDES

Monosaccharide *glycose* derivatives are the focus of this section, but the other categories of carbohydrates will also be discussed. Glycoses are subdivided into *aldoses*, where the carbonyl unit is an aldehyde, and *ketoses*, where the carbonyl unit is a ketone. This discussion begins with the characteristics and properties of glycoses, their structures, and their nomenclature.

Glycose

The basic unit of a glycose is shown, and there are several different monosaccharide glycoses. The carbon of each hydroxymethyl (CHOH) unit is a stereogenic center, so there are several diastereomers for most carbohydrates. Each time a CH_2OH unit is added, another stereogenic center is added, which increases the number of diastereomers. Each diastereomer will have a (+) and a (−) form (enantiomers). For the glycose where $n = 1$ (three carbons), there is one stereogenic center and there are two stereoisomers (the two enantiomers), following the 2^n rule (Section 5.18). When $n = 2$ (four carbons), there are 4 stereoisomers (two diastereomers, each with an enantiomer); when $n = 3$ (five carbons), there are 8 stereoisomers (four diastereomers, each with an enantiomer); and when $n = 4$ (six carbons), there are 16 stereoisomers (eight diastereomers, each with an enantiomer).

Glycoses are further classified by the total number of carbon atoms. The glycose with three carbon atoms ($n = 1$) is called a *triose*, and when there are four carbon atoms ($n = 2$), it is a *tetrose*. A *pentose* has five carbon atoms ($n = 3$), and a *hexose* has six carbon atoms ($n = 4$). If the carbonyl unit in a glycose is an aldehyde, it is an aldose (see above), so a three-carbon aldose is an *aldotriose* and a six-carbon aldose is an *aldohexose*. If the carbonyl unit in a glycose is a ketone, the molecule is a ketotetrose, a ketohexose, and so on.

A simple carbohydrate that fits the generic formula for a glycose is the aldotriose known as *glyceraldehyde*. This compound has one stereogenic center and two enantiomers, and the D- and L-designations are based on analogy to glyceraldehyde (see Figures 2.4 and 2.5 in Section 2.2). *Note that L-(+)-glyceraldehyde is* (*S*)-2,3-dihydroxypropanal *and D-(−)-glyceraldehyde is* (*R*)-2,3-dihydroxypropanal. There are two aldotetroses, D- and L-erythrose and D- and L-threose, as shown in Figure 13.1. There are four diastereomeric aldopentoses, and each has an enantiomer. The (*R*)- or (*S*)-configuration of the last two carbon atoms shown corresponds to the D or L nomenclature. With a focus on the D-diastereomers, the four diastereomers are D-ribose, D-arabinose, D-xylose, and D-lyxose. The absolute configuration of each stereogenic center is also provided.

Adding one carbon leads to the hexoses, and there are eight diastereomeric aldohexoses, each with an enantiomer, as shown in Figure 13.2. The D-compounds are D-allose, D-altrose, D-glucose, D-mannose, D-gulose, D-idose, D-galactose, and D-talose. The last –CHOH–CH_2OH unit (in violet) for all of the D-sugars shown has the same (*R*) absolute configuration. All the L-sugars will have the (*S*)-configuration at that carbon.

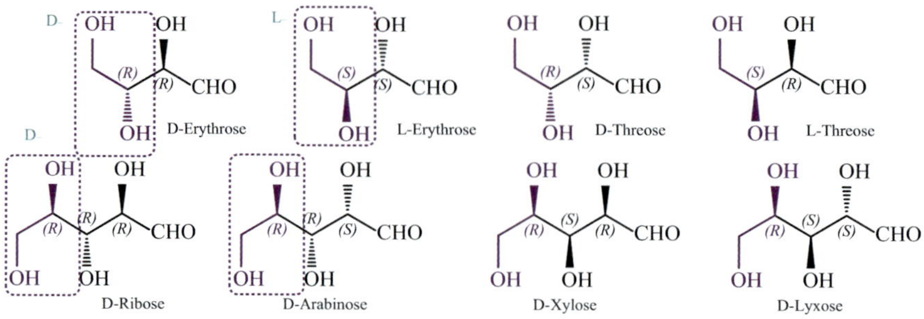

FIGURE 13.1 Aldotetroses and aldopentoses.

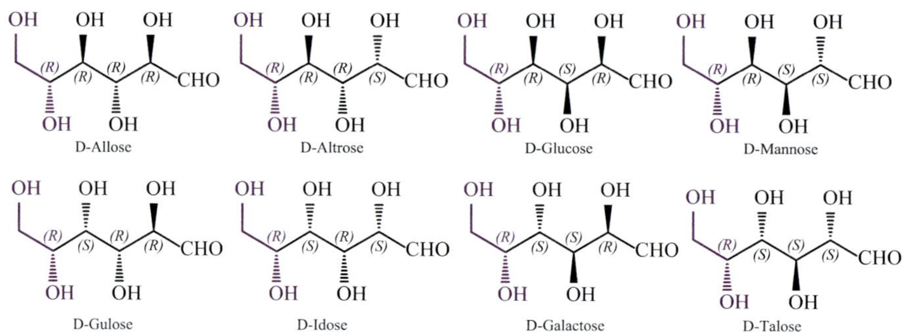

FIGURE 13.2 The eight D-hexoses.

TABLE 13.1
Three-Letter Codes Used to Represent Common Monosaccharides

Monosaccharide		Three-Letter Code
Allose		All
Altrose		Alt
Arabinose		Ara
Fructose		Fru
Fucose	(6-Deoxygalactose)	Fuc
Galactose		Gal
Glucose		Glc
Gulose		Gul
Idose		Ido
Lyxose		Lyx
Mannose		Man
Rhamnose	(6-Deoxy-L-mannose)	Rha
Talose		Tal
Xylose		Xyl

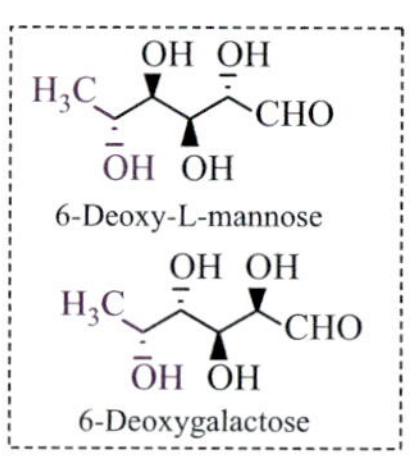

Carbohydrates are usually assigned a three-letter abbreviation (see Table 13.1)[1] that is used in conjunction with the letters *p* (for a pyranose) or *f* (for a furanose) to generate a shorthand name.

The aldoglycose compounds shown are hydroxy–aldehydes. The reaction of an aldehyde and an alcohol to give an acetal was presented in Section 6.8. A specific example is the reaction of 3,4-dimethylpentanal with ethanol, in the presence of an acid catalyst, to yield the acetal product 1,1-diethoxy-3,4-dimethylpentane. Formation of this acetal requires formation of a hemiacetal

[1] Kennedy, J.F.; White, C.A. *Bioactive Carbohydrates*, Ellis Horwood Ltd., Chichester, UK, **1983**, p. 41.

intermediate, 1-ethoxy-3,4-dimethylpentan-1-ol. However, in simple systems, the hemiacetal is usually a transient product and generally so unstable that it cannot be isolated. This does not apply to aldotetrose, aldopentose, and aldohexose derivatives, which are hydroxy-aldehydes. *Aldoglycoses do not form acetals because they form stable hemiacetals.*

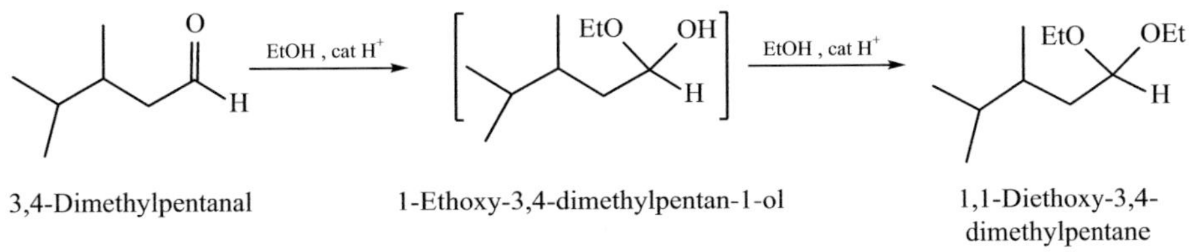

To form such a hemiacetal from a carbohydrate, a cyclic hemiacetal is formed. In the case of carbohydrates, the cyclic hemiacetal is stable and does *not* react further to form an acyclic acetal. The carbon bearing the acetal–OH unit is the hemiacetal carbon atom derived from the acyl carbon of the aldehyde unit, and it is called the *anomeric carbon* or an *anomeric center. Isomers that differ only in the configuration at the anomeric carbon are called anomers.* D-glyceraldehyde is arguably the simplest member of this family, so it is reasonable to ask if a hemiacetal can be formed. Attempts to form a cyclic hemiacetal from D-glyceraldehyde generate an epoxide, (3*R*)-(hydroxymethyl)-oxiran-2-ol, rather than the larger-ring oxetane derivative. In other words, D-glyceraldehyde does *not* easily form a cyclic hemiacetal.

Tetrose derivatives can form stable five- and six-membered ring hemiacetals. Both D-erythrose and D-threose form cyclic hemiacetals, as shown in Figure 13.3, but the terminal CH_2OH group is used to form the oxygen-containing ring. Cyclization favors the formation of five- and six-membered rings. D-erythrose cyclizes to yield tetrahydrofuran-2,3,(4R)-triol, and D-threose cyclizes to yield tetrahydrofuran-2,(3*R*,4*S*)-triol. Both are drawn in a hyperextended conformation in Figure 13.3 to emphasize which bond is formed and to which atoms, *but such a conformation cannot exist.* Therefore, the structures are drawn again in an envelope conformation to show a more normal conformation for the five-membered tetrahydrofuran rings.

The term *ose* is used for tetrahydrofuran-2,(3*R*,4*R*)-triol and tetrahydrofuran-2,(3*R*,4*S*)-triol because they are carbohydrates, but they are also derived from furan. This leads to the name *furanose* for a five-membered ring hemiacetal derived from an aldose. Thus, D-erythrose forms D-erythrofuranose, and D-threose forms D-threofuranose.

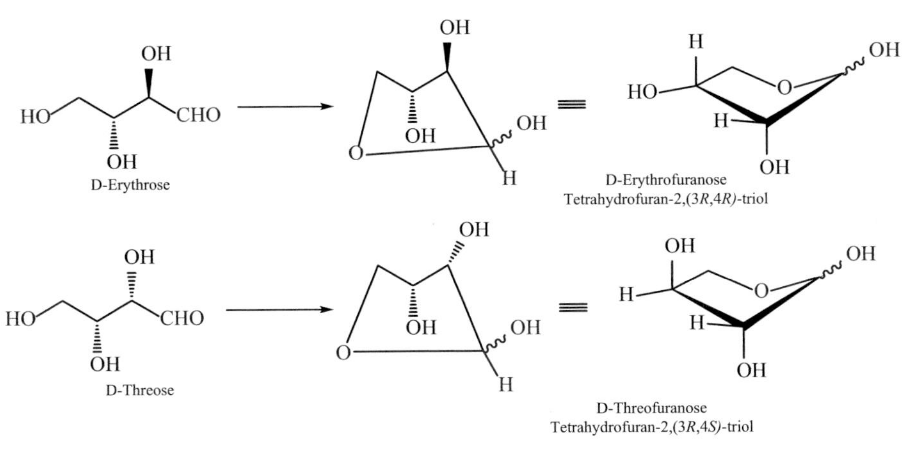

FIGURE 13.3 Hemiacetal forms of D-erythrose and D-threose.

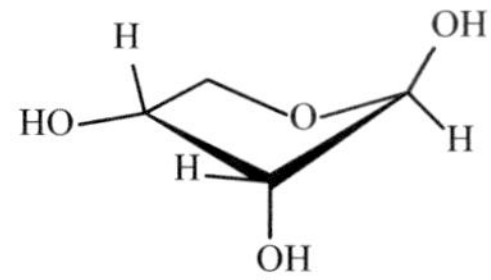

1

Tetrahydrofuran-(2*R*,3*R*,4*R*)-triol

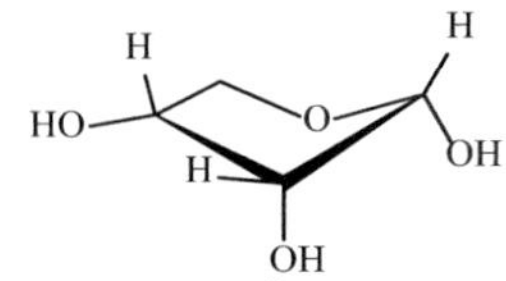

2

Tetrahydrofuran-(2*S*,3*R*,4*R*)-triol

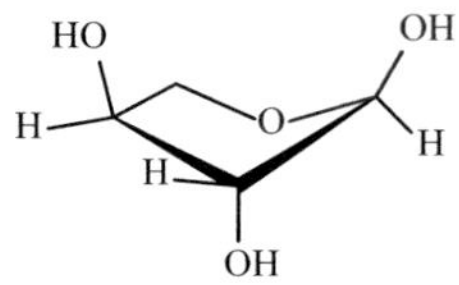

3

Tetrahydrofuran-(2*R*,3*R*,4*R*)-triol

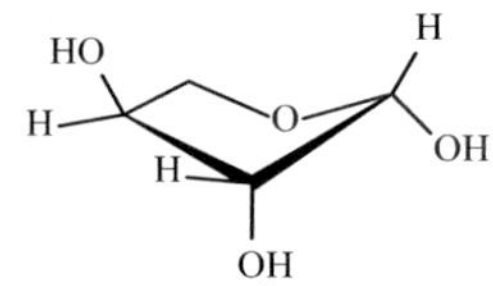

4

Tetrahydrofuran-(2*S*,3*R*,4*R*)-triol

As shown in Figure 13.3, D-erythrofuranose is actually a mixture of two compounds, indicated by the "squiggle" line to show undefined stereochemistry at the anomeric carbon. In this case, there are two stereoisomers at the anomeric carbon, C1, so there are two diastereomers. Similarly, D-threofuranose exists as two diastereomers. The four stereoisomers are **1** and **2** from D-erythrofuranose and **3** and **4** from D-threofuranose. The OH unit at C1 in **1** and in **3** is on the opposite side of the ring from the OH at C2. Indeed, the OH units are on the "top" of each ring as they are drawn. In **3** and in **4**, the OH at C1 is on the opposite side of the ring from the OH unit at C2, and on the "bottom" of each ring as they are drawn.

There is another nomenclature system that is used to identify the position of the OH unit on the anomeric carbon. In this system, the ring oxygen is drawn away from the viewer (to the rear) with the ring carbons forward (to the front), and the anomeric carbon is drawn on the right side (see Figure 13.4). In this representation, the OH on the upper bond of the anomeric carbon is identified as the β and the OH on the lower bond of the anomeric carbon is the α. The six-membered pyranose ring is similarly drawn with the ring oxygen to the rear and the anomeric carbon on the right, and both the α and β hydroxyl groups are shown. The two furanose anomers are therefore β-furanose and α-furanose, and the two pyranose anomers are β-pyranose and α-pyranose. Therefore, *α means down and β means up,* but only when the ring is drawn for the D-series, as in Figure 13.4. In the L-series, the definitions of α and β are reversed, in that the anomeric carbon is positioned to the left of the ring oxygen**.**

There is yet another way to represent furanose and pyranose derivatives that allows a focus on the sidedness of the ring, which can be useful in sorting out stereochemistry. The five- and six-membered rings are drawn as flat structures, with the OH substituents "up" or "down" relative to the "top" and "bottom" sides of the flat ring. This alternative representation is known as a *Haworth formula.*

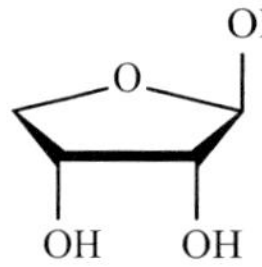

5

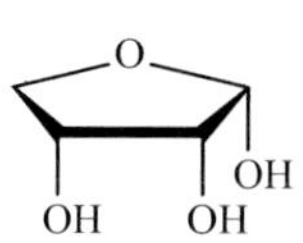

6

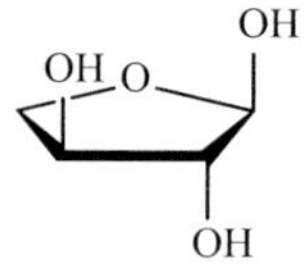

7

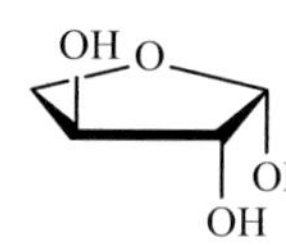

8

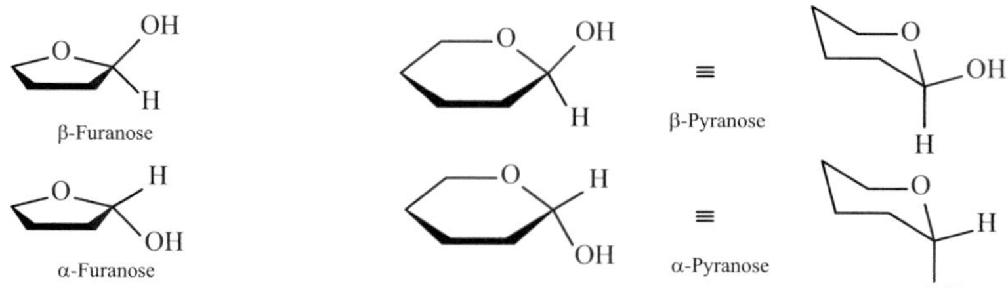

FIGURE 13.4 Convention for assigning α and β configurations for D-sugars.

Erythrofuranose derivatives **5** and **6** and threofuranose derivatives **7** and **8** are drawn again, but as Haworth formulas. Using these representations, **5** and **7** are α-anomers, and **6** and **8** are β-anomers. Therefore, **5** is α-D-erythrofuranose, **6** is β-D-erythrofuranose, **7** is α-D-threofuranose, and **8** is β-D-threofuranose.

Just as tetrose derivatives form cyclic hemiacetals, so do pentose and hexose derivatives. The hemiacetals formed by cyclization to the anomeric carbon in the pentose D-ribose are shown in Figure 13.5. D-Ribose exists in the cyclic, or furanose, form as α- and β-anomers. (5*R*)-(hydroxymethyl) tetrahydrofuran-(2*S*,3*R*,4*R*)-triol is identified as the β-anomer, or β-D-ribofuranose. Similarly, (5*R*)-(hydroxymethyl)tetrahydrofuran-(2*R*,3*R*,4*R*)-triol is the α-anomer for the D-series, or α-D-ribofuranose. There are other stereogenic centers, so cyclization produces two diastereomers.

Both β-D-ribofuranose and α-D-ribofuranose are drawn in Haworth projection in Figure 13.6, mostly for a comparison with the tetrose derivatives. Furanoses are five-membered ring compounds, so they will assume a conformation that approximates the envelope conformation observed for cyclopentane (Section 5.14). This result is essentially correct, and the envelope conformations are also shown in Figure 13.6 to provide a more realistic view of the conformational preferences of furanose derivatives.

The presence of α- and β-anomers for a furanose raises the question: Why is there a mixture at the anomeric carbon? The anomeric carbon yields a mixture because there is an equilibrium between the two hemiacetal forms and the open-chain aldehyde, as shown in Figure 13.7 for D-ribose. In an equilibrium process, the OH may attack the acyl carbon from one face (path A in cyan) or from the other face (see path B in green). Attack from these two faces (A and B) occurs because there is free rotation around the C1–C2 bond in the open-chain form of D-ribose (the aldehyde form), and cyclization can occur via each rotamer. Cyclization via path A leads to β-D-ribofuranose, and cyclization via path B leads to α-D-ribofuranose, as shown in Figure 13.7. Conversion of an acyclic aldoglycose to the cyclic hemiacetal occurs at the anomeric center, leading to a mixture of anomers.

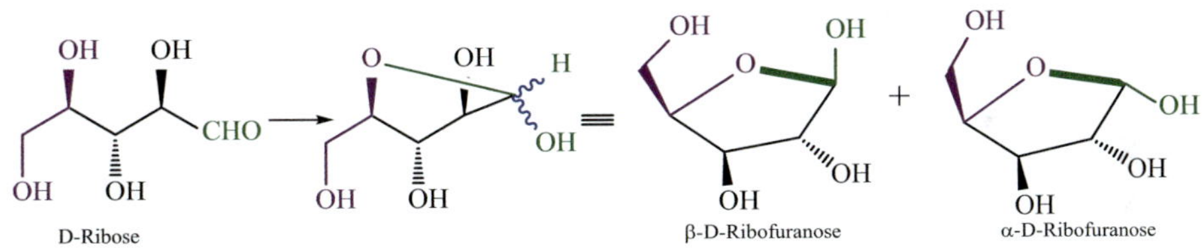

FIGURE 13.5 D-Ribose.

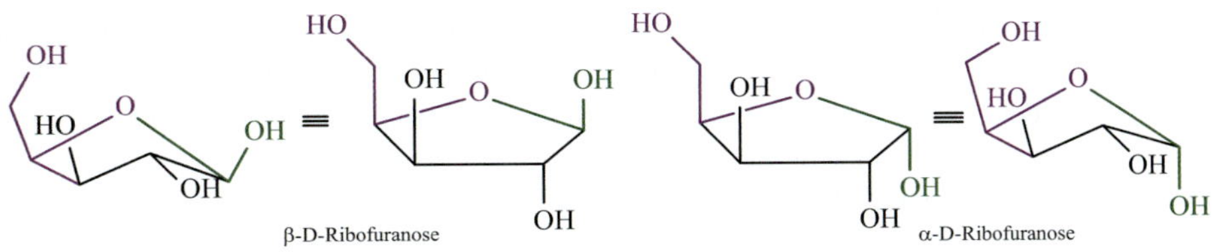

FIGURE 13.6 Envelope and Haworth formulas for α- and β-D-ribofuranose.

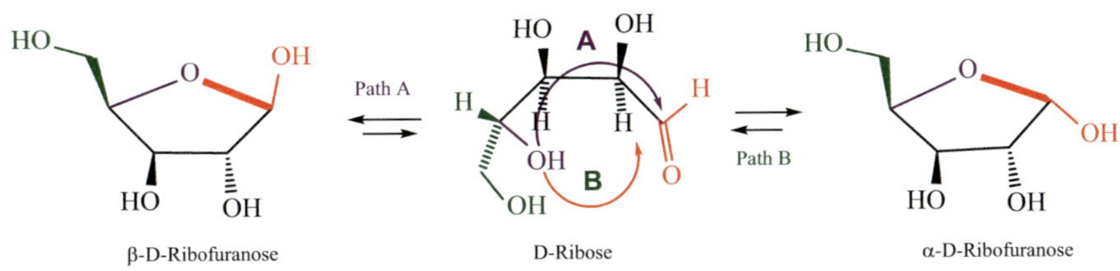

FIGURE 13.7 Equilibria for D-ribose.

13.3 MUTAROTATION

Cyclization of an aldopentose to the hemiacetal yields a five-membered ring, and, as noted, β-D-ribofuranose is in equilibrium with the acyclic aldopentose, which is in equilibrium with the α-D-ribofuranose. Similarly, an aldohexose can cyclize to both a five-membered ring (derived from *furan*, a *furanose*) and a six-membered ring. The six-membered ring derivative, derived from *pyran*, is called a *pyranose*. Analogous to the conformation of cyclohexane, a six-membered ring compound should assume a chair conformation as the lowest-energy form (Section 5.14). Using D-mannose as an example, in Figure 13.8, hemiacetal formation will yield two pyranose derivatives: the α-isomer (6*R*)-(hydroxymethyl)tetrahydro-2*H*-pyran-(2*S*,3*S*,4*S*,5*S*)-tetraol (α-D-mannopyranose) and the β-isomer (6*R*)-(hydroxymethyl)tetrahydro-2*H*-pyran-(2*R*,3*S*,4*S*,5*S*)-tetraol (β-D-mannopyranose). In α-mannopyranose and β-mannopyranose, the hydroxyl groups are axial or equatorial, which is the proper representation. Note that the anomeric OH is axial in β-mannopyranose and equatorial in α-mannopyranose, because the OH units are on opposite sides of the ring.

Hemiacetal formation from an aldehyde is a reversible process, and the formation of the anomers from the open-chain aldehyde of a furanose, or a pyranose, is also a reversible process. In other words, cyclization to the hemiacetals is an equilibrium. At equilibrium, D-mannose has three species in solution: α-D-mannopyranose, the open-chain aldehyde form, and β-D-mannopyranose, as shown in Figure 13.8. The position of this equilibrium can be measured, so it is possible to determine the relative percentages of these three species. This discussion must begin with an *assumption*, however. *Assume* that the open-chain aldehyde form of mannose accounts for $<1\%$ (for most sugars in this chapter, it is ~0.1%–0.5% or less) of the equilibrium, so the pyranose forms are the major conformations.

Since these are chiral compounds, one way to measure the equilibrium is to measure the specific rotation (see Section 5.17). The experimentally measured specific rotation for a pure sample of α-D-mannopyranose is +29.3°. The experimentally measured specific rotation of a pure sample of β-D-mannopyranose is −16.3°. The specific rotation of the mixture can also be measured, and once the specific rotations of each pure enantiomer are known, the relative percentages of α- and β-D-mannose may be determined. An interesting phenomenon known as *mutarotation* must be reckoned with, however.

When pure α-D-mannopyranose is allowed to stand in aqueous solution, the specific rotation changes, and it eventually stabilizes at +14.5°. Similarly, when a sample of pure β-D-mannopyranose is allowed to stand in solution, the specific rotation of the final mixture is also measured to be +14.5°. This observation indicates two things. First, the sign of the specific rotation is positive, which means that there is more α-D-mannopyranose than β-D-mannopyranose, indicating that there is more of the axial anomer than the equatorial anomer. Remember from Section 5.14 that a substituent in the equatorial position has less $A^{1,3}$-strain than when that substituent is in the axial position and is therefore expected to be lower in energy. *Since α-D-mannopyranose is the major conformation, a different effect must dominate the equilibrium.*

For the observed equilibrium to be established, a reaction must occur at the anomeric center of both isomers in aqueous solution. This change in the optical rotation of a carbohydrate at the anomeric carbon when it is dissolved in solution is the result of this reaction, and the phenomenon is called *mutarotation*. Mutarotation occurs at the anomeric carbon as a direct result of the

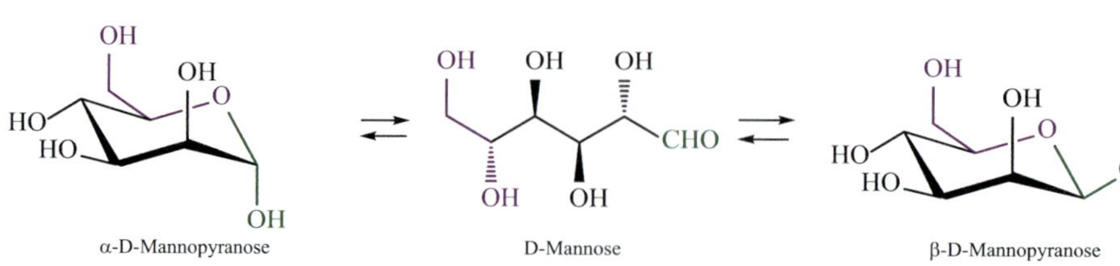

FIGURE 13.8 Equilibrium for D-mannose.

equilibrium between the open-chain aldehyde form and the two pyranose forms of mannose. When pure α-D-mannopyranose is in solution, an equilibrium is established with the open-chain aldehyde form of D-mannose, which can cyclize to β-D-mannopyranose, establishing an equilibrium. If pure β-D-mannopyranose is used, it also establishes an equilibrium with the open-chain aldehyde, which establishes an equilibrium with α-D-mannopyranose. This result means that adding one pure pyranose to the solution leads to equilibrium with the open-chain aldehyde and both α- and β-anomers. At equilibrium, both α- and β-D-mannopyranose are present in solution when the specific rotation of the mixture is measured, and the final value of the specific rotation will reflect the relative percentages of α- and β-mannose at equilibrium (*if the percentage of aldehyde is very small*). If this equilibrium value is +14.5° for D-mannopyranose, the percentage of each anomer can be calculated. At equilibrium, ~67.5% of α-D-mannopyranose (the axial anomer) and ~32.5% of β-D-mannopyranose, the equatorial anomer, are present. The equilibrium should favor the more stable isomer, which in this case has the OH in the axial position.

13.4 THE ANOMERIC EFFECT

Why is there more of the axial anomer for D-mannopyranose? In Section 5.14, an axial substituent on a six-membered ring should have more $A^{1,3}$-strain, which makes it less stable when compared to the chair conformation with that substituent in the equatorial position. However, as shown in Section 13.3, D-mannose has a higher percentage of the α-D-anomer at equilibrium, which indicates that the conformation with an axial OH group is preferred (see Figure 13.8). This issue is further complicated if mutarotation of D-glucose is examined.

D-Glucose has two cyclic hemiacetal forms (α-D-glucopyranose and β-D-glucopyranose), as shown in Figure 13.9. The specific rotation data lead to the calculation that there is ~36% of the α-D-glucopyranose and 64% of the β-D-glucopyranose, in which the OH is in an equatorial position. This result is contrary to the observation for D-mannopyranose, where the OH unit prefers an axial orientation, whereas in D-glucopyranose, the OH unit prefers an equatorial orientation.

To resolve this apparent conflict, all of the substituents on the six-membered ring must be considered for both molecules. For α-D-mannopyranose, there are two equatorial OH groups and two axial OH groups, whereas β-D-mannopyranose has three equatorial OH groups and one axial OH. For α-D-glucopyranose, there are four equatorial OH groups and zero axial groups, whereas β-D-glucopyranose has three equatorial OH groups and one axial OH. The preponderance of equatorial groups in β-D-glucopyranose is consistent with the behavior of cyclohexane derivatives in that equatorial substituents have less $A^{1,3}$-strain (see Section 5.14), which leads to a higher percentage of that conformation. However, $A^{1,3}$ strain does not explain the observations with α-D-mannopyranose. Simply looking at axial versus equatorial substituents leads to only a partial solution to the problem. Remember, however, that these are hydroxyl substituents and this is a pyran ring.

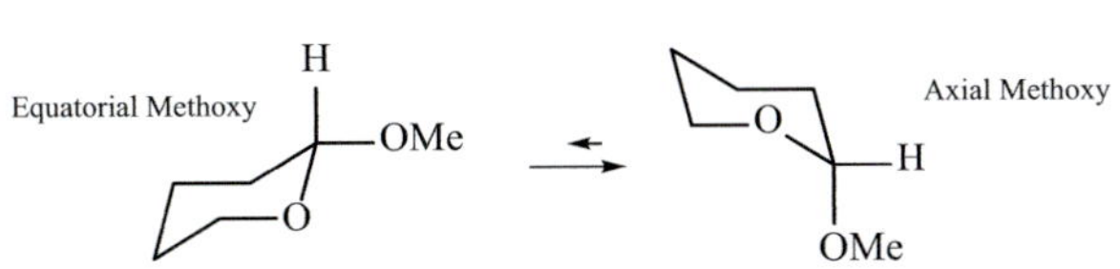

(*R*)-2-methoxytetrahydro-2*H*-pyran

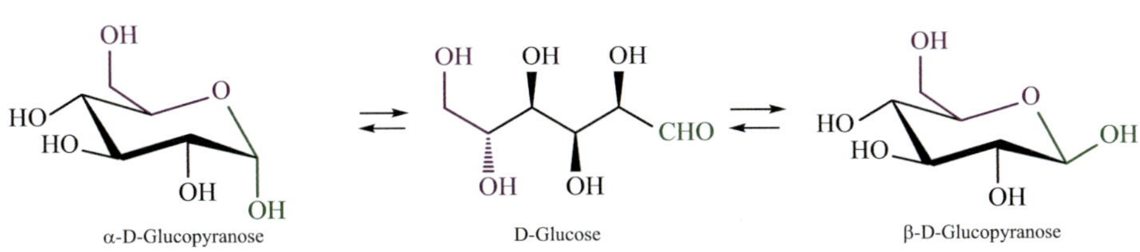

FIGURE 13.9 Mutarotation in D-glucose.

A phenomenon called the *anomeric effect* is the tendency for an electronegative substituent in a pyran ring to prefer an axial orientation when attached to an anomeric carbon. There are nonbonded interactions between the groups that play an important role in pyranose derivatives. The anomeric effect arises from the designation of the C1 carbon of a pyranose as the anomeric carbon. One explanation for the anomeric effect calls attention to the dipole moments of both heteroatoms. In an equatorial configuration, the dipoles of both heteroatoms are partially aligned, and they repel. In an axial configuration, these dipoles are opposed and represent a lower energy state.

In 1-methoxycyclohexane, the conformation with an equatorial methoxy is preferred to the conformation with an axial methoxy, as expected. In 1-methoxypyranose (2-methoxytetrahydro-2*H*-pyran), however, the anomeric effect makes the conformation with an axial OMe preferred to the conformation that has the equatorial OMe group. This anomeric effect is observed in structurally similar pyranose derivatives (e.g., mannopyranose and glucopyranose), so a pyranose with an axial OH group at the anomeric carbon is preferred in the pyranose–aldehyde equilibrium. While the anomeric effect explains why α-D-mannopyranose is preferred, it does not explain why β-D-glucopyranose is preferred. For glucose, the magnitude of the anomeric effect is small because of the equatorial OH at C2, and the lower conformational energy arises from having all-equatorial substituents. This observation leads to a preference for β-D-glucopyranose. It is important to note that in most cases, mutarotation in a pyranose derivative leads to *an equilibrium preference for the α-pyranose.* In most cases, experimental data are required to determine whether the α- or the β-anomer is preferred.

13.5 KETOSE MONOSACCHARIDES

There are obviously many aldose monosaccharides. There are also many ketose monosaccharides, but the discussion will focus only on the D-diastereomers, as with the aldoses. The triose is 1,3-dihydroxypropan-2-one (also called *glycerone*), and the tetrose is D-*glycerotetrulose*. There are two pentoses named D-*ribulose* and D-*xylulose*. There are four hexoses named D-*psicose*, D-*fructose*, D-*sorbose*, and D-*tagatose*. All of these compounds are ketoses and are further classified according to the number of carbon atoms, as noted in Section 13.1. For example, D-glycerotetrulose is a ketotetrose, D-ribulose is a ketopentose, and D-psicose is a ketohexose. In all cases, cyclization is possible to form a hemiketal. Just as aldehydes and alcohols react to form a hemiacetal, ketones and alcohols react to form a hemiketal. For D-ribulose, a six-membered ring (a pyranose) is formed if the terminal CH_2OH (in violet) unit reacts with the ketone carbonyl. The two anomers formed are α-D-ribulofuranose and β-D-ribulofuranose, as shown in Figure 13.10.

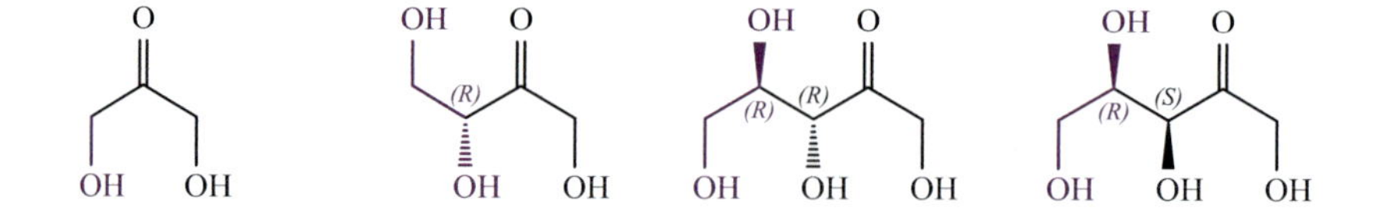

Glycerone
1,3-Dihydroxypropan-2-one
D-Glycerotetrulose
D-Ribulose
D-Xylulose

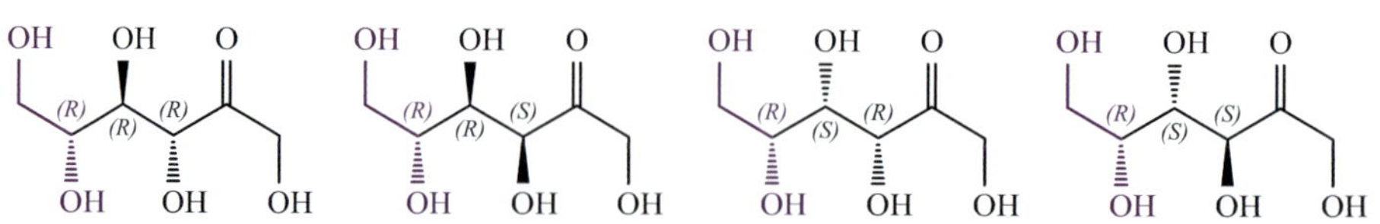

D-Psicose
D-Fructose
D-Sorbose
D-Tagatose

These furanose derivatives undergo mutarotation in a manner similar to the aldofuranose compounds. Ketohexoses form pyranose derivatives via the OH unit at C6 (in violet). D-Sorbose, for example, will form β-D-sorbopyranose and α-D-sorbopyranose, and D-fructose will form

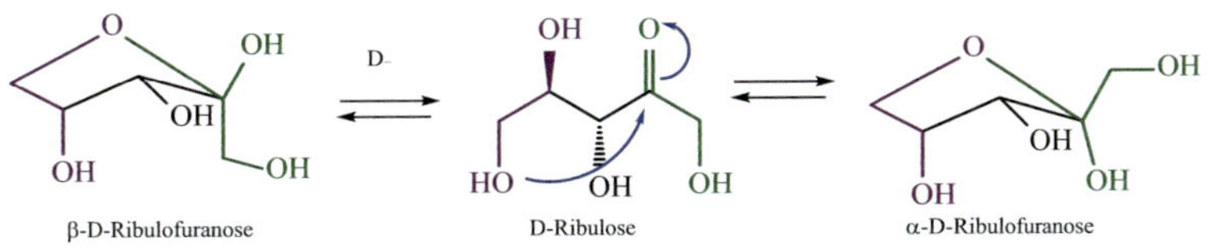

FIGURE 13.10 Mutarotation in D-ribulose.

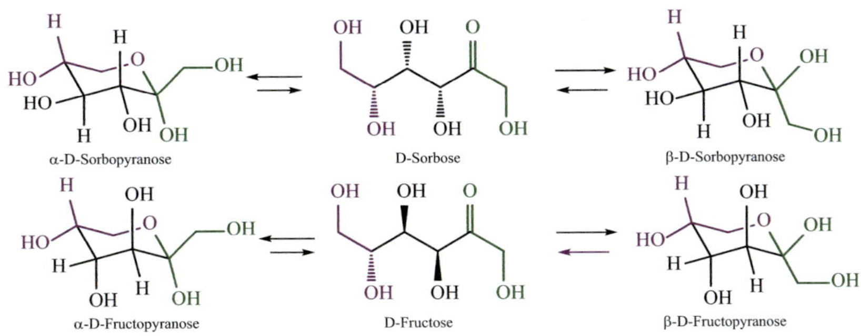

FIGURE 13.11 D-Sorbopyranose and D-fructopyranose.

α-D-fructopyranose and β-D-fructopyranose, as shown in Figure 13.11. For ketohexoses, however, the pyranose derivatives are *not* necessarily the major hemiketals that are formed.

Cyclization via the OH unit at C5 will lead to furanose derivatives as the preferred structure. In Section 5.14, cyclopentane was shown to be lower in energy than cyclohexane, and the tetrahydrofuran ring is indeed somewhat more stable than the tetrahydropyran ring. Stabilization of the hydroxyl groups and the hydroxymethyl groups contributes to the difference in stability in this case. Comparing D-sorbose and D-fructose, and using the OH at C5 of D-sorbose, α-D-sorbofuranose and β-D-sorbofuranose are formed, as are α-D-fructofuranose and β-D-fructofuranose from D-fructose, as shown in Figure 13.11. As mentioned, these compounds undergo mutarotation, but the equilibrium is more complex than that observed with most aldopyranoses.

HOMEWORK

13.1. What is an oligosaccharide?

13.2. Give one example of a tetrose, a pentose, and a hexose.

13.3. Examine the open-chain aldehyde form of D-galactose. How many stereogenic centers are there in this molecule? Draw the structure of the enantiomer of D-galactose.

13.4. Draw the structures of Alt, Fru, Gal, Lyx, and Tal.

13.5. Draw the product formed when cyclohexanone reacts with an excess of propanol in the presence of an acid catalyst.

13.6. Draw the furanose form of α-D-altrose. Give the correct name for this molecule.

13.7. Draw the Haworth formula of the β-form of D-xylofuranose.

13.8. Draw all three structures in the mutarotation equilibrium of (a) D-gulopyranose and also of (b) D-altropyranose. Look up the major structure found in this equilibrium and offer a brief explanation based on the anomeric effect.

13.9. What is the IUPAC name, with stereogenic centers identified as R or S, of β-D-threofuranose?

14 Glycosides

Carbohydrates were introduced in Chapter 13. This chapter will focus on carbohydrate derivatives, especially those that are functionalized at the anomeric carbon. A glycoside is a molecule in which a carbohydrate is bonded to another functional group by a glycosidic bond. A glycosidic bond usually refers to the bond of a functional group with the anomeric OH unit, but the IUPAC nomenclature for such compounds is *C*-glycosyl compounds. Glycosides can be linked by an *O*- (an *O-glycoside*), *N*- (a *glycosylamine*), *S*- (a *thioglycoside*), or *C*- (a *C-glycoside*) glycosidic bond. A Haworth projection (Section 13.2) is typically used for the functionalized carbohydrate, such as that used for the ribose unit in DNA and RNA structures. The carbohydrate group in a glycoside is known as the *glycone* and the non-sugar group as the *aglycone* or *genin* part of the glycoside. The glycone can consist of a single carbohydrate or several sugar groups.

14.1 MONOSACCHARIDES

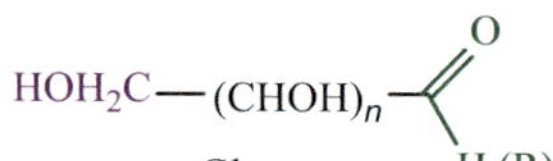

Glycose

A carbohydrate that contains one carbohydrate unit is categorized as a *monosaccharide*. Monosaccharaides were introduced in Section 13.2. Aldotetroses, aldopentoses, and aldohexoses are monosaccharides, as are ketotetroses, ketopentoses, and ketohexoses. The basic unit of a glycose is shown, and there are several different monosaccharide glycoses. The carbon of each hydroxymethyl (CHOH) unit is a stereogenic center, and each time a CH_2OH unit is added, another stereogenic center is added, which increases the number of diastereomers. Polyhydroxy aldehydes such as aldotetrose, aldopentose, and aldohexose derivatives form stable, cyclic hemiacetals (see Section 13.2). As noted in Section 13.2, carbohydrates are usually assigned a three-letter abbreviation[1] that is used in conjunction with the letters *p* (a pyranose) or *f* (a furanose) to generate a shorthand name.

Monosaccharides form cyclic hemiacetals, but an exception is D-glyceraldehyde, because the energy required to form a four-membered ring is too high and it does *not* form a cyclic hemiacetal. The carbon bearing the acetal–OH unit is the hemiacetal carbon atom derived from the acyl carbon of the aldehyde unit and is called the *anomeric carbon* or an *anomeric center.* If the ring is drawn as a flat structure, the OH unit can form on the 'top' of the ring or on the 'bottom' of the ring. These two compounds are isomers that differ only in the configuration at the anomeric carbon, and they are called anomers.

The fundamental features of monosaccharides can be described using tetrose derivatives as an example. Tetrose derivatives form both stable five- and six-membered ring hemiacetals, where cyclization favors the formation of five- and six-membered rings. Both D-erythrose and D-threose form cyclic hemiacetals, as shown in Figure 14.1. Both are drawn in an envelope conformation to show a more normal conformation for the five-membered tetrahydrofuran rings. There are two stereoisomers at the anomeric carbon (C1) of each compound, so there are two diastereomers for each: tetrahydrofuran-(2*R*,3*R*,4*R*)-triol and tetrahydrofuran-(2*S*,3*R*,4*R*)-triol for D-erythrose, and tetrahydrofuran-(2*R*,3*R*,4*S*)-triol and tetrahydrofuran-(2*S*,3*R*,4*S*)-triol for D-threose. Note that the anomeric OH unit is on the 'top' of the ring or on the 'bottom' of the ring in Figure 14.1. These derivatives are tetrahydrofuran derivatives, and the suffix ***"ose"*** is used to indicate they

[1] Kennedy, J.F.; White, C.A. *Bioactive Carbohydrates*, Ellis Horwood Ltd., Chichester, UK, **1983**, p. 41.

DOI: 10.1201/9781003605157-14

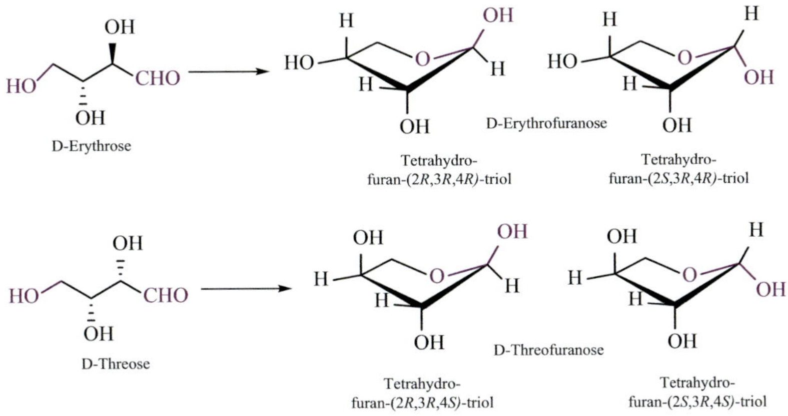

FIGURE 14.1 Hemiacetal forms of D-erythrose and D-threose.

are carbohydrates, so they are named as *furanose* derivatives. Therefore, D-erythrose is D-erythrofuranose (tetrahydrofuran-(2*R*,3*R*,4*R*)-triol and tetrahydrofuran-(2*S*,3*R*,4*R*)-triol), and D-threose is D-threofuranose (tetrahydrofuran-(2*R*,3*R*,4*S*)-triol and tetrahydrofuran-(2*S*,3*R*,4*S*)-triol). The "–ose" nomenclature will be used throughout.

14.2 DISACCHARIDES, TRISACCHARIDES, OLIGOSACCHARIDES, AND POLYSACCHARIDES

A *disaccharide* is a molecule that contains two sugar units, formed when two monosaccharides are coupled together. To prepare a disaccharide, two monosaccharides are coupled by the reaction of a hydroxyl unit of one saccharide with a carbonyl of the second saccharide to yield an acetal linkage or a ketal linkage. The anomeric partner is usually the "glycosyl donor" and the non-anomeric partner is the "acceptor." The glycosyl donor can, in principle, couple via any of the OH units of the acceptor.

The constituent carbohydrate moieties of disaccharides usually exist in the pyranose or furanose form. Most disaccharides are characterized by coupling at two positions, C1 and C4. In one example, the OH at C1 in β-D-allopyranose (from allose) was coupled to C1 (the anomeric carbon) of α-D-idopyranose (idose) to form the disaccharide (Figure 14.2)**.** This coupling is said to be a (1→1) linkage. If C1 in β-D-allopyranose couples to the C4 hydroxyl of α-D-idopyranose, the disaccharide has a so-called (1→4) linkage, as shown. The (1→4) linkage implies, according to the donor–acceptor model, that the anomeric partner is "1" and the non-anomeric partner is "4," and β-D-idopyranose is the anomeric partner. In both disaccharides, the acetal oxygen is α- for the allopyranose unit and the idopyranose unit.

Formally, the disaccharides in Figure 14.2 are named *O*-α-D-allopyranosyl-(1→1)-α-D-idopyran ose and *O*-α-D-allopyranosyl-(1→4)-α-D-idopyranose 5. In this name, the allopyranose unit is treated as a substituent, so *the -ose ending is dropped and replaced by -osyl.* In other words, an allopyranose is an allopyranosyl substituent, and an idopyranose unit is an idopyranosyl substituent. The two monosaccharide units are linked together via an oxygen, and this is indicated by *O*- in the name.

Trisaccharides contain three carbohydrate units linked via 1→1 or 1→4 linkages. Trisaccharides are prepared by sequential coupling reactions, and they are named in the same way as the disaccharides. The trisaccharide shown is *O*-α-D-ribofuranosyl-(1→4)-α-D-gulopyranosyl-(1→1)-α-D-allop yranose and is composed of ribofuranose (in violet), gulopyranose (in green), and allopyranose (in cyan). This nomenclature system certainly applies to large oligosaccharides and polysaccharides, but the names can be quite long and clumsy. Using a three-letter code (Table 13.1 in Section 13.2), the letters *p* (a pyranose) or *f* (a furanose) can be used to generate a shorthand name. Using this

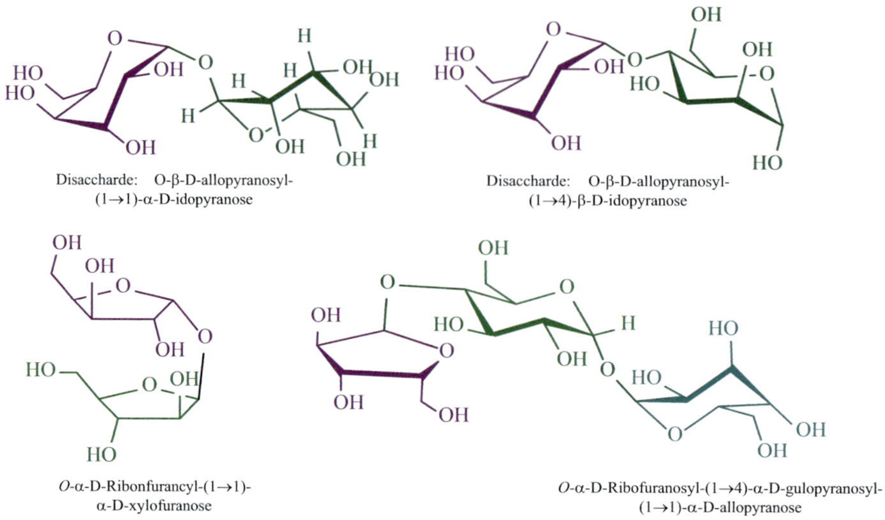

FIGURE 14.2 Disaccharides and a trisaccharide.

system, *O*-α-D-ribofuranosyl-(1→4)-α-D-gulopyranosyl-(1→1)-α-D-allopyranose becomes *O*-α-D-Rib*f*-(1→4)-α-D-Gul*p*-(1→1)-α-D-All*p*. Also, the structure for a hexasaccharide is *O*-α-D-Fru*f*-(1→4)-α-D-Gcl*p*-(1→1)-β-D-Gal*p*-β-D-Tal*p*-(1→1)-α-D-Ido*p*-(1→2)-α-D-All*p*.

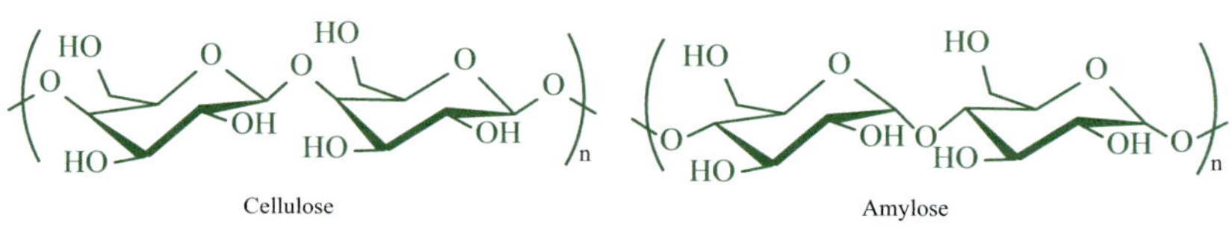

For polysaccharides, even the shorthand notation is unwieldy. Most of these compounds are given common names, and since they are usually polymeric, the repeating monosaccharide or disaccharide units are often shown. Many of these are *homopolymers*; that is, the polysaccharide is formed by coupling only one monosaccharide unit. *Cellulose*, for example, is a linear poly(glucopyranose), coupled in a (1→4)-β-D-manner, as shown, and it is a major constituent of plant cell walls. *Amylose* (a constituent of starch) is a linear poly(glucopyranose) coupled (1→4)-α-D-.

14.3 BIOLOGICALLY IMPORTANT GLYCOSIDES

Many plants store chemicals in the form of inactive glycosides that are activated by enzymatic hydrolysis. In addition to homopolymeric poly(saccharides), there are poly(saccharides) composed of more than one monosaccharide unit. For example, coniferous woods contain a poly(saccharide) that is a linear chain of D-glucopyranosyl and D-mannopyranosyl units. Plant cell walls (Figure 14.3a) also contain a polysaccharide that is a branched structure containing L-arabinofuranosyl and D-xylofuranosyl units.

All living things are composed of cells, but plant cells have a cell wall on the outer layer, whereas animal cells only have a cell membrane (Figure 14.3b). The cell wall is a protective layer outside the cell membrane that also provides support for the cell's structure. The primary walls surround growing cells, and secondary walls contain lignin and surround specialized cells. The primary cell wall of land plants is composed of the polysaccharides cellulose, hemicelluloses, and pectin. The cellulose in the cell wall is composed of D-glucose units linked by β-1, 4 glycosidic bonds; poly(1,4-β-D-glucopyranoside). Primary cell walls contain a protein, extensin, that is

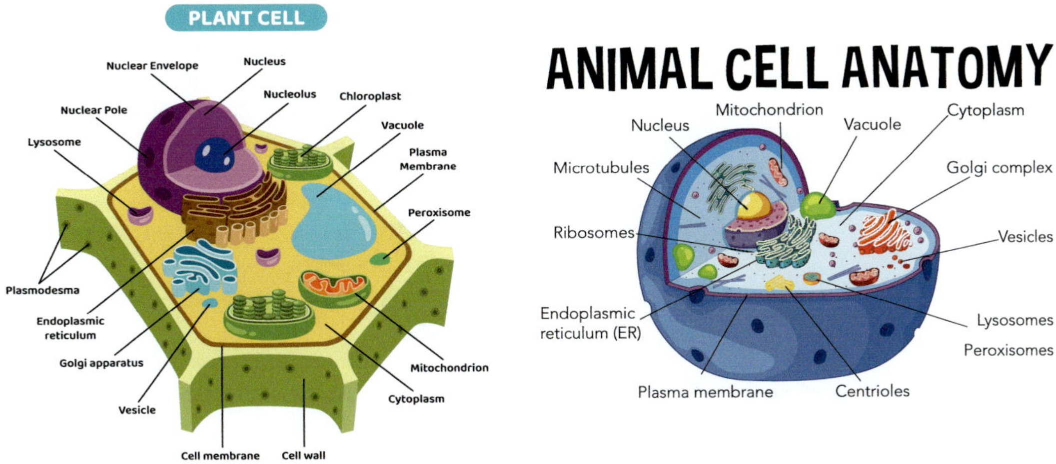

FIGURE 14.3 (a) The structure of plant cell. Stock Vector ID: 2465117249. (b) Diagram of animal cell anatomy illustration. Stock Vector ID: 1117954796.

rich in trans-4-L-hydroxyproline.[2] Extensins are rodlike, hydroxyproline-rich glycoproteins with many turns that are found in plant cell walls. Extensins probably crosslink with the wall polysaccharides by a glycosidic link through the hydroxyl group of hydroxyL-proline.[3] A short sequence of extensin that consists of two hydroxyproline residues and two or three other amino acid residues could be regarded as the crosslink between two polysaccharide chains. In bacteria, the cell wall is composed of peptidoglycan. Peptidoglycan (murein) is a polymer consisting of alternating β-linked *N*-acetylglucosamine and *N*-acetylmuramic acid residues and amino acid residues that form the cell wall.

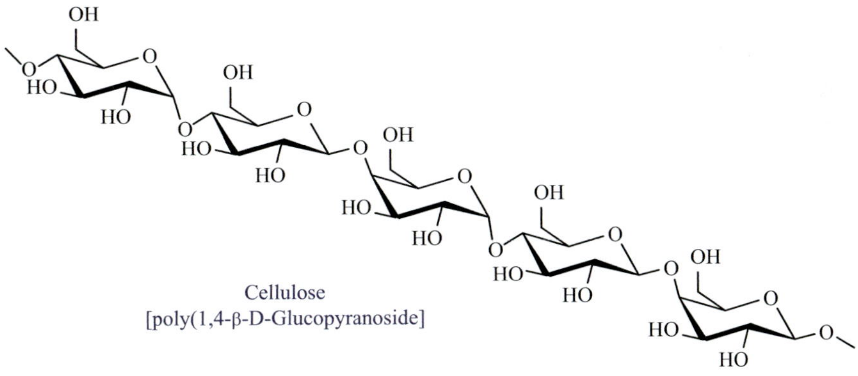

Cellulose
[poly(1,4-β-D-Glucopyranoside]

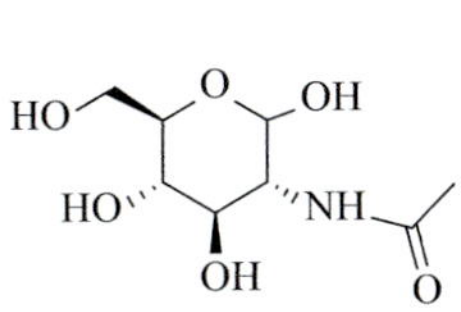

N-acetylglucosamine

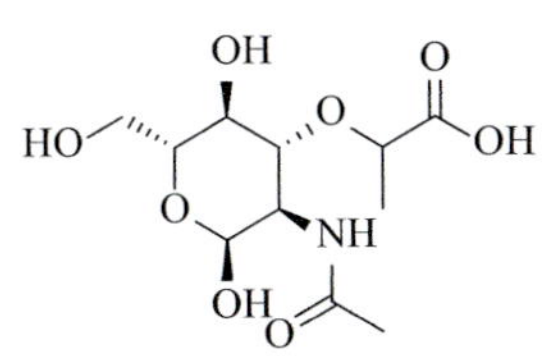

N-Acetylmuramic acid

---

[2] (a) Dougal, D.K.; Shlmbayashl, K. *Plant Physiol.* **1960**, *35*, 396–404; (b) Lamport, D.T.A.; Northcote, D.H. *Nature* **1960**, *188*, 665.

[3] Lamport, D.T.A. *Nature* **1967**, *216*, 1322.

In addition to cellulose, plant cell walls contain several matrix polysaccharides that include the pectic polysaccharides (homogalacturonan and rhamnogalacturonan I and II)[4] and the hemicellulosic polysaccharides (xyloglucans, glucomannans, xylans, and mixed-linkage glucans).[5] The biosynthesis of matrix polysaccharides and glycosylation of various cell wall glycoproteins occur in the Golgi membranes.[4,5,6] Many different enzymes are required for the biosynthesis of the polysaccharides found in cell walls.

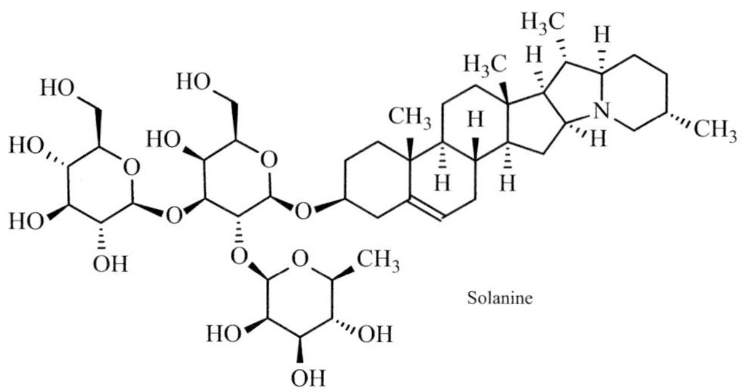

Solanine

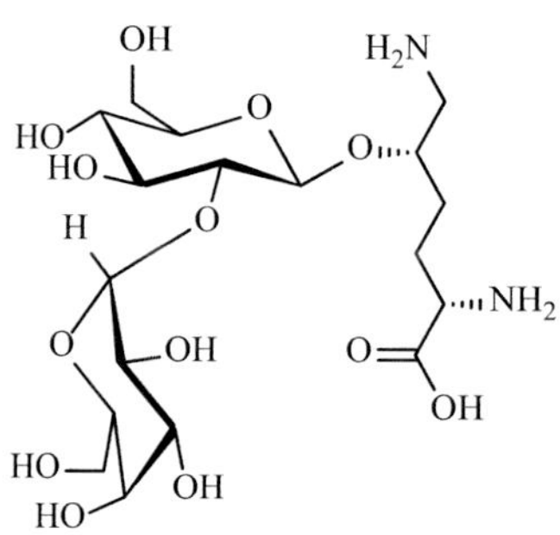

2-*O*-α-D-glucopyranosyl-*O*-β-D-galactosylpyranosyl hydroxylysine

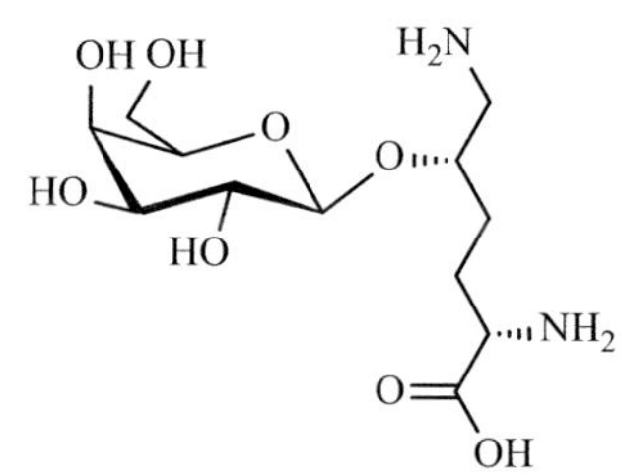

O-β-D-galactosylpyranosyl-hydroxylysine

Saponins, such as the glycoalkaloid solanine, are amphipathic glycosides (amphipathic molecules contain both hydrophilic and hydrophobic moieties). They are found in various plant species and are characterized by the presence of glycoside and terpene moieties. Saponins are found in plants where they may serve as anti-feedants and enhance nutrient absorption. They are known to lower the surface tension of water and form micelles that let them interact with biological membrane layers, which may perturb the membrane and its function, leading to membrane perforation or complete lysis.

Cervical spinal cord injury alters the collagen metabolism of affected patients,[7] which can be monitored by measuring the urinary excretion of collagen metabolites.[8] Two collagen metabolites are glucosylgalactosyl hydroxylysine and galactosyl hydroxylysine. The former metabolite is prevalent in skin collagen, and its increased presence in urine is associated with patients who have erythema multiforme (a hypersensitivity reaction associated with certain infections as an acute and sometimes recurring skin condition) or burns. The latter metabolite is prevalent in bone collagen, and its increased presence in patients is associated with bone diseases such as osteomalacia or Paget's disease.

[4] Harholt, J.; Suttangkakul, A.; Scheller, H.V. *Plant Physiol.* **2010**, *153*, 384.
[5] Scheller, H.V.; Ulvskov, P. *Annu. Rev. Plant Biol.* **2010**, *61*, 263.
[6] Ellis, M.; Egelund, J.; Schultz, C.; Bacic, A. *Plant Physiol.* **2010**, *153*, 403.
[7] Claus-Walker, J. *Int. J. Rehab. Res.* **1980**, *3*, 540.
[8] Rodriguez, G.P.; Claus-Walker, J. *J. Chromatogr.* **1984**, *308*, 65.

Many antibiotics and other medicines are glycosides, and the carbohydrate portion of the drug is often essential for full potency. The drug without the pendant carbohydrate is called the aglycone. An example is the cardiac glycoside digitoxin, isolated from foxglove (*Digitalis purpurea*). Streptomycin is an aminoglycoside antibacterial antibiotic produced by the soil actinomycete *Streptomyces griseus*. It acts by binding to the 30S ribosomal subunit of susceptible organisms, which disrupts the initiation and elongation steps in protein synthesis. Erythromycin is a broad-spectrum macrolide antibacterial that diffuses through the bacterial cell membrane to reversibly bind to the 50S subunit of the bacterial ribosome, which prevents bacterial protein synthesis.

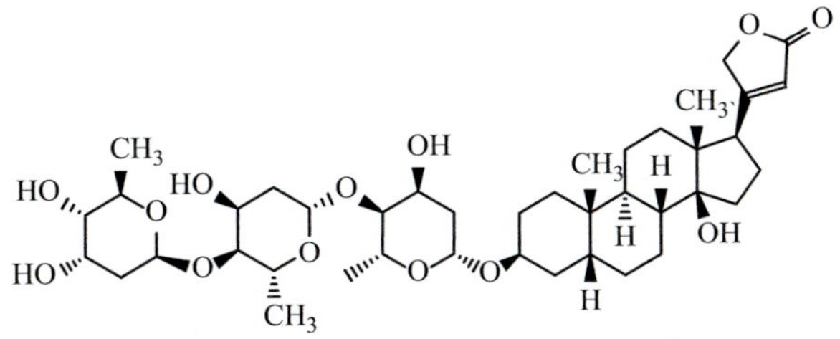

Digitoxin

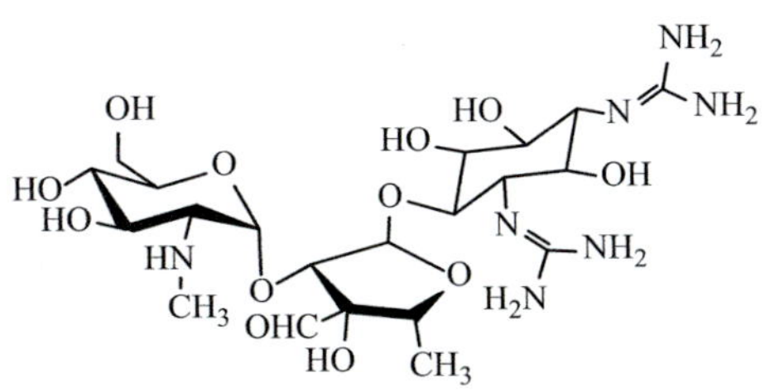

Streptomycin

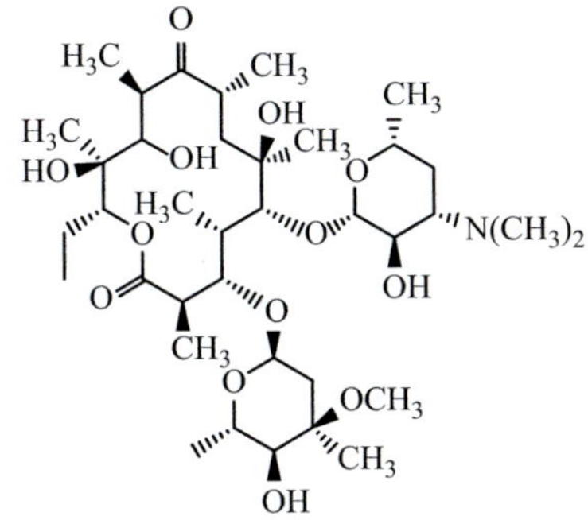

Erythromycin

14.4 BIOSYNTHESIS OF CARBOHYDRATES AND GLYCOSIDES

Catabolism is the breakdown of complex molecules to form simpler ones, with the release of energy. Anabolism is the process by which the body utilizes the energy released by catabolism to make complex molecules that are used to form cellular structures. Plants, for example, synthesize carbohydrates from CO_2 and water by photosynthesis. When animals consume plants, cellular respiration breaks down these carbohydrates, providing energy that is made available to cells.

Carbohydrates such as glucose are biosynthesized as shown in Figure 14.4.[9] Fatty acids are converted to acetyl-SCoA, which enters the *Krebs cycle*, where citrate and oxaloacetate are intermediates. Oxaloacetate is the source of pyruvate, which is then converted to glucose.[10] Figure 14.5[9a] shows the biosynthetic pathway that produces glucose from pyruvate. Glucose biosynthesis begins with the carboxylation of pyruvate to form oxaloacetate, which is converted to phosphoenolpyruvate. The addition of water yields D-2-phosphoglycerate, which is then converted to D-3-phosphoglycerate. A subsequent phosphorylation produces D-1,3-bisphosphoglycerate, which is reduced to D-glyceraldehyde-3-phosphate. Tautomerization gives dihydroxyacetone phosphate, which undergoes an aldolase-catalyzed reaction to form fructose-1,6-bisphosphate. Hydrolysis of the phosphate group at C1 gives fructose-6-phosphate, which tautomerizes to glucose-6-phosphate. A final hydrolysis step yields glucose.

[9] (a) Gabrielli, F. *J. Chem. Ed.* **1976**, *53*, 86. Also see (b) Utter, M.F.; Keech, D.B. *J. Biol. Chem.* **1963**, *238*, 2603; (c) Krebs, H.A. *Proceed. Royal Soc., Ser. B* **1963**, *159*, 545; (d) Berg, J.M.; Tymoczko, J.L.; Stryer, L. *Biochemistry*, 5th ed., section 16.3, W.H. Freeman, New York, **2002**. See Figure 16.24.

[10] (a) Weinman, E.O.; Strisower, E.H.; Chaikoff, I.L. *Physiol. Rev.* **1957**, *37*, 252. See (b) Landau, B.R.; Brunengraber, H. *Trends Biochem. Sci.* **1987**, *12*, 113; (c) Argiles, J.M. *Trends Biochem. Sci.* **1986**, *11*, 61.

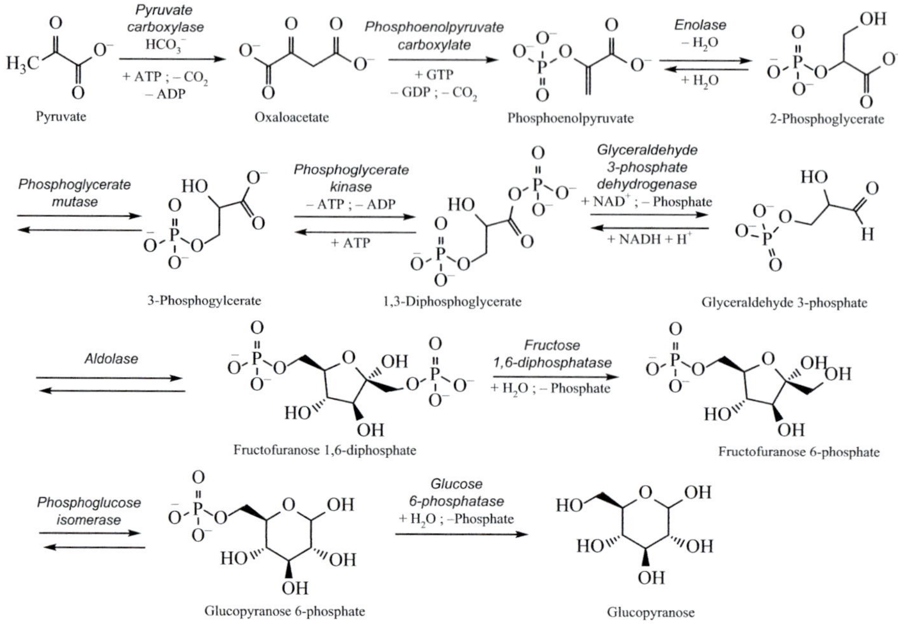

FIGURE 14.4 Glucopyranose biosynthesis. (Reprinted with permission from Gabrielli, F. *J. Chem. Educ.* **1976**, *53*, 86–91. Copyright 1976 American Chemical Society.)

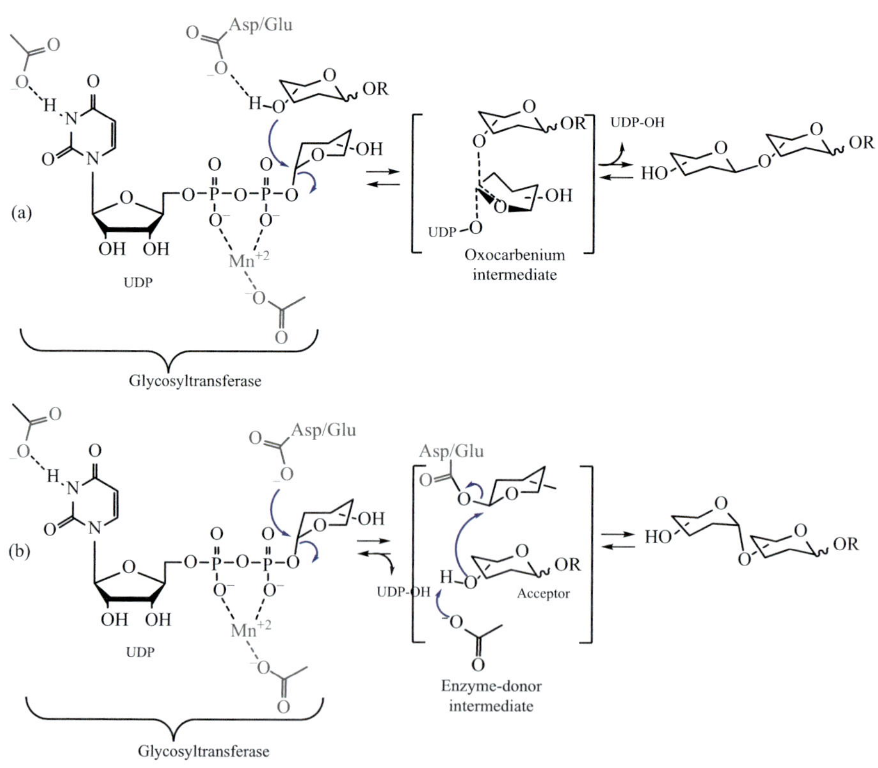

FIGURE 14.5 Mechanisms for inverting (a) and retaining (a) *glycosyltransferases* that involve uracil diphosphate. (Weijers, C.A.G.M.; Franssen, M.C.R.; Visser, G.M. *Biotechnol. Adv.* **2008**, *26*, 436–456, Glycosyltransferase-catalyzed synthesis of bioactive oligosaccharides. Copyright (2008), with permission from Elsevier.)

Disaccharides or oligosaccharides are prepared by enzymatic coupling of two monosaccharides or of a monosaccharide with an oligosaccharide or a glycoconjugate. Oligosaccharides (or glycans) are naturally occurring carbohydrates that consist of 3–10 monosaccharides, and they are biosynthesized in the endoplasmic reticulum and Golgi apparatus in the cell. In eukaryotes, bioactive oligosaccharides can commonly be found on the cell surface, where they have many functions. Enzymes known as *glycosyltransferases* (EC 2.4.1.X) and *glycosidases* (EC 3.2.1.X) are commonly used to make glycosidic bonds. Such enzymes can be segregated into 'retaining' or 'inverting' enzymes, according to whether the stereochemistry of the donor's anomeric bond is retained (α→α, β→β) or inverted (α→β, β→α). When the carbohydrate substrate reacts with a glycosidase enzyme, a glycosyl–enzyme intermediate is formed. Subsequent reaction with water gives the hydrolysis product, whereas reaction with a glycosyl acceptor gives a new glycoside or oligosaccharide.[11]

As noted, formation of the glycosidic bond can proceed either by inversion or retention of the anomeric carbon. For inverting *glycosyltransferases*, which involve a divalent cation such as Mn^{2+}, the reaction is initiated by a basic carboxylate residue (Asp or Glu), as shown in Figure 14.5.[11] Deprotonation of the glycosyl acceptor for in-line S_N2 nucleophilic attack at the C1 atom of the sugar of the nucleotide sugar donor leads to an oxocarbenium-like transition state, as shown in Figure 14.5.[11] The mechanism of retaining glycosyltransferases, which also uses Mn^{2+}, probably involves a 'double-displacement mechanism,' as shown.[11] An initial S_N2 nucleophilic attack by a carboxylate residue (from Asp or Glu) at the C1 atom generates a β-covalently-linked enzyme–glycosyl intermediate. Subsequent S_N2 nucleophilic attack on the C1 carbon of the intermediate by the glycosyl acceptor results in conversion to an α-linkage in the formed oligosaccharide product.

Mammalian *galactosyltransferases* (GalT; EC 2.4.1.X), with *β-1,4-galactosyltransferase* (β1,4GalT), use UDP-α-D-Gal as the donor, where UDP is uracil diphosphate. The glycosidic reaction generates β1–4, β1–3, α1–3, and α1–4 linkages to acceptor carbohydrates or glycoconjugates. As shown in Figure 14.6,[11] retaining GalT enzymes only recognize substrates with Gal (or Gal derivative) terminal structure, whereas all inverting β1,4GalT enzymes only accept Glc (or corresponding Xyl) terminal structures. Inverting β1,3GalT enzymes similarly accept terminal Glc structures, but more specific β1,3GalT4,6 enzymes prefer acceptors with a terminal Gal.[11] Note that Gal is galactosyl and Glc is glucosyl.

The pentose phosphate pathway (also called the phosphogluconate pathway and the hexose monophosphate shunt) converts glucose-6-phosphate to ribose-5-phosphate, which is a precursor for the synthesis of nucleotides. In the first step, *glucose-6-phosphate dehydrogenase* (EC 1.1.1.49) catalyzes the conversion of glucose-6-phosphate to 6-phosphogluconolactone, as shown in Figure 14.7.[12] The enzyme *gluconolactonase* (EC 3.1.1.17) opens the lactone ring to give 6-phosphogluconate, followed by *6-phosphogluconate dehydrogenase* (EC 1.1.1.43), which generates ribulose-5-phosphate. In the final step, *ribulose-5-phosphate isomerase* (EC 5.1.3.4) generates ribofuranose-5-phosphate.[12]

The enzymatic coupling of carbohydrates to other large molecules gives biologically important glycosides. Examples are glycolipids, glycopeptides, glycoproteins, and glyconucleotides. Relatively simple examples of a glycolipid are the sphingolipids. Galactosylceramide (GalCer), the principal glycosphingolipid in brain tissue, is an example of a glycolipid with the trivial name "cerebroside." The β-D-galactosylceramides are found in all nervous tissues, and they are major constituents of oligodendrocytes in brain tissue. Typically, sphingosine is the main long-chain base in cerebrosides of animal tissues.

Bacteroides fragilis, an anaerobic, Gram-negative, rod-shaped bacterium, produces an isoform of a galactosylceramide, which is a sponge-derived sphingolipid. *Bacteroides fragilis* lacks an endoplasmic reticulum (the site of eukaryotic sphingolipid synthesis) and biosynthesizes sphingolipids using a *serine palmitoyltransferase* (EC 2.3.1.50) in the first step, which forms

[11] Weijers, C.A.G.M.; Franssen, M.C.R.; Visser, G.M. *Biotech. Adv.* **2008**, *26*, 436.
[12] Kruger, N.J.; von Schaewen, A. *Curr. Opin. Plant Biol.* **2003**, *6*, 236.

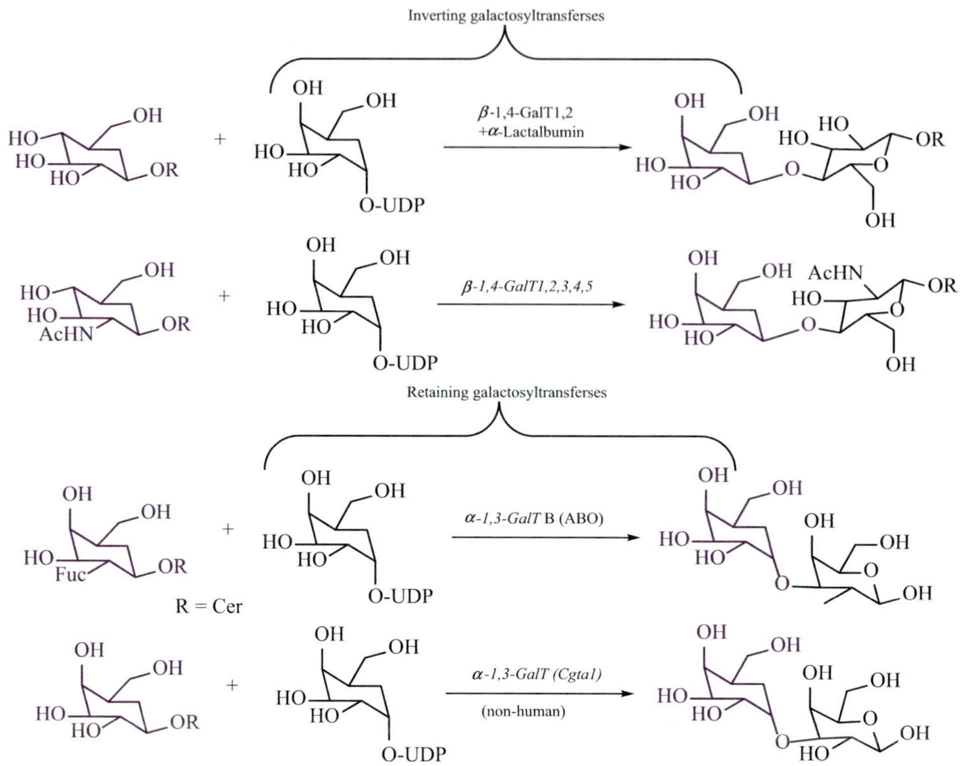

FIGURE 14.6 Substrates for mammalian *galactosyltransferase* superfamily (GalT) for natural glycoconjugates as acceptor substrates and UDP-Gal as sugar nucleotide donor. (Reprinted from Weijers, C.A.G.M.; Franssen, M.C.R.; Visser, G.M. *Biotechnol. Adv.* **2008**, *26*, 436–456, Glycosyltransferase-catalyzed synthesis of bioactive Oligosaccharides. Copyright (2008), with permission from Elsevier.)

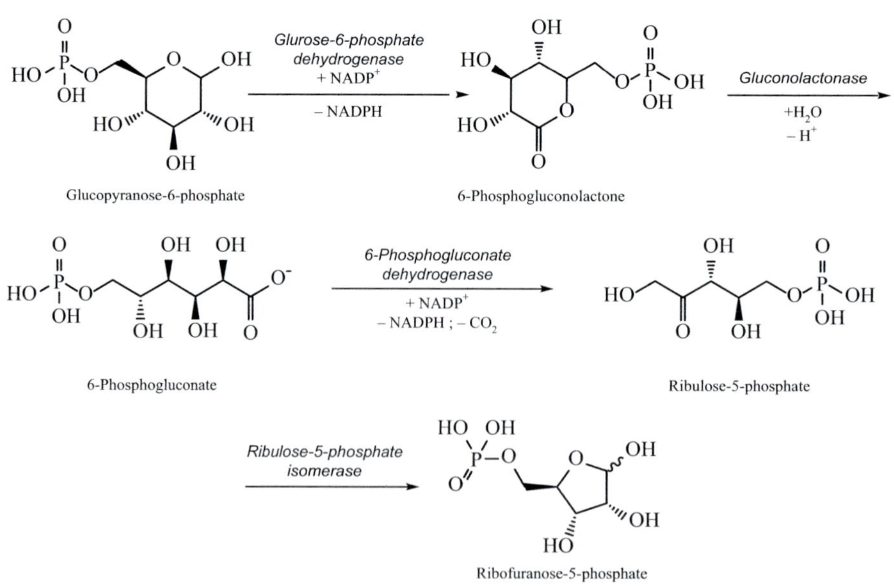

FIGURE 14.7 Biosynthesis of ribofuranse-5-phosphate. (Reprinted from Kruger, N.J; von Schaewen, A. *Curr. Opin. Plant Biol.* **2003**, *6*, 236–246, The oxidative pentose phosphate pathway: structure and organization. Copyright (2003), with permission from Elsevier.)

3-ketodihydrosphingosine from 13-methyltetradecanoic acid S-CoA (Figure 14.8).[13] A *ketoreductase* (EC 1.1.1.X) generates a dihydrosphingosine, and N-acylation gives a dihydroceramide. The glycosylation step forms the glycolipid using *α-GalCer synthase* (EC 2.4.1.XX) to form α-GalCer$_{Bf}$, the α-galactosylceramide.[13]

Glycopeptides contain carbohydrate units that are referred to as glycans, and they are attached to the side chains of amino acids in a peptide or (poly)peptide by covalent bonds. When the carbohydrate is attached to a nitrogen atom of an amino acid residue, such as asparagine, the covalent amide bond constitutes an N-linked glycan. When the carbohydrate is attached to an oxygen atom of an amino acid residue, such as serine or threonine, the covalent bond is an ether linkage, often, to the anomeric carbon. Such a covalent bond constitutes an O-linked glycan.

More than half of all human proteins are glycoproteins, and N-glycans constitute a major portion of glycoproteins.[14] N-glycans are delivered to a (poly)peptide chain as a lipid-linked oligosaccharide. Eukaryotic organisms use a multi-subunit *oligosaccharyltransferase*[15] on the lumenal face of the endoplasmic reticulum membrane to catalyze glycan transfer to an amino acid residue acceptor peptide site.[16] The active site of this enzyme probably contains a divalent metal ion, Mg^{2+} or Mn^{2+} (see Figure 14.9).[15] A lumen is the interior space of a tubular structure, such as an artery or the intestine. Oligosaccharyltransferase (EC 2.4.1.119) is a membrane protein complex that transfers an oligosaccharide and is a type of *glycosyltransferase*. Asparagine-linked (N-linked) glycosylation is very common, and the key step is the formation of an N-glycosidic linkage between the amide nitrogen of the acceptor asparagine and the C1 carbon of the first saccharide moiety of a lipid-linked oligosaccharide donor (see Figure 14.10).[15]

Many *hydrolases* that degrade plant cell walls consist of a single *glycoside hydrolase* (EC 3.2.1.169) domain and a single carbohydrate-binding module.[17] Catabolism is the breakdown of complex molecules in living organisms to form simpler ones, releasing energy.

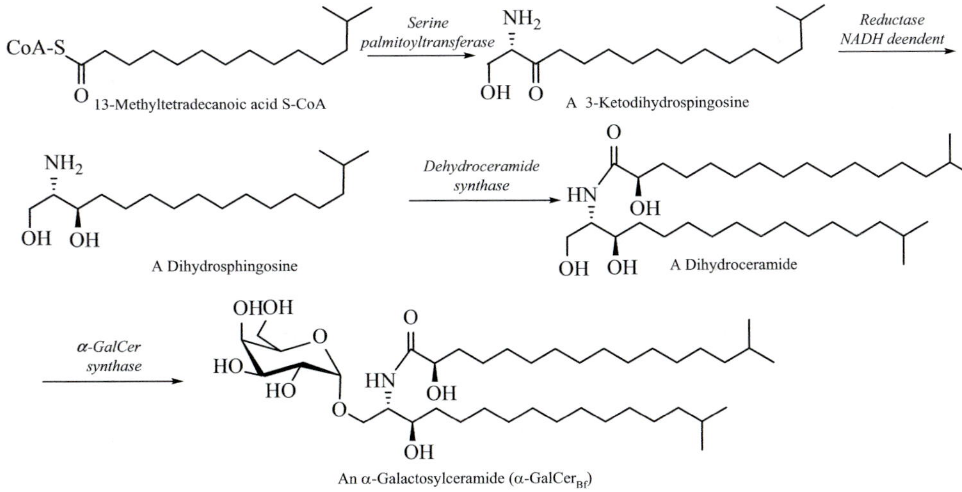

FIGURE 14.8 Proposed pathway for *Bacteroides* sphingolipid biosynthesis.

[13] Brown, L.C.W.; Penaranda, C.; Kashyap, P.C.; Williams, B.B.; Clardy, J.; Kronenberg, M.; Sonnenburg, J.L.; Comstock, L.E.; Bluestone, J.A.; Fischbach, M.A. *PLoS Biol.* **2013**, *11*, e1001610.
[14] Apweiler, R.; Hermjakob, H.; Sharon, N. *Biochim. Biophys. Acta* **1999**, *1473*, 4.
[15] Lizak, C.; Gerber, S.; Numao, S.; Aebi, M.; Locher, K.P. *Nature* **2011**, *474*, 350.
[16] Kelleher, D.J.; Gilmore, R. *Glycobiology* **2006**, *16*, 47R–62R.
[17] Yokoyama, H.; Yamashita, T.; Morioka, R.; Ohmori, H. *J. Bacteriol.* **2014**, *196*, 3784.

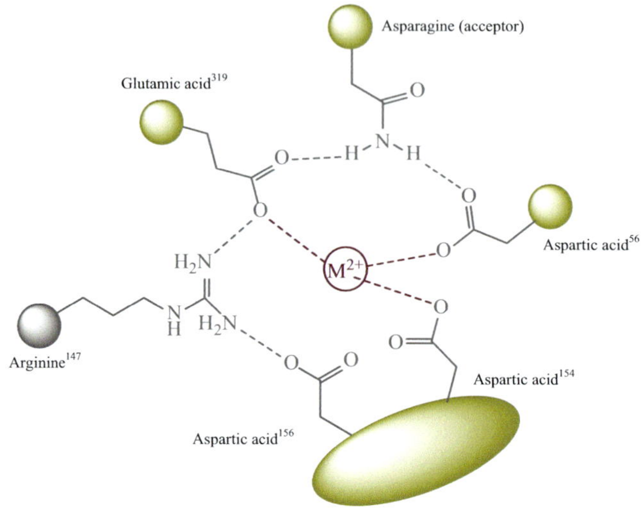

FIGURE 14.9 Active site of oligosaccharyltransferase. (Reprinted by permission from Springer Nature Customer Service Centre GmbH: Springer Nature. Lizak, C.; Gerber, S.; Numao, S.; Aebi, M.; Locher, K.P. *Nature* **2011**, *474,* 350–355. Copyright, 2011.)

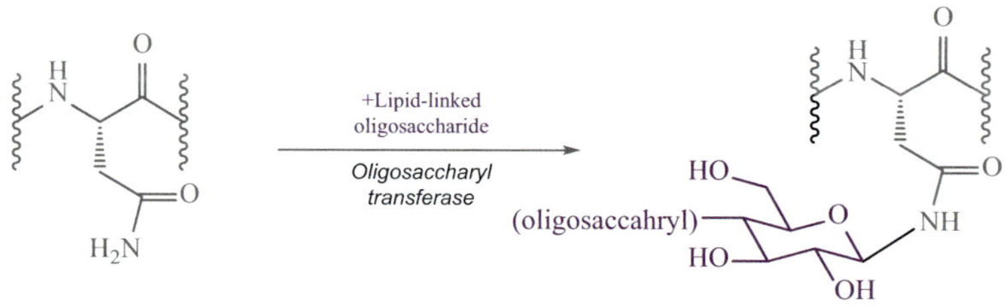

FIGURE 14.10 Reaction scheme of N-linked glycosylation, yielding an N-glycosidic bond. (Reprinted by permission from Springer Nature Customer Service Centre GmbH: Springer Nature. Lizak, C.; Gerber, S.; Numao, S.; Aebi, M.; Locher, K.P. *Nature* **2011**, *474,* 350–355. Copyright, 2011.)

14.5 BIODEGRADATION OF CARBOHYDRATES AND GLYCOSIDES

Glycolysis is a 10-step metabolic process used for aerobic and anaerobic respiration is which glucose in converted to two pyruvate molecules. In eukaryotes, glycolysis occurs in the cytoplasm, and all the glycolytic enzymes are found in the cytoplasm; since it is an anaerobic process, oxygen is not required. In the presence of oxygen, pyruvate is oxidized further to carbon dioxide, whereas in the absence of oxygen, pyruvate can be converted to lactate or ethanol. The standard eukaryotic glycolytic pathway is the *Embden-Meyerhof-Parnas (EMP) pathway* (see Figure 14.11),[18b] which consists of 10 enzymes, most of which catalyze reversible steps.[19] The *Entner-Doudoroff (ED) pathway*[20] is a related pathway for glycolysis that is active in bacterial primary metabolism.

The first step in glycolysis is the conversion of D-glucose into glucose-6-phosphate by *hexokinase* (EC 2.7.1.1), followed by the rearrangement of glucose-6-phosphate into fructose-6-phosphate catalyzed by *glucose phosphate isomerase* (*phosphoglucose isomer*ase; EC 5.3.1.9).[18] *Phosphofructokinase* (EC 2.7.1.11), with magnesium as a cofactor, converts fructose-6-phosphate

[18] Papagianni, M. *Conform. Struct. Biotechnol. J.* **2012**, *3*, e201210003. Also see Sańchez-Pascuala, A.; Ví, V.; Nikel, P.I. *ACS Synth. Biol.* **2017**, *6,* 793; (b) Romano, A.H.; Conway, T. *Res. Microbiol.* **1996**, *147,* 448. (c) Aryal, S. Microbiology Info.com **2015**. See https://microbiologyinfo.com/glycolysis-10-steps-explained-steps-by-steps-with-diagram/

[19] See Kresge, N.K.; Simoni, R.D.; Hill, R.L. *J. Biol. Chem.* **2005**, *280,* e46–e48.

[20] Entner, N.; Doudoroff, M. *J. Biol. Chem.* **1952**, *196,* 853.

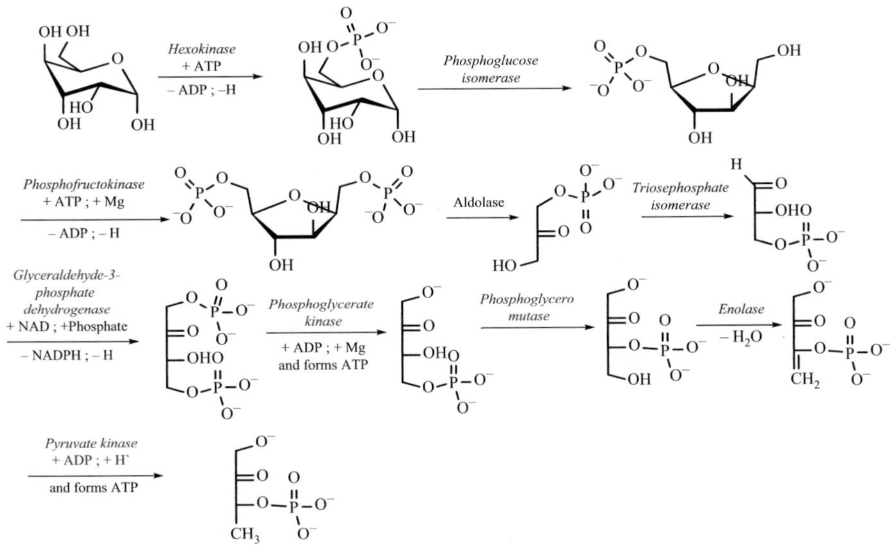

FIGURE 14.11 The Embden-Meyerhof-Parnas (EMP) pathway. (See Papagianni, M. *Conf. Struct. Biotechnol. J.* **2012**, *3*, e201210003. Open Access.)

into fructose-1,6-bisphosphate, and *aldolase* (EC 4.1.2.13) splits fructose-1,6-bisphosphate into dihydroxyacetone phosphate and the isomeric glyceraldehyde-3-phosphate (*triosephosphate isomerase* rapidly interconverts these isomers; EC 5.3.1.1). *Glyceraldehyde-3-phosphate dehydrogenase* (EC 1.2.1.9) dehydrogenates and then adds an inorganic phosphate to glyceraldehyde-3-phosphate to give 1,3-bisphosphoglycerate. *Phosphoglycerate kin*ase (EC 2.7.2.3) transfers a phosphate group from 1,3-bisphosphoglycerate to ADP to form ATP and 3-phosphoglycerate. The enzyme *phosphoglyceromutase* (EC 5.4.2.11) relocates the phosphate from 3-phosphoglycerate to form 2-phosphoglycerate, and then enolase generates phosphoenolpyruvic acid. The enzyme *pyruvate kinase* (EC 2.7.1.40) transfers a phosphate to ADP to form pyruvic acid and ATP. In humans, fructose is metabolized almost completely in the liver. Fructose can enter the glycolysis pathway, and there are enzymes that can add a phosphate group to fructose. Such pentose sugars enter the Embden-Meyerhof pathway of glycolysis via a shunt mechanism, as shown in Figure 14.11.[18b]

Gluconeogenesis converts non-carbohydrate molecules such as pyruvate, lactate, glycerol, alanine, and glutamine into glucose. Gluconeogenesis consists of a series of 11 enzyme-catalyzed reactions, and many of the reactions are the reverse of steps found in glycolysis. Glycogenesis refers to the process of synthesizing glycogen from excess glucose in humans. Glycogen is highly branched, with glucose-6-phosphate units. The enzymes *glycogen synthase* (EC 2.4.1.11) and glycogen branching enzyme (*1,4-alpha-glucan branching enzyme*; EC 2.4.1.18) catalyze the transfer of glucosyl residues from UDP-glucose for the synthesis of glycogen and the formation of α-1→4 and α-1→6 bonds.

Glycogenolysis is the breakdown of glycogen to glucose, which takes place in the cells of muscle and liver tissues in response to hormonal and neural signals. Glycogen is a branched polymer of glucosyl residues present in most mammalian cells, where it functions as a store of carbohydrate that can be degraded by enzymatic action, such as *glycogen phosphorylase* (EC 2.4.1.1). This enzyme catalyzes the glycogenolysis step that releases glucose-1-phosphate from a terminal alpha-1,4-glycosidic bond, as shown in Figure 14.12.[21] Pyridoxal phosphate is a cofactor for this reaction, which cleaves the glucose residues sequentially and yields glucose-1-phosphate (phosphorolysis). This process

[21] Agius, L. *Mol. Asp. Med.* **2015**, *46*, 34.

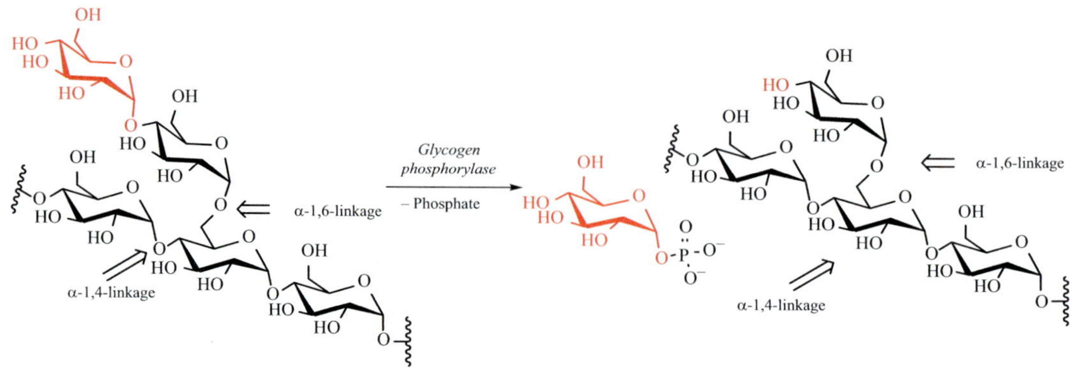

FIGURE 14.12 Phosphorylysis of glycogen.

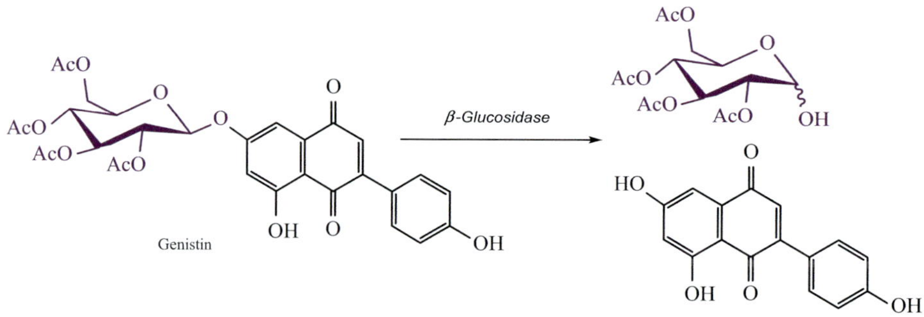

FIGURE 14.13 Hydrolysis of genistin.

continues until four residues are present on either side of a branching point, but the less-branched glycogen molecule cannot be further broken down by the enzyme.

Glycogen is polymeric, made up of repeated units of α-D-glucose that are linked to each other by α-1→4 glycosidic bonds, but there are branches linked by α-1→6 bonds.[22] Glycogen degradation is catalyzed by the actions of a phosphorylase (1,4-α-D-glucan: *orthophosphate α-D-glucosyltransferase*; EC 2.4.1.30), which releases the terminal glucosyl residue of the α-1→4 chains as glucose-1-phosphate. The debranching enzyme (*amylo-α-1,6-glucosidase, 4-α-glucanotransferase*; EC 3.2.1.3) acts as a transferase to transfer a terminal triose unit from an α-1→6 branch to an α-1→4 chain, and as a *glucosidase* to remove the remaining α-1→6 residue as glucose.[22]

Glycoside hydrolysis is, as the name implies, the enzymatic hydrolysis of the glycosidic bond of glycosides. In other words, the enzymes catalyze the hydrolysis of glycosidic bonds in complex carbohydrates. Flavonoids are polyphenolic compounds found in plants. An example is genistin, which is an isoflavone found in several dietary plants like soy and kudzu. A recombinant *β-glucosidase* (EC 3.2.1.21) from the hyperthermophilic archaeon *Pyrococcus furiosus* showed highly specific activity for the hydrolysis of genistin to glucose and the aglycone (see Figure 14.13).[23]

---

[22] Roach, P.J.; Depaoli-Roach, A.A.; Hurley, T.D.; Tagliabracci, V.S. *Biochem. J.* **2012**, *441*, 763.

[23] (a) Yeom, S.-J.; Kim, B.-N.; Kim, Y.-S.; Oh, D.-K. *J. Agric. Food Chem.* **2012**, *60*, 1535. Also see Sun, H.; Xue, Y.; Lin, Y. *J. Agric. Food Chem.* **2014**, *62*, 6763.

14.6 CHEMICAL REACTIONS OF CARBOHYDRATES

Reactions of carbohydrates are closely related to the normal reactions of alcohols, aldehydes, and ketones discussed in previous chapters. The presence of multiple functional groups is problematic, however, because one group may interact with another. The formation of furanose and pyranose rings in the previous sections, via a hemiacetal or a hemiketal reaction, and of disaccharides via an acetal or a ketal reaction, is one example where the functional groups interact with each other. There are many other reactions in which the hydroxyl units behave independently, or in which the aldehyde or ketone unit behaves independently. This section will explore several fundamental carbohydrate reactions.

A discussion of reactions will begin by introducing another classification scheme for carbohydrates: *reducing sugars and nonreducing sugars*. These sugars are generally categorized by whether they react with *Fehling's solution* [heating in aqueous copper (II) sulfate and sodium tartrate] or with *Tollens' solution* (silver nitrate in ammonia). Fehling's solution reacts with α-hydroxyaldehydes, α-hydroxyketones, and α-ketoaldehydes, so carbohydrates with these structural features will reduce Cu(II) to Cu(I) oxide, a red-brown solid. Due to mutarotation (Section 13.3), a carbohydrate has an aldehyde unit in equilibrium with the furanose or pyranose isomers, so it reacts with these reagents, which oxidize the aldehyde to a carbonyl compound. Carbohydrates that react with Fehling's or Tollens' reagents are known as *reducing sugars*. Ketoses do not react with these reagents. A disaccharide that is coupled 1→4 may have an aldehyde unit and be a reducing sugar, but a disaccharide coupled 1→1 cannot have an aldehyde unit available and is not a reducing sugar. In general, other aldehydes or ketones, including simple aliphatic or aromatic aldehydes or ketones, do not reduce Cu(II).

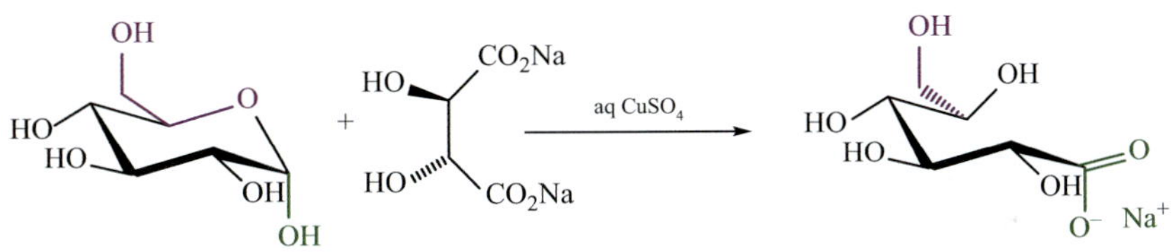

D-Glucopyranose   Sodium tartrate   (2*R*,3*S*,4*R*),5,6-Pentahydroxyhexanoate

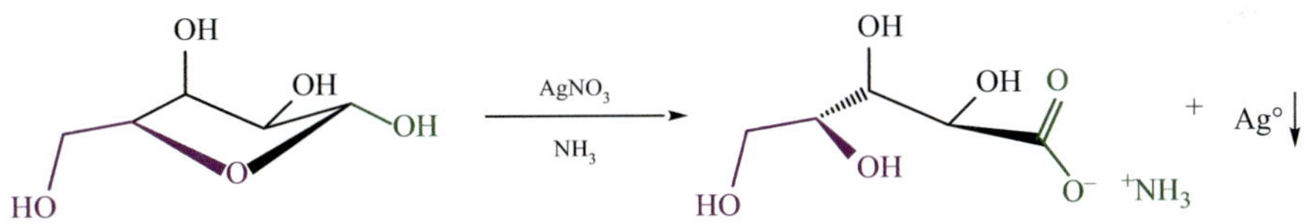

α-D-Ribofuranose   (2*R*,3*R*,4*R*),5-Tetrahydroxypentanoate

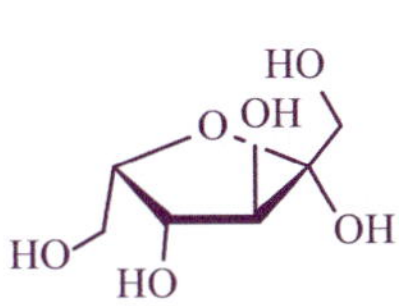

α-D-Fructofuranose

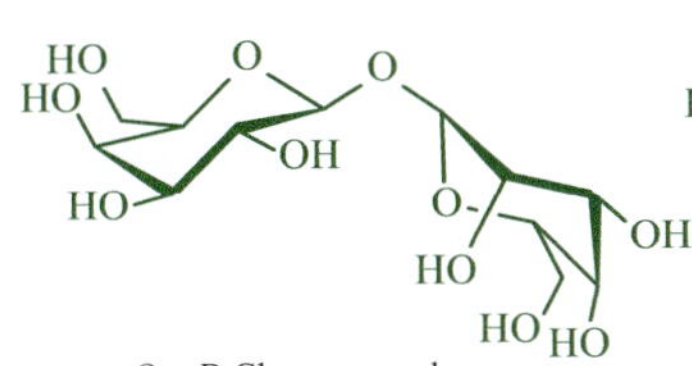

O-α-D-Glucopyranosyl-(1→4)-α-D-glucopyranose

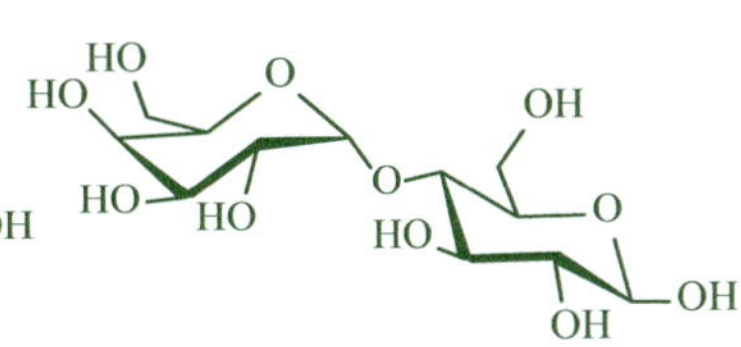

O-α-D-Glucopyranosyl-(1→1)-α-D-glucopyranose

One example of a reducing sugar is α-D-glucopyranose, which reacts with Fehling's solution to give 2,(3*S*,4*R*),5,6-pentahydroxyhexanoate and a brick-red precipitate of copper oxide (Cu_2O), which is the characteristic sign that the reaction worked. Another example is α-D-ribofuranose, which reacts with Tollens' reagent to give (2*R*,3*R*,4*R*),5-tetrahydroxypentanoate along with metallic silver, which coats the reaction vessel as a silver mirror and is diagnostic of a positive test. Note that the monosaccharide α-D-fructofuranose is a ketose, and it is not a reducing sugar.

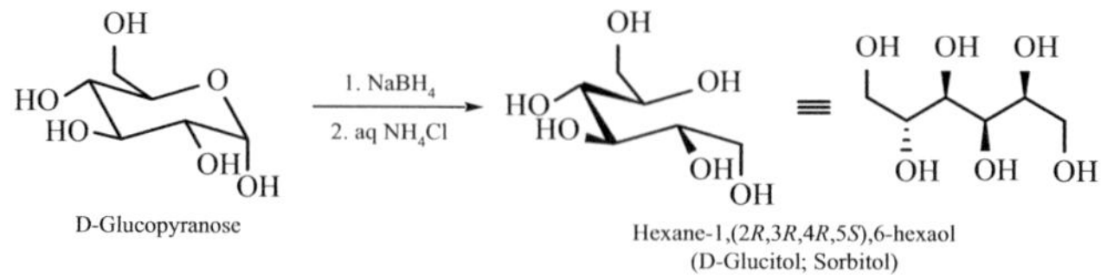

Disaccharides are potentially more difficult to classify. When *O*-α-D-glucopyranosyl-(1→4)-α-D-glucopyranose is mixed with Fehling's solution, the glucopyranose unit can undergo mutarotation, which means there is an aldehyde unit available, and it is a reducing sugar. When *O*-α-D-glucopyranosyl-(1→1)-α-D-glucopyranose is mixed with Fehling's solution, however, both anomeric carbons are tied up in the ketal linkage and mutarotation cannot occur. If no aldehyde unit is available to react, *O*-α-D-glucopyranosyl-(1→1)-α-D-glucopyranose is not a reducing sugar.

Carbohydrates are easily reduced under the proper conditions. The aldehyde unit is converted to a primary alcohol. When D-glucopyranose is reduced with sodium borohydride, the product is hexane-1,(2*R*,3*R*,4*R*,5*S*),6-hexol, known as D-glucitol (a glycitol) (also known as sorbitol). Sorbitol is a naturally occurring sweetening agent found in many berries, plums, apples, seaweed, and algae. It is sold commercially and added to many foods, particularly candy, as a natural sweetening agent. It has also been used as a sugar substitute for those who have diabetes. Most of the aldoses are similarly reduced, and *the name of the product is generated by dropping -ose and adding -itol.* Therefore, the reduction of D-mannose yields D-mannitol. The aldehyde unit of an aldose is reduced with sodium amalgam (Na/Hg), with Raney nickel in ethanol at reflux, or by catalytic hydrogenation to give a glycitol. Reduction of a ketose leads to a mixture of stereoisomers, and hydrogenation of D-fructose, for example, yields both the (*R*) and the (*S*) alcohol (D-glucitol and D-mannitol).

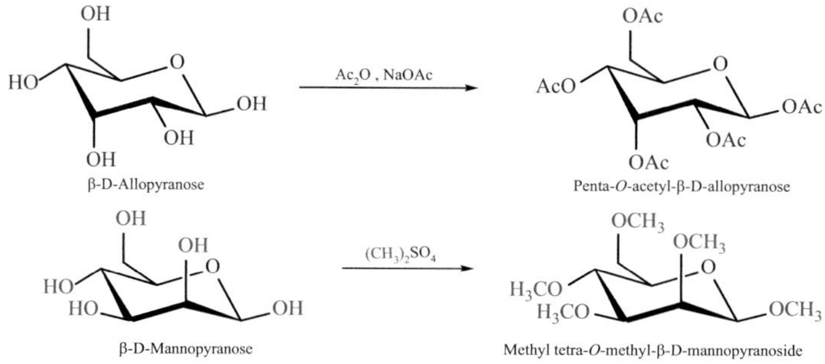

With many reagents, reaction with a carbohydrate occurs at all available hydroxyl units. When β-D-allopyranose reacts with acetic anhydride and sodium acetate, the product is the pentaacetate (penta-*O*-acetyl-β-D-allopyranose). Similarly, benzoyl chloride and pyridine react with all available hydroxyl units. When β-D-mannopyranose reacts with dimethyl sulfate [$(CH_3)_2SO_4$=Me_2SO_4], the product is the pentaether (methyl tetra-*O*-methyl-β-D-mannopyranoside). Dimethyl sulfate is the dimethyl ester of sulfuric acid (see Section 10.5), and it is a common reagent for converting an OH unit to an OCH_3unit. *The suffix -ide is used rather than -ose for the ether.*

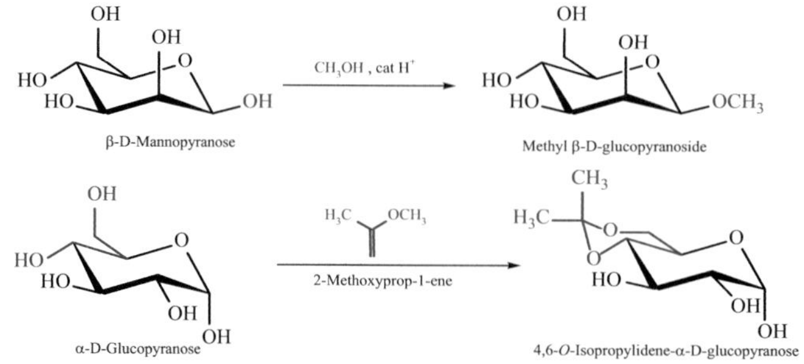

In reactions of carbohydrates, control for the reactivity for individual hydroxyl groups is essential. Although there are many hydroxyl units in a carbohydrate, using the proper reagents and reaction conditions allows one or two hydroxyl units to be selected for reaction, leaving the others untouched. If β-D-mannopyranose is treated with methanol and an acidic catalyst, for example, the product is the monoether, methyl β-D-glucopyranoside. When a carbohydrate reacts with a vinyl ether (e.g., 2-methoxyprop-1-ene), the product is a ketal, so α-D-glucopyranose reacts with this vinyl ether in the presence of an acid catalyst to give 4,6-*O*-isopropylidene-α-D-glucopyran ose. The isopropylidene group is the $C(CH_3)_2$unit, and the *O*-isopropylidene name refers to the acetone ketal unit.

Using an approach developed in an organic chemistry laboratory rather than a biosynthetic approach, the formation of a glycosidic bond is more complicated. The hydroxyl groups of the carbohydrate that are not targeted for reaction must be protected prior to the coupling reaction of the carbohydrates. In other words, the hydroxyl group must be converted to a different functional group that will allow the coupling reaction, and then a simple chemical reaction must convert the new functional group back to the hydroxyl unit (deprotected). In addition, one carbon of the carbohydrate must be converted to a reactive functional group (i.e., it is activated) to react with a hydroxyl group on the second carbohydrate that is left without a protecting group. Four examples are shown in Figure 14.14.[24] In all cases, hydroxyl groups are blocked (protected) as methoxy (OCH_3), benzyloxy (OCH_2Ph), silyloxy (R_3SiO), acetate ($OCOCH_3$, or pivaloyl ($OCOC(CH_3)_3$). In one case, a nitrogen substituent is blocked as the N-phthaloyl group. In all cases, these groups can be removed after the coupling to give the hydroxyl group or the amino group. In the first and

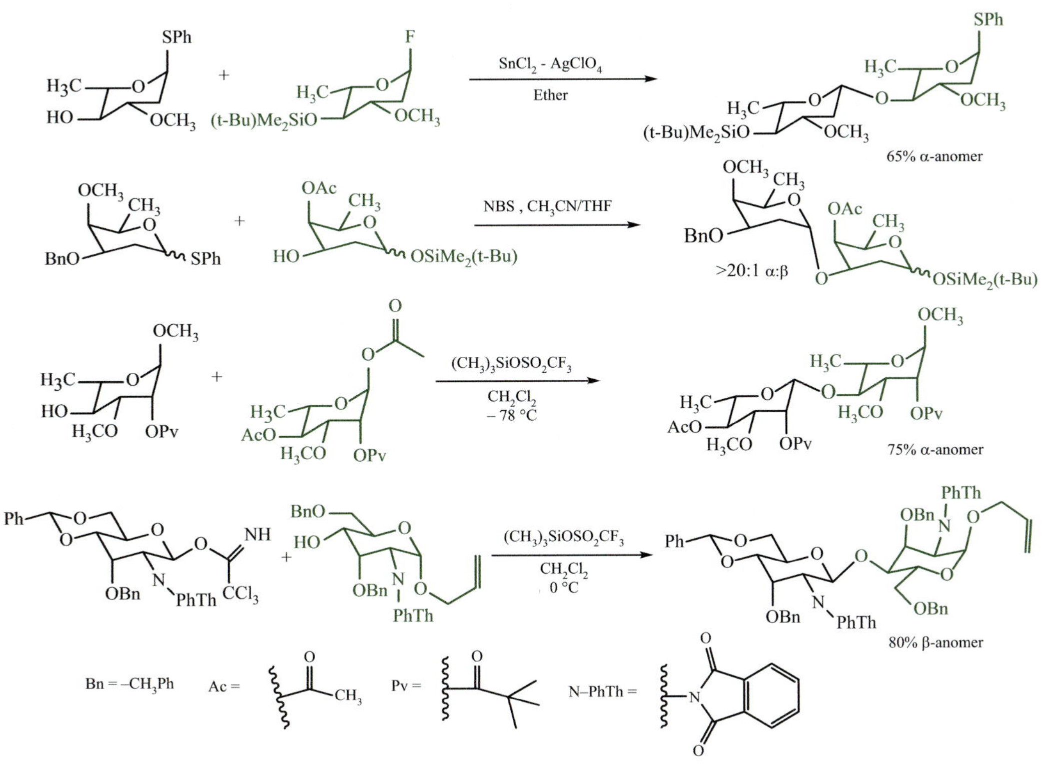

FIGURE 14.14 Chemical reactions that couple monosaccharides.

[24]Toshima, K.; Tatsuta, K. *Chem. Rev.* **1993**, *93*, 1503.

second cases, an SPh group at the anomeric carbon is labile, reacting with an anomeric fluorine in the first example or an anomeric silyloxy ether in the second. In the third example, an acetyl group at the anomeric position reacts with the unprotected hydroxyl group. In the last example, a trichloroimidate group [O(C=NH)CCl_3] reacts with an unprotected hydroxyl. The reagents used to facilitate the coupling are shown on the arrow in each case. Clearly, the protecting-deprotecting reactions that are required to form the 1→1, 1→4, 1→6, etc., glycan linkages consume time and chemical steps, making such reactions less efficient than the enzymatic reactions shown in the previous section.

HOMEWORK

14.1. Give one example of an aldotetrose, an aldopentose, and an aldohexose. Give one example of a ketotetrose, a ketopentose, and a ketohexose.
14.2. Give one example of a monosaccharide, one example of a disaccharide, and one example of a trisaccharide.
14.3. Draw both hemiacetal forms of D-ribofuranose and both hemiacetal forms of D-xylofuranose.
14.4. What is a homopolymer?
14.5. What is a reducing sugar?
14.6. Why is *O*-α-D-mannopyranose (1→1)-α-D-gulopyranose not a reducing sugar? Draw this disaccharide.
14.7. Give the product of the following reactions: (a) α-D-allopyranose + excess acetic anhydride and NaOAc, (b) α-D-talopyranose + excess dimethyl sulfate, (c) β-D-gulopyranose + hydrogen gas and a platinum catalyst.
14.8. Look up and draw a representation of the structure of pectin.
14.9. What is the definition of a glycoside?
14.10. Look up and draw the structure of paromomycin.
14.11. Look up and draw the structure of the glycoside azithromycin.
14.12. What is catabolism?
14.13. What is a *glycosyltransferase*?
14.14. What is a lumen?
14.15. What is a eukaryotic organism?
14.16. Draw the aglycone of (a) genistin, (b) digitoxin, (c) erythromycin.

15 Nucleic Acids, Nucleosides, and Nucleotides

Carbohydrates are essential components of deoxyribonucleic acid (DNA) and ribonucleic acid (RNA), particularly ribose. Nucleosides and their phosphate ester derivatives, called nucleotides, are ribose derivatives that have a purine or pyrimidine heterocycle attached to the anomeric carbon. This chapter will describe the formation and characteristics of nucleic acids, followed by the formation of nucleosides and then nucleotides.

15.1 NUCLEOSIDES AND NUCLEOTIDES

It is clear that carbohydrates are multifunctional compounds, as they contain hydroxyl, aldehyde, or ketone units, and sometimes carboxyl units. Carbohydrates may also contain an amine unit, yielding amino sugars. Examples include 2-amino-2-deoxy-β-D-mannopyranose (D-mannosamine) and 3-amino-3,6-dideoxy-β-D-glucopyranose. Carbohydrates that are missing an –OH unit are known as deoxy sugars. In some cases, an —OH unit is replaced with —NH_2 (as in 2-amino-2-deoxy-β-D -mannopyranose or 3-amino-3,6-dideoxy-β-D-glucopyranose), while in others the —OH unit is replaced with —H (as in 3-deoxy-β-D-idopyranose) or another group.

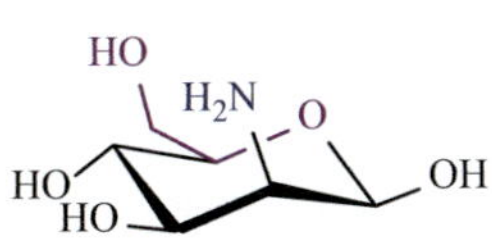

2-Amino-2-deoxy-β-D-mannopyranose (D-mannosamine)

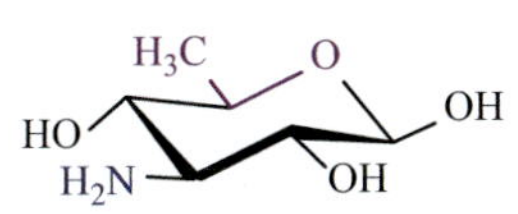

3-Amino-3,6-dideoxy-β-D-glucopyranose

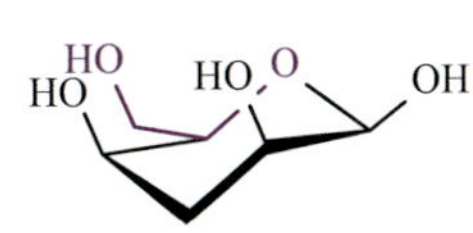

3-Deoxy-β-D-idopyranose

Ribose–purines and ribose–pyrimidines are particularly important because these structures are found in DNA and RNA. Both DNA and RNA are nucleic acids, which are biopolymers made up of repeating deoxyribofuranose or ribofuranose units. *Nucleic acids are oligo(sugar phosphates) because they contain a backbone of phosphorus linkers.* A monomeric ribofuranose with a purine or pyrimidine at C1 is called a *ribonucleoside*, or simply a *nucleoside*.

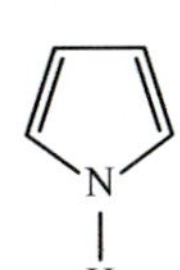

Pyrrole
1*H*-pyrrole

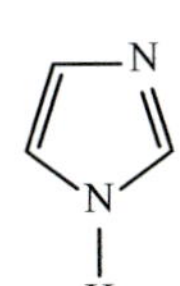

Imidazole
1*H*-imidazole

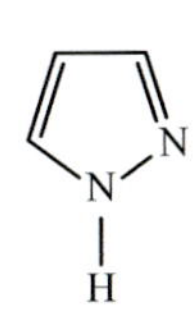

Pyrazole
1*H*-pyrazole

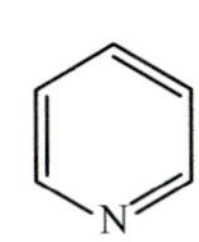

Pyridine

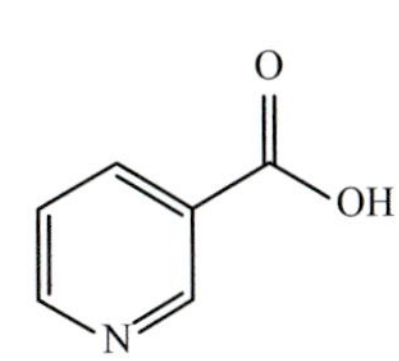

Nicotinic acid
(Niacin)

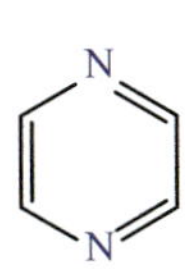

Pyrazine

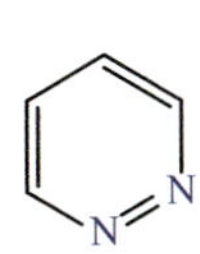

Pyridazine

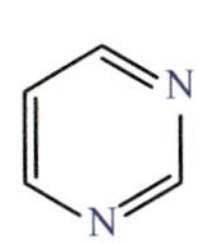

Pyrimidine

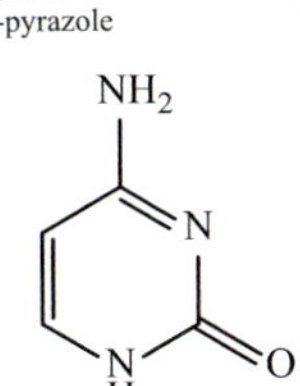

Cytosine

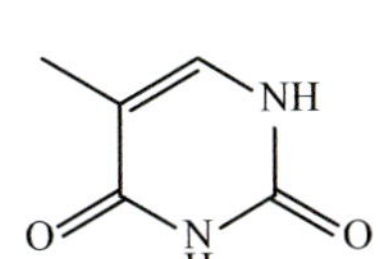

Thymine

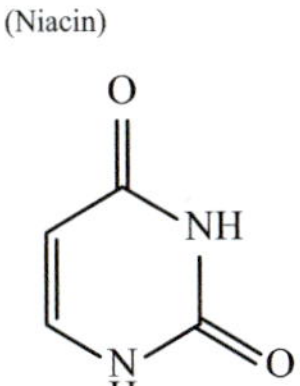

Uracil

DOI: 10.1201/9781003605157-15

Since heterocyclic compounds are part of nucleoside and nucleotide structures, a brief introduction to simple heterocycles is in order. Heterocyclic compounds were discussed in more detail in Chapter 11. Pyrrole is an aromatic compound that has one nitrogen atom in a five-membered ring. Although it is a secondary amine, the nitrogen atom is not very basic. The nitrogen electron pair must be parallel with the other π-electrons to maintain the aromatic character of the ring, which forces the hydrogen atom to be coplanar with the five atoms of the ring. Other aromatic five-membered ring amines have two nitrogen atoms in the ring. Molecules with the nitrogen atoms in a 1,3-relationship are called *imidazole*, and if the nitrogen atoms have a 1,2-relationship, the compound is called *pyrazole*.

The molecule that has one nitrogen atom in an aromatic six-membered ring is known as pyridine. Pyridine-3-carboxylic acid is commonly known as nicotinic acid, or niacin (vitamin B1). There are three different aromatic six-membered ring compounds that contain two nitrogen atoms in the ring: pyrazine, pyrimidine, and pyridazine. Both nitrogen atoms in these three heterocycles are basic. As with all aromatic compounds, the six-membered ring heterocycles are planar, with an aromatic π-cloud above and below the plane of the ring. The three pyrimidines that are found in nucleosides and nucleotides are cytosine, thymine, and uracil.

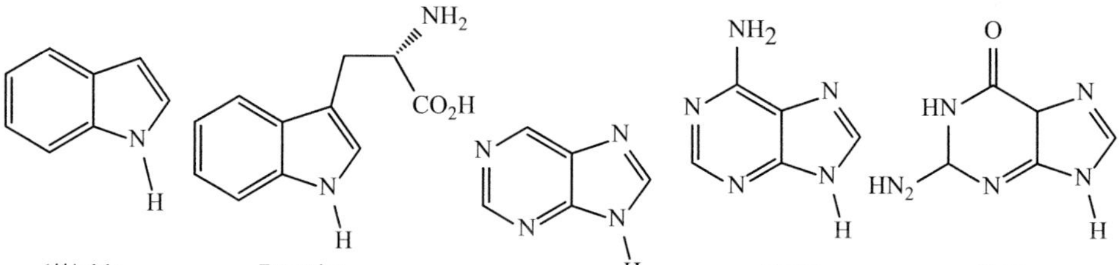

1*H*-indole Tryptophan 9*H*-Purine Adenine Guanine

There are derivatives that have one or two nitrogen atoms at different positions in bicyclic six–six, six–five, or five–five fused aromatic rings. It is also possible to incorporate three, four, or even more nitrogen atoms into these rings. An important six–five heterocyclic ring system that contains four nitrogen atoms is called purine, and derivatives of this fundamental heterocycle include adenine and guanine, which are components of deoxyribonucleic acid (DNA) and ribonucleic acid (RNA) (discussed in Section 15.2).

There are five particularly important nucleosides derived from D-ribofuranose. There are two purine derivatives: adenosine, which has an adenine unit on the ribose, and guanosine, which has a guanine unit on the ribose. There are three pyrimidine derivatives: uridine, with a uracil unit; cytidine, with a cytosine unit; and thymidine, with a thymine unit. There are also important deoxyribofuranose derivatives formed by using a ribose unit without an —OH at C2. Purine derivatives include 2′-deoxyadenosine and 2′-deoxyguanosine. Pyrimidine derivatives include 2′-deoxyuridine, 2′-deoxycytidine, and 2′-deoxythymidine. The bases found in DNA and RNA are listed in Table 15.1, along with their one-letter codes. The ribofuranose derivatives are the monomeric units found in RNA, and the 2′-deoxyribofuranose derivatives are the monomeric units found in DNA (deoxyribonucleic acids).

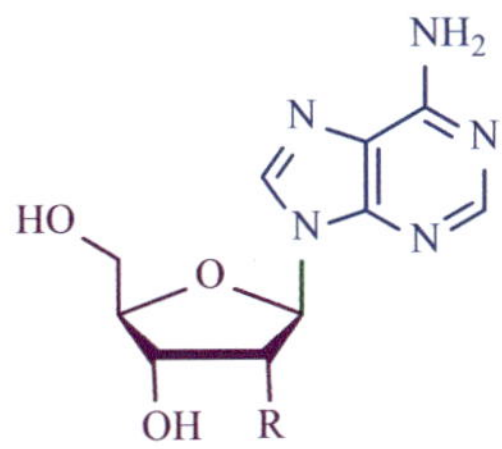

R = OH: Adenosine
R= H: 2'-Deoxyadenosine

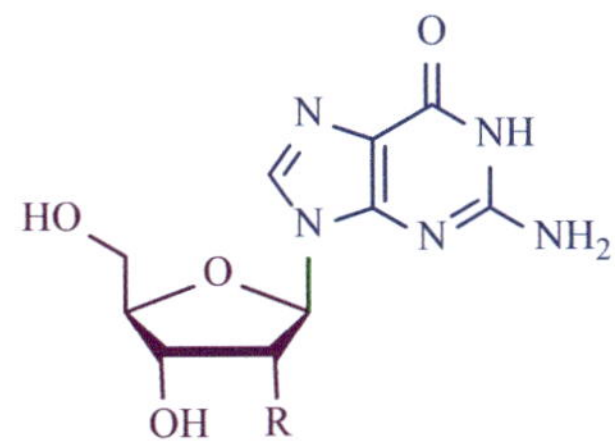

R = OH: Guanosine
R= H: 2'-Deoxyguanosine

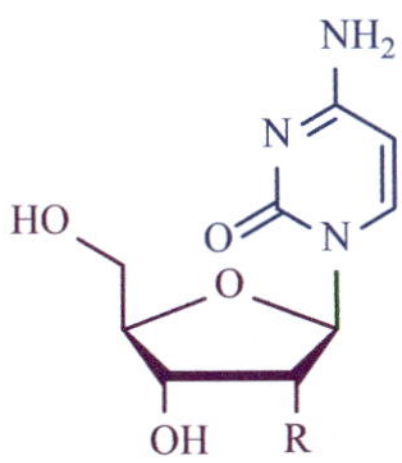

R = OH: Cytidine
R= H: 2'-Deoxycytidine

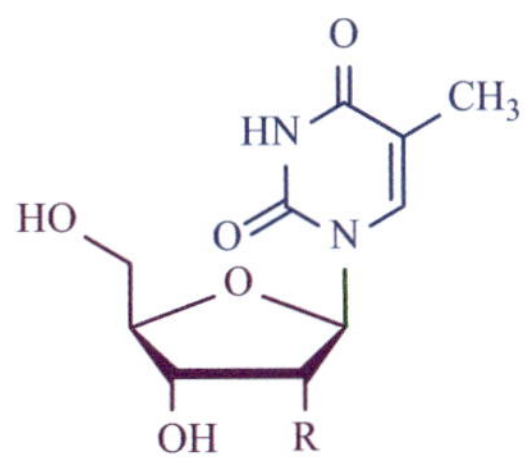

R = OH: Thymidine
R= H: 2'-Deoxythymidine

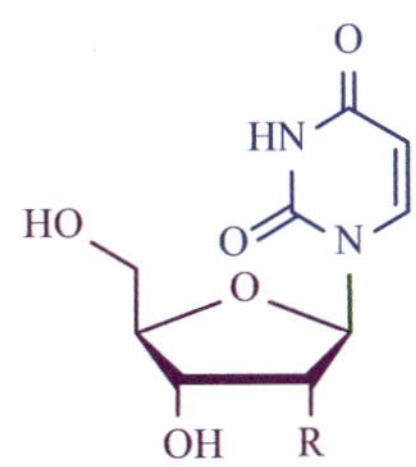

R = OH: Uridine
R= H: 2'-dexyuridine

TABLE 15.1
One-Letter Codes for Heterocyclic Amines

Heterocycles (Bases, B)	Structure	One-Letter Code
Adenine	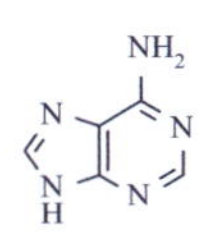	A
Guanine	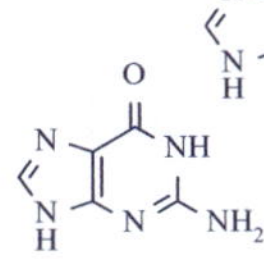	G
Uracil	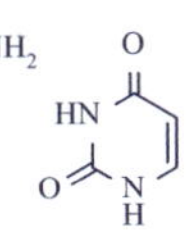	U
Cytosine	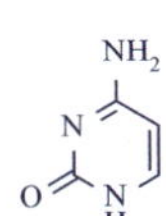	C
Thymine	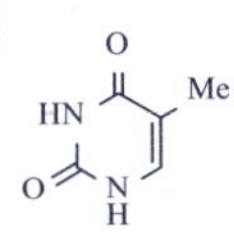	T

Each nucleotide can react with phosphoric acid or a phosphoric acid derivative to form a phosphate ester. *The phosphate ester of a nucleoside is called a nucleotide.* Using guanosine derivatives as an example, there are three possible monophosphate esters: guanosine 5′-monophosphate, guanosine 3′-monophosphate, and guanosine 2′-monophosphate.

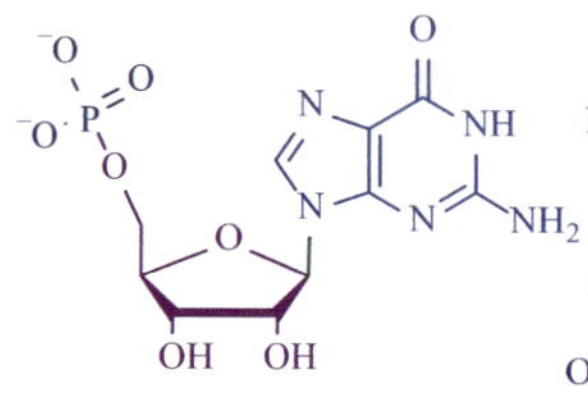

Guanosine 5'-monophosphate

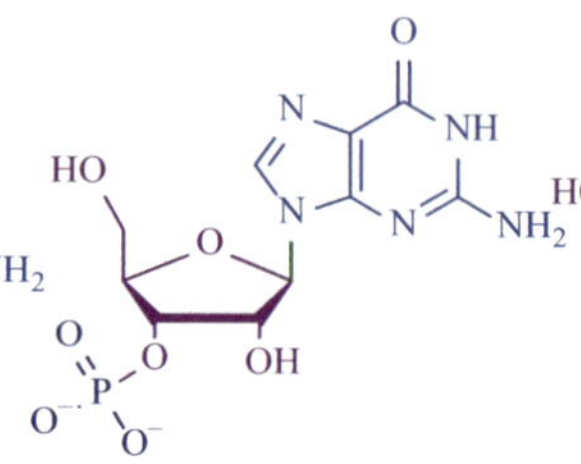

Guanosine 3'-monophosphate

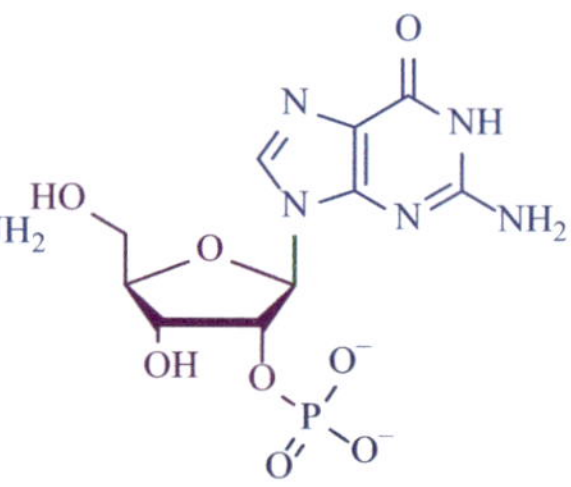

Guanosine 2'-monophosphate

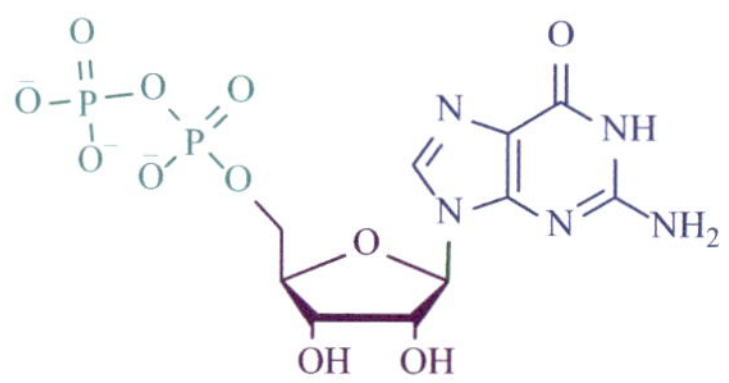

Guanosine 5'-diphosphate

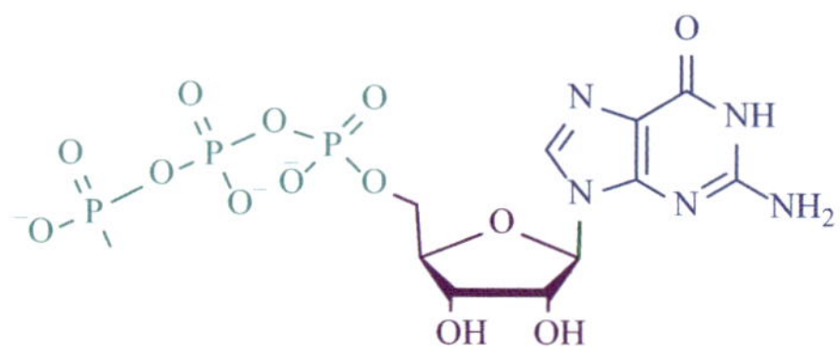

Guanosine 5'-triphosphate

A *nucleotide* with a second attached phosphate unit is a diphosphate (e.g., guanosine 5′-diphosphate), and a third phosphate unit is seen in adenosine 5′-triphosphate. These three compounds have commonly used abbreviations. The monophosphate is abbreviated GMP, the diphosphate is GDP, and the triphosphate is GTP. These abbreviations stem from a one-letter code for each heterocycle used to form the nucleotide; in this case, G stands for guanine (see Table 15.1). The code simply takes the first letter from each heterocycle.

15.2 POLYNUCLEOTIDES

Without question, the most important nucleotides are the purine and pyrimidine derivatives, as just shown. Nucleotides are linked together by phosphate linkages to form a long-chain nucleotide (a polynucleotide), resulting in the structures of deoxyribonucleic acids (DNA) and ribonucleic acids (RNA). DNA, of course, carries the genetic instructions used in the growth, development, functioning, and reproduction of all known living organisms. Both of these oligo(sugar phosphates) are formed by linking nucleotides together through phosphodiester linkages at the 3′–5′ positions of the nucleotides. In both DNA and RNA, a purine or pyrimidine heterocycle is attached to the anomeric carbon of the sugar, ribose. Both DNA and RNA are linear polymers that begin with a phosphodiester link at the 3′ position of one nucleotide, which is linked to the 5′-position of the next nucleotide. This 3′→5′ linkage continues as the polynucleotide chain grows. This structure results in a polymer with a phosphate backbone, a free 5′ terminus at the "beginning" of the polymer, and a free 3′ terminus at the "end" of the polymer.

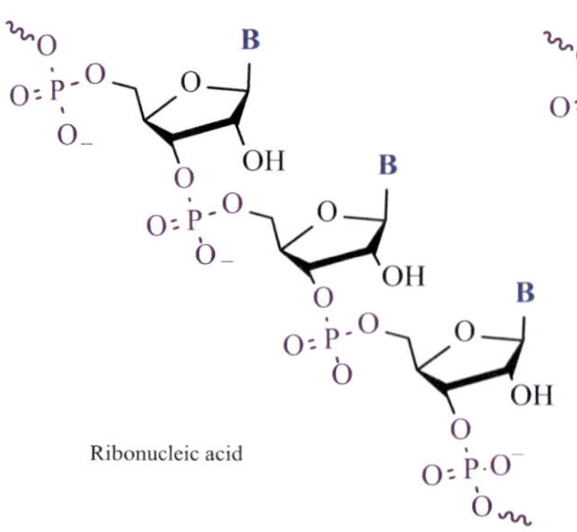

Ribonucleic acid

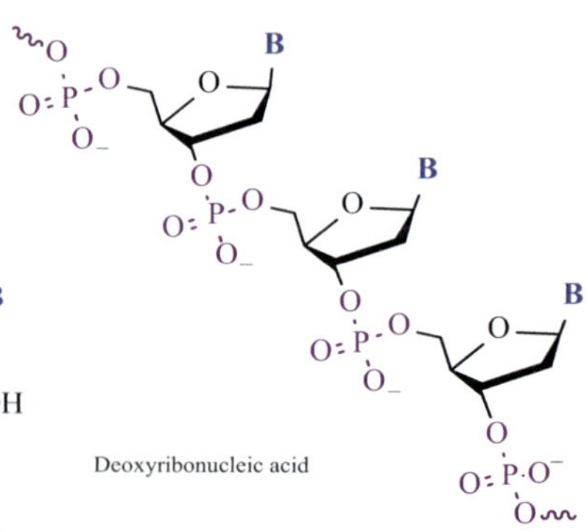

Deoxyribonucleic acid

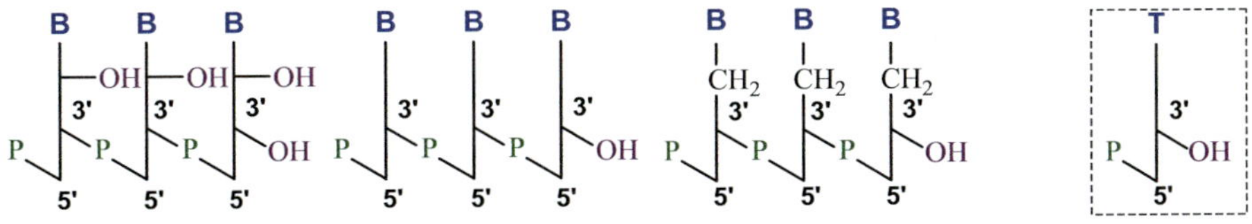

FIGURE 15.1 Shorthand representation of nucleic acids.

There is a shorthand notation that is sometimes used to represent the structures of nucleotides, roughly based on a Fischer projection, as shown in Figure 15.1. Using this model, a fragment of DNA is represented as **1,** and a fragment of RNA is represented as **2**, where the heterocyclic base is shown at the top of the structure, labeled as B (the letter code is used; T), and the 5′-OH unit is shown at the "bottom-left." The 2′ and 3′ OH units are shown on the right for the D-ribofuranose. Representation **2** has a methylene in the ribose ring rather than the CHOH unit, and **3** shows a –CH_2– unit to make the shorthand model **2** clearer, if there is any chance of confusion. Models **2** and **3** are meant to be identical. An individual nucleotide (e.g., thymidine-deoxyribofuranose) is represented as **4**, where T indicates the thymidine, and the one OH group that is shown in the structure is understood to be the 3′ OH.

The fundamental difference between DNA and RNA is the absence of a substituent at the C2′ position of the ribofuranose unit in DNA. The structure of DNA usually consists of nucleotides containing the purines adenine (**A**) and guanine (**G**), as well as the pyrimidines cytosine (**C**) and thymine (**T**). The same heterocycles are incorporated in RNA, but uracil (**U**) is usually found instead of thymine (**T**). Both DNA and RNA have a highly charged and polar sugar-phosphate "backbone" that is relatively hydrophilic (soluble in water). The heterocyclic bases attached to the ribofuranose or deoxyribofuranose units are relatively hydrophobic (relatively insoluble in water). These interactions lead to specialized secondary and tertiary structures of DNA and RNA. In other words, differences in the heterocyclic substituents result in different conformations for different nucleic acids.

As noted above, deoxyribonucleic acid is the repository of genetic information in cells. The major form of DNA under physiological conditions (those found in a normal cell) is the β-form, which consists of two anti-parallel strands: the *Watson-Crick double helix structure*, named after James D. Watson and Francis H.C. Crick. In earlier work, Rosalind Franklin obtained X-ray crystallographic images of DNA with the goal of producing images instructive for structure determination. She reportedly shared her images with Dorothy Crowfoot Hodgkin, who assisted in the interpretation of the X-ray images. There seems to be little doubt that X-ray crystallographic evidence such as this was critical for determining the structure of DNA, for which Watson and Crick shared the Nobel Prize in 1962. Dorothy Hodgkin was awarded the Nobel Prize in 1964 for her many contributions to understanding protein structure and the structure of important drugs such as penicillin. Rosalind Franklin died in 1958 of ovarian cancer and was not recognized with a Nobel Prize for her work.

The double strand in DNA is characterized by heterocyclic bases that have attached substituents capable of hydrogen bonding. When the heterocyclic base (simply referred to as the base) is adenine in an adenosine (A) nucleotide, it can hydrogen bond with a thymine of the other strand (an A–T base pair). A guanine base in a guanosine nucleotide on one strand can hydrogen bond with a cytosine base in a cytidine nucleotide on the other strand (a G–C base pair). The strands must be antiparallel (one strand is 3′→5′ while the other strand is 5′→3′) to maximize hydrogen bonding, because the stereogenic nature of the ribofuranose and deoxyribofuranose units leads to a "twist" in the polynucleotide backbone.

As noted, the hydrogen-bonding base pairs are called *Watson-Crick base pairs*, and the C–G and A–T base pairs are shown in Figure 15.2.[1] The inherent chirality of D-ribofuranose and

[1] Adapted from Figure 2 in Rajski, S.R.; Williams, R.M. *Chem. Rev.* **1998**, *98,* 2723.

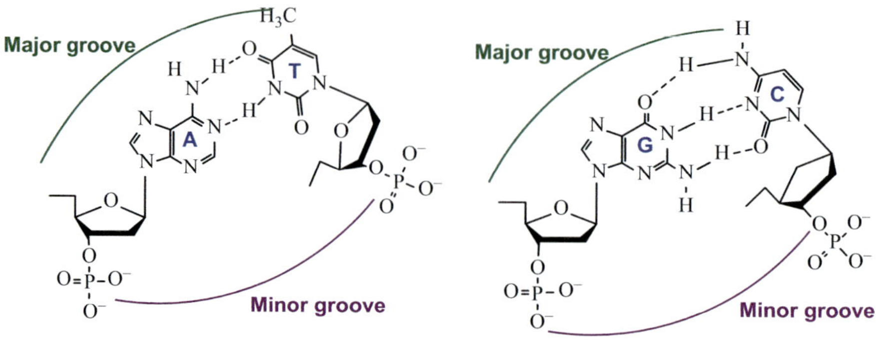

FIGURE 15.2 The C–G and T–A interactions. (Reprinted with permission and adapted from Figure 2 in Rajski, S.R.; Williams, R.M. *Chem Rev.* **1998**, *98,* 2723–2796. Copyright 1998 American Chemical Society.)

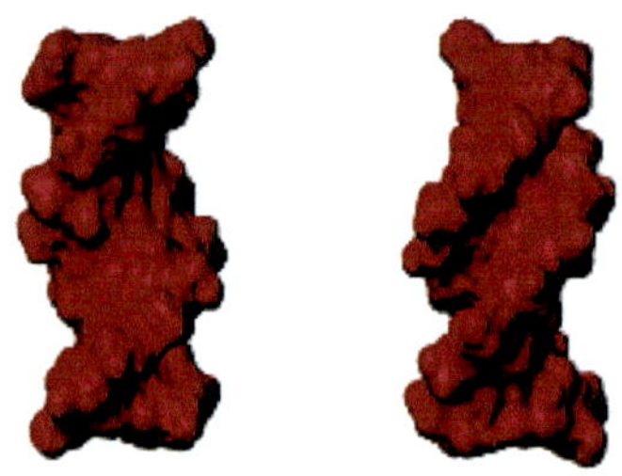

FIGURE 15.3 DNA: (a) major groove and (b) minor groove. (Reprinted with permission and adapted from Figure 6 in Liu, H.; Lynch, S.R.; Kool, E.T. *J. Am. Chem. Soc.* **2004**, *126,* 6900. Copyright 2004 American Chemical Society.)

D-deoxyribofuranose causes the β-form of DNA adopting a right-handed helix (see Figure 15.3),[2] and each base-pair plane is rotated ~ 36° relative to the one preceding it. The two strands are connected by hydrogen bonds between specific nucleotides on each strand (intermolecular hydrogen bonding). This leads to a complete right-handed turn for every 10 contiguous base pairs and a helical pitch of ~ 34 Å (34 pm).[3]

The β-form leads to the creation of two helical grooves: the *major groove* and the *minor groove,* as indicated in Figure 15.3. In the major groove, the nitrogen and oxygen atoms of the phosphodiester backbone point inward toward the helical axis. In the minor groove, the nitrogen and oxygen atoms point outward. These grooves are on opposite sides of the base pairs. The minor groove is narrow, while the major groove is wide because the backbones are far apart in the major groove and close together in the minor groove, but both are of approximately the same depth. When DNA interacts with small molecules or proteins, the major or minor groove provides a "microenvironment" for bonding and recognition of these compounds (called ligands).[2] As the specific nucleotides in the strands of DNA change, the "floor" of each groove also changes, leading to ligand-specific sites.

There are two other forms of DNA, the A-form and the Z-form.[4] If the relative humidity of the β-form of DNA decreases to 75% and the sodium chloride concentration drops to < 10%, the β-form is transformed into the A-form helix.[4] The A-form is a right-handed helix, while the Z-form of DNA adopts a left-handed helix, *but it is not simply the mirror image of the β-form or the A-form helices.*[4]

---

[2] Adapted from Figure 6 in Liu, H.; Lynch, S.R.; Kool, E.T. *J. Amer. Chem. Soc.* **2004**, *126,* 6900.

[3] Hecht, S.M. (Ed.), *Bioorganic Chemistry: Nucleic Acids*, Oxford University Press, New York, **1996**, p. 6.

[4] Hecht, S.M. (Ed), *Bioorganic Chemistry: Nucleic Acids*, Oxford University Press, New York, **1996**, see Figure 1–9, pp. 10–11.

Deoxyribonucleic acid stores genetic information, but RNA transcribes and translates this information for the cell. While DNA tends to form linear repeating structures, RNA generates diverse structures that have a unique function. There are several unique structures for RNA. There are three main classes of RNA found in a cell: tRNA, rRNA, and mRNA. Transfer RNA tends to be single strands of nucleic acid with 60–96 nucleotides. Its function is to transport an amino acid, covalently attached to a specific proton of the polynucleotide chain, to a ribosome where it can be incorporated in its proper sequence in a protein. This process is described in Section 3.2.

The primary and secondary structure of cysteine tRNA (the tRNA used to transport the amino acid cysteine) is shown as a "cloverleaf" structure (see Figure 15.4)[5] that contains double-stranded stems connected to single-stranded loops. The structure in Figure 15.4 consists of one nucleotide strand, but base-pairing in self-complementary regions of this strand leads to double-stranded loops. This strand is effectively held together by intramolecular hydrogen bonding. The 5′-terminus is phosphorylated, and the 3′-terminus of tRNA is where the amino acid is attached. This molecule always terminates in the sequence CCA-3′ with a free OH group, which can bind the amino acid. Almost all tRNAs have a seven-base pair structure called the *acceptor stem* (the seven base pairs prior to the CCA-3′ terminus). In Figure 15.4, the acceptor stem is U-U-G-C-A-A-A. The loop of nucleotides at the "bottom" of the "cloverleaf" has the three base pairs that will bind this tRNA to an mRNA ((G-C-A)A), which allows an amino acid to be transferred to the protein.

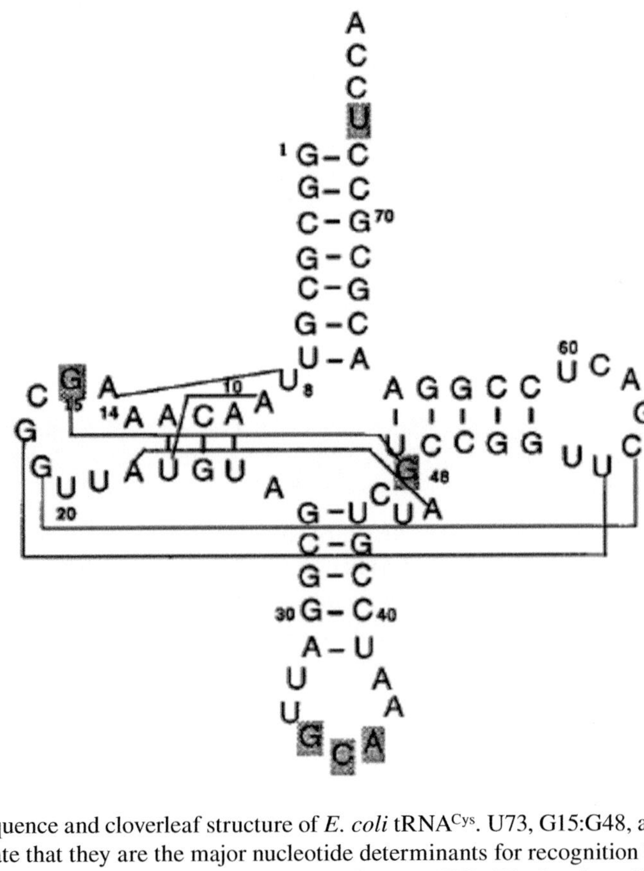

FIGURE 15.4 Sequence and cloverleaf structure of *E. coli* $tRNA^{Cys}$. U73, G15:G48, and the GCA anticodon are shaded to indicate that they are the major nucleotide determinants for recognition by cysteine tRNA synthetase. (Reprinted with permission from Hou, Y.-M.; Gamper, H.B. *Biochemistry* **1996**, *35*, 15340–15348. Copyright 1996 American Chemical Society.)

[5] Hou, Y.-M.; Gamper, H.B. *Biochemistry* **1996**, *35,* 15340.

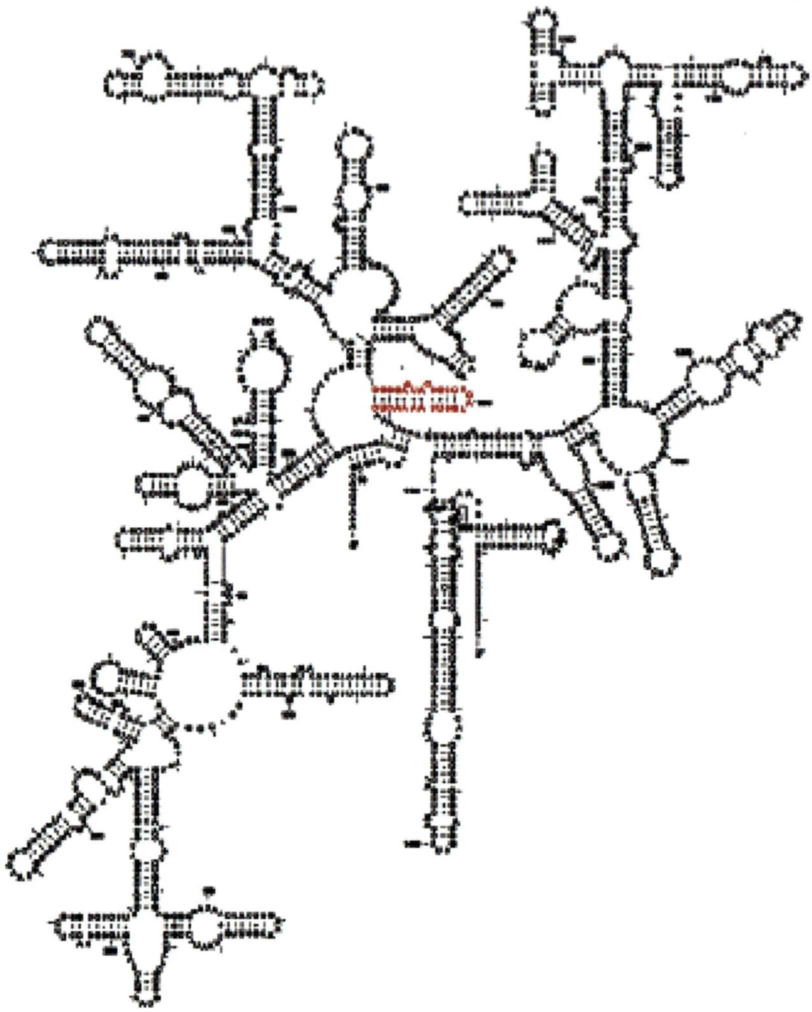

FIGURE 15.5 Secondary structure of the 16S RNA unit of *E. Coli*, for the 30S subunit of the 70S subunit. (Reprinted with permission from Hoerter, J.A.H.; Lambert, M.N.; Pereira, MJ.B.; Walter, N.G. *Biochemistry* **2004**, *43,* 14624. Copyright 2004 American Chemical Society.)

Ribosomal RNA usually has a complex structure, and rRNA is the main component of the ribosome. The ribosome of *Escherichia coli* (70S) has two rRNA units: the 50S and the 30S subunits. The 30S subunit has one large rRNA (16S RNA) and 21 individual proteins; the 50S subunit has two RNAs (5S and 23S) and 32 different proteins.[6] The secondary structure of the 16S rRNA is shown in Figure 15.5.[7] The *ribosome* is the body an organism uses to manufacture proteins. The ribosome binds mRNA, which then interacts with tRNA.

Messenger RNA is a single-stranded nucleic acid with unique secondary structures. It carries the genetic code from DNA to the ribosome, and this code is read via forming hydrogen bonds to the three base pairs of tRNA mentioned above (the triplet anticodon). To ensure this genetic code is easily "read," mRNA is generally less structured than tRNA or rRNA. A strand of mRNA may be categorized as having groups of three base pairs, where each group is the genetic code for a specific amino acid. These three base pairs (called codons) will match a complementary set of three base pairs on a tRNA (an anticodon). The anticodon is at the bottom loop of tRNA in Figure 15.4. The

[6] Hecht, S.M. (Ed.), *Bioorganic Chemistry: Nucleic Acids*, Oxford University Press, New York, **1996**, pp. 15–17.

[7] Hoerter, J.A.H.; Lambert, M.N.; Pereira, MJ.B.; Walter, N.G. *Biochemistry* **2004**, *43,* 14624.

codon and anticodon are complementary, which means that a G–C–C codon on mRNA will match a C–G–G anticodon on tRNA. Each amino acid will have a tRNA with a unique anticodon, and that amino acid will attach itself only to that tRNA. The ribosome will bind a strand of mRNA to build a specific protein, which has a specific sequence of amino acids (see Figure 3.10 in Section 3.2). The genetic code on the mRNA (the set of codons) matches the amino acid sequence of the protein. Therefore, the mRNA will bind tRNA molecules in a particular sequence, allowing that amino acid to be released to the growing protein sequence associated with the ribosome.

15.3 BIOSYNTHESIS OF NUCLEOTIDES

The biosynthesis of nucleotides is clearly a critically important process. One example is the biosynthesis of an adenine derivative, in which the purine moiety is constructed step by step on the ribose moiety, rather than being attached to the sugar unit as an intact heterocycle. Adenine monophosphate (AMP), adenine diphosphate (ADP), and adenine triphosphate (ATP) are examples. Clearly, nucleotides are ribose derivatives and perhaps the most common examples of glyconucleotides. The biosynthesis of AMP begins with the conversion of the diphosphate unit in 5-phosphoribosyl-1-pyrophosphate to an amine unit by transfer of an amido group from glutamine. All the steps in this sequence are enzymatically mediated, and the appropriate enzymes are shown in Figure 15.6.[8]

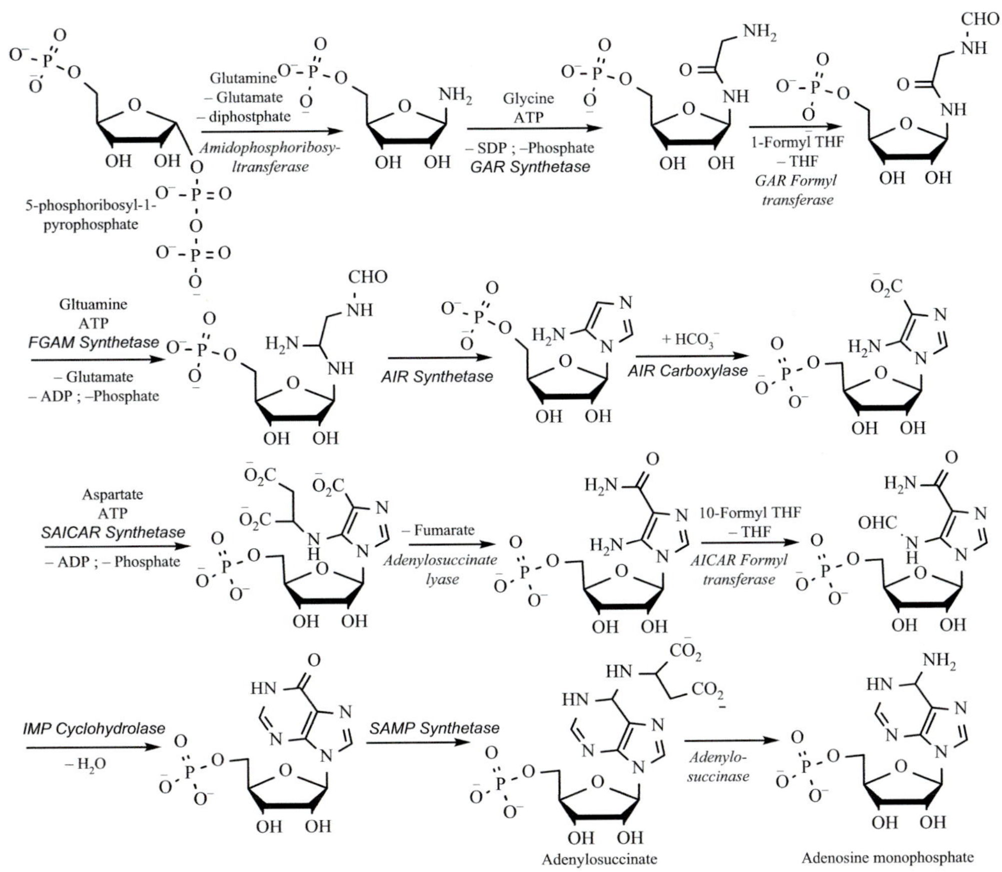

FIGURE 15.6 De novo biosynthetic pathway of purine nucleotides in plants. (Moffatt, B.A.; Ashihara, H. Purine and pyrimidine nucleotide synthesis and metabolism. *Arabidopsis Book*, **2002**;*1*, e0018. doi:10.1199/tab.0018. Open Access.)

[8] Moffatt, B.A.; Ashihara, H. *The Arabidopsis Book*, **2002**. https://doi.org/10.1199/tab.0018.

The next step is a reaction with glycine, followed by conversion to the corresponding formamide. The amide carbonyl is converted to an amino unit, and cyclization leads to an imidazole moiety. Carboxylation, followed by conversion to the aspartate derivative, allows formation of the imidazole with pendant amine and amide moieties. The primary amine moiety is converted to a formamide, allowing formation of the six-membered ring in a purine derivative. Conversion of the carbonyl to the aspartate derivative allows final formation of adenine monophosphate.[8] The enzymes are: *amidophosphoribosyltransferase* (EC 2.4.2.14); GAR synthetase, which is *glycinamide ribonucleotide synthetase* (EC 6.3.4.13); *GAR formyl transferase*, which is *formylglycinamide ribotide amidotransferase* (EC 6.3.5.3); *FGAM synthetase*, which is *formylglycinamide ribotide amidotransfer*ase (EC 6.3.5.3); *AIR synthetase*, which is *phosphoribosylformylglycinamidine cyclo-ligase* (EC 6.3.3.1); *AIR carboxylase*, which is *phosphoribosylaminoimidazole carboxylase* (EC 4.1.1.21); *SAICAR synthetase*, which is *phosphoribosylaminoimidazolesuccinocarboxamide synthase* (EC 6.3.2.6); *adenylosuccinate lyase* (EC 4.3.2.2); *AICAR formyl transferase*, which is *phosphoribosylaminoimidazolecarboxamide formyltransferase* (EC 2.1.2.3); *IMP cyclohydrolase* (EC 3.5.4.10); *SAMP synthetase*; *adenylosuccinase*, which is *adenylosuccinate synthetase* (EC 6.3.4.4).

The pyrimidine nucleotides are directly involved in plant carbohydrate metabolism. Pyrimidine nucleotides are also biosynthesized, and in the case discussed, the heterocyclic ring is assembled and then attached to the ribose moiety. The so-called *orotate pathway* of pyrimidine de novo synthesis involves the formation of the pyrimidine unit from glutamine and CO_2, and later coupling to 5-phospho-D-ribofuranose 1-diphosphate leads to uridine monophosphate (UMP). The biopathway is shown in Figure 15.7[9] and begins with the *carbamoyl phosphate synthase* (EC 6.3.5.5)-catalyzed

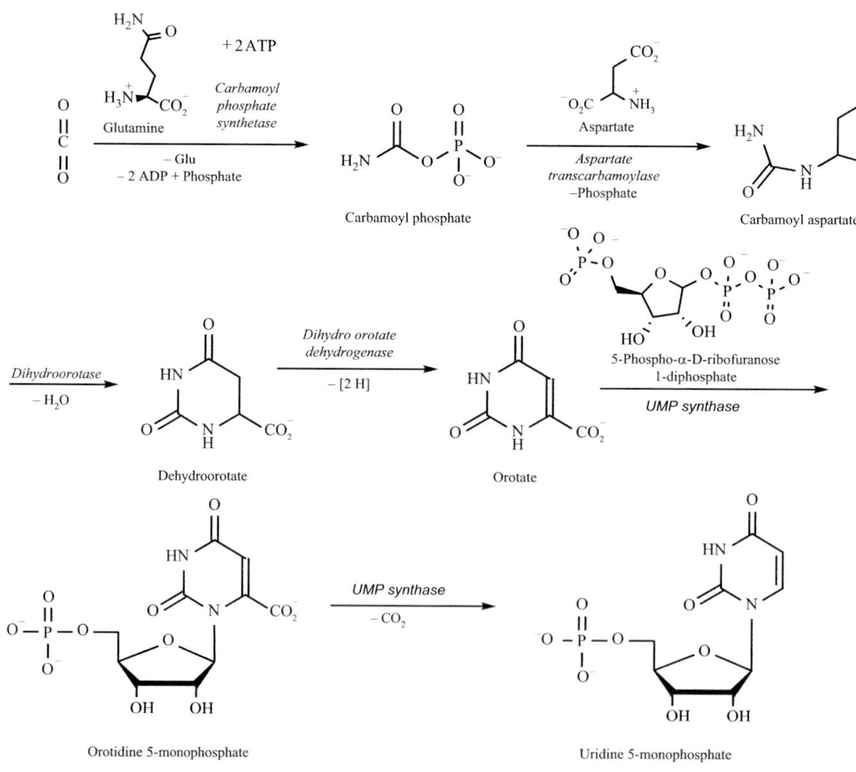

FIGURE 15.7 Biosynthesis of uridine 5-monophosphate. (Reprinted with permission from Christopheson, R.I; Lyons, S.D.; Wilson, P.K. *Acc. Chem. Res.* **2002**, *35*, 961–971. Copyright 2002 American Chemical Society.)

[9] Christopheson, R.I; Lyons, S.D.; Wilson, P.K. *Acc. Chem. Res.* **2002**, *35*, 961.

reaction of glutamine and carbon dioxide with adenosine triphosphate (ATP) to give carbamoyl phosphate, glutamine, adenosine diphosphate (ADP), and phosphate. Subsequent reaction with aspartate, catalyzed by *aspartate transcarbamoylase* (EC 2.1.3.2), gives carbamoyl aspartate and phosphate. The next step is cyclization with concomitant dehydration catalyzed by *dihydroorotase* (EC 3.5.2.3), to give dehydroorotate. Dehydrogenation is catalyzed by *dehydroorotate dehydrogenase* (EC 1.3.98.1) and gives orotate. Coupling to 5-phospho-D-ribofuranose 1-diphosphate, catalyzed by UMP synthase (EC 4.1.1.23), gives orotidine 5-monophosphate, and decarboxylation, also catalyzed by UMP synthase, gives uridine 5-monophosphate.

15.4 RIBOZYMES

A ribonucleic acid (RNA) enzyme that catalyzes a reaction is known as a ribozyme. An RNA enzyme is found in the ribozyme. The most common reaction, however, is the catalysis of cleavage or ligation of RNA and DNA, as well as formation of peptide bonds. Ribozymes can join amino acids to form a peptide bond, as occurs in a *peptidyl transferase* reaction. In ribozyme coupling, an esterified amino acid (N-blocked methionine) is coupled to the 3′(2′)-O of adenosine, as shown in Figure 15.8,[10] and the acceptor phenylalanine has a free amino group. In the ribozyme reaction, 5′-Phe-SS-RNA provides the amino group for an acyl substitution at the aminoacyl carbonyl carbon of adenosine monophosphate-Met-Bio to form a peptide bond. The wavy highlighted line represents the linker that joins guanosine-5′-monophosphorothioate and Phe: $CH_2C(O)NHCH_2CH_2SSCH_2CH_2NH$.[zhang]

Ribozymes are important for the synthesis of nucleotides used to make RNA polymers. In modern metabolism, there are various pathways by which nucleotides can be synthesized. 5-Phosphoribosylpyrophosphate (PRPP) is used by several *phosphoribosyltransferases* to synthesize purine, pyrimidine, and pyrimidine nucleotides.[11] An example is the ribozyme (a nucleotide synthase ribozyme) that facilitates the formation of 4-thiouridine at its 3′ terminus by coupling a tethered 5-phosphoribosyl 1-pyrophosphate (PRPP) with a free 4-thiouracil. Apparently, RNA uses a dissociative mechanism, as shown in Figure 15.9.[12] Loss of diphosphate

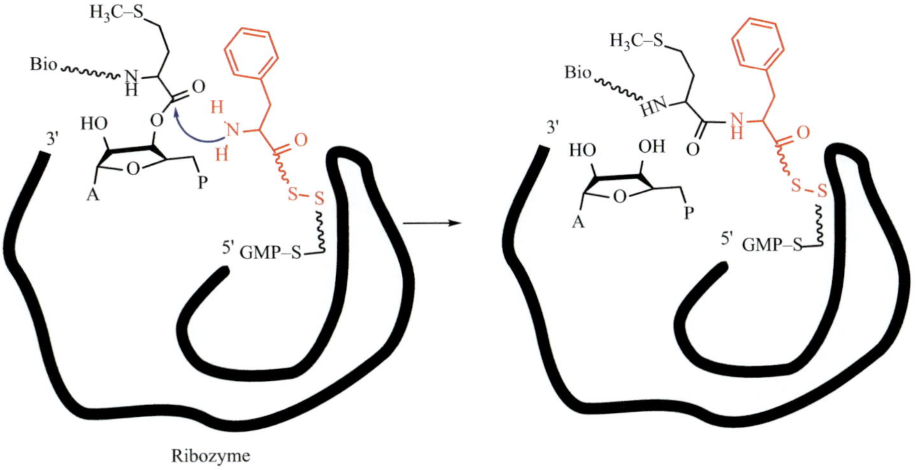

FIGURE 15.8 Peptide bond formation by a ribozyme. (Reprinted by permission from Springer Nature Customer Service Centre GmbH: Springer Nature. Zhang, B.; Cech, T.R. *Nature* **1997**, *390*, 96–100. Copyright 1997. Figure 1, p. 97.)

[10] Zhang, B.; Cech, T.R. *Nature* **1997**, *390*, 96.
[11] (a) Flaks, J.G. *Methods Enzymol.* ***1963***, 6, 136–158; (b) Unrau, P.J.; Bartel, D.P. *Nature* **1998**, *395*, 260.
[12] Unrau, P.J.; Bartel, D.P. *Proc. Nat. Acad. Sci. USA* **2003**, *100*, 15393.

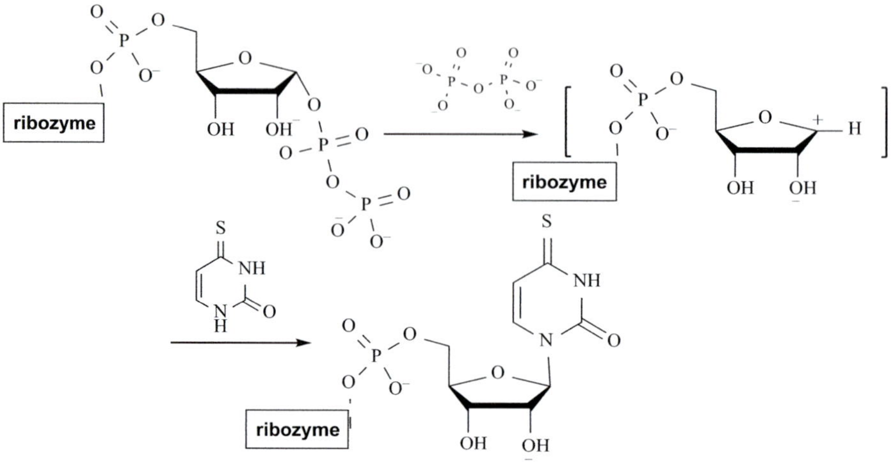

FIGURE 15.9 A stepwise dissociative mechanism for the displacement of pyrophosphate (PPi), formation of an oxocarbenium-ion intermediate, and nucleophilic addition of 4SUra to give 4-thiouridine. (Unrau, P.J.; Bartel, D.P. *Proc. Natl. Acad Sci. USA* **2003**, *100*, 15393–15397. Copyright (2003) National Academy of Sciences, U.S.A.)

leads to a carbocation intermediate, and subsequent reaction with 4-thiouracil gives 4-thiouridine attached to the ribozyme.

There are many examples of ribozymes, including the *GIR1 branching ribozyme*, which catalyzes cleavage of the RNA chain by transesterification and formation of a 2′,5′-phosphodiester bond between the first and the third nucleotide of the 3′-cleavage product.[13] The *Group II introns* are another example, and they catalyze RNA self-splicing by an autocatalytic two-step reaction in which the introns excise themselves from surrounding RNA (exons) and stitch the resulting pieces back together.[14] This form of RNA splicing is a foundation for RNA metabolism in plants, fungi, and yeast, and in many bacteria. The essential enzyme *RNase P* is present in all cells and cellular compartments that synthesize tRNA and is responsible for generating the mature 5’-end of tRNA.[15] All known *RNase P* enzymes are ribonucleoproteins and contain an RNA subunit essential for catalysis.

15.5 HYDROLYSIS OF RNA AND DNA

The carbohydrate-phosphodiester bond is broken in RNA cleavage. The hydroxyl group at the 2′ position is susceptible to this hydrolysis, which makes RNA more susceptible to hydrolysis relative to DNA. In this reaction, the 2′-hydroxyl is deprotonated and attacks the phosphorus of the phosphodiester moiety as a nucleophile, as shown in Figure 15.10[16] for a generic dinucleotide. The transition state for this reaction has a pentacoordinate phosphorus prior to loss of the phosphate and cleavage of the ester unit in the RNA backbone. A mononucleotide is cleaved, and a 2′,3′-cyclic phosphate is formed. If the RNA is double-stranded or involved in nucleotide base pairing, it is more stable, and cleavage is less likely, but enzymes are available for such cleavage, including *ribonuclease A* (*RNase A*; EC 3.1.27.5).

[13] Beckert, B.; Nielsen, H.; Einvik, C.; Johansen, S.D.; Westhof, E.; Masqida, B. *EMBO J.* **2008**, *27*, 667.
[14] Pyle, A.M. *Ann. Rev. Biophys.* **2016**, *45*, 183.
[15] Gopalan, V.; Vioque, A.; Altan, S. *J. Biol. Chem.* **2002**, *277*, 6759.
[16] (a) Breslow, R. *Proc. Nat. Acad. Sci. USA* **1993**, *90*, 1208; (b) Kirby, A.J.; Marriott, R.E. *J. Amer. Chem. Soc.* **1995**, *117*, 833.

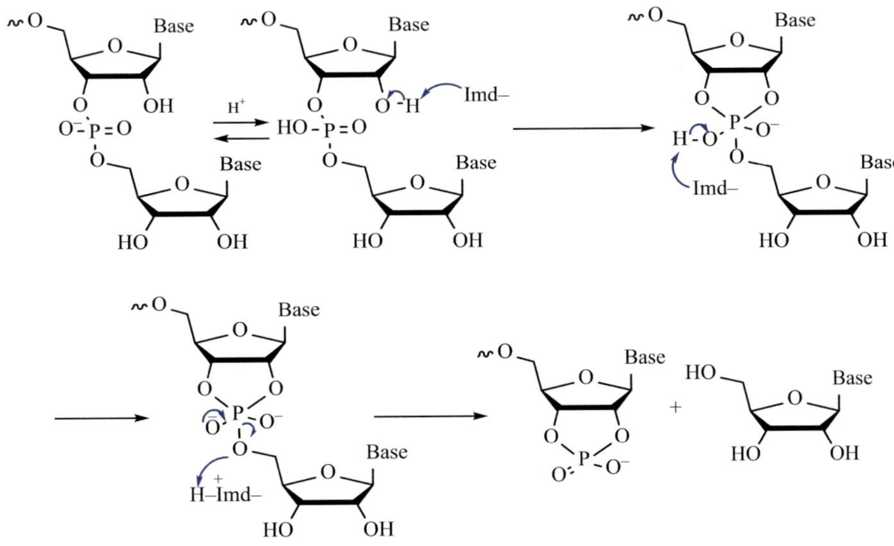

FIGURE 15.10 RNA cleavage. (Breslow, R. *Proc. Natl. Acad. Sci. USA* **1993**, *90*, 1208–1211. Copyright (1993) National Academy of Sciences, U.S.A.)

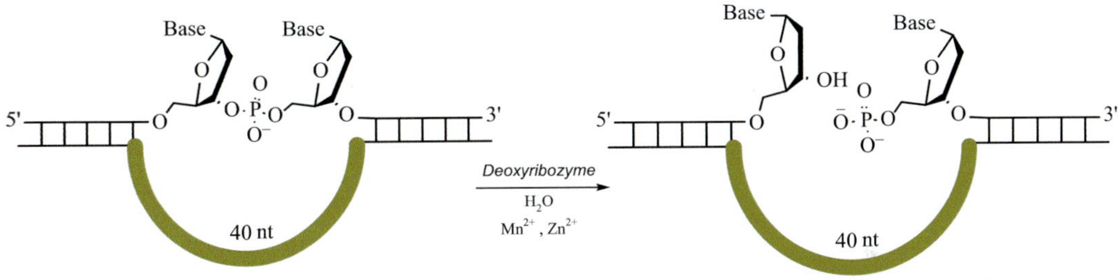

FIGURE 15.11 Hydrolysis of a DNA strand. (Reprinted by permission from Springer Nature Customer Service Centre GmbH: Springer Nature. Fekry, M.I.; Gates, K.S. *Nat. Chem. Biol.* **2009**, *5*, 710–711. Copyright 2009. Figure 1, p.711.)

The hydrolysis of DNA is an important process, but DNA is quite stable to hydrolysis. Indeed, it is estimated that the half-life of hydrolytic cleavage of the DNA backbone is about 30,000,000 years.[17] Nonetheless, it is estimated that 10,000 nucleotide glycosidic bonds spontaneously hydrolyze per cell each day.[18] This dramatic increase in rate is due to deoxyribozymes, which are short segments of DNA capable of catalyzing the hydrolytic cleavage of RNA phosphodiesters. This catalytic reaction allows the hydrolysis of the glycosidic bonds holding guanine residues to the DNA backbone[19] and the oxidative cleavage of DNA.[20] The hydrolysis of a DNA strand is shown in Figure 15.11,[21] with cleavage of the phosphate linkage. The hydrolysis is catalyzed by deoxyribozymes that use Zn^{2+} and Mn^{2+} to accelerate the reaction. Note that 40 nt is a peptide substrate segment from a 40-nucleotide stretch of randomized DNA sequence.[21]

[17] Schroeder, G.K.; Lad, C.; Wyman, P.; Williams, N.H.; Wolfenden, R. *Proc. Nat. Acad. Sci. USA* **2006**, *103*, 4052.
[18] Lindahl, T.; Nyberg, B. *Biochemistry* **1972**, *11*, 3610.
[19] Sheppard, T.L.; Ordoukhanian, P.; Joyce, G.F. *Proc. Nat. Acad. Sci. USA* **2000**, *97*, 7802.
[20] Chandra, M.; Sachdeva, A.; Silverman, S.K. *Nat. Chem. Biol.* **2009**, **5**, 718.
[21] Fekry, M.I.; Gates, K.S. *Nat. Chem. Biol.* **2009**, *5*, 710.

15.6 RNA-MEDIATED PROGRAMMABLE DNA CLEAVAGE

Genome engineering has led to endonucleases that are involved in bacterial adaptive immunity and can be reprogrammed with customizable small, noncoding RNAs. *Clustered regularly interspaced short palindromic repeats* (CRISPR) are essential components of nucleic-acid-based adaptive immune systems, widespread in bacteria.[22] Note: a palindromic sequence on double-stranded DNA or RNA has the 5′ to 3′ sequence on one strand match the 5′ to 3′ sequence on the complementary strand that forms a double helix. CRISPR-mediated immune systems have small RNAs that are sequence-specific and detect and silence foreign nucleic acids. Note: gene silencing is the regulation of gene expression in a cell to prevent the expression of a certain gene. In other words, an RNA-based adaptive immune system to target and destroy genetic parasites by integrating short fragments of foreign nucleic acid into the host.

The Cas9-crRNA ribonucleoprotein complex mediates specific DNA cleavage for CRISPR-encoded immunity in bacteria.[23] Note that Cas9 is *CRISPR-associated protein 9*, an RNA-guided *DNA endonuclease* enzyme, and crRNA is a CRISPR RNA. Cas9 is a DNA endonuclease guided by two RNAs. The Cas9 HNH (histidine–asparagine–histidine) nuclease domain cleaves the complementary strand, whereas the Cas9 RuvC-like domain cleaves the noncomplementary strand.[23] This system provides adaptive immunity against viruses and plasmids. The Cas9 endonuclease can be programmed with guide RNA engineered as a single transcript to target and cleave any double-stranded DNA sequence of interest.[23] As shown in Figure 15.12,[24] Cas9 DNA cleavage specificity can be reprogrammed to target a green fluorescent protein gene using a chimeric RNA fusion of the crRNA 3′-end with the tracrRNA 5′-end (tracrRNA is trans-activating crRNA). In Figure 15.12, the Cas9 protein (marked in light blue) combines with crRNA (marked in red) and tracrRNA (marked in orange) to form a ribonucleoprotein interfering complex. This complex coordinates with a complementary sequence in the target DNA (marked in dark blue). The complementary and noncomplementary strands are nicked, respectively, 3 nt away from the PAM sequence.

15.7 RESTRICTION ENZYMES

A restriction enzyme recognizes a specific sequence of nucleotides and cleaves DNA into fragments at or near that recognition site (restriction site). Restriction enzymes are endonuclease enzymes. To cleave DNA, all restriction enzymes must cleave DNA by breaking each strand at a

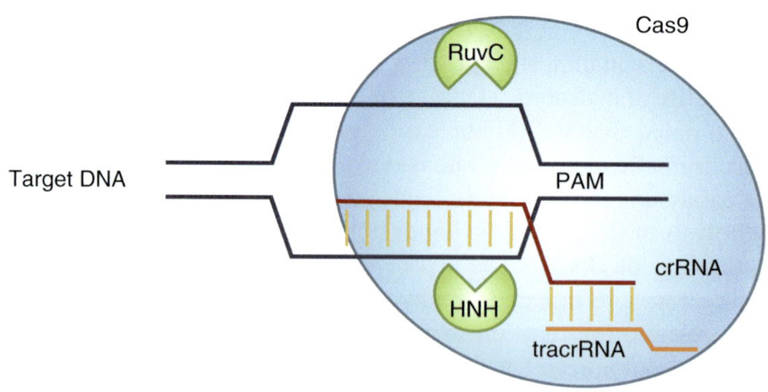

FIGURE 15.12 crRNA-Guided DNA cleavage by Cas9. (Reprinted by permission from Springer Nature Customer Service Centre GmbH: Springer Nature. Barrangou, R. *Nat. Biotechnol.* **2012**, *30,* 836–838. Copyright 2012. Figure 1a, p. 837.)

[22] Wiedenheft, B.; Sternberg, S.H.; Doudna, J.A. *Nature* **2012**, *482,* 331.
[23] Jinek, M.; Chylinski, K.; Fonfara, I.; Hauer, M.; Doudna, J.A.; Charpentier, E. *Science* **2012**, *337,* 816.
[24] Barrangou, R. *Nat. Biotechnol.* **2012**, *30,* 836.

carbohydrate-phosphate linage of the DNA double helix.[25] The recognition sequences can be classified by the number of bases in their recognition site, usually between 4 and 8 bases. There are five types of restriction enzymes that differ in their structure, whether they cut a DNA substrate at a recognition site, or whether the recognition site is separate from the cleavage site. Type I enzymes (EC 3.1.21.3) cleave at sites remote from a recognition site. Type II enzymes (EC 3.1.21.4) cleave within or at short specific distances from a recognition site. Type III enzymes (EC 3.1.21.5) cleave at sites a short distance from a recognition site; they require ATP but do not hydrolyze it. Type IV enzymes modify DNA with different substituents. Type V restriction enzymes (e.g., *cas9-gRNA*) utilize guide RNAs to target specific non-palindromic sequences found in invading organisms.

Most restriction endonucleases use Mg^{2+} to hydrolyze phosphodiester bonds at specific DNA sites. Under certain conditions, the terminal 3′-OH of one DNA strand can attack the target phosphodiester bond in the other strand to create a DNA hairpin. In the presence of Mg^{2+}, type IIS restriction *endonucleases (REases)* hydrolyze specific phosphodiester bonds in DNA to yield 5′-phosphate and 3′-hydroxyl termini. An example of a type II *endonuclease* is *BfiI*, a metal-independent restriction enzyme from the *phospholipase D* superfamily.[26] The *BfiI endonuclease* cuts DNA to give staggered ends with 1-nt 3′-extensions, and it cleaves DNA downstream of its recognition sequence:

$$
\begin{array}{c}
5' - \text{A-C-T-G-G-G-n-n-n-n-n} \downarrow \text{n} - 3' \\
3' - \text{T-G-A-C-C-C-n-n-n-n} \uparrow \text{n-n} - 5'
\end{array}
$$

[where n is any nucleotide (nt) and the site of cleavage is marked by an arrow.][27]

Like other *phospholipase D* enzymes, *BfiI* catalyzes not only phosphodiester hydrolysis but also transesterification. During DNA cleavage in the presence of ethanol or glycerol, *BfiI* generates transesterification products with an alcohol moiety covalently attached to the 50-terminal DNA phosphate. The putative mechanism is shown in Figure 15.13,[27] where transesterification leads to a hairpin product (see Figure 15.14),[26] and hydrolysis leads to the cleavage products. For the reaction of *BfiI* on 30T/30A DNA, with thymidine, the two strands of the duplex are marked as horizontal blocks, with the recognition sequence highlighted in grey; the chemical natures of the scissile phosphodiester bonds are given in full. The *BfiI* endonuclease is shown as a C-terminal DNA-binding domain connected by a linker to the N-terminal catalytic domain. After cutting the bottom strand, *BfiI*-N makes a nucleophilic attack on phosphate, forms a covalent intermediate, and the 30-leaving group is released. Nucleophilic attack of the 3′-OH group on the covalent intermediate joins the thymidine to the 3′-terminus and displaces the enzyme from the DNA. The active-site residue is His^{105}, which probably attacks the phosphorus at the scissile bond first to form a covalent intermediate and release the 3′-DNA.[26]

15.8 CHEMICAL SYNTHESIS OF NUCLEOTIDES

There are several synthetic methods to prepare purines and pyrimidines. The examples shown here date from the early work in this area, and some are more than 100 years old. These examples were chosen for two reasons. The first is to show that relatively well-known organic reactions can be used to prepare the heterocycles, and that the chemistry is not difficult or unusual. The second reason is to reinforce the idea that these heterocyclic bases are readily available.

The synthesis of thymine is shown in Figure 15.15, and it can be prepared from ethyl propionate by an initial reaction with sodium ethoxide.[28] The condensation reaction gives diethyl β-methyloxosuccinate, and catalytic hydrogenation reduces the ketone unit to an alcohol in diethyl

[25] (a) Arber, W.; Linn, S. *Ann. Rev. Biochem.* **1969**, *38*, 467; (b) Krüger, D.H.; Bickle, T.A. *Microbiol. Rev.* **1983**, *47*, 345.
[26] Sasnauskas, G.; Connolly, B.A.; Halford, S.E.; Siksnys, V. *Proc. Nat. Acad. Sci. USA* **2007**, *104*, 2115.
[27] Sasnauskas, G.; Connolly, B.A.; Halford, S.E.; Siksnys, V. *Nucleic Acids Res.* **2008**, *36*, 3969.
[28] Henderson, R.B.; Fink, R.M.; Fink, K. *J. Amer. Chem. Soc.* **1955**, *77*, 6381.

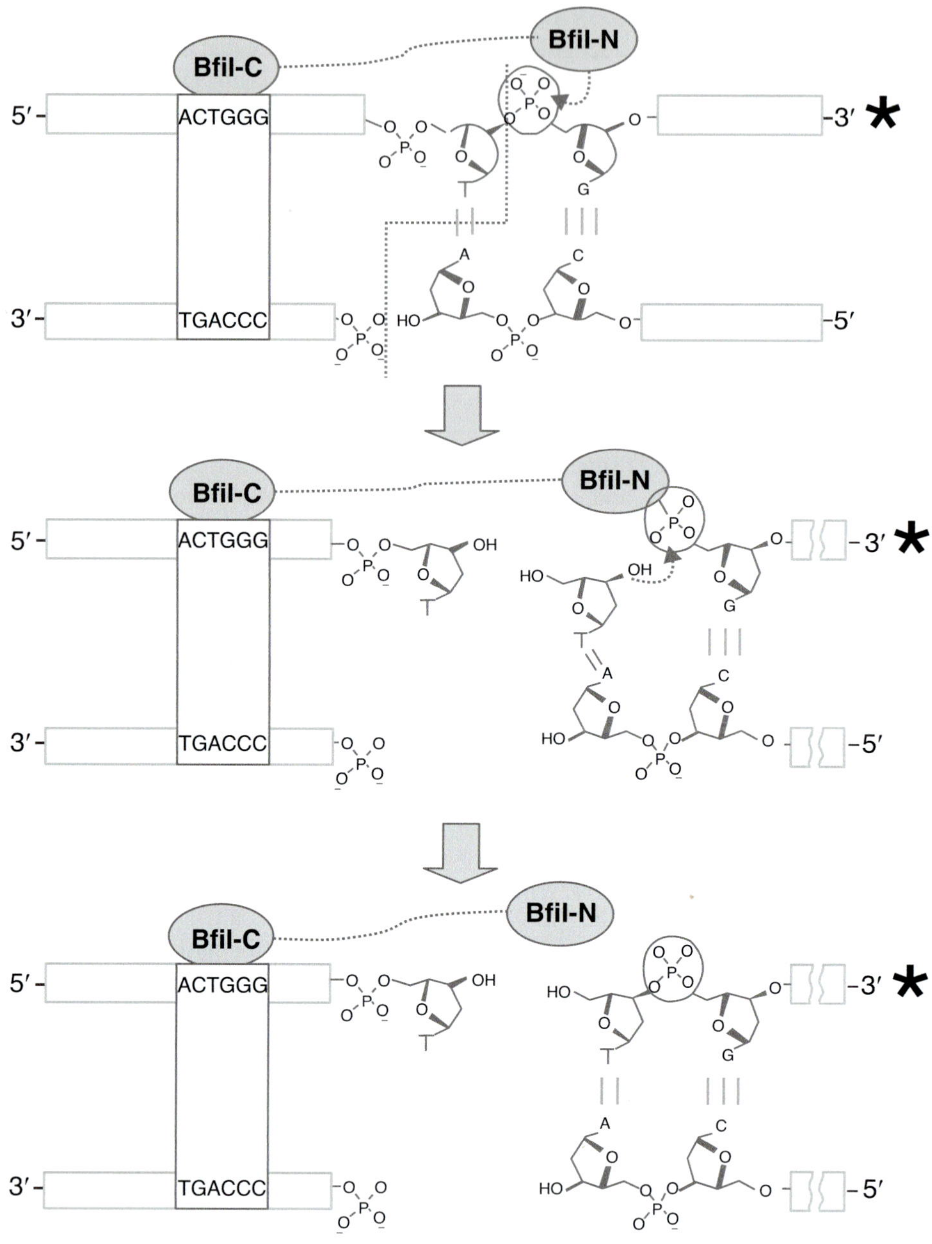

FIGURE 15.13 Mechanism of *BfiI*. (Sasnauskas, G.; Connolly, B.A.; Halford, S.E.; Siksnys, V. *Nucleic Acids Res.* **2008**, *36*, 3969–3977. Reproduced with permission from Oxford University Press. Figure 3, p. 3973.)

2-methyl malate. Saponification of the ester groups, followed by heating with fuming sulfuric acid, leads to cyclization, and neutralization with Amberlite-IRA 400 (hydroxide form) allows the isolation of thymine.

Uracil can be prepared by the reaction of malic acid and urea with fuming sulfuric acid, as shown in Figure 15.16. Loss of CO, CO_2 and SO_3 from the reaction medium leads to the formation of uracil.[29] A synthesis of cytosine involves reacting 1,1,3,3-tetraethoxypropane with hydroxylamine to give isoxazole, and heating with diethyl sulfate and NaOH in aqueous ethanol gives

[29] Davidson, D.; Baudisch, O. *J. Amer. Chem. Soc.* **1926**, *48*, 2379.

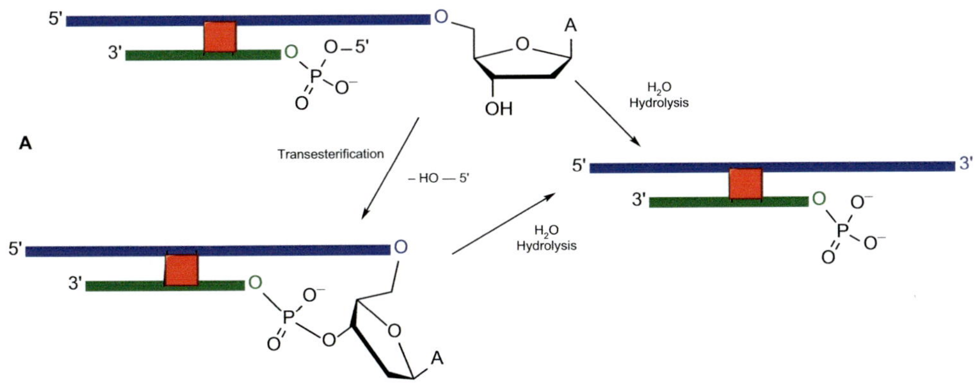

FIGURE 15.14 DNA hairpin formation. (Sasnauskas, G.; Connolly, B.A.; Halford, S.E.; Siksnys, V. *Proc. Nat. Acad. Sci.* **2007**, *104*, 2115. Copyright (2007) National Academy of Sciences, U.S.A.)

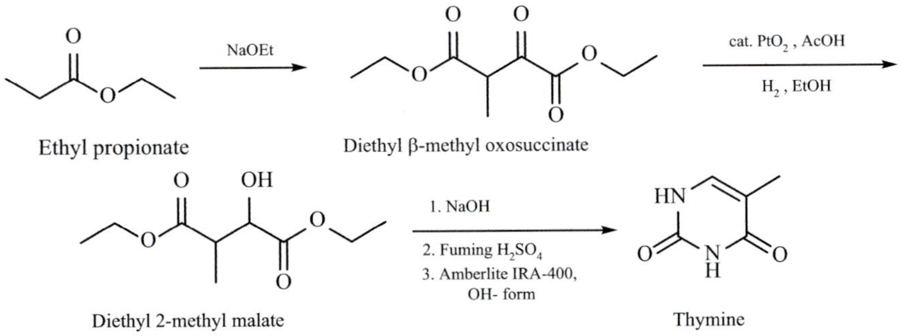

FIGURE 15.15 Synthesis of thymine.

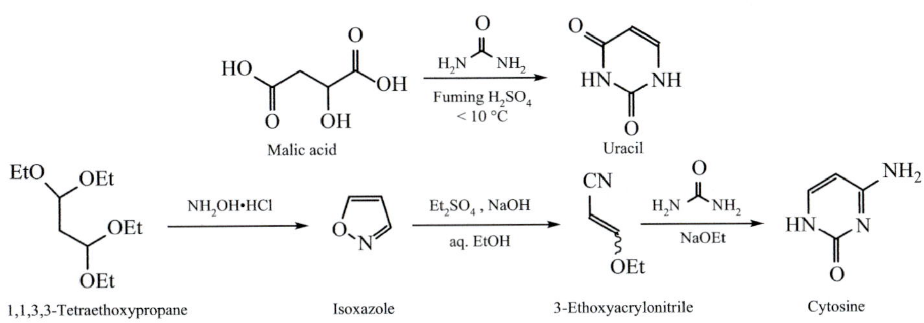

FIGURE 15.16 Syntheses of uracil and of cytosine.

3-ethoxyacrylonitrile. When this conjugated nitrile is treated with urea and NaOEt, cytosine is produced, as shown in Figure 15.16.[30]

Purines can be prepared from relatively simple precursors. Adenine is relatively straightforward to prepare, albeit in low yield. Heating 2,3-diaminomaleonitrile (also called HCN tetramer) in molten ammonium formate (110°C) gives an 18% yield of adenine (see Figure 15.17).[31] A synthesis of guanine is a bit more difficult, but still relatively straightforward. Traube's synthesis of guanine by the basic condensation of ethyl 2-cyanoacetate and guanidine is an example, giving

[30] Tarsio, P.J.; Nicholl, L. *J. Org. Chem.* **1957**, *22*, 192.
[31] Hill, A.; Orgel, L.E. *Origins Life Evol. Biosphere* **2002**, *32*, 99–102.

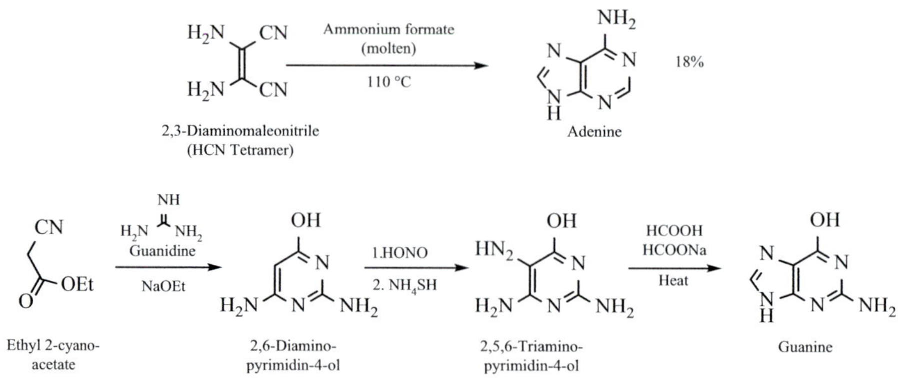

FIGURE 15.17 Syntheses of adenine and of guanine.

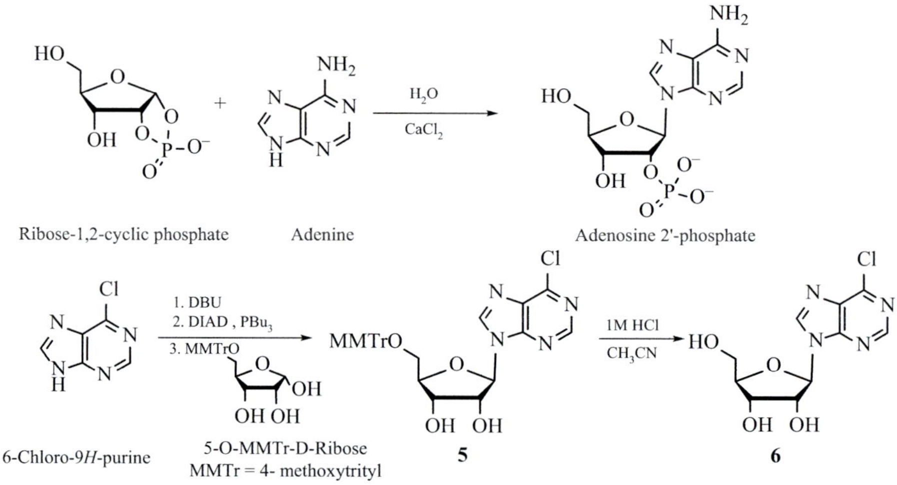

FIGURE 15.18 Syntheses of adenosine 2′-phosphate and of 2-(6-chloro-9*H*-purin-9-yl)-5-(hydroxymethyl) tetrahydrofuran-3,4-diol, **6**.

2,6-diaminopyrimidin-4-ol, as shown in Figure 15.17.[32] Subsequent treatment with HONO (nitrous acid) generates an oxime, and reduction with NH_4SH gives 2,5,6-triaminopyrimidin-4-ol. Heating with formic acid and sodium formate gives guanine.

With purine and pyrimidine bases in hand, several techniques exist to attach the heterocycle to a ribose molecule, in which a nitrogen of the heterocycle acts as a nucleophile and the anomeric carbon of the ribose is electrophilic.

Adenosine 2′-monophosphate was prepared by the reaction of adenine in an aqueous solution with ribose-1,2-cyclic phosphate in the presence of pre-dried $CaCl_2$ in water (see Figure 15.18).[33] Another coupling method involves the reaction of 6-chloro-pH-purine with diazabicycloundec-9-ene (DBU), followed by diisopropylazodicarboxylate (DIAD) and tributylphosphine (Mitsunobu conditions),[34] which is then coupled with 5-*O*-(4-methoxytrityl)-D-ribose to give **5**, as shown in Figure 15.18.[35] Deprotection with aqueous HCl gives **6**.

[32] Traube, W. *Ber. Deutsch. Chemisch. Gesell.* **1900**, *33,* 1371.
[33] Kim, H.-J.; Benner, S.A. *Proc. Nat. Acad. Sci. USA* **2017**, *114*, 11315.
[34] Mitsunobu, O. *Synthesis* **1981**, 1.
[35] Downey, A.M.; Richter, C.; Pohl, R.; Mahrwald, R.; Hocek, M. *Org. Lett.* **2015**, *17,* 4604.

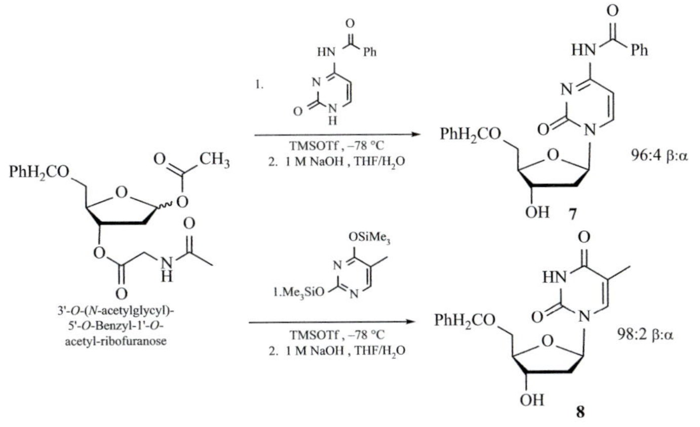

FIGURE 15.19 Syntheses of cytosine derivative (**7**) and thymine derivative (**8**).

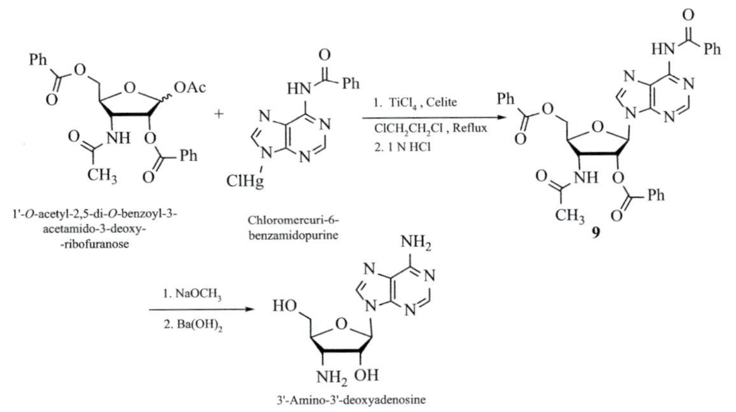

FIGURE 15.20 Synthesis of 3′-amino-3′-deoxyadenosine.

A useful leaving group for coupling the heterocyclic base to a ribose derivative is acetate. When 3′-*O*-(*N*-acetylglycyl)-5′-*O*-Benzyl-1′-*O*-acetyl-ribofuranose was reacted with N^4-benzoylcytosine and trimethylsilyl triflate (triflate is trifluoromethanesulfonate) at low temperatures, coupling gave the nucleoside (see Figure 15.19).[36] In this case, the *N*-acetylglycine protecting group at the 3′-hydroxyl group provided assistance by displacing the acetate intramolecularly to form a cyclic intermediate; that reacted with the cytosine derivative to give primarily the β-anomer **7**, in accord with *Baker's 1,2-trans rule.*[37] Deprotection with sodium hydroxide gave the protected nucleoside. A similar reaction with silylated thymidine (5-methyl-2,4-bis((trimethylsilyl)oxy)pyrimidine) gave primarily the β-anomeric nucleoside, and reaction with aqueous NaOH not only deprotected the 3′-hydroxyl but also converted the silylated thymine to the thymine nucleoside **8** (Figure 15.19).

An early method for preparing nucleoside derivatives used the mercuric derivatives of heterocyclic bases, which were coupled to ribose derivatives using a transition metal additive. The coupling reaction of 1′-*O*-acetyl-2,5-di-*O*-benzoyl-3-acetamido-3-deoxy-D-ribofuranose and chloromercuri-6-benzamidopurine,[38] in the presence of $TiCl_4$ for example, gave **9** (Figure 15.20).[39] Deprotection

[36] Liu, Z.; Li, D.; Yin, B.; Zhang, J. *Tetrahedron Lett.* **2010**, *51*, 240.

[37] Baker, B.R.; Joseph, J.P.; Schaub, R.E.; Williams, J.H. *J. Org. Chem.* **1954**, *19*, 1786.

[38] (a) Davoll, J.; Lowy, B.A. *J. Amer. Chem. Soc.* **1951**, *73*, 1650; (b) See Sundaralingam, M.; Carrabine, J.A. *Biochemistry* ***1971***, *10*, 292.

[39] Baker, B.R.; Schaub, R.E.; Kissman, H.M. *J. Amer. Chem. Soc.* **1955**, *77*, 5911.

using first sodium methoxide and then barium hydroxide gave 3′-amino-3′-deoxyadenosine. Initial formation of the β-nucleoside with a C1–C2 trans-configuration isomerizes to the product, the α-nucleoside with a C1–C2 cis-configuration, by interaction with titanium chloride[39] (the Baker 1,2-trans rule).[39] Note that several palladium-catalyzed coupling reactions have been developed.[40]

HOMEWORK

15.1. Draw the structures of (a) 2-amino-2-deoxy-α-D-altropyranose, (b) 3-amino-3,6-dideox y-β-D-allopyranose, and (c) 3,4-dideoxy-α-D-talopyranose).
15.2. Draw the structures of adenine and guanine.
15.3. Draw the nucleoside of cytidine and of 2′-deoxyadenosine.
15.4. Draw the structures of the 3′-monophosphate of uridine and the 5′diphosphate of thymidine.
15.5. Draw the shorthand structures of uracil, 2′-deoxycytosine, and adenosine.
15.6. What is the major groove of a DNA strand?
15.7. Why is the major groove wider than the minor groove in DNA strands?
15.8. Draw a generic structure of a tRNA and identify the region where the three anticodons are found.
15.9. What is a ribozyme?
15.10. What is a restriction enzyme?

[40] Agrofoglio, L.A.; Gillaizeau, I.; Saito, Y. *Chem. Rev.* **2003**, *103*, 1875.

16 Answers to Homework Problems

CHAPTER 1

1.1. If two dipeptide units are brought into proximity, the oxygen of one carbonyl can form a hydrogen bond with the proton on the amide nitrogen of the second dipeptide. For hydrogen bonding to occur, however, one dipeptide must be in an anti-orientation.

1.2. In a long-chain peptide composed of L-amino acid residues, intramolecular hydrogen bonding leads to a right-handed helix (called an α-helix), in which the hydrogen atom on the amide nitrogen is hydrogen-bonded to the oxygen of the carbonyl on the fourth amino acid residue. A random coil structure does not assume a regular conformation because hydrogen bonds are not easily formed. The β-pleated sheet involves intermolecular hydrogen bonds between two different peptide chains rather than intramolecular hydrogen bonds within a single peptide chain.

1.3. Four are alanine, valine, leucine, or isoleucine. Hydrophilic amino acids include arginine, lysine, asparagine, histidine, proline, and aspartic acid.

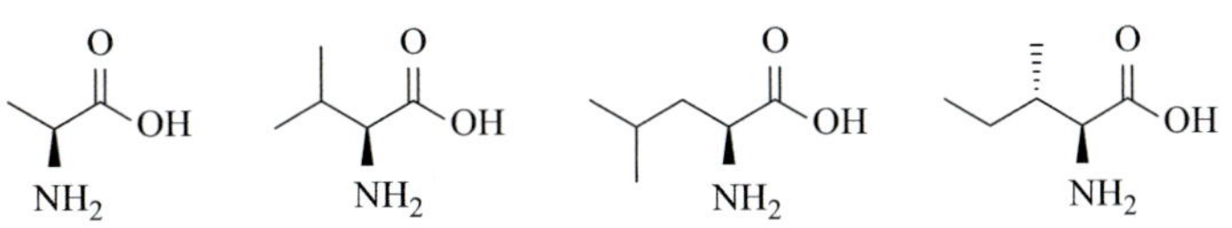

alanine valine leucine isoleucine

1.4. 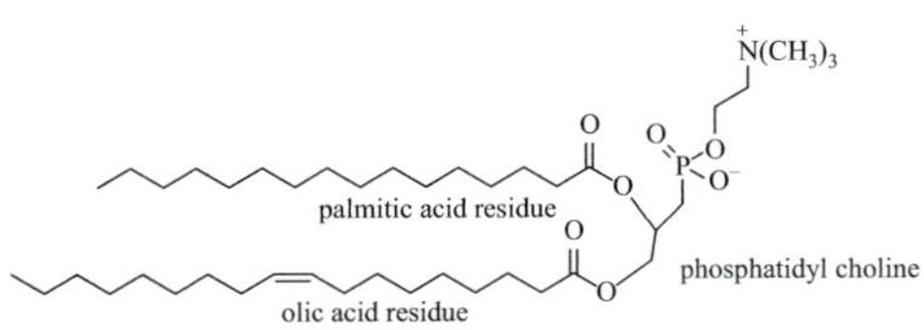

a phospholipid

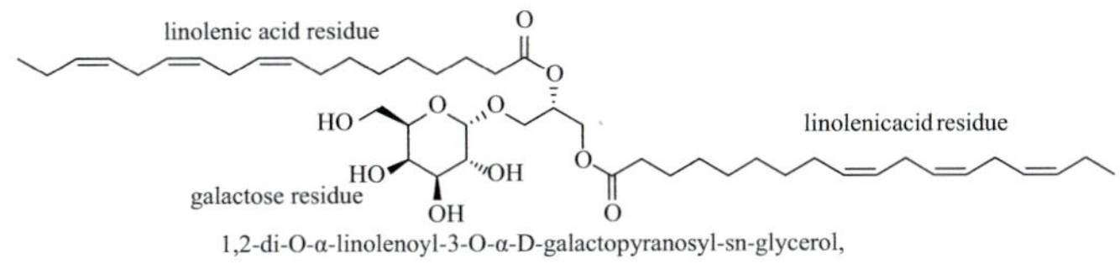

1,2-di-O-α-linolenoyl-3-O-α-D-galactopyranosyl-sn-glycerol,

a glycoplipid

1.5. The hydration layer is a layer of water molecules that covers a biomolecule.

1.6. An increase in the entropy of water molecules increases the aggregation of nonpolar side chains in the interior of a protein.

1.7. If an excess of sodium acetate is added, the equilibrium shifts to the left, mitigating the excess acidity due to the formation of H_3O. In other words, the excess sodium acetate acts as a buffer by reacting with the hydronium ion to prevent the solution from becoming too acidic.

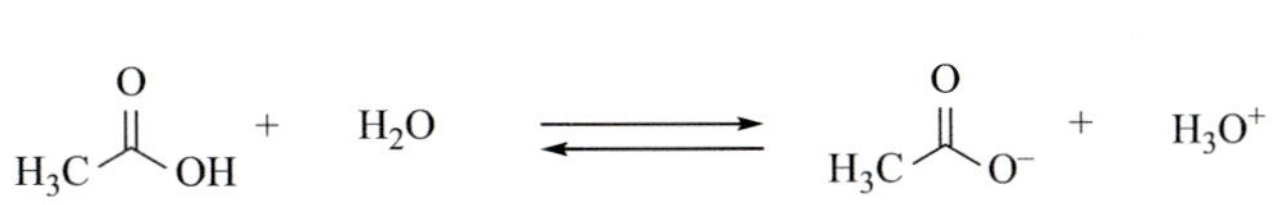

DOI: 10.1201/9781003605157-16

1.8. Acetic acid reacts with a base to form the resonance-stabilized acetate ion as its conjugate base. Ethanol reacts with a base to form the ethoxide anion, which is not resonance-stabilized. The greater stability of the resonance-stabilized acetate ion shifts the acid–base equilibrium to ward acetate, so the p*K*a of acetic acid is smaller, indicative of a more acidic compound.

1.9. 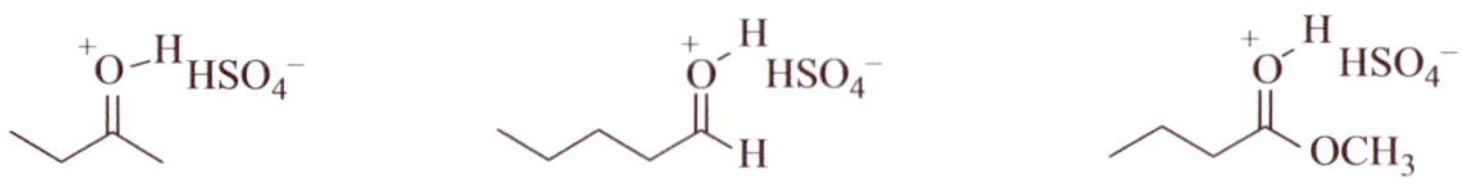

1.10. To calculate p*I*, use $pI = \frac{pK_{a_1} + pK_{a_2}}{2}$: glycine=(2.34+9.60)/2=5.97; valine=2.32+9.62/2=5.97; cysteine=1.96+10.28/2=6.12.

1.11. Give the complete mechanism for the acid-catalyzed esterification of butanoic acid and propan-2-ol.

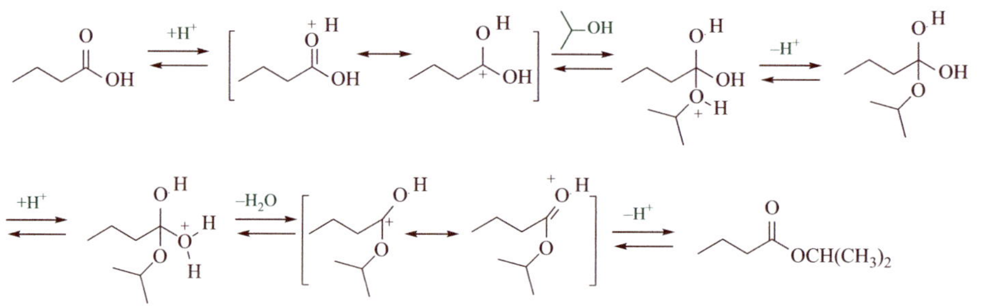

1.12. The structure is Me_3N+$CHCH_2OH$. Choline is an essential nutrient present in some foods and is a dietary supplement. Choline is a source of methyl groups in myriad metabolic sequences. Choline is required for phosphatidylcholine and sphingomyelin, two major phospholipids vital for cell membranes.

1.13. 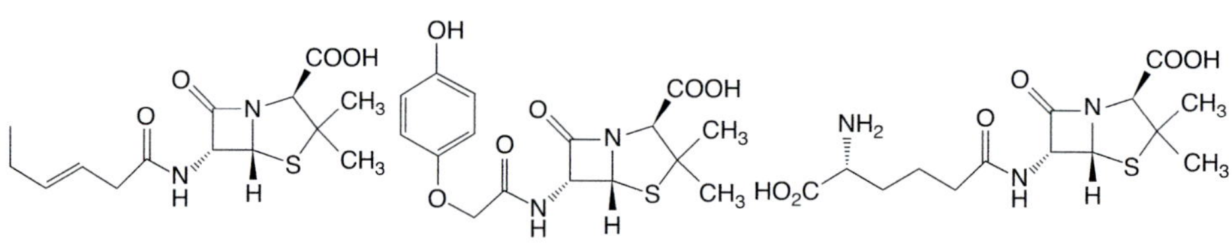

Penicillin F
2-Pentenylpenicillin

Penicillin F
p-Hydroxybenzylpenicillin

Penicillin N

CHAPTER 2

2.1. 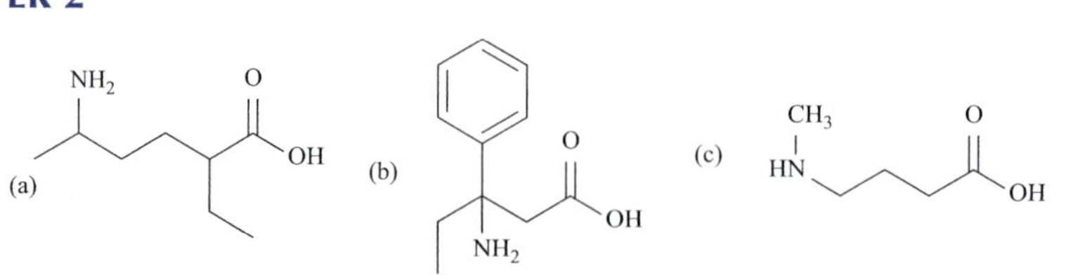

5-amino-2-ethylhexanoic acid

3-amino-3-phenylpentanoic acid

N-methyl 4-aminobutanoic acid

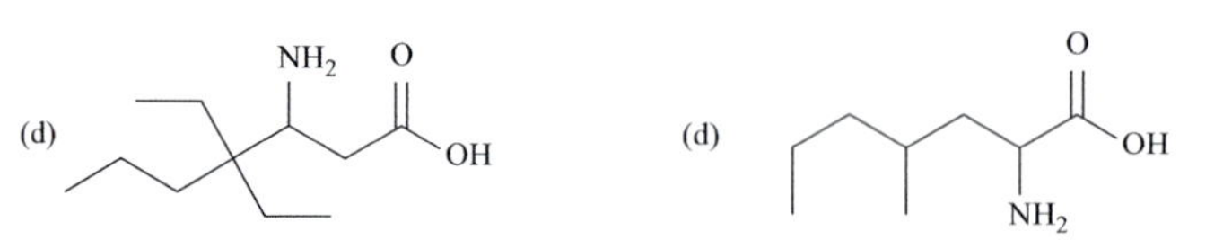

3-amino-4,4-diethylheptanoic acid

2-amino-4-(2-methylethyl)pentanoic acid

2.2.

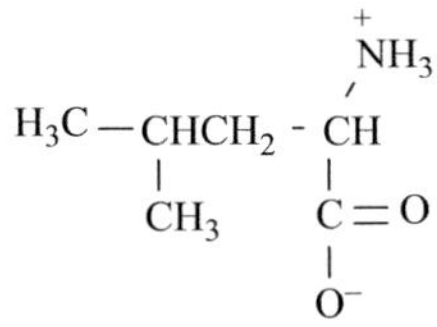

2.3. To calculate pI, use $pI = \dfrac{pK_{a_1} + pK_{a_2}}{2}$: tyrosine=(2.20 +9.11)/2=5.655; asparagine=(2.20+8.80)/2=5.5.

2.4. Using the one-letter codes, L is leucine, F is phenylalanine, D is aspartic acid, and W is tryptophan.

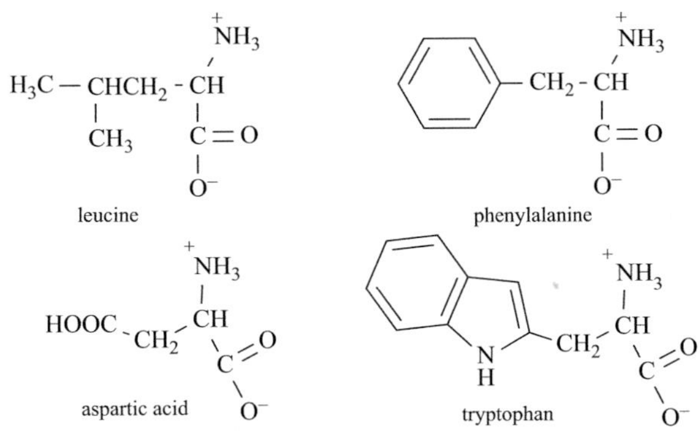

2.5. The enzyme *3-phosphoglycerate dehydrogenase* mediates the oxidation of an alcohol unit to a carbonyl in a ketone unit. The enzyme *3-phosphoserine aminotransferase* converts a ketone unit to an amine unit, so it mediates a variation of an acyl substitution reaction.

2.6. The final step to give L-lysine is a decarboxylation reaction, so the enzyme that mediates this transformation must be a decarboxylase rather than an epimerase.

2.7. It is formally a substitution reaction where an amino group replaces an alcohol unit.

2.8.

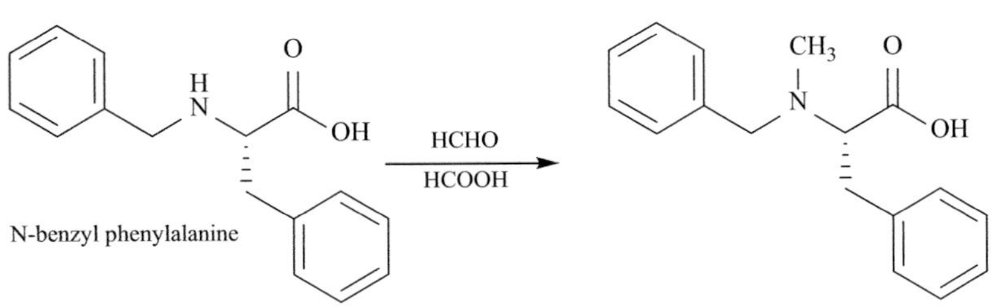

2.9.

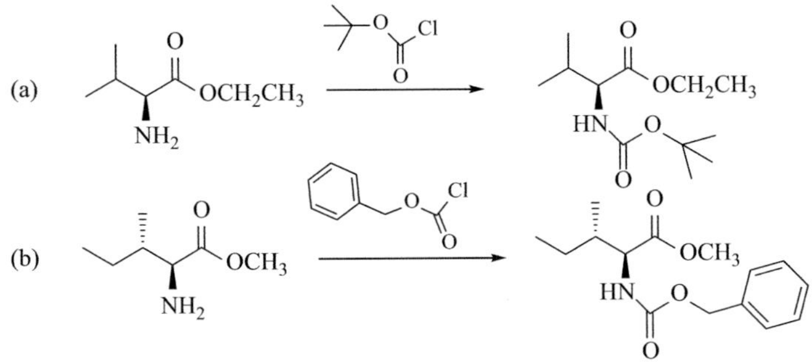

CHAPTER 3

3.1. The amide bonds within the peptide that connect the amino acid residues are called peptide bonds.

3.2. The amide unit of a peptide bond is essentially planar. The O=C–N bond of an amide shows two resonance structures, one with the carbonyl unit and the other with an alkoxy–iminium salt unit, so the C–N unit has a "partial double-bond character," which is normal for the C–N unit in simple amides. For this reason, the amide bond is essentially planar.

3.3. The amide bonds in a peptide are taken to have a planar geometry, so there is hindered rotation about the C–N bond due to resonance. The groups on carbon and nitrogen are typically opposite or trans to each other. The trans relationship is lower in energy than the arrangement of groups in conformations resulting from the other rotational angles.

3.4. The pitch of the α-helix is defined as the "rise" in the amino acid residue in the helix for each turn. Since there are 3.6 residues per turn and the rise per amino acid is measured to be ~ 1.5 Å (150 pm), the pitch of each turn is ~ 5.4 Å (540 pm).

3.5. In isoleucine, the side chain is a bulky sec-butyl group, whereas in alanine, the side chain is a much less bulky methyl group. Due to the increased steric hindrance, it is more difficult for isoleucine to form an a-helix when compared to alanine.

3.6. In a Richardson drawing of an enzyme, β-strands are shown as "thick" arrows, an α-helix is a spiral ribbon, and non-repetitive structures are shown as ropes

3.7. The *quaternary structure* of a polypeptide or protein refers to two or more peptide units (called subunits) that combine to form a protein structure. A peptide chain folds and coils into a very complex structure known as its *tertiary structure*. Using Figure 03.11, UUC codes for phenylalanine, AUA codes for isoleucine, CAC codes for histidine, CCA codes for proline, and AGA codes for arginine.

3.8. See *Journal of Agricultural and Food Chemistry* ***2004***, *52*, 385–406.

3.9.

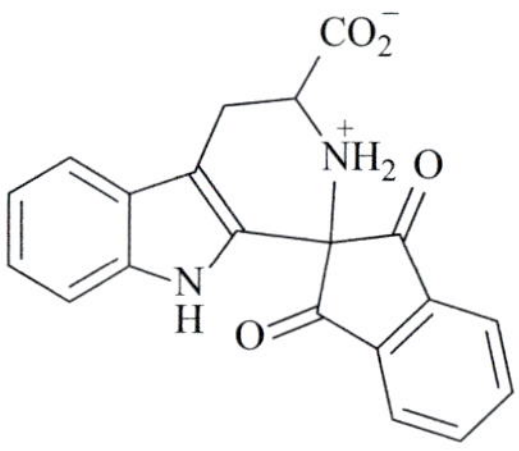

tryptophan ninhydrin adduct

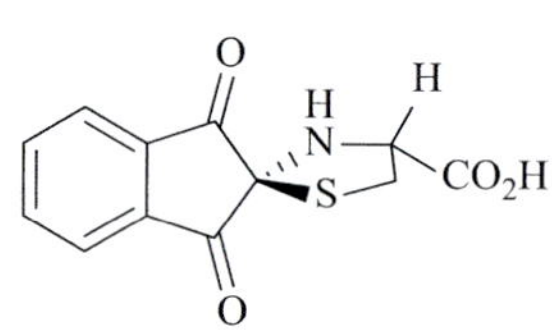

cysteine ninhydrin adduct

3.10.

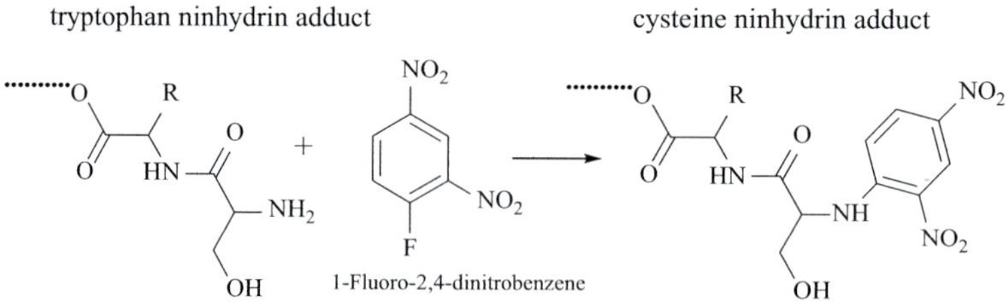

3.11.

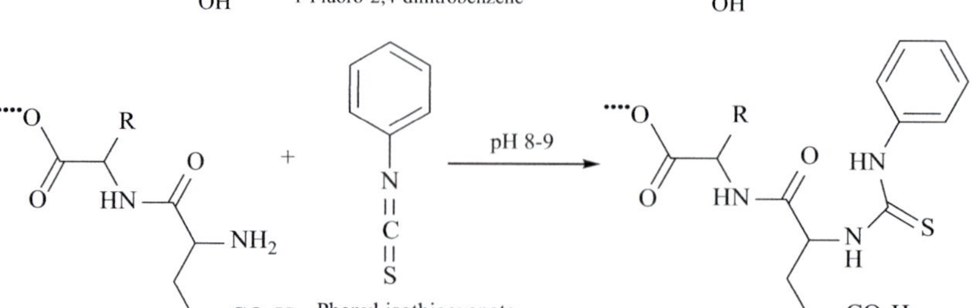

3.12.

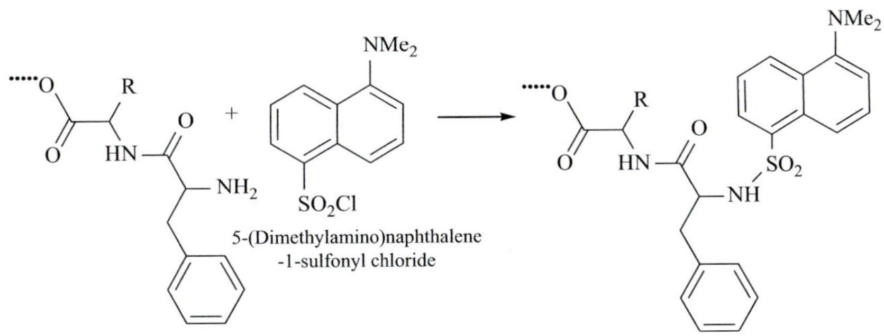

CHAPTER 4

4.1 The purpose of binding into a reactive complex is to weaken key substrate bonds to facilitate the desired chemical reactions.

4.2 For a zero-order reaction, the rate appears to be independent of the reactant concentration. The rate does not vary with increasing or decreasing reactant concentrations, so the rate of the reaction is equal to the rate constant, *k*, of that reaction.

4.3 A competitive inhibitor is typically a structural analog of the substrate that competes for the binding site of an enzyme. The competitive inhibitor forms a complex between the enzyme and inhibitor that effectively occupies the active site, which blocks the binding of the substrate.

4.4 Penicillin G (benzylpenicillin) is a competitive inhibitor for the last step in the formation of the cell wall, in which *glycopeptide transpeptidase* (EC 2.4.1.41) catalyzes the cross-linking reaction between peptidoglycan molecules. Penicillin, therefore, acts as an antibacterial agent that inhibits bacterial growth by inhibiting cell-wall synthesis.

4.5 Enzyme kinetics for a single substrate with a hyperbolic rate–substrate concentration relationship are described by the *Michaelis–Menten equation*, where *Km* is the *Michaelis constant*, [*S*] is the substrate concentration, *V* is the rate of the reaction, and *V*max is the reaction rate when the substrate concentration approaches infinity.

4.6 The B-vitamins that are cofactors include B1 (thiamine), B2 (riboflavin), and B3 (niacin).

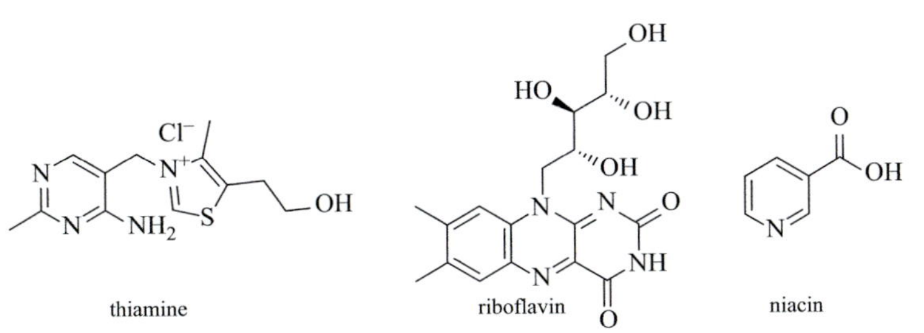

4.7 Globular proteins have a spherical shape because their tertiary structure has the polar, or hydrophilic, amino acids arranged on the outside and the nonpolar, or hydrophobic, amino acids on the inside of the three-dimensional shape. Due to this tertiary structure, globular proteins are typically soluble in water.

4.8 A prosthetic group is bound to an enzyme and acts as a cofactor. It is a non-polypeptide unit required for the biological function of some proteins. The group may be organic or inorganic, such as a metal ion, but is not composed of amino acids

4.9 The EC number is a numerical classification scheme based on the chemical reactions that are catalyzed

4.10 An oxidoreductase catalyzes oxidation/reduction reactions. A transferase catalyzes the transfer of a specific group from one molecule to another (donor to acceptor). A hydrolase catalyzes the hydrolysis of a chemical bond. An isomerase catalyzes the isomerization of one molecule to an isomer of that molecule.

4.11 An EC 2 enzyme is a transferase. An EC 3 enzyme is a hydrolase. An EC 5 enzyme is an isomerase. An EC 6 enzyme is a ligase.

4.12

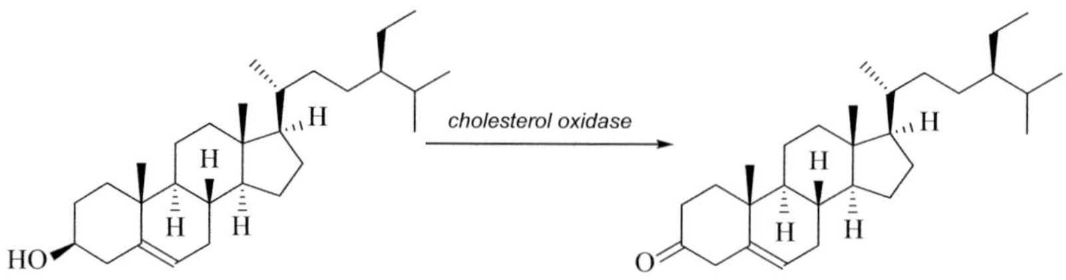

4.13 A catalyst is a substance that increases the rate of a chemical reaction without itself undergoing any permanent chemical change.

4.14 One half-life is the amount of time required for one-half of the starting material to be consumed. Assuming the starting material is present at 100% at the beginning of the reaction: 100%×0.5=50% remaining after one half-life; 50%×0.5=25%; 25%×0.5=12.5%; 12.5%×0.5=6.25%; 6.25×0.5=3.125%. If only 3.125% of the starting material remains, 96.875% has been consumed. Therefore, one additional half-life is required: 3.125%×0.5=1.5625%, where 98.375% of the starting material has been consumed. *Therefore, 6 half-lives are required.*

4.15 The reaction shown has two steps. First is the ionization of the carbocation, characterized by the rate constant K_{ion}. The second is substitution of the bromide ion with the highly reactive intermediate, the carbocation, and this reaction is characterized by the rate constant k_{sub}. The overall reaction rate is therefore $k_{rxn}=k_{ion}+k_{sub}$. Ionization is a very slow reaction, whereas the collision of the nucleophilic bromide with the highly reactive carbocation is a very fast reaction. Another way to phrase this statement is that *k*ion is very small and k_{sub} is very large. Since the first reaction is very slow and the second very fast, ignoring k_{sub} will have no significant effect on the overall rate, and the reaction is therefore classified as first order, which means that the rate depends only on the ionization of the substrate.

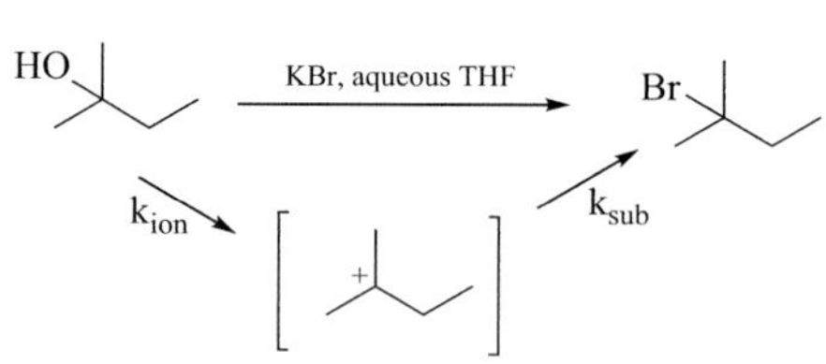

4.16

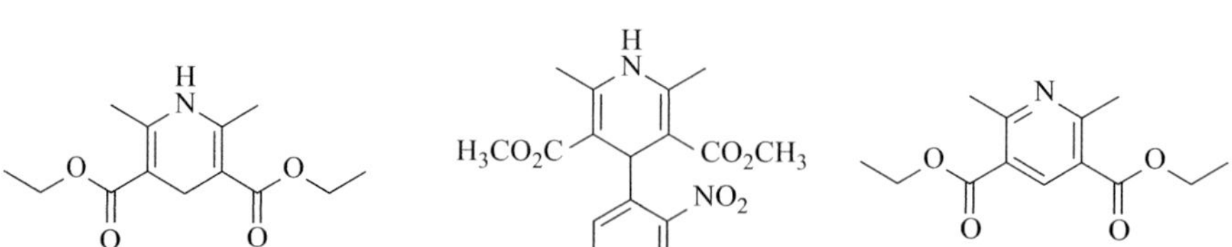

Hantzsch ester

nifedipine

oxidized form of Hantzsch ester

4.17 The Hantzsch ester is the reduced dihydropyridine form of the aromatic pyridine derivative. NAD+ is the oxidized pyridine form, whereas NADH is the reduced dihydropyridine form. In a transfer hydrogenation, hydrogen is transferred from the reduced dihydropyridine form to the substrate, which oxidizes the dihydropyridine to the pyridine form.

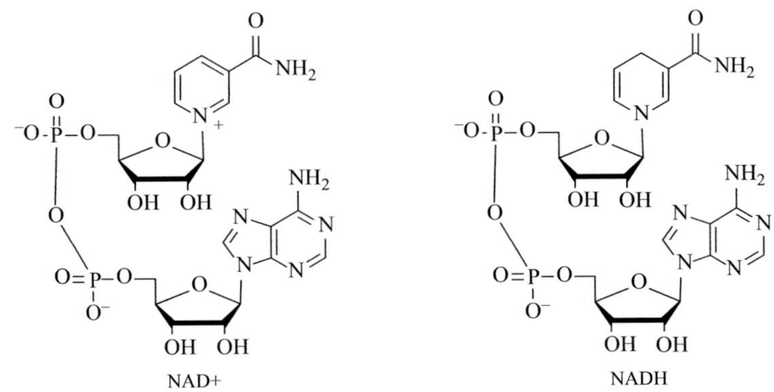

4.18 This transformation is a dehydrogenase reaction because the two hydrogen atoms marked in red are lost to form the carbonyl. This reaction is formally an oxidation.

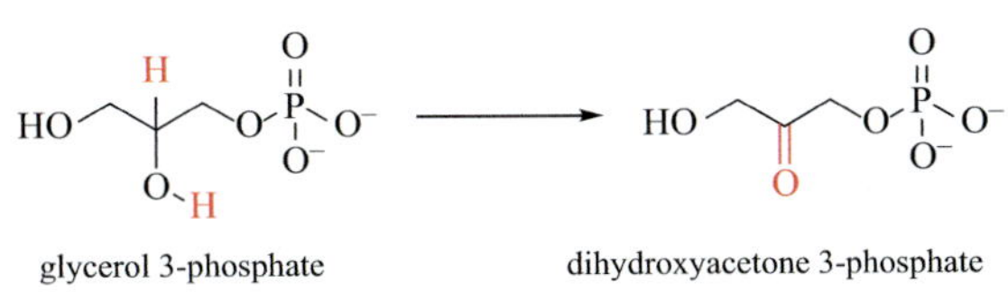

4.19

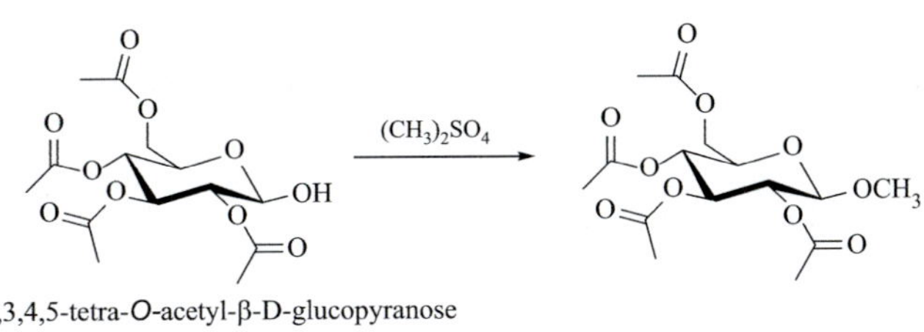

4.20

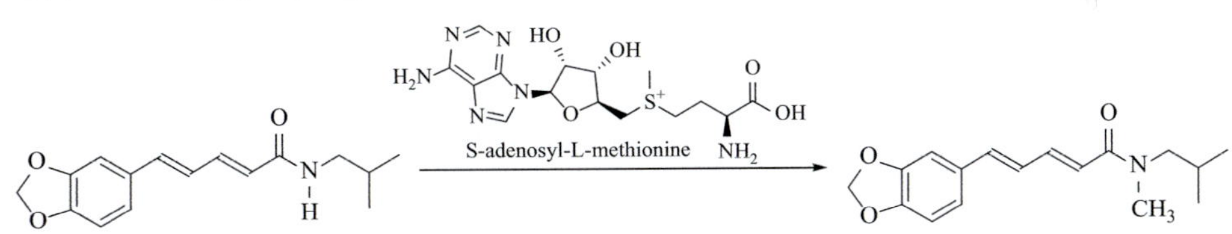

4.21

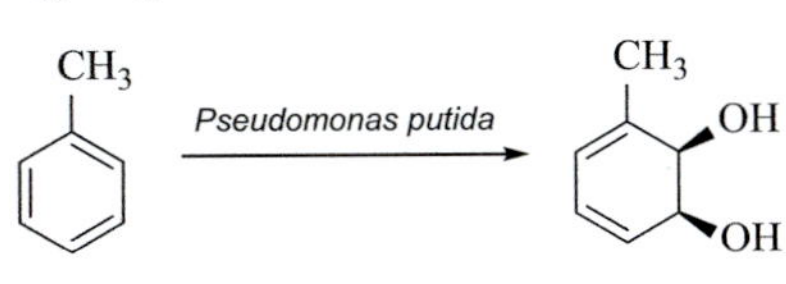

4.22

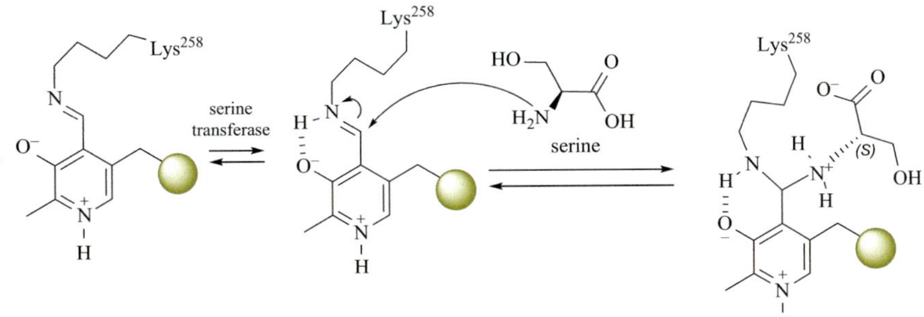

4.23 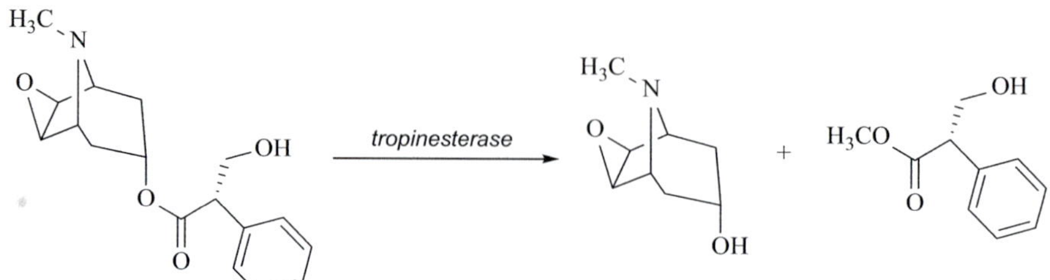

4.24 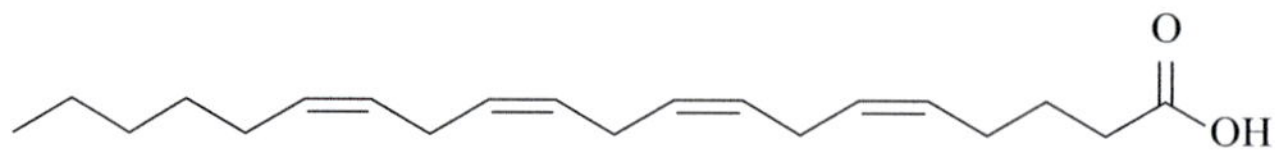

4.25 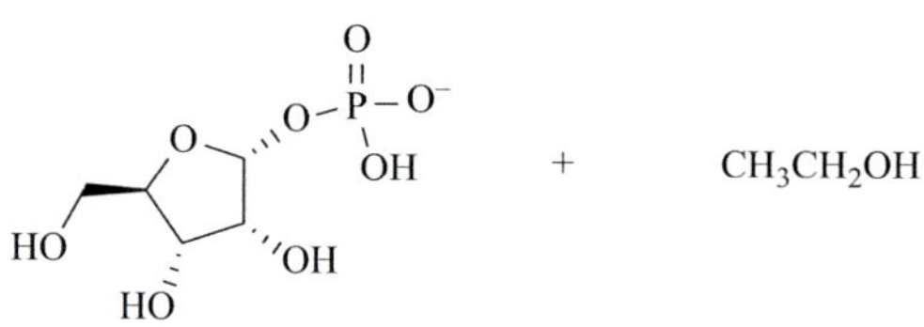

4.26 This is a DNA repair enzyme. The significance is that nicks in the DNA are repaired.

4.27 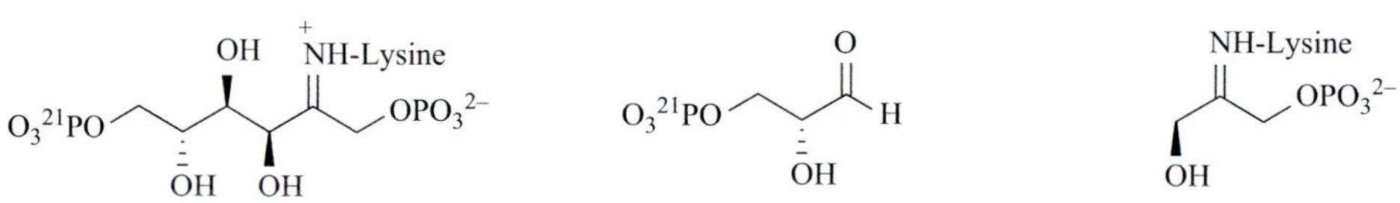

fructose   glyceraldehyde 3-phosphate

4.28 Note that 2,3-dimethylhexa-1,5-diene has two monosubstituted alkene units, whereas octan-2,6-diene has two disubstituted alkene units. The disubstituted alkenes are thermodynamically more stable than the monosubstituted alkene moieties, and this energy difference shifts the equilibrium toward octan-2,6-diene.

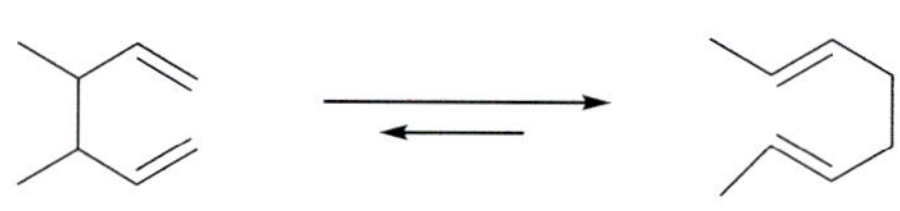

4.29 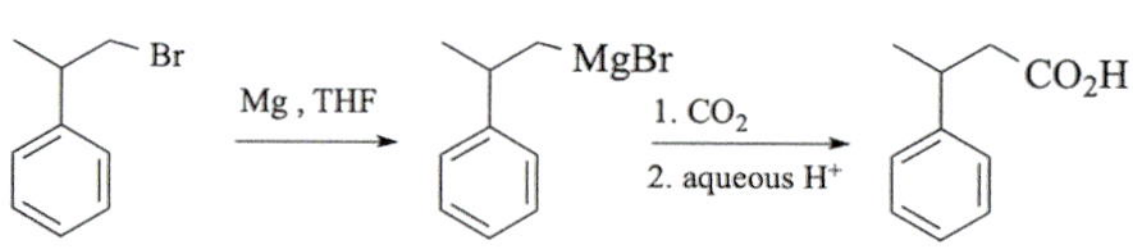

CHAPTER 5

5.1. This reaction is an S_N2 reaction, characterized by heterolytic bond cleavage.

5.2. 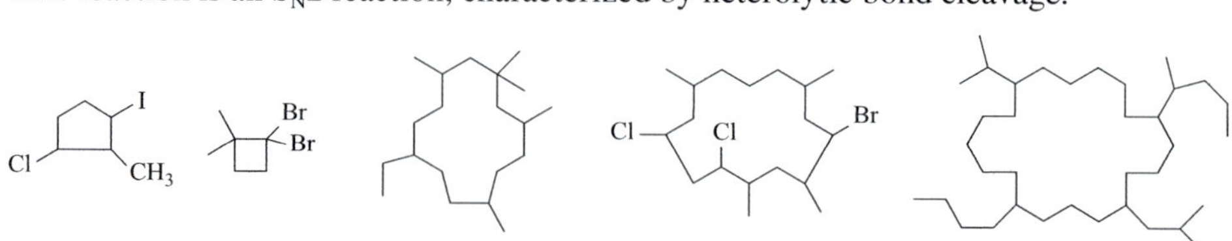

1-chloro-3-iodo-2-methylcyclopentane   1,1-dibromo-2,2-dimethylcyclobutane   6-ethyl-1,1,3,9,12-pentamethylcyclotridecane   1-bromo-5,7-dichloro-2,4,9,13-tetramethylcyclotetradecane   1-butyl-5-isobutyl-15-isopropyl-9-(pentan-2-yl)cycloicosane

5.3. 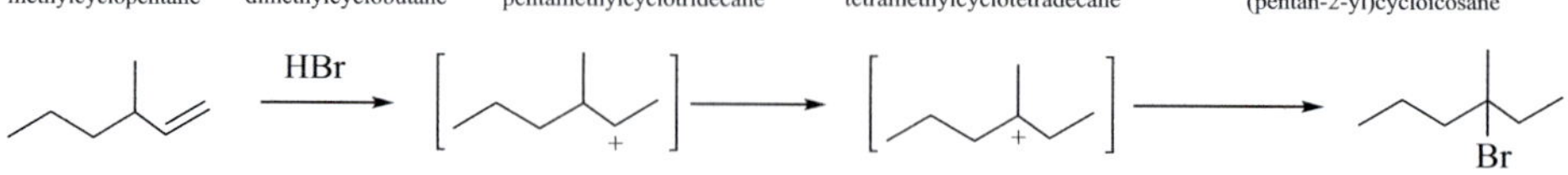

The intermediate is a carbocation. A 1,2-hydride shift to the more stable tertiary carbocation, followed by trapping of the bromide ion, gives the product.

5.4. Alcohols, thiols, amines, and carboxylic acids are all capable of hydrogen bonding.

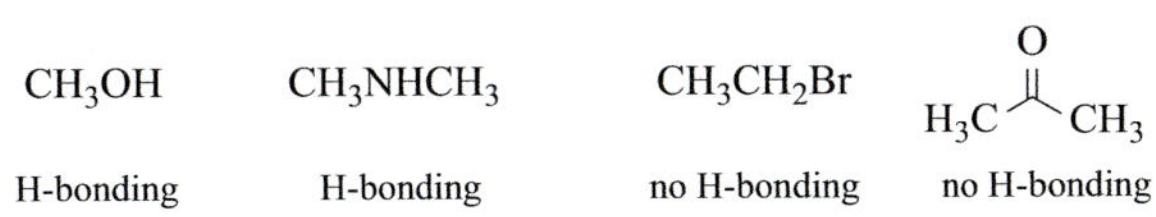

5.5. The conformation with the fewest axial groups has the least amount of A-strain and is therefore lower in energy, and it is the major chair conformation.

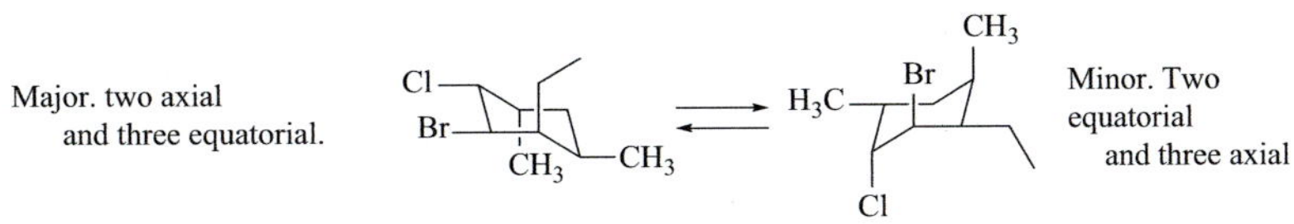

5.6.

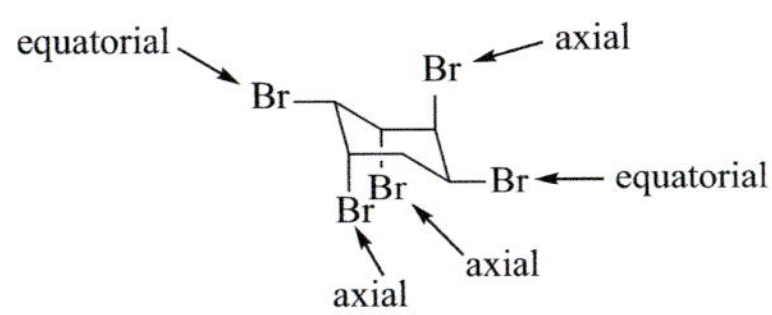

5.7.

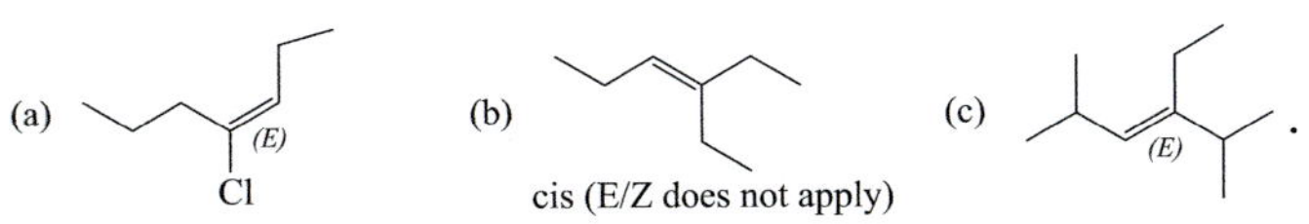

5.8. Al is $1s^22s^22p^63s^23p^1$ He is $1s^2$ Be is $1s^22s^2$ Mg is $1s^22s^22p^63s^2$
Cl is $1s^22s^22p^63s^23p^5$ Br is $1s^22s^22p^63s^23p^63d^{10}4s^24p^5$
Ti is $1s^22s^22p^63s^23p^63d^24s^2$ Cu is $1s^22s^22p^63s^23p^63d^{10}4s^1$

5.9. Both are group 1 elements. Potassium is a larger atom, which means that the $4s^1$ electron is held less tightly than the $3s^1$ electron on sodium. Using a very simple rationale, it should be easier to lose an electron from potassium, and the resulting K^+ ion is larger than the Na^+ ion and will be more stable.

5.10.

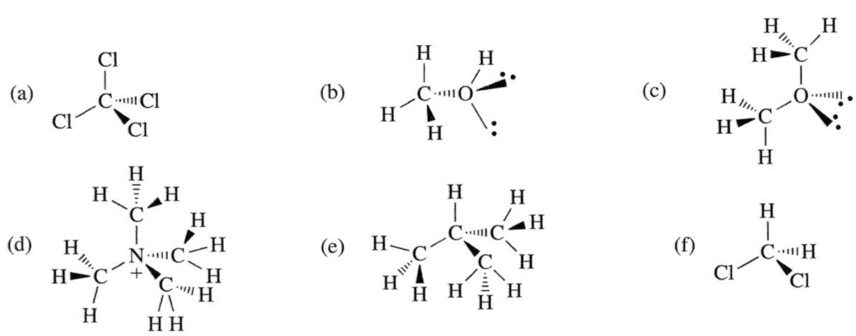

5.11. The dipole moment for the molecule is the vector sum of the two C–Cl bond moments, so the dipole moment for the molecule bisects the Cl–C–Cl bond as shown.

(a) 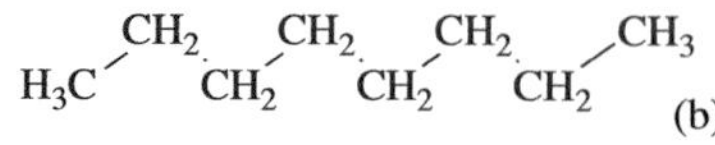

Octane

(b) 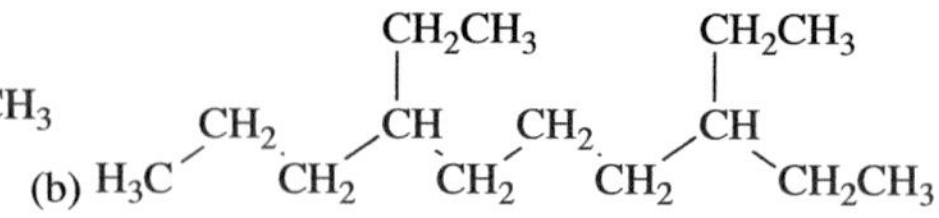

3,7-Diethyldecane

5.12.

(c) 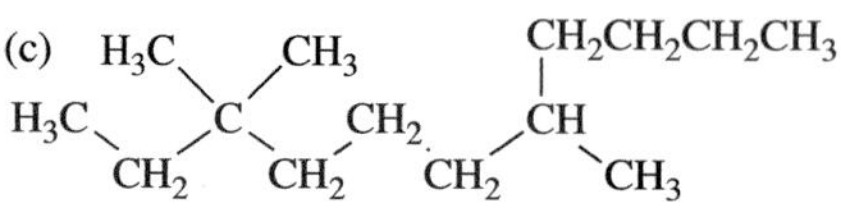

3,3,7-Trimethylundecane

(d) 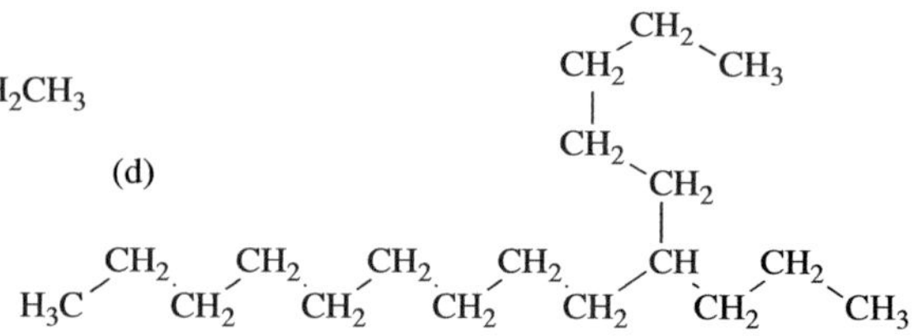

6-Propylpentadecane

(a) 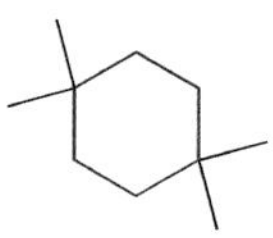

(b) 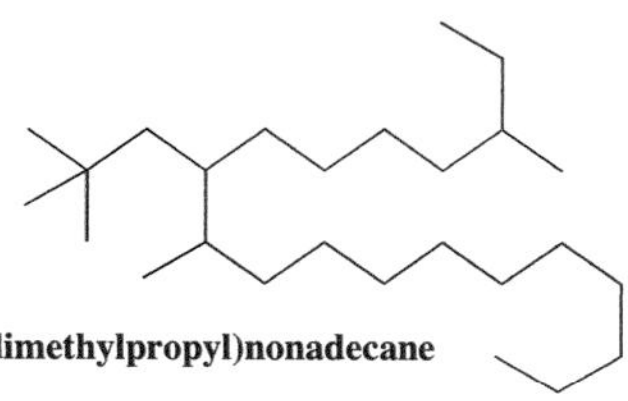

5.13. 1,1,4,4-Tetramethylcyclohexane 3,9-Dimethyl-8-(2,2-dimethylpropyl)nonadecane

(c) 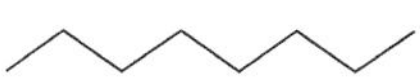

Octane

(d) 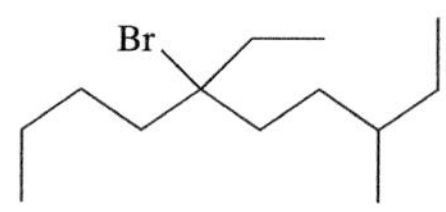

6-Bromo-6-ethyl-3-methyldecane

5.14. (a) Isomers of 1-bromooctane will have the formula $C_8H_{17}Br$. Several are shown, but there are others.

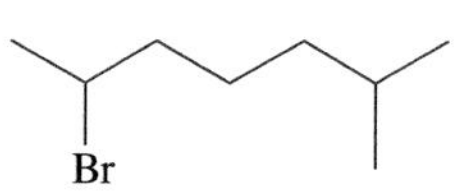

2-Bromo-6-methylheptane

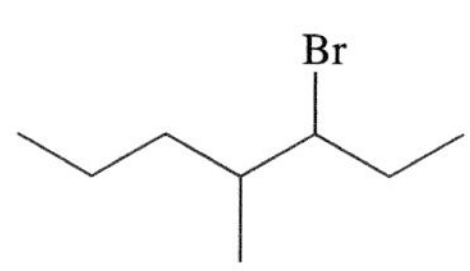

3-Bromo-4-methylmethylheptane

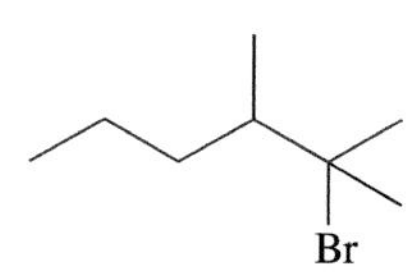

2-Bromo-2-3-dimethylhexane

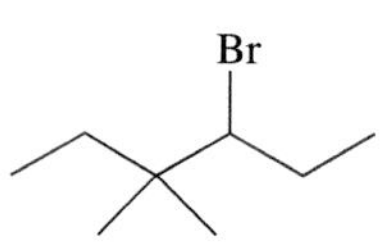

3,3-Dimethyl-4-bromo-hexane

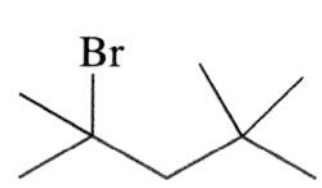

2-Bromo-2,4,4-trimethylpentane

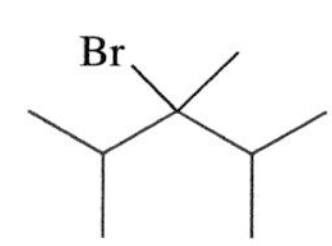

3-Bromo-2,3,4-trimethylpentane

(b) With the formula C_6H_{12}, assume they are cyclic alkanes. Several are shown, but there are others.

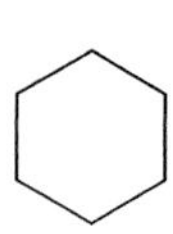

Cyclohexane

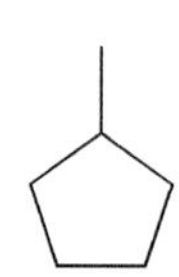

Methylcyclopentane

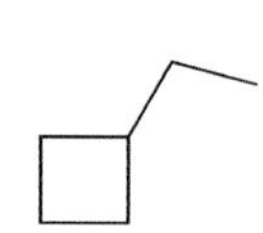

1-Ethylcyclobutane

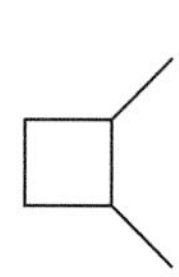

1,2-Dimethylcyclobutane

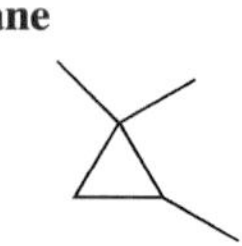

1,1,3-Trimethylcyclopropane

5.15. (a) 3,5-Dimethylheptane has the formula C_9H_{20}.

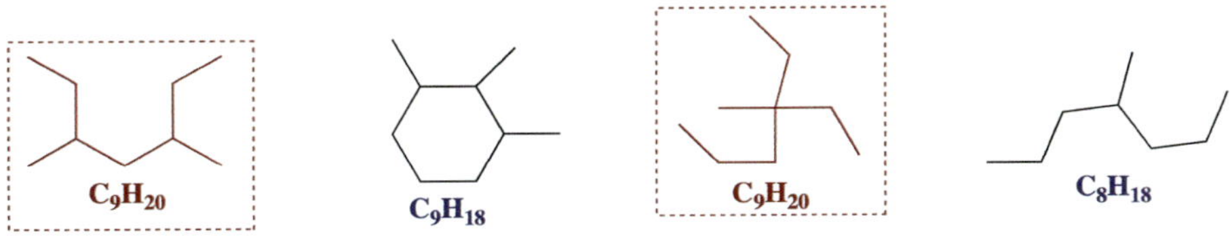

(b) Isomers of cycloheptane will have the formula C_7H_{14}.

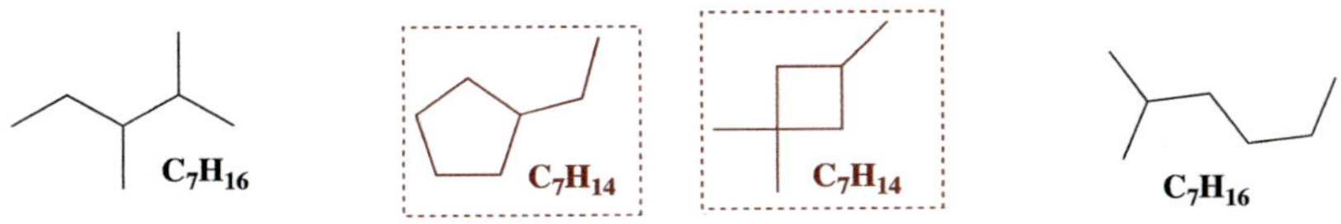

5.16. The transition state is that portion of the energy curve that represents the point at which bonds begin to break in the reaction and begin to form in the product. The difference in energy between the energy of the reactants and the energy of the transition state is the activation energy. The activation energy is the energy that must be provided to a reaction to begin the bond-making and bond-breaking process in the transition state.

5.17. No! Energy equal to the activation energy (E_{act}) must be added to initiate the reaction, and this has nothing to do with whether the overall reaction is endothermic or exothermic.

5.18. Calculate the half-life, given the following data. First-order half-life$=0.693/k$. Second-order half-life$=1/k[A_o]$, where for second-order reactions, assume $[A]_o=0.5$.

(a) A first-order reaction where $k=1.2\times 10-6$: half-life$=5.78\times 10^5$

(b) A second-order reaction where $k=4.5$: half-life$=0.44$

(c) A first-order reaction where $k=5.8\times 103$: half-life$=11.95\times 10^{-5}$

(d) A second-order reaction where $k=9.25\times 10-4$: half-life$=2162$

5.19. 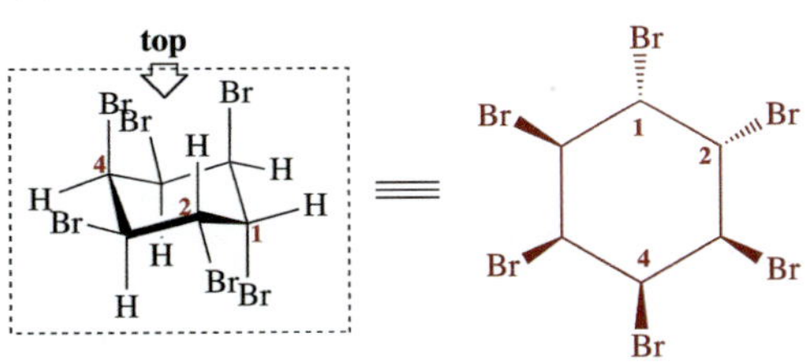

5.20. The flat structures are the highest in energy, and the boat conformations are the lowest. The boat conformation with the two chlorine atoms in flagpole positions has the highest transannular strain, and the boat conformation that has hydrogen atoms in the flagpole positions (circled) has the least amount of transannular strain and is the lowest energy conformation.

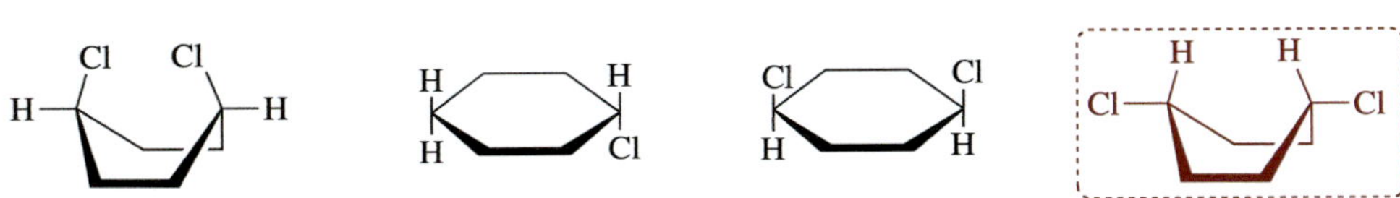

5.21. Conformation A has three axial groups (Cl, Br, Me) and B has two axial groups (Br, Cl). Since the relative steric demands of the substituents are unknown, simply count the number of axial groups, and B has only two. Therefore, the equilibrium should favor more B and less A.

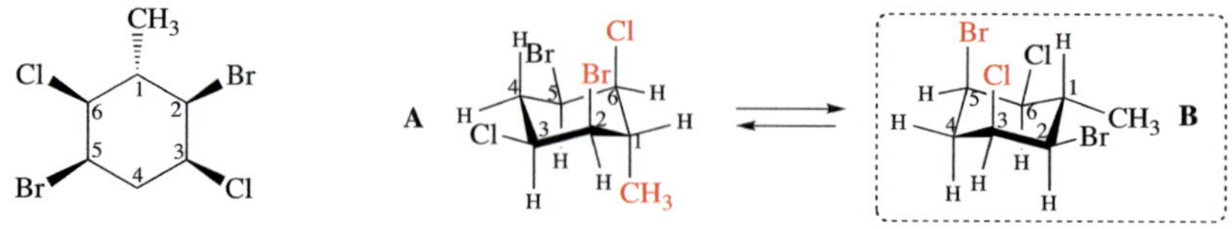

5.22. The covalent radius of Cl is 99 pm, and a methyl group has a covalent radius of 200 pm. The methyl group is larger. However, each chlorine atom has three unshared electron pairs that contribute to the Cl–Cl repulsion, which effectively raises the energy of the steric repulsion.

5.23. In the gas phase, the molecule should exist primarily in the gauche conformation shown due to stabilization via internal hydrogen bonding. In the hydrogen-bonding solvent methanol, intermolecular hydrogen bonding effectively makes the OH groups larger, and the minimal conformation will have the OH groups *anti* to minimize steric hindrance.

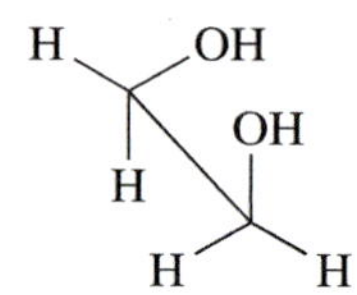

5.24. The circled solvents each have a stereogenic center, and therefore, they cannot be used for determining [α] of a compound in a polarimeter.

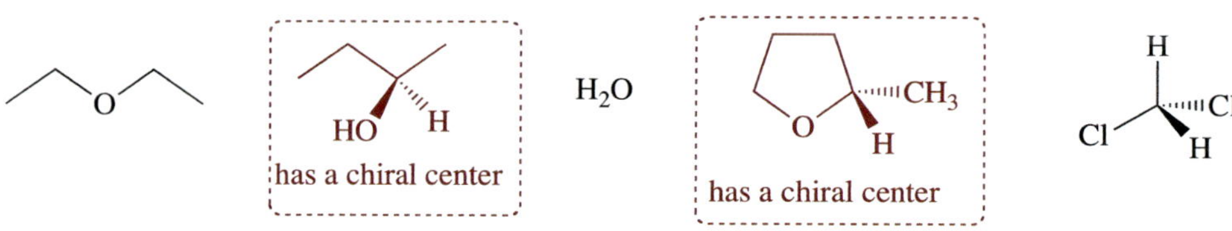

5.25.

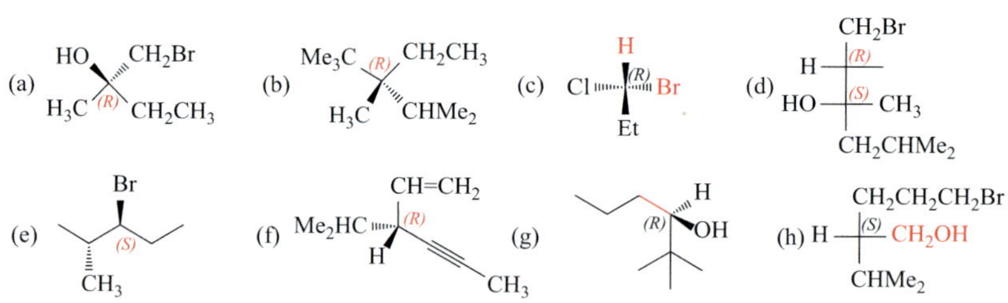

5.26. In all cases, convert centimeters (cm) to decimeters (dm), 1 cm = 0.1 dm.

(a) +18°/(1.1 × 2.5) = +6.55°

(b) (b) −176°/0.3 × 5.0) = −117.3°

(c) (c) −1.4°/(5.4 × 2.5) = −0.10°

(d) +94°/(2.3 × 3.0) = +13.6°

5.27.

(a) 3,3-Diphenyl-(4*E*)-nonen-1-ol

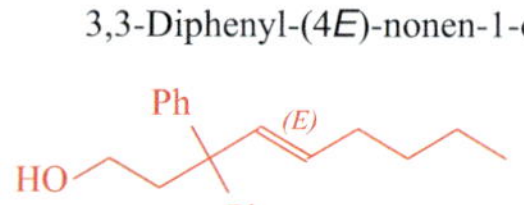

(b) 2,3,4,5-Tetrachlorohex-(2*Z*)-ene

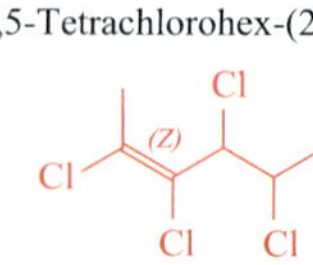

(c) 3-Bromo-6-fluorodeca-(3*Z*,6*E*)-dien-2-one

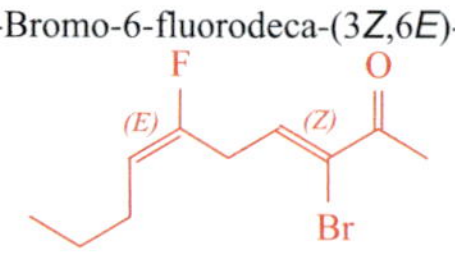

(d) 3-Ethylhept-(2*E*)-ene

5.28. 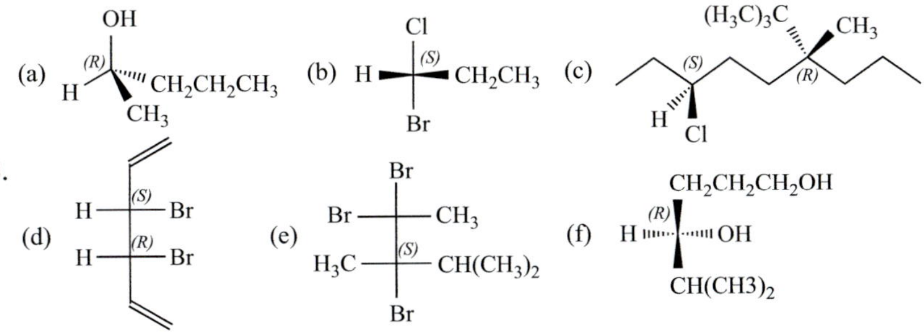

CHAPTER 6

6.1. 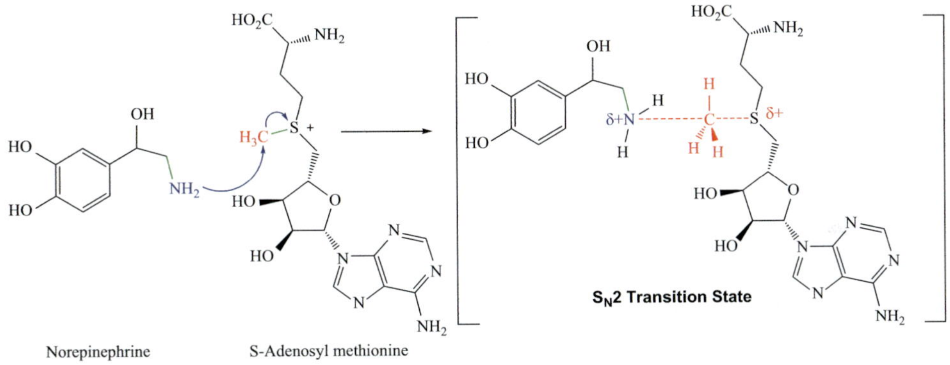

6.2. 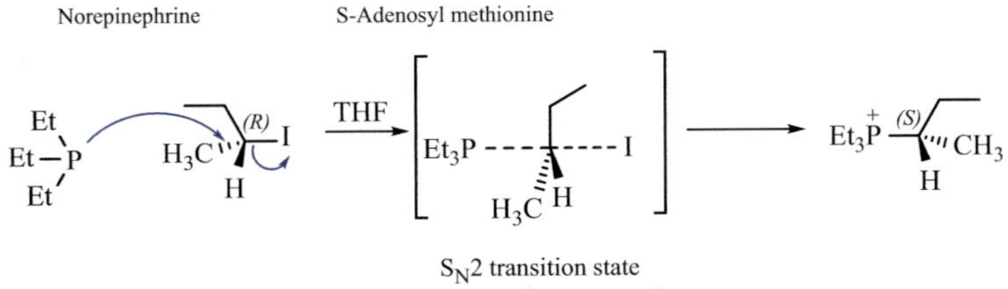

6.3. 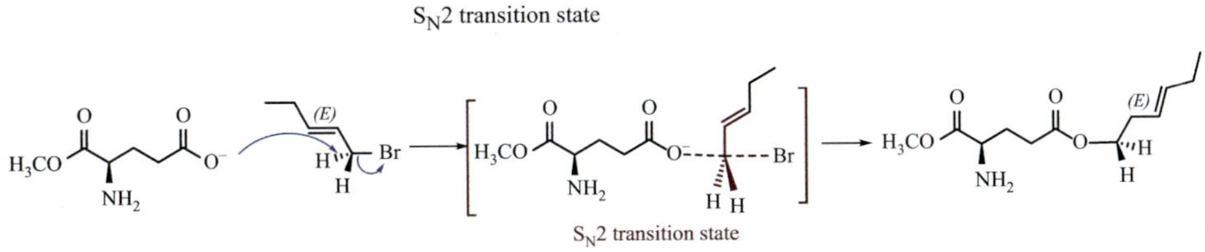

6.4. 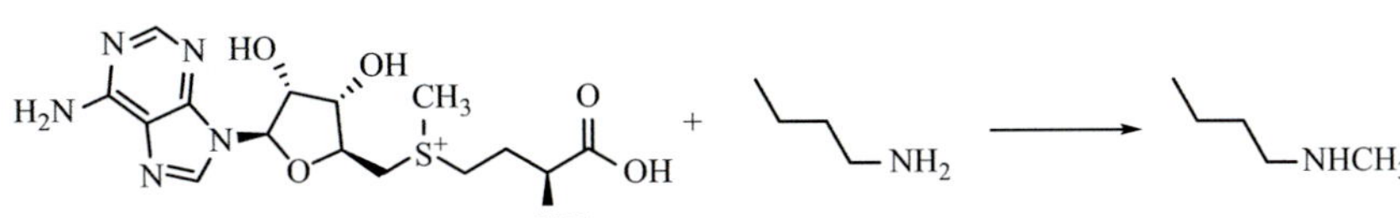

6.5. 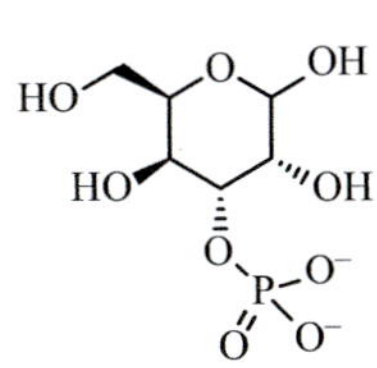

6.6. 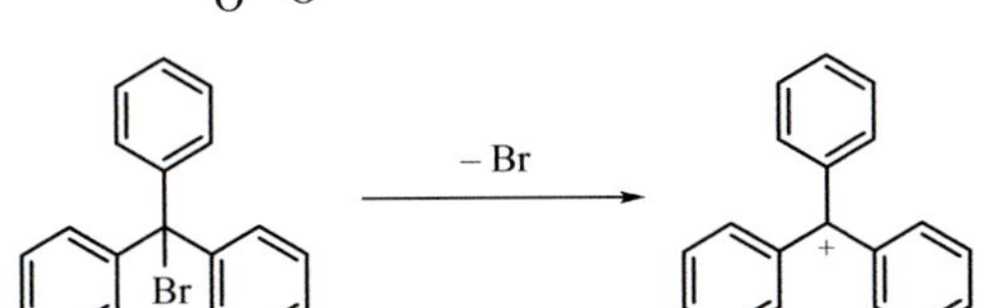

6.7. 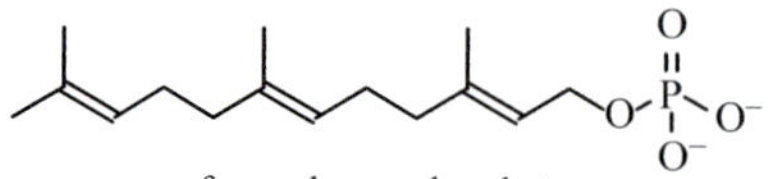

farnesyl monophosphate

6.8. 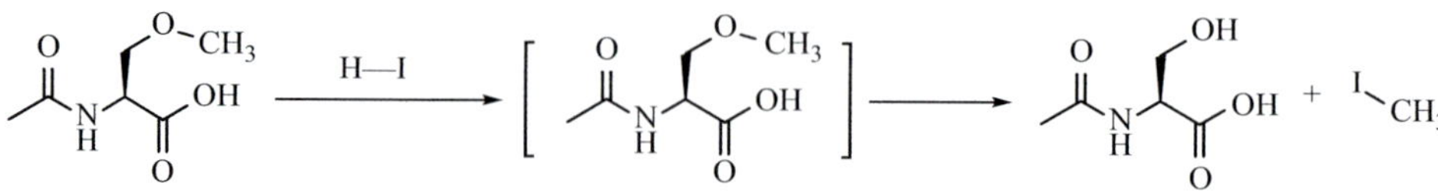

6.9. 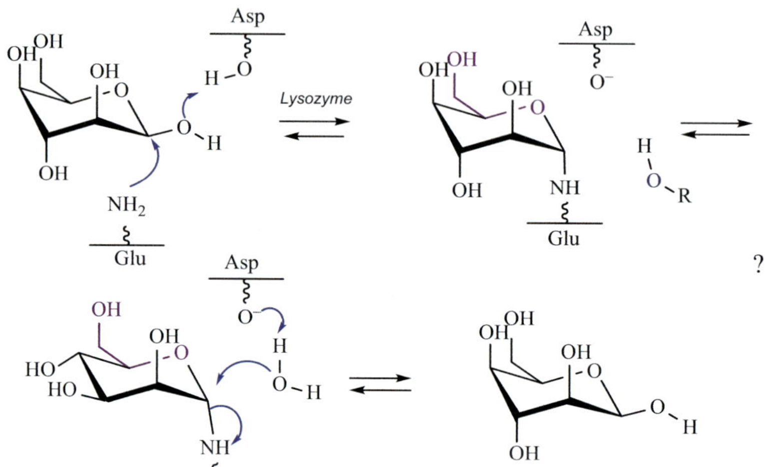

β-D-idopyranose

6.10. 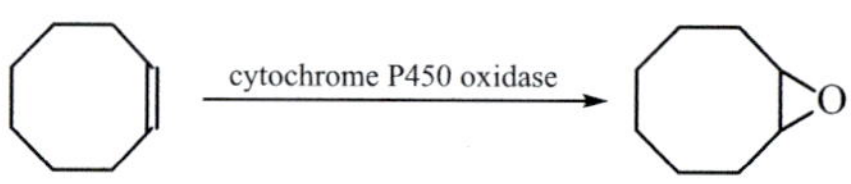

6.11. 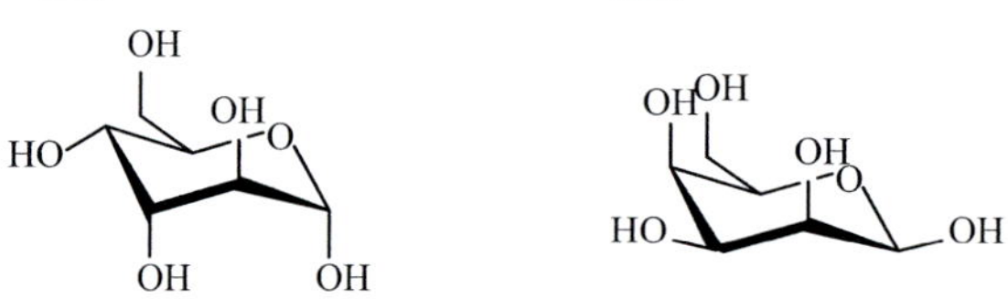

α-D-altropyranose β-D-alopyranose

6.12. 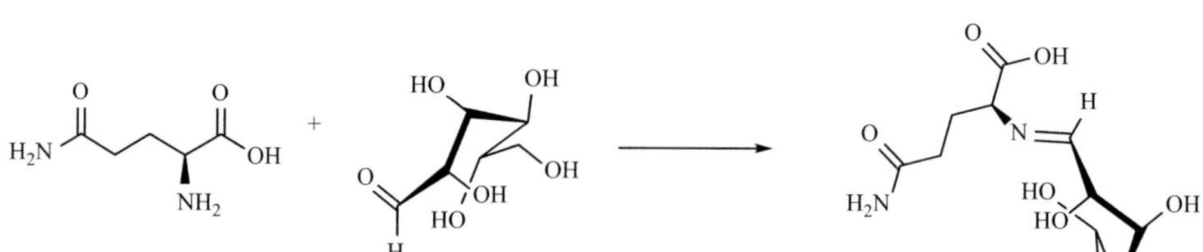

glutamine α-D-altropyranose

6.13. 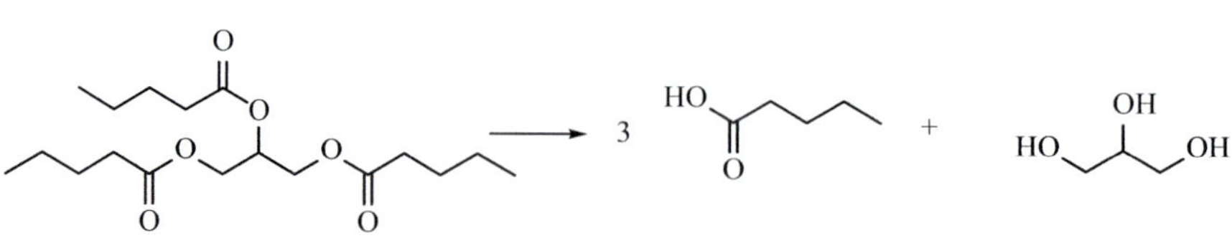

CHAPTER 7

7.1. 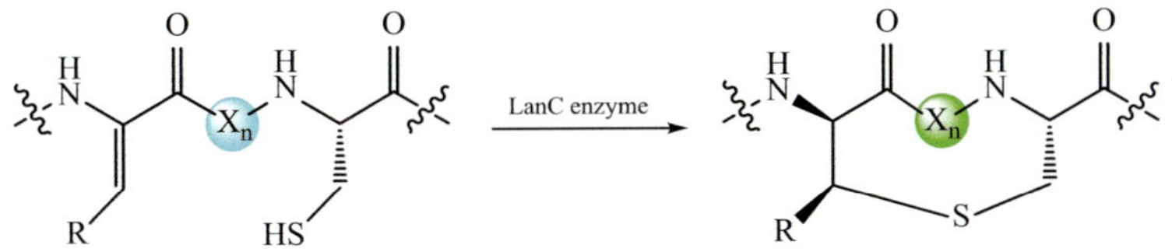

This is an internal Michael addition to give the cyclized product. *See Proceedings of the National Academy of Sciences*, **2023**, *120*, e2217523120.

7.2. The enzyme AbyU is a Diels-Alderase, and the internal [4+2]-cycloaddition gives the product shown. See *Chem. Sci.* **2024**, *15*, 11572.

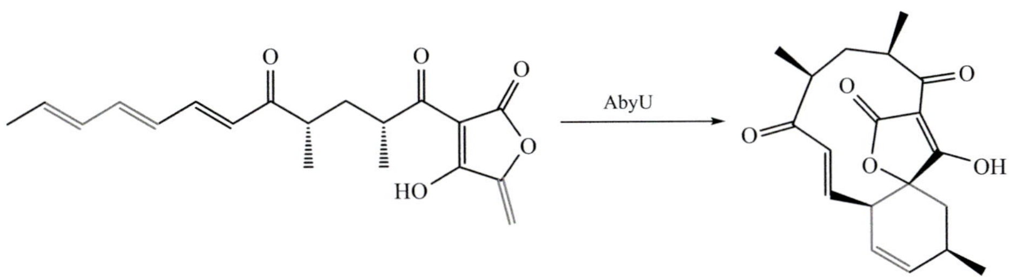

7.3. 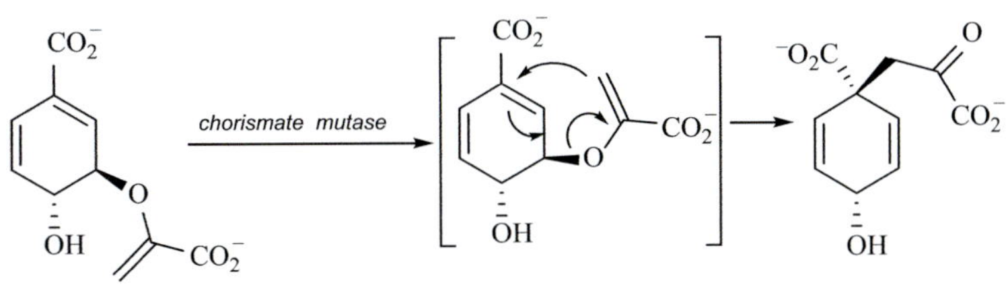

This conversion is a biological [3,3]-sigmatropic rearrangement. See *Angew. Chem. Int. Ed.* **2016**, *55*, 4711–4715.

7.4. Both reactions give the Michael addition product.

(a) 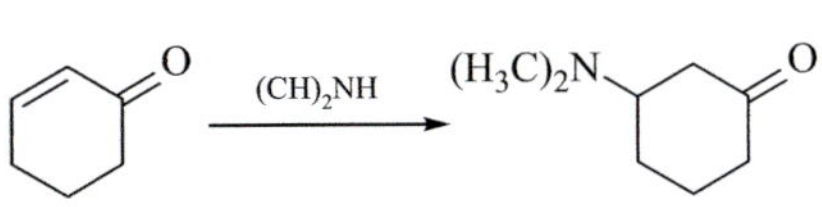

(b) 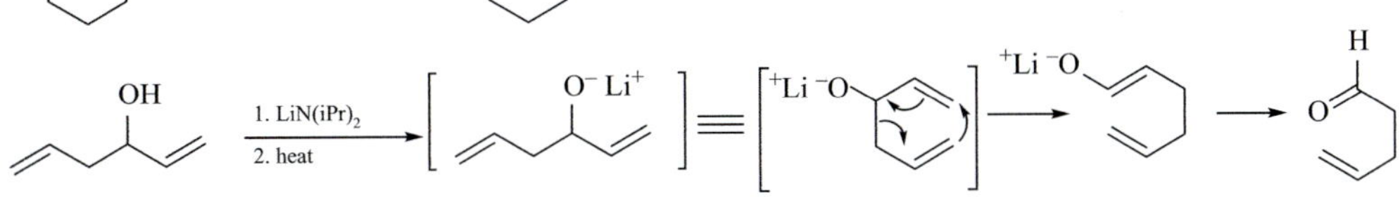

7.5. 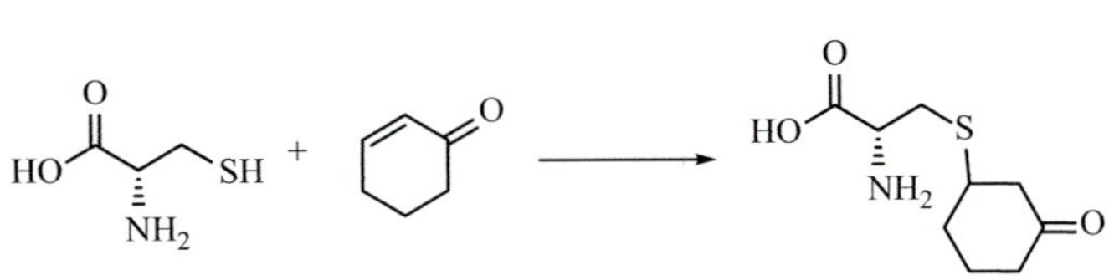

7.6. 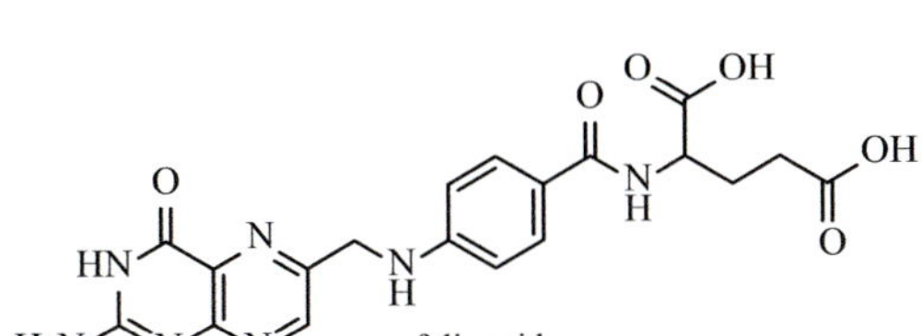

folic acid

7.7.

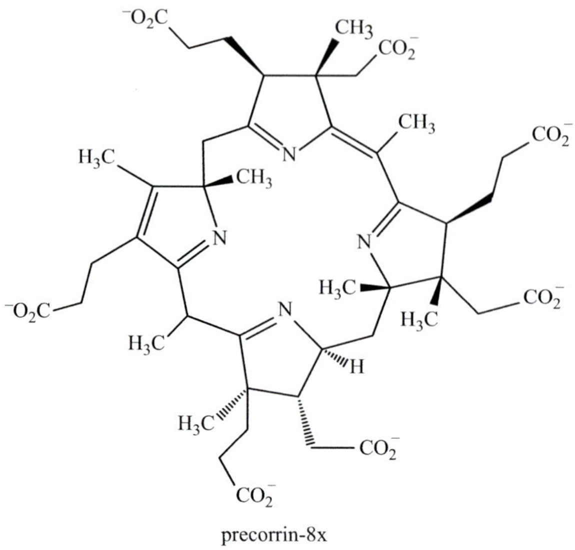

precorrin-8x

CHAPTER 8

8.1.

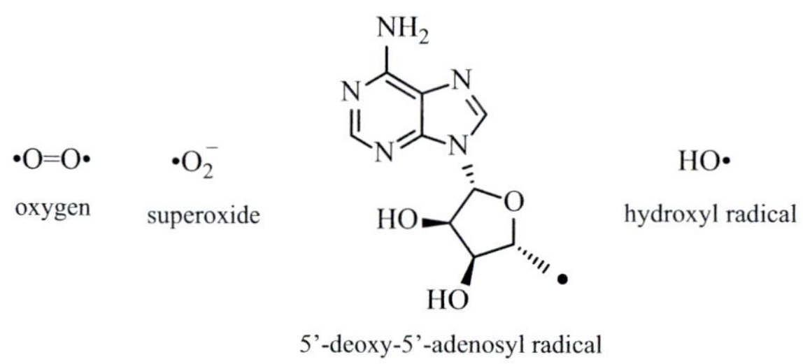

8.2.

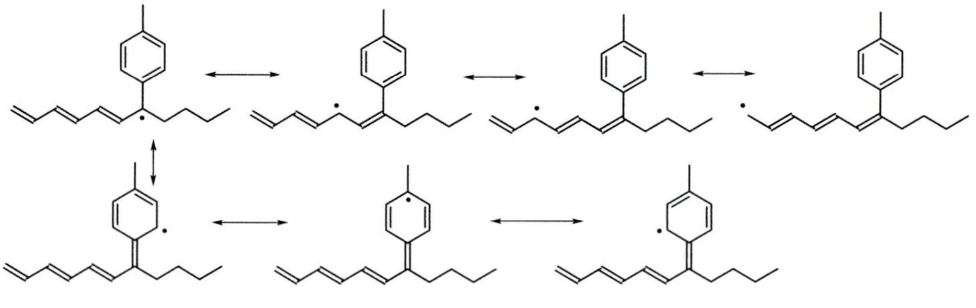

8.3. The cleavage of the C–H bond is an example of homolytic cleavage.

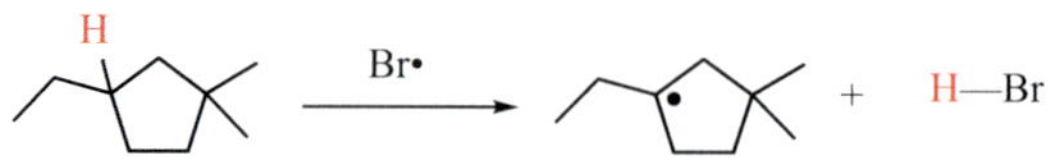

8.4.

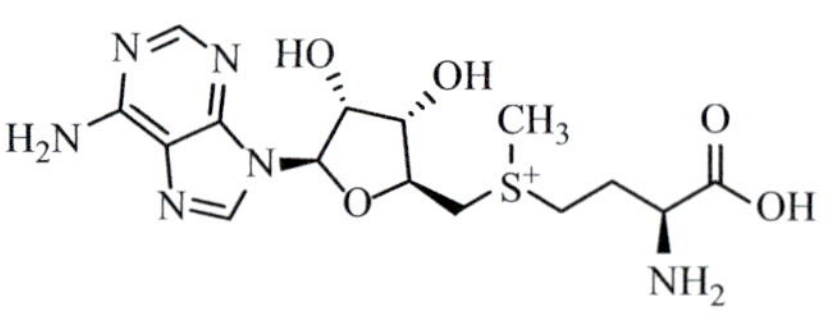

8.5. 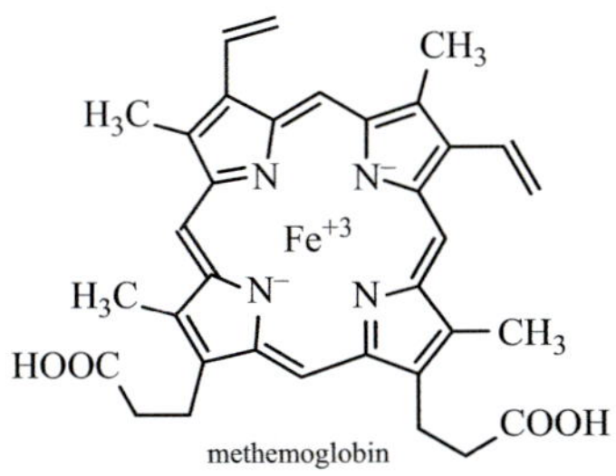

8.6. The disproportionation of superoxide gives O_2 and H_2O_2
8.7. Three scavengers are glutathione, α-tocopherol, and β-carotene.

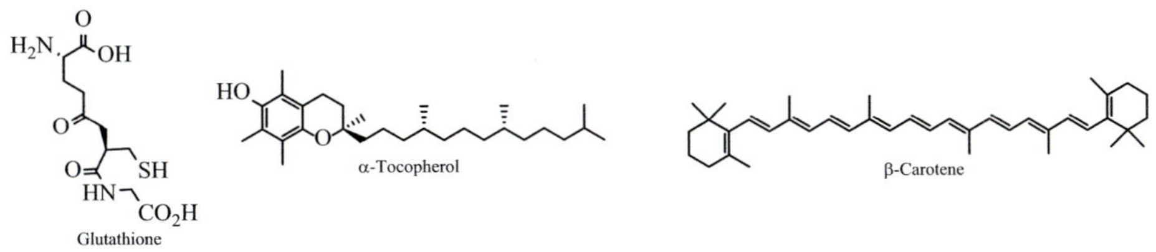

8.8. Non-enzymatic antioxidants include tocopherol and tocotrienol, L-Ascorbic acid (vitamin C; in citrus), carotenes such as lycopene and β-carotene, xanthophylls such as lutein (from green leaves) and zeaxanthin (from corn), phenolic compounds (ArOH) that are common in plants, lipoic acid, biliverdin, and bilirubin.
8.9. Captodative radicals contain both an electron-withdrawing group and an electron-releasing group. The captodative effect stabilizes radicals by the synergistic effect of an electron-withdrawing substituent and an electron-donating substituent.

8.10. 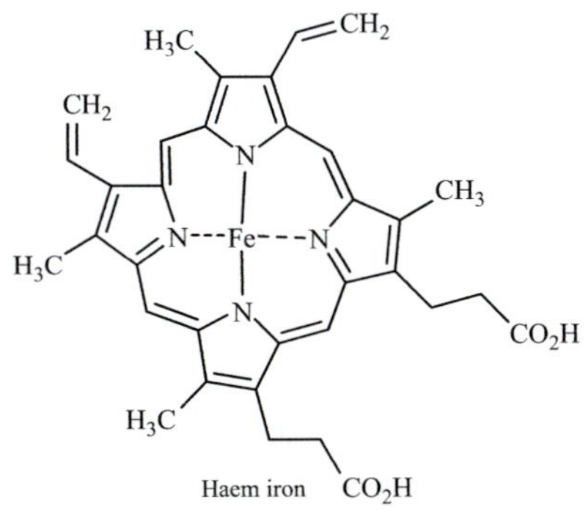

CHAPTER 9

9.1. Deprotonation of cyclohexan-1,3-dione gives a resonance-stabilized enolate anion with three resonance contributors. Deprotonation of cyclohexane gives an enolate anion with only two resonance contributors. The extra resonance stability of the enolate anion of cyclohexan-1,3-dione makes that conjugate base more stable, so the diketone is more acidic; i.e., it has a lower pK_a.

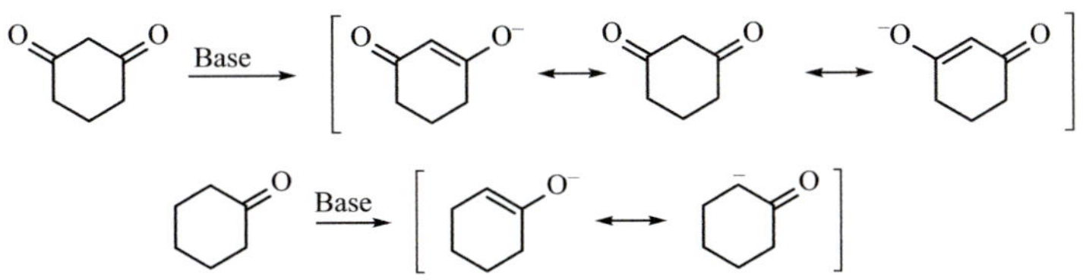

9.2. Since water is lost to form an alkene moiety, the conversion would be categorized as a dehydration.

9.3. Glycolysis involves reactions that convert glucose to pyruvate or lactate. Glycolysis is also called the Embden-Meyerhof pathway. It occurs in the cytoplasm of all living cells, aerobic as well as anaerobic.

9.4. *Enolase* enzymes can catalyze the thermodynamically difficult step of proton abstraction, which is important for epimerization transformations.

9.5. The histidine residue acts as a base to deprotonate the aspartic acid CO_2H moiety to generate a carboxylate anion, which is typically the unit that initiates deprotonation in an aldol-type condensation reaction mediated by an aldolase.

9.6. A racemase is an isomerase enzyme that catalyzes the inversion of stereochemistry in biological molecules.

9.7. It is called a tetrahedral intermediate, which is common in acyl substitution reactions.

9.8.

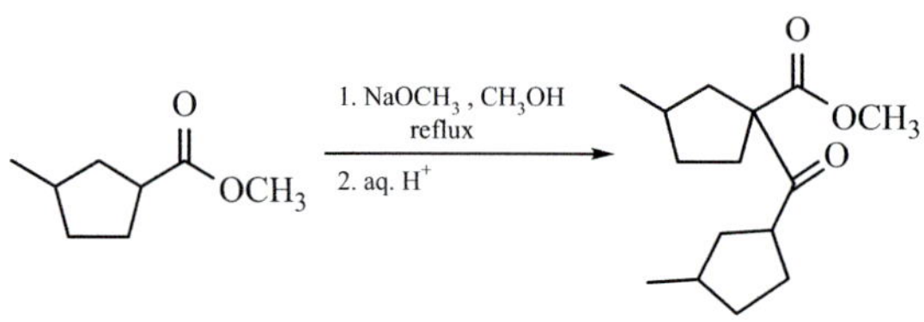

9.9. Acetyl-CoA is a molecule that participates in many biochemical reactions in protein, carbohydrate, and lipid metabolism and delivers the two-carbon acetyl group. In other words, it functions as a two-carbon unit in enzymatic synthesis.

9.10. Give the major product of the following reactions.

(a)

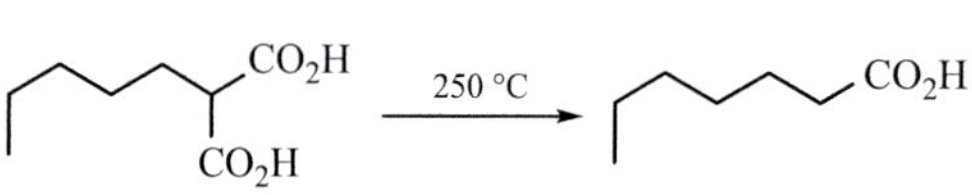

(b)

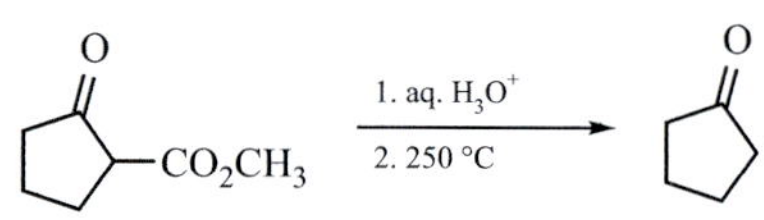

(c)

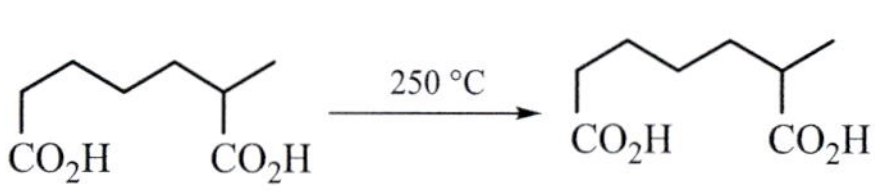

(d)

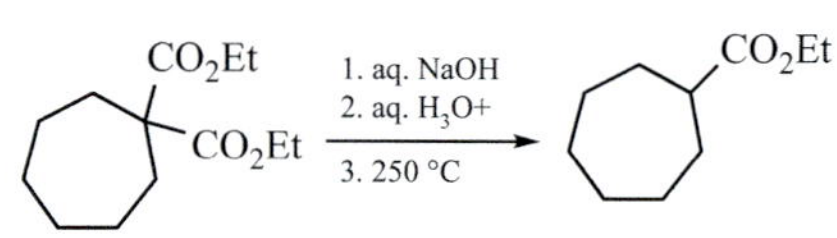

9.11.

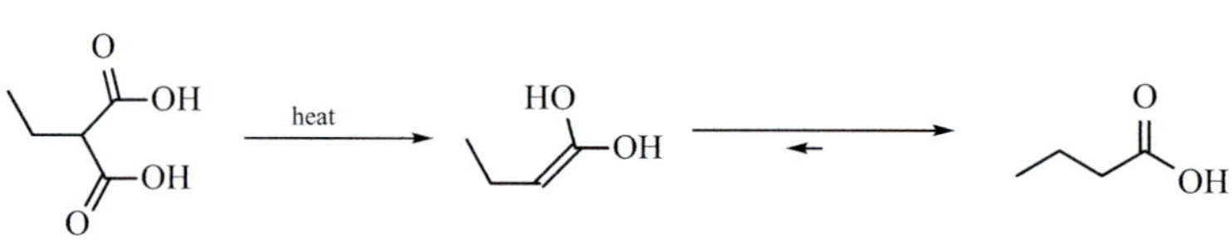

CHAPTER 10

10.1.

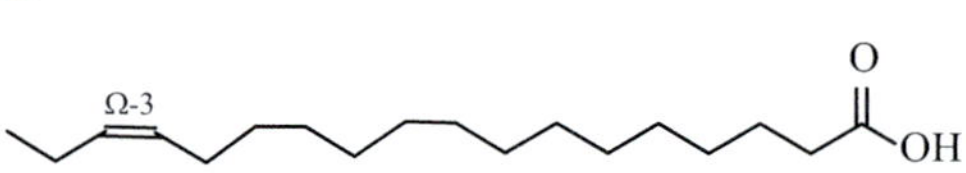

(*Z*)-octadec-15-enoic acid

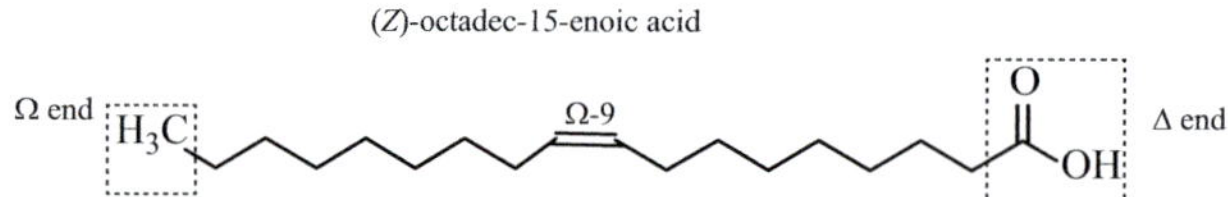

Oleic acid

10.2. Essential fatty acids are fatty acids that are required by humans and other animals for normal physiological function but cannot be synthesized in the body. They must be obtained from food or from a dietary supplement.

10.3. Choline is $Me_2N^+CH_2CH_2OH$

10.4. 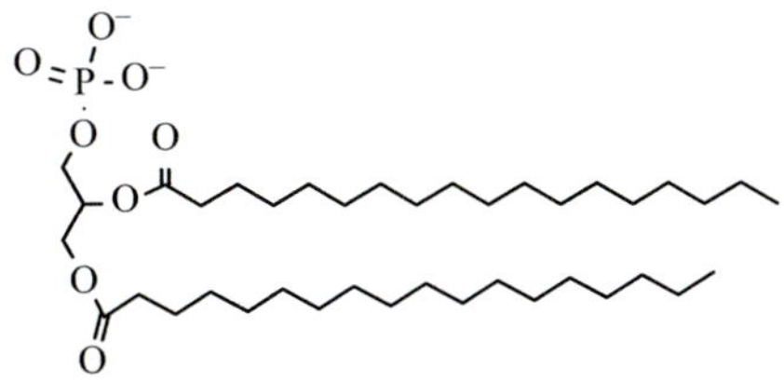

Phospholipid (2,3-bis(stearoyloxy)propyl phosphate)

10.5. 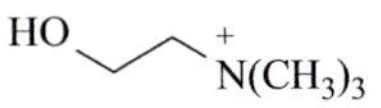

10.6. When a nucleophile reacts with a methyl group of dimethyl sulfate, the CH_3OSO_3– unit is a good leaving group. Therefore, dimethyl sulfate readily reacts as a methylating reagent.

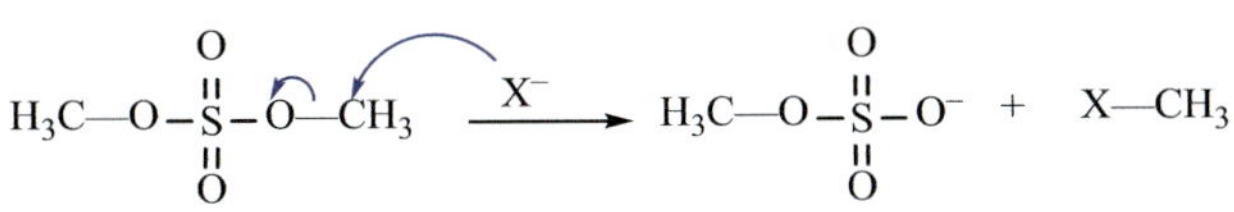

10.7. 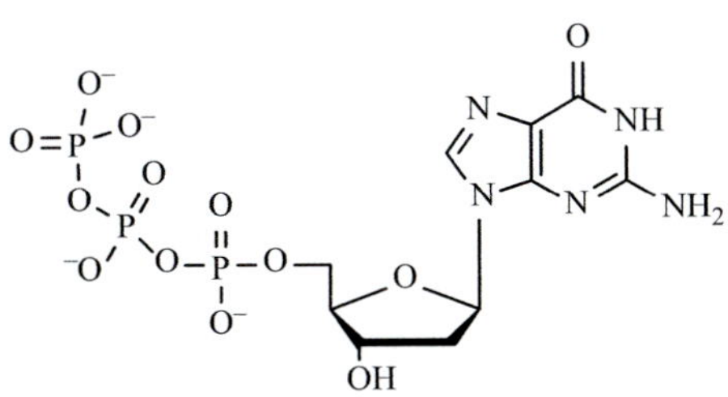

guanosine 5'-triphosphate

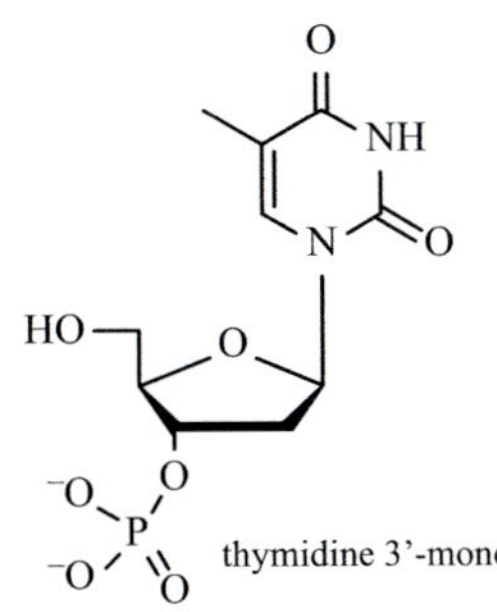

thymidine 3'-monophosphate

10.8. 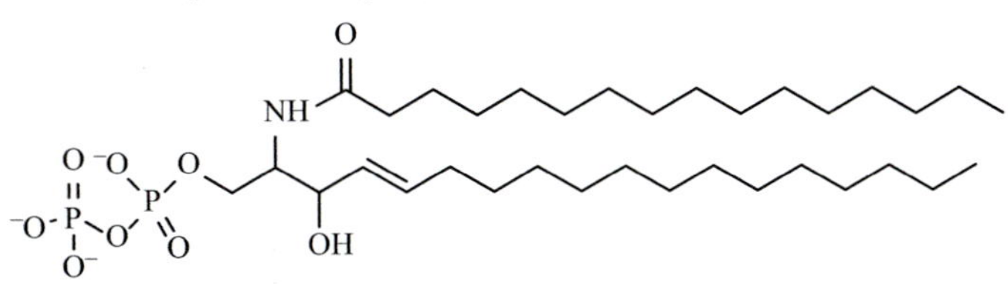

10.9. 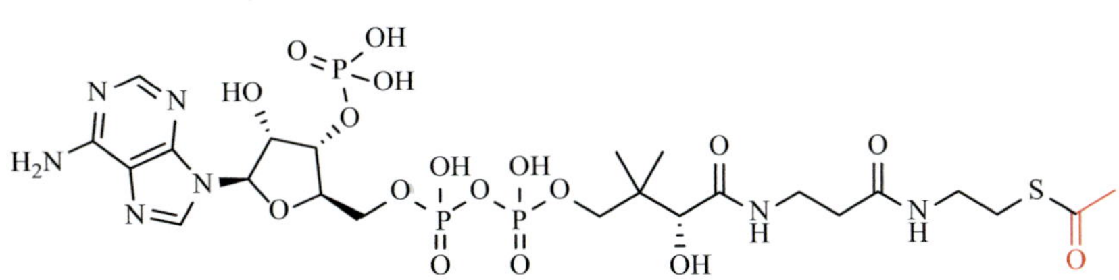

acetyl Co-A

CHAPTER 11

11.1. The reaction of a benzene ring attacking a carbonyl group is like the Friedel-Crafts acylation reaction.

11.2. The aromatic character of benzene means that benzene is a poor electron donor since electron donation would disrupt the aromaticity of the molecule. If benzene is a very poor electron donor, it is a very weak Bronsted-Lowry base, and it is so weak that it will not react with HBr.

11.3. In order to form catechol, cyclohexa-3,5-diene-1,2-diol must lose the two hydrogen atoms shown; hence, the enzyme that facilitates this reaction is a dehydrogenase. A dehydrogenase has an EC number of EC 1, and EC 1.3 is typical.

cyclohexa-3,5-diene-1,2-diol catechol

11.4. Although the actual answer may not be known for sure, one explanation may be that there is no radical anion intermediate that isomerizes to the 1,4-product. The aromatic ring must coordinate with the active site of the enzyme so there is 1,2-reduction by transfer of H atoms to give the 1,5-cyclohexadiene.

11.5. The definition of ipso- indicates that two substituents share the same ring position in an intermediate compound. In aromatic substitution, the ipso carbon is the site of the substituent attachment. In the case of anisole in question 09-4, the ipso carbon is the carbon on benzene to which the OCH_3 group is attached.

11.6. The biological function is to transfer a halogen atom to a substrate.

11.7. The N-methyl derivative of norepinephrine that is shown would be the expected product.

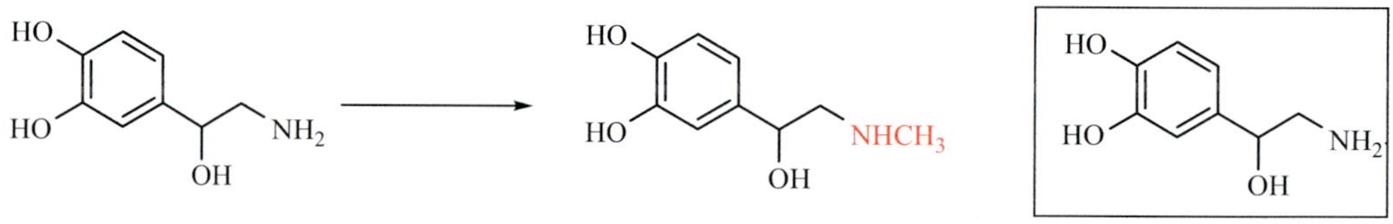

norepinephrine

11.8. 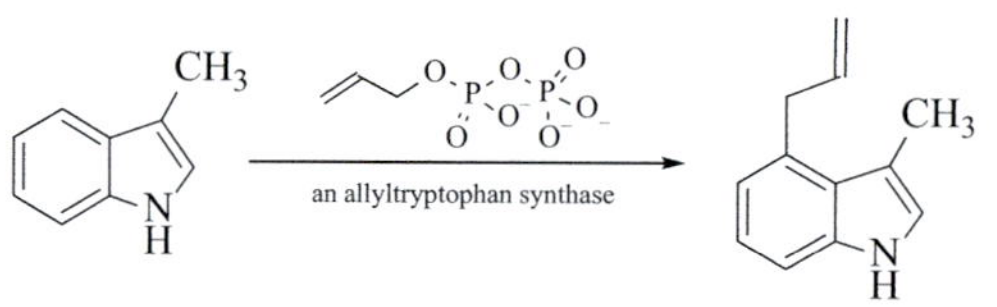

3-methyl-1H-indole 4-allyl-3-methyl-1*H*-indole

11.9. The reaction is completed by nucleophilic acyl substitution at the ester carbonyl to give a tetrahedral-type intermediate. Loss of the alkoxy group completes the nucleophilic substitution.

11.10. 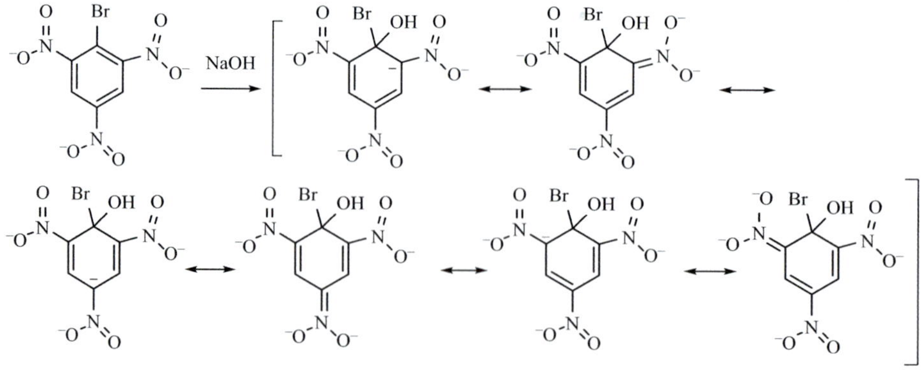

11.11. This reaction is an aromatic substitution, an S_EAr reaction.

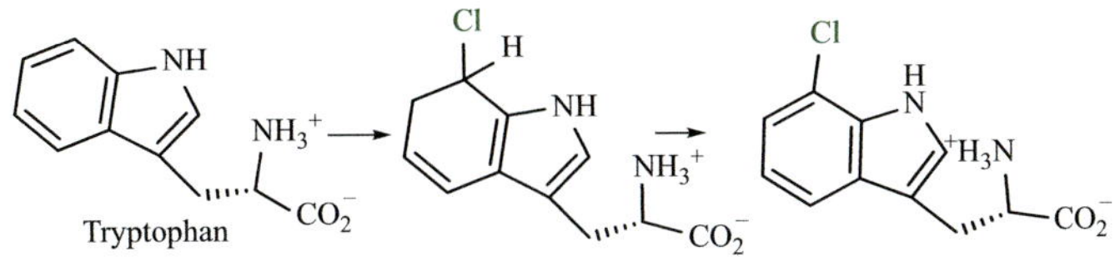

11.12. 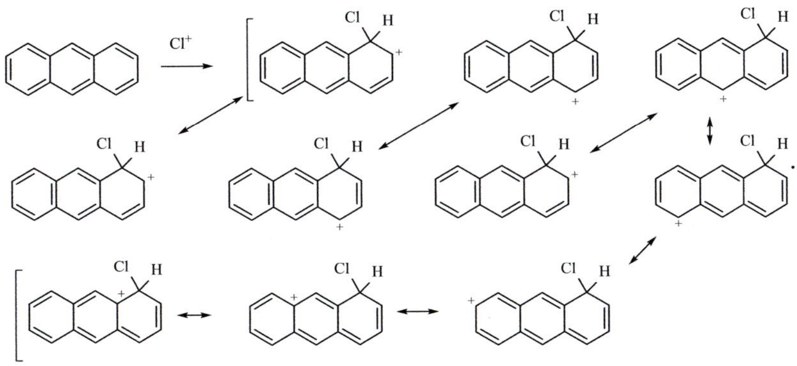

11.13. The five-membered heteroaromatic ring is more active and therefore more reactive than the six-membered ring.

11.14. In quinoline, one ring is an activated aromatic ring (Ar–N), whereas the other is a deactivated pyridine ring (N is in the ring). Therefore, an S_EAr reaction will occur preferentially at the activated aromatic ring.

11.15. The product would be a pyrrolidine product.

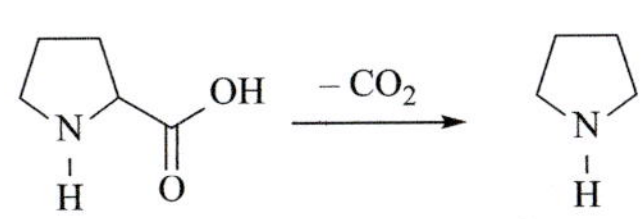

CHAPTER 12

12.1. Mg^{2+} (79 pm) and Ca^{2+} (99 pm) interact differently with an enzyme due to the difference in ionic radius. This size difference leads to large-scale structural effects induced by the metal.

12.2. 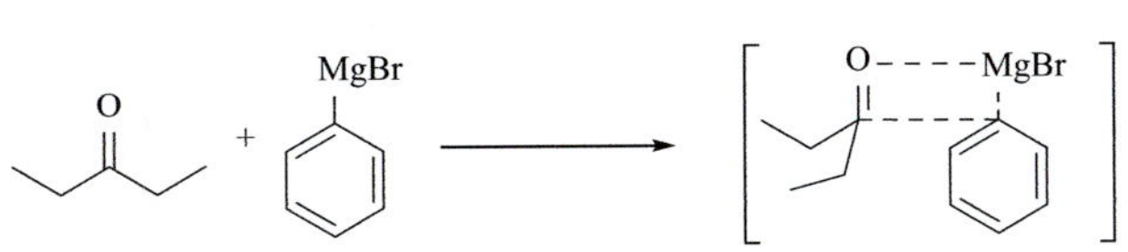

12.3. The phytol tail of chlorophyll is a long, hydrophobic hydrocarbon chain that anchors the molecule to membranes.

The most common metals in metalloenzymes are Fe, Zn, Cu, and Mn. Metalloenzymes contain a metal ion that is key to the activity of that enzyme, and tightly bound metal ions include Fe^{2+}, Cu^{2+}, Zn^{2+}, Mn^{2++}, Co^{3+}, Ni^{3+}, Mo^{6+}.

12.4. 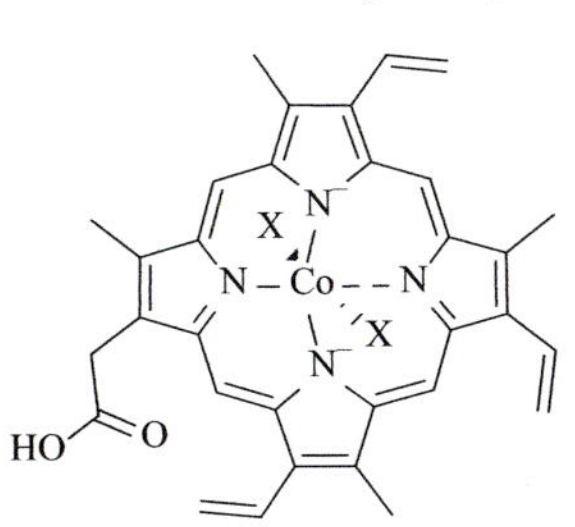

12.5. Molecules such as carbon monoxide (CO), amines such as methylamine (CH_3NH_2), phosphines such as trimethylphosphine ($P(CH_3)_3$), alkenes ($CH_2{=}CH_2$), thiols such as methanethiol (CH3SH), or halides such as bromide ion (Br^-).

12.6.

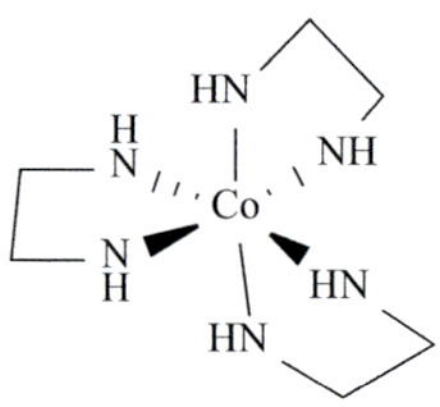

CHAPTER 13

13.1. An oligosaccharide is a carbohydrate that links 5–15 monosaccharides.

13.2.

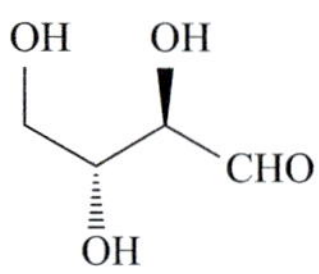

D-Erythrose
(tetrose)

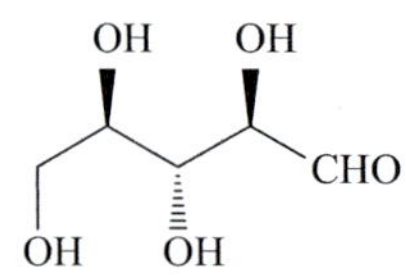

D-Ribose
(pentose)

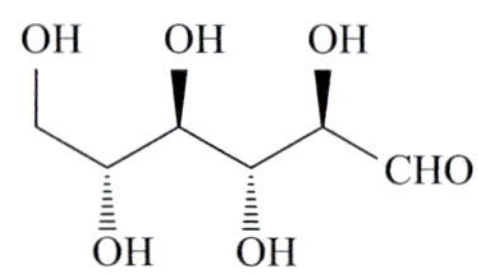

D-Allose
(hexose)

13.3.

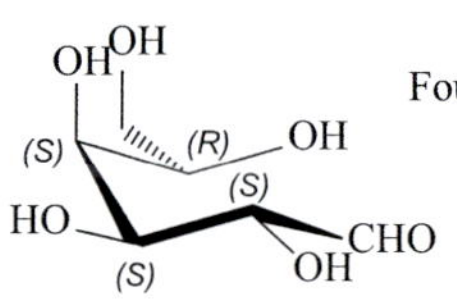

D-galactose

Four stereogenic centers

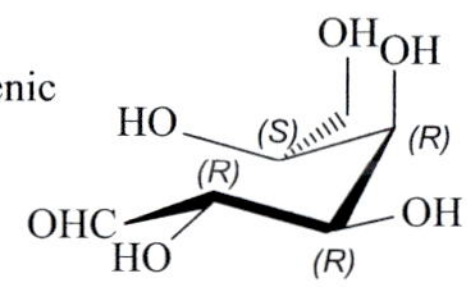

L-galactose

13.4.

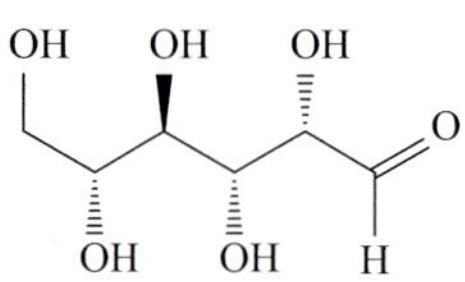

altrose

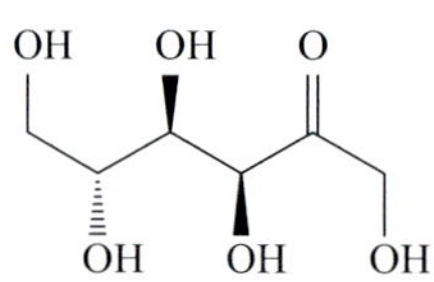

fructose

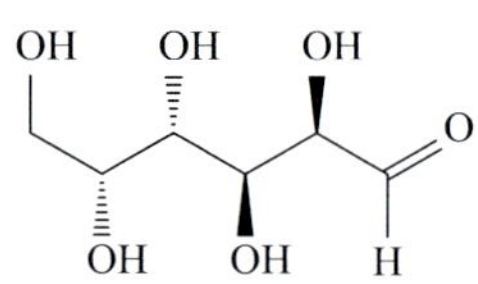

galactose

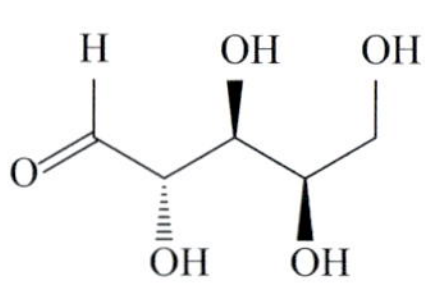

lyxose

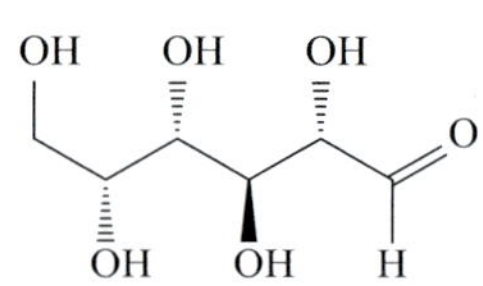

talose

13.5.

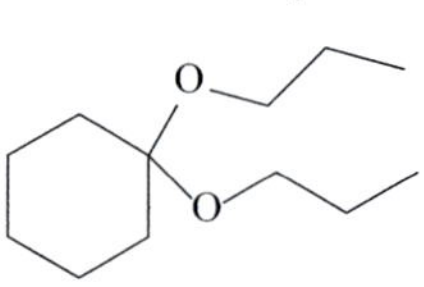

13.6.

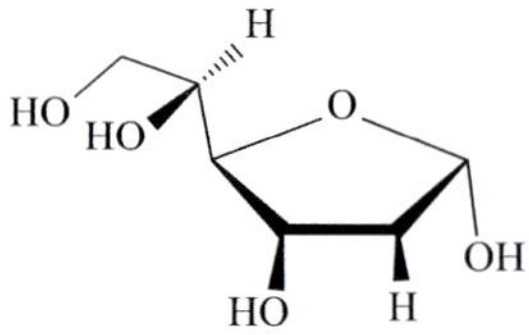

α-Altrafuranose

(2*S*,5*S*)-5-((*R*)-1,2-Dihydroxyethyl)tetrahydrofuran-2,4-diol

13.7. 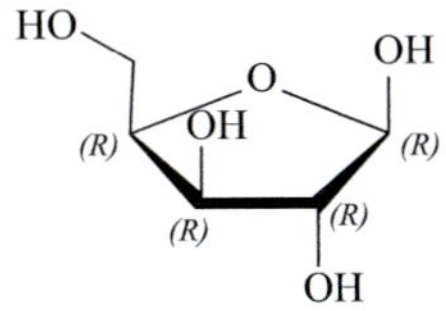

β-D-xylofuranose

13.8. The anomeric effect occurs when the OH at the anomeric carbon (the hemiacetal carbon) is in the axial position. In the axial position, the electron pairs on the oxygen atoms repel, which leads to the lower energy conformation, in both cases the α-isomer.

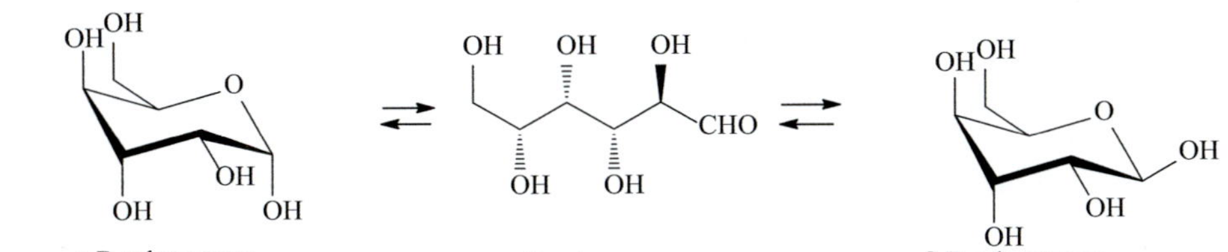

α-D-gulopyranose D-gulose β-D-gulopyranose

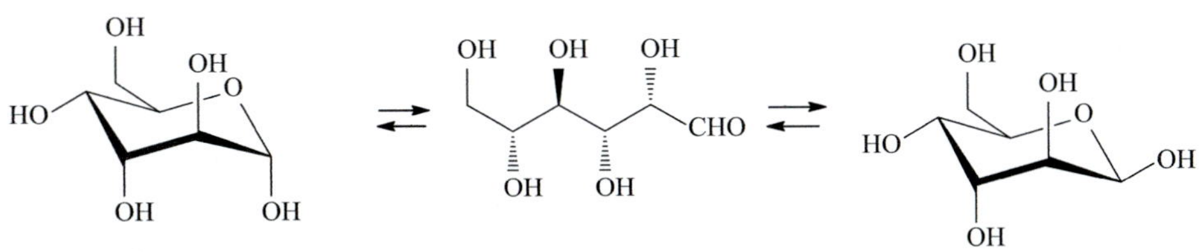

α-D-altropyranose D-altrose β-D-altropyranose

13.9. 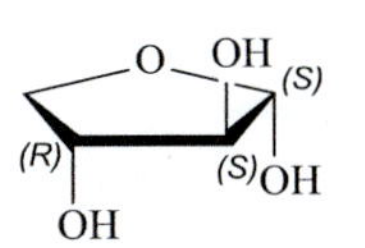

β-threofuranose
(2*S*,3*S*,4*R*)-tetrahydrofuran-2,3,4-triol

CHAPTER 14

14.1.

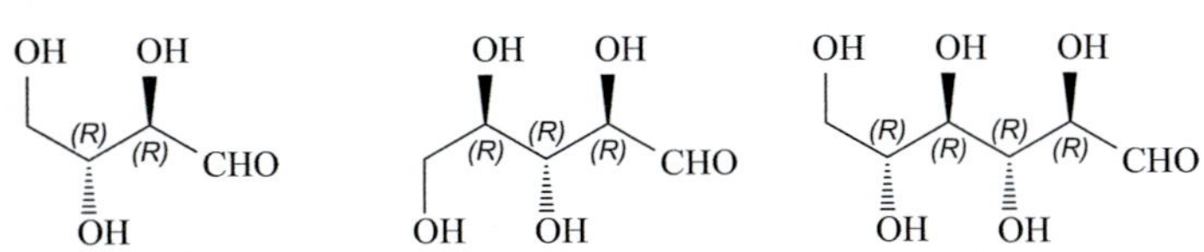

D-Erythrose (aldotetrose) D-Ribose (aldopentose) D-Allose (adohexose)

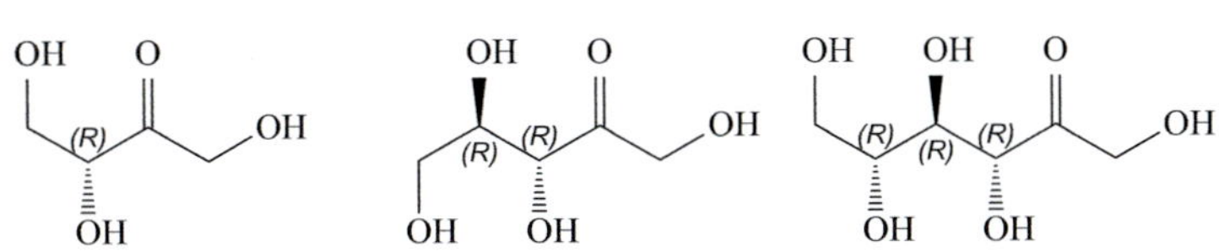

D-Erythrolose (ketotetrose) D-Ribulose (ketopentose) D-Psicose (ketohexose)

14.2. 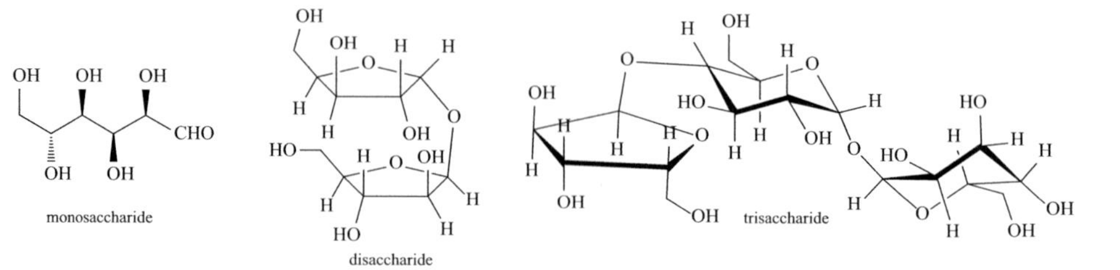

14.3. 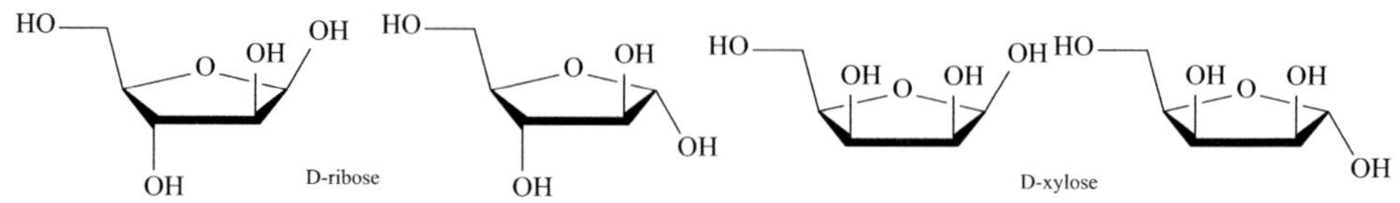

14.4. A homopolymer is a polymeric molecule that has only one type of chemical repeat unit or monomer.

14.5. A reducing sugar will react with Fehling's solution [heating in aqueous copper (II) sulfate and sodium tartrate] or with Tollens' solution (silver nitrate in ammonia).

14.6. With a (1 →1) linkage, no aldehyde moiety will be generated by mutarotation, and with no aldehyde, no reaction is possible with Fehling's solution or Tollens' solution.

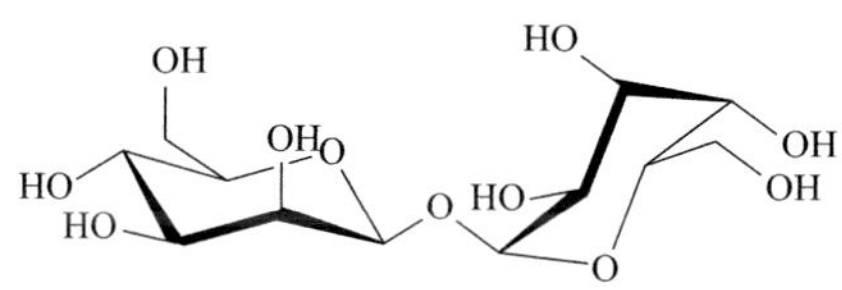

14.7. 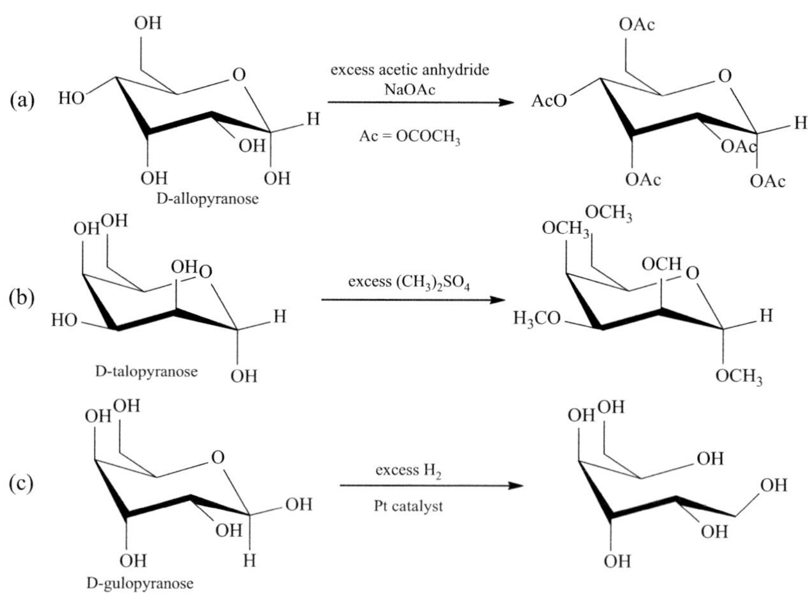

14.8. 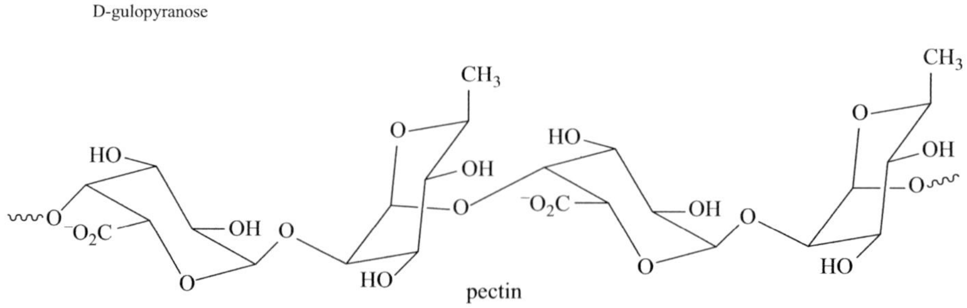

14.9. A glycoside is a compound formed from a carbohydrate and another compound by the replacement of a hydroxyl group in the carbohydrate.

14.10.

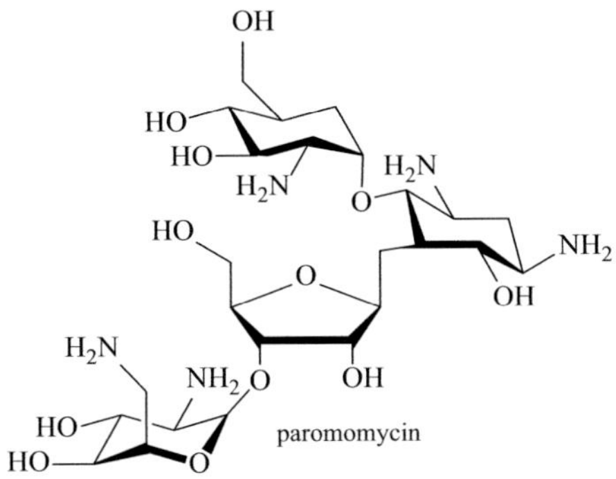

paromomycin

14.11.

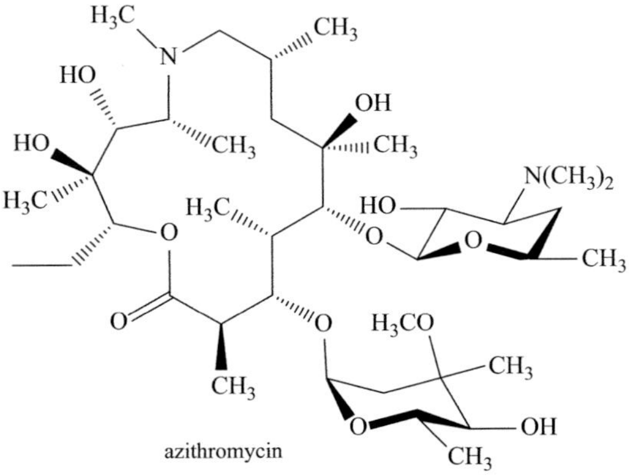

azithromycin

14.12. Catabolism is the breakdown of complex molecules to form simpler ones, with the release of energy.

14.13. A glycosyltransferase is an enzyme that catalyzes the formation of the glycosidic linkage to form a glycoside.

14.14. A lumen is the inside space of a tubular structure, such as an artery or intestine.

14.15. A eukaryotic organism has cells with a nucleus enclosed within membranes, whereas prokaryotes have no membrane-bound organelles.

14.16.

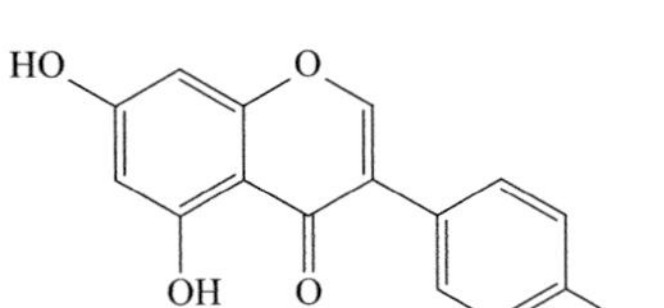

genistin aglycone

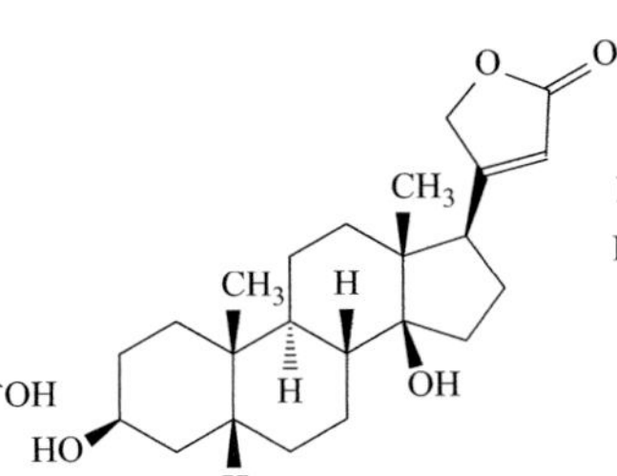

digitoxin aglycone

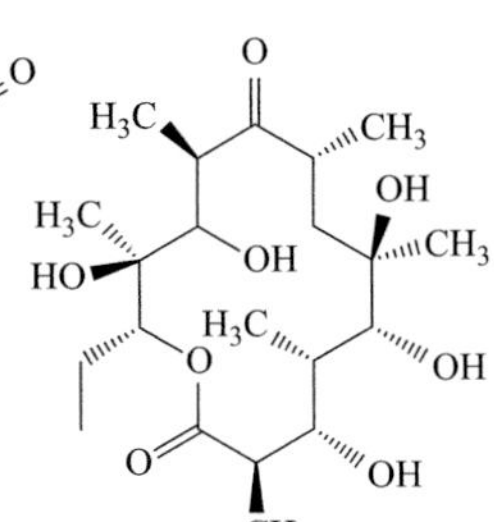

erythromycin aglycone

CHAPTER 15

15.1.

(a)

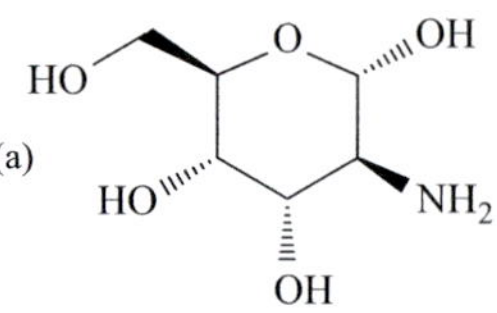

2-amino-2-deoxy-
α-D-altropyranose

(b)

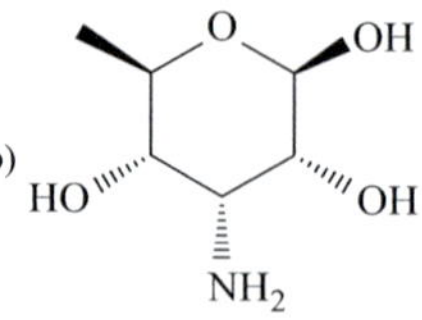

3-amino-3,6-dideoxy-
β-D-allopyranose

(c)

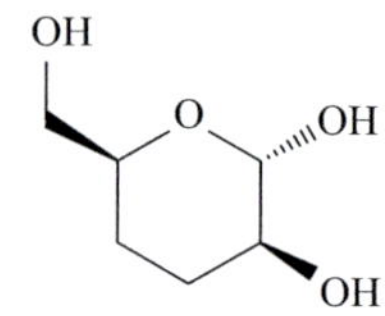

3,4-dideoxy-
α-D-talopyranose

15.2.

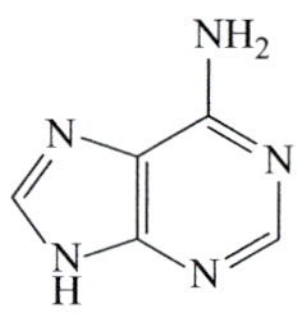

adenine

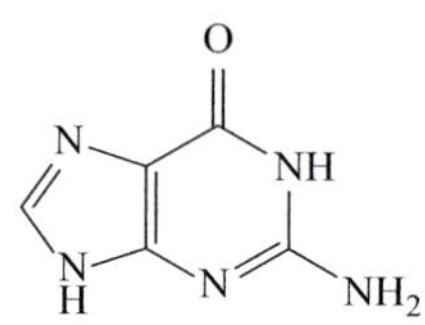

guanine

15.3.

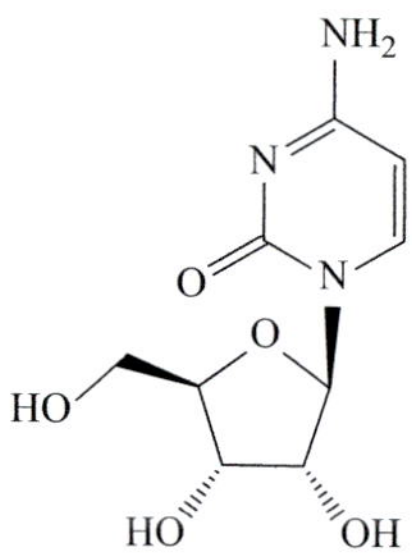

cytidine

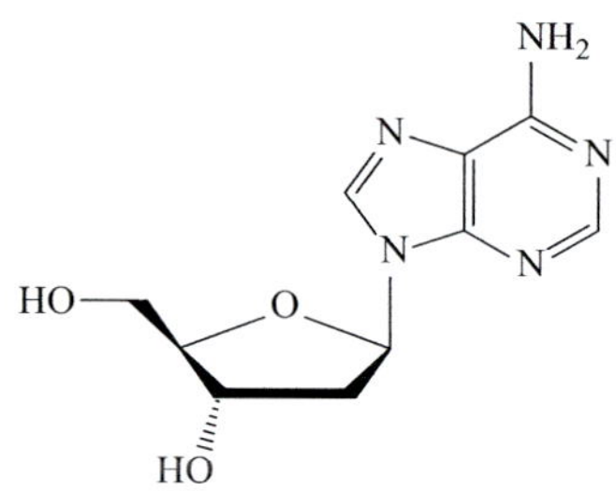

2'-deoxyadenosine

15.4.

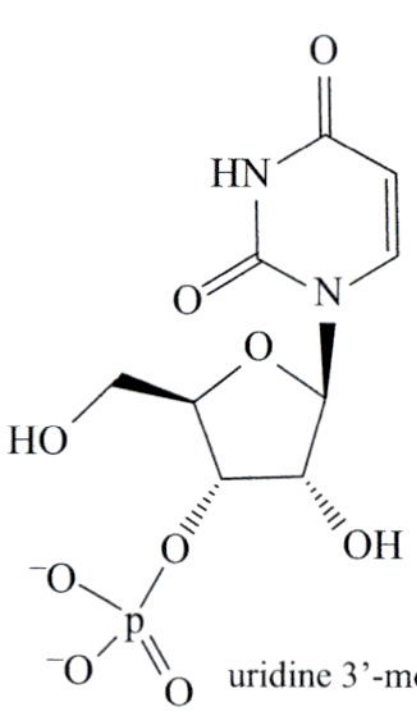

uridine 3'-monophosphate

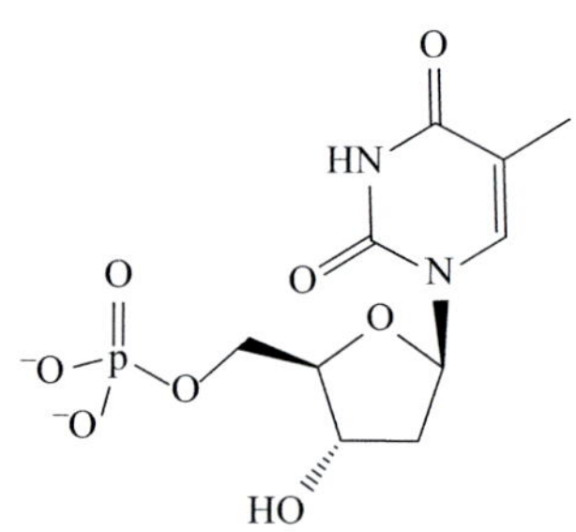

thymidine 5'diphosphate

15.5.

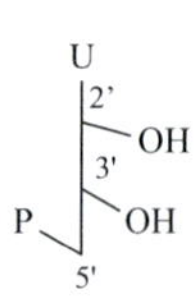

urasil

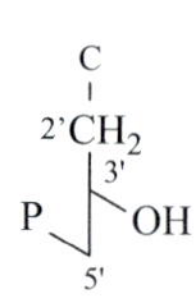

2'-deoxycytosine

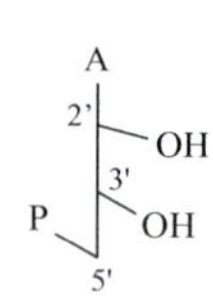

adenosine

15.6. In the DNA double helix, the strands of the phosphodiester backbone are closer together on one side of the helix than they are on the other. In the major groove, the nitrogen and oxygen atoms of the phosphodiester backbone point inward toward the helical axis, and the phosphodiester backbones are farther apart.

15.7. Because the nitrogen and oxygen atoms of the phosphodiester backbone point inward toward the helical axis

15.8. 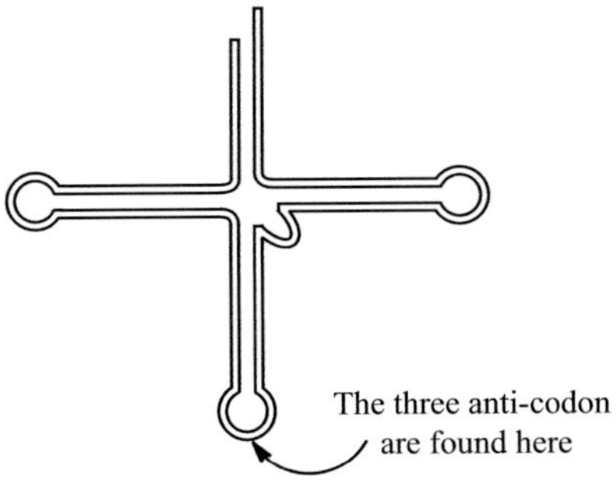

15.9. A ribozyme is a ribonucleic acid (RNA) enzyme that catalyzes a reaction.

15.10. A restriction enzyme recognizes a specific sequence of nucleotides and cleaves DNA into fragments at or near that recognition site (restriction site).

Index

D

E

F

G

H

I

J

K

L

M

N

O

P

Q

R

S

For Product Safety Concerns and Information please contact our
EU representative GPSR@taylorandfrancis.com Taylor & Francis
Verlag GmbH, Kaufingerstraße 24, 80331 München, Germany